INTERNAL COMBUSTION ENGINEERING: SCIENCE & TECHNOLOGY

INTERNAL COMBUSTION ENGINEERING: SCIENCE & TECHNOLOGY

Edited by

John H. Weaving

Technical Director, ERA Limited,
London Road, Dunstable, UK

ELSEVIER APPLIED SCIENCE
LONDON and NEW YORK

ELSEVIER SCIENCE PUBLISHERS LTD
Crown House, Linton Road, Barking, Essex IG11 8JU, England

Sole Distributor in the USA and Canada
ELSEVIER SCIENCE PUBLISHING CO., INC.
655 Avenue of the Americas,
New York, NY 10010, USA

WITH 24 TABLES AND 518 ILLUSTRATIONS

British Library Cataloguing in Publication Data
Internal combustion engineering.
1. Internal combustion engines
I. Weaving, John
621.43

ISBN 1-85166-410-6

Library of Congress Cataloging-in-Publication Data
Internal combustion engineering: science & technology/ edited by
John H. Weaving.
p. cm.
Includes bibliographical references.
ISBN 1-85166-410-6
1. Internal combustion engines—Design and construction.
I. Weaving, John H.
TJ785.I49 1990
621.43—dc20 89-16991
CIP

Printed in Northern Ireland by The Universities Press (Belfast) Ltd

Foreword

Sir Diarmuid Downs, CBE, FEng, FRS

Engineering is about designing and making marketable artefacts. The element of design is what principally distinguishes engineering from science. The engineer is a creator. He brings together knowledge and experience from a variety of sources to serve his ends, producing goods of value to the individual and to the community. An important source of information on which the engineer draws is the work of the scientist or the scientifically minded engineer. The pure scientist is concerned with knowledge for its own sake and receives his greatest satisfaction if his experimental observations fit into an aesthetically satisfying theory. The applied scientist or engineer is also concerned with theory, but as a means to an end. He tries to devise a theory which will encompass the known experimental facts, both because an all embracing theory somehow serves as an extra validation of the facts and because the theory provides us with new leads to further fruitful experimental investigation.

I have laboured these perhaps rather obvious points because they are well exemplified in this present book. The first internal combustion engines, produced just over one hundred years ago, were very simple, the design being based on very limited experimental information. The current engines are extremely complex and, while the basic design of cylinder, piston, connecting rod and crankshaft has changed but little, the overall performance in respect of specific power, fuel economy, pollution, noise and cost has been absolutely transformed. Incentives to improvement in all these areas have been provided by market forces and latterly by regulatory pressures from Government. The information on which these performance improvements have been based has been provided by the kind of detailed experimental and theoretical studies described in the various chapters of this book. To provide the necessary depth of study the subject is commonly divided, as this book has been divided, into a number of specialities, and engineers have frequently devoted their whole professional careers to the detailed investigation of such subjects as combustion, noise, pollution and the dynamic behaviour of engines. By such means is progress made. But, at the end of the process, the manufacturer has to sell and the customer wishes to buy, not just a high specific output device or a low fuel consuming device, let alone a low polluting or quiet device, but an engine which gives a performance in all these respects which will satisfy the marketplace. At the end of the process we need an engineer—a designer—who will put together the complete package, making use of the information provided by his specialist colleagues.

This book is addressed to specialists, but the chapters are also designed so that one specialist can understand the work of another specialist, and relate it to his own work and to the complete engine they both aim to produce. All are written by experts, most known to me personally. Between them they provide a guide to the present state of our knowledge of the various important aspects of the internal combustion engine design and performance. One difference which it is interesting to observe, as compared with a similar book which might have been compiled some 20 years ago, is the greater use of mathematical theory which has been made possible by the availability of powerful computers. A general trend in engineering is to calculate more, for our designs to become more quantitative, with less being left to inspired guesswork, or merely based on past experience. Additionally, the creation of mathematical models, trust in which will increase as they are more fully validated by experiment, not only enables more of the engine design to be calculated quantitatively, but also enables variables to be explored within the mathematical model rather than on the test bench, thus reducing the time and cost required to produce a new engine. In the ultimate, it may be possible to look forward to the time when experimental work will be a thing of the past: all will be subject to exact calculation. This, however, is far into the future, and this book wisely gives a balanced picture of experimental work and of calculation, mutually dependent and interrelated. I have no doubt that the picture it presents and the information it contains will be of great value to all engineers and scientists working in the internal combustion engine field for many years to come.

Preface

After so many years of development of the internal combustion (IC) engine since the days of Otto's first four-stroke engine (1876) it might have been thought that the IC engine would have been perfected. The contrary is the case, as every decade seems to throw up new challenges. The most recent challenge has been the bogey of air pollution, which started from the intolerable conditions often experienced in Los Angeles, where the climatology, combined with vast numbers of vehicles, produces the worst conditions of smog. Attention being called to this situation invited the monitoring of pollution worldwide and few cities were found to be free from pollutants. The results of these findings were severe regulations restricting the emission of carbon monoxide, oxides of nitrogen, hydrocarbons and lead. Are these regulations too severe or not severe enough?

What is the designer of a new vehicle to do to meet this challenge? Here a time factor intervenes: What can he do to meet legislation that is already promulgated or just round the corner and secondly, given time, can he design a clean economic and socially acceptable engine? It is hoped that this book will help on both counts.

For the immediate future the designer still has several alternatives. First he may take the easy way out, to accept that IC engines are polluters and decide to clean the exhaust gases with a catalyst, but there are disadvantages. Secondly he may say, 'Is a diesel engine to be preferred?' The answer is in the affirmative for commercial vehicles, and many think the same for private cars. Again, 'Can the high-compression lean-burn engine meet the challenge?' Looking to the future, 'Has the two-stroke engine advantages?' and 'What about the stratified charge type of engine?'

It will be seen that the designer has many difficulties in reaching a decision. It would be nice to say we will lay out the required specifications, build each of these types of engines and choose the best. Unfortunately it is not as easy as this because the conventional petrol and diesel engines have undergone decades of development, while others have been largely neglected. A more fundamental approach is necessary, and it is the object of this book to examine in depth the position of each type and its potential to meet the requirements of pollution, economy and power.

The major challenge for the passenger vehicle as indicated above is between the diesel engine and the spark-ignition engine.

The situation at present, as is well known, is that the diesel engine has proved itself for the commercial vehicle due to its reliability, economy and comparatively low pollution, but although it has made inroads into the passenger car field it is still only the choice of a minority. The reasons for this are apparent. The diesel, as a slower, heavier, noisier and more costly power unit has its major appeal to those who do high mileages, because of its economy; this includes most taxis. However, careful refinements and noise insulation have made it acceptable to a wider market. With the use of a turbo-blower the weight disadvantages may be ameliorated.

The reasons for the improved fuel consumption of the diesel engine compared with the spark-ignition (SI) engine are well known; namely, the higher compression ratio and the ability—in fact, desirability—to run unthrottled. Designers and research workers are endeavouring to reduce this gap by designing lean-burn running SI engines.

Two approaches present themselves, the stratified-charge engine and the so-called high-compression lean-burn (HCLB) engine. Both these engines can potentially reduce the gap but both need further research; however, the potential is good because the reward should be a lighter engine with lower pollutants and, one hopes, a less expensive solution than the diesel.

The success of the HCLB engine is related to the octane number of the fuel available for, as most researchers will be aware, the octane number reduces the propensity of the fuel to knock or detonate, and this limits the compression ratio achievable. The higher the number the greater the resistance to knock; there are two ratings which are determined in a standardized variable-compression single cylinder CFR (Co-operative Fuel Research) engine[1]—the research method

(*RON*), the value quoted at the pump (present 1989 four star fuel), is a minimum of 97. The second method is the motor method (*MON*), which is rated on the same CFR engine except that tests are conducted with a high inlet temperature and high speed; this rating gives some guidance to 'high speed knock'. The *MON* is usually some 10 numbers lower than the *RON*. The difference ($RON - MON$) is called the sensitivity. Neither of these tests correlate accurately with the performance of a multi-cylinder engine on the road. An approximate road octane number may be taken as $(RON + MON)/2$.

Tetra-ethyl lead and tetra-methyl lead are the best known anti-knock additives; they are being largely reduced or eliminated as they are air pollutants and also poison catalysts.

For the reasons outlined above, it will be seen that the internal combustion engine is still an exciting subject due to the need for personal door-to-door transport. Cars are made in the millions and success or failure is multiplied by a large factor.

The various chapters of this book address all these areas. Internationally well known and competent experts in each field have been chosen to give the clearest pictures of a most complex science.

As will be immediately apparent, this book is not another textbook on internal combustion engines *per se*; several such exist.[2–4] It is intended to put before the reader the position of the most advanced research and technology in the field of IC engines so that the reader, if an industrial designer or researcher, will himself be able to incorporate some of this work to improve his product or advance his research. It will give him the status of the work being performed in the universities and other research establishments.

For the academic graduate, if just starting on his studies for a higher degree, the book will give him the up-to-date state of the art. The undergraduate wishing to examine an area in depth for a project will find much that is useful.

The lecturer will find that each subject, treated by some of the leading experts in the field, will give a basis for lectures and tools and methodology for future research.

Finally, it is with much trepidation that the editor outlines the research and development requirements that will enable a detailed specification for a clean engine for the 21st century. Clearly, as is normal practice, the performance requirements of the vehicle are specified and this decides the power of the engine. Let us assume that it will be a spark-ignition (SI) piston engine. The questions that then have to be answered are:

(1) What size of engine and speed limitations are appropriate?
(2) Will it be blown or normally aspirated?
(3) What combustion chamber and valve layout is to be preferred?
(4) What compression ratio is optimum for the fuel available?
(5) How will it meet the pollution legislation?

To answer these questions a very detailed knowledge of the combustion process is required, and it is considered that in order to obtain the largest amount of information in the shortest time that computation and experiment need to go hand in hand.

If the engine concept is a new one where fundamental knowledge is lacking, an initial computation by mathematical model of the gas flow and fuel distribution will be most helpful, both in saving time in comparison with trial and error experiments and in giving a target for an achievement and a final specification. This will not eliminate experiments, as assumptions still have to be made of the physical and chemical processes, which are inordinately complex, but nevertheless it will give a clear concept of what is happening so that instruments may be positioned to give the maximum information. This information in turn can be used to refine the model. Again, computation gives an orderly discipline to a full research or development investigation, throwing up questions that need to be answered. This combination of mathematical modelling has been found to be very useful in diesel engine research and design, so the tracing of the air path and fuel spray is vital to a good engine. It has proved equally valuable in the SI field and with stratified charge engines.

The editor wishes to acknowledge and sincerely thank all the contributors to this book. The personal effort and time spent, both in writing the chapters and in the research that lies behind them, is gratefully appreciated. A not-inconsiderable proportion of this research work has been provided by grants from the Science and Engineering Research Council (SERC), under a Specially Promoted Programme on Research in Combustion Engines.

REFERENCES

1. American Society for Testing and Materials, ASTM-D2699 and ASTM-D2700. ASTM, Philadelphia, USA.

2. Benson, R. S. & Whitehouse, N. D., *Internal Combustion Engines*. Oxford, Pergamon Press, 1979.
3. Heywood, J. B. *Internal Combustion Engine Fundamentals*. McGraw-Hill, 1988.
4. Lilly, L. C. R. (ed.)., *Diesel Engine Reference Book*. Butterworth, 1984.

John H. Weaving

Contents

Foreword v

Preface vii

List of Contributors xv

1 Combustion in Spark-ignition Engines 1
M. T. Overington

2 Applied Research into Combustion in Small Diesel Engines 33
H. K. Oetting

3 The Two-Stroke Engine: Crankcase Compression Type . 65
G. P. Blair

4 The Two-Stroke Engine: The Blowdown and Uniflow Scavenge Process 103
F. J. Wallace

5 Stratified Charge Engines 137
J. H. Weaving

6 Mixture Preparation for Spark-ignition Engines 173
C. J. E. Nightingale

7 Diesel Engine Fuel Injection Processes and Spray Diagnostic Methods 213
A. A. Hamidi & J. Swithenbank

8 Turbulent Flows in Reciprocating Internal Combustion Engines 243
R. J. Tabaczynski

9 Combustion in Gasoline Engines 287
D. Bradley

10 Combustion in the Diesel Engine 333
S. Matsuoka

11 Computer Simulation of Fluid Flow and Combustion in Reciprocating Engines 385
D. B. Spalding

12 The Theory of Wave Action Approaches Applied to Reciprocating Engines 445
D. E. Winterbone

13 The Application of Wave Action Techniques to Reciprocating Engines 501
D. E. Winterbone

14 The Turbocharger 615
B. E. Walsham & D. E. Winterbone

15 Atmospheric Pollution 707
J. H. Weaving & J-P. Pouille

16 Instrumentation for Engine Flows 795
C. Arcoumanis, M. Yianneskis, D. R. Williams, C. A. Baker & D. A. Greenhalgh

Index 857

List of Contributors

C. Arcoumanis
Department of Mechanical Engineering, Imperial College of Science, Technology and Medicine, Exhibition Road, London SW7 2BX, UK

C. A. Baker
Harwell Laboratory, UK Atomic Energy Authority, Chemical Physics Group B551 Harwell Laboratory, Oxfordshire OX11 0RA, UK

G. P. Blair
The Queen's University of Belfast, Stranmillis Road, Belfast, Northern Ireland BT9 5AH, UK

D. Bradley
Department of Mechanical Engineering, University of Leeds, Leeds, LS2 9JT, UK

D. A. Greenhalgh
Harwell Laboratory, UK Atomic Energy Authority, Chemical Physics Group B551 Harwell Laboratory, Oxfordshire OX11 0RA, UK

A. A. Hamidi
Department of Mechanical and Process Engineering, University of Sheffield, Mappin Street, Sheffield S1 3JD, UK

S. Matsuoka
The Science University of Tokyo, 7-6-12 Seijo Setagaya-KU, Tokyo, Japan

C. J. E. Nightingale
University College London, Torrington Place, London WC1E 7JE, UK

H. K. Oetting
Volkswagen AG, Postbox 3180, Wolfsburg 1, FRG

M. T. Overington
Ricardo Consultants, Ricardo Consulting Engineers Ltd, Bridge Works, Shoreham-by-Sea, West Sussex BN4 5FG, UK

J-P. Pouille
Renault, 67 Rue des Bons-Raisins, 92508 Rueil Malmaison Cedex, France

D. B. Spalding
CHAM Ltd, Bakery House, 40 High Street, Wimbledon, London SW19 5AU, UK

J. Swithenbank
Department of Mechanical and Process Engineering, University of Sheffield, Mappin Street, Sheffield S1 3JD, UK

R. J. Tabaczynski
Ford Motor Co., PO Box 2053, Dearborn, Michigan 48121–2053, USA

F. J. Wallace
School of Mechanical Engineering, University of Bath, Claverton Down, Bath BA2 7AY, UK

B. E. Walsham
Holset Engineering Co. Ltd, PO Box A9, Turnbridge, Huddersfield HD1 6RD, UK

J. H. Weaving
ERA Ltd, London Road, Dunstable, Bedfordshire, LU6 3UR, UK *and* 150 Chessetts Wood Road, Lapworth, Solihull, B94, 6EN, UK

D. R. Williams
Harwell Laboratory, UK Atomic Energy Authority, Chemical Physics Group B551 Harwell Laboratory, Oxfordshire OX11 0RA, UK

D. E. Winterbone
UMIST, PO Box 68, Manchester Road, Manchester M60 1QD, UK

M. Yianneskis
School of Engineering, King's College London, University of London, Strand, London WC2R 2LS, UK

1

Combustion in Spark-ignition Engines

M. T. Overington

Ricardo Consulting Engineers Ltd., Shoreham-by-Sea, West Sussex, UK

1. Introduction

2. Research engines

3. Measurement and analysis methods
   3.1. Pressure data analysis
   3.2. Optical measurement methods
   3.3. Ionization probes

4. Combustion system characteristics
   4.1. The basic process
   4.2. Dilution with air
   4.3. Increased air motion
   4.4. Ignition retard
   4.5. Dilution with exhaust gas

5. Effects of air motion characteristics
   5.1. Swirl
   5.2. Tumble

6. Combustion chamber design effects

7. Characterization of in-cylinder air motion

1. INTRODUCTION

Later chapters in this book will discuss the origins of the fluid motions that occur within combustion chambers of spark-ignition engines, the fundamentals of the combustion process itself and some of the mathematical modelling of this process that has been carried out—all essentially looking at combustion from the inside. The aim of this chapter is to look at combustion from the outside; to take the position of the researcher or developer faced with trying to understand and develop the engine. It will try to illustrate some of the characteristics of the combustion process in spark-ignition engines and show how these may be measured. There is now a vast body of information on this subject: this chapter can only describe a selection which will illustrate the author's views on the important features.

2. RESEARCH ENGINES

Much development work on gasoline engines is carried out on multi-cylinder production versions. This type of approach is most appropriate where the interaction between one cylinder and the next is important and also where the behaviour of the whole system is being judged, e.g. where durability targets must be met. For combustion system investigations, however, the single-cylinder engine is often better suited to the work in hand. The absence of the variability between one cylinder and the next, which is often considerable, is removed and thus direct links can be established between the combustion process and observed characteristics such as fuel economy and exhaust emissions. This also enables programmes involving a number of variables to be carried out with a lower manufacturing cost for the prototype parts. However, for any particular experiment the judgement must always be made between the realism of the multi-cylinder engine and the experimental convenience and accuracy of the single cylinder.

Over the years many special single-cylinder engines have been made to meet the needs of different development programmes. Today the field is served by a number of single-cylinder research engines, two of which are illustrated in Figs 1 and 2. The Hydra engine (Fig. 1), described in Ref. 1, covers both gasoline and diesel engines for sizes used in passenger cars and light trucks, while the Proteus engine (Fig. 2) is used for the development of heavy duty truck diesel engines. These two are in addition to the many single-cylinder conversions of production engines, usually made by their manufacturer. The purpose-built engines share a number of basic characteristics. Each is robustly built to withstand long research programmes, many rebuilds, and

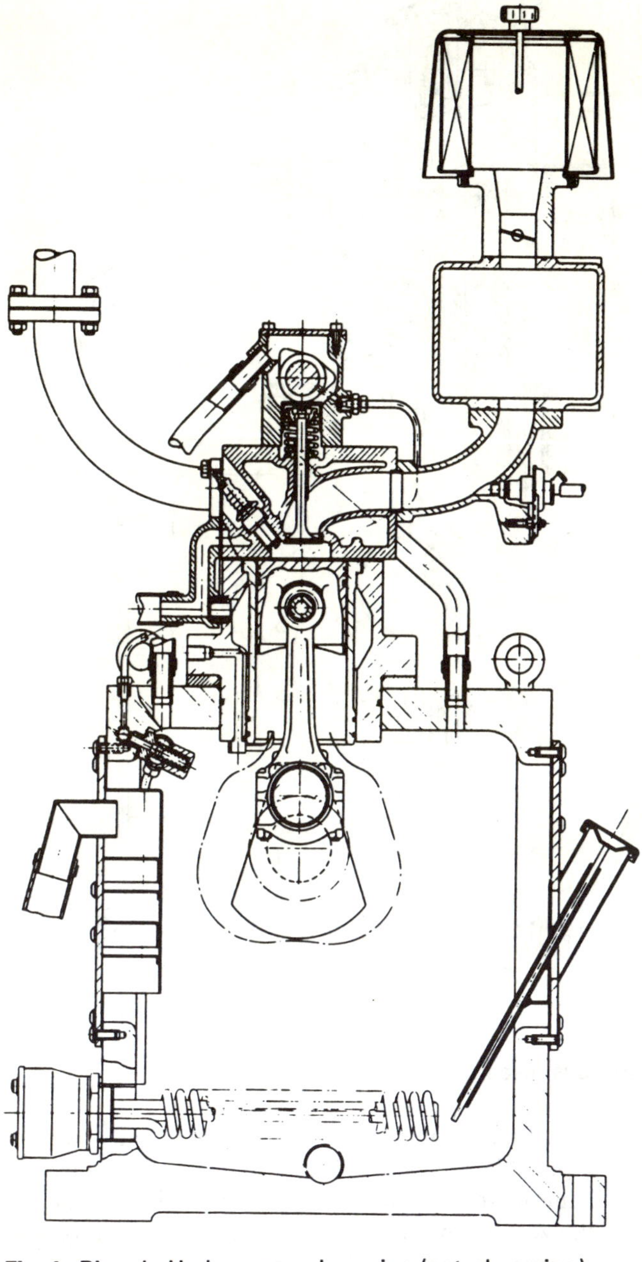

Fig. 1. Ricardo Hydra research engine (petrol version).

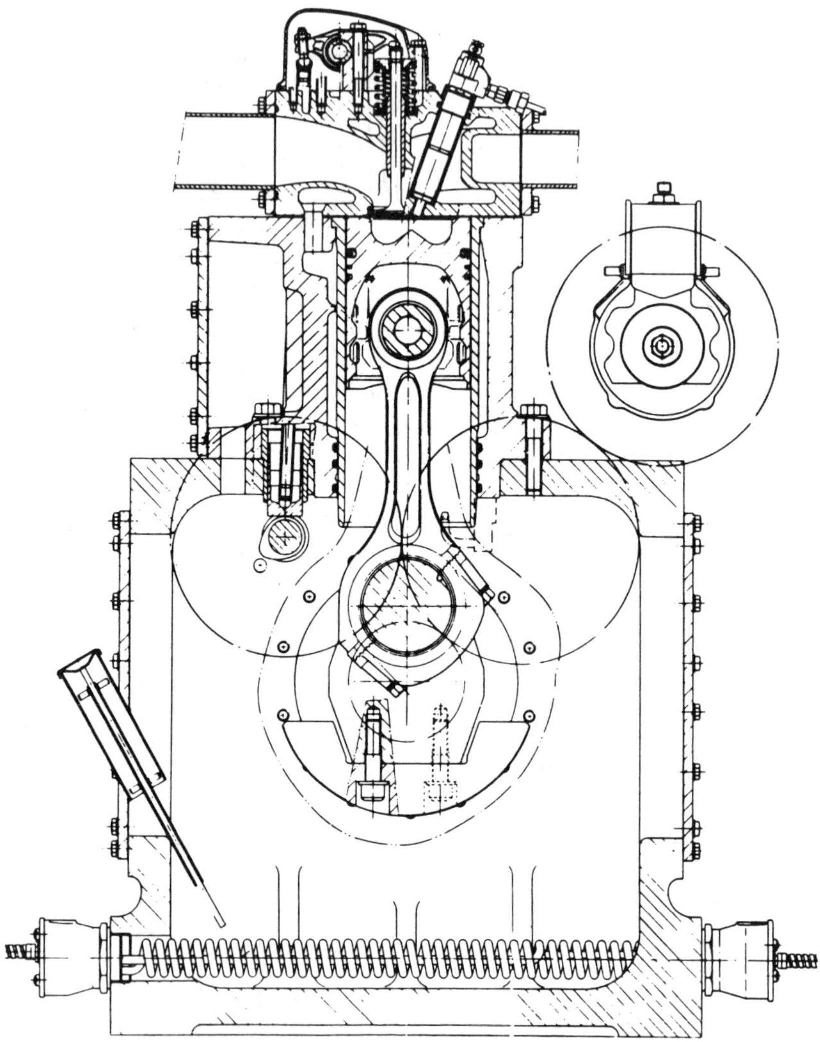

Fig. 2. Ricardo Proteus research engine (diesel version).

operation under heavy knock conditions. Bearing sizes are, however, made as similar as possible to comparable multi-cylinder engines so that specific mechanical friction levels are increased as little as possible. This minimizes the problem of operating at the very low loads that are so important for automotive engines.

These engines are normally fitted with comprehensive instrumentation, and measurement of exhaust emissions and cylinder pressures are made during nearly all testing in order to allow detailed analysis of the combustion process. For specific studies, these specially designed research engines can be easily modified to allow combustion photography or a number of laser-based measurement methods to be used.

3. MEASUREMENT AND ANALYSIS METHODS

For most experimental programmes an engine would be equipped as shown in Fig. 3. This set-up allows most basic studies to be carried out. Air supply to the engine would be measured using a viscous flow or similar meter, and supplied to the engine via a damping chamber with volume of five to ten times the engine swept volume. As an alternative, the air flow can be calculated from the air : fuel ratio of the exhaust stream, as deduced from exhaust emission measurements, and from an accurate determination of the fuel flow. This latter measurement would normally be carried out using a gravimetric fuel meter rather than the previously almost universal volumetric device. However, flow rates in experiments at low speeds and loads can be very low, and under these conditions, and in order to maintain measurement accuracy, volumetric meters can be preferable in some circumstances.

Measurement methods for exhaust emissions are now more or less standardized. Flame-ionization detectors (FID) are used for hydrocarbon analysis, non-dispersive infra-red (NDIR) analysers for CO and CO_2, paramagnetic analysers for O_2, and chemiluminescent analysers for oxides of nitrogen. One point that should be remembered is that although FID analysers have been developed to give a more or less uniform response to most hydrocarbon types, there are particular species which give a relatively low response. This point is addressed in Chapter 14.

The pressure within the combustion chamber should be measured using a purpose-designed flush-mounted piezo-electric transducer, incorporating a ceramic heat shield to reduce errors caused by radiation effects. The transducer, when used in conjunction with a suitable charge amplifier, produces a signal which can be logged by a high-speed data acquisition system. An indication of crankshaft position is required and this can be provided by a rotary encoder. The

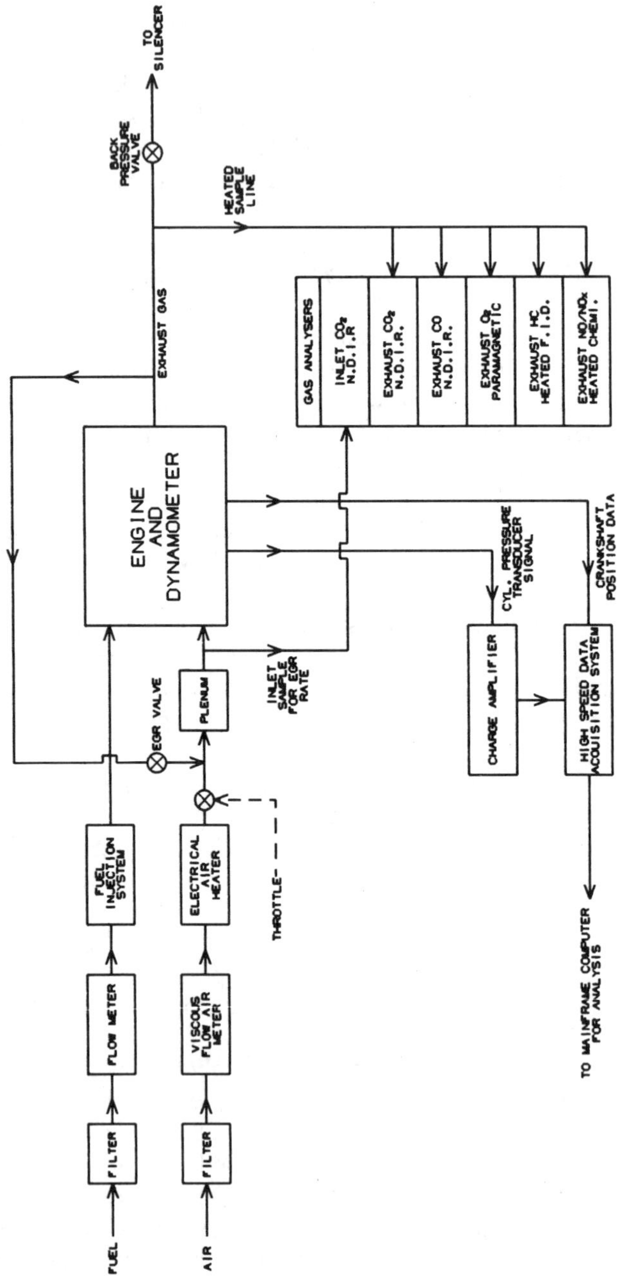

Fig. 3. Experimental engine set-up.

signal from this is also fed to the data acquisition system. Pressure data would then normally be logged at one degree intervals with 12 bit accuracy.

Phase errors can introduce significant errors into the analysis of pressure data and a number of authors have studied this subject in detail.[2] Suffice it to say here that phasing errors have very significant effects on IMEP calculations. A phase error of 1 degree gives an IMEP error of 4%: thus special care should be taken in setting up encoder phasing and also minimizing phase lags in the charge amplifier and data acquisition system.

3.1. Pressure Data Analysis

Two main methods for the analysis of cylinder pressure data are available. The first is based on a 'First Law' approach.

Heat release rates can be calculated from measured cylinder pressure diagrams by solving the First Law of Thermodynamics energy equation in conjunction with empirical heat transfer data. An estimate is made of the cylinder contents and their properties at the start of combustion, and also of the way in which these change as the combustion proceeds. The change in internal energy and the work output are then calculated from the pressure diagram and piston motion characteristics. The First Law can then be used to calculate the net heat addition once a correction for heat transferred to the combustion chamber walls has been made. In addition it is generally assumed that there is no leakage from the combustion space via valves or piston rings. Both this assumption and the lack of instantaneous heat-transfer data can give rise to significant errors, but the method does predict total heat release which can be used to estimate the proportion of fuel burnt.

The second method is based on a proposition made by Rassweiler & Withrow in 1938[3] that when combustion occurs in a fixed volume container, the fraction of the total pressure rise occurring at any time is proportional to the fraction of the charge burnt. Using this proposition, the pressure data from the combustion process in an engine is reduced to constant volume using the relationship $PV^n = K$, where n is calculated from the pressure diagram for the compression and expansion strokes, and during combustion it can be assumed that the value changes linearly from the compression to the expansion value. The

Fig. 4. Typical burn curves.

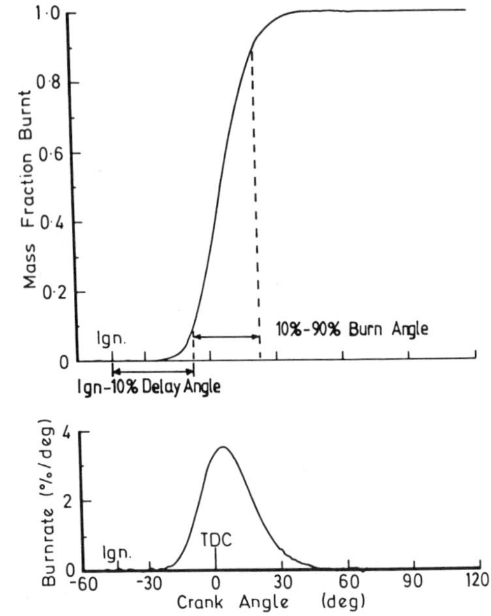

total pressure rise occurring during the process can then be calculated. The fraction of the charge burnt at any crank angle is then equal to the fraction of the total pressure rise that has occurred at that point. There is always debate over the precise value of n that should be used during the combustion process, but it turns out that the total calculation is fairly insensitive to the choice of n. The main advantage of this method is that the pressure data is sufficient information for the calculation to be carried out. Hence this method is the one most frequently adopted by workers in the field.

Cyclic pressure data is usually averaged over a number of cycles, between 300 and 400, to obtain a 'mean' pressure diagram, and then this is analysed using the technique described above. The results appear as shown in Fig. 4. The main parameters deduced from this analysis to characterize the combustion process are the so-called delay and burn angles. The delay angle is that from the point of ignition defined by the spark timing to that when a small proportion of the charge, usually 10% but latterly 3% or even 1%, is burnt. The choice of the smaller values to define the end of the delay period more accurately represents the concept of delay but makes greater demands on the pressure data acquisition and analysis methods used.

The main burning period is usually characterized by the crank angle required for the fraction from 10% to 90% of the charge to be burnt. The use of this centre portion reduces the effects of calculation uncertainties at the beginning and end of combustion on the results of the analysis.

The above analysis routine can also be carried out on individual firing cycles, and the gradually reducing cost and increasing 'power' of small computers is making this approach more frequently used. The analysis of each pressure diagram is carried out in the usual way and then the peak pressure, delay angle and burn angle values can each be analysed statistically to give mean values and standard deviations of peak pressure, delay angle etc. As vehicle engines become gradually more refined, these indicators of variation between cycles are more often used to indicate the direction in which development should be carried out.

One of the most universal indicators of combustion instability which is used is the coefficient of variation (COV) of $IMEP$.This is calculated very straightforwardly as:

$$COV(\%) = \frac{\sigma(IMEP)}{IMEP} \times 100, \qquad \text{where} \qquad IMEP = \int_{0^\circ}^{720^\circ} P\,\mathrm{d}V$$

and where $\sigma(IMEP)$ is the standard deviation of $IMEP$.

This analysis can be carried out on the individual pressure diagrams used for the burn rate analysis, but the use of a greater number of cycles (*c.* 1000) increases the statistical accuracy.

3.2. Optical Measurement Methods

Single-cylinder research engines are particularly suitable for the use of various optical techniques to study the initiation and progress of the combustion process. Many methods have been developed by workers in the field, only a few of which can be described here. The most straightforward of these is high-speed photography. Transparent windows usually made of quartz or Perspex can be fitted in either the cylinder head or to form a transparent piston crown. The former method is illustrated in Fig. 5. Suitable mirrors are used to create images of part of the combustion chamber and also an indication of crank angle within the frame. For 'through-the-piston' photography, the cylinder head is mounted on a raised cylinder bore which houses a

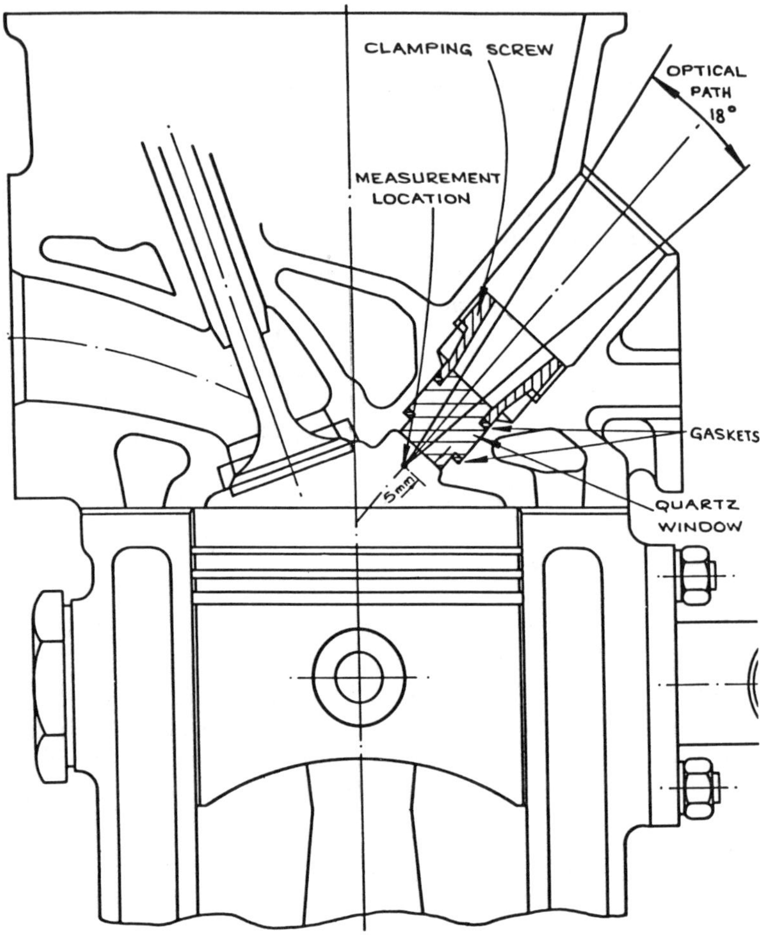

Fig. 5. Optical access through cylinder head.

piston with a transparent crown. This is connected to the normal piston which acts as a cross-head. The layout is shown in Fig. 6. An angled mirror immediately below the transparent crown provides a view over a large proportion of the combustion chamber area. The photographic techniques used for both of these applications are very similar. The flame front is usually barely visible except at rich mixtures, and thus special means are required to make the flame

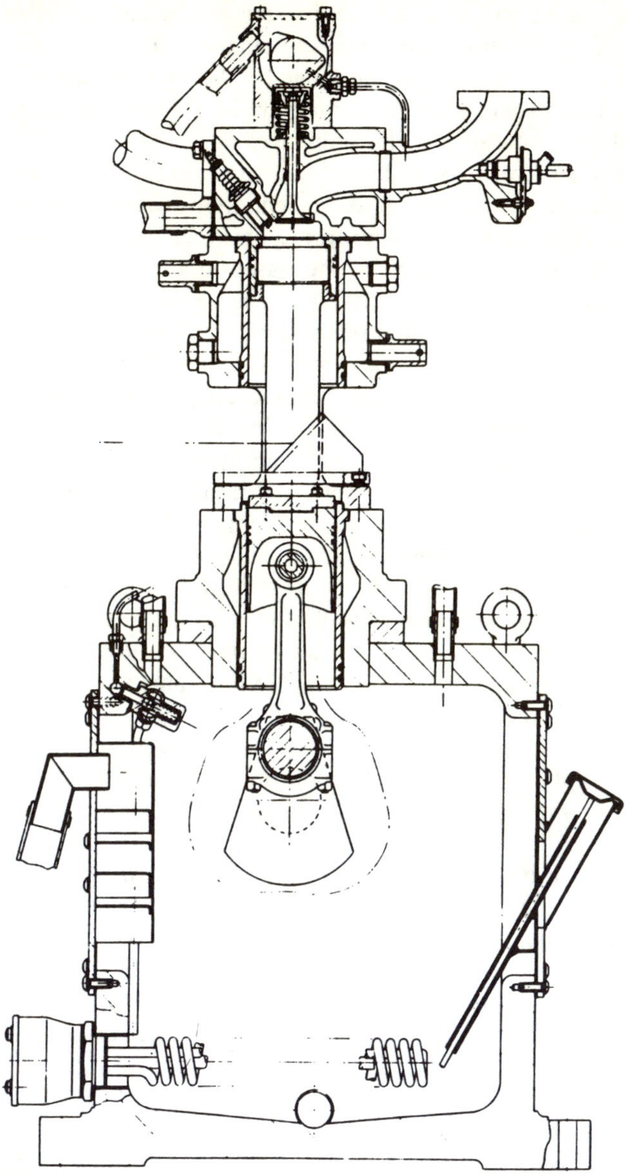

Fig. 6. Through-the-piston optical engine.

luminous. A solution of common salt can be sprayed into the intake manifold near the intake valve, or a metal oxalate can be added to the fuel. The camera speed used depends on engine speed and the resolution required, but is usually in the range 8000–16 000 frames per second. Reference 4 provides more details about the experimental techniques and shows some of the film sequences obtained.

The basic engine structures described above can also be used for a number of laser-based techniques. A laser Schlieren process may be used to study the structure of the flame front. This technique is based on the changes in refractive index of the gases in the combustion chamber during the combustion process. The engine set-up is shown diagrammatically in Fig. 7. A parallel beam from an argon ion laser is directed through a window in the cylinder head and reflected from a mirrored piston. The deflection of the beam caused by the density gradients is converted into light information by the Schlieren filter. The information is recorded on high-speed ciné film where the flame front appears light against a dark background. With this technique it is possible to trace the progress of the flame front in detail and also obtain a visualization of the detailed structure of the flame front.[5] The techniques described above rely on light reflected from internal surfaces of the engine, but it is also possible to use a similar technique with straight-through optical access to obtain similar information.[6]

For laser doppler anemometry (LDA), two laser beams are focused at a point within the combustion chamber, either through a window in the cylinder head or through the piston. In the latter case the position of the crossing point can be varied by altering the attitude of the mirror and moving the position of the laser along the beam axis. A fringe pattern is set up at the crossing point, and when seeding particles introduced into the air intake pass through the crossing point a doppler burst is produced. This is detected by a photomultiplier positioned either beyond the measuring volume or near to the laser—forward or back scatter. The frequency of the ripple on the photomultiplier output is then measured, and this is proportional to the velocity of the particle at right angles to the axis of the beams. By altering the beam orientation and the position of the crossing point, measurements can be made over a large proportion of the combustion chamber area and also at various distances along the cylinder axis. Typically, about 5×10^5 measurements are made at each point, and these are processed into vectors defining the mean velocity magnitude and direction, and also the r.m.s. value of the velocity. As

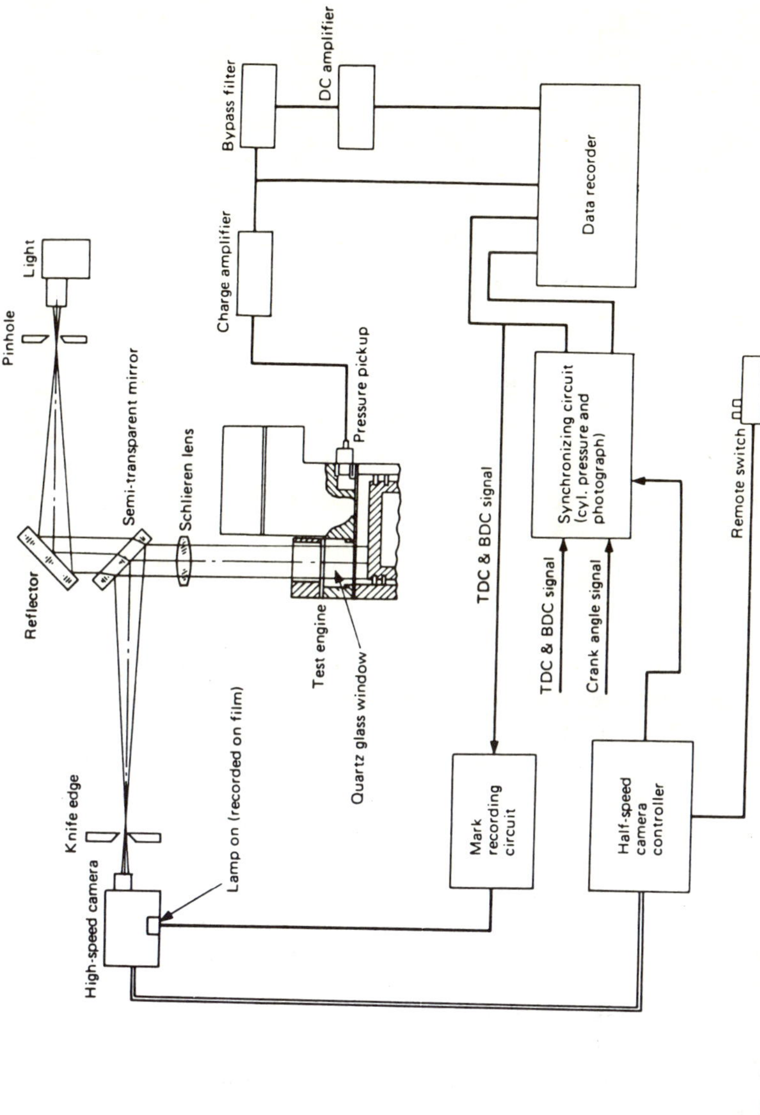

Fig. 7. Schlieren photography test equipment.

Fig. 8. Turbulence intensity and length scale measurements.

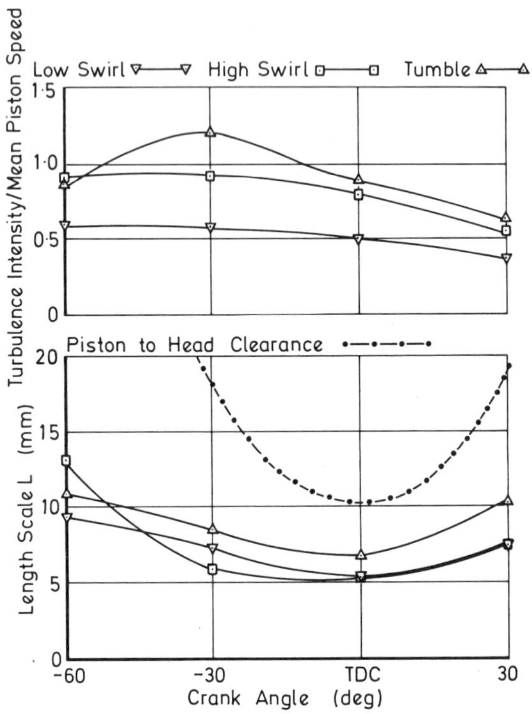

discussed in Chapter 8 this latter value, often known as the turbulence intensity, contains effects of cycle-to-cycle variability as well as true turbulence. Many workers have proposed methods to deduce the turbulence level without the cyclic variability effects. A number of these are based on the assumption that velocity variation of low frequency originates from variability from cycle to cycle, whereas variations with a high frequency content are true turbulence. A better approach is that suggested by Glover,[7] although the analysis requires very large data sets. In a recent variation[8,9] of traditional LDA, the measuring volume is scanned across the combustion chamber in a few crank degrees (*circa* 5°). By this means several velocities along a chord of the combustion chamber can be measured at one time; but, more importantly, data from repeated scans can be used to deduce a characteristic length scale. A typical result is shown in Fig. 8 from Ref. 9. Techniques of this sort are still very much in the development process and are therefore changing rapidly.

3.3. Ionization Probes

One of the properties of a flame is that the gases within and near to it are conducting. This property forms the basis of the ionization probe technique to measure the position of the flame front in a burning mixture. A number of probes are placed in the combustion space, usually near the walls, and each probe is formed such that there is a gap between the electrodes. A voltage of about 100 V is applied across the gap, and when the flame front passes a current flows in the circuit containing the electrodes. The phasing of the ionization signal is deduced from an oscilloscope display or by means of a data logging system. In a practical experiment a number of such probes may be used in various parts of the combustion chamber and from this data the progress of the flame front can be mapped.[10]

Unfortunately, there are a number of problems associated with this basically simple technique. These stem from the basic characteristics of the process, and its variability from cycle to cycle. This variability means that there is significant variation not only in phasing but also in the shape of the flame from cycle to cycle, and thus in order to form a view of the flame progression a statistical approach must be used and data from many cycles obtained. A further problem is that flames are wrinkled, and in some cases very wrinkled. Since the ionization probe only measures in a small zone this further adds to the problem of identifying the true flame front progression. For these reasons, this method is not popular for flame studies and optical methods are more frequently used.

4. COMBUSTION SYSTEM CHARACTERISTICS

4.1. The Basic Process

The process by which the charge in a spark-ignition engine is burnt has often been described in general terms. The more that has been learnt about the detail of the process, the more the basics have been confirmed. The spark at the plug initiates burning in a small region close to the plug electrodes. This flame grows slowly in a more or less spherical shape. The rate of energy release is low and its effect on the in-cylinder pressure is barely discernible. This phase of the process is generally referred to as the delay period. A number of models have been suggested for these processes. This initial burning occurs at near the laminar flame speed. As the flame grows and its size approaches and exceeds the turbulent length scales, the flame speed increases until it has a value that is much higher than the laminar flame speed. Its magnitude is a function of the intensity of the turbulence.

The flame is now still more or less spherical, but the flame front begins to take on a wrinkled appearance as it is affected by the turbulent structure of the air flow within the combustion chamber. This wrinkled structure has been shown up well by Schlieren photography by workers such as Namazian.[6] The rate of energy release has now increased to a high value, but soon the flame begins to encounter the walls of the combustion chamber and this leads to a slowing down of the process. The area of the flame front gradually decreases as a greater proportion of it reaches a wall, until the whole of the charge is consumed.

As the flame front progresses, the pressure in the chamber increases and so does the temperature of the unburned charge. Its temperature further increases due to radiation from the flame front. If a portion of the charge spends sufficient time at a high temperature it may auto-ignite. This process is known as detonation (or knocking) and results in a sudden increase in pressure and the familiar high-pitched ringing sound.

The above description is somewhat idealized; however, most practical combustion systems, even those with high levels of air motion, follow the stages described above. The shape of the flame does, however, tend to be distorted by the gross air motion patterns present.

It is evident from the foregoing description that the characteristics of the flows within the cylinder of the gasoline engine have fundamental effects on the rate at which the charge is burnt and hence on the pressure development within the cylinder. This process can also be affected by its phasing (because of the effect this has on the ambient temperature), by the shape of the combustion chamber and by the chemical constituents of the working fluid.

Many studies have been carried out to examine the effects of these variables. They can be illustrated with results taken from a number of studies on single-cylinder engines, showing the effects of a number of control and design parameters on the combustion process and the performance of the engine.

4.2. Dilution with Air

An examination of the effects of variables is best carried out by studying an engine initially running at part load conditions, stoichiometric mixture strength and with the ignition timing set for best torque (MBT). Air is often added to the intake under these conditions, to produce a mixture leaner than the chemically correct one, to improve fuel economy. But this has other effects as well, as shown in Fig. 9. The fuel consumption does indeed improve steadily as air is added, until it reaches a minimum at an 18:1 air:fuel ratio (A/F). During this period the hydrocarbon emissions also tend to decrease. However, NO_x emissions rise to a peak at 17 A/F and then tend to decrease to low levels as the mixture is further weakened. This peak occurs because the reactions under which the NO_x is formed are highly temperature dependent, and the combustion temperature reaches a maximum at about 17 A/F. The combustion process itself slows down gradually as the mixture becomes weaker and the delay period increases. For both of these reasons the ignition timing needs to be steadily advanced to maintain MBT. At very lean mixtures the process becomes very slow, and beyond a certain point the late burning of the charge within the cycle adversely affects the efficiency, and the fuel consumption rises. This late burning can also mean that some small residues of charge can remain unburnt, and thus unburnt hydrocarbon emissions also tend to rise as the lean limit of operation is reached. This trend is exaggerated by the fact that the exhaust gas temperature reduces and thus further oxidation of the emissions within the exhaust system is suppressed. Also, as the delay period in particular increases the instability of the process increases and eventually the engine becomes too unstable to run. In practice neither the turn-up in HC nor the reduced stability are desirable, but the limit is usually set on the basis of the stability measure. A limit value of *circa* 5% variability is often used, although the acceptable value in a vehicle varies with engine speed and load, and also with vehicle installation.

4.3. Increased Air Motion

One of the most effective means of improving the lean-running ability of a combustion system is to increase the turbulence intensity within the charge so that the burning rate at any mixture strength is

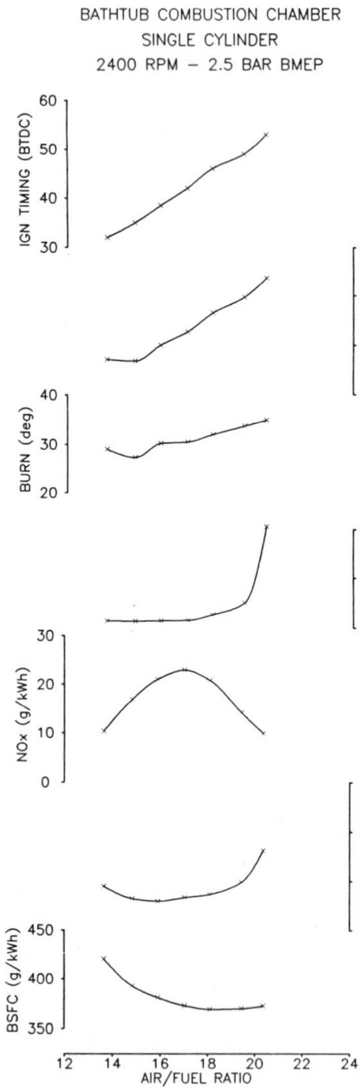

Fig. 9. Effect of air dilution on engine performance.

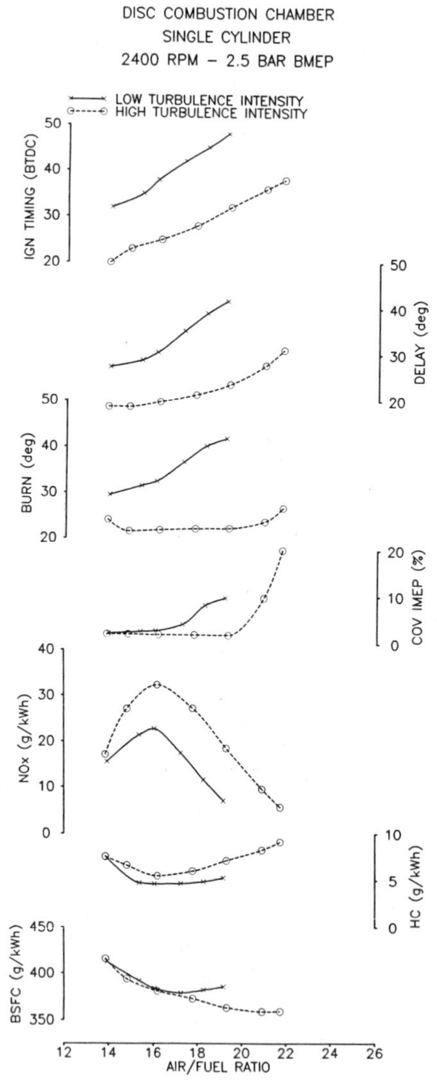

Fig. 10. Effect of increased turbulence on engine performance.

increased. Thus the critical mixture at which slow burning has an impact on efficiency is made leaner. The results shown in Fig. 10, with the level of air motion increased, demonstrate this well. The lean running limit is considerably extended from 19·5:1 to 21·8:1 A/F. The ability to run at the leaner mixtures enables better minimum fuel economy to be achieved. At the leaner running conditions lower levels of NO_x emissions are possible and HC emissions are maintained at a reasonably low level.

These comments on the effects of burning rate presuppose that the flame is initiated successfully by the spark. There is the possibility that the initial flame kernel does not grow but becomes extinguished because of locally adverse conditions. These conditions could be very high velocity and/or turbulence levels or a velocity direction which causes the kernel to be quenched on a cold surface, e.g. the spark plug body. During normal engine testing it can be difficult to deduce the cause of a lean limit being approached in any particular case. A method relying on the enhanced flame propagation conditions at retarded ignition timings due to the higher charge temperatures is described in Ref. 11, and this can be used to separate cases when the flame kernel fails to grow from the cases where burning becomes very slow.

4.4. Ignition Retard

The foregoing discussion has assumed that the ignition timing has been set at MBT under all conditions. However, there are advantages in using timings more retarded than this in many cases. The effect of retard is shown in Fig. 11. It can be seen that both HC and particularly NO_x emissions are reduced as the ignition timing is retarded. The effect on NO_x is due to reduced peak temperatures resulting from the greater expansion occurring during the burning process, and the reduced hydrocarbons result from the higher combustion temperatures which occur later in the cycle and the greater level of post-oxidation in the cylinder and exhaust system. It should be noted that at retarded timings the delay period is decreased as a result of higher temperature at the time of the spark, and thus the combustion event itself is not retarded to the extent that the change in ignition timing would suggest. It is also interesting that, in spite of these shorter delays, stability levels deteriorate at retarded conditions. The greatest drawback,

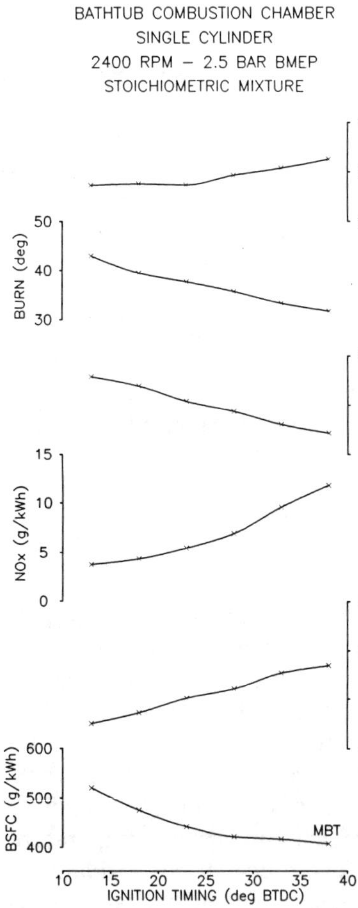

Fig. 11. Effect of ignition retard on engine performance.

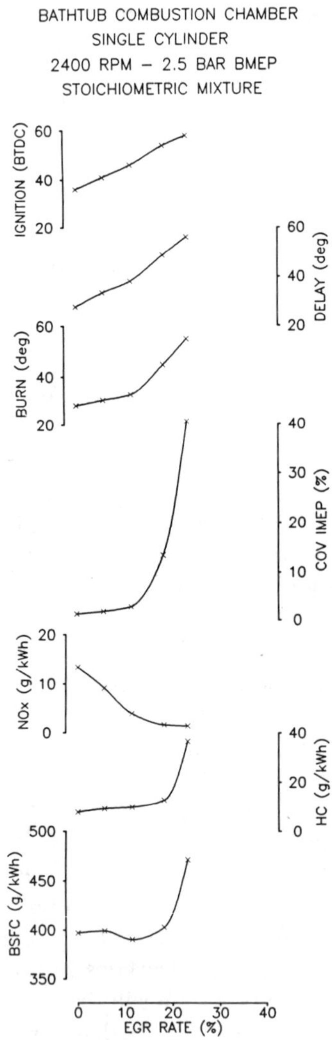

Fig. 12. Effect of EGR dilution on engine performance; slow burn engine.

though, is the deterioration in fuel economy which occurs, and these last two factors limit the extent to which this means can be used to achieve reduced emissions.

4.5. Dilution with Exhaust Gas

A further method which is often used to reduce NO_x emissions is to return a proportion of the exhaust gas to the engine intake. This is termed exhaust gas recirculation (EGR). This is analogous to running with a lean mixture, but the dilution is carried out with nearly inert exhaust rather than air. The addition of this inert gas to the working fluid has the effect of reducing peak temperatures, both because of the heat-sink effect of the gas and also because of the slower burn rate which results. The effect on NO_x emissions is significant, as is shown in Fig. 12, but this reduction occurs at the cost of some increase in HC emissions and also a deterioration in engine stability. These two factors place a limit on the level of EGR that can be used.

EGR was first used seriously in the mid-1970s and was applied, in the main, to engines with low rates of burning. In this type of combustion system, fuel economy deteriorates as soon as EGR is applied. The negative effect on vehicle fuel economy was immediately noticed by the public. More recent combustion systems developed to have higher burn rates and better tolerance to EGR do not respond in the same way. For these systems fuel economy tends to improve as EGR is applied, and only deteriorates at high EGR levels where the burning process is much degraded. This is illustrated in Fig. 13, where best fuel economy occurs at about 20% EGR. The gain relative to the zero EGR, $\lambda = 1$ value is 5%. The corresponding figure for dilution with air is 10%, and thus in a fast-burning engine the use of EGR is attractive for emissions reduction and can have a beneficial effect on fuel economy. It also can be used with a three-way catalyst system, which can only operate at stoichiometric conditions, to achieve very low exhaust emission levels.

5. EFFECTS OF AIR MOTION CHARACTERISTICS

It is now widely accepted that the flows within the combustion chamber can have a very significant effect on the combustion process.

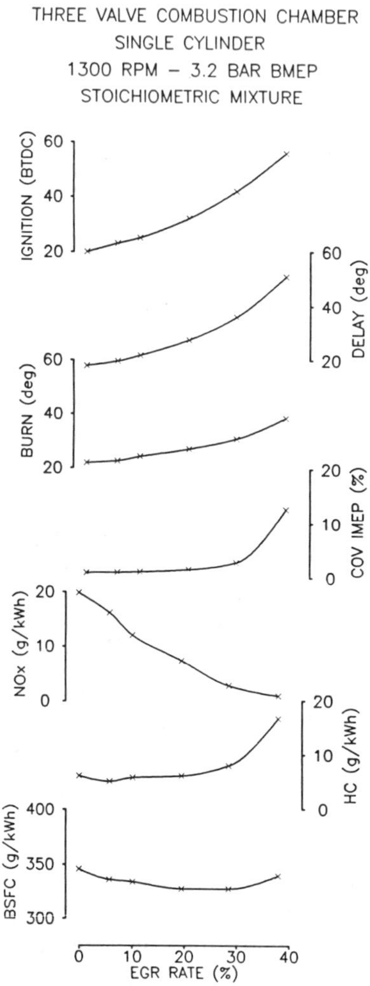

Fig. 13. Effect of EGR dilution on engine performance; fast burn engine.

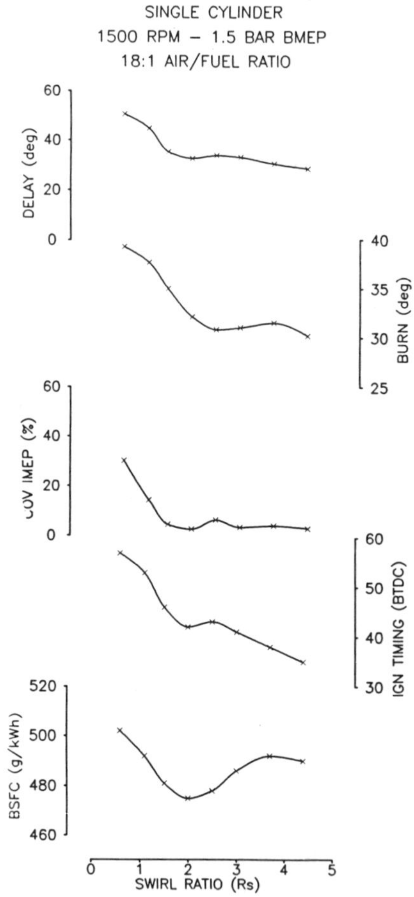

Fig. 14. Effect of intake swirl on engine performance.

Indeed, much of the current development effort on automotive combustion systems is directed at modulating these flows to give the best compromise performance characteristics.

5.1. Swirl

Many workers have used swirl to enhance the combustion process in various ways. The term 'swirl' is used to describe a more or less solid-body rotation of the contents of the cylinder about its centreline. The results from an experiment to investigate its effects are shown in Fig. 14, where improvements were noted in stability and fuel consumption up to a swirl ratio of 2. This measure of swirl, obtained from a steady flow rig test, represents the ratio between the expected solid-body angular velocity of the charge about the cylinder axis and the crankshaft angular velocity. This measure originates from the diesel engine area, where it is commonly used in the optimization of direct injection combustion systems. Here, good correlations between swirl and combustion parameters can be demonstrated for a wide variety of engines. However, in the gasoline engine no similar correlation exists. A recent study,[12] carried out on a single-cylinder engine where a number of swirl ratios were generated by different means, showed no correlation between swirl ratio and rate of burning (Fig. 15) although this combustion characteristic is the one that would be expected to be linked with swirl if it were an important parameter.

Thus while swirl can certainly influence the combustion process, this influence is more indirect than direct. The swirling motion is attenuated during the compression process by friction with the cylinder walls, and also by interaction with the surfaces of the combustion chamber and the piston crown. These interactions affect the rate of swirl decay and also the structure of the resulting flow field.

It is the characteristics of the final flow field, at the time of ignition or shortly thereafter, which prescribe the way in which the flame will subsequently propagate.

5.2. Tumble

Much has been made recently of a form of bulk air motion usually called tumble. This is similar to swirl but the axis of rotation is at right

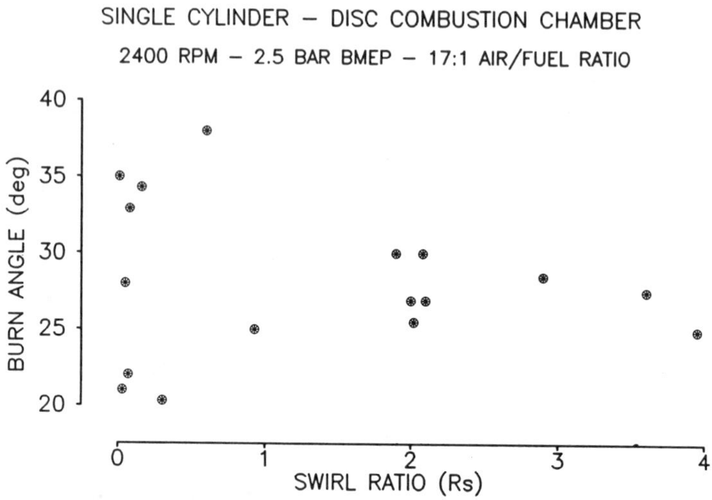

Fig. 15. Burn angle versus swirl ratio.

angles to the cylinder axis. The magnitude of its motion can be characterized in much the same terms as swirl, and this characteristic can be measured approximately using a suitable rig. Similar claims as to its beneficial effects have been made as in the case of swirl.[13] This particular work concentrated on its use in a four-valve per cylinder engine, but the basic phenomenon can be equally used in more conventional two-valve per cylinder engines. The same conclusion that was reached about swirl may well also prove to be true of tumble, i.e. that there is no direct correlation between a tumble ratio and a combustion parameter such as burn rate.

There is, however, one fundamental difference between swirl and tumble. Many measurements, both by photography and LDA means, have shown that swirl is not completely attenuated by the compression and combustion process but survives into the exhaust stroke.[9] Thus the proportion of the swirl energy that is converted into smaller scale turbulence around top dead centre (TDC) varies from one engine design to another. However, in the case of tumble, the evidence collected so far suggests that this type of motion is almost completely destroyed at TDC.[9,12] The reason is the more or less obvious geometric one that at TDC the minor diameter of the flow pattern is only approximately 14% of the major diameter. In the swirling case the two diameters remain constant irrespective of crank angle.

6. COMBUSTION CHAMBER DESIGN EFFECTS

An examination of engines used in current production cars shows that a wide variety of combustion chamber designs are used. The geometric changes that these designs embody can provide additional characteristics which can benefit the performance of the engine. These possibilities can be illustrated by considering three widely differing combustion chamber types, viz. the bathtub, the four-valve and the high ratio compact chamber (HRCC). The design features of each of these are shown in Fig. 16.

The bathtub is very commonly used, although often with minor variations, because it produces a design which has a low manufacturing cost. The four-valve design allows a significant increase in valve area for any given bore size, and thus higher power output, and is now used in many saloon cars as well as in engines for racing cars. The HRCC shape was developed to maximize fuel economy with high-octane gasoline. A design of this type reached production in the form of the May Fireball chamber in the Jaguar V12 engine.

A comparison of the full-load performance of these designs is shown in Fig. 17. Results from multi-cylinder engines would not necessarily show the same comparison as the results in Fig. 17 since intake tuning is often used to enhance the torque to improve driving characteristics. However, the figure does illustrate some of the characteristics of these combustion chamber designs. The superiority of the four-valve engine at high speed is clearly shown, but this necessarily means that the speed at which peak torque is generated is also increased and this can have an adverse effect on subjective driveability. The impression given is that low-speed torque is poor, whereas in fact the levels are similar to the other two designs. It is also apparent that the brake specific air consumption (BSAC) of the four-valve engine is also better than the other two types at high speed, and this acts in addition to the higher volumetric efficiency.

Of the other two types, the HRCC engine tends to produce higher *BMEP* levels because of its lower *BSAC*, which results from the higher compression ratio used. This design resulted from a systematic study of the possible design variables with the aim of maximizing fuel economy on high-quality fuels ($RON > 97$). The design shown has short flame travel lengths from the spark plug to the edge of the combustion chamber, and the air motion within the small chamber is very turbulent due to the flows generated by the large squish area. The

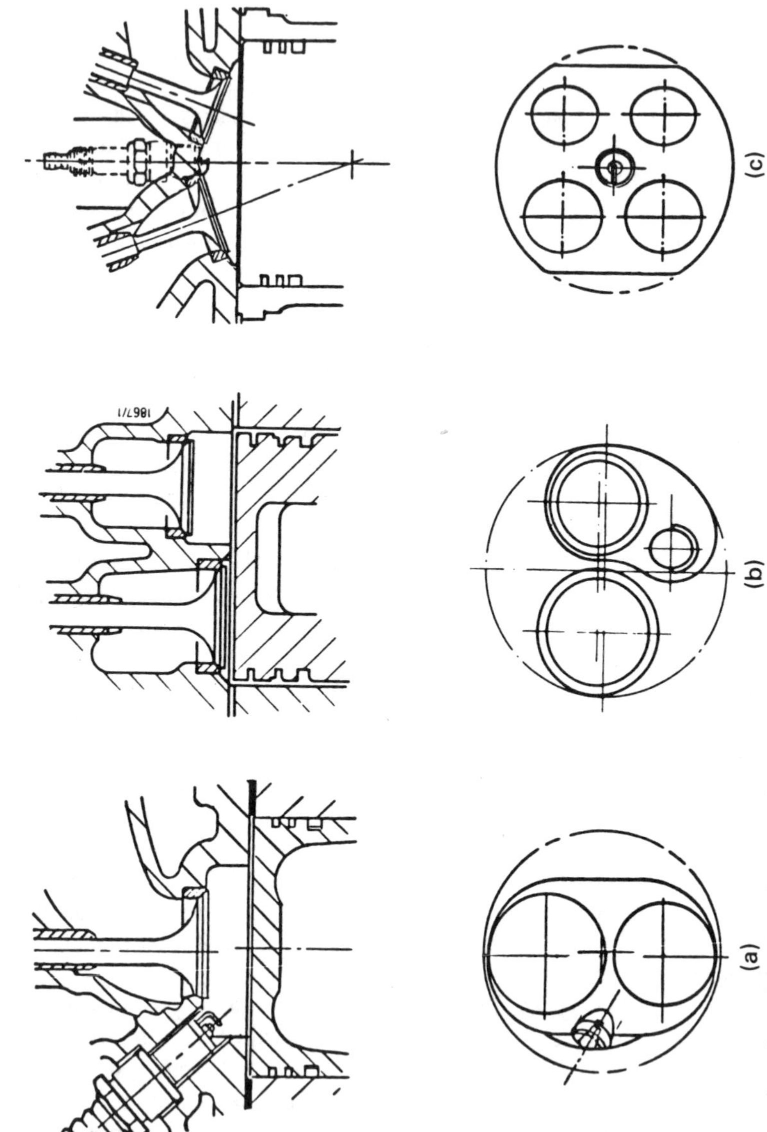

Fig. 16. Combustion chamber designs. (a) Bathtub combustion chamber. (b) HRCC combustion chamber. (c) Four-valve pent roof combustion chamber.

high turbulence produces very fast rates of burning of the charge and this, together with the short path lengths, tends to reduce the tendency for detonation to occur and allows high compression ratios to be used. This high compression ratio gives the good lean-running ability and fuel economy at part load which are the main objectives of the design.

A comparison of part-load performance is shown in Fig. 18. The improved lean running of the HRCC is immediately apparent. It also gives good fuel economy under these conditions but, unfortunately, the hydrocarbon levels are high and this increases the difficulty in meeting exhaust emission limits. The high HC levels come from the

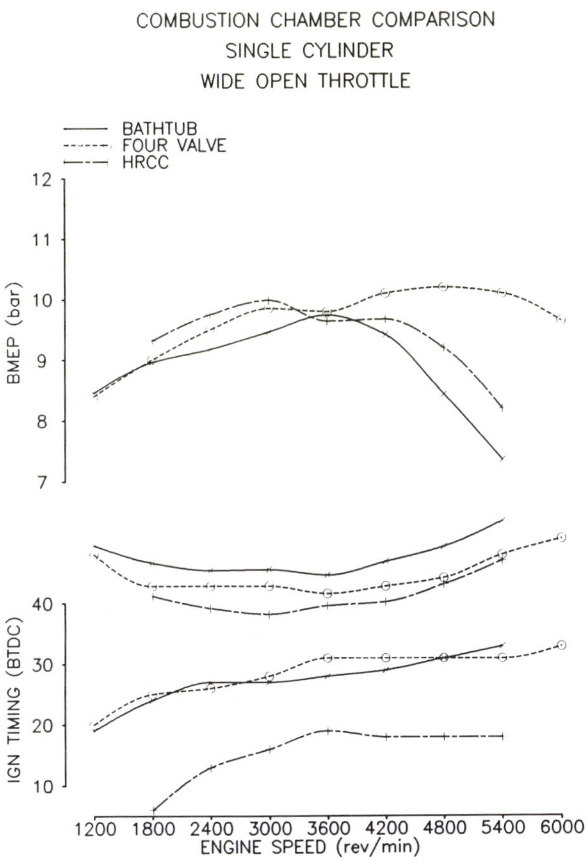

Fig. 17. Full-load performance.

Fig. 18. Part-load performance.

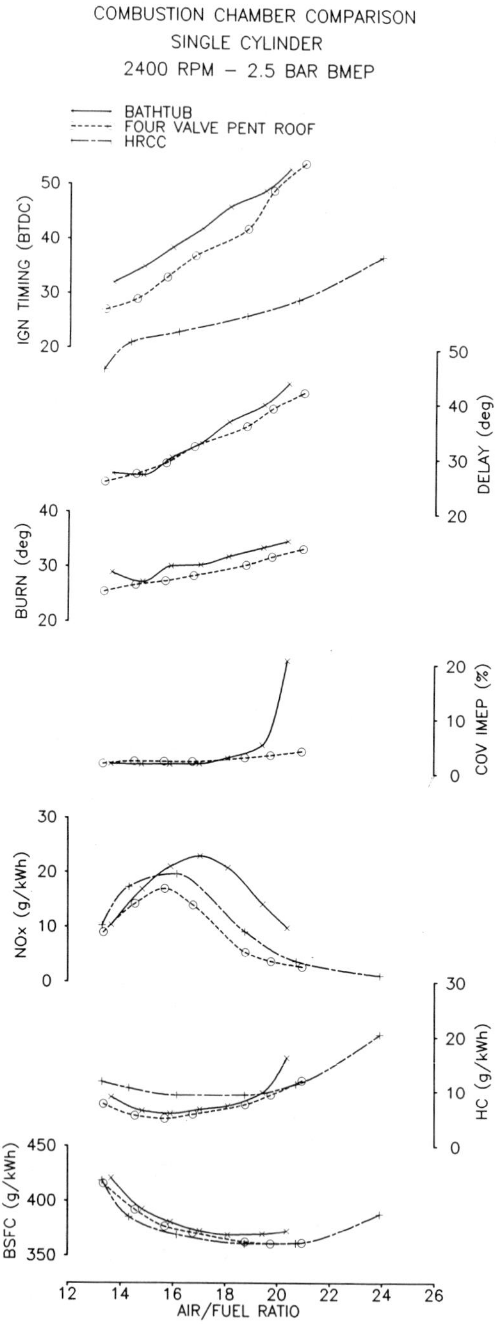

fact that the higher pressures generated by the higher compression ratio trap a greater mass of charge in the crevices in the combustion chamber, and some of this trapped charge fails to oxidize later in the cycle due to the lower exhaust temperature. This lower temperature is also a direct result of the high compression ratio used.

The four-valve design has also been developed with a tumbling air motion regime to have good lean-running characteristics, and its burn angle is shorter than that of the bathtub engine. An engine incorporating this design is shown in Fig. 19. Another contributing factor is that the spark plug is almost centrally located in the four-valve engine, which shortens the longest flame path length and thus tends to shorten the burn time.

It is clear from the ignition timing characteristics of the HRCC engine that, although burn rate data for this engine was not available, its combustion process is more rapid than in either of the other two

Fig. 19. The Austin Rover K16 engine—1·4 litre, 16 valve, twin overhead camshaft.

engines and this results in the ability to run at very lean mixtures. It is also apparent that the NO_x levels produced by the four-valve and HRCC designs are lower than the traditional bathtub, and this feature is an advantage in meeting low emissions limits.

7. CHARACTERIZATION OF IN-CYLINDER AIR MOTION

The sections above have described two flow regimes, swirl and tumble, both of which can have beneficial effects on combustion but where the measured values of parameters measured on rigs to characterize the major features of the flow do not necessarily correlate with combustion parameters.

The question then remains as to which characteristics of the in-cylinder flow field, as opposed to the geometry of the combustion chamber itself, directly influence the burning process. Many workers are tackling this problem at present and thus any answer must at this stage be speculative. However, previous experience and logic may be used to speculate as to which parameters are important.

A logical first choice would be the mean velocity magnitude and direction at the location of the spark plug electrodes. It is clearly important to avoid quenching of the flame in the early stages and thus a mean velocity towards the centre of the chamber should be beneficial. Also, a higher velocity should help to increase the rate of burning. Limited evidence exists to support both of these links.[14,15]

A number of combustion models have been created to predict the characteristics of the burning process. Most of these use characteristics of the turbulence structure as inputs to the model. Much evidence[16] supports the connection between these parameters and the way in which the charge burns. The two main parameters are the turbulence intensity (excluding the effects of cyclic variability) and a turbulence length scale. The most recent developments in measurement techniques now mean that these parameters can be measured in realistic engine geometries and at realistic speeds under motoring conditions. Many workers have measured mean flow vectors using either hot-wire or LDA techniques. In the most recent developments turbulence length scales are being measured. Perhaps the most advanced is that described in Refs 8 and 9, in which full account is also taken of the effects of cycle-to-cycle variability. The latter work, interestingly, seems to indicate that a single length scale may be sufficient to characterize the turbulent structure rather than the microscale and integral scale more commonly used. These techniques, however, are relatively new and thus little data yet exists to show whether these four parameters—mean velocity at the plug, its direction, the turbulence intensity and its scale—will be sufficient to correlate a flow field with the resulting combustion process.

REFERENCES

1. French, C. C. J., A universal test engine for combustion research. SAE paper 830453, 1983.
2. Lancaster, D. R., Krieger, R. B. & Lienesch, J. H., Measurement and analysis of engine pressure data. SAE paper 750026, 1975.
3. Rassweiler, G. H. & Withrow, L., Motion pictures of engine flames correlated with pressure cards. *Trans. SAE,* **42**(5) (1938) 185–204.
4. Muller, M. & Oblander, K., Analysis of the combustion process in gasoline engines with the aid of high speed flame photography and indicator diagrams. *MTZ,* October 1975, 267–74.
5. Witze, P. O., The effect of spark location on combustion in a variable swirl engine. SAE paper 820044, 1982.
6. Namazian, M., Hansen, S., Lyford-Pike, E., Sanchez Barsse, J., Heywood, J. & Rife, J., Schlieren visualization of the flow and density fields in the cylinder of a spark-ignition engine. SAE paper 800044, 1980.
7. Glover, A. R., Towards bias-free estimates of turbulence in engines. Proceedings of the Third International Symposium on the Application of Laser Doppler Anemometry to Fluid Dynamics, Lisbon, July 1986.
8. Glover, A. R., Hundleby, G. E. & Hadded, O., The development of a scanning LDA technique for the measurement of turbulence in engines. SAE paper 880378, 1988.
9. Glover, A. R., Hundleby, G. E. & Hadded, O., An investigation into turbulence in engines using scanning LDA. SAE paper 880379, 1988.
10. Witze, P. O. & Mendes-Lopes, J. M. C., Direct measurement of the turbulent burning velocity in a homogeneous charge engine. SAE paper 861531, 1986.
11. Quader, A. A., What limits lean operation in spark ignition engines—flame initiation or propagation? SAE paper 760760, 1976.
12. Kyriakides, S. C. & Glover, A. R., A study of the correlation between in-cylinder air motion and combustion in gasoline engines. Instn. Mech. Engrs. paper C55/88, 1988.
13. Benjamin, S. F., The development of the GTL 'barrel swirl' combustion system with application to four valve spark ignition engines. Instn. Mech. Engrs. paper C54/88, 1988.
14. Swords, M. D., Kalghatgi, G. T. & Watts, A. J., An experimental study of ignition and flame development in a spark ignition engine. SAE paper 821220, 1982.

15. Keck, J. C., Heywood, J. B. & Noske, G. Early flame development and burning rates in spark ignition engines and their cyclic variability. SAE paper 870164, 1987.
16. Tabaczynski, R. J., Turbulent flame propagation and detonation in spark ignition engines. Proceedings of the 19th International Symposium on Heat and Mass Transfer in Gasoline and Diesel Engines, Dubrovnik, Yugoslavia, 24–28 August 1987.

2

Applied Research into Combustion in Small Diesel Engines

H. K. Oetting

Volkswagen AG, Wolfsburg, FRG

1. Introduction
2. The influence of design
   2.1 Capacity and stroke: bore ratio
   2.2. IDI and DI processes
3. Swirl generation
4. Ignition delay
   4.1. Influence of fuel
   4.2. Thermodynamics of ignition delay
   4.3. Ignition delay and engine speed
   4.4. Influence of droplet size and swirl
   4.5. Ignition delay and engine starting
   4.6. The glow plug influence
   4.7. The influence of exhaust gas recirculation (EGR)
5. Main combustion
   5.1. Particle emission
   5.2. NO_x emission
   5.3. Heat transfer to the walls

1. INTRODUCTION

In a spark-ignition (SI) engine, the mixture formation is performed partly in the intake manifold and partly in the cylinder. At worst, a period of 360° crankshaft is available. In a diesel engine, on the other hand, all fuel injection, mixture formation, ignition delay and combustion must be commenced and completed within only 30–50° crankshaft. Thus it can be seen that the diesel engine can be primarily understood in terms of the special features of its mixture formation and its ignition delay, which takes from 2 to 20° CA from that time.

Mixture formation is performed both by the injectors (dealt with in Chapter 7), which are of primary importance for spray pattern formation, and by those measures which create a high level of air turbulence at the instant of mixture formation in the combustion chamber of a diesel engine (see Chapter 8), or which at least provide for a high relative speed between fuel and air. This high degree of fuel atomization and mixing—unlike the case of an SI engine—is indispensable in a diesel engine.

2. THE INFLUENCE OF DESIGN

2.1. Capacity and Stroke: Bore Ratio

Unlike the SI engine, the diesel engine is highly sensitive with regard to size. The greater the capacity, the easier it is to realize a high mean effective pressure (m.e.p.) for a given expenditure. On the other hand, the oscillating masses and the long stroke of the high-capacity diesel engine place an upper limit on its useful speed range. Nevertheless, a large capacity is helpful because with favourable surface: volume ratios (for a given compression and stroke: bore ratio) improved air utilization can be achieved. Additionally, a high level of air utilization is not necessary if there is sufficient capacity. This is a major step towards limiting the emission of soot.

For a given capacity in a diesel engine, the largest possible stroke should be striven for in order to produce high m.e.p. This is usually limited by the engine height, but also by the piston speed. A long stroke is even more desirable the smaller the capacity, because, as will be shown for example by the ignition delay (see pp. 46 and 47), it requires a particularly high compression ratio. However, if such compression ratios are to be provided with an acceptable ratio of combustion chamber surface to volume and with the smallest possible residual spaces for fresh air, which only participates in combustion to a limited degree, the greatest possible stroke is required. These relationships are shown in Fig. 1 by the example of an idealized combustion chamber of a direct injection diesel engine (DI). As one will take all the parameters in Fig. 1 to the limits of that which is technically feasible in order to keep the residual spaces small, the need

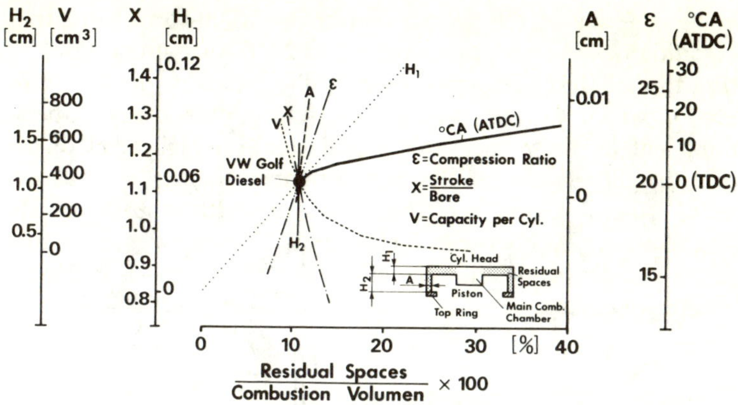

Fig. 1. Geometrical constraints on air utilization.

to have a long stroke can therefore only be avoided by the highly problematical abandonment of a high compression ratio (see p. 45).

A modern passenger vehicle diesel engine—generally designed in such a way that at least a large number of parts can be taken over from the corresponding SI engine, or vice versa—is limited in its realizable stroke by the conrod length and the cylinder block height, because of the crank pin diameter and compression height which must be greater than is the case for an SI engine, indeed generally considerably so. The heavier demands made on the cylinder head gasket also mean that the same bore dimension cannot normally be realized. This is, however, favourable for the longer stroke : bore ratio of a diesel engine.

The small diesel engine is therefore subject to severe capacity restrictions. This means that the residual spaces indicated in Fig. 1 are relatively large, with the effect that the air in main volume V_K must really be used almost completely in order that a tolerable m.e.p. may be achieved. This is only possible if the greatest attention is devoted to the generation of air turbulence and of a high relative speed between fuel and air (see p. 38).

2.2. IDI and DI Processes

In modern small vehicle diesel engines, four different combustion processes are currently being examined. Only two of these have gone

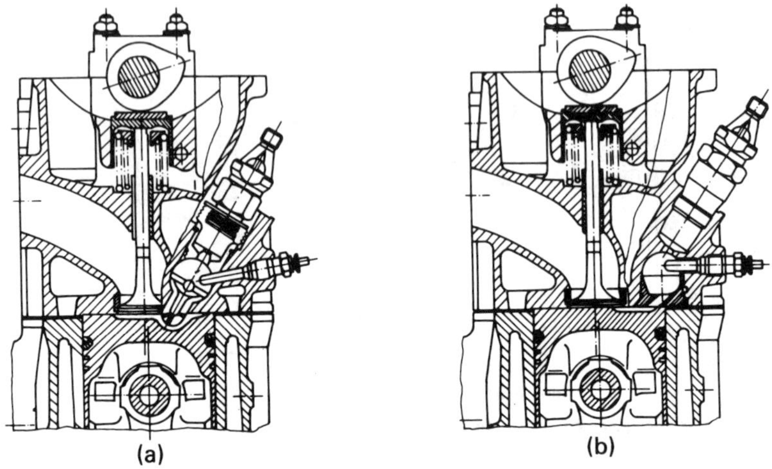

Fig. 2. IDI diesels. (a) IDI precombustion chamber. (b) IDI swirl chamber.

into large series production, the other two are about to be presented or have just gone into small series production. The processes used in production are the swirl chamber and the precombustion chamber processes (Figs 2(a) and (b)), that is to say those which have subdivided combustion chambers. They are also known as indirect injection (IDI) processes because they introduce the fuel into a part of the combustion chamber which is separated from the piston crown, on which the burning and expanding gas is to exert pressure, by a throat.

Those processes which are about to be or have been presented do not have subdivided combustion chambers. They are also known as DI processes, because they introduce the fuel directly into the chamber located in the piston crown (Figs. 3(a) and (b)). A distinction is made between the two DI processes depending on the way in which the fuel is distributed in the air. The more important of the two, which has already reached widespread use in truck and large diesel engines, employs a system which divides the fuel quantity injected upon each power stroke into four or more sprays, in order to cover the air in the combustion chamber as evenly as possible. The other (wall-wetting or MAN 'M' System) employs a spray which should not reach the combustion chamber wall at low load ranges if possible, while, at high loads, it is applied to a wall in a thin film, as evenly as possible, where it is then vaporized by the passing air. This wall is normally the wall of the main chamber inside the piston.

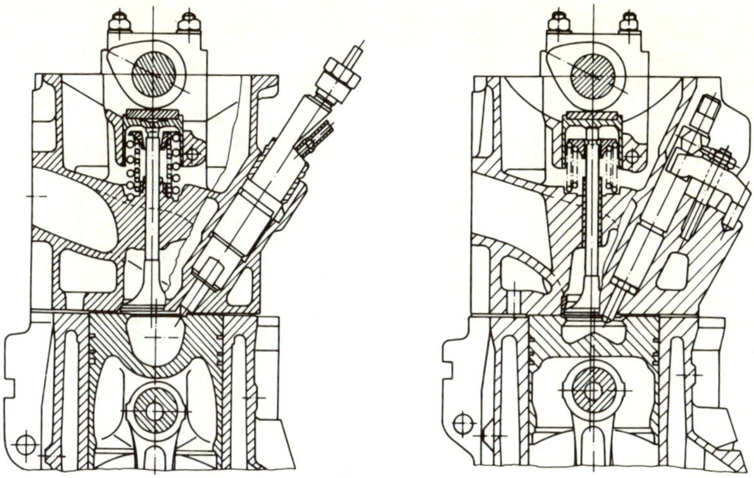

Fig. 3. DI diesels. (a) single-spray DI. (b) multi-spray DI.

Experience shows that the IDI processes are easier to develop, in particular for the purpose of the small passenger vehicle engine. Thus they are still favoured, although there is no doubt that the DI processes can achieve a 10–20% more favourable minimum fuel consumption for this size of engine also (see p. 42).

The IDI processes are easier to develop, particularly for small engines, because they generate the air movement necessary for diesel engine combustion in a very short period immediately before and during the fuel injection. The DI processes, on the other hand, are to a large extent dependent upon the air movement which arises as air flows into the cylinder (approx. 360 to 180° crankshaft), i.e. a long time before injection of the fuel has taken place.

3. SWIRL GENERATION

The typically calculated speed profile as it arises in the throat of an IDI swirl chamber is shown in Fig. 4.[1] Its diameter is determined as 32% of the engine bore. The throat is arranged so that a directed twisting flow (swirl) arises in the usually spherical chamber. The spray direction is normally tangential to the swirl.

Fig. 4. Throat velocities in IDI diesels.

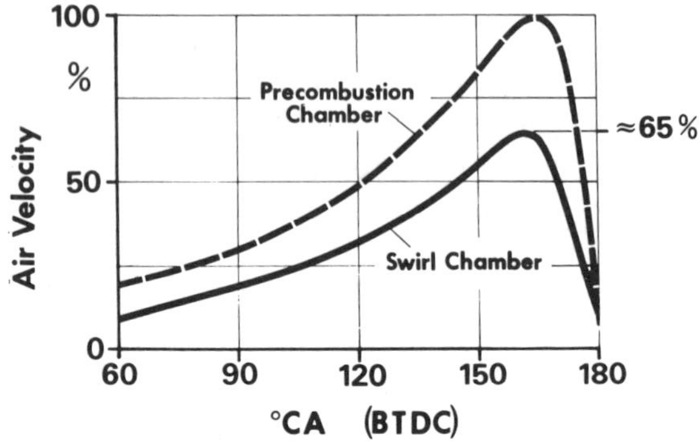

The typically calculated speed profile as it arises in the several throats of the precombustion chamber is also shown in Fig. 4. Here the flow speeds are greater than in the IDI swirl chamber. The heat losses are also correspondingly greater. Speeds up to 60 $m\,s^{-1}$ can be expected. They manifest themselves inside the chamber as almost uncontrolled high turbulence into which the fuel spray is introduced, and is quickly broken up and distributed. In Fig. 5[1] it is shown how the swirl energy develops in a precombustion chamber depending on the engine speed. The reduction in the maximum swirl energy at speeds above 3000 rpm can be put down to an overcritical pressure ratio between the main and precombustion chambers.

It is true of all modern inlet ports, regardless of whether they are

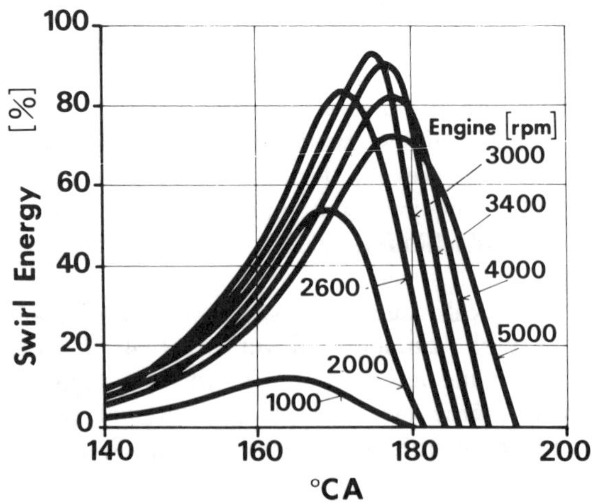

Fig. 5. Swirl energy in an IDI diesel.

used in a spark-ignition or a diesel engine, that they are designed for optimum air throughput. For peak performance, this requires a cross-sectional profile with the narrowest part of the passage located close to the entrance of the cylinder head and gradually expanding towards the inlet valve. For high-torque engines the valve cross-sectional area is normally smallest at its outlet from the cylinder head.

The cross-sectional profiles indicated above could, in principle, also be realized with helical ports, but flow separation could be expected in all probability because of the complicated shape resulting from the production tolerances. Therefore, in order to avoid these flow separations, helical ports are preferred for use with a cross-sectional profile which tapers in to the valve cross-section.

All facilities for the generation of rotational air movement in the cylinder have a throttling effect, so that they involve charge losses; the more so, the higher the engine speed. In order that additional cross-sectional constrictions are minimized, solutions are now generally employed in which the swirl is produced by the cross-sectional profile of the inlet port alone, the size of the port cross-sections being almost unchanged. Typical types of such modern helical ports are shown in Fig. 6.

Measurements of air flow and swirl ratio produced with a helical port are generally performed with steady state throughput for different inlet valve lifts using an anemometer or, increasingly, with the rectifier method. The rectifier method is shown in Fig. 7(a). The throughput and swirl values of a modern swirl port are compared (Fig. 7(b)) with those of a standard port that is a port without any intentional swirl effect, for an IDI swirl chamber engine.[2] In spite of the swirl rate, which approximately corresponds to a circumferential air speed of more than 0·6 times the engine speed, it was possible to reach very nearly the throughput characteristics of the swirl-free port.

For the swirl measurements it is normally assumed that all secondary flows die out during the compression stroke leaving the main swirl arising from the inflow process in the cylinder. For the swirl profile which arises at the point when injection commences, only the main swirl is significant which is measured by the anemometer speed or by the torque at the rectifier. Although this is certainly a valid theory, the question still remains open as to the degree of swirl flow which can be expected at the beginning of injection once the inlet-induced swirl has survived the wall friction during the compression stroke and, additionally, the squish process, in which, near TDC, the charge in the main combustion chamber is highly compressed.

Fig. 6. Helical ports (casting cores).

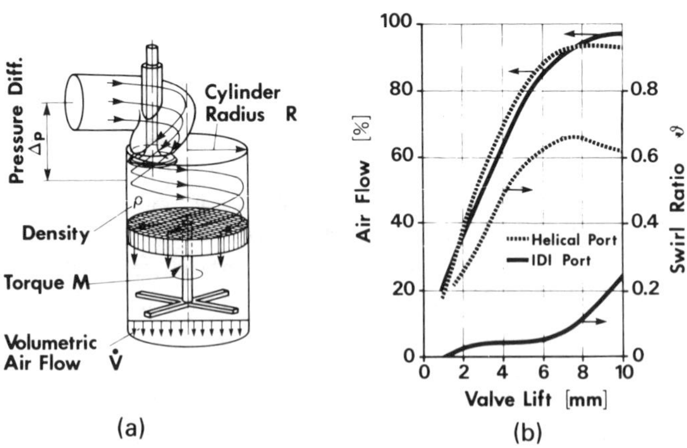

Fig. 7. Swirl measurement by rectifier method.

The squish process has a threefold effect: it increases the turbulence in the charge; it disturbs the orderly gas movement but also increases the swirl speed because it has compressed the swirl, which had formerly expanded to the diameter of the cylinder, to the smaller diameter of the main combustion chamber. The maximum swirl energy levels versus speed for a DI engine are shown in Fig. 8, those for an IDI swirl chamber engine also being given for purposes of comparison. The difference is almost two orders of magnitude.

This is one of the reasons why the IDI chamber engine must perform much more work in order to produce the charge movement. Its pumping work is of the same order of magnitude as its losses incurred by the increased heat exchange between gas and walls in comparison to the DI engine. The higher heat losses are caused by

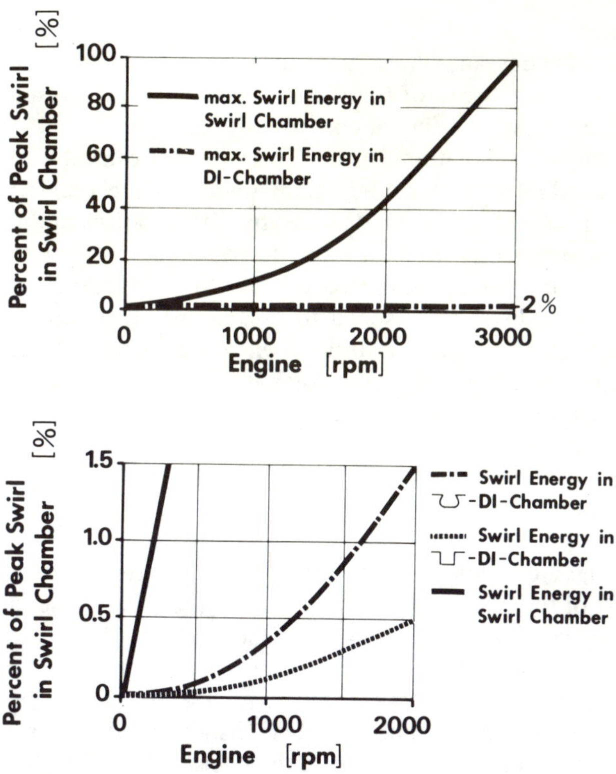

Fig. 8. Swirl energies in comparison.

Fig. 9. Swirl character.

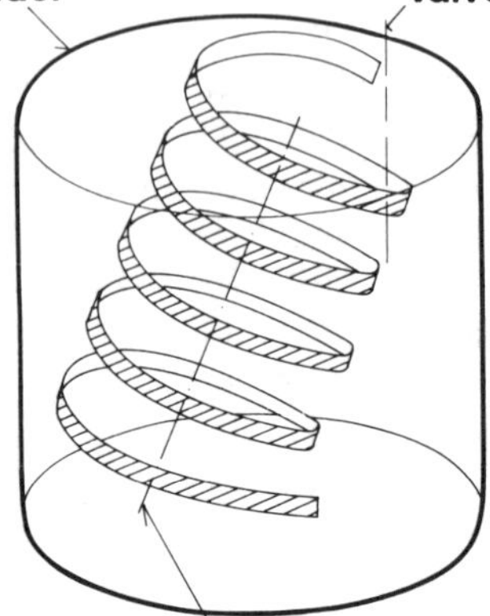

those higher gas velocities and the larger surfaces of its walls. The residual part of the higher losses of the IDI chamber engine is caused by the fact that there is less burning at constant volume.

From more recent 3-D simulations of the gas movement caused by a helical port in the cylinder[3] it is evident that, under all the realistic inlet conditions, a flow field at bottom dead centre (BDC) arises in the cylinder, which is sufficiently exact to be described as solid-body rotation (Fig. 9). For the compression stroke it was found that even at a swirl ratio of 2:1 (ratio of swirl speed to engine speed) it can be assumed that the circumferential constituent will predominate at the time at which injection commences, and that it is up to four times larger than the radial or axial constituent.

4. IGNITION DELAY

In practice, the ignition delay is the time between the start of injection, the start of the nozzle needle stroke and the first increase in pressure caused by combustion; that is the first deviation from the polytropic compression. Next to the injection system and the air movement in the combustion chamber, this manifestation of ignition delay has the most lasting influence on combustion in the diesel engine. In particular, ignition delay is one of the decisive factors in the complex starting process.

4.1. Influence of Fuel

First of all, ignition delay is a characteristic of the fuel. In Fig. 10[4] it is shown with vaporized fuel, depending upon the temperature and cetane number for five reference fuel blends of *N*-cetane and HMN (heptamethylnonane); that is to say, those fuels whose mixtures are used to determine the cetane number of any fuel. It can be seen how the ignition delay increases with decreasing temperature and falling cetane number, but also that, for vaporized fuel with normal cetane numbers and compression temperatures, the ignition delay is only slightly influenced by temperature.

Pressure profiles and the needle stroke for a DI engine with two fuels of differing cetane numbers are shown in Fig. 11.[5] The larger ignition delay (pressure increase is noted later for same injection process) produces a sharper increase in pressure and a higher peak pressure. Therefore, in order to keep the load on the engine (peak combustion pressure) and the engine noise (the first and second derivatives of the pressure profile are a measure of this in the first approximation) low, the smallest possible ignition delay must be striven for, particularly after cold starting of IDI engines and in the acceleration and full throttle ranges of DI engines. This can be achieved by selecting fuels with a high cetane number, which is, however, particularly in the case of alternative fuels, not always a guarantee of good ignition behaviour.[5]

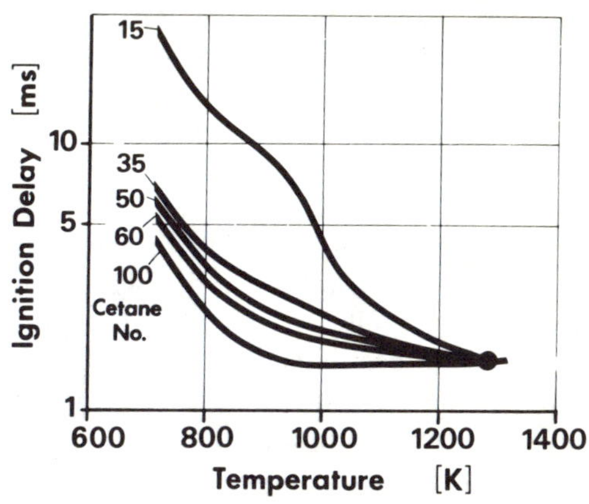

Fig. 10. Influences on ignition delay.

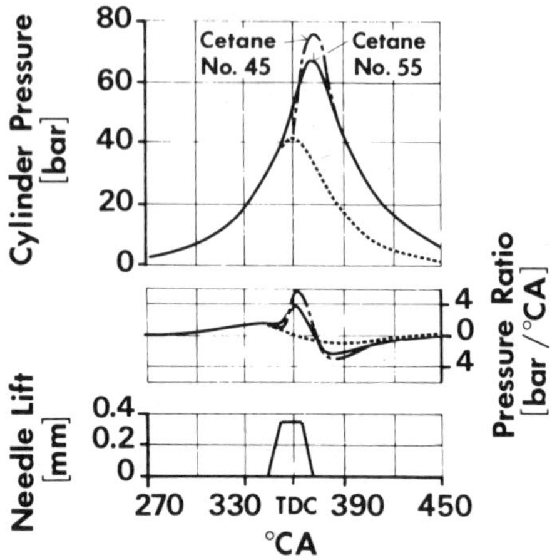

Fig. 11. Effects of ignition delay.

4.2. Thermodynamics of Ignition Delay

During the ignition process, mixture preparation processes take place, which can be summarized as follows:

(a) physical processes (atomization, vaporization and mixing, until a mixture capable of ignition at individual points is formed);
(b) chemical processes (preliminary reactions in the ignitable mixture and finally self-ignition).[6]

These processes may partly overlap in time and take place in all zones. The time of ignition and thus the ignition delay is determined by the mixture zone in which the mixture preparation has been initially completed. The later this point is, that is to say the longer the ignition delay, the larger the charge which is prepared for combustion and the larger the increase in pressure and peak pressure which it causes.

As the reactions become increasingly sluggish with an increasingly rich mixture, which can also be partly explained by the more pronounced mixture cooling through vaporization of the larger fuel quantities, the air ratio λ at the place of ignition is generally above 0·7. Because the ignition delay is dependent upon this local air ratio, the overall ratio has practically no effect on the ignition delay as it could only influence it via changes in the local mixture process.

As the physical and the chemical ignition delay times are dependent upon the air temperature T_2 and the pressure p_2 in the combustion chamber at the beginning of injection, this also applies for the ignition delay t_z itself.

The relevant quantifiable equation is as follows[7]:

$$t_z = C_1 \beta \frac{e^{E/T_2}}{p_2^{C_2}} \sqrt{T_2}$$

where C_1 and C_2 are fuel constants, E is the activation energy, and $\beta = f(T_2, E)$.

As $T_2 = T_1 \varepsilon^{n-1}$ and $p_2 = p_1 \varepsilon^n$, where ε is the compression ratio, p_1 and T_1 are the initial conditions, and n is the polytropic exponent, the pressure and temperature at start of compression, the compression ratio and the polytropic exponent are immediately seen to be further influencing factors.

The supercharging of diesel engines, which produces higher pressures and temperatures at the start of compression, thus enables a considerable reduction in the ignition delay which, in turn, produces advantages with regard to noise emission (less fuel present in the combustion chamber at the start of combustion). This is of special importance for DI engines because they suffer noise problems under acceleration and full load where turbochargers are in action, while IDI engines, which suffer noise problems during idle after cold starts, benefit less.

For its part, the polytropic exponent is dependent upon the engine speed and the surface:volume ratio of the combustion chamber, as shown in Fig. 12.[5] As Fig. 12 also shows the influence of typical capacities of smaller to medium DI engines, it can be seen that, in

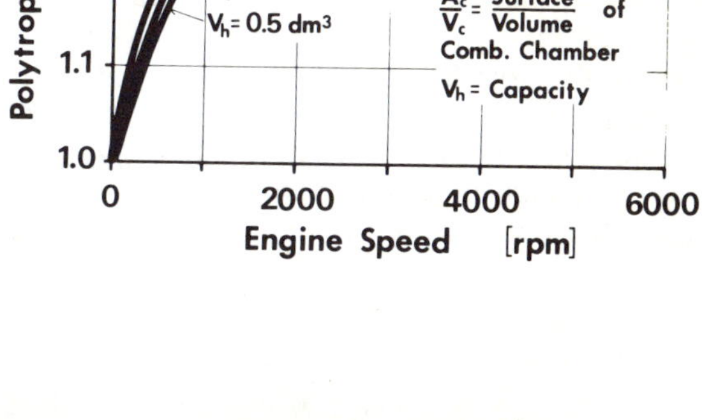

Fig. 12. Polytropic exponent.

accordance with the A_c/V_c ratio, the ignition delay problem increases with smaller volumes of cylinder (see also p. 34).

The polytropic exponent also takes the blow-by at the piston into account. If this increases, the polytropic exponent decreases and, consequently, the pressure and temperature at the point of injection also drop. This is the main reason why older engines are more difficult to start and are often louder after cold-starting.

4.3. Ignition Delay and Engine Speed

The minimization of the ignition delay is also one of the most important preconditions for the operation of diesel engines over a wider speed range. Only a sufficiently short ignition delay guarantees the acceleration of the combustion processes necessary at high speeds. From this it can be deduced that high-speed engines require a higher compression ratio than slow-running ones (see p. 34). Fortunately, on the one hand the polytropic exponent describing the heat and air mass losses occurring during compression is, as Fig. 12 shows, dependent upon the engine speed in such a way that it achieves higher levels with increasing speed as the time-dependent heat and mass losses decrease, although the heat transfer between the gas and the wall increases through greater combustion chamber turbulence. On the other hand, the higher turbulence in the combustion chamber also leads to a clear reduction in the ignition delay.

4.4. Influence of Droplet Size and Swirl

In addition to the gas temperature and pressure at the time of fuel injection, the size of the droplets (dependent upon injection pressure and nozzle-hole geometry) and the relative speed between air and droplets have a great influence on ignition delay. A schematic description of the effects of the air speed and droplet-size parameters is given in Fig. 13.[5]

For the purposes of simplification, a chemically uniform fuel (*N*-heptane) was used and piston movement was not considered. The temperatures of the compressed air of the fuel vapour and of the droplets are given on the *y* co-ordinate, the abscissa being the time axis. After one part of the physical ignition delay t_M has taken place,

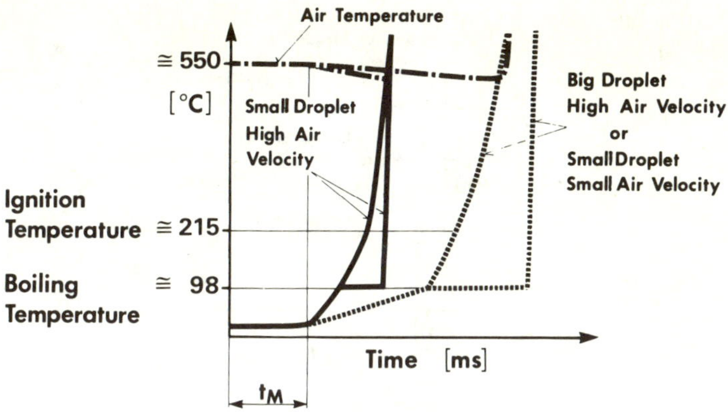

Fig. 13. Droplet size and air velocity effects.

during which the fuel temperature can be assumed constant as a first approximation because the spray disperses during this time; subsequently the temperature of the droplets climbs rapidly to boiling point with simultaneous cooling of the compressed combustion air. The vaporized fuel quickly heats up to ignition temperature in the warmer surrounding air and this, in turn, then heats the combustion chamber contents, after ignition has taken place. In the case of practical fuels, which have a boiling point curve (fuels vaporize within a specified temperature range) the conditions are indeed more complicated, but are similar in principle. The differences between the continuous and the broken curves indicate the influence of the droplet size and the relative speed between air and spray. Thus, small droplets and a high relative speed are desirable.

4.5. Ignition Delay and Engine Starting

The ignition delay plays its most important role in the stabilizing phase immediately after the starting of the diesel engine. The cold-starting process for a truck engine is shown in Fig. 14.[8] These curves depict the end of the electric starting process through the declining starting torque. The frictional torque of the engine, the torque resulting from the gas pressure and the resulting engine speed are plotted against time. The entire starting process is run at the full-acceleration pedal

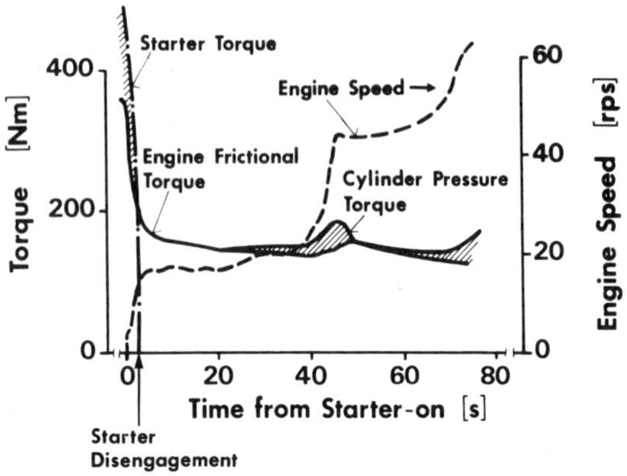

Fig. 14. Engine behaviour during starting.

position, and it can be seen that the engine only attains stable operating conditions after approximately 70 s.

The cause of the two highly typical stagnation phases which are particularly noticeable in the speed profile lies in the ignition delay (Fig. 15).[8] Here, as an example, the first acceleration phase after the cold start is presented with the transition into the first stagnation phase. The picture starts with the first ignition occurring. It is observed in the fourth expansion cycle which takes place in cylinder 4. At 10° CA BTDC, injection starts and the first ignition follows 5° CA later; that is to say, there is an ignition retard of 5° CA, corresponding to approximately 4 ms. The result of this first ignition process is a powerful acceleration of the crankshaft which, as the ignition delay time remains constant at approximately 4 ms, leads to an increase in the ignition delay in °CA to 8° in the next cycle. This continues in this manner until, in the ninth cycle, the ignition only occurs at 15° CA after TDC; that is to say, the ignition delay has reached an extent of 25° CA. However, even this cycle accelerates the crankshaft once again, so that in the tenth cycle the cooling down which occurs as a result of the expansion of the piston overtakes the ignition delay and ignition no longer occurs. This condition continues until the piston speed is so reduced by the losses that the pre-reaction process of the ignition delay is no longer interrupted by the expansion, and ignition once again occurs. However, as this is all happening at an engine speed level, there are only very few ignitions, one or two after each other, causing the piston speed to once again become so great that cut-outs once again occur. That is to say the engine has ceased to accelerate, the speed is almost constant and the first stagnation phase (Fig. 14) has been reached. However, one has to appreciate that these ignition cut-outs are distinctly audible and that the engine is running roughly.

During this stagnation phase the bearings now heat up so that the oil viscosity is reduced to one half, the frictional torque declines and, after a further cut-out phase, there is suddenly such a surplus of torque as ignition recommences that a higher speed level is achieved. As a rule, the stagnation phase is repeated at this level until the combustion chamber walls are sufficiently warm to cause a lasting reduction in the ignition delay so that the engine now receives combusted gas from every operating point without delay.

Essentially, two measures are taken in order to control the ignition delay, namely the selection of a high compression ratio and the installation of a heating device. The choice of the most suitable compression ratio represents a compromise between consumption,

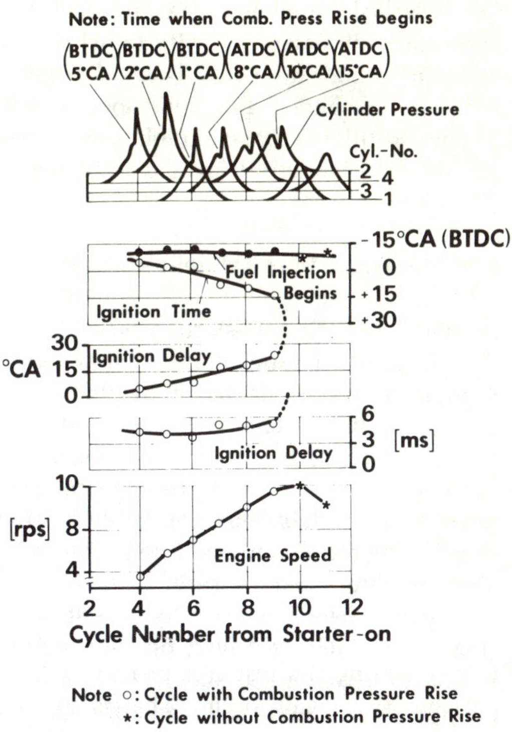

Fig. 15. Ignition delay during engine starting.

Fig. 16. Glow plug.

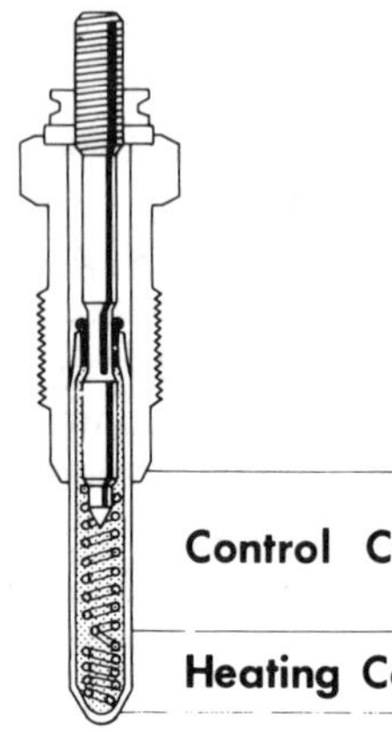

emissions and ignition behaviour, the ignition delay being the most important parameter. Thus, in the first place, a high compression ratio is required for good starting characteristics down to normal passenger motor vehicle minimum starting temperatures of −25°C. This is particularly necessary in order to achieve the pressures and temperatures important for short ignition delay at the commencement of injection. Secondly, the high speeds, which are typical for small engines, with their unfavourable surface: volume ratio conditions also require short ignition delay and thus high compression ratio.

4.6. The Glow Plug Influence

A solution to the cold-starting problem down to −25°C by means of the compression ratio alone is, however, unthinkable. Facilities for heating the intake air and glow plugs are necessary as support. The glow plugs have a dual advantage: firstly, they represent a particularly highly heated surface in the combustion chamber, to which fuel is applied in order to vaporize particularly rapidly, and this quickly provides an ignitable mixture. Secondly, they lend heat to the process even during the compression stage, which leads to a further increase in air temperature and pressure.

Modern glow plugs basically consist of a heating coil in a glow tube (Fig. 16)[9]. After switching on the engine, these glow elements are heated to between 850 and 1200°C. In the most modern fast-heating plugs, a short heating coil made of a Cr–Al–Fe alloy is used, together with a control coil made of Ni. The control coil ensures that the temperature limit of 1200°C is never exceeded. The progress which has been made with these plugs just recently is shown in Fig. 17.[10]

4.7. The Influence of Exhaust Gas Recirculation (EGR)

It has already been suggested several years ago[11] that EGR can be used to reduce the effect of ignition delay. It was assumed that EGR does not influence the ignition delay itself, but suppresses the following reaction so that neither such sharp increases in pressure nor extremely high pressure peaks occur.

Corresponding measurements undertaken on an IDI engine showed that with an EGR rate of, for example, 40% a reduction in noise emissions of about 4 dB(A) occurred when idling, with no measurable worsening in fuel consumption having to be accepted. Indeed, as expected, an improvement in the NO_x emissions was observed.

The fact that this idea has nevertheless not yet gone into mass production lies in its effect on the other emissions, which tend to increase with increasing EGR (Fig. 18),[12] so an extremely sensitive EGR control system—limiting high quantities to the lower load range—is required in order to maintain the noise advantages with unchanged emissions. Only in the future will diesel engines with

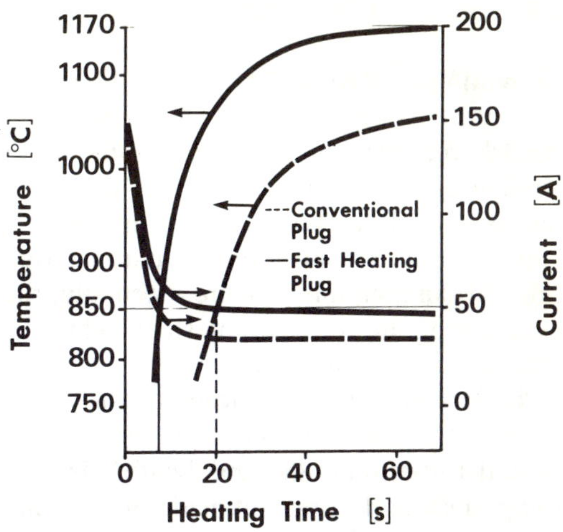

Fig. 17. Heat plug warm-up.

Fig. 18. Emission effects of EGR.

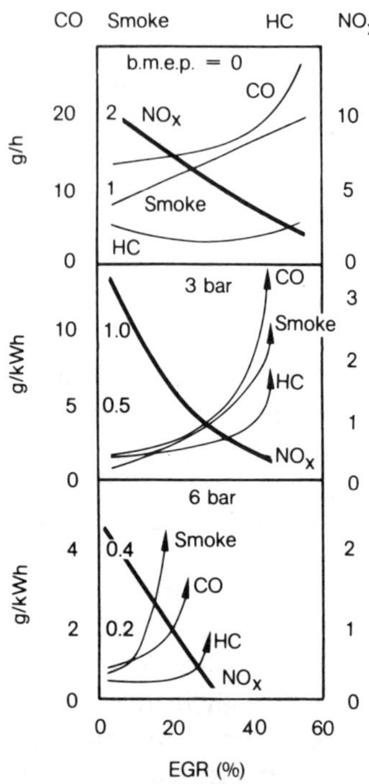

electronic injection control be able to realize this sensitive control system, and consequently the added advantage of noise reduction.

5. MAIN COMBUSTION

Good combustion in a diesel engine is distinguished from bad combustion by a low noise level, as well as low NO_x and particulate emissions. Naturally, in the range of the full throttle curve, there should be as low an air:fuel ratio as possible in order to extract the highest possible torque from a given capacity.

5.1. Particulate Emission

The particulate emission of the diesel engine, which tends to be two to ten times as large as that of spark-ignition engines, arises during combustion through reaction and cracking processes in orders of magnitude of 200 times that which can later be found in the exhaust. That is to say, the majority of the particles originally arising disappear again through oxidation during the course of combustion. Oxidation is all the more likely to occur the higher the oxygen content in the gas. Amongst other things, the particle emission drops with declining load and tends to be lower with supercharged engines.

Sizes of the sampled material vary widely, in the range 0–30 μm, but the large soot particles have been shown to consist of individual spherical units of diameter around 300 Å, typical of gas-phase soot formation.[13] These primary units exhibit an almost log-normal size distribution which varies hardly at all with the engine type or mode of operation.

The particles start to arise in a diesel engine from pyrolysis reactions in the spray cone, which is very fuel-rich but also continue into the mixture formation phase, as the fuel is allowed to go from the fluid to the vapour state under high temperature and pressure conditions. As the beginning of this process is automatically tied to an air:fuel ratio of zero, the particulate formation is initially extremely strong. In order to achieve a low formation rate for particles during, and particularly at the beginning of, the combustion an endeavour must be made to maintain a low fuel temperature in the combustion chamber for as long as possible, and yet also bring the fuel, which vaporizes to an immediate intimate contact with the air to produce as near as possible a stoichiometric mixture and thus cause it to oxidize. In principle, this functions particularly well with wall-wetting injection processes such as are found in IDI chamber engines and in the single-spray DI process. The wall temperature is between 200 and 300°C, which allows practically no pyrolysis processes in pure fuel. The extent to which fuel evaporates from the wall governs the addition of oxygen which is brought about by the extreme turbulence in IDI chamber engines. In principle, the same also applies for the single-spray DI engine which therefore requires an air swirl two to four times higher than in the multiple-spray DI engine.

Such a high air swirl is inappropriate for a multiple-spray DI engine because it would carry oxidizing areas of a spray into the exhaust which has already arisen from a neighbouring spray. As a rule, resulting from the inert gas effect there are premature oxidation cut-offs from the fuel spray and also from the produced particulate products—and consequently an increase in particle emission. The

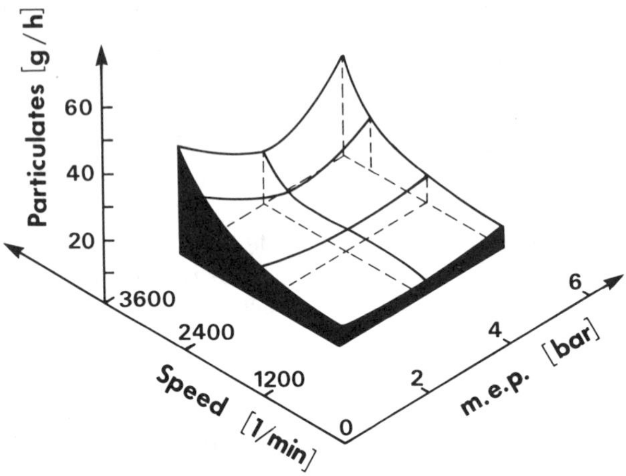

Fig. 19. Particulates in engine map.

multiple-spray DI engine can therefore only be helped by a solution from the injection system. It requires high opening pressures right down to the lowest load fractions, in order to achieve small fuel droplets as quickly as possible. The demands which it makes on the injection system are correspondingly high, while those of the IDI chamber engines and of the single-spray DI engine are correspondingly low.

However, if improved atomization is accompanied by lower spray penetration because of the reduced momentum of the smaller particles, a very rich primary zone can be created which may actually encourage particulate formation. Therefore atomization of the spray has an upper limit, but this is not normally achievable in small diesel engines.

The net exhaust particulate emissions vary in a complex fashion with engine speed and load. This is shown for a four-cylinder passenger car diesel engine in Fig. 19. At low speeds, the long particle combustion time available results in low particulate emissions. At high speeds, the particulate emission is increased as the burnout time is reduced. At these speeds, an increase in load raises chamber temperatures and promotes particle burnout so that the emissions actually decrease. However, above a certain load, the increased formation of particles in the primary zone due to higher fuel concentrations outweighs this temperature effect and the net emissions climb to very high values for heavy loads at high speeds.

5.2. NO_x Emission

Fundamentally, the NO_x emission in a diesel engine follows the same laws as in an SI engine. The Zeldovich mechanism[14] also applies here; that is to say, NO principally arises within the engine, the greater part of it only oxidizing to form NO_2 in the exhaust pipe or when entering the air.

In Fig. 20(a)[15] the calculated NO concentrations, with chemical equilibrium, are plotted over the range of reaction temperatures for the combustion of octane with air at 20 bar. The air : fuel ratio λ is the parameter. The strong influence of temperature on the reaction of atmospheric nitrogen and atmospheric oxygen can be seen from the figure. The effect of the air : fuel ratio λ can also be seen; NO_x still increasing, however, at air : fuel ratios over 1 . 4. This does not correspond to the experience with engines, and evidently can be put down to the equilibrium reaction upon which the illustration is based, that is to say without consideration of the reaction kinetics, and to the thermal pretreatment of the air in the reaction chamber. For this reason, a typical scatter pattern (established with SI engines) of

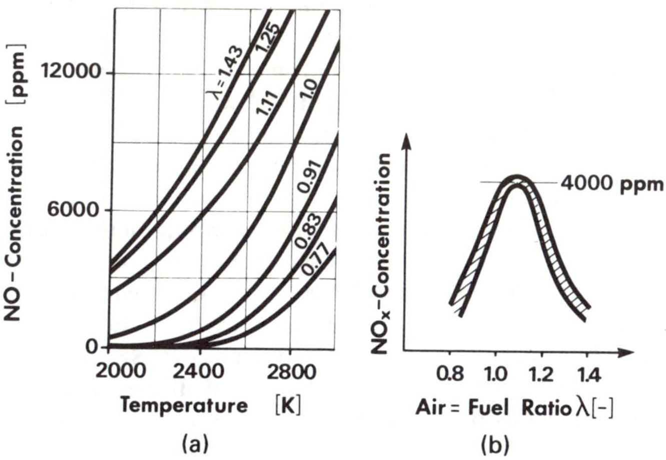

Fig. 20. NO_x emission.

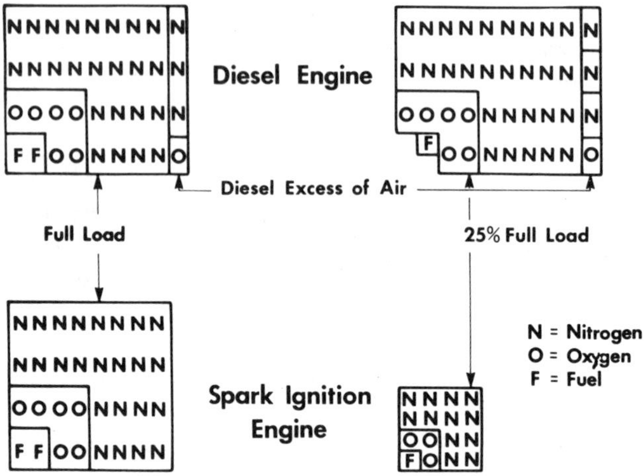

Fig. 21. NO formation conditions.

measured NO_x concentrations over an air:fuel ratio of around 1, is added in Fig. 20(b).

The principal differences between diesel and SI engines as regards NO_x emissions are shown in Fig. 21—the areas represent gas masses. In the figure, it can be assumed that the engines compared offer approximately the same performance at full and part load. For a modern diesel the intake air mass must be approximately 15% larger than for the SI engine.

It can be clearly seen from Fig. 21 that, at full load, there is an approximately equally high probability for SI and diesel engines that a nitrogen and an oxygen molecule are in the vicinity of an oxidizing fuel molecule and are thus brought to the oxidation temperature of N_2. At part load, however, this probability is much lower in the case of the diesel. For the diesel, in the assumed part-load case, there is one quarter of the fuel molecules for the same, if not an even higher, number of nitrogen and oxygen molecules as at full load, whereas this ratio remains roughly the same for the SI engine. That is to say, the probability that a NO molecule arises at part load in a SI engine is approximately six times as great as in a diesel engine. If one considers that in the SI engine at this part-load point only about one-quarter of

† λ = (actual A:F ratio)/(stoichiometric A:F ratio).

the O_2 and N_2 molecules of the diesel engine are utilized, one arrives at the result that the ratio of the performance-rated NO_x emissions of SI and diesel engines at such part-load points must be approximately 1.5.

Only at first sight is it astonishing that this simple declaration comes very close to the conditions observed in practice. The fact of far higher peak pressures in the diesel, resulting from the high compression, is largely compensated for by the diffusion combustion, so that the peak temperatures which occur in fact remain comparable, even at full load. If this were otherwise, different materials would certainly have to be chosen for the combustion chamber walls of the diesel. At part load, the temperatures in the diesel tend to be lower because of the ballasting effect of the gas masses.

It naturally also applies for the diesel engine, as for the SI engine, that exhaust gas recirculation (EGR), or the introduction of water vapour, leads to an increase in the quantity of ballast gases (water and carbon dioxide have high specific heats) and thus lower temperatures and consequently emissions.

Let us consider the quantitative aspects of EGR. We will assume:

(1) Air consists of nitrogen and oxygen, the ratio generally being 79:21 parts.
(2) If the air:fuel equivalence ratio is higher than 1, combustion in diesel engines is complete as a rule, so that the exhaust gas will contain nothing but N_2, O_2, CO_2 and H_2.
(3) N_2, CO_2 and H_2O are inert gases.
(4) The formation of black smoke begins as soon as the percentage of oxygen E at point (2) (top of Fig. 22) is equal to A, i.e. to the percentage of the air which is consumed by the combustion process.

A different assumption is sometimes made, namely that the black smoke begins at an equivalence air:fuel ratio of 1.15, as in the VW Diesel Rabbit. The black-smoke limit, in this case, is of course $E = 1.15A$.

With reference to Fig. 22, we now define the following variables: S is the percentage of oxygen at (1) (normally 21%); E is the percentage of oxygen at (2); EGR is the EGR percentage at (2) and (3); X is the percentage of oxygen at (3); $A = E - X$ is the percentage of oxygen in

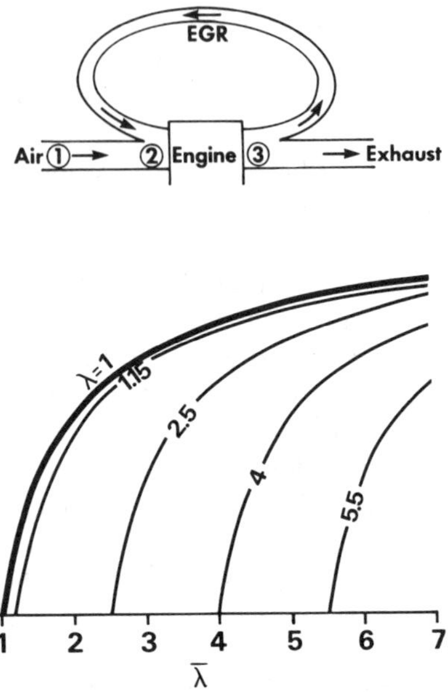

Fig. 22. Oxygen dilution by EGR.

the engine consumed for combustion. Then:

$$E = EGR \times \frac{X}{100} + \frac{(100 - EGR) \times S}{100} \tag{1}$$

and

$$X = E - A \tag{2}$$

Thus:

$$E = S - \frac{A \times EGR}{100 - EGR} \tag{3}$$

Now if we remember that the apparent equivalence air : fuel ratio, i.e. the one which stems from the assumption that the quantity of fuel injected is burned in pure air, can be expressed as

$$\bar{\lambda} = S/A \tag{4}$$

or

$$A = S/\bar{\lambda} \tag{5}$$

we have

$$\bar{\lambda} = \frac{S \times EGR}{(S - E)(100 - EGR)} \tag{6}$$

Taking S as 21% we obtain the thick black enveloping line for EGR

plotted against $\bar{\lambda}$ for the condition when E equals A namely for stoichiometric combustion (true $\lambda = 1$).

This condition would, of course, produce black smoke. A typical IDI engine would be set for $\lambda = 1.15$. Figure 22 thus shows the diluting effect of the EGR over a wide range. It will be noted that there is indeed still an excess of oxygen, even at high EGR rates. However, EGR is limited (see Fig. 18) due to the generation of particle emissions.

There are, of course, differences in the NO_x emissions between different diesel engine processes. In principle, it can be assumed that the direct injection processes, that is to say those with the better efficiency, reach higher gas temperatures and that their NO_x emission is therefore higher than in the IDI chamber process, the single-spray DI engine having a slightly better position. In practice, this means that today's IDI chamber engines are generally able to fulfil even strict emission regulations with the conventional means of parameter optimization, while the DI engines already require EGR for this.

5.3. Heat Transfer to the Walls

One of the most interesting questions of modern diesel engine research is that concerning the role of heat transfer to the wall during combustion and expansion. This arose recently out of the belief that, particularly with new ceramic materials, the means were available to reduce drastically the heat loss to the walls, if not indeed to completely eradicate it. These hopes were fostered by the results of calculations which predicted excellent levels of efficiency for the 'adiabatic engine'; that is, an engine in which the walls do not extract any heat from the gas.

However, these calculations were based on erroneous assumptions, insofar as there is hardly a material the thermodynamic and tribological characteristics of which make the adiabatic engine feasible. The transfer of heat from a gas to a wall depends only upon the heat transfer coefficient, the heat transfer surface and the temperature difference. As, in a real engine, neither the heat transfer surfaces nor the heat transfer coefficient—there are great gas movements—can be reduced to zero, the adiabatic engine would have to be achieved by the wall temperature following the temperature of the gas without a temperature drop. A wall capable of this would either have to consist of a material of extremely low conductivity, so that only the topmost layer of this material follows the gas temperature, or it would have to be of very low thermal capacity, for instance of one-tenth of the thermal capacity of the cylinder charge. The second solution cannot be considered for two technical reasons: firstly, the thermal capacity of the cylinder charge of a passenger vehicle diesel engine is about $0{\cdot}5\ \mathrm{J\,K^{-1}}$ (joules per degree kelvin). The wall would therefore have to have a thermal capacity of $0{\cdot}05\ \mathrm{J\,K^{-1}}$, which, in the case of steel, would require an absolute wall mass of 0·1 g and, in the case of the most favourable ceramic material (ZrO_2), an absolute mass of about 0·1 g also. The second reason is that neither ceramic nor metallic materials are known which can withstand temperatures of at least 2000 K without failing, particularly with regard to the pressures and the tribological loads placed on the walls of a diesel engine combustion chamber.

This, in fact, also renders the first solution impossible. However, it should be briefly considered in view of its thermal conductivity characteristics. There are solid materials the conductivity of which is about $0{\cdot}1\ \mathrm{W\,(mK)^{-1}}$ (watts per metre kelvin): they include slag, peat and soot. However, their thermal conductivity is still greater than the thermal conductivity of static air or static xenon, by factors of five and 25 respectively. Taking the factor of five, this layer must be five times as thick as the layer of air in order to work off the same temperature drop. In order that it does not absorb more heat than the air layer, its mass would have to be about one-fifth of the mass of the air layer, because the specific thermal capacities of air and the materials mentioned are the same with regard to order of magnitude. This, however, is hopeless, as even the bulk density of flaky soot is at least 100 times the mass of air. To summarize, this means that the adiabatic engine seems technically unrealizable.

Although it has, in the past, been possible to make the engine superficially cooler by using ceramic materials, this only means that, in the first place, the inclusion of these materials in the cylinder wall made a much higher temperature possible, so that the natural convection of the air could remove sufficient heat. It is unimportant whether this happened with a higher or lower temperature of the engine external wall than is found in a conventional engine. At any rate, it must be assumed that, with such engines, less cooling heat finally reaches the surrounding air. Nevertheless, and this has only been known for certain for a short time although practicians long suspected it, the tendency of these engines is to work with a lower level of efficiency than conventional ones.

The cause lies in the heat transfer coefficient, α, between the working gas and the wall. This depends upon the thermal conductivity in the layer of the gas bordering directly onto the wall:

$$\alpha = 2/d$$

where d is the thickness of the boundary layer. The thermal conductivity coefficient λ results, in accordance with the kinetic theory of ideal gases, from

$$\lambda \sim k\bar{c}/r^2$$

where k is the Boltzmann constant, $\bar{c}$ is the mean speed of the gas molecules, and r^2 is the dimension of the molecule cross-section. The mean molecule speed $\bar{c}$ immediately next to the wall results from the wall temperature T_W and the molecule mass m_0 as

$$\bar{c} \sim \sqrt{kT_W/m_0}$$

As the Boltzmann constant k, the molecule mass m_0 and the molecule radius r do not depend upon the temperature, the heat transfer coefficient increases with T_W if the thickness of the boundary layer is unchanged. If we take, as realistic temperatures, 523 K for a metal wall and 973 K for a ceramic wall, the heat transfer coefficient increases by a factor of 1·4 through this alone. Because the temperature difference between the gas and the wall has reduced by a factor of 1·4 in this case, this would just compensate for the effect of the increased heat transfer coefficient. That is to say, the ceramic wall would absorb just as much heat as the metallic one. In reality it absorbs noticeably more, as is shown in the most recent publications,[16–19] all of which are based on measurements made by different researchers in different places.

Figure 23[16] has typically resulted from heat flow and temperature and pressure measurements, the latter for calculations of the mean gas temperature in accordance with the state equation for gaseous matter. It can be seen that the heat transfer coefficient does not only have a much higher peak, but also that it remains, for a longer time (up to 90° after TDC, indeed until BDC), larger by more than 100% in the case of those insulating materials than for the conventional aluminium alloy.

The kinetic gas theory has already predicted greater heat transition

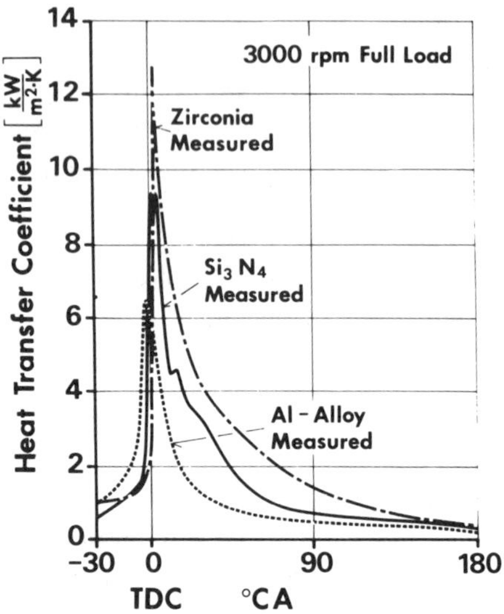

Fig. 23. Effects of hot walls.

at higher wall temperatures, provided, however, that there is the same boundary layer thickness (see p. 61). In Ref. 16 it is suspected that this very condition is no longer correct at higher wall temperatures under the thermodynamic conditions in the engine, and rather that flame quenching takes place in closer proximity to the wall. At a fixed temperature T_q, the flame is extinguished at the wall. The position at which T_q is reached is much closer to the wall at a high wall temperature than at a lower temperature which, on the one hand, increases the temperature gradient in the gas and, on the other hand, has the additional effect of also carrying the turbulence closer to the wall, thus making the boundary layer thinner.

If then, more heat generally enters the ceramic wall than the metallic one, the question arises as to where this heat goes if it is not conducted away by coolants. The answer is simple, for this heat is carried away by higher exhaust heat out of necessity, and is thus lost from the process of the naturally aspirated engine; for it cannot be assumed that the wall temperature becomes so high that heat from the wall flows back to the gas during the expansion cycle.

However, this is not the whole picture, as the higher wall temperature naturally also leads, on the one hand, to more heat transfer to the fresh gas, which is why its mass becomes smaller. At part loads, this is not generally a problem, but at full load it results in reduced m.e.p.

REFERENCES

1. Pischinger, A. & Pischinger, F., *Gemischbildung und Verbrennung im Dieselmotor*. Springer-Verlag, Berlin, 1957.
2. Schindler, K. P., Bestimmung von Drall und Durchsatz nach der Gleichrichtermethode. VW-Forschungsbericht, 1982 (unpublished).
3. Schäpertöns, H. & Thiele, F., Three-dimensional computations for flowfields in DI piston bowls. SAE paper 860463.
4. Siebers, D. L., Ignition delay characteristics of alternative diesel fuels: implications on cetane number. SAE paper 852102.
5. Hardenberg, H. & Wagner, W., Der Zündverzug in direkteinspritzenden Dieselmotoren. *MTZ*, **32** (197) 240–8.
6. Pischinger, F., *Verbrennungsmotoren. Band II. Vorlesungsmanuskript*, Aachen, 1980.
7. Schmidt, F. A. F., *Verbrennungskraftmaschinen*. Springer-Verlag, Berlin, 1967.
8. Kobayashi, A., Kurashima A. & Endou, S., Über den Kaltstart von Dieselmotoren mit Direkteinspritzung. *Jidosha Gijutsu*, **38** (1984) 1059–66.
9. N. N., *Starthilfsanlagen für Dieselmotoren*. Robert Bosch GmbH, 1982.
10. von Puttkamer, E., Immer schneller, immer bequemer. MOT 2/1985.
11. Oetting, H. & Papez, S., Reducing diesel knock by means of exhaust gas recirculation. SAE paper 790268.
12. Oetting, H. & Schweimer, G. W., Diesel NO_x control: air pollution by nitrogen oxides. US–Dutch International Symposium, Maastricht, 1982.
13. Haynes, G. S. & Wagner, H. Gg., Soot formation. *Progr. Energy Combustion Sci.*, **7** (1981).
14. Zeldovich, Ya. B., The oxidation of nitrogen in combustion explosions. *Acta. Phys.-chim. URSS*, (1946) 21.
15. Heitland, H. & Bernhardt, W., Möglichkeiten zur Verminderung der Stickoxidbildung in Ottomotoren, insbesondere durch Schichtladungsbetrieb und katalytische Konvertierung. 14. Fisita-Kongress, 1972.
16. Woschni, G., Untersuchung des Verhaltens eines Dieselmotors mit wärmedichtem Brennraum. *FVV-Abschlußbericht* No. 311, December 1985.
17. Enomoto, Y., Furuhama, S. & Musashi, M. T., Heat transfer to wall of ceramic combustion chamber of internal combustion engine. SAE paper 865022.
18. Langer, M., Distler, D., Giesen, K., Heinrich, H. & Klarhoefer, C., Wärmegedämmter PKW-Dieselmotor mit Abgasenergienutzung. 12. Statusseminar des Bundesministeriums für Forschung und Technologie (BMFT), Bad Ems, 1986.
19. Schroer, A., Stingl, P., Steiner, M., Zernig, N., Zloch, N. & Lingenauber, R., Herstellung und Erprobung von keramischen Bauteilen für Otto- und Dieselmotoren (KEBOD). 12. Statusseminar des Bundesministeriums für Forschung und Technologie (BMFT), Bad Ems, 1986.

3

The Two-stroke Engine: Crankcase Compression Type

G. P. Blair

The Queen's University of Belfast, Belfast, Northern Ireland, UK

1. Historical introduction
2. Later development: wave action
3. Two-stroke petrol engines: general description
4. Engine performance and validation of theory
5. Induction control
6. Noise prediction
7. Scavenging efficiency
8. Design method for scavenging
9. Mathematical modelling of the scavenging process
10. Concluding discussion

'Is there anyone so wise as to learn by the experience of others?'
(Voltaire)

1. HISTORICAL INTRODUCTION

It is assumed that the reader of this chapter has a basic understanding of the principles of operation of the two-stroke cycle engine and of the many mechanical variations which can occur on this simplest of all engine themes. It cannot be assumed that the reader, unless he is actively involved in the design, development or research of two-stroke engines, will be aware of the many problems which beset the progress of this type of power unit. For example, as the working cylinder has a compression and combustion process taking place within it upon every crankshaft revolution, it follows that some 50% of the cycle will be occupied by that firing process. It follows equally that, as combustion produces exhaust gas from the original air and fuel, the remaining half of the cycle must be used to remove the exhaust gases and replace them in the working cylinder by fresh air and, either simultaneously or eventually, fuel. As the working cylinder pressure varies from about one to many tens of atmospheres during this action, clearly the cylinder filling and emptying process is very dynamic and is conducted in an unsteady manner. Rather than use my own words at this point, I will quote the quite classic words of Hopkinson[1] given in 1914:

> 'The performance of 2-cycle internal combustion engines is determined very largely by the efficiency or otherwise of the process of charging. In the course of less than one quarter of a revolution the products of combustion resulting from the previous explosion have to be replaced as far as possible by the fresh charge of air, or of gas and air, which is blown in through the inlet valves and drives before it through the open exhaust ports the exhaust gases. Inevitably some mixing occurs and some of the fresh charge or of the scavenging air passes away to the exhaust and is wasted. On the amount of this waste depends very largely the performance of the engine. Its economy suffers to the extent of the waste of fuel.'

Hopkinson went on in this paper to derive the classic mixing equation for scavenging:

$$SE = 1 - e^{-SR}$$

where *SE* is the scavenging efficiency and *SR* is the scavenge ratio.

Following up this work, Bird[2] made a contribution on port design using classical fluid dynamic steady flow theory.

In 1954 Professors Taylor and Rogowski of MIT presented a paper[3] on 'Scavenging the 2-stroke engine'. This paper illustrates a very modest technological advance in thinking, considering that some 40 years and two World Wars had elapsed in the intervening period since Hopkinson's address.

In Bird's work no experimental verification was made, and it was not obvious that steady flow theory would be valid; indeed in 1936 this quasi-steady flow theory was challenged by Kadenacy. In his patent[4] he made extravagant claims, stating that the exhaust gases in his engines left the cylinder with ballistic velocities, leaving behind a vacuum that could be used for inducing the fresh charge. These claims inspired many research workers in turn to challenge the Kadenacy theory and, as a result, the theory of the dynamic fluid flow was put on a far better footing.

Weaving[5] demonstrated experimentally that without exhaust pipes no vacuum pressures occurred in the cylinder and that discharge was consistent with quasi-steady fluid flow theory. This work was confirmed and expanded considerably later by Johnson.[6] When exhaust pipes are present wave action becomes very relevant; indeed, tuned exhaust pipes had been used in Crossley–Carter engines since about 1946.[7]

2. LATER DEVELOPMENT: WAVE ACTION

The pioneering work in wave action was performed by Giffen.[8] Mucklow & Bannister at Birmingham University[9] simulated, with a 'rig', the exhaust 'blow-down' and subsequent flow and wave action in a pipe by bursting a diaphragm from a pressure source. They used these experimental results to validate a theory using unsteady gas dynamics. This work showed that there was no substance in Kadenacy's claims. In Switzerland, Jenny[10] and De Haller[11] made further contributions to the theory.

Building on this background, a very informative set of papers appeared, based on the use of the method of characteristics to analyse unsteady gas flow in the pipes, ducting and ports or valves of a reciprocating IC engine; typical of such papers are those by Wallace & Mitchell,[12] Wallace & Boxer[13] and Wallace & Nassif.[14] Unfortunately, while they were indeed informative, they added little to the design capability of the engineer, because the horrendous arithmetic complexity of the method of characteristics, in the slide rule and log-tables days of the mid-1950s, quite literally prohibited their usefulness. As one who carried out research under Professor F. J. Wallace's supervision in this area in the late 1950s, the author still grows quite weary as he leafs through Wallace & Boxer, thinking of the man–hours put into the production of that epic publication. Relief was at hand, for at the crucial point in history there appeared the digital computer, and it was through the classic paper by Benson, Garg &

Woollatt[15] that the real advantages of the computation of wave action in the ducting of an IC engine could be clearly seen—a design method had arrived.

The discussion, comments and references have so far been about two-stroke engines in general, and they would apply equally well to spark-ignition (petrol) or compression-ignition (diesel) power units. From this point forward the author will concentrate on petrol-burning engines and their technology; however, much of the discussion has also applicability to two-stroke diesel engines.

3. TWO-STROKE PETROL ENGINES: GENERAL DESCRIPTION

Two-stroke spark-ignition engines are usually of the crankcase-compression type and by virtue of their very simplicity are used for the production of inexpensive power units for agricultural implements such as lawnmowers, chainsaws and water pumps. Where a good specific power output is essential they find favour in outboard marine units, snowmobiles, motorcycles and for RPV (remotely piloted vehicle) applications. In the recent past they have been used in small numbers for automobiles, but high hydrocarbon emissions have caused their demise at this time, although this situation could change in the future for reasons which will be discussed later. Most engines today are of the Schnurle or loop-scavenge type with a flat-topped piston, but cross-scavenged units with deflector pistons are still used for some applications: see Schweitzer[16] for various types. There is evidence to be discussed later that the deflector-piston engine, the original design, could make a comeback for the best of technical reasons.

It is in the motorcycle field that the first of the modern high specific output two-stroke engines began to appear in the late 1950s and early 1960s. Prior to that time, virtually all racing motorcycles used four-stroke power units, and trade names such as Norton, AJS and Gilera will stir memories for many a reader. From 1935 to 1960 the engineers in such companies were able to produce racing four-stroke engines with bmep levels of 190–200 lb in^{-2} (13–13·5 bar). The equivalent two-stroke could manage but 75–90 lb in^{-2} (5–5·5 bar) and, at equal engine speeds, it can be seen that the two-stroke racing motorcycle in any given class was at a 20% power disadvantage. However, from East Germany there appeared a virtually unknown machine, the MZ (Motorradwek Zschopau), in the 125 cm^3 class, and it was clearly superior in power to the four-stroke competition; these machines were a 'revelation' and used a new type of exhaust system called an expansion chamber, not invented or designed by any theoretical procedures, which provided a 20–25% power boost to the engine to about 100–110 lb in^{-2} bmep. That device not only stimulated tremendous interest into research in two-stroke engines which has continued to this day, but in the field of Grand Prix racing motorcycles it is the four-stroke engine which has disappeared. The naturally aspirated four-stroke racing engine today produces 220–230 lb in^{-2} BMEP (15–15·5 bar) and the two-stroke equivalent is 150–160 lb in^{-2} (10–11 bar). A good discussion of the design of modern racing two-stroke engines is given by Naito & Taguchi,[17] Naitoh & Nomura[18] and Blair.[19]

4. ENGINE PERFORMANCE AND VALIDATION OF THEORY

It is important to consider the basic design philosophy of an expansion chamber for a ported two-stroke cycle engine, for it is worth remembering that in 1960 when it appeared on the MZ racing motorcycle it was not technologically understood, by either the MZ engineers or the engineers who first examined it in Europe. The fact, and that was clearly demonstrable on the racing circuits, that it practically doubled the engine power output seemed more miraculous than thermodynamic: In short, even those engineers whose business it was to develop engines had a sufficiently poor knowledge of unsteady gas dynamics as to be unable to comprehend its mechanism of operation. When it is explained now, the reader may well wonder why it took until the sixth decade of this century before it appeared experimentally. (Those who find difficulty in understanding the unsteady gas dynamics referred to in the text below should first study Refs. 20 and 21.)

The dimensions of an expansion chamber fitted to the QUB 500 racing motorcycle engine described by Blair[19] are shown in Fig. 1; the piston in the illustration is at BDC so that the exhaust port and the transfer ports are fully open and scavenging flow rate of the fresh charge is at a maximum. While the theory says that the fresh charge pushes the exhaust gas ahead of it, even Hopkinson in those early days, was under no delusion that some mixing of the incoming fresh charge with the burnt gas in the cylinder takes place, and that some of

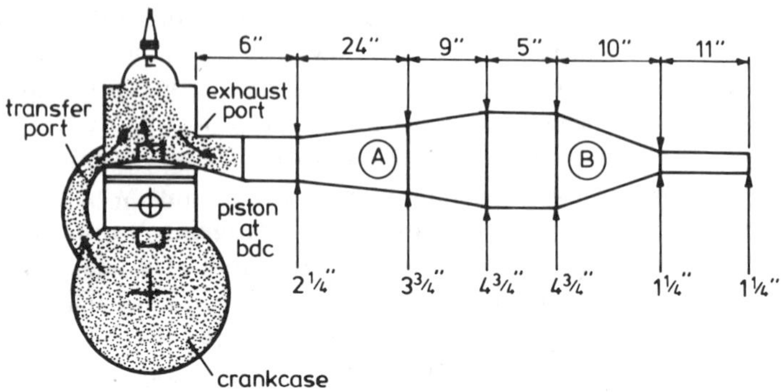

Fig. 1. Expansion chamber exhaust pipe attached to two-stroke engine with piston at BDC during the scavenging process.

this mixture enters the exhaust pipe at the position shown in Fig. 1. This process is occurring rapidly, the entire 'open' cycle period from exhaust open to exhaust closure taking some 0·004 s at 7500 rpm. The truly dynamic nature of the scavenging process and of the exhaust pipe wave action is best illustrated in Fig. 2, which is reproduced from

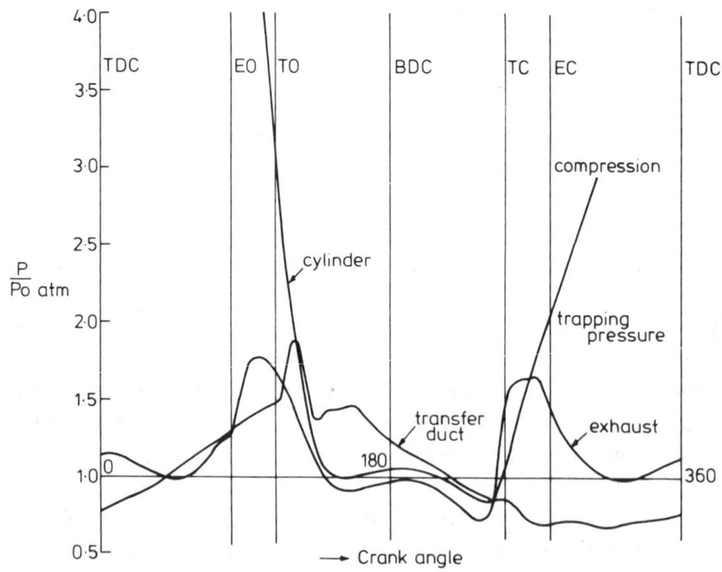

Fig. 2. QUB 500 cylinder, exhaust and transfer duct pressure–time history at 7500 rpm.

Blair.[19] The figure shows the pressures taken in the cylinder near the exhaust port but in the exhaust pipe and in the transfer duct between the cylinder and the crackcase; all these as a function of crank angle at 7500 rpm where the peak power of 64·5 hp from the engine occurred. The bmep was 110 lb in^{-2} (or 7·6 bar). The crank-angle axis commences and finishes at TDC for one complete cycle, and the porting events are distinguished by vertical lines of 'EO' for exhaust port opening, 'EC' for exhaust port closing, and 'TO' and 'TC' for transfer port opening and closing respectively. The first event of interest in Fig. 2 is the creation of the exhaust pulse after EO, from a cylinder pressure initially at about 8 bar at EO. Near the exhaust port, the exhaust pulse is seen to rise to about 1·7 bar and this massive 'pop' is propagated at supersonic velocity down the exhaust pipe, firstly to the expanding area section marked A and then to the converging section marked B. It can be seen also that the cylinder pressure at transfer port opening TO is some 3·0 bar, whereas the transfer duct pressure is only 1·5 bar. Not unnaturally, another exhaust pulse is sent down the transfer duct, in the 'wrong direction' and some 10° CA occur before the flow is reversed to the 'proper' direction. It will now be observed that for virtually all of the remainder of the scavenge period between TO and TC, the forcing pressure situation is from the transfer duct to the cylinder and thence to the exhaust pipe.

Also, the exhaust pipe port pressure has fallen below the atmospheric pressure value of 1·0 atm: this is because the exhaust pulse, having arrived at section A in the pipe, sends back a reflection of the opposite 'sign' to the cylinder at a sub-atmospheric pressure value. This valuable suction pressure helps to increase the flow rate of fresh charge from the crankcase, but it also helps to pull a proportion of it into the exhaust pipe. It is at this point that the critical action of the converging section B of the exhaust pipe takes place. The ongoing exhaust pulse is 'echoed' from the converging section B as a wave of the same 'sign' as the exhaust pulse and it is this wave, often called the plugging pulse, which is effectively sealing off the exhaust port and preventing further fresh charge loss. It can be seen that the plugging pulse is actually higher in pressure than the cylinder pressure and this forces some lost or short-circuited fresh charges back into the cylinder. The resulting trapping pressure at exhaust port closing is now 2·0 bar, whereas if the tuned expansion chamber had not been connected to the exhaust port then that trapping pressure, P_t, would have been 1·0 bar, or about 1·2 bar at best. This has vital implications regarding trapped charge mass, M_t, irrespective of its proportions of fresh charge and exhaust gas. From a simple application of the state equation,

$$M_t = P_t V_t / R T_t$$

it is clear that, as T_t, V_t and R are unlikely to vary substantially, the simple act of doubling trapping pressure P_t from 1·0 to 2·0 bar thence doubles the trapped mass M_t. If the scavenging efficiency remains constant, that doubled quantity of fresh charge, when combusted, will roughly translate into a doubling of the indicated work per cycle or twice the power at the same engine speed. The experimental proof of the importance of the tuned pipe is that when a split or crack appeared in the metal skin under racing conditions, the QUB 500 engine power fell off markedly from 65 hp to about 40 hp.

The explanation given above for the behaviour of the tuned expansion chamber was not available in 1960 when the tuned pipe motor-cycle first appeared, and so the research and interest which this device created was remarkable. In one racing year the two-stroke motorcycle underwent a metamorphosis from a loser to a winner on the Grand Prix circuits. The research stimulated Blair[19] at The Queen's University of Belfast to use the approach of Benson,[15] and this led to a series of publications illustrating that, for the first time, a true design procedure was available for the two-stroke cycle engine, based solely on the physical geometry of the engine, its porting and intake and exhaust ducting. From this computer calculation, based on the thermodynamics and unsteady gas dynamics of the flow in and through the engine's physical geometry detailed by the user, a prediction could be made at any desired engine rotational speed of the power, torque, mean effective pressure and specific fuel consumption, and of the pressure–time histories at any point in the engine ducting cylinder or crankcase. Among the most significant of these publications are those by Blair & Johnson,[22] Blair & Cahoon,[23] Blair & Ashe,[24] Blair[25] and Blair, Hinds & Fleck.[26]

To illustrate the usefulness of such design methods, the cylinder and exhaust pipe pressures recorded by fast-response pressure transducers at engine speeds of 4000, 6000 and 8000 rpm are shown in Figs 3, 4 and 5. At similar engine speeds and with the same exhaust system in place the crankcase and inlet duct pressures are shown in Figs 6, 7 and 8. The engine being used is a moto-cross racing engine, a 250 cm^3 Husqvarna single-cylinder piston-ported power unit with a tuned exhaust pipe, the dimensions of all of the relevant physical geometry

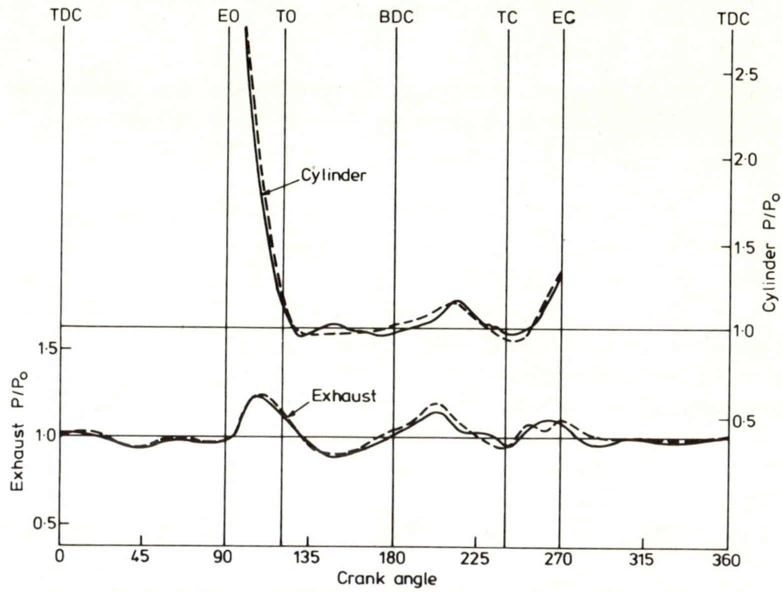

Fig. 3. Open-cycle cylinder and exhaust pressures of Mk 4 Husqvarna engine at 4000 rpm.

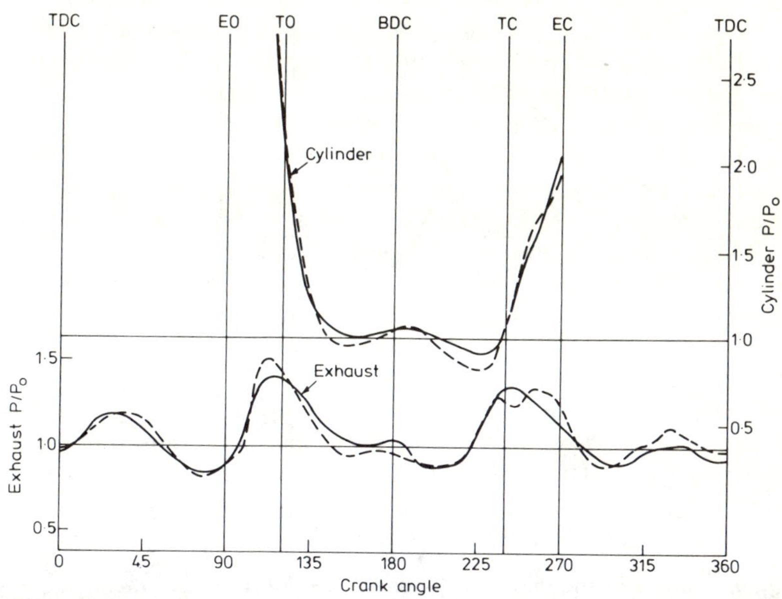

Fig. 4. Open-cycle cylinder and exhaust pressures of Mk 4 Husqvarna engine at 6000 rpm.

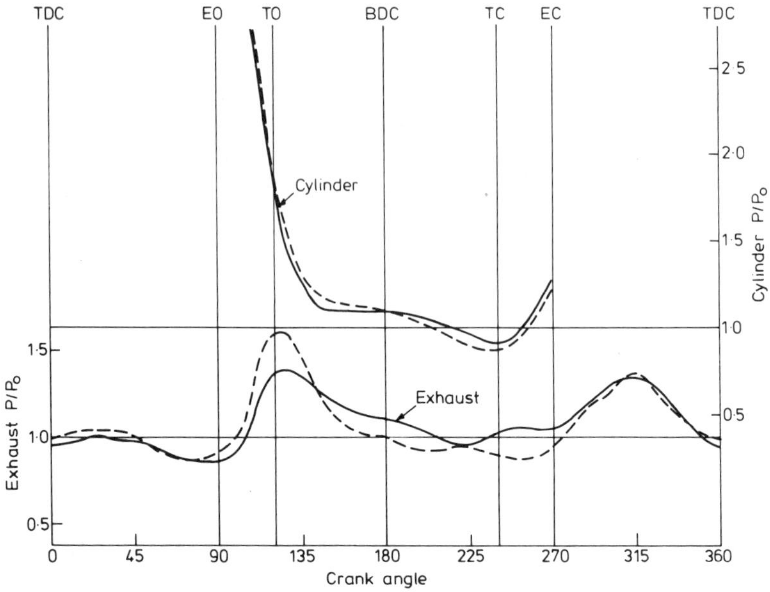

Fig. 5. Open-cycle cylinder and exhaust pressures of Mk 4 Husqvarna engine at 8000 rpm.

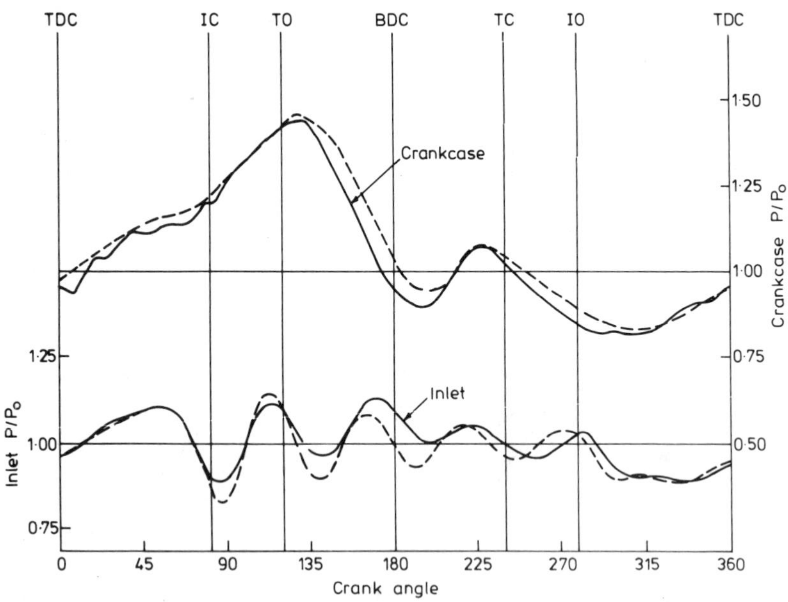

Fig. 6. Crankcase and inlet duct pressures for Mk 4 Husqvarna engine at 4000 rpm.

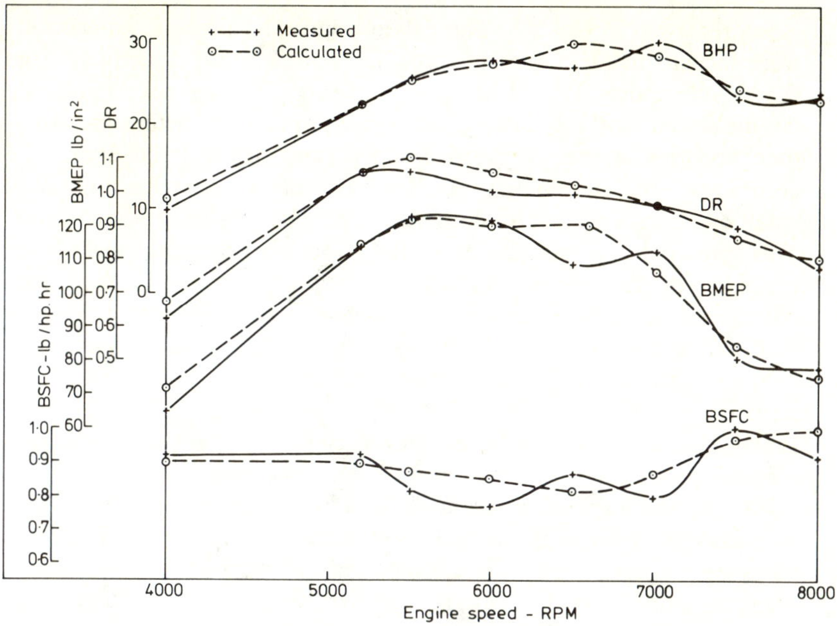

Fig. 9. Measured and calculated Mk 4 Husqvarna engine performance characteristics.

speeds selected are 65, 120 and 75 lb in^{-2}, giving ratios of 1·3 : 2·4 : 1·5, whereas the trapping pressure ratios are 1·3 : 2·1 : 1·25. To assist designers who would not have access to sophisticated computer programs for the design of tuned-exhaust pipes for two-stroke engines, Blair & Johnson[22] distilled much of their experimental and theoretical work into the form of simple empirical equations which permit the design of the physical dimensions of tuned expansion chambers to suit the varying requirements of engines with differing swept volumes, exhaust port timings and speed range requirements. As is common with all such simplistic approaches, it is far from a perfect design method, but it is definitely superior to having none at all, which was the situation before 1960.

5. INDUCTION CONTROL

The two-stroke engine has various forms of induction control not seen on the four-stroke cycle engine, where the poppet valve is the

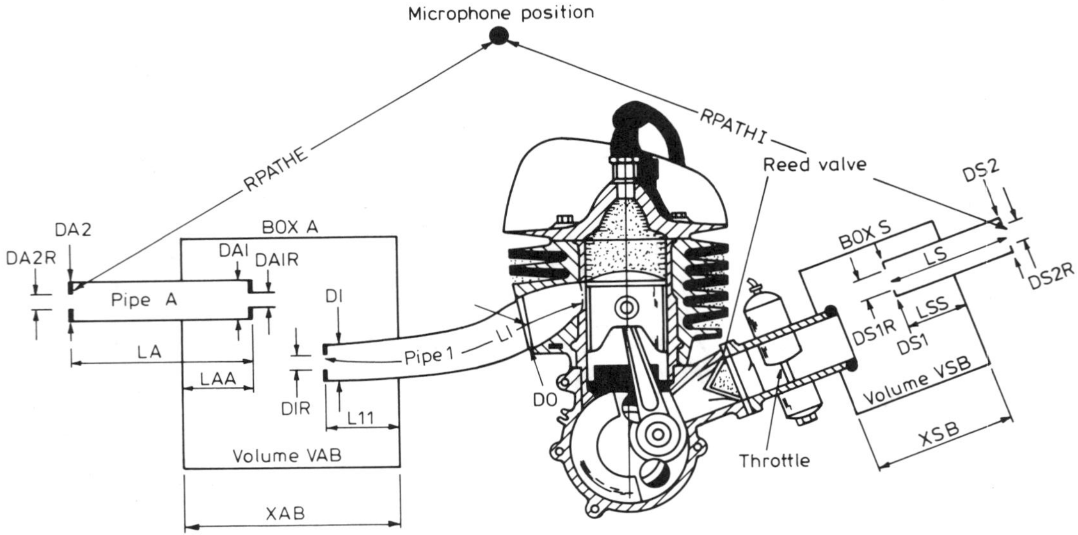

Fig. 10. Arrangement of reed block, intake and exhaust silencers for data insertion into computer program GRB1.

universally accepted device. On two-stroke engines the various forms utilized are piston-controlled inlet ports, rotary sleeve valves, rotary disc induction valves and reed valves. All of these devices have an application in particular designs for the marketplace. The piston-controlled port is the simplest and the cheapest, but suffers from the disadvantage of having symmetrical port timing. The other types mentioned above, sleeve, disc and reed, are more expensive to produce, but have the advantage that the timing of the opening and

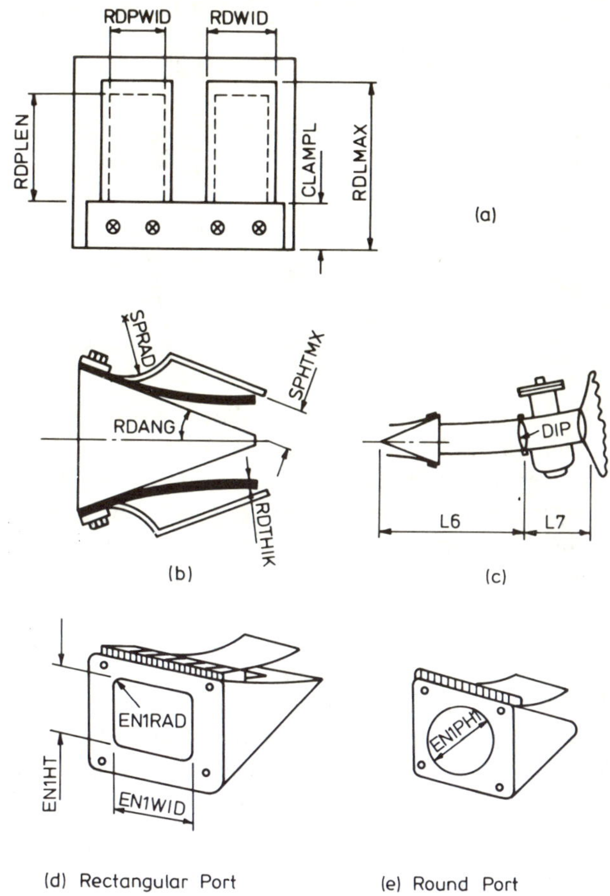

Fig. 11. Some reed block and petal data required for digital computation of unsteady gas flow and reed motion.

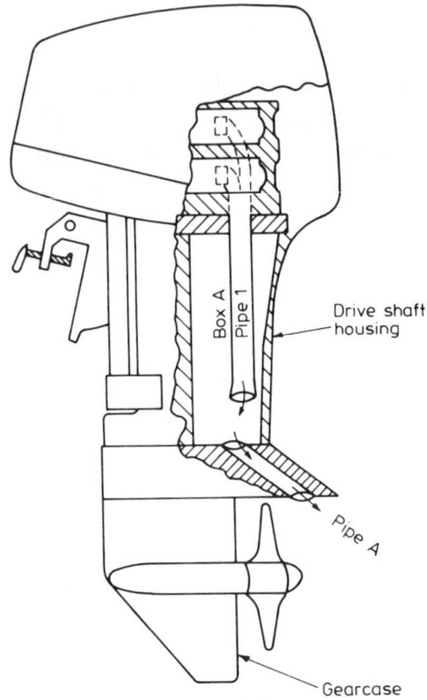

Fig. 12. Arrangement of exhaust ducting for a two-cylinder outboard marine engine.

closing point may be unsymmetrical and thus allow the engine to breathe a higher scavenge ratio into the crankcase. All of the devices mentioned, with the exception of the reed valve, provide the engineer programming the device into the computer programs with few problems, for their physical geometry is relatively easy to specify as a function of the engine's rotational movement. This is not the case with the reed valve where the spring steel (or plastic, carbon fibre or fibreglass) reed is opened and closed purely as a function of the pressure difference between the atmosphere and the varying crankcase pressure. A typical location of the reed valve is shown in Fig. 10. It is important to note that all of the physical dimensions listed there are used as data values for that particular program. The reed valve needs to be specified most carefully, and the geometry data required is shown in Fig. 11. One data item missing from Fig. 11 is the Young's modulus of the reed material.

Earlier, Hinds & Blair[27] had determined, under simulated dynamic flow and pressure conditions, that they could predict the unsteady motion of the reed in those circumstances. This was incorporated into the main engine simulation program. The program was expanded to deal with the more complex twin-cylinder outboard marine power unit configuration, as in Fig. 12.[26] The result is shown in Fig. 13, and the close correlation between the measured and calculated power, BMEP, air flow rate and brake specific fuel consumption is clearly evident. Another piece of the design jigsaw had been put into place, with the

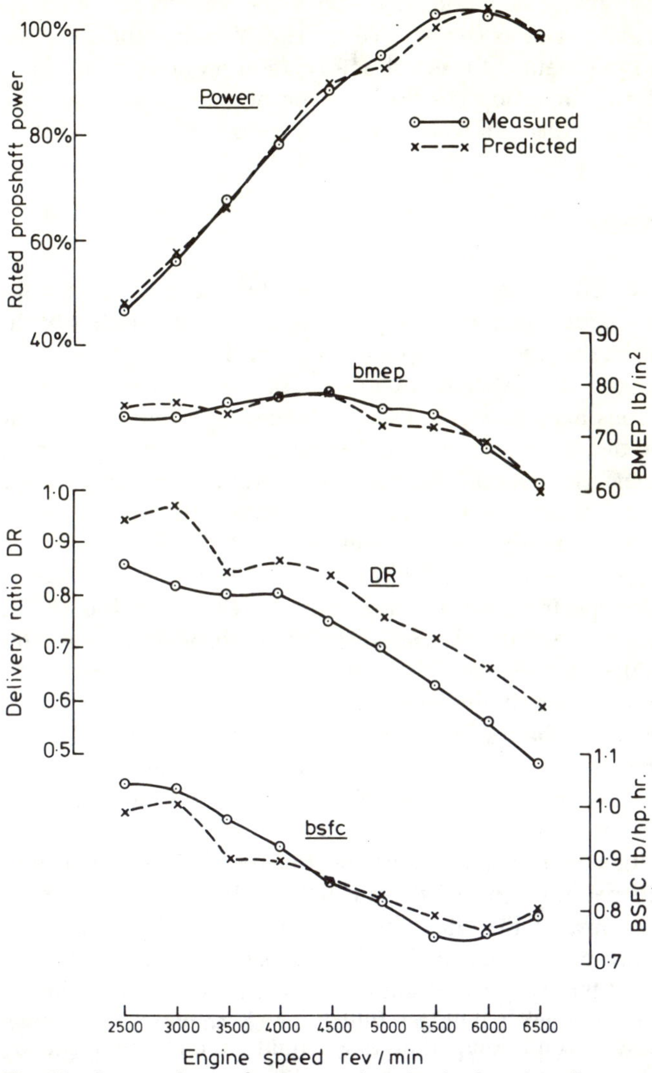

Fig. 13. Comparison of measured and calculated performance characteristics of engine X.

possibility being opened up for calculation work with multi-cylinder engines. This is important, for many large outboard marine units are of in-line three, V4, V6 or V8 construction, with the exhaust manifolds branched and collected into the gear-casing leg of the outboard motor and exhausted underwater near the propellor. The branching of exhaust, and intake, manifolds has been the subject of separate research studies by such as Benson, Woollatt & Woods[28] and Bingham & Blair,[29] and that detailed research information on pressure losses at the branch during unsteady gas flow is even now being programmed into multi-cylinder calculations for two-stroke engines.

6. NOISE PREDICTION

It will be observed in Fig. 10 that there are two data values for the computer calculation which refer to path length from the outlets of the inlet and exhaust ducts to a microphone position. It has been shown by Blair & Spechko[30] and Blair & Coates[31,32] that a Fourier analysis of the instantaneous mass flow rate leaving the end of a pipe system can be used to predict the noise level and spectrum at any point in space from that outlet. The overall sound pressure level and spectra are well predicted up to about 2 kHz, but a more complete calculation to include turbulence noise levels would be needed to obtain better accuracy at higher frequencies. Shown in Fig. 14 is a typical set of predicted noise spectra from a system used by Coates & Blair[32] and the correlation between the measured spectra (unbroken line) and the calculated values (broken line) is quite good. The four diagrams for $\theta = 0$ to $\theta = 90°$ refer to the microphone location, varying from an in-line position to that at right angles to the axial particle flow direction. The most important aspect of this useful addition to the unsteady gas dynamics computer calculations is that it allows the designer to (a) observe the effect of placing various silencer combinations to the inlet and exhaust ducts, on the trade-off between noise reduction and deterioration of the engine's performance characteristics, and (b) to observe the effect of the silencers on the production of noise separately by the intake and exhaust systems and be able to balance them so that they are equally quiet (or noisy!). This ability of the computer program to detect which noise source is the more voluble is invaluable to the designer, for it is extremely difficult and tedious work to determine such effects experimentally. There have

Fig. 14. One-third octave noise spectrograms.

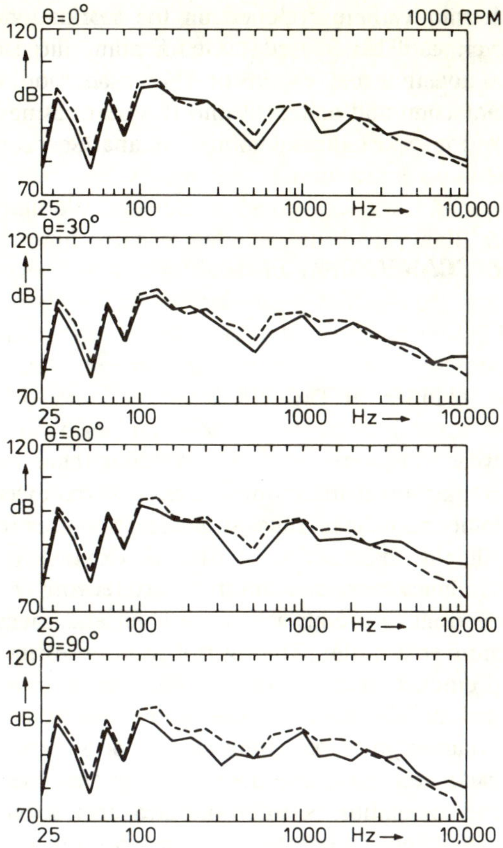

been other works in this area of research into silencing such engines, notably those of Roe[33] and Jones & Brown.[34]

It will be reasonably clear to the reader, at this stage of the discussion, that these unsteady gas dynamics calculations effected by computer provided a tremendous design capability for all types of two-stroke engines. Naturally, this computation work continues to this present day and is being expanded to deal with more difficult calculation regimes, and with more complex types of engines and their ducting. Throughout the discussion it is apparent that the calculation operates by predicting mathematically the instantaneous state conditions, velocity and mass content of the cylinder, crankcase, and at all sections of the intake and exhaust ducting. There are two parts to the thermodynamic cycle within the working cylinder, namely the open cycle and the closed cycle. During the closed cycle, compression, combustion and expansion take place and it is possible to deal with that computationally by the simple treatment described by Blair[25] or by more complex combustion analyses such as those proposed by Mattavi & Amman.[35]

7. SCAVENGING EFFICIENCY

For further consideration of events during the open cycle it is necessary to return to the earliest statements in this chapter by Hopkinson[1] or Taylor & Rogowski[3] and to examine the progress made in scavenging analysis since that period of history. During the open cycle, when the cylinder is being charged with fresh mixture, as Hopkinson pointed out, some displaces exhaust gas from the cylinder, some mixes with exhaust gas and some of it may short-circuit straight out the exhaust port into the exhaust pipe. In the unsteady gas dynamics computer program, one is forced to use some mathematical relationship which will tell the ensuing calculation what proportion of the fresh charge entering the cylinder either (1) displaced exhaust gas, (2) mixed with it, or (3) was short-circuited. In other words, the scavenging process is intimately connected with the unsteady gas dynamic processes going on around it, and indeed is implemented by these processes. The actual flow of that fresh charge in various modes and directions is one to be analysed using fluid mechanics. The significance of the above statements can be most easily seen from the experiments conducted by Sweeney, Kenny, Swann & Blair[36] in which a single-cylinder 250 cm^3 Yamaha DT250 engine was provided with several cylinders which had identical geometries of exhaust, transfer and inlet ports but the directions of those transfer ports induced differing regimes of loop scavenging which resulted in the engine having different trapped scavenging efficiencies when supplied with *identical* scavenge ratios. The measured scavenging efficiencies at full throttle over the speed range of the engine, and the BMEP and BSFC characteristics over that same speed range, are shown in Figs 15 and 16. It can be clearly seen that the highest scavenging efficiency results in the highest power and torque and the lowest specific fuel consumption. As Hopkinson[1] would have pointed out, his statements were well proven.

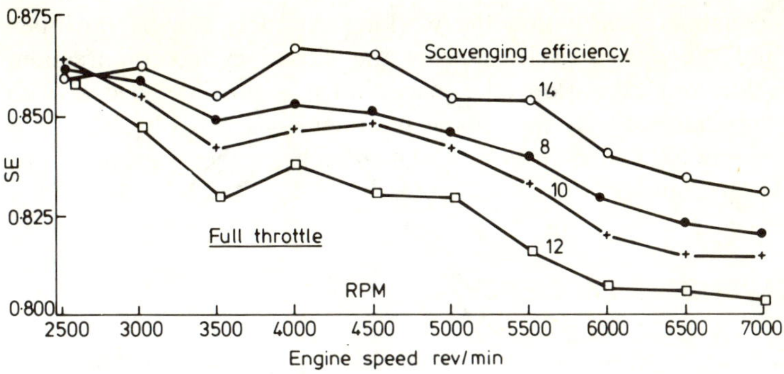

Fig. 15. Scavenging efficiency–speed behaviour at full throttle for engine cylinders numbered 8, 10, 12 and 14.

It is worthwhile, for the remainder of this discussion, to recapitulate that,

$$\text{scavenging efficiency, } SE = \frac{\text{mass of fresh charge trapped}}{\text{total trapped charge mass}}$$

$$\text{scavenge ratio, } SR = \frac{\text{mass of fresh charge supplied}}{\text{mass of fresh charge to fill swept volume at STP}}$$

It is also rather interesting to note, from Figs 15 and 16, that a mere 4% variation in scavenging efficiency produces some 15% variation in torque and specific fuel consumption. Scavenging is clearly a very important factor in the design of such engines.

It should also be remembered that, while this information reminds the engine designer of the vital importance of scavenging, it does little to tell him what he must do to avoid *his* engine from being a cylinder type 12 rather than a cylinder type 14. In Fig. 17 are the transfer plan drawings of cylinders 12 and 14, originally published by Blair & Kenny,[37] and the designer's natural query is at what angles he should set the horizontal plan and vertical angles of the ports in his engine for maximized scavenging efficiency. No one in history has as yet satisfied this totally logical question, but many have postulated testing methods to allow the designer to develop his porting systems so as to produce good scavenging.

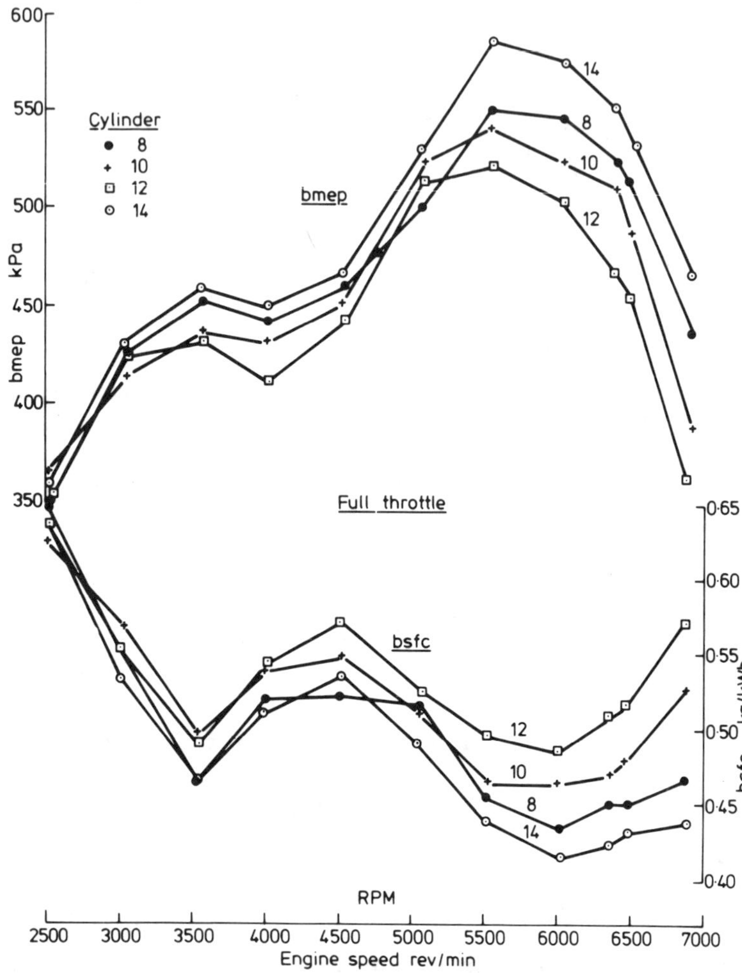

Fig. 16. BMEP–BSFC–speed characteristics at full throttle for engine cylinders numbered 8, 10, 12 and 14.

8. DESIGN METHOD FOR SCAVENGING

While there have been many papers published on this subject, one of the papers which has had the most impact in recent times is that by Jante.[38] Jante proposed the use of a comb of pitot tubes mounted at cylinder head level over the open bore of a motored two-stroke

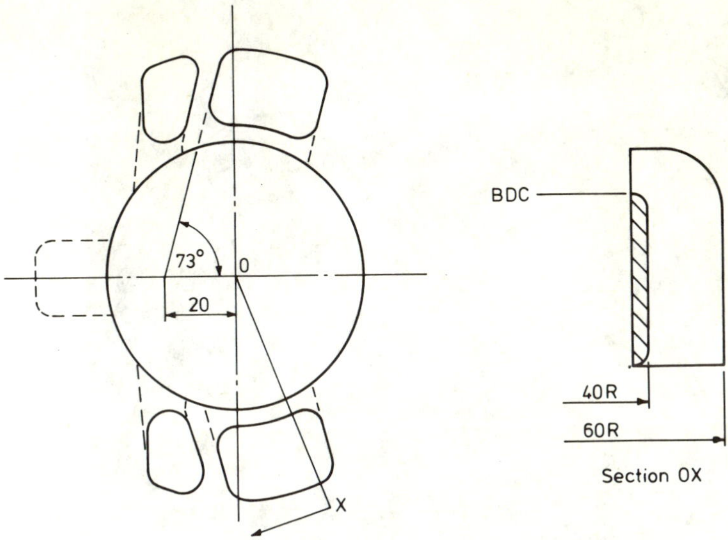

(a)

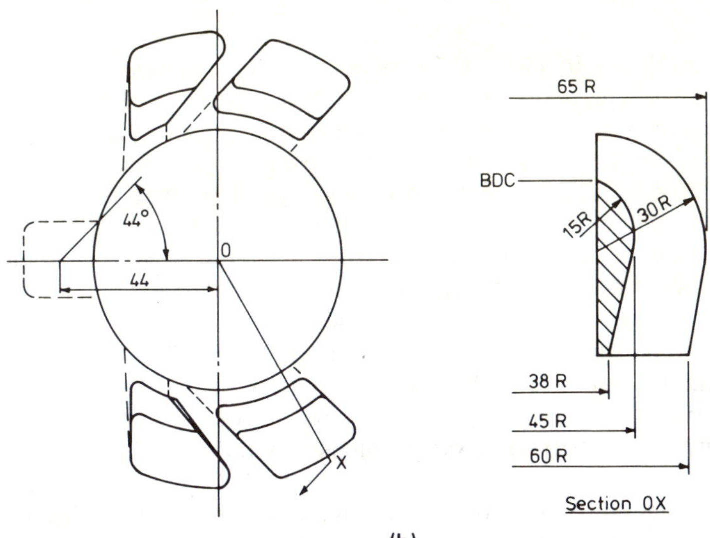

(b)

Fig. 17. Transfer port geometrical layout for cylinders numbered (a) 12 and (b) 14.

Fig. 18. An engine on the test-bed for scavenging flow assessment with pitot tubes in position.

engine, as shown in Fig. 18. From the velocity patterns which one obtained from this array of pitot tubes when indexed over the entire bore area, it was possible to deduce whether the fresh air emanating from the transfer ports and ducts was providing a 'good' or 'bad' scavenging flow. The flow patterns for such 'good' or 'bad' flows were described by Jante[38] in his paper and are summarized in Fig. 19.

Jante defined 'good' and 'bad' scavenge diagrams by the shape and position of the zero velocity line with respect to the plane of symmetry of the scavenge flow. The basic definitions are presented in Fig. 19, where type (a) shows the 'perfect' scavenge diagram. The arrow at the top of the diagram shows the axis of symmetry of the scavenge flow and inherently, for most engines, the direction of exhaust flow from the cylinder. The perfect scavenge velocity contour map has the zero velocity line across the bore diameter, with lines of increasing velocity extending smoothly towards the scavenge flow side, with the highest velocity at the opposite side of the cylinder from the exhaust port. The negative 'velocity' on the exhaust half of the bore area shows the entrained flow and that caused by piston motion; inevitably any diminution of that negative flow area by positive flow pockets would indicate stray flow emanating from the fresh air or scavenging stream and would denote deteriorating scavenging. Considering a diametrical line across the bore, the flow symmetry ratio (*FSR*) is the ratio of the flow up the bore on the inlet port side to the flow down the bore to the exhaust ports. If the flow is one-sided, then the zero velocity line will be shorter than the bore diameter. With the following definitions:

Flow Symmetry Ratio, *FSR*

$$= \frac{\Sigma \text{ (upward positive pitot velocity)} \times \text{(pitot subtended area)}}{\Sigma \text{ (downward negative pitot velocity)} \times \text{(pitot subtended area)}}$$

Mixing Length Ratio, *MLR*

$$= \frac{\text{Length of the zero velocity across cylinder}}{\text{Cylinder bore diameter}}$$

Scavenged Area Ratio, *SAR*

$$= \frac{\Sigma \text{ (all pitot subtended areas recording positive velocity)}}{0{\cdot}5 \times \text{Bore area}}$$

Derived numerical data in the form of these dimensionless parameters from the pitot tube mean velocity measurements can be obtained. Thus the 'perfect' scavenge diagram would (as shown in Fig. 19) have *FSR, MLR* and *SAR* values of unity. The 'bad' diagrams of scavenging were described by Jante as being unsymmetrical (Fig. 19(b)), as having

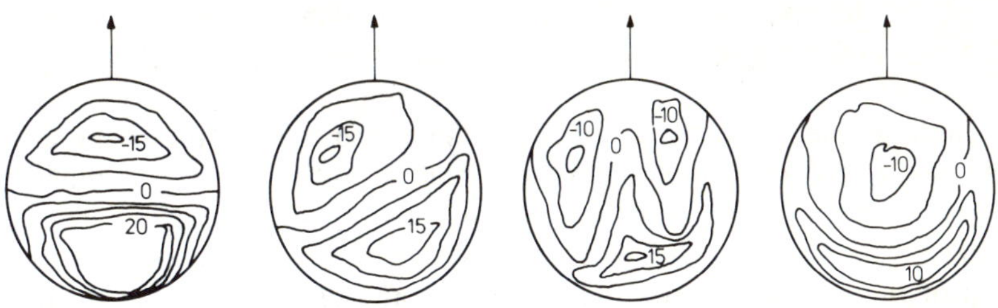

	A	B	C	D	
FRS	— 1·0	1·4	1·15	1·1 —	FSR
MLR	— 1·0	1·1	2·3	2·2 —	MLR
SAR	— 1·0	0·6	0·55	0·5 —	SAR
Pattern	-"Perfect"	"Unsymmetrical"	"Tongue"	"Wall"	

Fig. 19. The several basic types of scavenge flow patterns produced by testing using Jante's method. Contour velocities are in m s^{-1}.

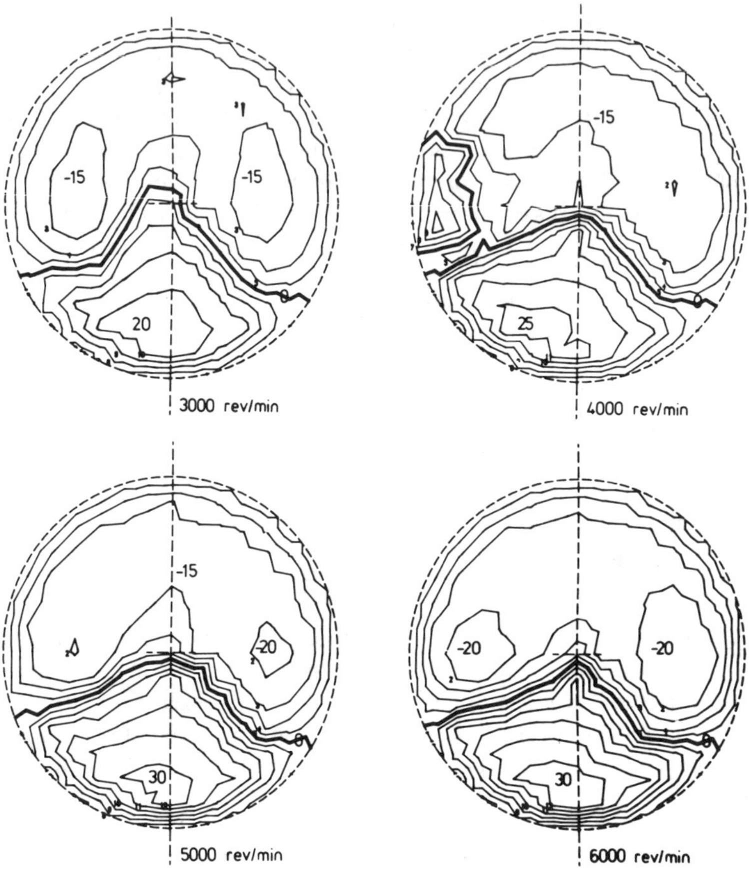

Fig. 20. Scavenging patterns with Jante's method at 'full throttle' for cylinder numbered 12. Contour velocities in m s^{-1}.

a 'tongue' pattern (Fig. 19(c)), or forming a 'wall' pattern (Fig. 19(d)). Some typical values for *FSR, MLR* and *SAR* are shown on Fig. 19(c) beneath each diagram, indicating how these derived values may quantify the inferior nature of the scavenging flow and how patterns (b), (c) and (d) have at least one derived value straying significantly away from the perfect value of unity.

Blair[39] showed that the somewhat tedious measurement method of Jante could be automated and then he, Kenny and others in their papers[36–40] systematically tested some 15 different Yamaha 250 cylinders under both firing and Jante motoring test conditions for the effect of scavenging flow variations. The basic conclusion reached was that the Jante method of testing was extremely effective in determining 'good' and 'bad' scavenging cylinders but was somewhat indeterminate in differentiating between the 'good' and 'nearly good' designs. A good example of this will be noted from the Jante scavenge patterns for the cylinders 12 and 14 already mentioned; these are shown in Figs 20 and 21. It would be very difficult, on the basis of Jante's own

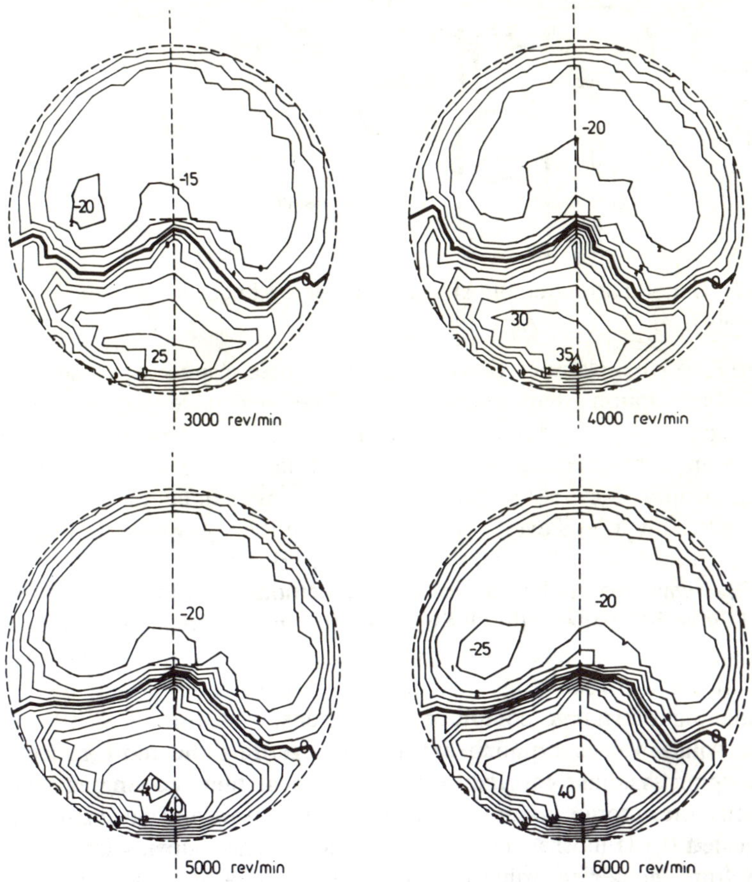

Fig. 21. Scavenging patterns with Jante's method at 'full throttle' for cylinder numbered 14. Contour velocities in m s^{-1}.

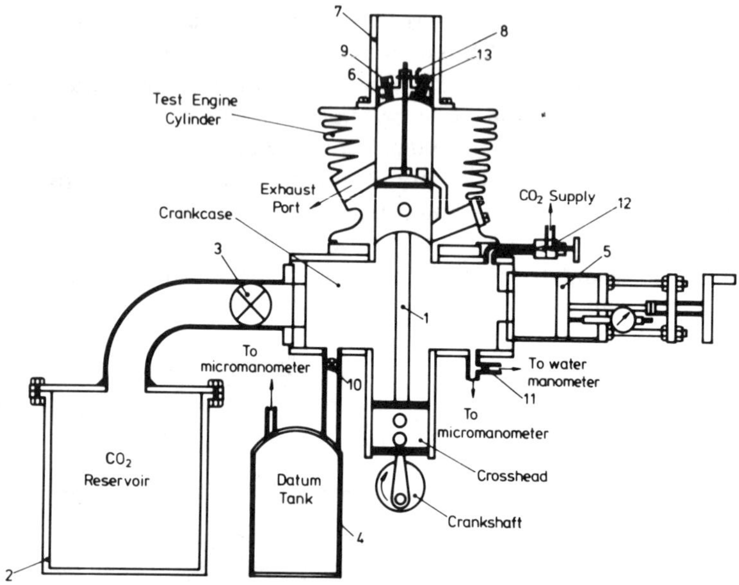

Fig. 22. General arrangement sketch of single-cycle gas simulation apparatus.

criteria, to label cylinder 12 as being 15% worse in power and specific fuel consumption than cylinder 14. The real clue to the poor scavenging flow of cylinder 12 is contained in the 4000 rpm picture where an unsymmetrical flow appears at left-centre of the picture which would inevitably short-circuit, but it would take a very experienced worker's eye to spot it, and even then he would be unable to put a number on the quantity of that deterioration.

What was required was an absolute testing method, one which would give feedback to the designer in the same practical development terms as the Jante test did in part, as well as to aid the theoretician with his need for scavenging efficiency–scavenge ratio models for the unsteady gas dynamic computer calculations, and also to assist with the resolution of continuing arguments regarding the most effective scavenging systems, namely loop, cross- or uniflow scavenging techniques. After some initial experiments with a single-cycle testing rig suggested by Sanborn,[40] the research team at The Queen's University of Belfast decided to bring up-to-date an old idea of Sammons[41]; this was published in Ref. 36. The basic layout of the apparatus is shown in Fig. 22 and features a test engine cylinder mounted on a constant-volume crankcase and with a double piston arrangement in the working cylinder which permits a constant-volume scavenging process. The crankcase is filled with carbon dioxide to represent the fresh charge and air fills the working cylinder to simulate exhaust gas. The 'engine' turns over once at an appropriately high speed for one cycle and is then stopped by an electromagnetic clutch. The reduction in crankcase volume determines the scavenge ratio and the trapped cylinder contents are delivered to an oxygen analyser to deduce scavenging efficiency. Repeating the process at varying crankcase pressures gives a series of tests at varying scavenge ratios. The results of such tests can be plotted and compared on an absolute basis irrespective of cylinder size, type or scavenging system employed. Typical of such results from this apparatus are those plotted in Figs 23 and 24. On Fig. 23 can be seen the pure displacement scavenging line (where $SE = SR$), the Hopkinson perfect mixing line (where $SE = 1 - e^{-SR}$) and the test results for the already mentioned Yamaha cylinders numbered 12 and 14, as well as for others numbered 7, 9 and 15. As only an 8% variation in measured scavenging efficiency covers the lot, it looks a rather complex picture, and hence the expanded and magnified picture for a narrower SR band given in Fig. 24 makes the cylinder ranking order very much clearer. It will be observed that cylinder 14 is best followed by 15, 7, 9 and lastly, as before, cylinder

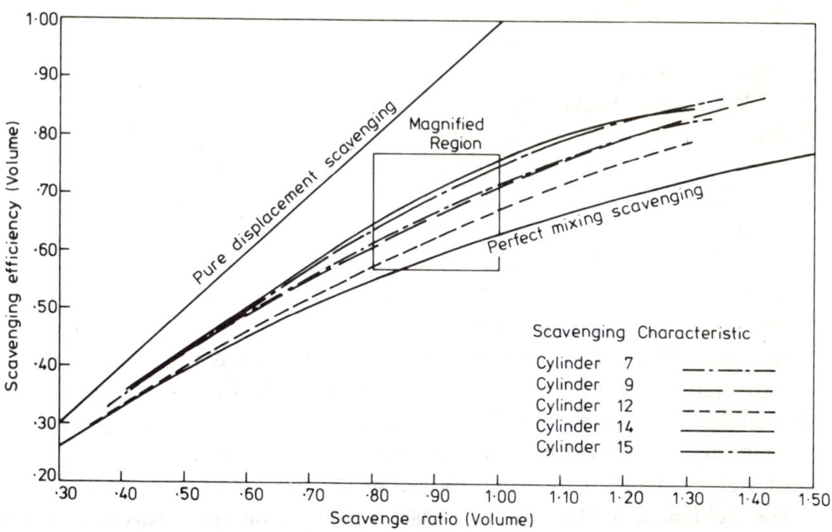

Fig. 23. Measured isothermal scavenging efficiency–scavenge ratio characteristics for cylinders numbered 7, 9, 12, 14 and 15.

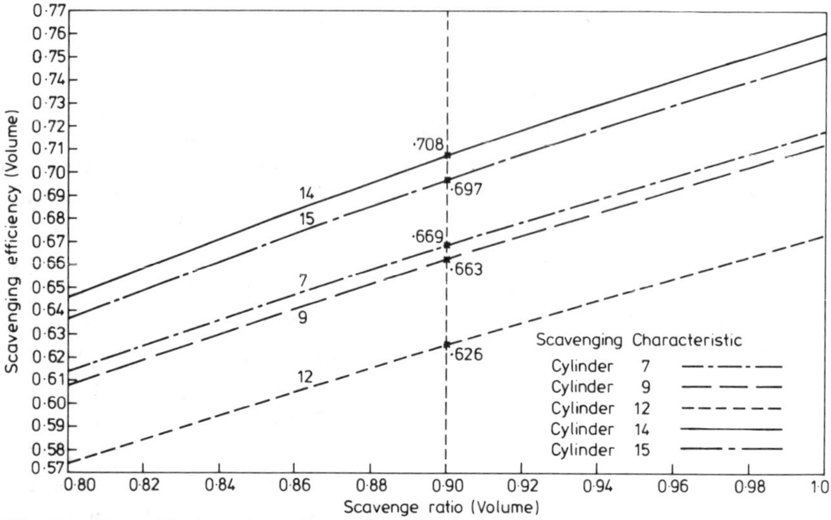

Fig. 24. Magnified region of measured scavenging characteristics for cylinders 7, 9, 12, 14 and 15.

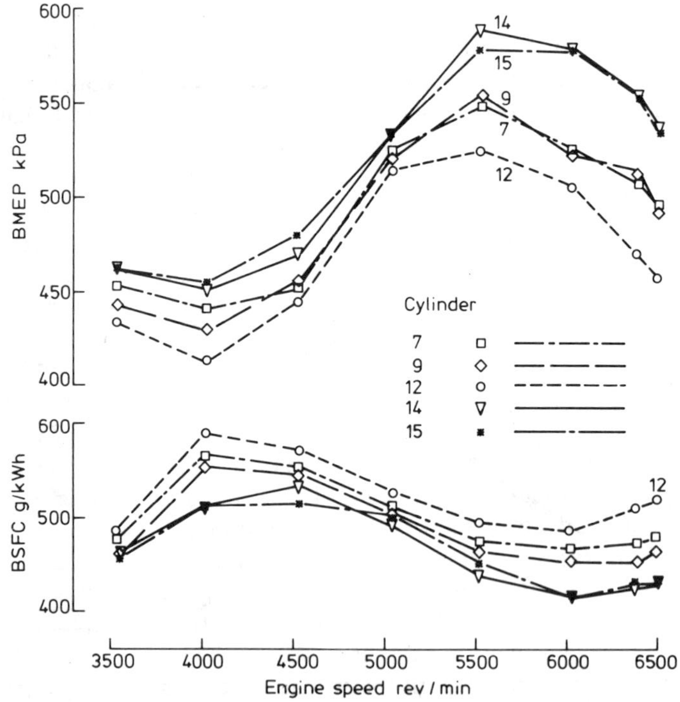

Fig. 25. Measured full throttle BMEP–BSFC–speed characteristics for cylinders numbered 7, 9, 12, 14 and 15.

number 12. The reader's natural question as to the effectiveness of this experimental assessment technique will be answered by an examination of Fig. 25, which shows the torque (as BMEP) and brake specific fuel consumption characteristics over the speed range at full throttle for the named cylinders 14, 15, 9, 7 and 12. It can be seen that the single-cycle test results in Fig. 24 rank the cylinders in the correct order when compared with the performance characteristics for those same cylinders. From the designer's standpoint it is possible to develop the porting of any engine, and compare the effectiveness of loop, cross or uniflow scavenging on such a rig. The test cylinder can be made in a plastic, wood, or a low melting point alloy and the ports so modified as to produce the highest scavenging efficiency–scavenging ratio characteristic prior to the expensive procedure of pre-production casting die investment. This is totally feasible with a single-cycle test apparatus using 'cold' gases for the simulation.

9. MATHEMATICAL MODELLING OF THE SCAVENGING PROCESS

In the future it may be possible to use a computational fluid dynamics program to calculate the movement of the gases during the scavenging process, based on the actual geometry of the exhaust and transfer ports. Sweeney *et al.*[42] used the PHOENICS software developed by Spalding[43] (see Chapter 11) to compute the scavenging flows in the cases of the various cylinders used on their single-cycle simulation rig. Cylinders numbered 12 and 14 were used as the examples presented in their paper[42] and an illustration of the future potential of that calculation technique is given in Figs 26 and 27. In Figs 26 and 27 are shown the concentration plots of fresh charge for cylinders 14 and 12 respectively at a particular crank angle of 9° BBDC. Other figures in that paper[42] similarly cover the complete open cycle. In each case the individual figures (a), (b), (c) and (d) show elevation sections, through the cylinder vertically and transversely in (a) and (b) and plan sections at the piston crown in (c) and at half-stroke in (d). Examination of Figs 26 and 27 explains why, for the first time since Hopkinson[1] raised the issue, one cylinder has superior scavenging behaviour to another. The fresh charge concentrations by 9° BBDC in front of the exhaust port range from 0·1 to 0·3 for the 'good' cylinder 14 and 0·1 to 0·6 for

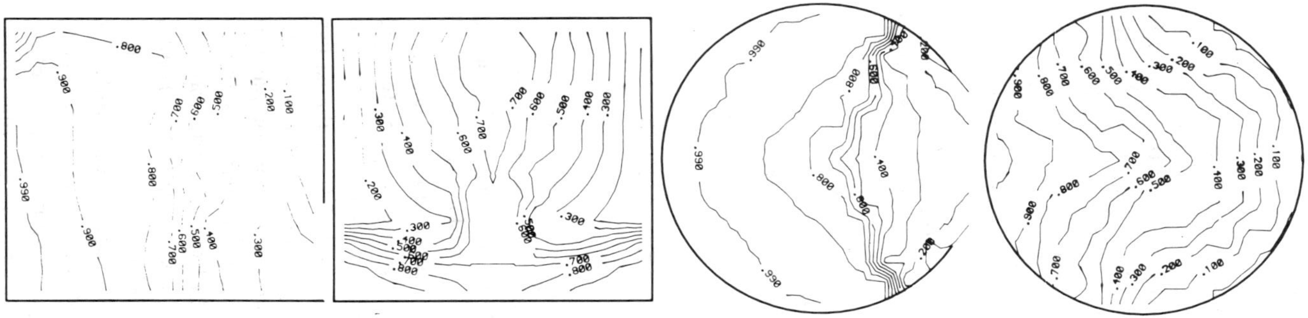

Fig. 26. Scavenging efficiency concentration plots at 9° BBDC for cylinder numbered 14.

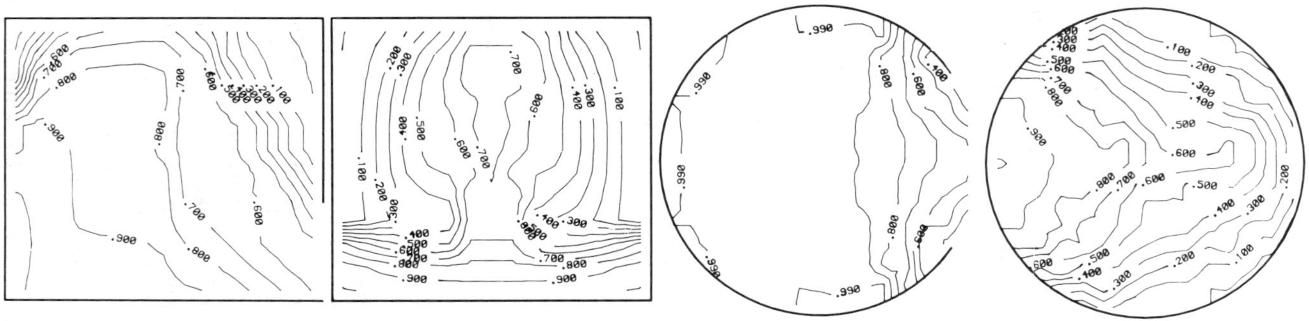

Fig. 27. Scavenging efficiency concentration plots at 9° BBDC for cylinder numbered 12.

the 'bad' cylinder 12. Not surprisingly, cylinder 14 has better scavenging than cylinder 12. The overall cycle effects coincide well both by calculation and experiment, as reported in that paper.[42] The designer's questions may well be finally answered—he can have the fine detail of his scavenge port's geometry detailed by computer to obtain a maximized scavenging efficiency, as well as having the overall physical geometry of the engine's port timing, area and ducting also optimized by computer so as to produce a desired set of performance characteristics.

As an addendum to the above discussion on scavenging, the reader may consult a paper by Blair *et al.*,[44] which uses the single-cycle testing rig to compare experimentally loop and cross-scavenging of two-stroke cycle engines. The conclusion drawn is that loop scavenging is superior to the conventional method of cross-scavenging, but that an alternative method of cross-scavenging has better scavenging than the best known loop scavenging values. The loop scavenging versus cross-scavenging debate is one which has stimulated many a technical conversation among designers of two-stroke engines.

10. CONCLUDING DISCUSSION

By now the reader will have gathered that if the best scavenging system yet seen has a scavenging efficiency of 0·7 at a scavenge ratio of 1·0 then some 30% of the fresh charge is leaving the cylinder of the engine. If the engine is carburetted then 30% of the fuel is also lost. This contributes to an automatic 30% deterioration in specific fuel consumption and a horrendously high unburned hydrocarbon count in the exhaust emissions. On the other hand, due to the 30% exhaust gas retention, and the fact that the engine is firing on every cycle, there is a reduced nitrogen oxide emission after combustion by comparison with the four-stroke cycle power unit. Many researchers have, naturally, looked at scavenging the engine with fresh air and supplying the fuel by injection after the exhaust ports have closed. The results reported can only be described as 'mixed', and to obtain a fuller understanding of the problems involved the reader should study the papers by Yamagishi *et al.*,[45] Pichard,[46] Vieilledent[47] and Douglas & Blair.[48] In general, the results quoted show specific fuel consumption figures reduced by some 20–30%, but not to the levels one could expect if all of the fuel short-circuiting is avoided. The fundamental reason seems to be that the mixing time for the fuel and air from exhaust port closure is too short, so that to obtain good mixing/evaporation of the fuel before combustion the injection starting point must be brought before exhaust port closure. The result is then fuel short-circuiting. It would appear that fuel injection of the two-stroke engine awaits two things, a rapid and effective injection system which will operate at engine cycling speed and a combustion system which is sufficiently turbulent as to promote the pre-combustion mixing process. Should these 'inventions' be forthcoming then the two-stroke engine may become the automobile engine of the future, for it has other excellent characteristics, such as the following: good combustion of a lead-free fuel; good specific power output by weight, volume/bulk or capacity; good low-speed fuel economy due to reduced pumping/friction losses; low NO_x exhaust emission due to the built-in EGR (exhaust gas recirculation) process due to lower scavenging efficiency as compared to a four-stroke.

However, not all traditional two-stroke engine markets can sustain the application of expensive fuel injection systems. Yet for outboards, lawnmowers, motorcycles and chainsaws there is a need to reduce fuel consumption and exhaust emissions. Hence designs have appeared which lay stress on the effectiveness of the scavenging process to improve the engine's performance characteristics. Notable among these efforts are those by Onishi *et al.*[49] and Blair *et al.*[50,51] Onishi *et al.*[49] conclude that by paying careful attention to controlling the streams of liquid fuel in the transfer ducts during scavenging flow it is possible to have a carburetted two-stroke engine return fuel consumption and exhaust hydrocarbon emission figures which are little different from those regarded as conventional for four-stroke engines. Blair *et al.*[50] and Hill *et al.*[50,51] use a stratified charging approach to keep a rich air–fuel mixture well away from the exhaust port during scavenging, and hence they report fuel consumption figures similar to those declared by Onishi *et al.*[49] These are significant results, for in both cases the reported fuel consumption values would be some 25–40% less than those regarded as conventional for two-stroke motorcycle power units.

REFERENCES

1. Hopkinson, B., The charging of two-cycle internal combustion engines. *Trans. N E Coast Inst. Engrs Shipbldrs*, **30** (1914) 433.

2. Bird, A. L., *Oil Engines*. Methuen, London, 1923.
3. Taylor, C. F. & Rogowski, A. R. Scavenging the two-stroke engine. *Trans. SAE*, **62** (1954) 487.
4. Kadenacy, M., British Patents 431856, 431857 etc.
5. Weaving, J. H., Two-strokes and turbines. *Proc. Inst. Mech. Engrs.* **183** (October 1968); Discharge of exhaust gases in two-stroke engines. *Proc. Inst. Mech. Engng*, **161** (1949) 98.
6. Johnston, S. C., A characterisation of unsteady gas distilled from a vessel. Sandia Laboratories, Jan. 1975.
7. Carter, H. D., Loop scavenged diesel engine. *Proc. Inst. Mech. Engrs* **154** (1946) 386.
8. Giffen, E., Rapid discharge of a gas from vessel into atmosphere. *Engineering*, **150** (1940) 139.
9. Mucklow, G. F. & Bannister, F. K. Wave action following sudden release of compressed gas. *Proc. Inst. Mech. Engrs.* **159** (1948) 269.
10. Jenny, E., Unidimensional transient flow with consideration of friction, heat transfer and change of section. *Brown Boveri Review*, **37**(11) (1950) 447.
11. De Haller, R., The application of a graphic method to some dynamic problems in gases. *Sulzer Tech. Rev.* **1** (1945) 6.
12. Wallace, F. J. & Mitchell, R. W. S., Wave action following the sudden release of air through an engine port system. *Proc. Inst. Mech. Engrs*, **1B** (1952–3) 343–63.
13. Wallace, F. J. & Boxer, G., Wave action in diffusers for exhaust pipe systems, with special reference to the scavenging of two-stroke engine. *Proc. Inst. Mech. Engrs*, **170**(39) (1956) 1131–56.
14. Wallace, F. J. & Nassif, M. H., Air flow in a naturally aspirated two-stroke engine. *Proc. Inst. Mech. Engrs*, **168**(18) (1954) 515–41.
15. Benson, R. S., Garg, R. P. & Woollatt, D., A numerical solution of unsteady flow problems. *Int. J. Mech. Engng Sci.*, **6**(1) (1964) 117–44.
16. Schweitzer, P. H., *Scavenging of Two-stroke Cycle Diesel Engines*. Macmillan, New York, 1949.
17. Naitoh, H. & Taguchi, M., Some development aspects of two-stroke cycle motorcycle engines. *Trans. SAE* 660394, 1966.
18. Naitoh, H. & Nomura, K., Some new development aspects of two-stroke cycle motorcycle engines. *Trans. SAE* 710084, 1971.
19. Blair, G. P., Further developments of a 500 cc single cylinder 2-cycle motorcycle engine for motorcycle racing and moto-cross applications. *Trans. SAE* 740745, 1974.
20. Bannister, F. K., Pressure waves in gases in pipes. Ackroyd Stuart Memorial Lectures, University of Nottingham, 1958.
21. Annand, W. J. D. & Roe, G. E., *Gas Flow in the Internal Combustion Engine*. G. T. Foulis, Yeovil, England, 1974.
22. Blair, G. P. & Johnston, M. B., Simplified design criteria for expansion chambers for two-stroke gasoline engines. *Trans. SAE* 700123, 1970.
23. Blair, G. P. & Cahoon, W. L., A more complete analysis of unsteady gas flow through a high specific output two-cycle engine. *Trans. SAE* 720156, 1972.

24. Blair, G. P. & Ashe, M. C., The unsteady gas exchange characteristics of a two-cycle engine. *Trans. SAE* 760644, 1976.
25. Blair, G. P., Prediction of two-cycle engine performance characteristics. *Trans. SAE* 760645, 1976.
26. Blair, G. P., Hinds, E. T. & Fleck, R., Predicting the performance characteristics of two-cycle engines fitted with reed induction valves. *Trans. SAE* 790842, 1979.
27. Hinds, E. T. & Blair, G. P., Unsteady gas flow through reed valve induction systems. *Trans. SAE* 780766, 1978.
28. Benson, R. S., Woollatt, D. & Woods, W. A., Unsteady flow in simple branch systems. *Proc. Inst. Mech. Engrs,* **178** (Pt. 31 iii, No. 10) (1963–4) 21.
29. Bingham, J. F. & Blair, G. P., An improved branched pipe model for multi-cylinder automotive engine calculations. *Proc. Inst. Mech. Engrs,* **199** (D11) (1984–5) 65–77.
30. Blair, G. P. & Spechko, J. A., Sound pressure levels generated by internal combustion engine exhaust systems. *Trans. SAE* 720155, 1972.
31. Blair, G. P. & Coates, S. W., Noise produced by unsteady exhaust efflux from an internal combustion engine. *Trans. SAE* 730160, 1973.
32. Coates, S. W. & Blair, G. P., Further studies of noise characteristics of internal combustion engine exhaust systems. *Trans. SAE* 740713, 1974.
33. Roe, G. E., An empirical approach to motorcycle silencing. *Trans. SAE* 770188, 1977.
34. Jones, A. D. & Brown, G. L., Determination of two-stroke engine exhaust noise by the method of characteristics. *J. Sound Vib.,* **82**(3) (1982) 305–27.
35. Mattavi, J. N. & Amman, C. A., *Combustion Modelling in Reciprocating IC Engines, Proceedings of a Symposium Held at General Motors.* Plenum Press, New York, 1980.
36. Sweeney, M. E. G., Kenny, R. G., Swann, G. B. & Blair, G. P., Single cycle gas testing method for two-stroke engine scavenging. *Trans. SAE* 850178, 1985.
37. Blair, G. P. & Kenny, R. G., Further developments in scavenging analysis for two-stroke engines. *Trans. SAE* 800038, 1980.
38. Jante, A., Scavenging and other problems of two-stroke cycle spark-ignition engines. *Trans. SAE* 680468, 1968.
39. Blair, G. P., Studying scavenge flow in a two-stroke cycle engine. *Trans. SAE* 750752, 1975.
40. Sanborn, D. S., Blair, G. P., Kenny, R. G. & Kingsbury, A. H., Experimental assessment of scavenging efficiency of two-stroke cycle engines. *Trans. SAE* 800975, 1980.
41. Sammons, H., A single-cycle test apparatus for studying loop scavenging in a two-stroke engine. *Proc. Inst. Mech. Engrs,* **161** (1949) 233–42.
42. Sweeney, M. E. G., Swann, G. B., Kenny, R. G. & Blair, G. P., Computational fluid dynamics applied to two-stroke engine scavenging. *Trans. SAE* 851519, 1985.
43. Spalding, D. B., A general purpose computer program for multi-dimensional one and two-phase flow. *Mathematics and Computers in Simulation,* **23** (1981) 267–76.

44. Blair, G. P., Kenny, R. G., Smyth, J. G., Sweeney, M. E. G. & Swann, G. B., An experimental comparison of loop and cross-scavenging of the two-stroke cycle engine. *Trans. SAE* 861240, 1986.
45. Yamagishi, G., Suto, T. & Iwasa, W., A study of two-stroke cycle fuel injection engine for exhaust gas purification. *Trans. SAE* 720195, 1972.
46. Pichard, J., Two-cycle stratified charge engine. Fourth International Symposium on Automobile Propulsion, 17–22 April 1977, Washington DC, by ERDA.
47. Vieilledent, E., Low pressure electronic fuel injection system for two-stroke engines. *Trans. SAE* 780767, 1978.
48. Douglas, R. & Blair, G. P., Fuel injection of a two-stroke cycle spark ignition engine. *Trans. SAE* 820952, 1982.
49. Onishi, S., Jo, S. H., Jo, P. D. & Kato, S., Multi-layer stratified scavenging (MULS)—a new scavenging method for two-stroke engine. *Trans. SAE* 840420, 1984.
50. Blair, G. P., Hill, B. W., Miller, A. J. & Nickell, S. P., Reduction of fuel consumption of a spark ignition two-stroke cycle engine. *Trans. SAE* 830093, 1983.
51. Hill, B. W. & Blair, G. P., Further tests on reducing fuel consumption with a carburetted two-stroke cycle engine. *Trans. SAE* 831303, 1983.

4

The Two-stroke Engine: the Blowdown and Uniflow Scavenge Process

F. J. Wallace

School of Mechanical Engineering, University of Bath, Bath, UK

1. Introduction
2. Fundamental considerations
   2.1. Port or valve arrangements for uniflow scavenged engines
   2.2. Formulation of flow equations for the blowdown process
   2.3. Incorporation of blowdown equations in general cycle simulation schemes
3. Practical implications
   3.1. Exhaust lead calculations
   3.2. Exhaust pulse calculations
4. Uniflow scavenging of two-stroke engines
   4.1. Introduction
   4.2. The isobaric–isochoric scavenging model
   4.3. Phenomenological (jet-based) scavenging models
5. Concluding remarks

1. INTRODUCTION

Chapter 3 has dealt with the loop scavenged system used principally for spark-ignition two-stroke engines and has dealt mainly with that part of the gas exchange process when the inlet and exhaust ports are simultaneously open, allowing replacement of the products of combustion by fresh air. Diesel engines, on the other hand, more often use the uniflow system of scavenging which will be considered in the second half of this chapter. A comparison will also be made with the loop and cross-scavenge systems.

The blowdown process only briefly alluded to in the previous chapter will be treated in detail in the first part of this chapter. It is the important phase of the gas exchange process preceding inlet port opening, viz. the period which occurs between exhaust opening and inlet opening and which allows the rapid escape of a large part of the products of combustion from the cylinder into the exhaust manifold, such that at 'inlet port opening' the cylinder pressure has ideally dropped to the pressure in the inlet manifold. The analysis in the second half of this chapter is largely concerned with the degree to which the process of replacement of the products of combustion succeeds. This is based on a greatly simplified model in which dynamic effects in the external manifolding are completely neglected and in which, furthermore, the in-cylinder process is assumed to proceed in an isobaric, isochoric manner; i.e. at constant pressure and constant volume. Finally, the analysis proceeds on the assumption that there are three separate zones of continuously changing volume:

(a) a diminishing residual zone;
(b) a growing pure air zone;
(c) a zone consisting of a mixture of residual gas and pure air.

By making assumptions about the flows of air or gas into and out of, as well as between, these zones, it is possible to model most types of scavenging systems and to assess their relative merits.

For full simulation of the two-stroke cycle, whether for diesel or spark-ignition engines, much more complex modelling is required, taking into account the full complexity of the in-cylinder processes during the scavenging phase, and of the flows through valves or ports of continuously varying area and with continuously varying pressures both upstream and downstream. These latter boundary conditions have to be combined with the conditions in the inlet and exhaust manifolds which may be evaluated by one of two basic methods:

(a) The so-called filling and emptying method[1–3] which treats the manifolds as single volumes subject to inflows and outflows and experiencing continuous changes in pressure, temperature and, in the case of exhaust manifolds, also of composition.
(b) The wave action method which takes into account the propagation of pressure pulses induced by the intermittent nature of the flows into and out of the cylinder, and therefore models not only the temporal, but also the spatial variations of pressure,

temperature and composition in the manifolds. The literature on the various methods of calculating wave action effects is extremely extensive.[4–6]

It is not the object of this short chapter to deal comprehensively either with the filling and emptying or the wave action technique, but rather to concentrate on the essentials of the blowdown process insofar as they affect the timing of inlet and exhaust opening under different engine operating conditions, and with either constant or variable exhaust manifold pressure.

2. FUNDAMENTAL CONSIDERATIONS

2.1. Port or Valve Arrangements for Uniflow Scavenged Engines

In Figure 1 is shown a typical geometric area schedule for the exhaust and inlet ports of an opposed-piston two-stroke diesel engine, as a function of crank angle for the exhaust piston. The timing of the exhaust and inlet events is strongly asymmetric, with the exhaust opening some 20–30° CA before the inlet and closing slightly before inlet closing. The flat top of the exhaust opening profile is due to the fact that the exhaust piston, after completely uncovering the exhaust ports, continues to travel towards its own BDC while, in the case of the inlet ports, these are fully uncovered only when the inlet piston has reached its own BDC position.

The reason for asymmetric timing is that, as already stated, it is essential to allow sufficient time for the blowdown process to proceed to the point at which the cylinder gas pressure has fallen to the level of the inlet manifold pressure when the inlet ports open, since otherwise blowback of cylinder gases into the inlet manifold will occur, with harmful effects on the subsequent scavenging process. In the opposed-piston engine asymmetric timing is achieved by a phase displacement between the exhaust and inlet piston, of the order of 10–15° CA depending on the required blowdown angle, which is primarily a function of the 'release' conditions at exhaust opening and of engine speed.

The opposed-piston engine represents the optimum solution for both the blowdown and the scavenge process, in that both the rate of

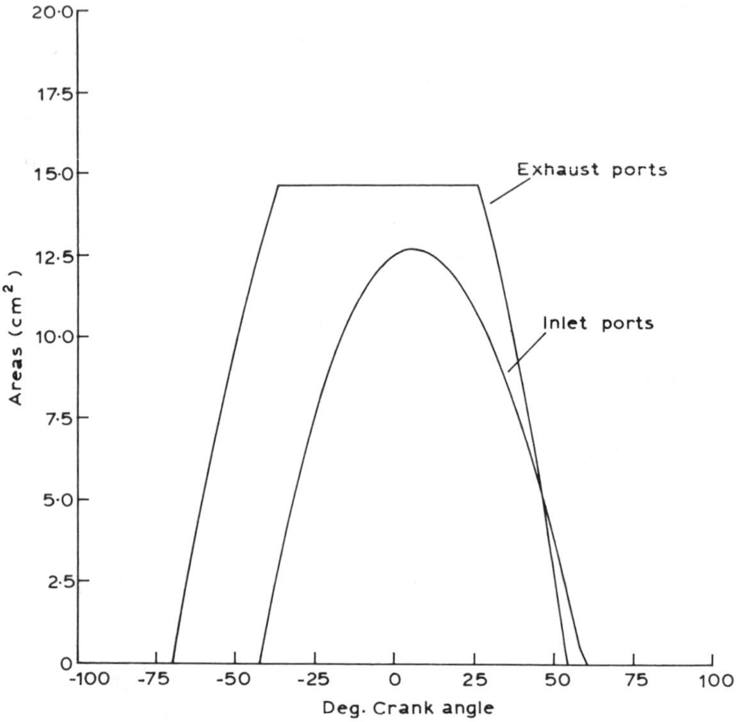

Fig. 1. Porting diagram of optimum design.

port opening and the available flow area can be maximized, resulting in minimum flow resistance and the maximum degree of scavenging. On the other hand, it is mechanically complex and hence expensive to produce. It has found application in high-performance military engines and in the marine propulsion field.

The solution more generally adopted is that of the exhaust-valve-in-head, piston-controlled inlet port engine, of which the Sulzer marine engine is a leading example (Fig. 2). With the exhaust valve situated opposite the inlet ports, the flow of scavenging air through the cylinder is again unidirectional, thus encouraging the maximum degree of displacement scavenging. Both inlet port and exhaust valve areas can be made large, and timing of exhaust and inlet opening can once again be made asymmetric. Hence both the opposed-piston and the exhaust-valve-in-head engines represent highly desirable, albeit expensive, solutions of the scavenging problem.

Reference must also be made to the simplest type of two-stroke engine, in which a single piston controls both the inlet or transfer and the exhaust ports (Fig. 3). Depending on the port arrangement, these engines are either of the loop or cross-scavenged type, with inherently symmetrical timing. The chief disadvantage of the latter is that provision of the necessary exhaust lead for blowdown purposes inevitably involves an equal lag between inlet and exhaust closing. The latter results in loss of trapped charge from the cylinder to the exhaust after inlet closing, this being particularly damaging to efficiency in the case of spark-ignition engines, in which loss of trapped charge implies loss of air–fuel mixture rather than pure air.

2.2. Formulation of Flow Equations for the Blowdown Process

The blowdown process commences at exhaust opening when the cylinder gas conditions are assumed to be known.

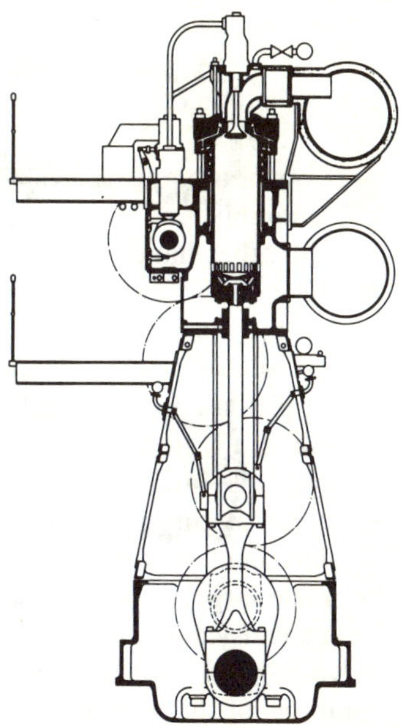

Fig. 2. Exhaust-valve-in-head engine.

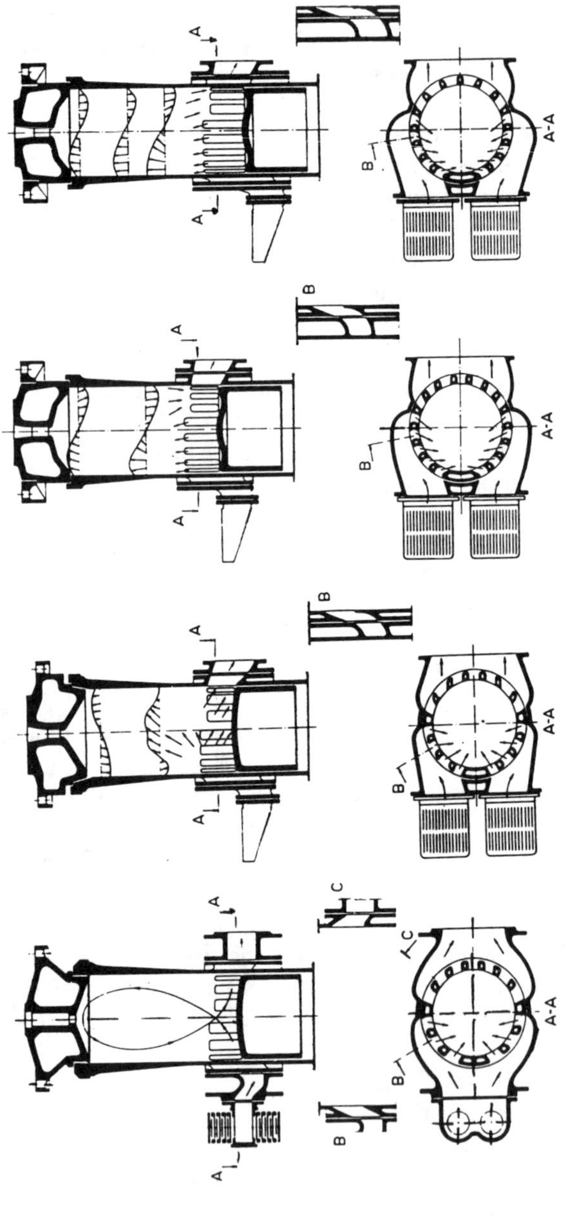

Fig. 3. Loop scavenged engine.

$$p_{cyl} = p_{release} \qquad \text{(pressure)}$$

$$T_{cyl} = T_{release} \qquad \text{(temperature)}$$

$$f_{cyl} = f_{release} \qquad \text{(equivalence ratio)}$$

In addition to these upstream conditions, the downstream pressure p_{exh} in the exhaust manifold is also assumed to be known, particularly when dealing with naturally aspirated or constant-pressure turbocharged engines or, in the case of pulse-turbocharged or wave action assisted naturally aspirated engines, where instantaneous exhaust manifold pressure is obtained step-by-step from appropriate calculations extending over the whole cycle.[7]

The other information required concerns the variation of the geometric exhaust flow area with crank angle (see Fig. 1) which must be combined with an experimentally obtained coefficient of discharge schedule, such that

$$(A_{exh})_{effective} = (A_{exh})_{geometric} \times COD = f(\theta) \tag{1}$$

where θ is the crank angle. While the actual flow is extremely complex, being both three-dimensional and unsteady, i.e. involving the three spatial variables x, y, z as well as time t as independent variables, calculations are invariably based on a drastic simplification in which the flow is treated as one-dimensional and quasi-steady (i.e. during any one calculation step, steady state conditions are assumed to prevail) and in which $(A_{exh})_{effective}$ is taken as the throat area A_t of a simple orifice type restriction that is concentric with the exhaust pipe of area A_{exh}. This simplified flow configuration is shown in Fig. 4.

Finally, a distinction must be made between the *sonic* and the *subsonic* flow regimes. The former occurs in the early part of blowdown, when the pressure ratio p_{cyl}/p_{exh} is sufficiently large to cause choking at the throat. Eventually the transition takes place from sonic to subsonic flow, with the mass discharge rate now becoming a function of downstream as well as upstream conditions.

2.2.1. Sonic Discharge Phase

The equations used are the equation of state, and the conservation equations for mass and energy. In addition, the flow from the cylinder to the throat is assumed to be isentropic, while that from the throat to the pipe is irreversible, but has to satisfy mass continuity.

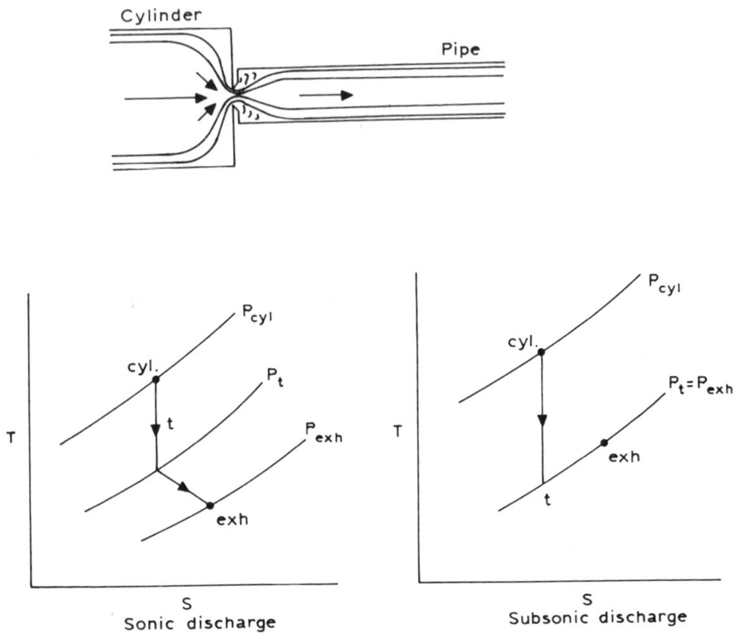

Fig. 4. Simplified exhaust flow configuration.

Energy equation—cylinder to throat. Throat velocity U_t

$$U_t^2 = a_t^2 = 2C_p(T_{\text{cyl}} - T_t) \tag{2a}$$

where suffix t refers to conditions at the throat, and where

$$\frac{T_t}{T_{\text{cyl}}} = \left(\frac{p_t}{p_{\text{cyl}}}\right)^{(\gamma-1)/\gamma} = (PR)_{\text{crit}}^{(\gamma-1)/\gamma}$$

PR is the pressure ratio, and $(PR)_{\text{crit}} = [2/(\gamma+1)]^{\gamma/(\gamma-1)}$, i.e.

$$U_t^2 = a_t^2 = a_{\text{cyl}}^2 \frac{T_t}{T_{\text{cyl}}} = a_{\text{cyl}}^2\left(\frac{2}{\gamma+1}\right) \tag{2b}$$

Mass continuity—cylinder to throat. The rate of discharge

$$\dot{m}_t = A_t\rho_t U_t$$
$$= A_t\rho_{\text{cyl}}a_{\text{cyl}}\left(\frac{2}{\gamma+1}\right)^{(\gamma+1)/2(\gamma-1)} \tag{3}$$

Although the downstream conditions do not influence the rate of discharge when the flow is sonic, relationships have to be established between the conditions in the pipe and at the throat, under now non-isentropic state relationships.

Energy equation—throat to pipe. Exhaust pipe velocity U_{exh}

$$U_{\text{exh}}^2 - U_t^2 + 2C_p(T_{\text{exh}} - T_t) = 0$$

or

$$\begin{aligned} U_{\text{exh}}^2 &= U_t^2 + 2C_p(T_t - T_{\text{exh}}) \\ &= a_{\text{cyl}}^2\left(\frac{2}{\gamma+1}\right) + 2C_p\left[T_{\text{cyl}}\left(\frac{2}{\gamma+1}\right) - T_{\text{exh}}\right] \end{aligned} \tag{4}$$

where suffix 'exh' refers to conditions in the exhaust manifold.

Mass continuity—throat to pipe:

$$\dot{m}_t = \dot{m}_{\text{exh}}$$

i.e.

$$\begin{aligned} A_t\rho_t U_t &= A_{\text{exh}}\rho_{\text{exh}}u_{\text{exh}} \\ &= A_{\text{exh}}(p_{\text{exh}}/RT_{\text{exh}})u_{\text{exh}} \end{aligned}$$

or

$$A_t\rho_{\text{cyl}}a_{\text{cyl}}\left(\frac{2}{\gamma+1}\right)^{(\gamma+1)/2(\gamma-1)} = A_{\text{exh}}\left(\frac{p_{\text{exh}}}{RT_{\text{exh}}}\right)u_{\text{exh}} \tag{5}$$

With p_{exh} known, eqns (4) and (5) may be solved simultaneously for the pipe velocity U_{exh} and the temperature T_{exh}.

Finally, the rate of discharge through the throat area is equated to the rate of change of mass contents of the cylinder. Therefore,

$$\begin{aligned} \frac{\mathrm{d}}{\mathrm{d}t}(V_{\text{cyl}}\rho_{\text{cyl}}) &= -A_t\rho_t U_t \\ &= -A_t\rho_{\text{cyl}}a_{\text{cyl}}\left(\frac{2}{\gamma+1}\right)^{(\gamma+1)/2(\gamma-1)} \end{aligned} \tag{6a}$$

Since the expansion in the cylinder during the very rapid blowdown process can also be assumed to occur isentropically, cylinder density and acoustic velocity can be simply related to the known release conditions. Thus:

$$\rho_{\text{cyl}} = \rho_{\text{rel}}(p_{\text{cyl}}/p_{\text{rel}})^{1/\gamma} \qquad \text{and} \qquad a_{\text{cyl}} = a_{\text{rel}}(p_{\text{cyl}}/p_{\text{rel}})^{(\gamma-1)/2\gamma} \tag{i,ii}$$

Substituting from the auxiliary (i) and (ii) in eqn (6a) and putting

$$\frac{\mathrm{d}}{\mathrm{d}t}=\frac{\mathrm{d}}{\mathrm{d}\theta}(6N_e)$$

where N_e is the engine speed (in rpm), we have

$$6N_e\frac{\mathrm{d}}{\mathrm{d}\theta}\left[V_{\mathrm{cyl}}\rho_{\mathrm{rel}}\left(\frac{p_{\mathrm{cyl}}}{p_{\mathrm{rel}}}\right)^{1/\gamma}\right]=-A_t(\rho_{\mathrm{rel}}a_{\mathrm{rel}})\left(\frac{p_{\mathrm{cyl}}}{p_{\mathrm{rel}}}\right)^{(\gamma+1)/2\gamma}\left(\frac{2}{\gamma+1}\right)^{(\gamma+1)/2(\gamma-1)} \tag{6b}$$

With cylinder volume and throat, i.e. effective exhaust port or valve area, both known as functions of crank angle θ (see eqn (1)), the cylinder pressure p_{cyl} throughout the sonic phase of discharge may be determined from eqn (6b), applied step-by-step using suitable numerical techniques.

2.2.2. Subsonic Discharge Phase

Under subsonic conditions, the throat velocity U_t falls below the acoustic velocity a_t, and the rate of discharge becomes dependent on the downstream pressure p_{exh}.

Two flow models may be adopted for the flow from the throat to the pipe:

(a) the simple constant pressure model, i.e. $p_t=p_{\mathrm{exh}}$, which neglects any possible pressure recovery between throat and pipe;
(b) the pressure recovery model which, by the combined application of the conservation equations for momentum and energy, makes it possible to evaluate the notional pressure rise $(p_{\mathrm{exh}}-p_t)$ between throat and pipe.

Only the simple model (a) will be considered here (see also Fig. 4).

Transition from sonic to subsonic discharge. Since $p_t=p_{\mathrm{exh}}$ and with p_{exh} known, a simple test for sonic flow may be applied:

$$\text{sonic:}\ \left(\frac{p_{\mathrm{exh}}}{p_{\mathrm{cyl}}}\right)<(PR)_{\mathrm{crit}}=\left(\frac{2}{\gamma+1}\right)^{\gamma/(\gamma-1)} \tag{7a}$$

$$\text{subsonic:}\ \left(\frac{p_{\mathrm{exh}}}{p_{\mathrm{cyl}}}\right)>(PR)_{\mathrm{crit}} \tag{7b}$$

$$\text{transition:}\ \left(\frac{p_{\mathrm{exh}}}{p_{\mathrm{cyl}}}\right)=(PR)_{\mathrm{crit}}=\left(\frac{2}{\gamma+1}\right)^{\gamma/(\gamma-1)} \tag{7c}$$

Subsonic discharge equations. The subsonic discharge equations may again be solved by the simple application of the conservation equations for energy and mass, first between cylinder and throat and then between throat and pipe.

Energy equation—cylinder to throat:

$$U_t^2 = 2C_p(T_{\mathrm{cyl}} - T_t)$$

and with $p_t = p_{\mathrm{exh}}$ and isentropic expansion up to the throat

$$T_t = T_{\mathrm{cyl}}(p_{\mathrm{exh}}/p_{\mathrm{cyl}})^{(\gamma-1)/\gamma}$$

whence

$$U_t^2 = 2C_pT_{\mathrm{cyl}}[1 - (p_{\mathrm{exh}}/p_{\mathrm{cyl}})^{(\gamma-1)/\gamma}] \tag{8}$$

Mass continuity—cylinder to throat. The rate of discharge,

$$\begin{aligned}\dot{m}_t &= A_t\rho_tU_t\\ &= A_t\rho_{\mathrm{cyl}}(p_{\mathrm{exh}}/p_{\mathrm{cyl}})^{1/\gamma}\{2C_pT_{\mathrm{cyl}}[-(p_{\mathrm{exh}}/p_{\mathrm{cyl}})^{(\gamma-1)/\gamma}]\}^{1/2}\end{aligned} \tag{9}$$

Energy equation—throat to pipe ($p_t = p_{\mathrm{exh}}$):

$$U_{\mathrm{exh}}^2 - U_t^2 + 2C_p(T_{\mathrm{exh}} - T_t) = 0$$

i.e.

$$\begin{aligned}U_{\mathrm{exh}}^2 &= U_t^2 + 2C_p[T_{\mathrm{exh}} - T_{\mathrm{cyl}}(p_{\mathrm{exh}}/p_{\mathrm{cyl}})^{(\gamma-1)/\gamma}]\\ &= 2C_p\{T_{\mathrm{cyl}}[1 - 2(p_{\mathrm{exh}}/p_{\mathrm{cyl}})^{(\gamma-1)/\gamma}] + T_{\mathrm{exh}}\}\end{aligned} \tag{10}$$

Mass continuity—throat to pipe:

$$A_t\rho_tU_t = A_{\mathrm{exh}}\rho_{\mathrm{exh}}U_{\mathrm{exh}}$$

i.e.

$$\begin{aligned}&A_t\rho_{\mathrm{cyl}}(p_{\mathrm{exh}}/p_{\mathrm{cyl}})^{1/\gamma}\{2C_pT_{\mathrm{cyl}}[1 - (p_{\mathrm{exh}}/p_{\mathrm{cyl}})^{(\gamma-1)/\gamma}]\}^{1/2}\\ &\quad = A_{\mathrm{exh}}(p_{\mathrm{exh}}/RT_{\mathrm{exh}})(2C_p)^{1/2}\\ &\quad\quad \times \{T_{\mathrm{cyl}}[1 - 2(p_{\mathrm{exh}}/p_{\mathrm{cyl}})^{(\gamma-1)/\gamma}] + T_{\mathrm{exh}}\}^{1/2}\end{aligned} \tag{11}$$

Equation (9) directly determines the discharge rate, while eqns (10) and (11) may be solved for the pipe velocity U_{exh} and temperature T_{exh}.

The overall calculation scheme will continue to include the cylinder discharge, eqn (6), now in its subsonic form, i.e.

$$\begin{aligned}\frac{\mathrm{d}}{\mathrm{d}t}(V_{\mathrm{cyl}}\rho_{\mathrm{cyl}}) &= 6N_e\frac{\mathrm{d}}{\mathrm{d}\theta}[V_{\mathrm{cyl}}\rho_{\mathrm{rel}}(p_{\mathrm{cyl}}/p_{\mathrm{rel}})^{1/\gamma}]\\ &= A_t\rho_tU_t\\ &= A_t\rho_{\mathrm{cyl}}(p_{\mathrm{exh}}/p_{\mathrm{cyl}})^{1/\gamma}(2C_pT_{\mathrm{cyl}})^{1/2}[1 - (p_{\mathrm{exh}}/p_{\mathrm{cyl}})^{(\gamma-1)/\gamma}]^{1/2}\end{aligned} \tag{12}$$

Equations (9)–(12) continue to be applied step-by-step, to evaluate the change of cylinder pressure for the remainder of the blowdown period, i.e. up to exhaust opening.

2.3. Incorporation of Blowdown Equations in General Cycle Simulation Schemes

2.3.1. General Comments

Cycle simulation uses the conservation equations for energy, mass and chemical species for the various control volumes—cylinders, manifolds etc.—making up the system, in conjunction with equations for

(a) mass discharge through ports, valves etc.,
(b) heat transfer from gas to wall in cylinders, manifolds, pipes etc., and
(c) fuel burning and/or heat release rates.

The literature on cycle simulation is extremely extensive and a detailed discussion in this chapter would be inappropriate. However, in any two-stroke engine simulation, the equations of Section 2.2, which have been presented in slightly simplified form for greater clarity, would form part of the complete calculation process, with due allowance for variable gas properties—as functions of pressure, temperature and composition—and for departure from isentropic conditions for the expansion process in the cylinder (eqns (6) and (12)).

The cylinder discharge equations (6) and (12) would then appear in the form of a general energy conservation equation together with the differential form of the state equation, to be solved simultaneously for $\mathrm{d}p_{\mathrm{cyl}}$ and $\mathrm{d}T_{\mathrm{cyl}}$.

Energy:

$$\frac{\mathrm{d}}{\mathrm{d}\theta}(mu)_{\mathrm{cyl}} = -p\frac{\mathrm{d}V_{\mathrm{cyl}}}{\mathrm{d}\theta} - \frac{\mathrm{d}Q_{\mathrm{cyl}}}{\mathrm{d}\theta} - \frac{\mathrm{d}m_{\mathrm{cyl}}}{\mathrm{d}\theta}h_{\mathrm{cyl}} \tag{13}$$

where $u_{\mathrm{cyl}} = C_v T_{\mathrm{cyl}}$ is the specific internal energy, $h_{\mathrm{cyl}} = C_p T_{\mathrm{cyl}}$ is the specific enthalpy, and $\mathrm{d}Q_{\mathrm{cyl}}/\mathrm{d}\theta$ is the rate of heat transfer (+ ve when from wall to gas).

State:

$$\frac{\mathrm{d}m_{\mathrm{cyl}}/\mathrm{d}\theta}{m_{\mathrm{cyl}}}+\frac{\mathrm{d}R_{\mathrm{cyl}}/\mathrm{d}\theta}{R_{\mathrm{cyl}}}+\frac{\mathrm{d}T_{\mathrm{cyl}}/\mathrm{d}\theta}{T_{\mathrm{cyl}}}=\frac{\mathrm{d}p_{\mathrm{cyl}}/\mathrm{d}\theta}{p_{\mathrm{cyl}}}+\frac{\mathrm{d}V_{\mathrm{cyl}}/\mathrm{d}\theta}{V_{\mathrm{cyl}}} \tag{14a}$$

where the rate of change of mass in the cylinder $\mathrm{d}m_{\mathrm{cyl}}/\mathrm{d}\theta$ is given by

$$\mathrm{d}m_{\mathrm{cyl}}/\mathrm{d}\theta = -\dot{m}_t/6N_e \tag{14b}$$

in which $\dot{m}_t$, the throat rate of discharge, is in turn given by eqn (3) for sonic, and eqn (9) for subsonic conditions.

In eqns (13) and (14a) all gas properties, i.e. the specific heats C_p and C_v and the gas constant R, as well as the specific heat ratio γ in all previous equations, become functions of pressure, temperature and composition.

2.3.2. Allowance for Pressure Fluctuations in Exhaust Manifold

In the equation of Section 2.2 it was assumed that the exhaust pressure p_{exh} is constant or, at any rate, known. Unless the engine discharges direct to the atmosphere or, in the case of constant-pressure turbocharging, into a large plenum, this is not the case. Under such circumstances blowdown calculations must include a method for simultaneously evaluating exhaust pressure variations by one of two methods:

(a) the so-called filling and emptying method,[1–3] or
(b) the method of characteristics for pressure wave action in the manifold.[4–6]

Only the former will be briefly described, involving the use of the energy and state equations not only for the cylinder, but also for the exhaust manifold

$$\frac{\mathrm{d}}{\mathrm{d}\theta}(mu)_{\mathrm{exh}}=\frac{-\mathrm{d}Q_{\mathrm{exh}}}{\mathrm{d}\theta}+\sum_{i=1}^{u}\left(\frac{\mathrm{d}m_{\mathrm{cyl}}}{\mathrm{d}\theta}h_{\mathrm{cyl}}\right)-\frac{\mathrm{d}m_{\mathrm{exh}}}{\mathrm{d}\theta}h_{\mathrm{exh}} \tag{15}$$

$$\frac{\mathrm{d}m_{\mathrm{exh}}/\mathrm{d}\theta}{m_{\mathrm{exh}}}+\frac{\mathrm{d}R_{\mathrm{exh}}/\mathrm{d}\theta}{R_{\mathrm{exh}}}+\frac{\mathrm{d}T_{\mathrm{exh}}/\mathrm{d}\theta}{T_{\mathrm{exh}}}=\frac{\mathrm{d}p_{\mathrm{exh}}/\mathrm{d}\theta}{p_{\mathrm{exh}}} \tag{16}$$

where the symbols have the same meaning as in eqns (13) and (14), but now apply to the exhaust manifold rather than the cylinder. The summation sign $\sum_{i=1}^{u}$ in eqn (15) allows for simultaneous inflow into the manifold from more than one cylinder, while $\mathrm{d}m_{\mathrm{exh}}/\mathrm{d}\theta$ represents outflow from the manifold.

3. PRACTICAL IMPLICATIONS

3.1. Exhaust Lead Calculations

Although the literature on scavenging of two-stroke engines is very extensive (see Chapter 3), the blowdown period itself has not received very wide attention and there is virtually no information in the literature on practical blowdown results for uniflow engines. Instead, reference will be made to a simplified 'design' approach put forward by Wallace & Nassif[8] in 1954, which illustrates both the simple blowdown theory of Section 2.2 for a constant exhaust pressure system and a method for calculating exhaust pulse amplitudes in the absence of reflections from a turbine.

The required exhaust lead angle in terms of a dimensionless parameter derived directly from eqn (6),

$$N_e V_{\text{cyl}} \bigg/ \left(a_{\text{cyl}} \frac{\mathrm{d}A_t}{\mathrm{d}\theta} \right)$$

is given in Fig. 5 for various values of the release pressure ratio $p_{\text{rel}}/p_{\text{exh}}$. Evidently, high engine speed, large cylinder volume and high values of the release pressure ratio all lead to large values of the required exhaust.

Two widely differing examples will be used, a low-speed exhaust-valve-in-head marine engine (engine A) and a high-speed, opposed-piston, automotive engine (engine B) with the following characteristics:

	Engine A	Engine B
Bore (m)	0·380	0·058
Stroke (m)	1·100	0·125
V_{swept} (m^3)	0·125	$0{\cdot}33 \times 10^{-3}$
Max. valve or port area (m^2)	0·02	0·001 54
Engine speed (rpm)	196	4000
Release pressure ratio	4:1	2·8:1
$N_e V_{\text{cyl}} \bigg/ \left(a_{\text{cyl}} \frac{\mathrm{d}A_t}{\mathrm{d}\theta} \right)$	48·0	49·7
Exhaust lead (°CA) (from Fig. 5)	29	25

These values of exhaust lead correspond closely to those adopted in the actual design of engines A and B.

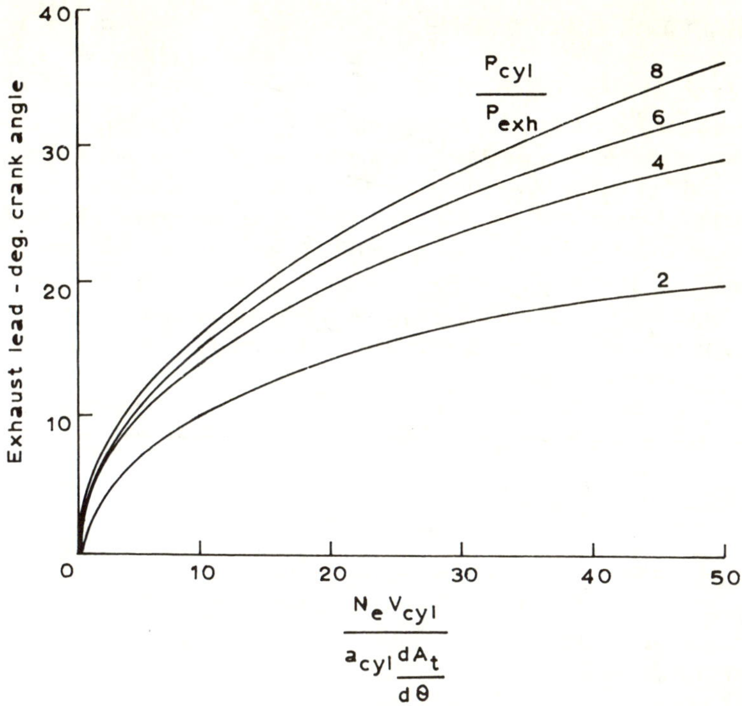

Fig. 5. Exhaust lead chart. n = Engine speed (rpm); a_{cyl} = acoustic velocity of undisturbed air.

3.2. Exhaust Pulse Calculations

A design chart for exhaust pulse calculations, based on a simplified pressure wave approach, is shown in Fig. 6. The dimensionless parameters used are $N_e V_{cyl}/A_{pipe}a_{cyl}$ along the abscissa, and the non-dimensional rate of port or valve area opening $(dA_t/d\theta)/A_{pipe} = b$, as the applicable value for each of the three 'carpets', which in turn contain lines of constant release pressure ratio p_{rel}/p_a and of crank angle θ at which the peak amplitude, represented by the ordinate, p_{max}/p_0, occurs.

It is evident that high engine speed and a high rate of port opening lead to large pulse amplitudes, with a real amplitude p_{max}/p_0 of 2·7 predicted for $N_e V_{swept}/a_{cyl}A_{pipe} = 2{\cdot}6$ and $p_{rel}/p_0 = 8{\cdot}0$.

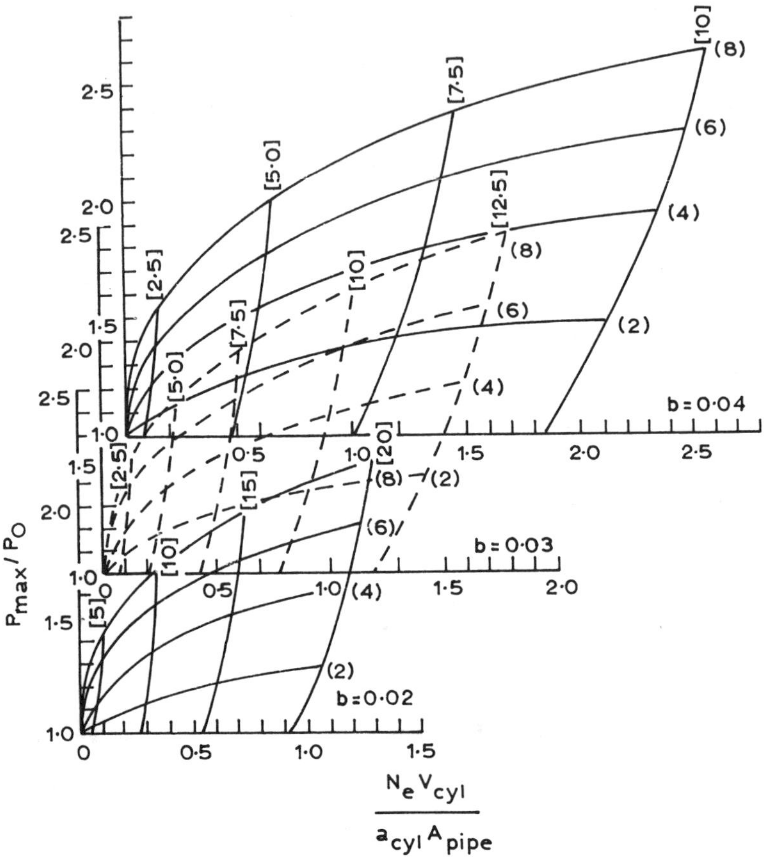

Fig. 6. Exhaust pulse chart. Values of P_{REL}/P_0 are shown in parentheses, and values of θ (degrees) in square brackets.

Although such calculations are now invariably performed by complex cycle simulation programs, a simple design chart such as Fig. 6 still provides useful information.

4. UNIFLOW SCAVENGING OF TWO-STROKE ENGINES

4.1. Introduction

Three main scavenging systems are employed in two-stroke engine design; (a) loop scavenging, (b) cross-scavenging and (c) uniflow scavenging. Types (a) and (b), considered in Chapter 3, are popular in small engines as they are the least expensive to produce but uniflow scavenging is the most efficacious in terms of achieving the highest degree of charge purity for a given air throughput per working cycle. In practice, this system has been limited in its application to the larger type of two-stroke engine, mainly because of the greater degree of mechanical complication generally associated with it in either of its two forms; namely the exhaust-valve-in-head single-piston engine (Fig. 7) or the opposed-piston engine, either with separate crankshafts for the exhaust and air pistons, or with a single crankshaft and siderods. The outstanding feature of all uniflow scavenging systems is that, unlike the simpler loop and cross-scavenging designs, the timing of exhaust and inlet opening, instead of being symmetrical about bottom dead centre, can be asymmetric. As a result, the very necessary exhaust lead, i.e. the interval between exhaust and inlet opening, usually of the order of 20–30° CA, no longer has to be associated with a similar inlet lag, i.e. the interval between exhaust and inlet closing,

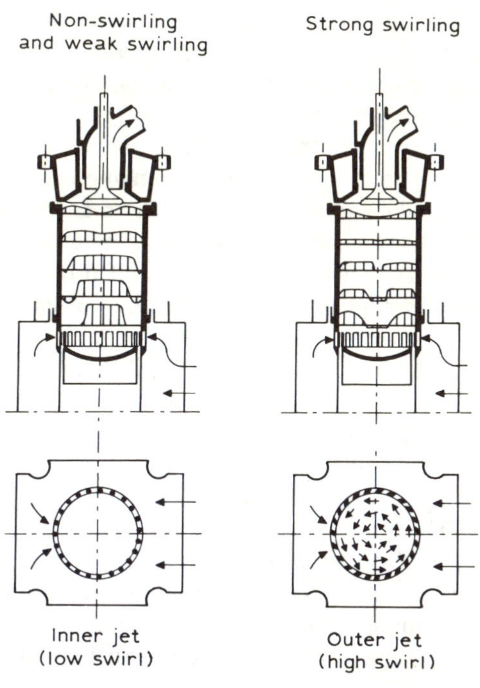

Fig. 7. Uniflow scavenging.

which can lead to a substantial loss of fresh charge through the exhaust, and thus a loss of scavenging efficiency.

Recognizing the crucial importance of scavenging in relation to the performance of two-stroke engines, it is not surprising that the research literature dealing both with experimental and analytical studies is very extensive indeed. Starting with the seminal work of Hopkinson,[9] who was the first to make the fundamental distinction between the two limiting cases of perfect mixing and perfect scavenging and who formulated the original perfect mixing scavenging equation

$$SE = 1 - e^{-SR}$$

where SE is the scavenging efficiency and SR the scavenge ratio, there have been many important contributions.

On the experimental side two groups of studies must be mentioned; namely those concerned with flow visualization, usually based on water models of the scavenging system giving limited dynamic similarity (Refs 10 and 11, and others), and effecting visualization by dye injection or similar means, and a second group based on the use of gases under isothermal conditions, in which the experiments are performed on an engine simulation rig, and employing two different and chemically identifiable gases representing scavenging air and products of combustion.[12,13] In yet a further subgroup, the engine itself is used, and a tracer technique is employed to measure the final charge purity at the end of the scavenging process.[14]

Analytical work has progressed from the simple isothermal, isobaric Hopkinson analysis[9] through a series of important contributions concerned with the utilization of pressure wave effects in the external pipe system based on the method of characteristics, and considered in the previous chapter, to the sophisticated present-day approach based on the use of computational fluid dynamics (CFD) to solve the dynamic flow distribution within the cylinder in terms of temperature, density, concentration, velocity etc.[15] However, this latter technique requires large computing resources and is still in its early days (see Chapter 11).

The present section deals with the progressive development of a general-purpose scavenging model of intermediate complexity. The first of these treatments retains the simplification of isobaric–isochoric conditions, but allows the formulation of scavenging expressions for multi-zone models by the use of flow proportion coefficients into and out of particular zones. The second treatment extends this approach to the 'real' engine situation in which pressure, volume and temperature changes all occur simultaneously. It may be regarded as a phenomenological model based on the use of simplified jet models, recognizing that the entry of scavenging air through the inlet ports of a uniflow scavenged engine, and its subsequent history within the cylinder is essentially a jet mixing phenomenon.

4.2. The Isobaric–Isochoric Scavenging Model

4.2.1. Introduction

We are basically dealing with a three-zone model, these zones consisting respectively of pure air, products of combustion, and a mixture, in varying proportions, of air and products. The individual volumes occupied by these zones vary continuously, both spatially and temporally, during the scavenging process. Coefficients of 'intake proportion' and 'discharge proportion' are assigned to the flows into and out of each of these zones. By applying the conservation equations for mass and energy, together with a composition equation and the perfect gas equation of state, all under the overall constraint of constant pressure and constant overall volume, closed solutions for the condition of the cylinder contents at various stages of the scavenging process may be obtained.

The main phases of the scavenging process may be defined as follows:

Phase A: a displacement scavenging phase during which only residual gas is discharged from the cylinder, with some residual gas remaining in the cylinder at the end of phase A.

Phase B: a mixing scavenging phase in which an air-residual mixture of varying composition, and pure air are discharged respectively from the mixing and the air zone until the mixing zone disappears completely.

Phase C: a short-circuiting phase during which fresh air is discharged straight from the exhaust ports.

4.2.2. Simplified Derivation of Governing Equations (Boundary Conditions)

For the inlet ports, the mass of air entering for any incremental step is Δm^a_{in}. Part of this air enters the air zone, and part the mixture zone, i.e.

$$\Delta m_{\text{in}} = \Delta m_{\text{in}}^{a} + \Delta m_{\text{in}}^{m} \tag{17}$$

Similarly for the exhaust ports, assuming that these discharge a mixture of residuals, g, a mixture of variable composition, m, and air, a,

$$\Delta m_{\text{out}} = \Delta m_{\text{out}}^{g} + \Delta m_{\text{out}}^{m} + \Delta m_{\text{out}}^{a} \tag{18}$$

Defining coefficients of intake proportion, in eqn (17), as

$$I^{a} = \Delta m_{\text{in}}^{a}/\Delta m_{\text{in}} \qquad \text{and} \qquad I^{m} = \Delta m_{\text{in}}^{m}/\Delta m_{\text{in}} \tag{19,20}$$

$$I^{a} + I^{m} = 1 \tag{21}$$

Similarly, defining coefficients of discharge proportion, in eqn (18), as

$$D^{a} = \Delta m_{\text{out}}^{a}/\Delta m_{\text{out}}, \qquad D^{g} = \Delta m_{\text{out}}^{g}/\Delta m_{\text{out}}, \qquad D^{m} = \Delta m_{\text{out}}^{m}/\Delta m_{\text{out}} \tag{22–24}$$

$$D^{a} + D^{g} + D^{m} = 1 \tag{25}$$

For each zone we may now formulate the conservation equations for mass, species and energy, as follows.

(a) *Mass conservation*:

For the air zone, $$\Delta m^{a} = \Delta m_{\text{in}}^{a} - \Delta m_{\text{out}}^{a} \tag{26}$$

For the gas zone, $$\Delta m^{g} = -\Delta m_{\text{out}}^{g} \tag{27}$$

For the mixture zone, $$\Delta m^{m} = \Delta m_{\text{in}}^{m} - \Delta m_{\text{out}}^{m} \tag{28}$$

(b) *Species conservation*: using the notation C_{y}^{x} to denote the concentration of species y in zone x, we have the following. For the air zone,

$$C_{a}^{a} = 1, \qquad C_{g}^{a} = 0 \tag{29,30}$$

For the residual zone,

$$C_{a}^{g} = 0, \qquad C_{g}^{g} = 1 \tag{31,32}$$

For the mixing zone,

$$C_{a}^{m} = m_{a}^{m}/(m_{a}^{m} + m_{g}^{m}), \qquad C_{g}^{m} = m_{g}^{m}/(m_{a}^{m} + m_{g}^{m}) \tag{33,34}$$

(c) *Conservation of energy*:

For the air zone,

$$\Delta(m^{a}u^{a}) = \Delta m_{\text{in}}^{a}h^{a} - \Delta m_{\text{out}}^{a}h^{a} - \Delta Q^{a} - p\Delta V^{a} \tag{35}$$

For the gas zone,

$$\Delta(m^g u^g) = -\Delta m^g_{\text{out}} h^g - \Delta Q^g - p\Delta V^g \tag{36}$$

For the mixture zone,

$$\Delta(m^m u^m) = \Delta m^m_{\text{in}} h^a - \Delta m^m_{\text{out}} h^m - \Delta Q^m - p\Delta V^m \tag{37}$$

In addition, we have the perfect gas state equation in the form, for the general zone i, and for the constant (isobaric) pressure p,

$$pV^i = m^i RT^i \tag{38}$$

and the overall volume constraint, for the isochoric assumption,

$$V_{\text{cyl}} = \text{constant}, \qquad V^a + V^g + V^m = V_{\text{cyl}} = \text{constant} \tag{39}$$

By manipulation of the three sets of conservation equations, together with the state equations for each zone (38) and the volume constraint equation (39), and using the auxiliary equations for intake and discharge proportion, we arrive at five simple difference equations which may be solved numerically for the following variables, all as functions of an incremental inflow, Δm_{in}:

(1) increment of outflow, Δm_{out};
(2) increment of air mass in air zone, Δm^a;
(3) increment of gas mass in gas zone, Δm^g;
(4) increment of air mass in mixing zone, Δm^m_a;
(5) increment of gas mass in mixing zone, Δm^m_g.

4.2.3. Consideration of Individual Phases

Phases A (displacement scavenging), B (mixing scavenging) and C (short circuiting) have already been defined. The different phases are illustrated in the scavenging model (Fig. 8).

Phase A. At the end of phase A there is an air zone A, in addition to the enlarged mixing zone M and the reduced gas zone G. While the residual content of the gas zone has been reduced, that of the mixing zone has remained constant at its original value, $\eta_0 m^g_0$, where $m^g_0 = p_a V_{\text{cyl}}/RT_g$, i.e. the total residual content of the cylinder at inlet opening.

Defining a delivery ratio λ_A at the end of phase A, $\lambda_A = (m_{\text{in}})_A/M$, where M is a reference mass, $M = p_a V_{\text{cyl}}/RT_a$, as well as a specific volumetric fraction η_A for the residual zone, at the end of phase A, we

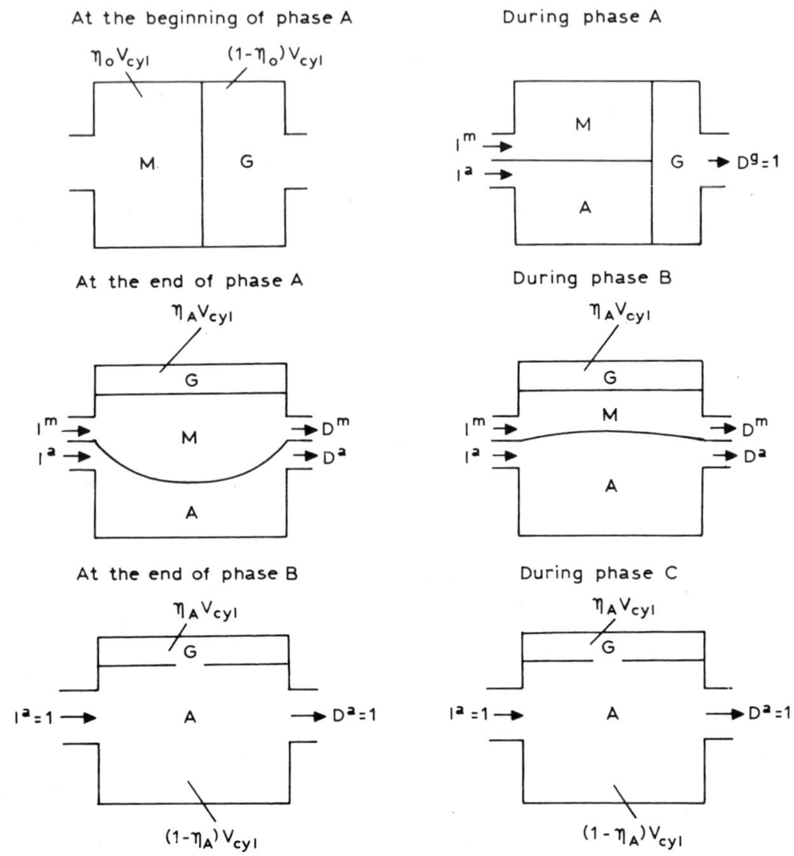

Fig. 8. The simplified analytical model.

may show that the volumetric ratios of the residual, mixing and air zones are given by

$$V^g/V_{\mathrm{cyl}} = \eta_{\mathrm{A}} = 1 - \eta_0 - \lambda_{\mathrm{A}}, \qquad V^m/V_{\mathrm{cyl}} = \eta_0 + \lambda_{\mathrm{A}} I^m, \qquad V^a/V_{\mathrm{cyl}} = \lambda_{\mathrm{A}} I^a \tag{40–42}$$

(i.e. all air entering the air zone is retained), giving a charging efficiency

$$(\eta_{\mathrm{ch}})_{\mathrm{A}} = \lambda_{\mathrm{A}} \tag{43}$$

Phase B. During phase B the residual zone remains at its final value for phase A, i.e. η_{A} (eqn (40)), as discharge through the exhaust ports now switches from the residual to the mixing zone, which vanishes completely at the end of phase B (see Fig. 8). The air zone continues to grow, but there is now simultaneous inflow as well as outflow. Solutions of the relevant difference equations depend on the assumptions concerning the relative magnitudes of the assumed coefficients of intake and discharge proportion for the mixing zone, I^m and D^m, respectively.

However, in all cases, the charging efficiency at the end of phase B approaches the limit, for very high delivery ratios,

$$(\eta_{ch})_B = 1 - \eta_A \tag{44}$$

Phase C. During the short-circuiting phase C (unlikely to occur in uniflow engines), any inflow of air Δm^a_{in} results in an equal discharge of air Δm_{out}. As a result, there is no further change in the properties of the air and residual zones.

4.2.4. Results of Scavenging Analysis

It is clear that the above treatment is not a predictive one, but is wholly analytical. It lends itself particularly well to parametric studies, due to its inherent flexibility, and can therefore be used both to compare various empirical scavenging models and to assess the relative merits of alternative scavenging systems, particularly the traditional systems of loop, cross- and uniflow scavenging. However, since the model is not predictive, empirical values have to be assigned to the coefficients of intake and discharge proportion, I^a, I^m, D^a, D^g and D^m (eqns (19)–(25)), the initial volumetric fractions occupied by the mixing and gas zones η_0 and $(1 - \eta_0)$, as well as that occupied by the residual zone at the end of the initial displacement scavenging phase A, η_A. Thus, with appropriate substitutions all the well known empirical scavenging models can be represented from the simple Hopkinson perfect mixing model[9] with $I^m = 1$, $D^m = 1$, $\eta_0 = 1$, $\eta_A = 0$, to the most sophisticated three-zone models such as that of Baudequin & Rochelle[16] with $I^m = 0{\cdot}9$, $D^m = 0{\cdot}9$, $\eta_0 = 0{\cdot}5$ and $\eta_A = 0{\cdot}1$. Three of these models, those of (a) Benson & Brandham,[17] (b) Maekawa[18] and (c) Baudequin & Rochelle,[16] are illustrated in Fig. 9. By comparing each with the optimistic perfect scavenge and generally pessimistic perfect mixing models in terms of charging efficiency vs. delivery ratio, good insights can be gained into the relevance of such models to particular real systems, thus the Benson & Brandham and Maekawa models[17,18] predict relatively high charging efficiencies and are thus

Main hypothesis: Two zones—the residual gas zone and the mixing zone.
No air zone.
Two phases—phase A and phase B.
No rèmaining residual gas after the end of phase A.

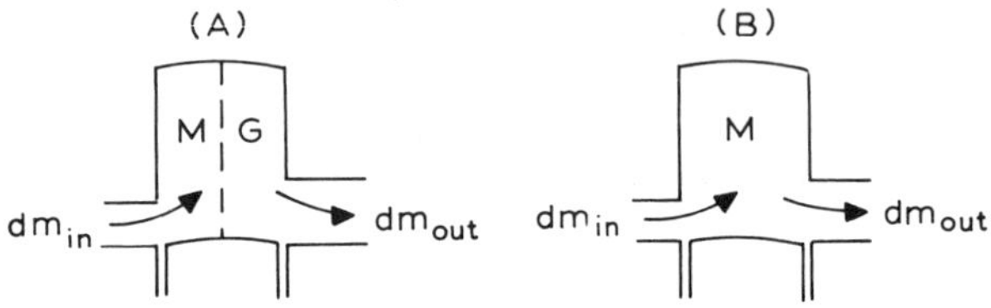

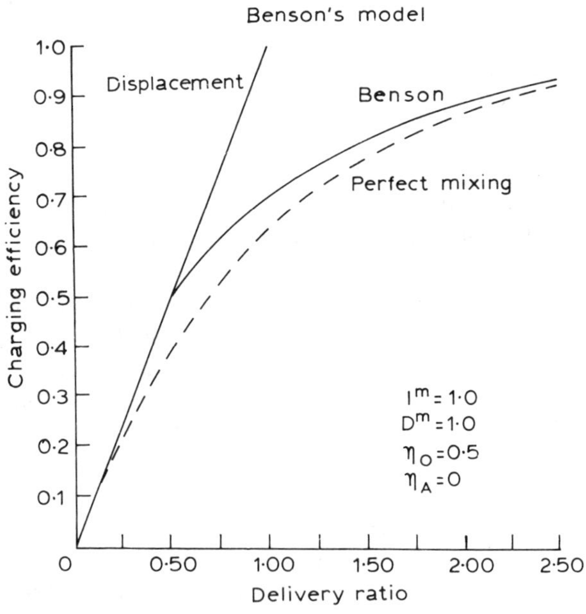

$\eta_{ch} = \lambda$ $\qquad (\lambda \leqslant \lambda_A = 1 - \eta_0)$

$\eta_{ch} = 1 - \eta_0 e^{-(\lambda - (1-\eta_0))}$ $\qquad (\lambda > \lambda_A)$

Remarks: No short-circuiting of fresh air.
No 'dead space' of residual gas.
No curves under $\eta_{ch} = 1 - e^{-\lambda}$.

Fig. 9a. Benson's model.

Main hypothesis: Two zones.
No phase A.
No remaining residual gas.
Isothermal process
$D^m > I^m$

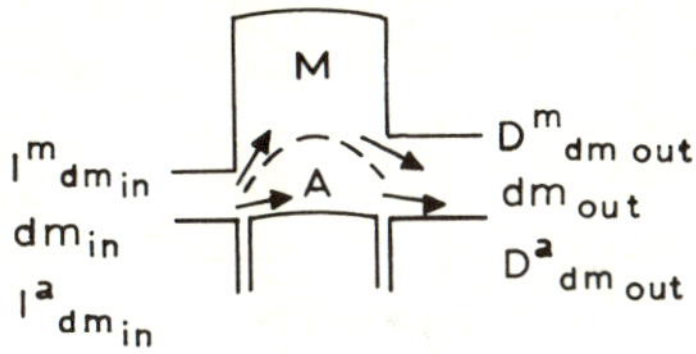

Maekawa's model

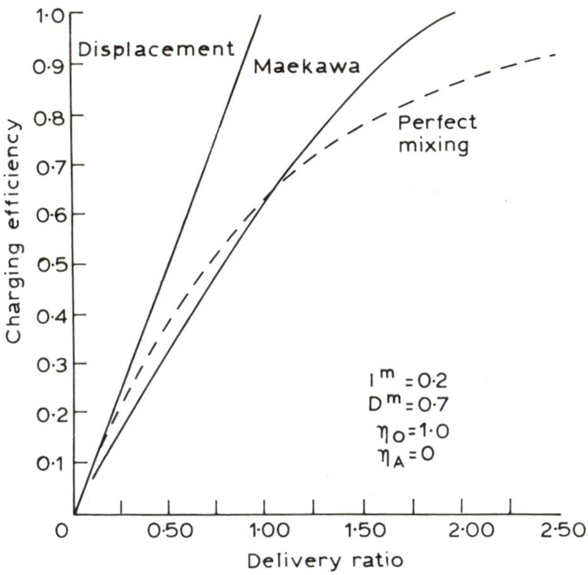

$$\eta_{ch} = 1 - (1 - (D^m - I^m)\lambda)\frac{D^m}{(D^m - I^m)} \qquad \left(D^m > I^m \text{ and } \lambda \leqslant \lambda_B = \frac{1}{(D^m - I^m)}\right)$$

$$\eta_{ch} = 1 \qquad (\lambda > \lambda_A)$$

Remarks: No initial displacement phase.
No 'dead space' of residual gas.
Isothermal idealization.

Fig. 9b. Maekawa's model.

Main hypothesis: Three zones.
Two phases—phase A and phase B.
Remaining residual gas.
Isothermal process.

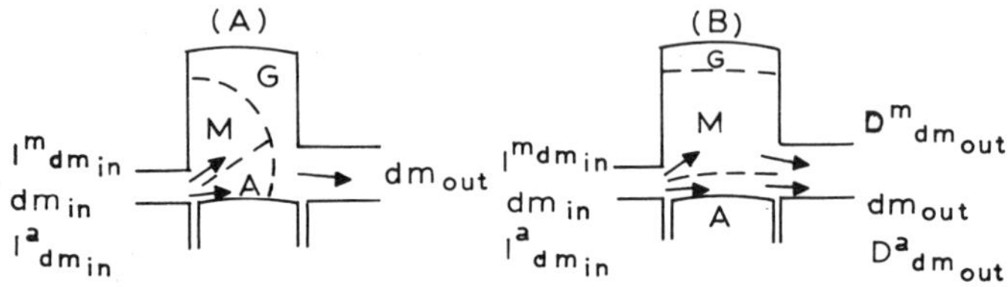

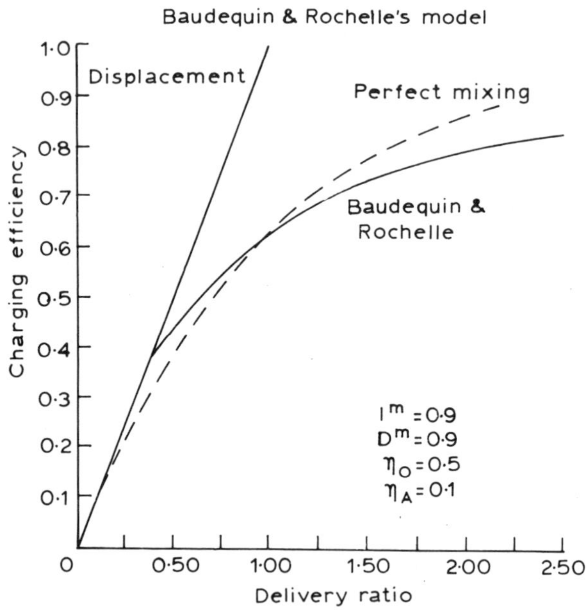

$$\eta_{ch} = \lambda \qquad (\lambda < \lambda_A = 1 - \eta_0 - \eta_A)$$

$$\eta_{ch} = (1 - \eta_A) - \eta_0 \exp\left(-\frac{D^m(\lambda - (1 - \eta_0 - \eta_A))}{D^m(1 - \eta_0 - \eta_A) + \eta_0}\right) \quad (\lambda > \lambda_A)$$

Remarks: Isothermal idealization.

Fig. 9c. Baudequin & Rochelle's model.

appropriate to 'good' scavenging systems, e.g. uniflow, while the Baudequin & Rochelle model predicts poor charging efficiencies due to a high degree of short-circuiting, and would thus be appropriate to 'poor' scavenging systems, e.g. loop. In Fig. 10 is presented a comparison of the three main scavenging systems, i.e. uniflow, loop and cross-scavenging, based on experimental work by List,[19] for air and residual gas temperatures of 300 K and 800 K respectively, and with the iteratively adjusted values of volumetric ratios and flow coefficients summarized in Table 1. The differences in the characteristics of the different systems are clearly shown in Fig. 10, the uniflow systems appearing as by far the best.

It may thus be claimed that the isobaric–isochoric model, while in

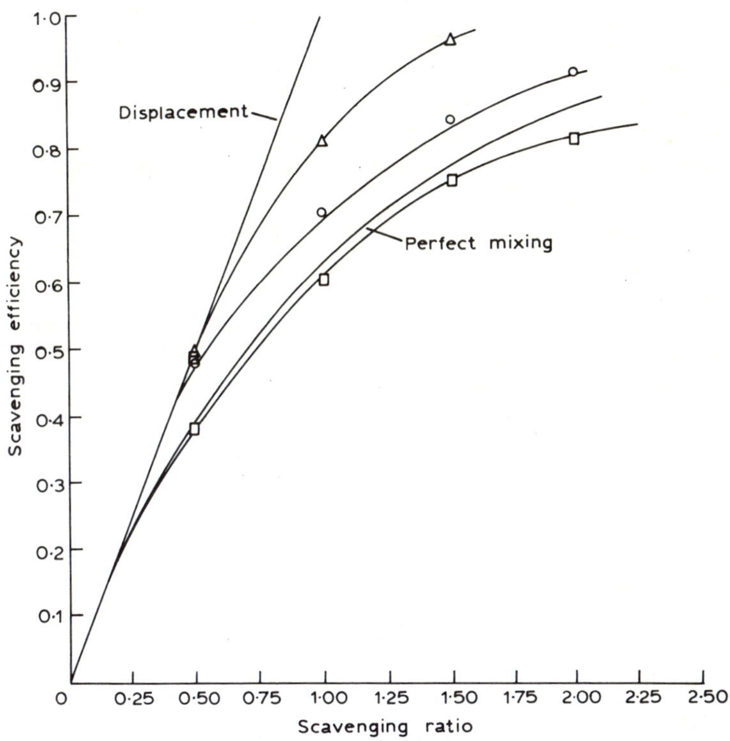

Fig. 10. Comparison between experimental and computational results using the isobaric–isochoric model. Experimental results: △, uniflow; ○, loop; □ cross.

Table 1. Parameters and Coefficients Used in Computations

Type of scavenging system	Initial volumetric ratio of mixing zone	Volumetric ratio of displacement scavenging	Volumetric ratio of remaining residual gas
Uniflow	0·55	0·45	0·0
Loop	0·60	0·32	0·08
Cross	0·63	0·12	0·25

Type of scavenging system	Coefficients of intake proportion		Coefficients of exhaust proportion	
	Air zone	Mixing zone	Air zone	Mixing zone
Uniflow	0·58	0·42	0·10	0·90
Loop	0·65	0·35	0·44	0·56
Cross	0·65	0·35	0·41	0·59

no sense a substitute for, as yet, undeveloped predictive techniques, still provides valuable insights into the complex scavenging process.

4.3. Phenomenological (Jet-based) Scavenging Models

4.3.1. Introduction

Brief reference has been made to the computational fluid dynamic approach of Spalding[15] as providing the ultimate basis for a truly predictive scavenging model. However, at present this method still suffers from major uncertainties, particularly concerning turbulence modelling, and absorbs excessive amounts of computing time. It is appropriate, therefore, to explore simpler approaches to the predictive problem. Such an approach is provided by the recognition that, particularly in the uniflow engine which is almost invariably equipped with piston-controlled inlet ports, the scavenging process involves a form of jet flow through the inlet ports, accompanied by progressive entrainment of the residual gas content of the cylinder, as well as possible recirculation in the later stages. By building a relatively simple one-dimensional, but multi-zone and unsteady jet model into the conventional step-by-step thermodynamic model of the gas exchange process (e.g. Streit & Borman[20]), it is possible to build up a 'quasi-predictive' model taking into account some of the features of the relatively simple geometry of the uniflow scavenge engine, e.g.

inlet port timing, area and orientation, as well as the running conditions of the engine. Such a model therefore represents a significant advance on the parametric isobaric–isochoric model described in the previous section.

The model, in its simplest form, is illustrated in Fig. 11, which is divided into two parts. The upper half illustrates the two basic types of jet flow which are assumed to exist depending on port orientation:

(a) inner jetting flow for port orientations between radial and an arbitrary 40° to the radial direction, and

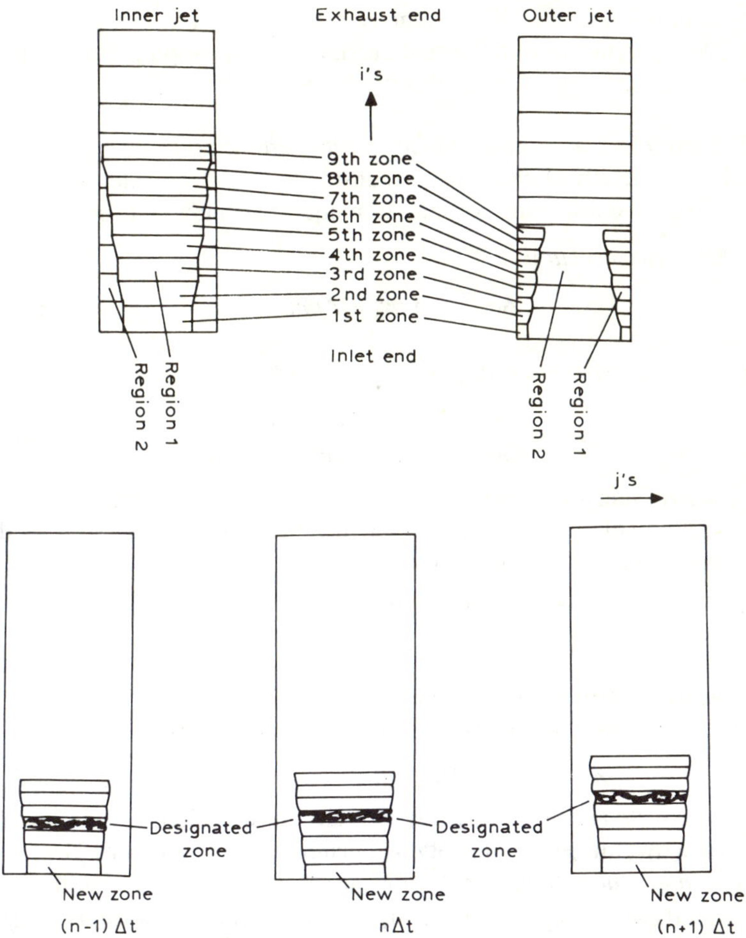

Fig. 11. Simplified flow patterns and divided jet zones: version 1.

(b) outer jetting flow for port orientations in excess of 40° to the radial direction.

The transition from 'inner' to 'outer' jetting flow is somewhat arbitrary, but is related to a theory of vortex breakdown. The lower half of Fig. 11 illustrates the progressive development of the jet (for inner jetting flow), as well as the ith shaded jet zone of mass $m^{j,i}$ to which a calculation scheme based on conservation of mass, species, axial and angular momentum and energy is applied. Additionally, Fig. 11 also shows how for each calculation step a new jet zone is created. In the scheme shown, the jet is represented by a series of radially undivided disc-shaped zones; in a more refined model the jet is further subdivided also in the radial direction. Even in its simplified form the model is capable of yielding useful predictive results.

4.3.2. Simplified Derivation of Governing Equations

Considering the ith element of the unsteady jet, j, i.e. element $E^{j,i}$ (see Fig. 11), the following conservation equations may be formulated.

(a) *Mass conservation*:

$$\Delta m^{j,i} = \Delta m^{j,i}_{\mathrm{in}} - \Delta m^{j,i}_{\mathrm{out}} \tag{45}$$

(b) *Species conservation* (equivalence ratio F):

$$F^{j,i} = \frac{m^{j,i}_f}{f_s m^{j,i}_a} = \frac{m^{j,i}_{f,0} + m^{j,i}_{f_{\mathrm{in}}} - m^{j,i}_{f_{\mathrm{out}}}}{f_s[m^{j,i}_{a,0} + m^{j,i}_{a_{\mathrm{in}}} - m^{j,i}_{a_{\mathrm{out}}}]} \tag{46}$$

where suffixes f and a refer respectively to fuel and air, the second suffix 0 to conditions at the beginning of the step, and the remaining symbols are obvious.

(c) *Conservation of axial momentum*:

$$\Delta[m^{j,i}w^{j,i}] = \Delta[m^{j,i}_{\mathrm{in}}w^{j,i}_{\mathrm{in}}] - \Delta[m^{j,i}_{\mathrm{out}}w^{j,i}_{\mathrm{out}}] \tag{47}$$

where w represents axial velocity.

(d) *Conservation of angular momentum about the centreline*:

$$\Delta[m^{j,i}r^{j,i}v^{j,i}] = \Delta[m^{j,i}_{\mathrm{in}}r^{j,i}_{\mathrm{in}}v^{j,i}_{\mathrm{in}}] - \Delta[m^{j,i}_{\mathrm{out}}r^{j,i}_{\mathrm{out}}v^{j,i}_{\mathrm{out}}] \tag{48}$$

where $r^{j,i}$ and $v^{j,i}$ refer, respectively, to the effective radius and effective tangential velocity associated with the element.

(e) *Conservation of energy*:

$$\Delta[m^{j,i}u^{j,i}] = \Delta[m^{j,i}_{\mathrm{in}}h^{j,i}_{\mathrm{in}}] - \Delta[m^{j,i}_{\mathrm{out}}h^{j,i}_{\mathrm{out}}] - \Delta Q^{j,i} - p\Delta V^{j,i} \tag{49}$$

where u and h refer respectively to specific internal energy and specific enthalpy. $\Delta Q^{i,j}$ is the heat transferred to the element (actually ignored) and $p\Delta V^{i,j}$ is the work done by the element.

In addition to the above conservation equations, use is made of the perfect gas equation of state for each zone and an entrainment equation due to Ricou & Spalding[21] in the form

$$\Delta m = 0{\cdot}28\rho_s F_0 \Delta Z \, \Delta t \tag{50}$$

where ρ_s is the density of the surrounding gas, F_0 is the initial injected momentum, ΔZ is the axial length of given jet zone and Δt is the time interval under consideration.

Since later zones may travel faster than earlier zones, the model also allows for coalescence of zones which takes place under conditions of conservation of axial and angular momentum.

Finally, there is the overall condition that the sum of all jet volumes and the surrounding volume must equal the cylinder volume, coupled with the condition of constant pressure throughout, at any one instant.

The resultant system of equations, together with the overall step-by-step calculation scheme, may now be solved, degree by degree, for incremental changes of mass m, equivalence ratio f, axial velocity w, radial velocity v, internal energy u, within the jet and, in virtue of entrainment, also within the recirculation zone. As a result, the complete scavenging history may now be determined analytically. This technique has been applied successfully to simulate earlier experimental work, using a tracer gas technique, on the Rootes TS 3 opposed-piston two-stroke engine.[14]

4.3.3. Results of Scavenging Analysis

The results of the analysis, together with a set of eight experimental conditions for the Rootes TS 3 engine, are shown in Table 2 for both the radially undivided (version 1) and radially divided (version 2) forms of the analysis. It is clear that, in general, both versions of the program give remarkably good predictions of overall performance, as well as of airflow and charging efficiency. While, clearly, the model is incapable of predicting detailed spatial distributions of density, velocity and composition within the cylinder, the postulated notional distribution scheme nevertheless results in good overall predictions.

Table 2. Comparison between Experimental and Computational Results

	Case no.							
	1	2	3	4	5	6	7	8
Operating conditions								
Engine speed, rpm	1010	1126	1510	1510	1505	1515	1500	2010
Boost pressure, bar	1·99	2·50	1·52	1·60	1·99	2·50	3·00	1·99
Exhaust pressure, bar	1·84	2·26	1·39	1·38	1·59	2·15	2·55	1·54
Trapped air : fuel ratio	30·5	39·2	30·5	29·8	30·5	41·2	56·7	31·7
Measured values								
Output power, kW	48·61	52·56	54·52	59·57	74·85	68·88	58·86	94·03
Brake thermal efficiency	0·381	0·398	0·391	0·398	0·414	0·412	0·398	0·403
Charging efficiency	0·88	0·86	0·80	0·83	0·85	0·78	0·72	0·82
Air mass flow, kg min^{-1}	9·88	13·65	8·28	10·63	16·66	17·96	21·53	17·83
Calculated values from version 1								
Output power, kW	48·24	53·22	52·77	60·06	73·50	67·59	56·97	90·33
Brake thermal efficiency	0·377	0·389	0·383	0·390	0·397	0·400	0·393	0·404
Charging efficiency	0·87	0·87	0·77	0·83	0·85	0·84	0·79	0·79
Air mass flow, kg min^{-1}	10·31	14·26	8·53	11·14	16·07	17·02	20·61	16·78
Calculated values from version 2								
Output power, kW	49·56	54·18	54·78	61·89	74·19	68·52	58·20	92·64
Brake thermal efficiency	0·380	0·391	0·387	0·393	0·398	0·402	0·395	0·406
Charging efficiency	0·89	0·89	0·85	0·85	0·86	0·86	0·86	0·81
Air mass flow, kg min^{-1}	10·35	14·06	8·53	11·05	15·93	16·99	20·36	16·79

5. CONCLUDING REMARKS

The outline of considerations appertaining to uniflow scavenging given in this chapter indicates that much remains to be done, both analytically and experimentally, before the subject can be said to be fully understood.

The isobaric–isochoric approach, while being very far from a predictive tool, nevertheless lends itself very well to parametric investigations and can be used to calibrate different scavenging systems against each other, as well as providing useful insights into the various mass and energy transfers.

The phenomenological jet-based model is intended to give at least partial insights into momentum, as well as mass and energy transfers and, as part of complete cycle simulation programs, can be described as an approximate predictive tool.

A full understanding of the complexities of the scavenging process will only be attained when refined three-dimensional turbulent flow models and sophisticated experimental techniques involving temporally and spatially resolved velocity measurements as well as gas sampling, can be applied simultaneously.

ACKNOWLEDGEMENT

Much of the material in Section 4 is based on work by Chen.[22,23]

REFERENCES

1. Janota, M. S., Hallam, A. K., Brock, E. K. & Dexter, S. G., The prediction of diesel engine performance and combustion chamber component temperatures using digital computers. *Proc. Instn. Mech. Engrs,* **182** (1967–68) 3L.
2. Wallace, F. J., Vergleich des Gleichdruck—und Stossaufladerverfahrens bei der Abgasturboaufladung von Dieselmotoren mit hohem Aufladedruck. *MTZ,* **25**(5) (1965) 196.
3. Horlock, J. H. & Winterbone, D. E., In *The Thermodynamics and Gas Dynamics of Internal Combustion Engines,* Vol. II; Chapter 18, Engine simulation models with filling-and-emptying methods. Clarendon Press, Oxford, 1986.
4. Benson, R. S., A new gas dynamic model for the gas exchange process in two-stroke loop and cross scavenged engines. *Int. J. Mech. Sci.,* **19** (1977) 693–711.

5. Benson, R. S., Baruah, P. C. & Whelan, B., Simulation model for a crankcase compression two-stroke spark-ignition engine including intake and exhaust systems. *Proc. Instn. Mech. Engrs,* **189** (1975) 167–75.
6. Blair, G. P. & Ashe, M. C., The unsteady gas exchange characteristics of a two cycle engine. SAE paper 760644, 1976.
7. Benson, R. S., In *The Thermodynamics & Gas Dynamics of I. C. Engines,* vol. I; Chapter 7 Numerical solutions of non-steady flow problems with simple boundary conditions. Clarendon Press, Oxford, 1982.
8. Wallace, F. J. & Nassif, M. H., Airflow in a naturally aspirated two-stroke engine. *Proc. Instn. Mech. Engrs,* **168**(18) (1954) 515–35.
9. Hopkinson, B., The charging of two-cycle engines. *Trans. NE Coast Inst. Engrs Shipbldrs,* **XXX** (1914).
10. Rizk, W., Experimental studies of the mixing process and flow configurations in two-cycle engine scavenging. *Proc. Instn. Mech. Engrs,* **172** (1958) 417–37.
11. Dedeoglu, N., Modelluntersuchungen zur Spülung-und Gemischbildung in Einem Verbrennungsmotor. Ph.D. thesis, ETH Zurich, 1968.
12. Sammons, H., A single cycle test apparatus for studying loop scavenging in a two-stroke engine. *Proc. Instn. Mech. Engrs,* **174** (1960) 409.
13. Blair, G. P., An experimental comparison of loop and cross scavenging for two-stroke engines. SAE paper 861240, Milwaukee, 1986.
14. Wallace, F. J. & Cave, P. R., Experimental and analytical studies on a two-cycle opposed-piston diesel engine. SAE paper 710175, 1971.
15. Spalding, D. B., A general purpose computer program for multi-dimensional one and two-phase flow. *Mathematics and Computers in Simulation,* **23** (1981) 267–76.
16. Baudequin, F. & Rochelle, P., Some scavenging models for two-stroke engines. *Proc. Instn. Mech. Engrs,* **194**(12) (1980).
17. Benson, R. S. & Brandham, P. T., A method for obtaining a quantitative assessment of the influence on two-stroke engine. *Int. J. Mech. Sci.,* **11**(3) (1969) 303–12.
18. Maekawa, M., Text of course. *J. Soc. of Mech. Eng.* **636** (1957) 23 (in Japanese).
19. List, F., The charging processes of internal combustion engines. *Proc. Instn. Mech. Engrs, Automobile Div.* (1953–4).
20. Streit, E. E. & Borman, G. L., Mathematical simulation of a large turbocharged two-stroke engine. SAE paper 710176, 1971.
21. Ricou, F. P. & Spalding, D. B., Measurement of entrainment by axisymmetrical turbulent jets. *J. Fluid Mech.,* **11** (1961) 21–32.
22. Chen, C. & Wallace, F. J., A generalized isobaric acid isochoric thermodynamic scavenging model. SAE paper 871657, Milwaukee 1987.
23. Chen, C. & Wallace, F. J., A phenomenological unsteady jet model for uniflow scavenging. SAE paper 871658, Milwaukee 1987.

5

Stratified Charge Engines

J. H. Weaving
ERA Ltd, Dunstable, UK

1. Introduction
   1.1. Theoretical considerations

2. Methods of achieving stratified charge
   2.1. Pre-chamber designs
   2.2. Open-chamber stratified charge engines

3. Design and research with stratified charge engines
   3.1. The pre-chamber combustion system
   3.2. Research with the Ford Proco open combustion chamber engine
   3.3. Mathematical model and experimental investigation of combustion in the BL stratified charge engine

1. INTRODUCTION

The idea of the stratified charge engine (SCE) is old. Sir Harry Ricardo is credited with first demonstrating the pre-chamber type,[1] and there is also an early GM patent dating back to 1922. The SCE is capable of bridging the gap between the diesel engine and the spark-ignition engine. The diesel has excellent fuel consumption and is almost the standard type of engine for commercial vehicles, while the spark-ignition (SI) engine with the lower compression ratio is smoother in running, quieter and free from objectionable odour. Because of its lower maximum cylinder pressures the SI engine is lighter and may be run at higher speeds, giving higher specific power.

The fundamental difference between the two engines is that the power in the diesel engine is varied by reducing the fuel without throttling the air, while with the conventional SI engine power is reduced by throttling. The SCE endeavours to get the best of both worlds.

In the stratified charge engine the mixture ratio of petrol to air of the charge in the vicinity of the spark plug has to be kept close to stoichiometric in order that it is ignitable and capable of producing a robust kernel of flame to ignite the rest of the charge, which is kept at a weak air:fuel ratio. As noted, the potential of the SCE has long been realized, and experimental engines were evaluated. However, such engines are inevitably more complex than the conventional SI engine and no engine was put into quantity production until Honda produced the CVCC engine to meet severe pollution regulations in Japan and the USA in 1974.

It is fair to say that the only type of four-stroke engine that gives lower nitric oxide production, without loss of efficiency, is the spark-ignition SCE. However, it does have problems with hydrocarbon emissions.

1.1. Theoretical Considerations

At full load little is to be gained by stratification, as both the conventional SI engine and the SCE will be unthrottled, but at part load considerable gains are possible. Schweitzer & Grunder[2] demonstrated this clearly by calculating the heat balance at part load for the throttled and unthrottled situation. This has been updated for a modern type of engine and recalculated† using assumptions that are as realistic as possible (Fig. 1). The height of the bars represents the total heat in the fuel.

The SI engine (Otto cycle) works on a cycle approximating to the constant-volume cycle. The theoretical ideal efficiency is therefore

$$\text{thermal efficiency} = \eta = 1 - \frac{1}{r^{\gamma - 1}}$$

where r is the compression ratio and γ is the ratio of the specific heats. On part load, with a stratified charge engine, γ is nearer to 1·4 than in a throttled engine because more air is present and the temperatures

† Using a mathematical model, by Dr H. James of Loughborough University.

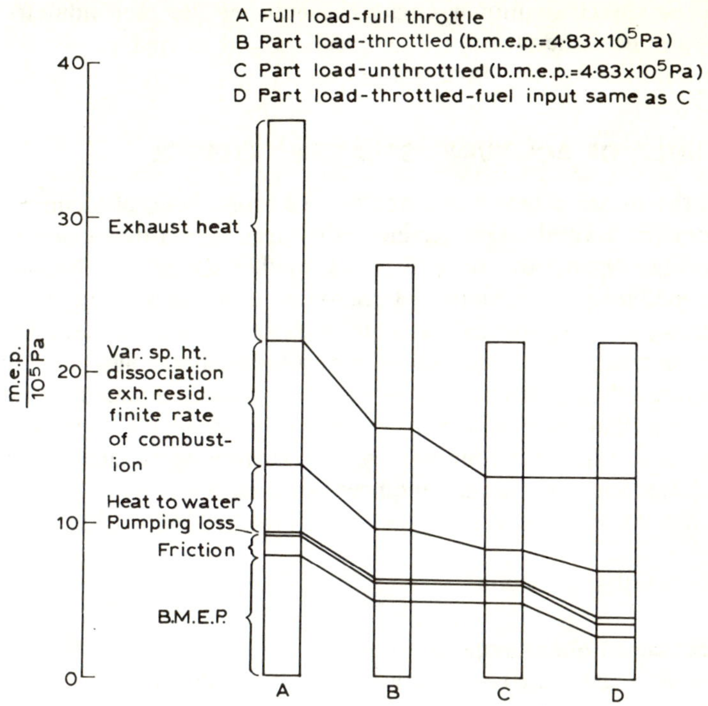

Fig. 1. A comparison of the losses involved in running both throttled and unthrottled at part load.

are lower. For example:

at 10:1 compression ratio for $\gamma = 1{\cdot}4$, $\eta = 60\%$

for $\gamma = 1{\cdot}3$, $\eta = 50\%$

Additionally, pumping losses to charge the cylinder are lower and losses associated with high-temperature dissociation of the combustion gases are reduced. Heat losses to coolant are also reduced. Comparison of bars B, C and D demonstrates these points. These calculations are for the best estimate of a practical situation, as distinct from the ideal efficiency conditions.

Reference to the pollution characteristics of an SI engine (Fig. 1, Chapter 15) shows that if an engine can be run near 20:1 air:fuel ratio then nitric oxide emissions will be low. This is not as difficult with an

SCE as with a conventional engine because of the stoichiometric mixture ratio near the plug.

2. METHODS OF ACHIEVING STRATIFIED CHARGE

As with the diesel engine there are two basic methods of achieving stratified charge combustion, namely the pre- or divided chamber method (Fig. 2) and the open-chamber method (Fig. 3). The pre-chamber method is the easier practical solution, but as with the diesel the efficiency may be slightly less than with the open chamber. This, however, depends very much on how successful the designs are, as stratification in an open chamber is not easily accomplished. Fuelling with the pre-chamber engine may be by carburation or low-pressure fuel injection, but open-chamber engines require moderately high-pressure injection to obtain the required atomization.

2.1. Pre-chamber Designs

2.1.1. BL† Stratified Charge Engine

It will be seen that the BL engine utilizes a third valve (Fig. 2) to feed the pre-chamber, with a separate small carburettor linked to the main carburettor. The third valve is actuated from the same overhead camshaft through a small rocker.

The pre-chamber is similar to that in an IDI diesel engine, except for its size, which is much smaller, and the design of the nozzle between the two chambers, which is orientated to direct a jet of burning mixture to the far side of the main combustion chamber.

2.1.2. Honda CVCC Engine

This is the now well known engine fitted in Honda CIVIC vehicles in quantity production. The combustion chamber arrangement is shown in Fig. 4 (for details see Refs 3 and 4). The pre-chamber with the third valve is some 12% of the swept volume and acts as a torch igniter for the main charge. The two inlet valves are fed by a two-barrel carburettor of complex design.

The total air–fuel mixture supplied to the CVCC engine is fairly

† British Leyland, now Austin Rover.

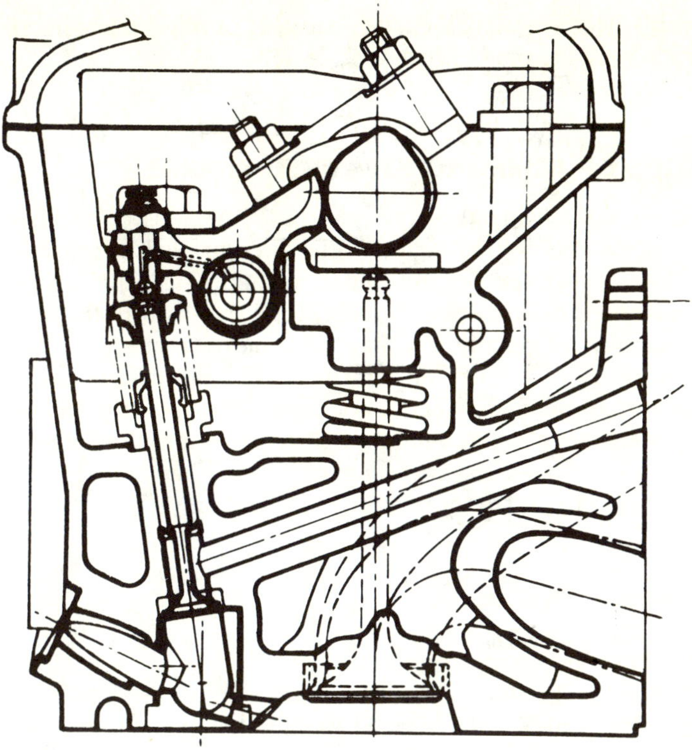

Fig. 2. A cross-section of the BL stratified charge engine cylinder head.

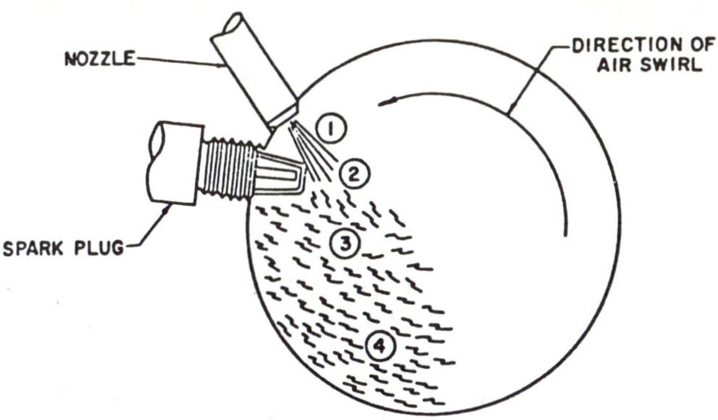

Fig. 3. The Texaco combustion process. 1. Fuel spray. 2. Fuel–air mixing zone. 3. Flame front area. 4. Combustion products. (Bore: 4.00; Stroke: 3.50; Comp. ratio: 11 : 1; C.I.D./CYL: 44.)

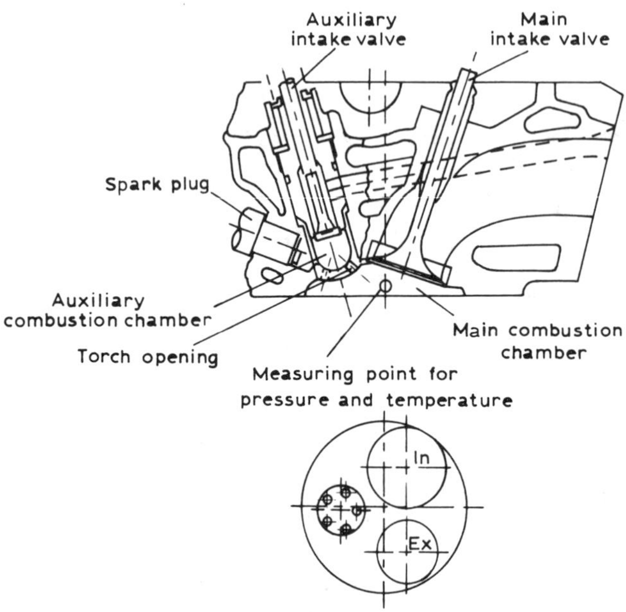

Fig. 4. The Honda CVCC engine cylinder head.

lean, leaner than stoichiometric. However, means are provided so that just before ignition (at the end of compression stroke) a comparatively rich mixture that is easy to ignite is present in the auxiliary combustion chamber, where the spark plug is located. The mixture in the main chamber is rather lean and thus more difficult to ignite. But by proper geometry and proportioning of air and fuel, a mixture of intermediate richness forms near the torch opening connecting the respective chambers. This mixture, which is termed ‘mixture cloud’, is ignited by the flame initiated in the auxiliary chamber and ensures positive combustion of the lean mixture in the main chamber.

Because the CVCC engine operates on lean mixtures and in accordance with a carefully controlled cycle, the peak temperature in the main combustion chamber is lower than that found in conventional engines—thereby reducing NO_x emissions. In later designs the combustion chamber is nearer to the axis of the cylinder and has five holes so that the torch flames spread through the whole combustion chamber, with the object of reducing the tendency to knock. However, the compression ratio of 9:1 used is quite modest.[5]

2.1.3. Porsche Stratified Charge Engine SKS

This is a pre-chamber experimental engine but without a third auxiliary valve in the pre-chamber. A third valve is not essential as residual gas in the pre-chamber receives adequate fresh air from the main chamber during compression. The charge in the pre-chamber is enriched by means of a small fuel injector injecting into the pre-chamber. This has advantages and disadvantages over the third valve construction. Because of the residual burnt gas and rich air:fuel (A/F) ratio in the pre-chamber, nitric oxide generation will be lower than with a third valve. This was the Porsche finding by a direct comparison using a single-cylinder research engine.[6] Disadvantages are that fuel has to be injected at a moderately high pressure to give reasonable atomization and that it is not easy to make reliable injection equipment to pump petrol, which is a poor lubrication. Porsche also found slightly worse hydrocarbon emissions.

Experimental engines of similar designs to those described above have been made by VW and Daimler–Benz in Germany and GM in the USA. All of these engines have added to knowledge, and the significant result will be considered below in Section 3.

2.2. Open-chamber Stratified Charge Engines

In 1934 Hesselman of Sweden reported to the SAE[7] experiments with an open-chamber experimental spark-ignited diesel engine, having a combustion chamber in the piston crown and air induced through a shrouded inlet valve producing high air swirl. By suitably locating the spark plug and the fuel injector he was able to produce an ignitable mixture at the plug over a large range of running conditions; however, some throttling was required at low loads.

2.2.1. Texaco Engine

Witzky[8] and Barber (Texaco)[9] also investigated similar ideas which ultimately resulted in the Texaco engine, which was actually produced in small numbers.[10] The design of the Texaco engine is shown in Fig. 3, which shows the combustion process. A high air swirl is produced from a helical inlet port which, as in DI diesel engines, is enhanced by squeezing the induced air into a cylindrical bowl in the piston. Fuel is injected tangentially into this swirling air and, immediately after injection, a continuous series of sparks are produced from a spark plug

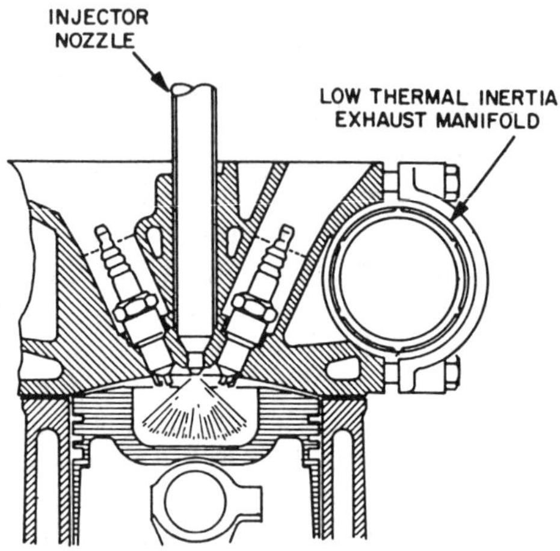

Fig. 5. The Ford Proco combustion chamber.

ahead of the injection point. This forms a flame zone from the burning fuel droplets which is fed with air from the rotating swirl. When fuel delivery is cut off the remaining evaporated fuel continues to burn until used up. It will be appreciated therefore that, as with the diesel engine, its power output is 'smoke-limited'. However, this system has no highly compressed end-gas and it does not suffer from 'knock'. Because it has spark ignition it is not sensitive to cetane number, and will burn a large variety of fuels from alcohol through the petroleum range, from petrol through wide-cut fuels to diesel.

2.2.2. Ford Proco

A very considerable R&D programme has been conducted by the Ford Motor Co., in the USA. The engine is named the Proco and a diagram of an engine, the outcome of this R&D, is shown in Fig. 5. Air, with a considerable degree of swirl, is induced through the inlet valve and is compressed into a small combustion bowl in the piston crown, not unlike the DI diesel engine. The fuel injector is centrally orientated and by careful design a spray gives a stoichiometric air : fuel ratio in the vicinity of two spark plugs, with a weaker mixture in the rest of the chamber. As there is need for evaporation, injection has to be earlier than the spark and has to be advanced with load. A high-energy spark is also needed. For a variety of reasons, not related to efficiency, this engine has not reached the production stage.

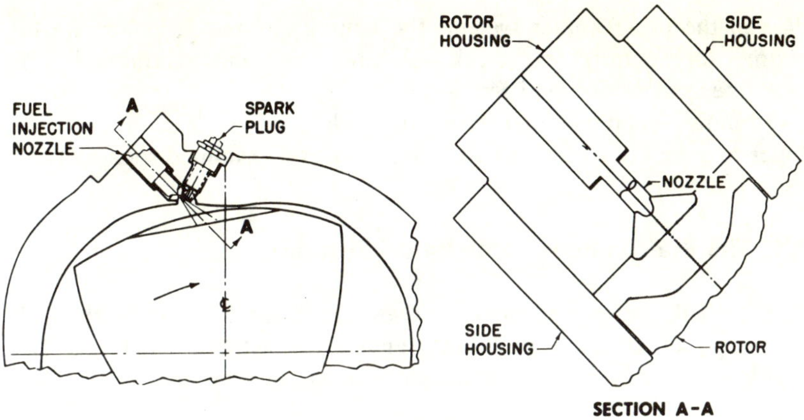

Fig. 6. The Curtis Wright stratified charge combustion chamber (Wankel type rotary engine).

2.2.3. Curtis Wright Rotary Engine

Curtis Wright have for many years been experimenting on applying the principles of charge stratification to the Wankel type engine. They point out that the rotor moves the air past the wasp waist of the trochoidal housing once every revolution and that this lends itself to a stratified system.

Numerous schemes have been evaluated,[11] one promising scheme involving injection of fuel in the vicinity of the spark plug (Fig. 6). A second scheme involves injection of a pilot quantity of fuel for ignition in the proximity of the spark plug, and supplies the remainder through a carburettor in the normal way.

3. DESIGN AND RESEARCH WITH STRATIFIED CHARGE ENGINES

As it is a spark-ignition engine, the SCE works basically with the same type of combustion processes. However, it is a lean-burn engine with the consequence or slower flame speeds. In the conventional lean-burn engines (Chapter 1) this is catered for by higher compression ratios and higher small-scale turbulence. Similar principles are necessary for the SCE; however, it has in-built facilities to promote fast burn. In the case of the pre-chamber systems the issuing burning jet is a powerful source for igniting the weak mixture zone and, additionally, it generates gas motion and small-scale turbulence.

With the open-chamber type the high swirl is available for turbulence generation. These two systems will be examined separately.

3.1. The Pre-chamber Combustion System

In the 1970s, several companies investigated the SCE in depth, with the hope of meeting pollution regulations without catalysts. The highlights of some of these investigations will be examined.

3.1.1. GM Investigations

The size and orientation of the jet passage connecting the pre-chamber to the main chamber is of considerable significance. Krieger of GM conducted a useful research project with a comparatively large pre-chamber orifice.[12] He defined a dimensionless orifice size, D_0:

$$D_0 = \frac{\pi d_0^2 \times \text{bore}}{4V}$$

where d_0 is the orifice diameter and V is the volume of the pre-chamber. In general, when the orifice is large the jet does not penetrate completely across the main combustion chamber, so that when the pre-chamber flame emerges from the orifice the main charge is ignited and spreads spherically across the chamber. The jet itself, however, is rich and as it entrains weak mixture from its surroundings it expands into a growing flame. This is illustrated in Fig. 7. The Krieger orifice number was 1·3.

3.1.2. Porsche Investigations

Porsche[6] investigated the effect of several variables in this type of engine. In the first place they compared the difference in mixture preparation and effectiveness of feeding the pre-chamber with a fuel injector in comparison with a small third valve feeding in a premixed charge. Results of their tests are shown in Fig. 8, where 'λ_{overall}' is the average mixture ratio number, defined as λ = (actual air:fuel ratio)/(stoichiometric A/F ratio) and t_E is the exhaust gas temperature.

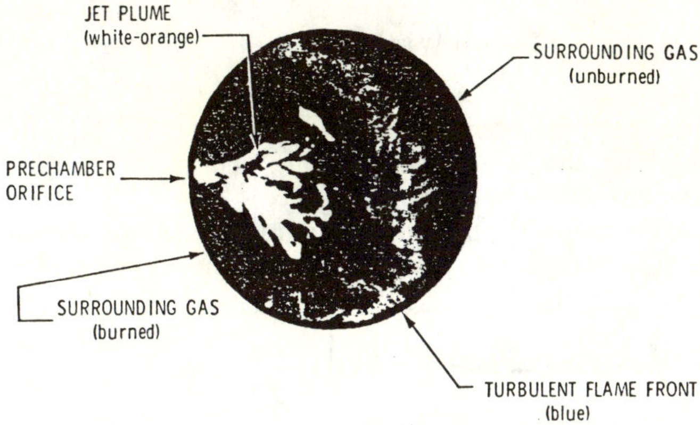

Fig. 7. Single frame from high-speed film of jet ignition combustion taken through transparent piston. Conditions: 600 rpm; $\phi_{OA} = 0{\cdot}83$, $\phi_{PS} = 4{\cdot}83$, $FR = 10\%$; $\beta = 3{\cdot}6$; spark advance = 5° BTDC; gasoline.

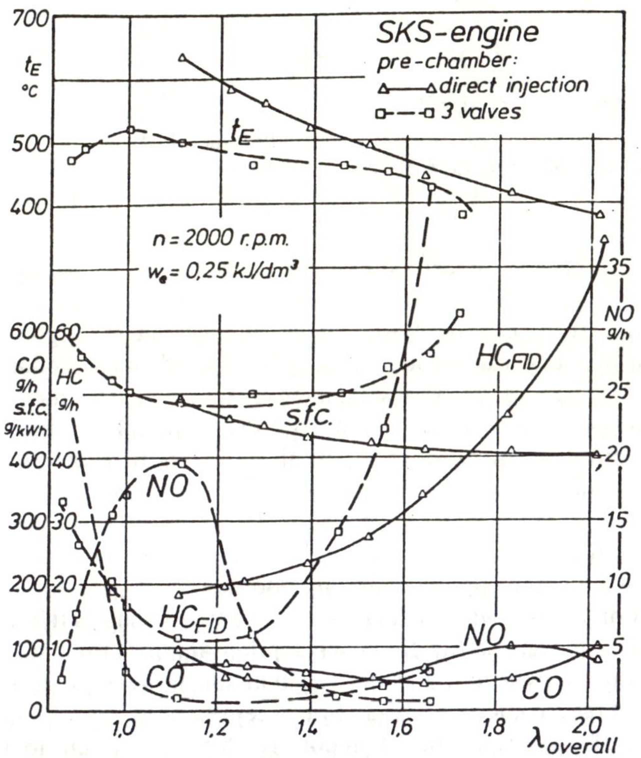

Fig. 8. Porsche: comparison of stratified charge combustion processes.

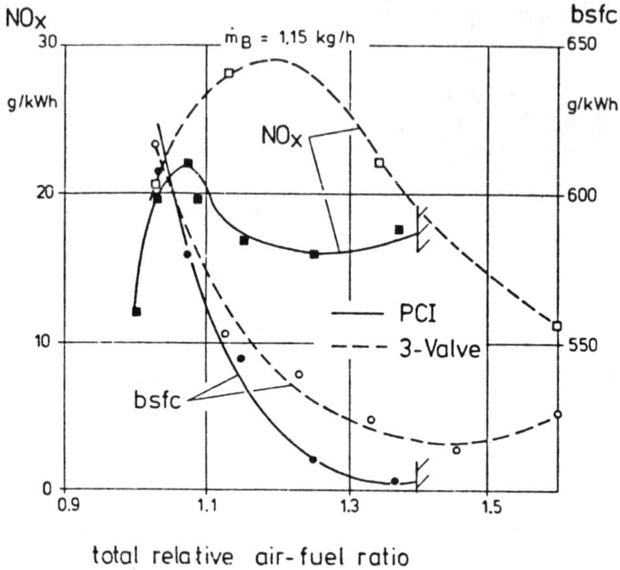

Fig. 9. VW single-cylinder engines: comparison between PCI and three-valve type at 2600 rpm and MBT spark advance.

It will be noted that the injection into the pre-chamber gives a better fuel consumption, is capable of running at a weaker air:fuel ratio and gives lower nitric oxide, but the 'third-valve' scavenged chamber gives lower hydrocarbons and carbon monoxide. VW carried out a similar investigation (Fig. 9) and found a similar advantage of pre-chamber direct injection (PCI)—for NO reduction—but the 'third-valve' engine was capable of burning satisfactorily at a weaker mixture than the PCI giving, at lean ratios, even lower nitric oxide. Porsche, using direct injection to the pre-chamber, investigated the influence of pre-chamber size, and from Fig. 10 it will be seen that 10% of the total combustion space volume appears to be the optimum.

The effect of orifice size was also evaluated, and Fig. 11 shows that a small orifice gives a reduction in fuel consumption and in nitric oxide and hydrocarbons. It was found, however, that the power output is also reduced so that a compromise has to be accepted. A value of 6 on the horizontal scale gave a loss of 3% in power. This is equivalent to a 'Krieger' orifice number of 0·41 and would give high penetration. It will be noted also that injection timing (IT) has to be advanced, indicating slower burning. Compared with the GM research engine the orifice size is smaller, giving greater penetration of the jet.

3.1.3. BL Investigations

The BL project was pursued in two parallel programmes.[13] The first used a converted four-cylinder IDI diesel engine of 1·5 litre capacity, while the parallel programme followed a systematic research with a single-cylinder engine. From the results of this work a four-cylinder 1·855 litre engine was designed and evaluated on the test-bed and in a medium-sized passenger vehicle. The design function was assisted by a mathematical model produced by CHAM, coupled with an experimental validation programme using hot-wire and laser-doppler anemometry. This latter work is detailed in Section 3.3.

The single-cylinder investigation resulted in the selection of an

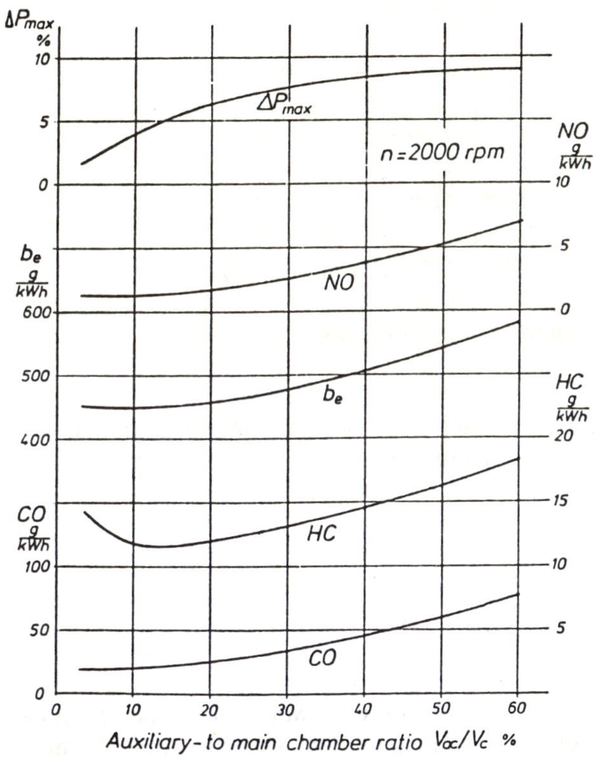

Fig. 10. Porsche engine: influence of auxiliary chamber volume.

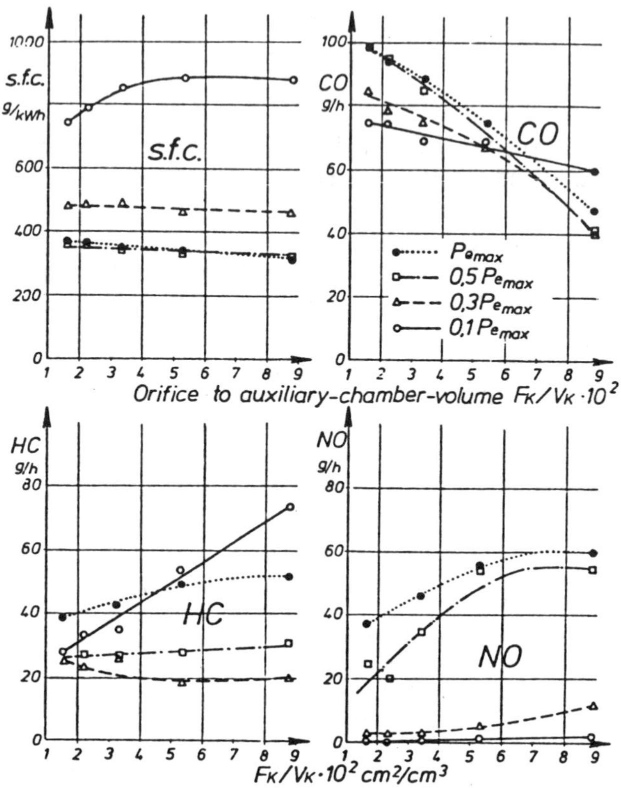

Fig. 11. Porsche engine: influence of orifice size.

optimum configuration of the many variables possible with this type of engine. The combustion chamber design for this optimum was as follows:

Pre-chamber volume = 5 cm^3 (nominally 10% of total clearance volume)
Throat diameter = 8 mm
Pre-carburettor airflow = 6% total
Compression ratio = 9·5 : 1

It was found that some throttling was necessary for part-load running, as is shown in Fig. 12. Mixture loops are shown here, with the main throttle set in stages to give reduced load from the fully opened position. The throttle position is indicated arbitrarily by the inlet manifold depression, in inches of mercury.

It will be noted that maximum fuel consumption is attained by running with fully open throttle from full load to half load, after which instability occurs due to the mixture in the main chamber being too weak. For lower loads progressive throttling is clearly desirable.

A comparison of several stratified charge engines with a conventional engine is shown in Fig. 13.

The effect of throttling on nitric oxide emissions (plotted as NO_2) is shown in Fig. 14, from which it will be noted that the full-throttle characteristic starts with peak NO from what is a slightly weaker than stoichiometric homogeneous situation and, as the mixture weakens, the NO drops rapidly to approximately half load. By following the minima on part throttle a very low NO_x characteristic may be attained.

The comparison of several stratified charge engines with a conventional engine for NO_x emissions is shown in Fig. 15.

When we consider the unburnt hydrocarbon situation the picture is

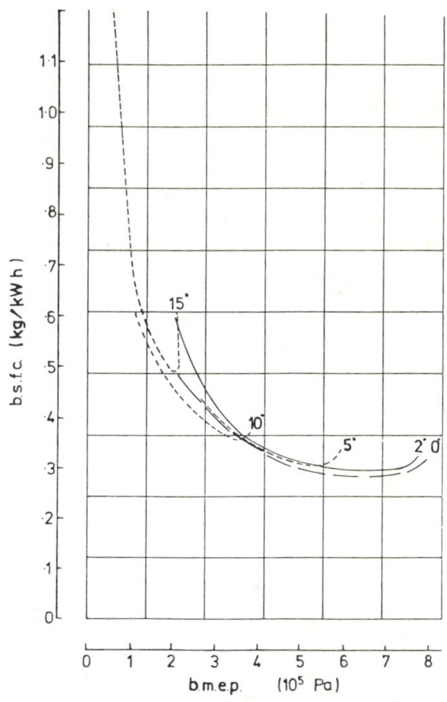

Fig. 12. BL stratified charge: effect of throttling on specific fuel consumption.

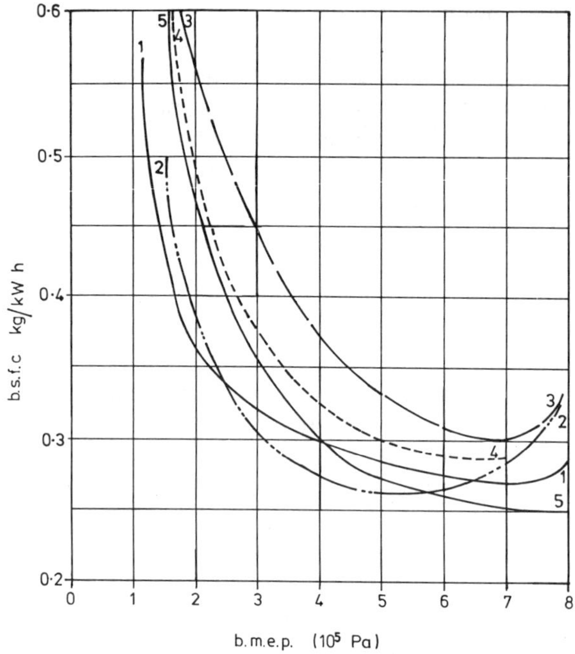

Fig. 13. A comparison of the specific fuel consumption of some BL engines with contemporary stratified engines, at the closest speeds for which information is available. 1, BL 1·85 litre stratified charge engine, 3000 rpm; 2, Texaco TCCS stratified charge engine, 2500 rpm; 3, Honda CVCC, 3000 rpm; 4, Porsche, 2000 rpm; 5, BL 1·8 litre 'B' series 'conventional' engine, 3000 rpm.

not good, and this is probably the major reason why there has only been limited application of the SCE in production. The fitting of an oxidizing catalyst in the present state of the art would be necessary, as will be seen from the following tests.

A four-cylinder 1·85 litre SCE fitted in a BL passenger car and tested according to US emission test procedure gives the following results:

	HC (gm/mile)	CO (gm/mile)	NO_x (gm/mile)	Mpg (US gallons)
Without catalyst	8·0	6·0	0·7	24·0
With oxidizing catalyst	0·35	1·00	0·80	24·2
US Federal Standard	0·4	3·4	1·0	—
US Californian Standard	0·4		0·4	—

It will be seen, therefore, that BL experience showed that it is possible to meet the US Federal Standards but not the Californian Standards. It should be possible to meet the proposed European Standards. The same vehicle gives a 10% improvement in fuel consumption over the standard vehicle. The design of the combustion chamber was optimized with the aid of a mathematical model (Section 3.3 below).

3.2. Research with the Ford Proco Open Combustion Chamber Engine

Very high quality research work has been performed by US Ford with their Proco concept. The basic work was performed by Bishop on a

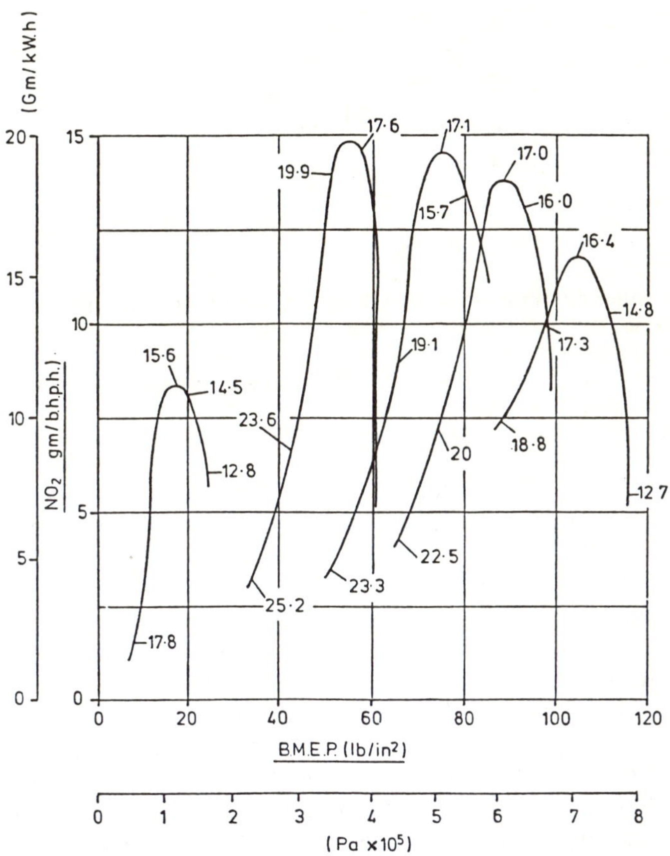

Fig. 14. BL 2 litre stratified charge engine: effect of throttling on oxides of nitrogen. Figures indicate overall air–fuel ratios.

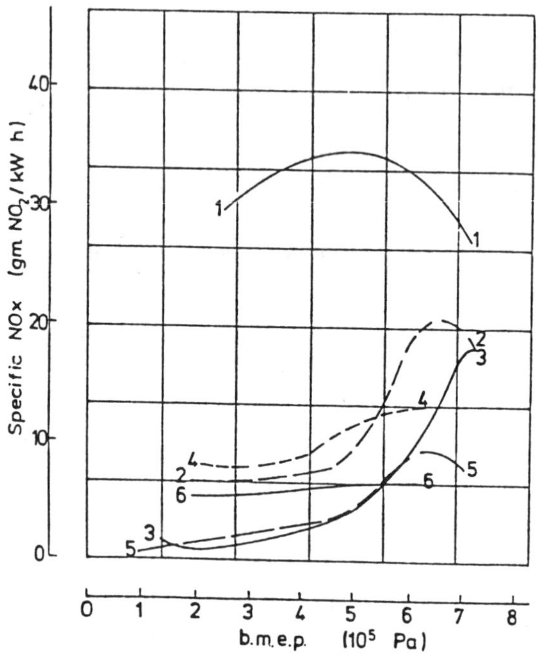

Fig. 15. A comparison of the specific NO_x emissions of the BL engines with the Porsche, VW and Texaco engines. 1, A BL 1·8 litre 'conventional' engine; 2, BL 1·85 litre stratified charge (economy tuned); 3, BL 1·85 litre stratified charge (tuned for low NO_x); 4, VW stratified charge engine; 5, Porsche; 6, Texaco TCCS engine.

single-cylinder engine.[14] The fundamental problem with this type of engine, as observed above, is to organize a spray pattern to give a near-stoichiometric air:fuel ratio near the plug. To investigate this area a carefully designed injector nozzle working at some 400 lb in.$^{-2}$, giving a soft well distributed spray, was designed. Elaborate tests using high-speed sampling valves on the combustion chamber enabled them to plot contour lines of the air:fuel ratio over the whole chamber.

The results from this research were incorporated in V8 engines of 5 and 6·6 litre capacity which were assessed in a passenger car. The compression ratio was 11:1 but the research octane number requirement was below 90. It was also insensitive to fuel composition, and gave a fuel consumption some 15% better than a standard-type engined vehicle.[15,16]

3.3. Mathematical Model and Experimental Investigation of Combustion in the BL Stratified Charge Engine

As little was known with certainty of the combustion process in the divided-chamber stratified charge engine, it was decided, as noted earlier, to use both high-speed photography and a two-dimensional mathematical model to elucidate, if possible, the gas motion in the combustion chamber and the subsequent burning. Professor Spalding and Dr Glynn via the company CHAM (Concentration, Heat and Momentum Limited) were commissioned to develop the model.

Although, ideally, a three-dimensional (3-D) model would be desirable, it was considered at the time that such complexity would require too long a time for the results to be useful to guide this research. Additionally, the amount of computer time required in either case is bound to be large, and a two-dimensional (2-D) model would allow better resolution. Fortunately, the pre-chamber stratified charge engine being investigated lends itself quite well to a 2-D approach, as the main combustion chamber is confined between two parallel surfaces. The accuracy of the 2-D approach does, of course, fall off as the piston descends. However, the most important period for combustion is from 30° BTDC to 30° ATDC, where little piston movement is involved. The combustion chamber modelled therefore consisted of two cylinders of the same height connected by a short throat (Fig. 16).

Because of the complexity of the fluid motion in this type of engine it was essential that the model should represent, spatially and in time, the motion of the gas and other physical phenomena taking place during compression and combustion.

3.3.1. Mathematical Representation of the Stratified Charge Process

The following are the dependent variables for which conservation equations are solved: the axial and lateral velocity components, u and v ($\mathbf{u}$ is the corresponding vector value) the pressure, p; the stagnation enthalpy, $\tilde{h}$; the turbulence energy, k, and its dissipation rate, ε; the mixture fraction, f; and the concentration of reactant fuel and other gaseous species, m_{fu}, m_{N}, m_{NO}.

The control volumes must satisfy conditions for continuity and

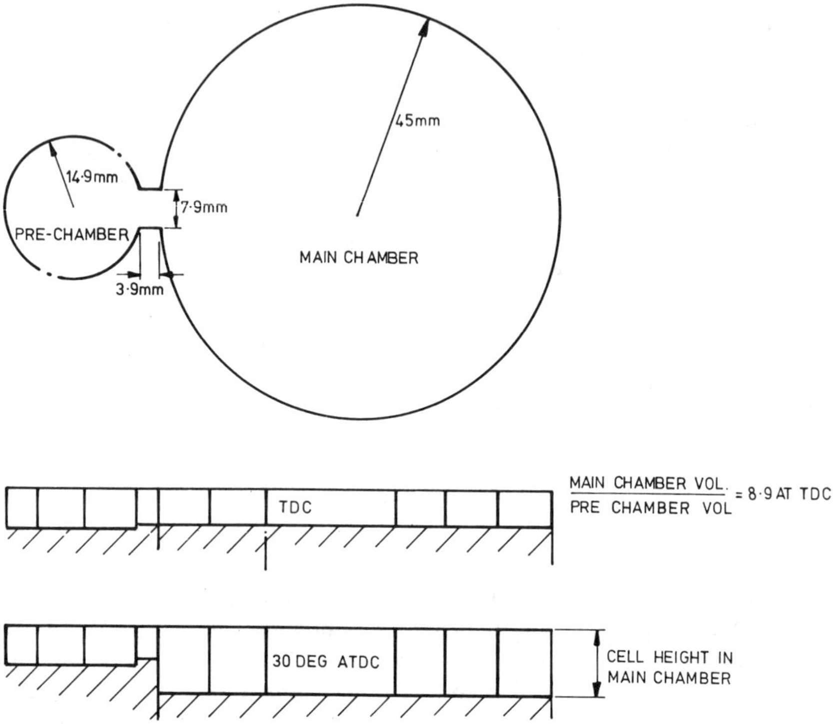

Fig. 16. The simulated divided combustion chamber.

conservation. For continuity,

$$\frac{\mathrm{D}\rho}{\mathrm{D}t} + \rho \operatorname{div} \mathbf{u} = 0$$

$$\mathbf{u} \equiv (u, v), \qquad \operatorname{div} \mathbf{u} \equiv \frac{\partial u}{\partial x} + \frac{\partial v}{\partial y} \tag{1}$$

The other conservation equations for the various factors involving momentum, energy, gas concentrations, etc. may be expressed in the form:

$$\frac{\partial}{\partial t}(\rho\phi) + \operatorname{div}(\rho \mathbf{u} \phi - \Gamma_\phi \boldsymbol{\nabla} \phi) = S_\phi$$

$$\boldsymbol{\nabla}\phi \equiv \left(\frac{\partial \phi}{\partial x}, \frac{\partial \phi}{\partial y}\right) \tag{2}$$

Table 1. Diffusion Coefficients and Source Items

ϕ	Diffusion coefficient, Γ_ϕ	Source term, S_ϕ
u	μ_{eff}	$-\partial p/\partial x$
v	μ_{eff}	$-\partial p/\partial y$
$\bar{h}$	$\frac{\mu}{\sigma_{\bar{h}}}+\frac{\mu_t}{\sigma_{t,\bar{h}}}$	$\partial p/\partial t$
k	$\frac{\mu}{\sigma_k}+\frac{\mu_t}{\sigma_{t,k}}$	$\mu_t - \rho\varepsilon$
ε	$\frac{\mu}{\sigma_\varepsilon}+\frac{\mu_t}{\sigma_{t,\varepsilon}}$	$(C_1 G_k - C_2\rho\varepsilon)\frac{\varepsilon}{k}$
f	$\frac{\mu}{\sigma_f}+\frac{\mu_t}{\sigma_{t,f}}$	0
m_j	$\frac{\mu_j}{\sigma_{m_j}}+\frac{\mu_{t,j}}{\sigma_{t,m_j}}$	$-R_j$

Here, Γ_ϕ is the diffusion coefficient. As is conventional in turbulence modelling an augmented coefficient is used for turbulent flow. S_ϕ is the source term of the dependent variable ϕ. Expressions for Γ_ϕ and S_ϕ are shown in Tables 1 and 2. Here σ is the laminar Prandtl number, and the symbols μ, μ_t and μ_{eff} stand, respectively, for the laminar, turbulent and effective values of viscosity. G_k is the generation rate of turbulence kinetic energy. The turbulent viscosity is evaluated from the local computed values of k and ε. The quantities $\sigma_{t,\phi}$ are turbulent Prandtl numbers, and the values which must be determined empirically as input to the turbulence-model constants are given in Table 2; these values are based on experience of calculating a wide range of turbulent flows. A detailed description of the turbulence model is given by Launder & Spalding.[17]

Table 2. Constants for the Turbulence Model

$\sigma_{\bar{h}}$	0·7	$\sigma_{t,\bar{h}}$	0·86
σ_k	1·0	$\sigma_{t,k}$	1·0
σ_ε	1·0	$\sigma_{t,\varepsilon}$	1·3
σ_f	0·7	$\sigma_{t,f}$	0·86
$\sigma_{m_{fu}}$	0·7	$\sigma_{t,m_{fu}}$	0·86
C_1	1·43	C_D	0·09
C_2	1·92		

The following formula and values close the equation set. The effective viscosity is the sum of the laminar viscosity, μ, and a turbulent viscosity, μ_t, evaluated from k and ε:

$$\mu_{\text{eff}} = \mu + \mu_t = \mu + C_D \frac{k^2}{\varepsilon} \tag{3}$$

where C_D is a constant, which takes the value given in Table 2.

The generation term, G_k, is evaluated from:

$$G_k = 2\left[\left(\frac{\partial u}{\partial x}\right)^2 + \left(\frac{\partial v}{\partial y}\right)^2\right] + \left(\frac{\partial u}{\partial y} + \frac{\partial v}{\partial x}\right)^2 \tag{4}$$

The density is calculated from the equation of state:

$$\rho = p/nRT \tag{5}$$

where

$$n = \frac{m_{\text{fu}}}{w_{\text{fu}}} + \frac{m_{\text{ox}}}{w_{\text{ox}}} + \cdots$$

and w is the molecular weight.

The temperature is obtained from the enthalpy as follows:

$$T = [\tilde{h} - h_{\text{fu}} m_{\text{fu}} - h_{\text{H}_2} m_{\text{H}_2} - h_{\text{CO}} m_{\text{CO}} - \tfrac{1}{2}(u^2 + v^2)]/c_p \tag{6}$$

where c_p is the specific heat, and h_{fu} is the heat of combustion of the fuel. c_p is calculated as in Section 3.3.9 below.

The mathematical modelling of a particular engine also requires a statement of the conditions at its boundaries. There are two types of boundaries; (a) wall boundaries, and (b) the axis of symmetry. The symmetry axis has a simple zero-flux condition on all variables.

At a wall, the component of velocity normal to the wall is zero and the fluxes of chemical species, mixture fraction, turbulence fluctuations and their dissipation rates are also zero.

Wall friction and heat transfer are treated by specifying the values of the shear stress coefficient and the Stanton number. This approximation is consistent with the approximation implied by the two-dimensional assumption. Wall functions[18] are provided for k and ε.

3.3.2. The Finite-difference Grid

The geometry which has been considered is shown in Fig. 17, which also indicates the type of grid used. The grid is non-orthogonal. The velocity components and the other variables are stored in 'staggered' positions on the grid. These storage locations and the corresponding control volumes are shown in Fig. 18.

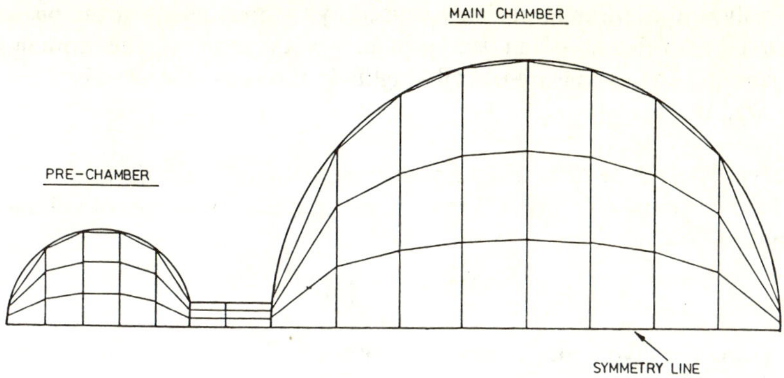

Fig. 17. Simplified geometry of the stratified charge engine and type of grid used.

3.3.3. Piston Movement

Piston movement is accounted for in two respects. Firstly, the volumes and face areas of the control cells in the main chamber are varied appropriately with time. Secondly, the change in volume of the control cells introduces an additional term into the continuity equation.

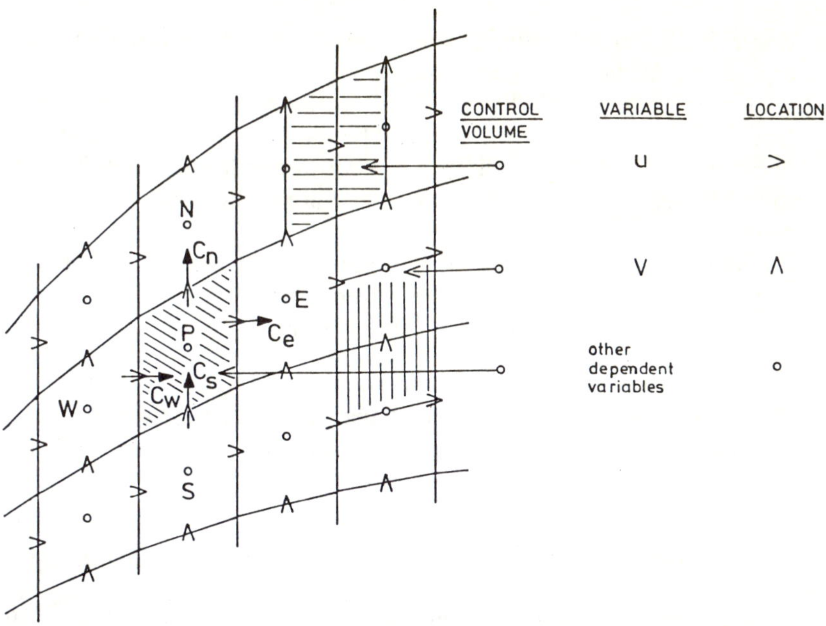

Fig. 18. Control volumes and storage locations.

Finally, the changes in pressure caused by the movement of the piston are taken account of in the pressure–work term of the enthalpy equation. The expansion of cell heights in the main chamber is shown in Fig. 16.

3.3.4. The Finite-difference Equations and Their Solutions

The differential equation (2) may be integrated over a control volume, giving:

$$\int_V \mathrm{div}(\rho \mathbf{u}\phi - \Gamma_\phi \nabla\phi)\,\mathrm{d}V = \int_V \left(S_\phi - \frac{\partial(\rho\phi)}{\partial t}\right)\mathrm{d}V \qquad (7)$$

Using the divergence theorem we obtain:

$$\sum_i (\rho \mathbf{u}\phi - \Gamma_\phi \nabla\phi)_i(A_i \mathbf{n}_i) = \left(S_\phi - \frac{\partial(\rho\phi)}{\partial t}\right)\Delta V \qquad (8)$$

where the sum is over the cell faces. A_i and $\mathbf{n}_i$ are, respectively, the cell face area and the unit normal to the cell face. Equation (8) can be approximated to give:

$$\phi_p = \frac{a\phi_N + b\phi_S + c + S_U}{d - S_p} \qquad (9)$$

where ϕ_p, ϕ_N and ϕ_S refer to the values of ϕ at the centre of the cell and at its 'north' and 'south' neighbours, as shown in Fig. 18. The quantities a, b, c and d are related to the convective and diffusive fluxes of ϕ through the cell faces. The values of ϕ at the 'east' and 'west' neighbours of the cell, which occur in eqn (8), are included into the coefficient c. The quantities S_u and S_p are related to the source and transient expressions.

The equation (9) for a line of cells at constant x forms a closed set of simultaneous equations which are solved to obtain the values of ϕ for these cells by the tri-diagonal matrix algorithm.

3.3.5. Pressure-correction Equation

After the velocity components have been calculated for a line of cells, the pressures are adjusted so that the continuity equation is satisfied. Suppose that R^* is the volumetric continuity error in a cell, so that

$$R^* = C_N - C_S + C_E - C_W - \frac{V}{\gamma \Delta t}\log(p/p_{\mathrm{old}}) - \frac{V}{L}u_{\mathrm{pis}}, \qquad (10)$$

where C is the volume flux through cell face (see Fig. 18), V is the volume of the cell, Δt is the time step, p_{old} is the pressure in the cell at the previous time step, L is the distance between the piston and the top of the cylinder, and u_{pis} is the piston velocity. The penultimate term on the right-hand side of this equation represents the volume flux out of the cell due to the effects of compressibility; the derivation of this term is given by Awn.[19] The last term is the change in volume of the cell due to piston movement.

It may be shown[18,20] that:

$$R^* = c_1 \Delta p_p + c_2 \Delta p_N + c_3 \Delta p_S \tag{11}$$

where Δp_p, Δp_N and Δp_S are the changes in pressure at the cell under consideration and at its north and south neighbours which are necessary for the continuity error to vanish, and c_1, c_2 and c_3 are coefficients which may be evaluated. Then:

$$c_1 \Delta p_p + c_2 \Delta p_N + c_3 \Delta p_S = C_N - C_S + C_E - C_W - \frac{V}{\gamma \Delta t} \log\left(\frac{p}{p_{\text{old}}}\right) - \frac{v}{L} u_{\text{pis}} \tag{12}$$

This is the pressure-correction equation. The set of equations (11) for a line of cells at constant x may be solved simultaneously for the pressure corrections Δp. The same tridiagonal matrix algorithm is used.

3.3.6. Solution Procedure

The calculation proceeds by determining the solution at successive time steps, the magnitude of the time step having been specified. At each time step the procedure is as follows. On each line of cells at constant x, in turn, the velocities and the other variables listed in Table 1 are found by solving the finite-difference equations (9), and the pressure-correction equation (12) is then used to find the values of pressure on the line which are consistent with the continuity equation. This 'sweep' of the domain gives an approximation to the solution of the equations. This approximation is then improved by making repeated sweeps of the domain until the required degree of accuracy is obtained. The calculation can then proceed with the next time step.

3.3.7. Modelling of the Combustion Process

The combustion chamber initially contains a lean mixture in the main chamber and a richer mixture in the pre-chamber. Following ignition the pre-chamber gas expands into the main chamber, preceded by a flame front. Behind the flame front a slower reaction takes place in which unburnt fuel emerges from the pre-chamber and burns in any excess oxygen remaining in the main chamber. The methods of modelling these two combustion processes are now discussed.

Flame front treatment is of vital importance to achieving the objectives. The flame front should reach the remote portions of the combustion chamber as quickly as possible, but without causing engine roughness or detonation.

It would theoretically be possible to treat the entire combustion process, including the flame front, by solution of the equations set out above. However, the flame front is thin and rapid changes in fluid properties occur across it. To prevent the flame front from becoming excessively smeared out due to numerical diffusion it would be necessary to use an extremely fine grid. A better approach is to track the progress of the flame explicitly and to ensure the correct variation in properties across it.

The position of the flame front is therefore stored as a set of coordinates of points p_n. At each time step, each of these points is moved with the local flame velocity, which is the vector sum of the burning velocity and the local unburnt gas speed. This flame tracking is a Lagrangian approach, which contrasts with the Eulerian solution of the finite-difference equations.

There are two components of velocity for each point on the flame front;

(a) the burning speed, normal to the flame (which may be constant or a function of local turbulence, pressure, etc.), and
(b) the local gas velocity, on the *unburnt* side of the flame.

The local gas velocity is found by interpolation; the (x, y) co-ordinates of the point are then incremented appropriately.

Note that the local expansion caused by the flame makes the local gas velocities on the two sides of the flame different; it is that on the unburnt side which convects the flame and hence should be used. When a cell is intersected by the flame front, an additional volume-source term must appear in the pressure-correction equation.

At each time step, the mass fraction of fuel burnt by the flame in each cell must be calculated and the other dependent variables adjusted accordingly. For the mixing-controlled combustion behind the flame front a simple one-step reaction is assumed:

$$1\ \text{kg fuel} + s\ \text{kg oxygen} \rightarrow (1+s)\text{kg products} \tag{13}$$

where s represents the stoichiometric ratio. No intermediate species are considered in this process and the products of the reaction are considered as a single entity.

3.3.8. Prediction of Flame Speed

The rate of combustion initiated from the spark depends on the turbulence state of the compressed mixture.

From Chapter 9 it will be seen that Bradley and co-workers delineate the sequential development of the flame propagation as follows:

(1) The mixture is ignited and the initial flame spread is laminar, with enhancement due to spark energy.
(2) As the flame propagates it is first wrinkled by the higher frequencies of the turbulent spectrum then, increasingly, by ever-lower frequencies leading to the attainment of the fully developed turbulent burning velocity.
(3) In the later stages the pressure increases at an increasing rate and, particularly at the high flame straining rates, there is appreciable after-burning behind the flame front.

The development of the turbulent burning velocity is given by

$$u_{t_k} = u_t\left[1 - \exp\left(-0{\cdot}2\frac{t_k}{t_a}\right)^{0{\cdot}75}\right]^{0{\cdot}5} \tag{14}$$

where u_{t_k} is the burning velocity at time t_k from ignition and t_a is the auto-correlation time. u_t is the fully developed turbulent burning velocity, and is found for a given value of R_2 at the point where $u'_k = u'$ on Figs 6 and 7 in Chapter 9.

At the time of this research and the application of the mathematical model, the above approach had not been sufficiently refined and not all the empirical data was available. In view of this, the propagation of the flame across the chamber was measured experimentally by high-speed photography and flame ionization gaps, and the resulting data was fed explicitly into the model.

3.3.9. Combustion Behind the Flame Front

Behind the flame front there may be a region of distributed combustion, where incompletely burnt fuel emerging from the pre-chamber burns in the oxygen remaining in the main chamber. This process is also modelled by the simple one-step reaction (eqn (13)); the transport equation (2) for mixture fraction and fuel mass fraction is solved to predict the concentrations of the various species and the heat release.

The solution scheme proceeds in the following manner. First, the finite-difference equations are solved for the velocity components, the turbulence quantities and the stagnation enthalpy. Then the equations for the mixture fraction, f, and the unburnt-fuel mass fraction are solved throughout the chamber. The mixture fraction is linearly related to the difference between the local fuel concentration and the stoichiometric value and indicates the mixing process between the two chambers. For example, since nitrogen is inert, the local mass fraction of nitrogen is obtained from:

$$m_{N_2} = f(m_{N_{2,pre}} - m_{N_{2,main}}) + m_{N_{2,main}} \qquad (15)$$

From the new values of f and m_{fu} the mass fraction of oxygen, m_{ox}, is found by the equation:

$$m_{ox} = \left[m_{fu} - \frac{f - f_{st}}{1 - f_{st}}\right]s \qquad (16)$$

where f_{st} is the stoichiometric value of f. The mixture specific heat C_p is now determined from the equation:

$$C_p = m_{fu}C_{fu} + m_{ox}C_{ox} + (1 - m_{fu} - m_{ox} - m_{N_2})C_{pr} \qquad (17)$$

where C_{fu}, C_{ox} and C_{pr} are the specific heats of fuel, oxygen and products respectively. Equation (17) is used together with eqn (6) to determine the local temperature of the gas mixture. The reaction rate is obtained from an Arrhenius expression.

3.3.10. Nitric Oxide Prediction

A chemical equilibrium analysis is performed of the products of the above reaction to estimate the concentrations of the individual species H, O, OH, H_2, H_2O, CO and CO_2. This analysis is based on the minimization of the Gibbs free energy, and details are given in Ref. 21.

The formation of nitric oxide is then determined by use of the

Zeldovich mechanism:

$$\begin{aligned}
&(1)\ N_2 + O \rightleftarrows NO + N \\
&(2)\ O_2 + N \rightleftarrows NO + O \\
&(3)\ OH + N \rightleftarrows NO + H
\end{aligned} \tag{18}$$

This is done by solving equations of the form (2) for the concentrations of N and NO, using specified reaction rates. A basic assumption made is that these reactions can be considered as decoupled from the main fuel-burning reaction, in the sense that at each time step (while the calculation of the latter reaction will determine the NO_x produced in that time step) the NO_x calculation has no influence on the fuel-burning reaction. Other reactions, involving N_2O and NO_2, might also be considered.

3.3.11. Model Validation

The approach taken was to divide the process into two parts: the first was to establish the gas dynamics in the combustion chamber without combustion; this represents the situation of a motored engine. Having established the pressure and velocity fields, the process was then modified by the addition of combustion. This has the advantage of allowing validation of the simpler motoring conditions before adding the complication of combustion, and allows more than one combustion theory to be explored. To accomplish this a rig was built to simulate as closely as possible the two-dimensional situation. This consisted of a single-cylinder 'engine' with a special cylinder head that had been simplified to two cylindrical chambers joined by a parallel circular passage. It is a closed system with no valves, as is shown in Fig. 16. The smaller cylinder has the same volume as the pre-chamber of the actual engine. Hot-wire anemometry (HWA) and laser–doppler anemometry (LDA) measurements were taken in several places in the cylinder head of this rig. Two experimental sets of LDA measurements were made. For the first experimental set HWA measurements were also made to provide a comparison between the two methods. A second experimental set of LDA measurements were made in order to obtain data near to the throat. For the simulation the flow domain was divided into a grid of 51 × 12 cells. One time step was taken every half degree of crank angle. The simulation began at TDC, assuming the fluid to be at rest. Although this is an approximation it is reasonable for the motored situation, as without combustion there should be little

Fig. 19. Location of measurement and prediction points in main chamber for the axial *u* and lateral *v* velocity components.

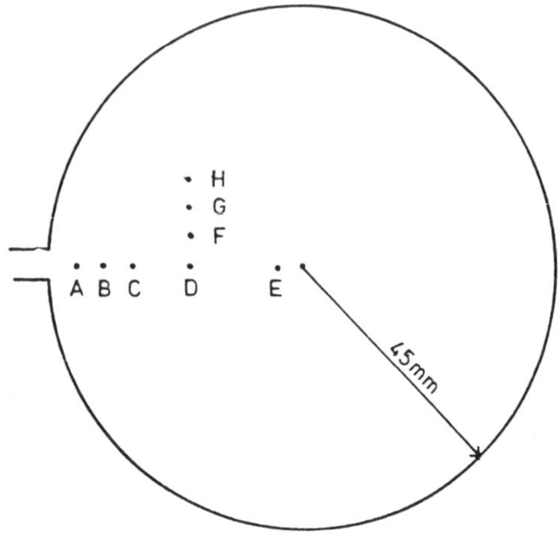

pressure difference between the chambers at TDC. The model predicted that as the pre-chamber fluid emerged into the main chamber a symmetrical pair of recirculating eddies would form either side of the throat in the main chamber.

Experimental set 1. The progression of the emerging jet across the main chamber and its associated velocity field are shown in Figs 20–22. The positions of the measuring points A–H are shown in Fig. 19. All measurements are by LDA, except for the HWA measurement at A, shown as A* in Fig. 20.

A comparison between the observations and predictions of the axial component of velocity (u) along the axis of the main chamber is shown

Fig. 20. LDA observations and predictions of the axial velocity component u: motored case. * HWA corrected according to free stream temperature.

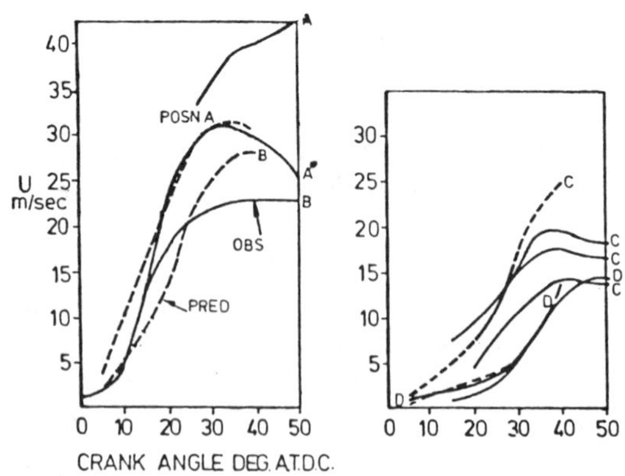

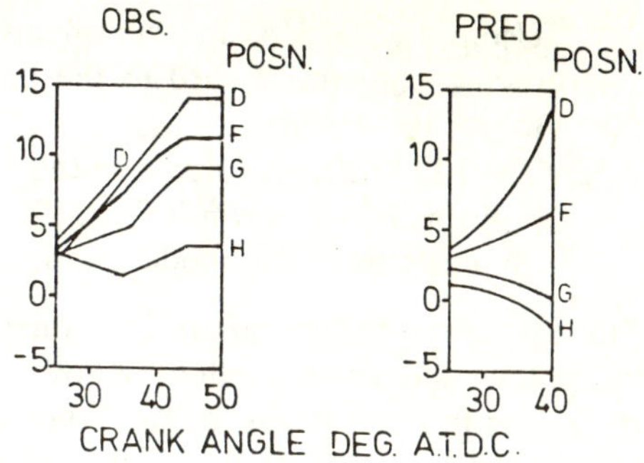

Fig. 21. LDA observations and predictions of the axial velocity component *u*: motored case.

in Fig. 20. A comparison across the chamber is shown in Fig. 21. The predicted negative values of *u* at the point furthest from the axis indicate the presence of recirculation. Comparisons of the lateral component of velocity (positive away from the axis of symmetry), both on and off the axis of symmetry, are shown in Fig. 22. The following points should be borne in mind when assessing model performance:

(a) Differences of the order of 30% are observed between measurements made at the same point in the chamber in different runs (see for example Fig. 20, point C, for which three runs were made).
(b) Significant differences were also observed between the laser–doppler and the hot-wire anemometry (HWA) measurements. For example, a phase difference of 25° between the peak velocity at position A between the two methods was observed.

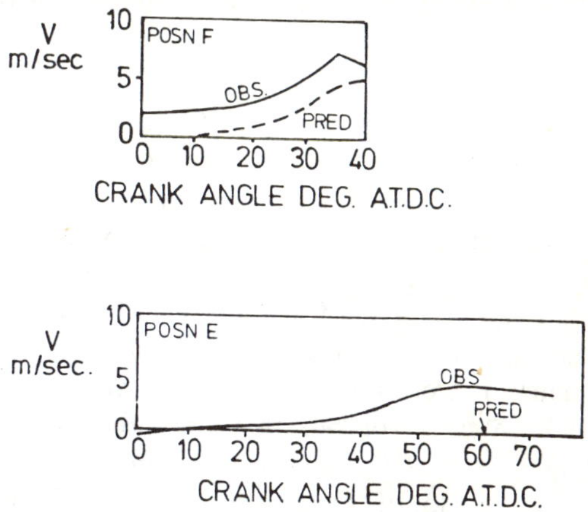

Fig. 22. LDA observations and predictions of the lateral velocity component *v*: motored case.

Using the HWA measurements, Fig. 20 shows that better correlation with regard to the timing of the peak velocity at A would be obtained.

(c) The LDA measurements of the lateral velocity component *V* at position E, as shown in Fig. 22, indicates that the flow may be asymmetrical. The model assumes the flow field is symmetrical.

This first set of results raised some questions concerning the symmetry of the flow and also the extent of any recirculation regime predicted by the model but not observed. A further set of LDA measurements were subsequently made, in which attention was focused on these points.

Experimental set 2. Predictions and observations of the velocity field at 30° ATDC and 40° ATDC are shown in Fig. 23. The predictions show the development and growth of a recirculating eddy either side of the throat. The LDA observations from both experimental sets 1 and 2 have been combined. It can be seen that observations also support the development of a recirculating regime. It would seem that the centre and shape of the predicted and observed eddies are not coincident insofar as they can be determined accurately. Allowance must be made for the fact that 3-D effects at 30 and 40° ATDC are progressively becoming important, but certainly the general features of the observed and predicted distributions suggest that encouraging predictions are being made. Also note that a test of symmetry was made by taking measurements at points symmetrically opposed to each other. These vectors are compared at point X on Fig. 23. Clearly there is asymmetry, and predictions must be viewed with regard to this.

3.3.12. SCAN Combustion Simulations

Extensive correlations under firing conditions on suitable chambers using the SCAN code are difficult to make. However, it was considered of interest to exercise the model to ascertain its viability. These simulations were started at ignition, which is assumed to take place on the axis of symmetry at the pre-chamber wall opposite the throat. For the purposes of demonstration it was assumed that the chambers were initially quiescent, and that the A/F ratio was constant in each chamber but that the pre-chamber was richer. A constant flame speed of 20 m s^{-1} was assumed, guided from high-speed photography. A plot of the progression of the flame front at three consecutive

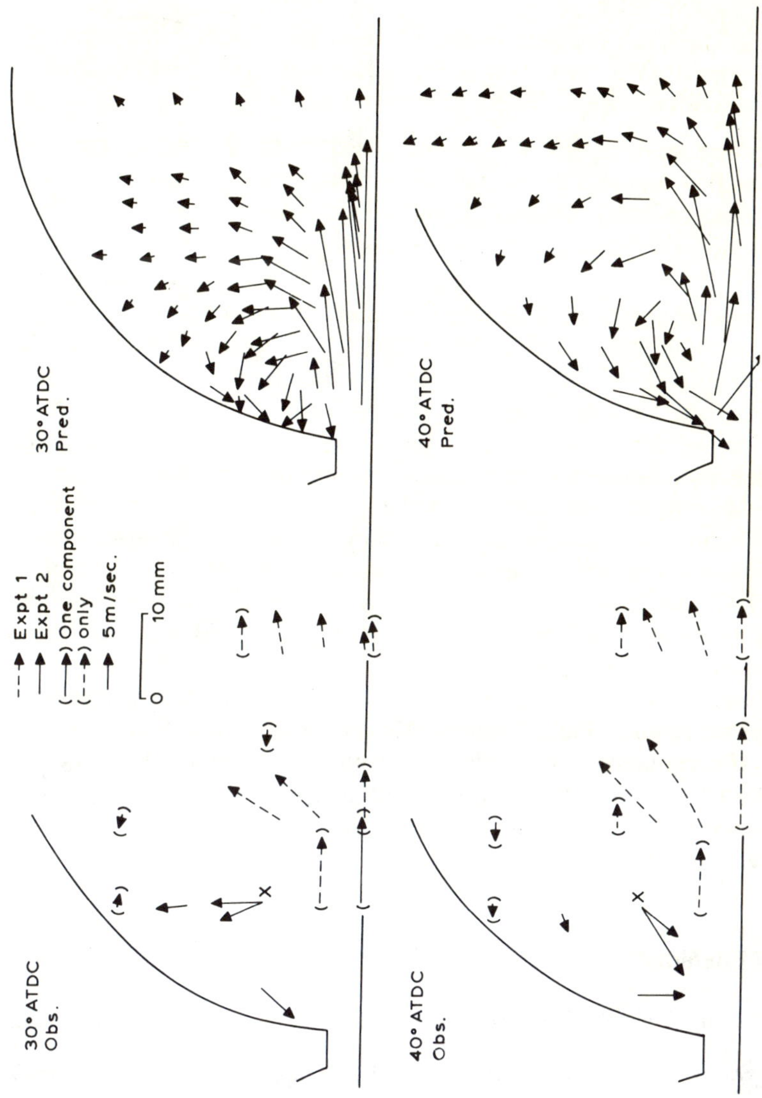

Fig. 23. Observed and predicted velocity fields in a motored divided chamber.

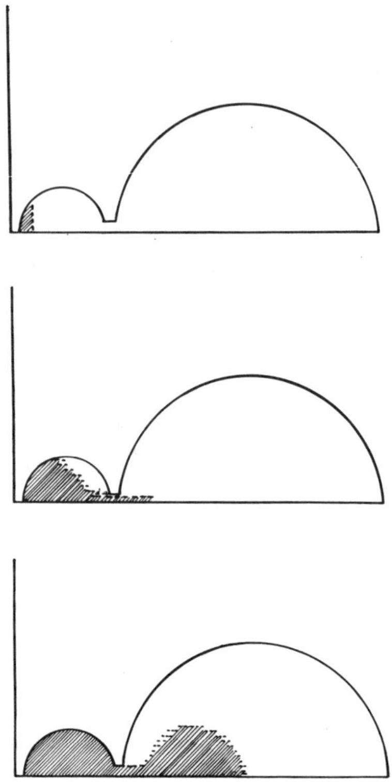

Fig. 24. Flame front progression.

times is shown in Fig. 24. Ignition is simulated by burning the column of cells including the spark cell at the left-hand side of the pre-chamber at 25° BTDC. It is recalled that the front is treated explicitly and is calculated and stored as a set of co-ordinate points within the program.

REFERENCES

1. Ricardo, H. R., Recent work on the internal combustion engine, *SAE Journal,* **10** (May 1922) 305–36.
2. Schweitzer, P. H. & Grunder, L. J., Hybrid engines, *Trans. SAE,* **71** (1963) 541.
3. Date, T., Yagi, S., Ishizuya, A. & Fujii, I., Research and development of the Honda CVCC engine. Honda Research and Development Co. Ltd, SAE paper 740605, August 1974.

4. Yagi, S., Date, T., Ishizuya, A. & Fujii, I., Research and development of the Honda CVCC Engine. Honda Research and Development Co. Ltd, SAE paper 741158, 1974.
5. Yagi, S., Fujii, M., Nishikawa, M. & Shirai, H., A newly developed 1·5 litre CVCC engine. SAE 800321, 1980.
6. Gruden, D., Stratified charge engines. Instn. Mech. Engrs Conference, Nov. 1976, paper C243.
7. Dillstrom, T., A high-power spark ignition fuel injection engine. *SAE Journal* (Dec. 1934) 431–8.
8. Witzky, J. A. & Clark, J. M. Jr, Stratification and combustion in reciprocating engines. ASME Diesel and Gas Engine Power Conference, Houston, Texas, 12–16 May 1968.
9. Barber, E. M., Reynolds, S. & Tierney, W. T., Elimination of combustion knock—Texaco combustion process. *Trans. SAE,* **59** (Jan 1951) 50.
10. Tierney, W. T., United Parcel Service applies Texaco stratified charge technology to power parcel delivery vans. SAE paper 801429, 1980.
11. Jones, C., A progress report on Curtis Wright's rotary stratified charge engine development. Curtis Wright Corp., SAE paper 741206, 1974.
12. Krieger, R. B. & Davis, G. C., The influence of the degree of stratification on jet-ignition engine emissions and fuel consumption. Instn. Mech. Engrs Conference, Nov. 1976.
13. Weaving, J. H. & Corkill, W. J., British Leyland experimental stratified engine. Instn. Mech. Engrs Conference on Stratified Charge Engines, Nov. 1976.
14. Bishop, I. N. & Simko, A. O., A new concept of stratified charge combustion. SAE paper 680041, 1968.
15. Simko, A. O., Scussel, A. & Wade, W., The Ford Proco engine update. SAE paper 780699, 1978.
16. Choma, M. A., Havstad, P. H., Simko, A. O. & Stockhausen, W. F., Fuel tolerance tests with the Ford Proco engine. Instn. Mech. Engrs Conference on Stratified Charge Engines, paper C398/80, 1980, pp. 41–7.
17. Launder, B. E. & Spalding, D. B., *Mathematical Models for Turbulence.* Academic Press, London 1972.
18. Patankar, S. V. & Spalding, D. B., A calculation procedure for heat, mass and momentum transfer in three-dimensional parabolic flows. *Inst. J. Heat Mass Transfer,* **15**(8) (1972) 1787–806.
19. Awn, A. E. G., Calculation of unsteady compressible flows. CHAM Report 1391/2, 1977.
20. Spalding, D. B., Mathematical modelling of fluid mechanics. Heat Transfer and Mass Transfer Processes: a lecture course. Imperial College, London, Heat Transfer Section. Report No. HTS/80/1, 1980.
21. Gordon, S. & McBride, B., Computer program for calculation of complex chemical equilibrium compositions. NASA SP-273, 1971.

6

Mixture Preparation for Spark-ignition Engines

C. J. E. Nightingale
University College London, London, UK

1. Introduction
2. Engine requirements
3. Droplet formation
   3.1. Basic principles
   3.2. Mean droplet diameter
   3.3. Droplet size distributions
   3.4. Prediction of mean and maximum droplet diameters
   3.5. Droplet formation in fuel systems
4. Evaporation of drop
   4.1. Motionless drop
   4.2. Moving drop
5. Transport of mixture through manifolds
   5.1. Theoretical investigations
   5.2. Experimental investigations
   5.3. Manifold design for transport of mixture
6. Measurement techniques
   6.1. Assessment of problem
   6.2. Measurement of mean droplet size and droplet size distribution
   6.3. Wall-film flow rate measurement
7. Future work
   7.1. Manifold investigations
   7.2. Inlet valve investigations.

NOTATION

b_1 Term in mean diameter expression
b_2 Term in mean diameter expression

c_1	Constant
c_2	Constant
d	Diameter of orifice
D	Diameter of drop
D_{32}	Sauter mean diameter of drops
D_{30}	Volume mean diameter of drops
$\bar{D}$	Characteristic diameter (Rosin & Rammler distribution)
$\mathscr{D}$	Coefficient of diffusion
k	Thermal conductivity
k_1	Constant of proportionality
k_v	Coefficient of mass transfer
L	Latent heat of vaporization
m	Mass
M	Molecular mass
n	Term in Rosin & Rammler distribution
N	Number of drops
P	Total pressure
p	Partial pressure
p_r	Partial pressure of vapour at surface of drop
p_∞	Partial pressure of vapour at infinite distance
q	Heat flow rate
Q	Volume flow rate
$\mathscr{R}$	Universal gas constant
t	Time
T	Temperature
VF	Volume fraction
Nu	Nusselt number (heat transfer)
Nu'	Nusselt number (mass transfer)
Pr	Prandtl number
Re	Reynolds number
Sc	Schmidt number
Γ	Gamma function
μ	Viscosity
v	Relative velocity
ρ	Density
σ	Surface tension

Subscripts

A	Air
L	Liquid
max	Maximum

i Initial
v Vapour

1. INTRODUCTION

The fuel system and inlet manifold are servants of the spark-ignition (SI) engine in that they are called upon to deliver to each cylinder the required amount of fuel mixed with an appropriate quantity of air, served in a digestible form. The objective of this chapter is to review current knowledge of the processes involved in the supply of this fuel/air mixture. This topic is one that has been researched continuously, but not always with the highest priority, during the evolution of the SI engine; it has demanded particular attention recently due to the development of lean-burn engines. Lean-burn engines not only require tighter tolerances on air:fuel ratio variations, both between cylinders and from one cycle to the next, but they are also sensitive to the mixture preparation (i.e. the percentage of fuel vaporized, droplet size and distribution, etc.) within the cylinder. The starting point of this chapter will be a discussion of the requirements of SI engines in terms of mixture preparation.

2. ENGINE REQUIREMENTS

Two of the fundamental requirements are for a fuel system and its manifold to supply a uniform distribution of fuel and air between cylinders, and for this supply not to vary on a cycle-to-cycle basis for as long as the engine condition is constant. For a multi-point fuel injection system (i.e. one injector per cylinder intake port), the success of this operation will depend on the uniformity and consistency of the injectors, and also on the ability of the inlet manifold to deliver equal quantities of air to each cylinder. When a central fuel preparation system (e.g. a carburettor) is used, the conventional approach is to apply heat to the manifold walls at the first really wet surface so that as much of the fuel as possible is vaporized. This tends to avoid the difficulties of dividing a two-phase flow, as this involves ensuring that each phase is shared equally.

Thus, a vaporized fuel and air mixture is desirable for good distribution, but the question arises as to whether having a large proportion of vaporized fuel at this stage is what is ideally required for overall engine performance considerations. A significant disadvantage is that vaporized fuel will reduce the full-throttle torque output of the engine. This is due to a loss of volumetric efficiency caused both by the increase in the volume occupied by the fuel in its vapour phase and by the increased mixture temperature. As an approximate guide, there would be a 13·5% loss in volumetric efficiency for an engine supplied with fully vaporized fuel and air mixture at 50°C compared with an 85% wet mixture at 20°C. In terms of combustion, it is less clear as to the ideal mixture for an engine to burn. Dodd[1] performed tests on a single-cylinder engine supplied with fuel by, in turn, a port-mounted injector and a vaporizing unit. In each case, the engine should have received equal amounts of fuel from one cycle to another, and so the observed extension of the part-throttle lean limit of the engine by an average of a 2·0 A/F ratio indicated that there had been an improvement in the combustion of lean mixtures from using pre-vaporized fuel.

There have also been examples of engines with good part-throttle combustion of lean mixtures that have not been supplied with pre-vaporized fuel. A pattern can be discerned and a convenient illustration is given in the work of Beale & Hodgetts.[2] Their work was performed on an inlet valve throttled engine again using, in turn, a port injector and pre-vaporized mixture. They found that with port injection the effect of valve throttling was to extend the lean limit considerably: when a pre-vaporized mixture was used, only a marginal further improvement was noted. However, when vaporized fuel was supplied in conjunction with conventional engine throttling, the lean limit was extended in the same way as with valve throttling. Beale & Hodgetts attributed the effects of valve throttling to the high velocity of flow through the inlet valve and the turbulence caused by the dissipation of this flow energy. This gas motion breaks down all large drops, distributes them around the combustion chamber and aids their evaporation: it will also have further repercussions on the initiation and propagation of the flame, but this is outside the scope of this section. Inlet valve throttling is difficult to implement on practical engine designs, but some of its benefits can be obtained in other ways. The engine for the Ford Eltec research vehicle[3] has two inlet valves and ports per cylinder (Fig. 1). The mixture is directed through the smaller diameter port at low air flows to keep a high velocity entering the cylinder, while at high flows the mixture is allowed down both ports to minimize the flow losses. A retractable swirl-generating device

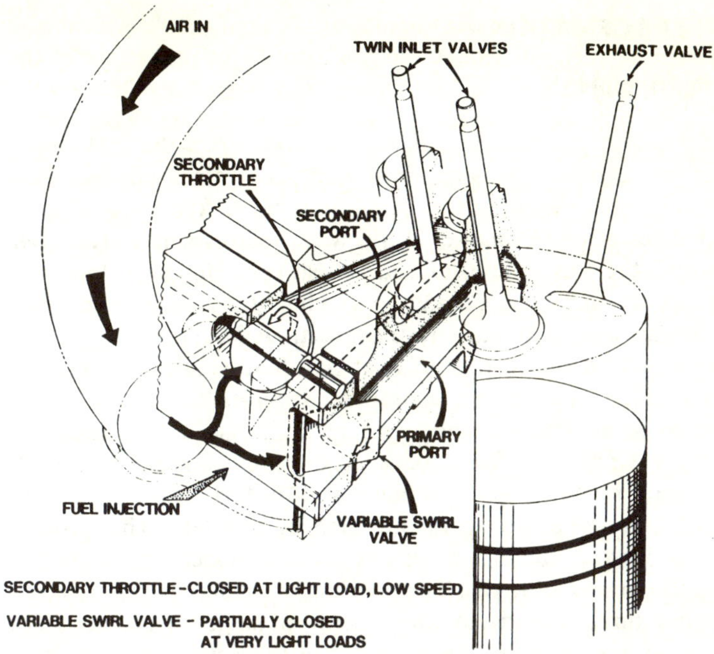

Fig. 1. Inlet port layout of Ford Eltec engine (reprinted with permission of the Ford Motor Company).

is positioned in the smaller port with the objective of increasing the swirl set up within the cylinder at the very low flow conditions. The enhanced mixture motion at low-flow conditions resulting from these measures will produce small droplets, help distribution and evaporation, and for these (and other reasons) extend the part-throttle lean limit.

Any attempt to generalize on mixture preparation requirements would have to say that different engine designs have different requirements. Some designs will only require uniform air/fuel distribution and will perform the final mixture preparation by such means as creating high-velocity flow through the inlet valve: other engine designs will prefer a well prepared mixture prior to the inlet valve, and this will probably be achieved by ensuring that a high proportion of the fuel is already evaporated.

Since gasoline will only burn as a vapour, the processes of droplet formation and evaporation are crucial for mixture preparation, and so these processes are reviewed individually in the following two sections.

3. DROPLET FORMATION

3.1. Basic Principles

Droplet formation is performed in carburettors when the fuel issues from a jet into a rapidly moving air stream. The velocity of fuel flow through the jet is of the order of $1\ \mathrm{m\,s^{-1}}$ and, in the terms that droplet formation is described in Chapter 7, this would correspond to Rayleigh jet break-up if the fuel was introduced into still air. However, the air is far from still: for instance, it could be moving at a velocity of around $60\ \mathrm{m\,s^{-1}}$ in an air-valve carburettor (Section 3.5.1), with further acceleration around the throttle disc at part-load conditions. The air flow tears the column of fuel apart into separate threads of liquid which try to form droplets under the action of surface tension. In many cases, the relative velocity between the drops and the surrounding gas is such that the aerodynamic forces distort them to such an extent that they break up again into smaller droplets. This process continues as long as there is sufficient relative velocity between the drops and the surrounding gas. This mechanism of successive break-up of the liquid threads leads to there being a range of droplet sizes present in any spray. Thus, to characterize the spray, one needs to specify the form of the droplet size distribution and also a mean diameter.

3.2. Mean Droplet Diameter

The situation is confused by there being a number of mean diameters that can be applied. The most commonly used mean diameter in the IC engine context is the *Sauter mean* diameter which represents the droplet diameter of a notional spray of uniform size, with the same overall volume as the actual spray, and also with the same overall surface area. In mathematical terms:

$$\text{Sauter mean diameter} = \sum ND^3 / \sum ND^2 \qquad (1)$$

where N is the number of drops in a size group of mean diameter D. The Sauter mean diameter is particularly relevant because the evaporation of fuel sprays is of prime interest and the rate of evaporation is a function of surface area.

Mean volume diameter is used in one of the spray formulae given later in this section; it is the diameter of a drop having a volume equal to the average volume per droplet of the entire spray. In mathematical terms:

$$\text{mean volume diameter} = (\Sigma ND^3/\Sigma N)^{1/3} = (\Sigma ND^3/N_{TOT})^{1/3} \quad (2)$$

where N_{TOT} is the total number of drops in the spray. There are at least five other mean diameters that have been used in different applications. A general equation can be formed to cover all these mean diameter definitions:

$$\text{mean diameter} = (\Sigma ND^{b_1}/\Sigma ND^{b_2})^{1/(b_1-b_2)} \quad (3)$$

where $b_1 = 3$ and $b_2 = 2$ for the Sauter mean diameter (termed D_{32}), and $b_1 = 3$ and $b_2 = 0$ for the mean volume diameter (termed D_{30}).

3.3. Droplet Size Distributions

A number of different droplet size distribution equations have been applied to describe the variation of droplet diameter throughout a fuel spray. The Rosin & Rammler distribution was put forward in the 1920s by two German researchers to describe the distribution of particle size after the crushing of coal. It has since been applied to droplet sprays. The equation takes the form:

$$VF = e^{-(D/\bar{D})^n} \quad (4)$$

where VF is the volume fraction of particles or drops with a diameter greater than D, $\bar{D}$ is the characteristic diameter of particles/drops, and n is a measure of the spread, such that $n = \infty$ for a spray with droplets of uniform size. In practice, $1{\cdot}5 < n < 3{\cdot}0$ for most fuel sprays. The characteristic diameter, $\bar{D}$, is related to the Sauter mean diameter, D_{32}, by:

$$D_{32} = \frac{\bar{D}}{\Gamma\left(1 - \frac{1}{n}\right)} \quad (5)$$

where Γ is the gamma function, the value of which may be determined from tables.[4]

Two typical Rosin & Rammler distribution curves, plotted on a volume fraction basis are shown in Fig. 2; the implication of this type of distribution is that most of the droplets in a spray are relatively small in size. For instance, in the case where $\bar{D} = 25\ \mu m$ and $n = 3$, around 65% of the droplets would have a diameter less than 4 μm.

Other frequently used distributions have been the Nukiyama and Tanasawa distribution and the log-probability distribution. All three

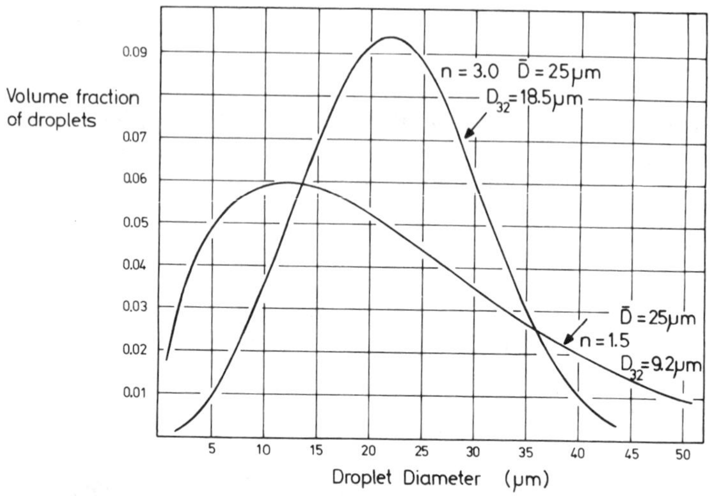

Fig. 2. Rosin & Rammler distribution curves.

distributions were reviewed critically by Mugele & Evans[5] who put forward a further alternative, the upper limit distribution. Mugele & Evans' appraisal of the Rosin & Rammler distribution was that it was usually possible to obtain a curve fit that was reasonably close to experimental data, but there was sometimes a difference in Sauter mean diameter between the value calculated from eqns (4) and (5) and that obtained from summation of experimental data. A more recent review by Elkotb[6] included two further possible distributions, the chi-square and the Tanasawa–Tesima distributions, and the conclusions were that either would be suitable for the specific case that they investigated (sprays injected through pintle-type nozzles).

3.4. Prediction of Mean and Maximum Droplet Diameters

It has been a focus of attention for researchers to investigate which are the predominant parameters that determine droplet size and to put forward equations to describe their effect. Two different approaches will be described in the following:

(1) The empirical approach of performing experiments, varying important parameters and measuring the resulting droplet sizes: from this data, correlations can be developed to predict mean droplet diameter. The work of Nukiyama & Tanasawa[7] and Ingebo[8] would fit into this category.

(2) The theoretical approach of studying the factors governing the stability of a drop and developing an equation to predict the maximum diameter of droplet that can exist under given conditions. The work of Prandtl[9] and Hinze[10] would fit into this category.

The following subsections include two examples of each of the two fundamental approaches.

3.4.1. The Nukiyama & Tanasawa Equation

$$D_{32} = \frac{585}{v}\sqrt{\frac{\sigma_L}{\rho_L}} + 597\left(\frac{\mu_L}{\sqrt{\sigma_L \rho_L}}\right)^{0.45}\left(\frac{1000 Q_L}{Q_A}\right)^{1.5} \qquad (6)$$

where D_{32} is the Sauter mean diameter (μm), v is the relative velocity between air and liquid streams (m s^{-1}), ρ_L is the liquid density (g cm^{-3}), σ_L is the liquid surface tension (dyne cm^{-1}), μ_L is the liquid viscosity (poise), Q_L is the volume flow rate of fuel, and Q_A is the volume flow rate of atomizing air. In the case of SI engine applications, the second term on the right-hand side of the equation may be neglected. Thus, this equation predicts the Sauter mean diameter to be mainly a function of relative velocity, surface tension and liquid density. The equation was put forward after work in the late 1930s on plain-jet air blast atomizers (Fig. 3): the geometry of the atomizing system and the pressures involved are obviously different from those

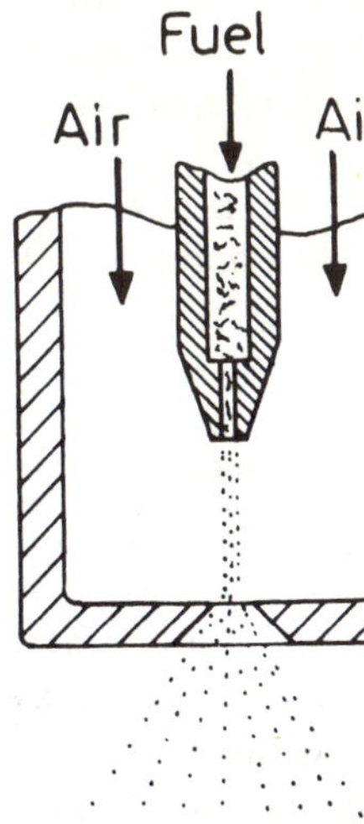

Fig. 3. Air-blast atomizer.

appertaining to a carburettor, but the equation has been applied to carburettors with moderately good agreement between the predictions and experimental results. The good agreement can be explained by the fact that it is a high-velocity air stream in each case that is causing the break-up of the fuel into small droplets. In the case of the air-blast atomizer, a number of more recent investigations have been performed[11]; many of these have been inspired by the need to improve fuel preparation in gas turbine combustion chambers.

3.4.2. The Ingebo & Foster Equation

Ingebo & Foster[8] performed a series of experiments in the mid-1950s on the injection of fuel through a plain orifice at right angles to a high-velocity air stream. This geometry is representative of the arrangement that exists within carburettors when fuel is introduced into the air stream. Their results could be described by:

$$D_{30} = 3{\cdot}9d\left(\frac{1}{Re}\right)^{0{\cdot}25}\left(\frac{\sigma_L}{\rho_A d v_i^2}\right)^{0{\cdot}25} \tag{7}$$

where d is the diameter of the orifice, σ_L is the surface tension, ρ_A is the density of air, v_i is the initial relative velocity between liquid and air, D_{30} is the mean volume droplet diameter, and Re is the Reynolds number (based on d). It should be noted that the grouping $\sigma_L/\rho_A d v^2$ appears later in Prandtl's and Hinze's equations. It is known as the Weber number and it is the non-dimensional grouping of parameters that are most significant in determining the size of drops formed during atomization.

3.4.3. Prandtl Equation

Prandtl's approach[9] was to estimate the maximum size of droplet that can exist when acted upon by aerodynamic forces:

$$15{\cdot}4 \approx D_{max}\rho_A v^2/\sigma_L \tag{8}$$

where D_{max} is the maximum diameter of droplet. The form of the equation was obtained from theoretical consideration of droplet forces, while the constant was obtained from experimental data. Prandtl pointed out that since the processes of deformation take time, one must insert an average value of the relative velocity v rather than just an instantaneous value. He also pointed out that an extra term would have to be included in the equation for viscous liquids to cover the damping effect.

3.4.4. Hinze Equation

The effects of viscous damping and velocity variation have been investigated in more detail by Hinze.[10] He pointed out that the application of dimensional analysis to droplet formation suggests that there is a non-dimensional viscosity group, $\mu_L/\sqrt{(\rho_L \sigma_L D}$, which will affect the critical Weber number for droplet shatter. Hinze states that, for values of viscosity number close to zero, the critical Weber number is 13 for the 'shock' condition. The 'shock' condition is defined as when the droplet is suddenly exposed to a parallel air stream, as is usually the case with SI engine fuel systems. Hinze's graph for the modifying effect of viscosity number on critical Weber number shows that, for a 50 μm diameter droplet of gasoline (viscosity group number = 0·02), the critical Weber number will be around 14·5. Thus, the maximum diameter of droplet that can exist without shattering is given by

$$14{\cdot}5 = D_{\max}\rho_A v_i^2/\sigma_L \tag{9}$$

A comparison between the predicted droplet diameters of the Nukiyama–Tanasawa and Hinze equations will be made later in the following subsection.

3.5. Droplet Formation in Fuel Systems

3.5.1. Carburettors

The salient points arising from the preceding discussion of the mechanisms involved in droplet formation can be applied to obtain an understanding of the form of spray that issues from a carburettor. A schematic drawing of a section through a piston-type air-valve carburettor is shown in Fig. 4. A qualitative description of the operation

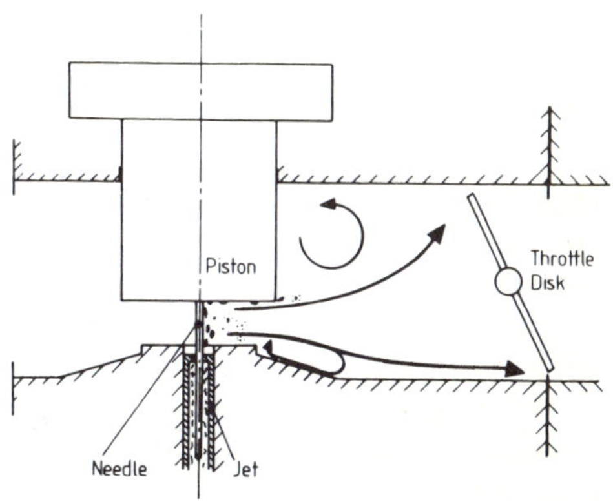

Fig. 4. Section of piston-type air-valve carburettor.

of this type of carburettor is given in Ref. 12, but for the purposes of this discussion it is sufficient to know that the mean velocity under the piston is, typically, of the order of $60\,\mathrm{m\,s^{-1}}$: it is into this air stream that the fuel is introduced. First, let us simplify the situation by considering steady flow through the carburettor. Applying Hinze's equation to this situation suggests that the Sauter mean diameter will be of the order of 40 μm. The droplets are carried past the piston into what is a region of confused flow: the efflux from the piston first enters a relatively large flow section and then has to divide between the flow areas of the leading and trailing edges of the throttle disc. The peak velocity of the mixture as it passes the throttle is dependent on the depression in the manifold and this is somewhat higher than the velocity under the piston at all but the wider open-throttle conditions. The acceleration of the gas flow as it approaches the throttle disc will produce drag forces on the droplets which are greater than they have previously experienced, so that all but the smallest droplets will break up further. At part-throttle conditions, it is therefore the atomization that takes place at the throttle disc which determines the characteristics of the carburettor spray, and not the initial atomization that takes place at the needle/jet. This is illustrated in Fig. 5, where the mean droplet diameters issuing from a piston-type air-valve carburettor are

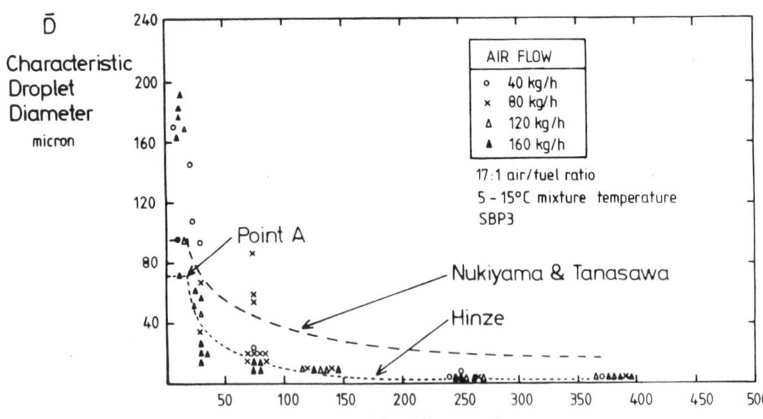

Fig. 5. Air-valve carburettor droplet sizes: comparison of results from test rig with theoretical predictions.

plotted against downstream depression (equivalent to manifold depression). The measurements[13] were taken on a steady flow rig using a Malvern Droplet Size Analyser (see Subsection 6.2.2 and Chapter 7): a substitute fuel was used (Shell–BP solvent SBP 3) which has a similar relative density and surface tension to gasoline. Tests were performed over a range of throttle angles and air flows, and it was apparent that the controlling variable was downstream depression: the downstream depression is related to the peak velocity achieved at the throttle disc and hence is related to drag forces experienced by the droplets as they approach the throttle. It can be seen that there was a considerable scatter of results at the low downstream depressions (i.e. at the near wide open throttle conditions). At these conditions, the fit of the Rosin & Rammler distribution curve to the observed data was relatively poor (but still within an arbitrary limit set for the Analyser). This difficulty stemmed from there being at least three mechanisms of droplet formation:

(1) Some of the droplets formed at the needle/jet would pass through the carburettor without further break-up.
(2) Some of the larger drops formed at the needle/jet would break up further due to the drag forces that they experienced as they passed through the carburettor.
(3) The remaining drops would impact, some on the piston and some on the throttle, and then be re-atomized by the passing stream.

The Nukiyama–Tanasawa and Hinze equations [eqns (6) and (9)] have been applied to predict mean droplet diameter and the results shown in Fig. 5. The two equations predict a similar trend in droplet size, but the Hinze equation shows better agreement with experimental results. The value of velocity used in each case was the theoretical maximum flow velocity in the carburettor, calculated assuming no velocity head recovery either after the piston or after the throttle. The prediction of maximum diameter given by the Hinze equation was converted to Sauter mean diameter using the relationship suggested by Lenz[14] for carburettor sprays:

$$D_{32} = 0{\cdot}532 D_{\max} \tag{10}$$

Sauter mean diameters, D_{32}, were converted to characteristic Rosin & Rammler mean diameters, $\bar{D}$, by assuming a value of $n = 2$ for the spray spread.

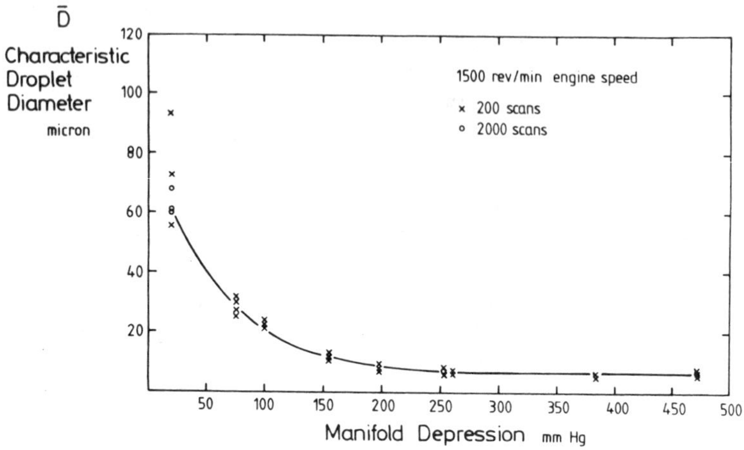

Fig. 6. Air-valve carburettor droplet sizes: measured on engine[15] (reprinted with permission of the Institute of Physics).

On an engine, the situation is complicated by the pulsating nature of the air flow. At high mean manifold depressions, the effect on the flow through the carburettor is limited by the fact that the flow through the throttle is choked for a significant proportion of an engine cycle. However, for lower depressions, the flow through the carburettor will be highly pulsating and so the velocity, and hence mean diameter of droplets formed, will vary throughout a cycle: the region of confused flow (referred to earlier) between the piston and the throttle will become even more confused. Measurements[15] of characteristic droplet diameter have been taken on a running engine directly downstream of a piston-type air-valve carburettor and the results are shown in Fig. 6.

The preceding section has referred to droplet formation in air-valve carburettors. A similar description would apply to fixed-choke carburettors apart from the fact that there would be more variation of the mean diameter of the droplets formed at the metering point in response to changing air flow. One consequent effect would be that the changeover from throttle plate atomization would usually occur at a higher depression than at point A in Fig. 5.

3.5.2. Central Injection Systems

Fuel systems, where either one or two injectors are mounted in a body behind a throttle disc, are known variously as throttle body injection (tbi) systems, central fuel injection (cfi) systems or central injection (ci) systems. The main difference between these systems and carburettors in terms of mixture preparation is that the fuel supply is pulsed instead of the fuel being induced by the depression resulting from the air velocity through the air valve or venturi. The latter is the preferable situation, since fuel flow is linked to air flow, rather than having the fuel introduced as a number of discrete 'packets' with the likelihood of subsequent inter-cylinder distribution problems. A further problem with central injection systems is that, for reasons of cost, relatively low fuel supply pressures are used, resulting in a correspondingly large mean droplet diameter: Toyota[16] have measured a Sauter mean diameter of 470 μm for a fuel supply pressure of 69 kPa. It is common practice with central injection systems to arrange for the injector spray to impact on the upstream side of the throttle disc. This is to ensure that re-atomization will take place over as wide a range of conditions as possible since the initial mixture preparation is so poor. If the spray angle is kept narrow (Fig. 7a), there will be a division of fuel between the leading and trailing sides of the throttle disc which will change with throttle angle and this will cause inter-cylinder distribution problems. If the spray angle is too wide (Fig. 7b), some impaction will occur on the flow passage walls, with the consequence that some fuel will pass the throttle as wall flow. However, Toyota[16] have found the narrow spray angle better in terms of engine transient response (because of less wall-wetting) and so their final solution used an offset narrow spray angle injector (Fig. 7c). The offset gave a

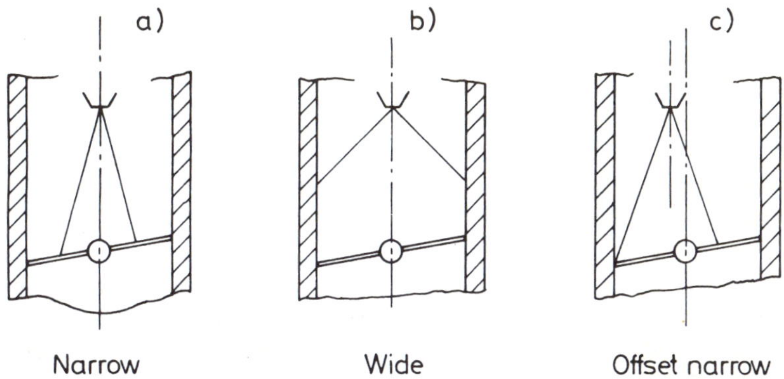

Fig. 7. Injector mounting positions for central fuel injection.

consistent, but uneven, split of fuel between the leading and trailing edges of the throttle disc.

3.5.3. Multi-point Fuel Injection Systems

The injectors are mounted in the intake ports and so, unlike the central injection systems, there can be no secondary atomization at the throttle disc. The fuel supply pressure for multi-point systems (2–8 bar gauge) is higher than for central injection systems so as to reduce the dependence of flow on the prevailing manifold pressure. The higher injection pressure leads to a higher fuel velocity ($20\,\mathrm{m\,s^{-1}}$ upwards) through the injector nozzle. The injection pressure is still way below typical diesel engine injection pressures. Conditions within the inlet manifold of an SI engine are far from quiescent and the liquid leaving the injection nozzle is immediately subjected to strong aerodynamic forces: even if drops survive immediate break-up, they are likely to be either shattered as they pass the inlet valve or deposited on the valve or passage walls and then reformed into drops.

4. EVAPORATION OF DROP

4.1. Motionless Drop

The most straightforward case is that of the evaporation of a single spherical drop, motionless in an infinite medium. The rate of evaporation can be determined by combining the equations governing diffusion and heat transfer. The rate of diffusion of vapour away from the droplet is given by Stefan's equation[17]:

$$\frac{\mathrm{d}m}{\mathrm{d}t} = \frac{-2\pi D \mathscr{D} M_v P}{\mathscr{R} T_v} \ln\left\{1 - \frac{(p_r - p_\infty)}{(P - p_\infty)}\right\} \tag{11}$$

where $\mathrm{d}m/\mathrm{d}t$ is the mass rate of evaporation, $\mathscr{D}$ is the coefficient of diffusion of vapour, D is the drop diameter, M_v is the molecular mass of vapour, $\mathscr{R}$ is the universal gas constant, T_v is the absolute temperature of vapour, P is the total pressure, p_r is the partial pressure of the vapour at the surface of the drop, and p_∞ is the partial pressure of the vapour at infinite distance. In the case of relatively non-volatile constituents, p_r and p_∞ will be small and it can be shown[18] that the Stefan equation will simplify to:

$$\frac{\mathrm{d}m}{\mathrm{d}t} \approx \frac{\partial \pi D \mathscr{D} M_v}{\mathscr{R} T_v}(p_r - p_\infty) \tag{12}$$

The above equation was obtained independently by Maxwell about the same time as Stefan's work (towards the end of the 19th century). When dealing with gasoline evaporating in air at around ambient temperatures, eqn (11) would be required for such constituents as butane, whereas eqn (12) would be an acceptable approximation for the less volatile octane.

For the normal situation that exists inside an SI engine manifold (fuel introduced at a similar temperature to the air, with the fuel above its wet bulb temperature), the temperature of the drop will initially fall relative to its surroundings due to the provision of latent heat for the evaporating fuel. The resulting temperature difference between the drop and its surroundings will mean that heat will flow from the surroundings to the drop as given by:

$$q = \pi D^2 h (T_A - T_L)\left(\frac{z}{e^z - 1}\right) \tag{13}$$

where

$$z = \left(\frac{\mathrm{d}m}{\mathrm{d}t} c_{pL}\right) \Big/ \pi D^2 h$$

and where q is the heat flow rate, h is the convective heat transfer coefficient, T_L is the temperature of the drop, T_A is the temperature of the surrounding air, and C_{pL} is the specific heat of the liquid. $[z/(e^z - 1)]$ is a correction factor to allow for the fact that some of the heat transferred from the surroundings does not reach the drop as it raises the temperature of the surrounding vapour. This correction was used by El Wakil *et al.*[19] Equation (13) ignores any heat transfer by radiation and assumes a constant temperature, T_L, throughout the drop. The subsequent rate of change of temperature by the drop is:

$$C_{pL}\left(\frac{\pi D^3}{6}\right)\rho_L \frac{\mathrm{d}T_L}{\mathrm{d}t} = -L\frac{\mathrm{d}m}{\mathrm{d}t} + q \tag{14}$$

where L is the latent heat of vaporization and ρ_L is the density of liquid. Equations (12), (13) and (14) can be solved for a series of steps in time to give mass–time and temperature–time histories. The results for a drop of heptane in air are shown in Fig. 8. It can be observed

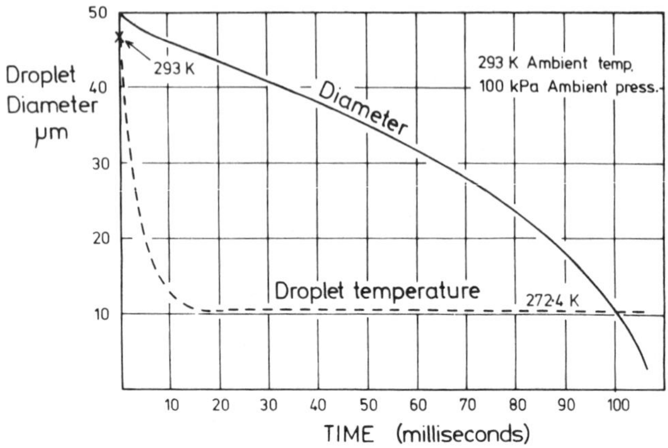

Fig. 8. Evaporation history of a heptane droplet.

that the temperature of the drop becomes steady after a short period of time: this corresponds to the wet bulb temperature of the drop. An appreciation of the factors that affect the rate of evaporation of a drop can be obtained from studying eqn (12):

(a) The rate of evaporation is proportional to the droplet diameter. This does not mean that a sample of fuel formed into N_1 drops of diameter D_1 will evaporate more slowly than the same volume of fuel formed into N_2 drops of diameter D_2 (where $D_2 > D_1$), because:

the rate of evaporation of sample 1,

$\mathrm{d}m_1/\mathrm{d}t$, will be proportional to $N_1 D_1$

the rate of evaporation of sample 2,

$\mathrm{d}m_2/\mathrm{d}t$, will be proportional to $N_2 D_2$

and if there is the same quantity of fuel in both samples,

$$N_1(\pi D_1^3/6) = N_2(\pi D_2^3/6)$$

$$\mathrm{d}m_1/\mathrm{d}t = k_1(N_2(D_2^3/D_1^3)D_1$$

$$\mathrm{d}m_2/\mathrm{d}t = k_1(N_2 D_2)$$

so

$$\frac{\mathrm{d}m_2}{\mathrm{d}t} = \frac{\mathrm{d}m_1}{\mathrm{d}t}\left(\frac{D_1}{D_2}\right)^2$$

Hence the benefit of having smaller drops to increase the rate of evaporation.

(b) $p_r - p_\infty$ may be considered as the 'driving force' of evaporation, so that a high evaporation rate will be obtained when the fuel concerned has a relatively high vapour pressure at the temperature of the drop, while at the same time the surroundings are kept relatively devoid of vapour.

(c) Equation (12) implies that a high temperature will reduce the rate of evaporation. The coefficient of diffusion is also temperature dependent, and the overall effect of increasing temperature is to make a number of changes to terms which, in total, result in the expected increase in evaporation rate.

(d) The coefficient of diffusion, $\mathscr{D}$, is an obvious factor in determining the evaporation rate. Values are available for some of the fuel constituents to describe the rate of vapour diffusion in still air. However, conditions within an SI engine manifold are such that the drops are continuously being accelerated and decelerated during their passage through the manifold. Both the coefficient of diffusion and the convective heat transfer coefficient will increase with motion of the surrounding fluid. The prediction of their values is covered in the following subsection.

4.2. Moving Drop

Considering the case of heat transfer first, it can be shown by dimensional analysis that the heat transfer between a sphere and its surrounding medium under conditions of relative motion can be given by an equation expressing the Nusselt number as a function of the Reynolds and Prandtl numbers. Nusselt, Reynolds and Prandtl numbers are all dimensionless groups: the Nusselt number is a measure of the rate of convective heat transfer; the Reynolds number describes the flow; and the Prandtl number is a grouping of the fluid properties relevant to heat transfer. Various correlations between these groups have been put forward, but Ranz & Marshall's equation[20] has become widely accepted:

$$Nu = 2{\cdot}0 + Pr^{1/3}Re^{1/2} \tag{15}$$

The value of the convective heat transfer coefficient calculated from the preceding equation can be used in equation (13) for moving drop calculations.

An analogy exists between the processes of heat transfer and mass transfer. It is such that similar equations can be used to cover relationships of corresponding terms. Thus, the equation for mass transfer is:

$$Nu' = 2{\cdot}0 + Sc^{1/3}Re^{1/2} \tag{16}$$

where Sc is the Schmidt number $(\mu/\rho\mathscr{D})$ which is a dimensionless group of fluid properties relevant to mass transfer, and Nu' is the Nusselt number for mass transfer $(k_v MDp/\mathscr{D}\rho)$. The value of the diffusion coefficient can be calculated from the above equation and used in eqn (11) for moving drop calculations.

5. TRANSPORT OF MIXTURE THROUGH MANIFOLDS

5.1. Theoretical Investigations

This section will concentrate on the transport of mixture through manifolds where there is some kind of central fuel preparation system supplying several cylinders. It was noted in the first section that the fundamental requirements were the supply of an equal distribution of fuel between cylinders without variation from one cycle to another for as long as the engine conditions were constant. These requirements could be served by either vaporizing a large proportion of the fuel before it has to be divided between cylinders, or by producing a stream of small, well mixed, droplets which would follow the air stream with a minimum of impaction on the manifold walls. If the first approach is adopted (which it usually is), the question arises as to how best to ensure the rapid evaporation of fuel: 'Is it best to pre-heat the fuel, or to have warm intake air, or to heat the surfaces of the manifold, or any combination?' If the latter approach is adopted, how small do the droplets have to be and what are the minimum permissible radii of manifold bends? Researchers have developed mathematical models to help answer these and other questions. Such is the complexity of the situation that it is inconceivable at the moment for a comprehensive model to be developed to cover all aspects of mixture transport, and so current models have been developed to concentrate only on specific aspects of the processes. These models can be split into the following three categories.

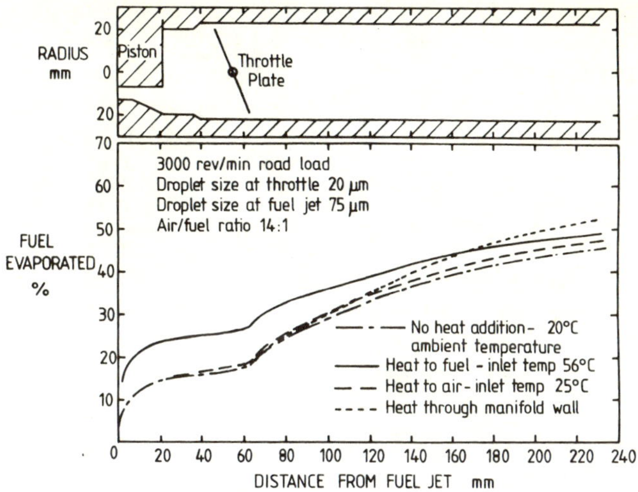

Fig. 9. Effect of different forms of heating on droplet evaporation[21] (reprinted with permission of Her Majesty's Stationery Office).

5.1.1. Modelling of Droplet Behaviour During Transport

Finlay *et al.*[21] have developed a model to describe the behaviour of fuel mixture from the point of formation at the needle/jet of a piston-type, air-valve, carburettor to when it has passed down a passage representative in length of the mixture path through a manifold. Salient features of the model are: the use of the Ranz & Marshall equations to help predict heat and mass transfer rates; a 16-component model for the fuel; assumed droplet sizes at the piston and throttle disc based on experimental results; and an assumption that the wall-film is built up at a constant rate after the throttle until it is 20% of the total initial flow and, thereafter, the Reynolds analogy is used to predict the rate of droplet deposition. The application of this model is illustrated in Fig. 9, in which the effects can be seen of different methods of heating on the evaporation of fuel.

All models of this type involve many simplifying assumptions. The most crucial concerns the proportion of fuel existing as a wall-film. At part-throttle conditions, the region just downstream of the throttle is in a highly confused state: there are two jets of high-velocity mixture, one from each of the leading and trailing edges of the throttle, and these issue into a region of relatively large cross-sectional area. The

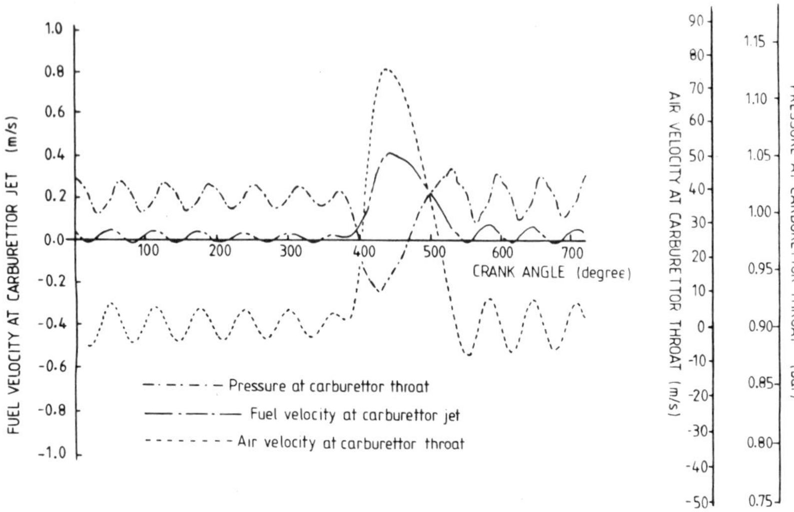

Fig. 10. Instantaneous air velocity, fuel velocity and pressure at carburettor[22] (reprinted with permission of the Society of Automotive Engineers).

highly turbulent nature of the flow is such that significant numbers of droplets are impacted on the flow passage walls. In addition, the layout of the manifold is often such that the mixture is required to turn sharply soon after leaving the fuel system and before the issuing jets have dissipated—further impaction will take place. It is because of this chaotic behaviour that it is very difficult to model representatively the behaviour of the wall-film.

One of the simplifications of Finlay's model was to assume steady flow, and Winterbone *et al.*[22] have subsequently developed a model for droplet behaviour in non-steady, non-homentropic, flow in which they used the method of characteristics to solve the resulting set of partial differential equations. The Ranz & Marshall correlation was used in the prediction of heat and mass transfer rates. The model was applied to a single-cylinder, four-stroke, engine fitted with an air-valve carburettor, and the results reported for full throttle conditions with a single-component fuel. It was assumed that no wall-film was present. One of the interesting predictions of this model (Fig. 10) is that there is likely to be a small oscillatory flow of air through the carburettor during the valve closed period and this continues to induce fuel, leading to an accumulation of liquid fuel in the carburettor flow passage during this period. Another prediction is that the fuel droplets maintain their movement through the air long after the closing of the inlet valve, eventually impacting on or near the valve.

5.1.2. Bend Impaction Models

The most comprehensive study was carried out as part of the Battelle investigation[23] in which the results of their theoretical investigation were checked using a pulsating flow test rig. The procedure adopted for the theoretical part of the study was to define the velocity field for the carrier gas and to perform a step-by-step calculation of the droplet trajectory from knowledge of its starting position, mass and drag coefficient. The conclusion was that the majority of droplets in the size range 10–20 μm should be able to pass through a typical manifold without impaction. However, this conclusion had to be modified in the light of results obtained from the experimental part of the investigation. This was carried out on a test rig in which various bend geometries were investigated under conditions of pulsating air flow. Droplets of known size were introduced upstream of the bend under test and the resulting wall-film was measured using a sampling section of the type described in Section 6.3.1. The conclusion of the experimental work was that there would be significant impaction, even with droplets of less than 20 μm diameter: in fact, considerably more impaction would occur than the theoretical work would suggest. However, the researchers still supported a fine fuel atomization approach as it would minimize impaction while helping evaporation. This is an important conclusion because it gives some hope of success to those trying to improve inter-cylinder distribution by developing a fuel system that produces small droplets, well distributed throughout the air stream.

5.1.3. Simple Wall-film Models

One of the problems of having wet manifold walls is that of the extra inertia during transients of the wall-film compared with droplet and vapour flow. A number of relatively simple models have been put forward both to aid understanding of the processes involved and to be the basis of fuel system control strategies for implementation during transients. The model of Hires & Overington[24] fits into the latter category. It assumes that a fraction of the fuel metered by the fuel

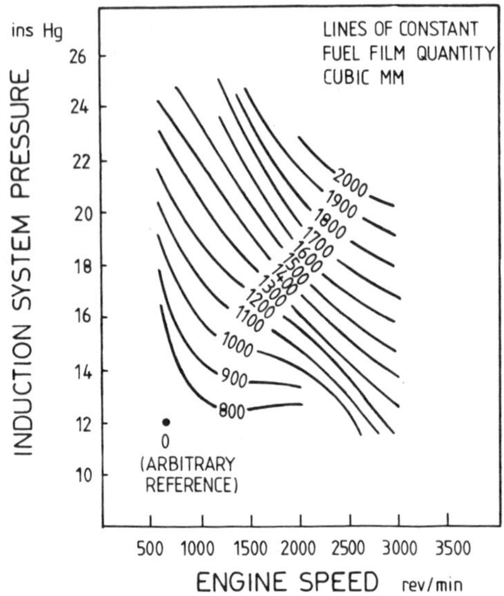

Fig. 11. Lines of constant mass of liquid fuel for a range of engine load and speed conditions[24] (reprinted with permission of the Society of Automotive Engineers).

system impacts on the manifold walls while the remainder is transported to the engine in the air stream. It further assumes that the amount of fuel leaving the wall-film is proportional to the total fuel film mass. From these two assumptions, Hires & Overington derived a simple first-order differential equation which needs to be solved to provide information concerning the corrections which should be made to the base fuel flow rate to compensate for wall-film inertia effects. The equation contains two terms the values of which change with engine condition. In Fig. 11 it is shown how one of the terms, which represents the total mass of liquid fuel present in the wall-film, varies with engine speed and load for a 5·0 litre, V-8 engine. The values are only relative, as an arbitrary zero point has been assumed, but the plot does give an interesting insight as to how wall-film does vary for the particular engine under investigation. The experimental values for the other term in the equation, representing a modified time delay, were not plotted because insufficient data was generated. The implementation of this model to modify the fuel system control strategy resulted in measured reductions of transient air:fuel ratio excursions of up to 85%. Wu *et al.*[25] have put forward a more complex two-puddle model, and this has been used to develop an understanding of engine/induction system behaviour.

5.2. Experimental Investigations

Researchers[24–26] have consistently confirmed the effect of wall-film inertia on the air/fuel mixture reaching the engine under transient conditions. Most researchers have also found that the wall-film is one of the most important factors in causing inter-cylinder distribution problems. Demel & Lenz[27] have quantified the inter-cylinder distribution problems for a particular multi-cylinder engine in the form shown in Fig. 12. The figure shows that the differences in air:fuel ratios between cylinders are generally below 2·5% in the region up to 50% engine load: at higher engine loads, the distribution worsens. The inlet manifold of the engine incorporated a water-heated hot spot, a locally heated region of the manifold wall, in order to evaporate as much of the wall-film as possible. The dashed lines in Fig. 12 illustrate the ratio of actual heat transfer from the hot spot to the heat required for complete fuel vaporization. It can be seen that there is excess heat

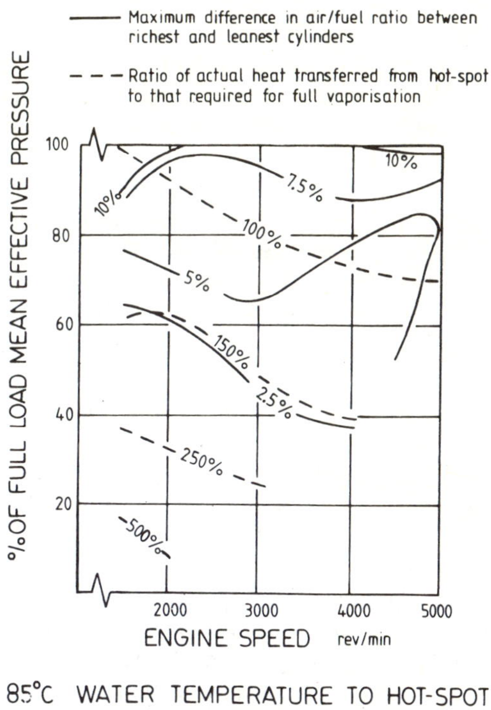

Fig. 12. Variation in mixture distribution for carburetted engine at normal operating temperature[27] (reprinted with permission of the Society of Automotive Engineers—Australasia).

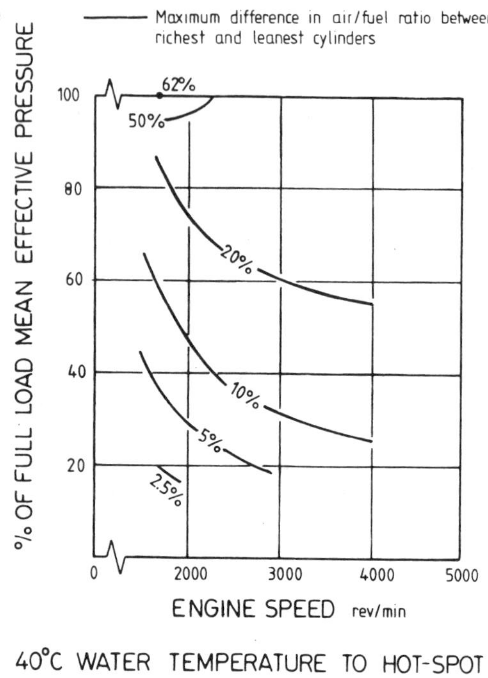

Fig. 13. Variation in mixture distribution for carburetted engine at reduced hot spot operating temperature[27] (reprinted with permission of the Society of Automotive Engineers—Australasia).

available at low load conditions, suggesting that much of the wall-film is evaporated, with the consequent improvement of inter-cylinder distribution. The importance of the hot spot is illustrated further by reference to Fig. 13, which is similar to Fig. 12 apart from a reduction in heating water temperature from 85°C to 40°C. The deterioration of distribution apparent between Figs 12 and 13 gives an appreciation of the distribution problems that engines suffer under starting and warm-up conditions. Inter-cylinder distribution problems caused by wall-film flow are at their worst during cold-cranking conditions (i.e. when the engine is completely cold and is being rotated by its starter motor). This is quantified to some extent in Fig. 14, where the proportion of wall-film flow to each cylinder (expressed as a percentage of the fuel supplied by the fuel system) is shown for extended engine cranking; the different distribution plots shown are for similar tests performed on different days. It can be seen that the wall-film is shared between cylinders 2 and 3 on rather an unpredictable basis, with negligible quantities reaching cylinders 1 and 4.

Demel & Lenz produced a comparable set of characteristics to Fig. 12 for an engine supplied by a multi-point fuel injection system. The plot was completely different in that there was even distribution at high load and poorer distribution at low load (7·5–10% variation below 50% load). The relatively poor distribution at low loads could be attributable to difficulties in matching injectors at low flows and also to some variation in air distribution.

As well as causing inter-cylinder air:fuel ratio variations, the presence of wall-film and large drops in manifolds will cause variations in the air:fuel ratio supplied to any one cylinder from one cycle to the next. These variations will lead to a reduction of the effective

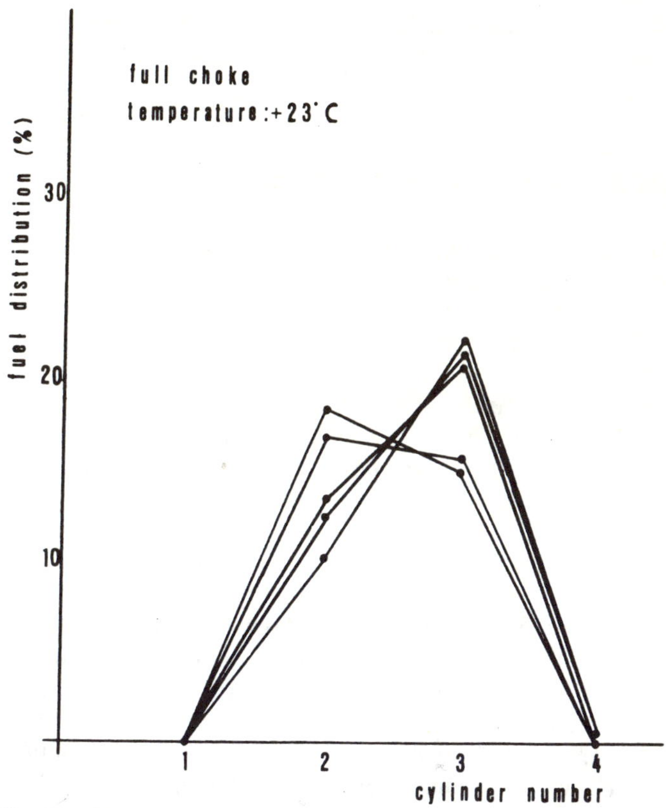

Fig. 14. Wall-film distribution under cranking conditions[13] (reprinted with permission of the Institution of Mechanical Engineers).

lean-burn limit of the engine, as the manufacturer will have to allow an increased safety band when tuning the engine for lean running. It is extremely difficult to quantify these variations due to problems in measuring either what enters or what leaves a cylinder during an individual cycle. Most attempts have used indirect ways of quantifying the effect. Hansel[28] noted the difference in terms of the cycle-to-cycle fluctuations in peak cylinder pressure between running on a well mixed vapour/air mixture (and hence with virtually no variation in air:fuel ratio from one cycle to the next) and running with the usual wet manifold. Hansel came to the conclusion that approximately one-third of the observed cycle-to-cycle fluctuations in peak cylinder pressure could be attributed to variations in the air:fuel ratio of the mixture supplied to a given cylinder between consecutive cycles. The remaining two-thirds of the fluctuations have been attributed to such causes as variations in the local flow field around the spark plug at the time of sparking.

5.3. Manifold Design for Transport of Mixture

It follows from the foregoing that engine performance problems associated with inter-cylinder or cycle-to-cycle distribution would be virtually eradicated if it were possible to eliminate wall-film and large droplets. There are a number of possible approaches to this objective.

5.3.1. Heating the Manifold

Manifolds can be designed to give greatly enhanced evaporation by being heated, particularly at the first potentially wet surface in the flow passage. This has been known and incorporated in manifold design since the earliest IC engines. The use of computer-based models (subsection 5.1.1) has now enabled researchers to investigate fully other sources of heat flow (e.g. intake air, fuel before metering, etc.). It was illustrated in the preceding section how high load conditions require large heat flows to improve distribution (Fig. 12) but, as pointed out in the first section, this is at the expense of volumetric efficiency. The designer must reach a compromise between high volumetric efficiency and good distribution. One further difficulty is that neither of the two main sources of heat flow (engine cooling water and exhaust gas) can be used during engine starting and the early stages of warm-up. An alternative source is the dissipation of electrical energy, and a convenient, self-controlling, method is to use PTC heaters.[29]

5.3.2. Improving Mixture Preparation

Ideally, the designer needs to develop fuel systems that produce small droplets throughout their required operating range: not only this, but the droplets must be distributed throughout the air stream. The Battelle investigation (subsection 5.1.2) gave weight to the argument that such an approach would significantly reduce droplet impaction. The sonic throat carburettor was one attempt to implement this policy, and Aquino[30] reported that the inter-cylinder distribution achieved with this device in conjunction with an unheated manifold matched that of a conventional carburettor with a heated manifold. The sonic throat carburettor is essentially a convergent–divergent nozzle with the fuel introduced at, or near, the point of highest air velocity. Idealized velocity and static pressure characteristics are shown in Fig. 15 for two different manifold depression cases, for both a sonic throat and an air-valve carburettor. The characteristics illustrate that the sonic throat carburettor achieves higher peak velocities and, hence, it produces smaller droplets. Engine throttling is achieved by variation of the throat area of the sonic throat carburettor.

5.3.3. Improving Manifold Design

An alternative approach is to design symmetrical path manifolds so that the mixture travels to each cylinder under similar conditions. If one considers the common 'rake-type' manifold (Fig. 16) with a cylinder firing order of 1–3–4–2, there are two main reasons why the outside two cylinders are likely to receive a different quantity of fuel from the inside two:

(1) the path lengths are longer to the outside cylinders;
(2) the inside two cylinders have to induce after the outside two of the opposing side, while the outside cylinders induce after the inside cylinders of the *same* side.

The problems arise from the inertia of the wall-film and large drops and to some extent the effects can be compensating, but a technically elegant approach would be to set up similar induction conditions for each cylinder. Jante[31] suggested the design shown in Fig. 16. However, there are two serious drawbacks to this approach: the manifold shape is often seriously constrained by available space, and symmetrical-path

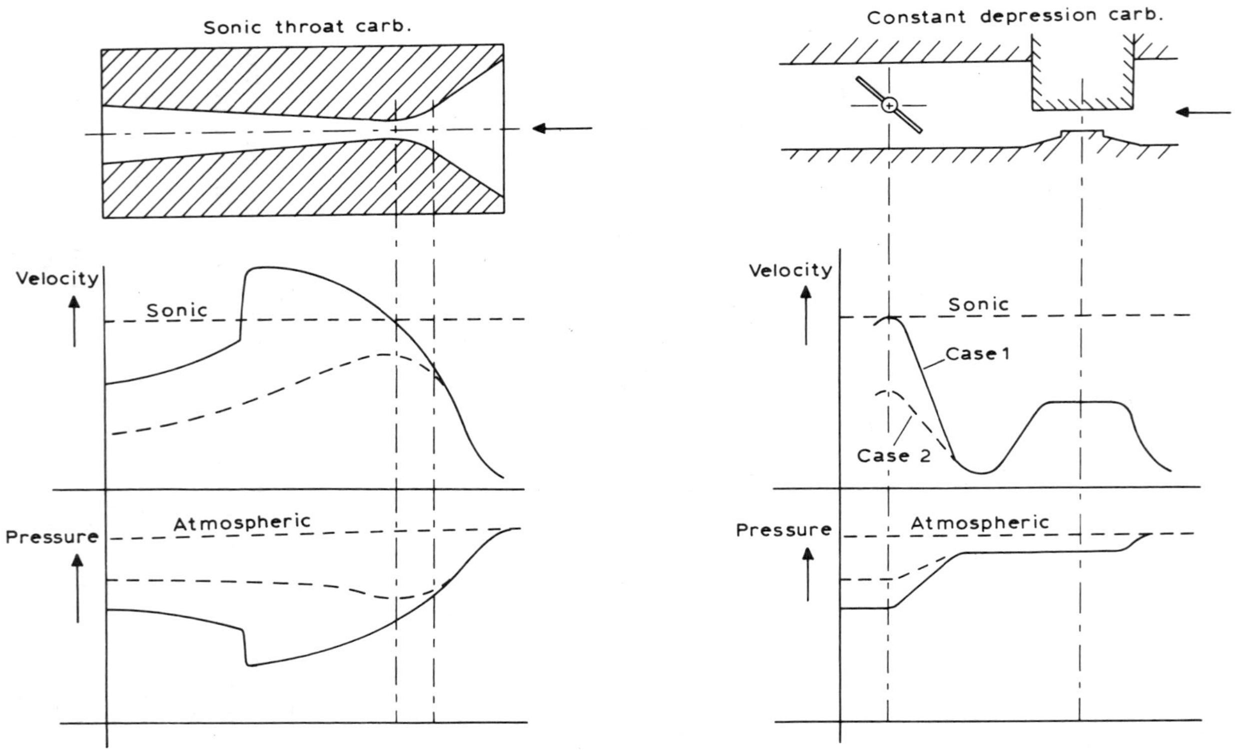

Fig. 15. Idealized velocity and pressure profiles for sonic throat and air-valve carburettors[13] (reprinted with permission of the Institution of Mechanical Engineers).

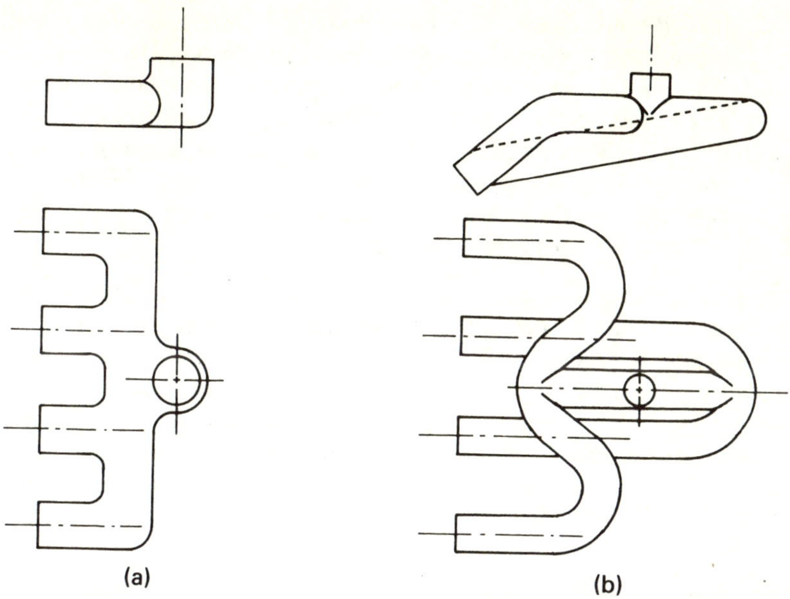

Fig. 16. (a) Rake-type and (b) symmetrical manifold arrangements.

manifolds tend to have larger volumes and surface areas, leading to poor transient response.

6. MEASUREMENT TECHNIQUES

6.1. Assessment of Problem

There are three possible forms for the fuel to be in once it has been introduced into the air stream: liquid droplets, liquid on the flow passage walls, and vapour. In order to provide a complete description of the mixture preparation at a particular section of the manifold, the following would have to be specified:

(a) Liquid droplets: the proportion of original fuel supplied; mean diameter; size distribution; velocity and direction of travel and variations across the section; temperature and knowledge of any variation between droplets; constituent hydrocarbons.
(b) Liquid wall-film: the proportion of original fuel supplied; thickness and its variation around the flow passage; flow rate and its

variation around the flow passage; temperature and its variation around the flow passage; constituent hydrocarbons.

(c) Air and fuel vapour: pressure; temperature and how it varies across the section; local concentrations of fuel vapour; constituent hydrocarbons in fuel vapour; vapour expressed as a proportion of original fuel supplied; velocity and direction of flow and how they vary across the section; details of turbulence and how it varies across the section.

The problem is compounded by the pulsating nature of the flow and the variations that are present from one cycle to the next in SI engines. The obvious consequence is that nobody has attempted a complete characterization; any attempt to do so would produce more data than could sensibly be assimilated. Nevertheless, it is important in some instances to be able to measure certain key parameters, and so this section will include a brief review of methods of measuring droplet size distribution and wall-film flow rate.

6.2. Measurement of Mean Droplet Size and Droplet Size Distribution

6.2.1. Impaction and Photographic Methods

Early attempts centred around either the magnesium oxide method or photographic techniques. With the magnesium oxide impaction method,[32] a glass plate covered in magnesium oxide smoke particles was exposed to the flow and subsequently the diameters of the indentations produced by droplet impaction were measured. This method suffered from a number of problems: the smaller droplets tended not to be captured; the larger drops broke up on impact; the flow was disturbed; and both testing and analysis were time-consuming. Photographic methods are discussed in Chapter 7 and the comment here will be brief. One of the main problems with photographing carburettor sprays is obtaining suitable optical access and sufficient illumination[33] to be able to resolve the movement of small, fast-moving, droplets within a flow passage. The latter difficulty can be overcome by using pulsed lasers for illumination, and this has recently revived interest in photographic techniques. A recent development[34] in the analysis of photographic results has been to transfer the images onto video frames and to make use of modern image enhancement and analysis techniques to help the counting and sizing of droplets.

6.2.2. Holographic, Diffraction-based and Laser-doppler Techniques

Some examples of these techniques are described in Chapter 7, and comment here will be limited to observing that the development of non-intrusive droplet sizing techniques such as that developed by Swithenbank *et al.*[35] has enabled experiments to be performed which have led to a much greater understanding of fuel system droplet sprays. An instrument operating on these principles is available commercially and is known as the Malvern Droplet Size Analyser (manufactured by Malvern Instruments Ltd, Malvern, UK).

6.3. Wall-film Flow Rate Measurement

Hewitt[36] lists five techniques for measuring wall-film flow rate in two-phase flow. It is apparent that three of these techniques could be, or have been, applied to the measurement of the flow rate of fuel on the walls of SI engine manifolds, and these three techniques will be discussed in the following subsection.

6.3.1. Film Removal Methods

The fuel can be removed by skimming it off using a concentric cylinder[13] of slightly smaller diameter than the flow passage (as shown in Fig. 17), or by using a porous section inserted into the side of the flow passage.[23] Other investigators[37] have sampled locally with the facility of being able to move the measurement area around the periphery of the flow passage so as to study the variation of flow rate with peripheral position. Measurements have shown that the wall-film flow rate can vary between around 50% of the total fuel supplied (under extended cold-cranking conditions) to zero (at part-load conditions with a heated manifold). Removal of the wall-film will affect engine running under conditions where the wall-film is a significant proportion of the total, and this has to be countered by re-introducing compensating quantities of fuel[38] downstream of the skimming section.

6.3.2. Tracer Mixing Methods

A dye or some other form of tracer can be added to the wall-film at a known rate and, after allowing mixing to take place, the concentration

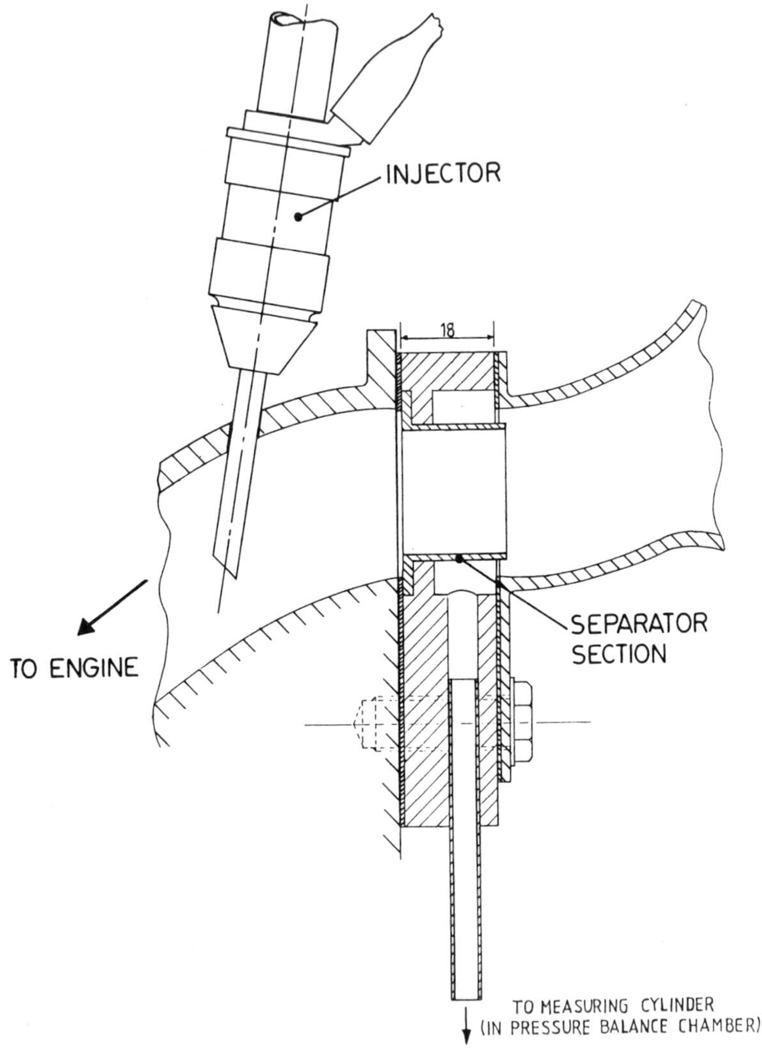

Fig. 17. Wall-film separator[13] (reprinted with permission of the Institution of Mechanical Engineers).

can be measured and the wall-film flow rate deduced from a mass balance. The author is not aware that this method has yet been applied to SI engines: inaccuracies would be caused by evaporation, re-entrainment and deposition effects. A related approach[39] is to add an agent to the fuel to make it conductive and to insert probes to measure the resistance between various points in the manifold. Gasoline is not suitable but methyl alcohol is. This technique does not measure flow rate directly but it does give an indication of wall-film thickness.

6.3.3. Photographic Methods

High-speed ciné photographs give an impression of wall-film flow rate from observation of the bulk flow of puddles of fuel. The impression can be misleading where surface waves are present, as the observer is tempted to link the movement of the wave with the bulk movement of the liquid. If the fuel is dyed, an impression of wall-film thickness can be obtained from observing the colour contrasts. The majority of photographic investigations have been performed using either wholly or partially transparent manifolds. This approach enables the observer to obtain an overall impression of the wall-film behaviour under engine running conditions, but it is not representative of actual manifolds because the transparent section will neither have corresponding wall temperatures nor surface finish. It is now possible to use endoscopes[38] to view wall-film behaviour from inside the manifold. Endoscopes take the form of a light collection lens or prism situated at the end of either a viewing tube or a bunch of fibre optic cables. The image is viewed through an objective lens. It is extremely difficult to use high-speed ciné photography under these conditions because of the low levels of light intensity available; this, in turn, is due to the small size of the light collection optics.

6.3.4. Miscellaneous

A completely different method of wall-film measurement has been proposed by Lenz *et al.*[40] It consists of making a simultaneous comparison between the amount of fuel supplied by the fuel system and the amount of fuel burnt under conditions of varying fuel supply. The application of control theory enables an assessment to be made of the quantity of fuel stored in the manifold. This method is related to the modelling approaches discussed in Section 5.1.3.

7. FUTURE WORK

7.1. Manifold Investigations

This chapter has discussed the role of the fuel system and inlet manifold in delivering mixture to the cylinders of SI engines. It was pointed out that a really comprehensive model of all aspects of mixture transport within manifolds would be an extremely complex undertaking, and so current models have tended to concentrate on specific aspects of the processes. These models, supported by some experimental data, have led both to better understanding of the processes and also to the provision of useful information for design and product engineers on which decisions concerning where and how to apply heat to manifolds, when and how much transient enrichment to supply, and so on, can be based. There is still room for development of current models, particularly in the case of wall-film behaviour; one of the problems is the difficulty in making comprehensive measurements of wall-film behaviour to validate models, because each of the techniques described in the preceding section has its shortcomings. The most useful future developments in manifold research work will be to improve methods of wall-film measurement and to refine models of wall-film behaviour.

7.2. Inlet Valve Investigations

The chapter began by discussing the mixture preparation requirements of engines, and it was apparent that the effect of the flow past the inlet valve was likely to be particularly important in the disintegration of drops. The inlet valve has another role to play in arrangements of multi-point fuel injection systems where the spray is directed onto the back of the valve; in this instance, it provides a hot surface to aid evaporation. A future area of research must therefore be to build up an understanding of the evaporation and droplet disintegration processes that occur at the inlet valve. This work would link with in-cylinder flow modelling work to move towards an understanding of how mixture preparation affects combustion in particular designs of engines. This linking of manifold and cylinder work should encourage designers to think in terms of an integrated fuel system/

manifold/engine design rather than the common practice of designing the engine, picking a fuel system and using the manifold and engine porting to link the two.

REFERENCES

1. Dodd, A. E. & Wisdom, J. W., Effect of mixture quality on exhaust emissions from single-cylinder engines. *Proc. Instn Mech. Engrs,* **183,** Part 3E (1969) 117–32.
2. Beale, N. R. & Hodgetts, D., Inlet valve throttling and the effects of mixture preparation and turbulence on the exhaust gas emissions of a spark ignition engine. *Proc. Instn Mech. Engrs,* **190,** 1/76 (1976) 13–21.
3. Bowman, T. J., Chopra, A. S. & Kunisch, K., Performance and powertrain development of an engine for integrated powertrain control. Proc. FISITA Congress, Belgrade, 1986, Paper 865028.
4. Weast, R. C., *Handbook of Chemistry and Physics,* 65th edition, CRC Press, Florida, 1984.
5. Mugele, R. A. & Evans, H. D., Droplet size distribution in sprays. *Indust. Eng. Chem.,* **43**(6) (1951) 1317–24.
6. Elkotb, M. M., Fuel atomization for spray modelling. *Proc. Energy Combust. Sci.,* **8** (1982) 61–91.
7. Wallis, G. B., *One-dimensional Two-phase Flow,* McGraw-Hill, New York, 1969, 377 pp.
8. Ingebo, R. D. & Foster, H. H., Drop size distribution for crosscurrent breakup of liquid jets in airstreams. NACA Technical Note 4087, 1957.
9. Prandtl, L., *Essentials of Fluid Dynamics.* Blackie, London, 1952, 328 pp.
10. Hinze, J. O., Fundamentals of the hydrodynamic mechanism of splitting in dispersion processes. *Am. Inst. Chem. Engrs J.,* **1**(3) (1955) 289–301.
11. Richards, G., Sojka, P. E. & Lefebvre, A. H., Dropsize studies in a radially-uniform fuel spray. SAE paper 852083, 1985.
12. Knight, P. G. G., The SU carburettor. *Proc. Instn Mech. Engrs, Automobile Div.,* No. 3 (1962–3) 128–40.
13. Nightingale, C. & Tsatsami, V., Improved mixture preparation for better cold starting. *Automotive Engr,* **9**(5) (1984) 97–100.
14. Lenz, H. P., Uber Grundlagen der Zerstaubung und ihre Anwendung bei Vergasern von Ottomotoren (Atomization and its application to four-stroke petrol engines) *VDI Zeitschrift,* **6,** No. 14 (December 1967).
15. Finlay, I. C., McMillan, T., Bannell, J. L. K. & Nightingale, C., The measurement of the sizes of droplets leaving the throttle plate of an air-valve carburettor, *J. Phys. D: Appl. Phys.,* **18** (1985) 1213–22.
16. Takeda, K., Shiozawa, K., Oishi, K. & Inoue, T., Toyota Central Injection (Ci) system for lean combustion and high transient response. SAE paper 851675, 1985.
17. Godsave, G. A. E., The combustion of drops in a fuel spray. Memorandum No. M. 95, National Gas Turbine Establishment, Pyestock, UK, 1950.

18. Fuchs, N. A., *Evaporation and Droplet Growth in Gaseous Media.* Pergamon Press, New York, 1959.
19. El Wakil, M. M., Uyehara, O. A. & Myers, P. S., A theoretical investigation of the heating-up period of injected fuel droplets vaporizing in air. NACA Technical Note 3179, 1954.
20. Ranz, W. E. & Marshall, W. R., Evaporation from drops. *Chem. Engng Prog.*, **48**(3) (1952) 141–6; **48**(4) 173–80.
21. Boam, D. J. & Finlay, I. C., A computer model of fuel evaporation in the intake system of a carburetted petrol engine. Instn. Mech. Engrs Conference on Fuel Economy and Emissions of Lean Burn Engines, C89/79, 1979.
22. Low, S. C., Baruah, P. C. & Winterbone, D. E., Transportation of liquid fuel droplets in the pulsative air flow within the s.i. engine intake manifold. SAE paper 810497, 1981.
23. Trayser, D. A., Gieseke, J. A., Fischer, R. D. & Creswick, F. A., A study of the influence of fuel atomization, vaporization, and mixing processes on pollutant emissions from motor-vehicle powerplants—Phases 1 & 2. Battelle Memorial Institute, Columbus, Ohio, April 1969 and January 1972.
24. Hires, S. D. & Overington, M. T., Transient mixture strength excursions—an investigation of their causes and the development of a constant mixture strength fueling strategy, SAE paper 810495, 1981.
25. Wu, H., Aquino, C. F. & Chou, G. L., A 1·6 liter engine and intake manifold dynamic model. Am. Soc. Mech. Engrs, paper 83-WA/DSC-39, 1983.
26. Aquino, C. F., Transient A/F control characteristics of the 5 liter central fuel injection engine. SAE paper 810494, 1981.
27. Demel, H. & Lenz, H. P., Kraftstoffeinsparung durch verbesserte Gemischverteilung an Motoren mit zentraler Gemischbildung. Proceedings of XIX FISITA Congress, Paper 82129, 1982. (Reduction of fuel consumption through improved mixture distribution in engines with central mixture formation, MIRA Translation 45/83.)
28. Hansel, J. G., Lean automotive engine operation—hydrocarbon exhaust emissions and combustion characteristics. SAE paper 710164, 1971.
29. Paganelli, J., PTC ceramic heaters in automotive controls. SAE paper 840143, 1984.
30. Aquino, C. F., The design and development of the upper-pivoted sonic carburettor. SAE paper 780078, 1978.
31. Jante, A., Uber Gemischverteilung an Ottomotoren, *Automobil Industrie,* 3/76 (1976) 57–67.
32. Giffen, E. & Muraszew, A., *The Atomization of Liquid Fuels.* Chapman & Hall, London, 1953.
33. Jones, A. R., A review of drop size measurements—the application of techniques to dense fuel sprays. *Prog. Energy Combust. Sci.,* **3** (1977) 225–34, Pergamon Press.
34. Lo, R. S., Matysiewicz, E. J. & Hotham, G. A., Fuel injector atomization using laser imaging techniques. SAE paper 851673, 1985.
35. Swithenbank, J. *et al.,* A laser diagnostic for the measurement of droplet

and particle size distribution. AIAA 14th Aerospace Sciences Meeting, Washington, DC, 1976.
36. Hewitt, G. F., *Measurement of Two-phase Parameters*. Academic Press, London, 1978.
37. Finlay, I. C., McMillan, T. & Bannell, J. L. K., Liquid fuel distribution in the region of the throttle plate of an air valve carburettor. National Engineering Laboratory report Y5/32, East Kilbride, Scotland, 1979.
38. Nightingale, C. & Tsatsami, V., Improved cold-starting prospects—by studying wall film flow. *Automotive Engr*, **11**(4) (1986) 28–34.
39. Hayashi, S. & Norihiro, S., Measurement of fuel liquid-film flow in intake pipe of two-stroke motorcycle engine using conductive probe. SAE paper 840554, 1984.
40. Demel, H., Diem, E. & Lenz, H. P., Quantitative Messung des Kraftstoffilms in Saugrohren von Ottomotoren. *Motortechnische Z.*, **44** (7/8) (1983).

7

Diesel Engine Fuel Injection Processes and Spray Diagnostic Methods

A. A. Hamidi & J. Swithenbank

Department of Mechanical and Process Engineering, University of Sheffield, UK

1. Introduction
2. Fuel injection components
3. Diesel spray visualization
4. Atomization and spray break-up
5. Aerodynamics of diesel spray
6. Drop size and velocity distribution diagnostics
   6.1. Holography
   6.2. Diffraction-based techniques
   6.3. Laser–doppler anemometry (LDA)
7. Size and velocity distribution measurement: results
8. Conclusion

NOTATION

C	Discharge coefficient
d	Liquid column diameter
d_0	Orifice diameter
D	Drop diameter
D_{32}	Sauter mean diameter
D_{10}	Linear diameter
G	Mass of fuel delivered per pulse
l	Break-up length
L	Penetration length

n Number of holes
P Pressure
ΔP Pressure drop
r Radial distance along spray
t Time
t_f Total fuel delivery time
u_{FO} Initial liquid discharge velocity
u Velocity
x Distance downstream of nozzle
δ x/d_0
θ Equivalent spray angle
θ_s Spray angle
λ Incident light wavelength
λ' Liquid column wavelength
μ Dynamic viscosity
ν Kinematic viscosity
ρ Density
σ Surface tension
ϕ Oxygen ratio
χ Density ratio

Subscripts
f Fuel
g Gas
L Liquid

1. INTRODUCTION

In diesel engines the processes taking place in the combustion chamber have been studied in detail, both experimentally and theoretically, in order to achieve a better understanding of the combustion and pollutant formation. The atomization and spray formation of the liquid fuel is one of the most important processes in diesel engine combustion. Detailed knowledge of the formation of the spray and its interaction with the combustion air is essential to satisfactory design of diesel engines and achievement of complete and efficient combustion and low pollutant emission.

The aim of this chapter is to highlight the main characteristics of the diesel sprays and their dependency on the various operating conditions. To this end some of the latest experimental work carried out in measurement of diesel spray characteristics and measurement techniques employed are outlined.

2. FUEL INJECTION COMPONENTS

The components of the fuel injection system in all diesel engines of almost any design and duty are similar and their required operation is essentially the same.[1,2] A diagrammatic representation of typical fuel injection equipment is shown in Fig. 1(a). The sequence of events are briefly outlined as follows. The fuel feed pump, which is cam driven, may be a high-pressure plunger type or a low-pressure diaphragm type. It draws the fuel from the service tank and delivers it to the high-pressure injector pump through a fuel filter. The high-pressure feed pump delivers the fuel through a suitable nozzle into the combustion chamber. This pump is cam driven at half engine speed for a typical four-stroke engine. The timing and duration of the injection is determined by the cam position and shape. For most diesel engines the injection is started when the piston is approaching its top dead centre. There are two types of injection pump used with diesel engines. The in-line pump, used with all medium and some high-speed engines, is of the plunger type and each injector requires a pump of its own. The rotary type injection pump is used with small high-speed engines and has the advantage that all injectors are supplied from the same pump (Fig. 1(b)).

The type of injector used is the needle-spring type. The fuel from the high-pressure pump is progressively pressurized, exerting an ever increasing force on the needle. At some predetermined pressure the needle is lifted against the spring pressure and the nozzle opens, starting the injection of the fuel. At the end of the pump stroke the pressure in the injector drops and the spring pushes the needle back on its seat, so that the injection of the fuel comes to an abrupt end.

There are a large number of factors and parameters that must be considered when the injection system of a diesel engine is being designed—pipe–nozzle match, cavitation erosion, loading figure, nominal mean injection velocity, trapped volume ratio and nozzle opening pressure are just some of these factors. The injection pressure–time

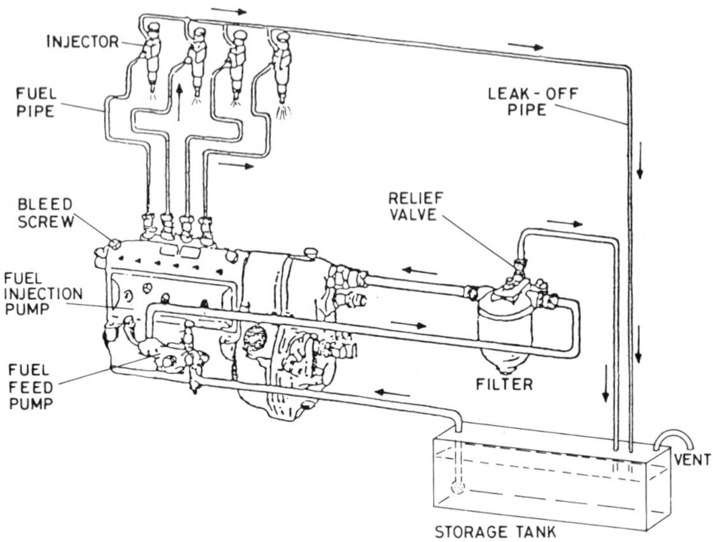

(a)

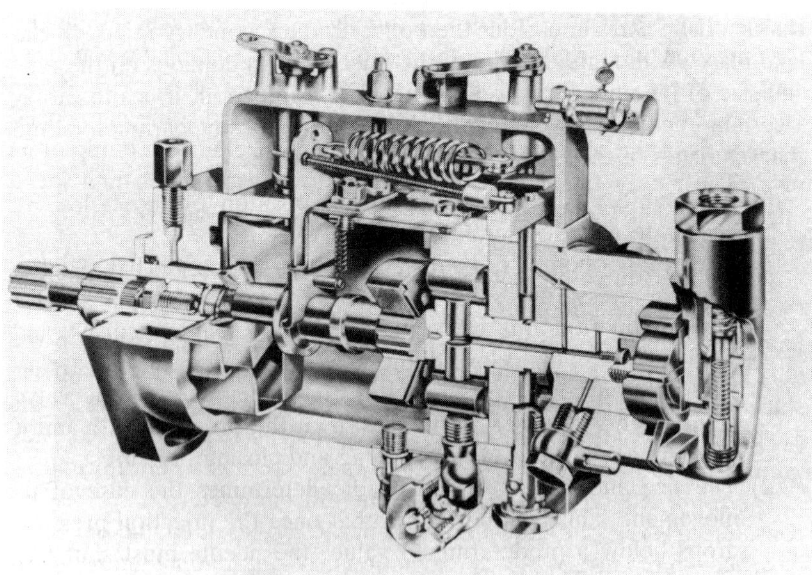

(b)

Fig. 1. Diesel engine fuel injection components.

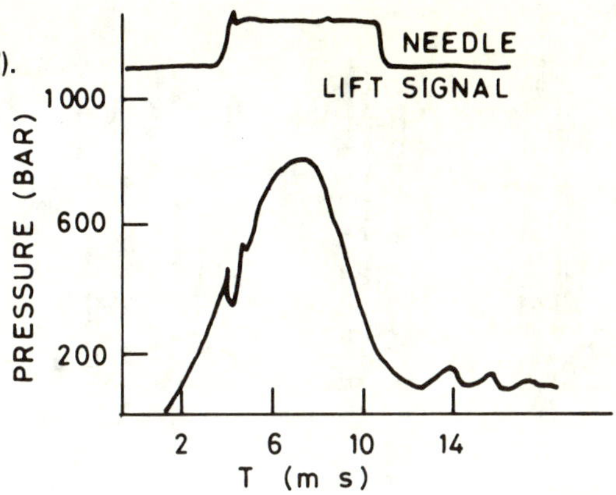

Fig. 2. Typical fuel injector line pressure trace (Ewan[30]).

history of the injector, which is predominantly determined by the pipe–nozzle match, is of great importance in the injection process.

A typical plot of a pressure–time history is shown in Fig. 2. There is an instantaneous drop in pressure as the needle lifts. If the length of the pipe from the injection pump to the injector is not of the correct characteristic and dimensions then undesirable secondary injection can take place. Although the type of the injector used depends on the type and size of the engines as well as the type of injection, it has to satisfy a certain engine requirement. Some of the important parameters and characteristics of the nozzle are briefly outlined here.

(a) The number, diameter, length and position of the holes are chosen to give best air/fuel mixing.
(b) The nozzle opening pressure ranges from 200 to 400 bar. Higher opening pressure results in better atomization and a shorter injection period but causes greater stress on the spring and seating, especially at low idling speeds.
(c) The differential ratio is the ratio of the diameter of the valve guide to the valve seat. This ratio is of the order 1·7–2·7 and it determines the ratio of the opening and closing pressures.
(d) The size and weight of the needle determines the ease of its movement. This is important in that once the injection pressure drops below a predetermined value, the needle must shut fast enough to avoid the flow back of combustion gases into the injector or dribble of the liquid fuel.
(e) The needle must lift up sufficiently to allow adequate fuel flow

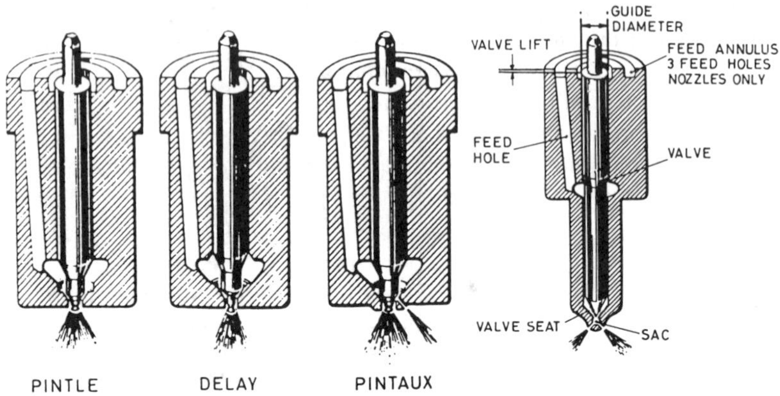

Fig. 3. Different types of nozzle used in diesel injection.

through the nozzle and to avoid excessive pressure drop across the nozzle hole.

(f) The relationship between the sac volume, indicated in Fig. 3, and the hydrocarbon emission is well established and hence this volume must be kept to a minimum.

There are several designs of injector nozzle used. Single-orifice hole type and pintle type nozzles are used in high-speed smaller diesel engines where direct injection is employed. Multi-hole nozzles are used for the heavier medium-speed engines. In indirect injection engines, the pintaux nozzle is used, whereby a portion of the spray is directed towards the centre of the chamber where the higher temperature and greater degree of turbulence enhance the liquid spray vaporization. A diagrammatic representation of these nozzles is given (Fig. 3).

3. DIESEL SPRAY VISUALIZATION

Although in recent years important advances have been made in the development of non-intrusive optical techniques for measurements of the liquid fuel spray characteristics, photography still remains as one of the most important techniques available in the study of diesel sprays. Photographs can be taken of the spray at different delay times after the start of liquid injection, and therefore changes in the spray behaviour as a function of time can be studied. Photographic examination of diesel sprays produces qualitative and quantitative data on the behaviour of the spray. Some characteristics of the diesel spray can only be measured by photographic studies. Yule *et al.*[3] discuss the cyclic variation in the structure of the diesel sprays, which they call the 'spray jitter', from photographic data. Air entrainment, liquid fuel and vapour concentration can all be studied from normal or Schlieren photography.

Single- and double-exposure photography[4,5] has been used to produce data on drop size and velocity distribution in the relatively unpopulated regions of the spray. A major disadvantage of the technique is the time-consuming nature of the analysis of the photographs. Therefore automatic and semi-automatic techniques have been

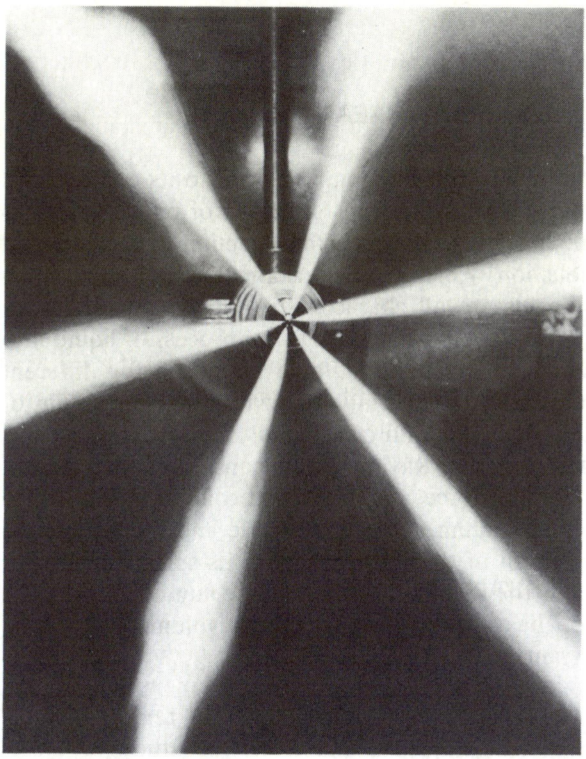

Fig. 4. Six-hole injector nozzle spray pattern, 3·5 ms after start of injection.

developed[6] to reduce this problem, and to eliminate the need for an operator who is prone to fatigue and hence misjudgement of the image dimensions. Those techniques which permit the measurement of drop size from photographic negatives must ensure that the drop is in focus before measuring and including it in the data.

In the more practical study of the diesel sprays, the injection is carried out inside a pressure vessel at high pressures and temperatures. Recent advances in fibre optics and light pipes[7] enable easy and satisfactory illumination of the spray for good quality photographs.

Ciné camera and high-speed photography have been employed extensively to study the aerodynamic behaviour of the spray. A very fast ciné camera, with up to a million frames per second, has been used[8] to record the spray formation and liquid atomization at the tip of the nozzle. A typical photograph of a multi-hole diesel injector is shown in Fig. 4.

4. ATOMIZATION AND SPRAY BREAK-UP

In diesel engines the liquid must be atomized prior to its combustion. Atomization, best described as the transformation of bulk liquid into drops, is the intermediate stage between the injection of the fuel and its subsequent combustion. Efficient combustion, low pollutant emission and low noise levels are all characteristics of effective atomization. Therefore an understanding of the break-up process of liquid into drops is of importance in designing suitable atomizers for different combustion requirements. In general, atomization is the result of growth of instabilities in the column of a liquid. This phenomena was studied mathematically by Lord Rayleigh[9] who considered the effect of inertial and surface tension forces on an inviscid column of liquid. He found that a free liquid column becomes unstable and breaks up into droplets if the wavelength of the disturbance on its surface is greater than $\sqrt{2}\,\pi d$, where d is the diameter of the liquid column. This implies that the largest drop has a volume equal to the volume of the one wavelength of the ligament[10]:

$$\frac{\pi}{6}D^3 = \frac{\pi}{4}d^2\lambda' \tag{1}$$

i.e.

$$D = 1{\cdot}89d$$

Other studies[11] have included the effect of viscous forces in the analysis, and similar conclusions have been reached. For a diesel injection system the fuel is normally forced through a small orifice at the tip of the injector causing atomization. For this type of atomizer three break-up regimes can be considered, depending on the liquid flow velocity. Rayleigh jet break-up, which takes place at low discharge velocities, is caused by the growth of axisymmetric oscillations on the jet surface, induced by surface tension. The break-up occurs many jet diameters downstream of the nozzle and the drop diameter usually exceeds that of the orifice. The limits for this region are established[12] by the Weber number for the liquid and gas such that:

$$We_L = \rho_L u^2 d_0/\sigma > 8{\cdot}0, \qquad We_g = \rho_g u^2 d_0/\sigma < 0{\cdot}4 \tag{2}$$

The wind-induced jet break-up which occurs at higher liquid discharge velocities is due to the unstable growth of short-wavelength surface waves on the jet surface, caused by the relative motion of the jet and surrounding media. The average drop diameter is smaller than the orifice diameter, and the break-up occurs some jet diameters downstream of the nozzle. The criteria for the existence of this break-up régime is given by:

$$0{\cdot}4 < \rho_g u^2 d_0/\sigma < 13 \tag{3}$$

Atomization, the break-up régime in diesel sprays, results in complete and virtually immediate disruption of the liquid jet as it leaves the nozzle. The average drop diameter is much smaller than the orifice diameter. This régime takes place when:

$$We_g = \rho_g u^2 d_0/\sigma > 13 \tag{4}$$

The break-up mechanism involved in atomization is as yet unknown, but it can be attributed to a combination of some of the following processes:

Aerodynamic jet break-up[13]
Liquid turbulence[14]
Liquid cavitation[15]
Velocity profile rearrangement[16]
Liquid supply pressure oscillation[17]

Figure 5 is a plot of Reynolds number ($\rho_L u d_0/\mu_L$) against Ohnesorge number ($\mu_L/\sqrt{\rho_L \sigma d_0}$). This diagram illustrates the boundaries

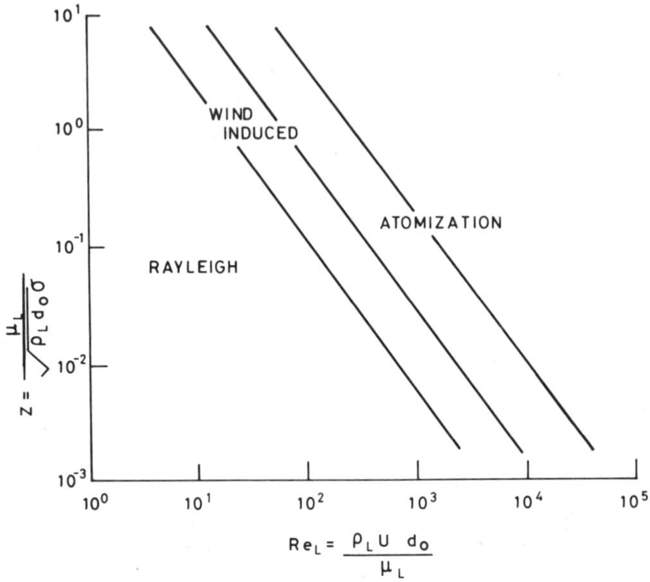

Fig. 5. Liquid jet break-up régime boundaries.

for the three régimes of jet break-up discussed above. On the basis of a large number of experimental and theoretical studies carried out on plain jet atomizers, the liquid break-up can be considered to go through the following stages[10]:

(1) spread of the liquid fuel into a conical jet;
(2) entrainment of air and subsequent break-up of fuel into ligaments—the outer ligaments are retarded faster than the core ligaments;
(3) the occurrence of small waves in the liquid surface;
(4) tearing of the surface into finer ligaments due to drag, viscous and surface tension forces;
(5) break-up of the fine ligaments into drops;
(6) shatter of the large fast-moving drops;
(7) agglomeration and coagulation of drops by collision.

The relative importance of these stages varies depending on nozzle geometry, discharge velocity, fuel and ambient properties.

Stage 6 occurs for large drops moving at high velocities. The criterion for drop shatter is given by:

$$We = \rho_g u^2 D / \sigma > 13 \tag{5}$$

The intact column of liquid at the nozzle exit, called 'the break-up', is a function of the injection parameters and ambient air conditions. The break-up length is a strong function of Reynolds number at lower values of this parameter ($Re < 5000$) but remains constant at high Reynolds number. A long break-up length results in a long combustion delay (Section 5). As a result of experimental studies several empirical equations have been derived for the break-up length. One such equation is:

$$l_{bu} = 372{\cdot}0 d_0 Z^{0{\cdot}308} We^{-0{\cdot}71} \rho_g^{-1{\cdot}2} \tag{6}$$

5. AERODYNAMICS OF DIESEL SPRAY

The dynamics of the transient diesel spray, its interaction with the air charge in the combustion chamber and some major aspects of the spray vaporization and combustion are presented in this section. A number of experimental findings regarding the spray tip penetration, angle and other characteristics are also presented. Drop size and velocity measurement techniques and some experimental findings for these two important parameters are discussed in later sections.

The diesel spray can be regarded as having two distinct regions, the main jet zone and the mixing zone. A diagrammatic representation of a diesel spray and its characteristic parameters are shown in Fig. 6. The mixing zone is where most of the liquid drop and charge–air interactions take place. In fact it is in this zone that the first visible flame appears. It can justifiably be assumed that no further atomization takes place in this region. Experimental study of this region yields valuable data in relation to spray dynamics and combustion. In the main jet region the fuel concentration is very high and the liquid jet is undergoing atomization at all times. Due to the very low air concentration in the 'initial part' of the jet, the flame never enters this region as long as injection is still taking place.

The spray penetration and angle are of prime importance in the combustion of the spray. These two spray characteristics can be affected by ambient conditions in the combustion chamber, as well as by other atomization characteristics such as the injection pressure.

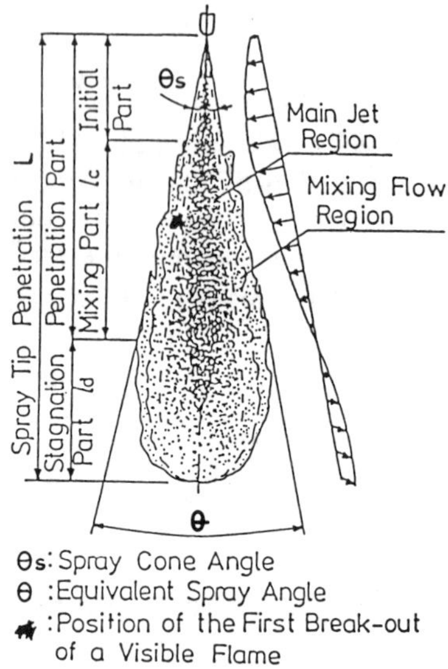

Fig. 6. Diesel spray characteristics (reprinted with permission of the Society of Automotive Engineers, 1984; Jong-Yul Ha[22]).

Wakuri *et al.*[18] showed that the motion of the drops in the dense diesel spray cannot be modelled by assuming independence of drops from each other and applying Stokes's law to the motion of spherical particles. Due to the small diameter of the drops, their high number concentration and the high gas density in the chamber, a new approach was considered. The theory is based on the idea that the air induced into the fuel jet stream produces a mixed gas and liquid drop flow with little velocity slip. The penetration of the diesel spray can therefore be calculated from a conservation of momentum study.

The application of the principle of the conservation of momentum to the liquid jet flow at the exit of the nozzle orifice and spray drops and entrained air mixture at some distance downstream of the nozzle results in an equation relating the penetration of the spray tip to the various injection parameters and the time after the start of liquid injection. Defining δ as the ratio of penetration length to orifice diameter and χ as the ratio of the fuel density to the air density, the simplified form of the equation is:

$$\sqrt{\frac{1}{\chi C}}\,\delta^2 \tan\theta = u_{FO}\frac{t}{d_0} \tag{7}$$

The dependency of the spray tip penetration on the gas density is clearly illustrated in eqn (7); this finding is consistent with experimental findings.

Experimental investigations[19,20] have indicated the applicability of this theory to dense diesel sprays, and a large number of empirical equations have been derived expressing the spray tip penetration as a function of various physical and geometrical properties of the injection process. One such equation has been derived by Fujimoto[19] as a result of extensive experimental study of diesel sprays:

$$L = 27{\cdot}1 u_{FO}^{0{\cdot}5} t^{0{\cdot}38} \rho_g^{(0{\cdot}092 \log t - 0{\cdot}41)} d_0^{0{\cdot}5} \tag{8}$$

u_{FO} is the mean discharge velocity of the liquid jet and is usually calculated from the following relationship:

$$u_{FO} = C\sqrt{2\Delta P/\rho_L} \tag{9}$$

Photographic techniques have been mostly used to measure the spray tip penetration at various ambient and injection conditions. It has been shown that the spray tip penetration decreases as the ambient pressure increases. Yule *et al.*[3] have shown that the relevant parameter is the gas density, with the gas temperature being of secondary importance. An increase in the charge temperature results in evaporation of the drops and therefore the spray tip penetration is decreased. However, the increased temperature causes a reduction in the gas density which results in an increase in the spray penetration. The observation that the gas density is the influencing parameter in spray penetration is reflected in the appearance of gas density in most empirical equations derived as a result of experimental investigations.

For almost all medium- and a large proportion of high-speed diesel engines, the injector nozzle is of the multi-hole type. The effect of neighbouring sprays and the presence of walls and obstructions in the vicinity of the nozzle have been studied by Fujimoto.[19,21] It was shown that the spray tip penetration was unaffected by the presence of other neighbouring sprays or the presence of the wall prior to impact. The inclination of the wall to the spray was also found to be of little importance to the spray tip penetration.

In Fig. 7 is shown a typical set of data on spray tip penetration. The dependence of the penetration on the ambient pressure (density) is clearly illustrated. Also indicated is the independence of penetration from the presence of other neighbouring sprays. It is clear from this diagram that the rate of spray tip penetration is a function of time.

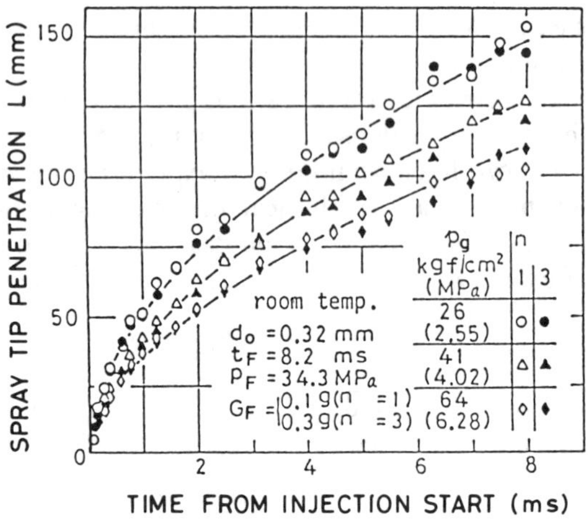

Fig. 7. Spray tip penetration (reprinted with permission, Jpn Soc. Mech. Engs., 1982; Fujimoto[19]).

A distinction is usually made between spray cone angle (θ_S) and the equivalent spray angle (θ), as illustrated in Fig. 6. The spray cone angle is usually of the order of 10 degrees at atmospheric conditions and increases with increasing chamber density.[3] The effect of ambient temperature is again of secondary importance. As illustrated in eqn (7), the spray cone angle and the spray tip penetration are related to each other. The equivalent spray angle is influenced by the degree of air movement around the spray, as well as the presence of walls and obstructions. In the study by Fujimoto[21] on spray and wall impingement it was concluded that the spray cone angle remains unaffected by the existence of a wall, but the equivalent spray angle increases as the wall is placed closer to the nozzle. These experimental findings are of course of great importance in understanding the behaviour of the spray as it approaches the cylinder wall or the piston crown.

The entrainment of the combustion air into the diesel spray is of prime importance due to its effect on the ignition delay, combustion efficiency and exhaust gas emission.

In Fig. 6 is illustrated a representation of the axial velocity distribution of the entrained air, based on experimental investigation.[22] The air is entrained into the spray in the initial and mixing part of the spray due to the lower static pressure in the spray at these regions. The boundary between these two regions has the maximum entrainment velocity. In the stagnation zone of the spray, part of the entrained air is expelled due to the higher stagnation pressure in this region. The experiments were carried out using a calibrated hot-wire anemometer inside a pressurized bomb. The probe could be traversed so that a complete map of air entrainment velocity could be obtained.

A typical entrainment velocity distribution for different locations around the spray is illustrated in Fig. 8. These measurement locations are shown in Fig. 9. Observations of smoke being entrained into the spray show that the entrainment takes place at right angles to the spray envelope.

The air flow pattern in the vicinity of the spray is greatly influenced by the presence of neighbouring sprays. Using the smoke tracer technique it has been shown[19] that the air located between the sprays moves towards the tip of the nozzle so that the air requirements of neighbouring sprays are in competition with each other.

The effect of the presence of walls and obstructions on the spray tip penetration and spray angle have already been indicated. The shape and structure of the spray are altered as a result of the impact on a surface. The spray tip is flattened depending on position and the relative tilt of the surface. Empirical equations have been derived[21] expressing the subsequent dispersion of the spray after impact as a function of position and the relative tilt of the surface.

Vaporization and combustion of the injected fuel are the final steps in the process. The process of evaporation of pure liquids is a well studied phenomenon. In general, the rate of evaporation of a spherical

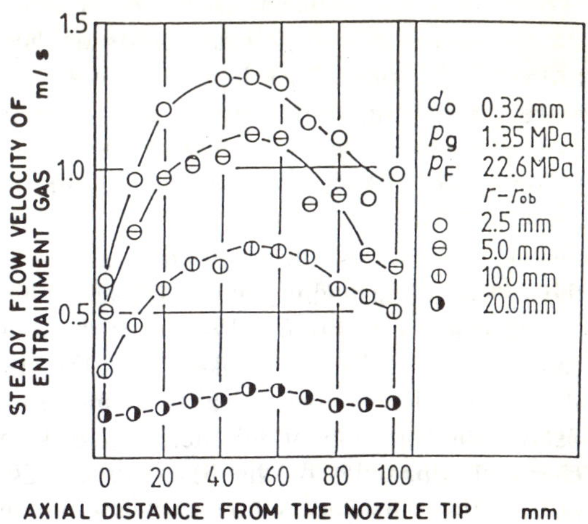

Fig. 8. Axial velocity distribution of the entrained air (reprinted with permission of the Society of Automotive Engineers, 1983; Jong-Yul Ha[20]).

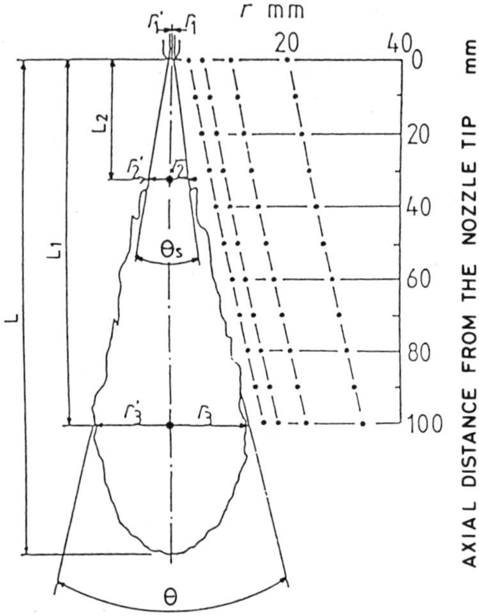

Fig. 9. Measurement locations for the axial air velocity distribution (reprinted with permission of the Society of Automotive Engineers, 1983; Jong-Yul Ha[20]).

liquid drop is proportional to the surface area of the drop. This relationship is well known as the d^2-law, the details of which can be found in standard texts on heat and mass transfer. Diesel fuel is a mixture of organic liquids having a boiling point in the range 180–360°C. The implication of this is that the conventional d^2-law cannot be applied to the evaporation of these fuels since, in addition to the evaporation of the liquid, a process of distillation is also taking place. This results in the variation of the physical characteristics of the fuel as a function of time. Therefore a modified system of equations based on the d^2-law, applicable to the pure liquids in the fuel mixture, have been solved to provide data on the rate of evaporation for different conditions.[23] This modified d^2-law has been evaluated against the experimental findings and good agreement has been achieved.

Evaporation of the liquid is one of the limiting steps before ignition can take place. The time taken from the start of the injection to the start of the ignition of the air/fuel mixture is known as the ignition delay. The duration of this delay time depends on the physical and chemical properties of the fuel and air, as well as on some of the engine operating parameters such as the compression ratio. Figure 10 is an experimental cylinder pressure–crank angle plot showing the start of injection, the ignition delay and start of the combustion, which take place just before top dead centre. The effect of injection timing is clearly illustrated. There have been a number of empirical equations derived[24] relating the ignition delay to various operating conditions. Jong-Yul Ha *et al.*[20] showed that the ignition delay becomes a constant beyond a certain ambient pressure and temperature. From photographic studies it was shown that the air:fuel mass or volume ratio was small for a period of time after the start of the injection so that no combustion could take place. However, the ratio increased after a delay time and the combustion was initiated. This delay period was found to be a function of injection pressure of the liquid, varying from 0·6 to 0·25 ms as the nozzle opening pressure was increased from 226 to 490 bar. The air:fuel mass and volume ratios are shown in Fig. 11 as functions of time for several different nozzle opening pressures; the arrows show the time at which the ignition starts.

The combustion of the fuel results in an increase in the chamber pressure and temperature. The rate of pressure rise and the decay due to the expansion is illustrated in Fig. 10. The rate of heat release (*ROHR*), which is an indication of the temperature rise in the chamber, is shown in Fig. 12. The rate of heat release, flame characteristics and propagation have been studied experimentally and

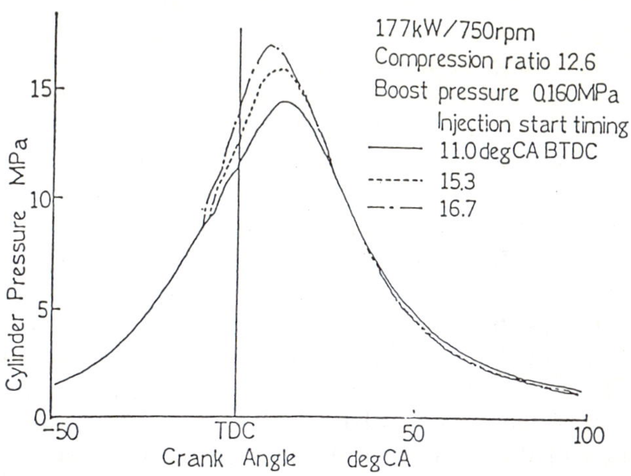

Fig. 10. Examples of cylinder pressure–crank angle diagrams as a function of injection start timing obtained from experiments.

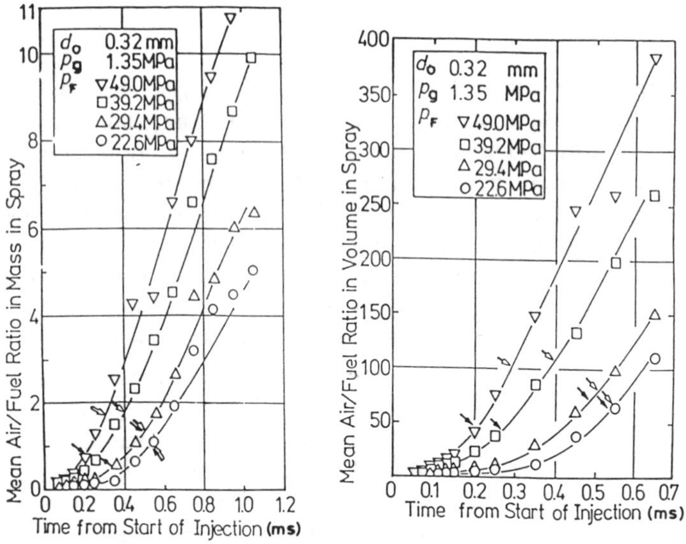

Fig. 11. Effect of nozzle opening pressure on mean air : fuel ratio (reprinted with permission of the Society of Automotive Engineers, 1983; Jong-Yul Ha[20]).

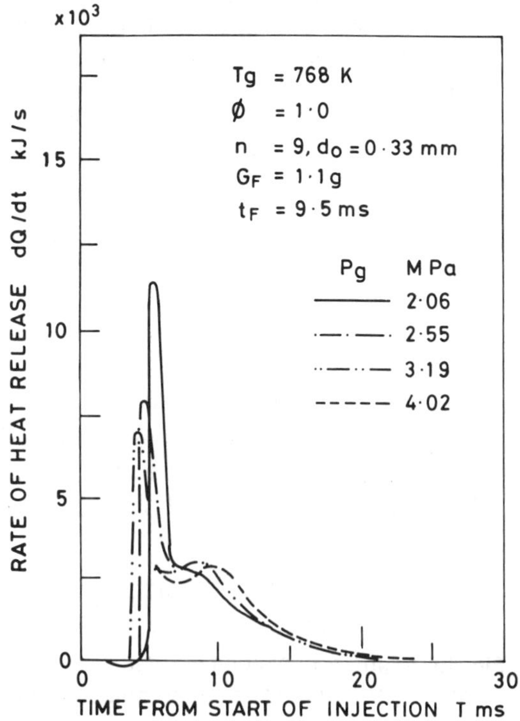

Fig. 12. Dependence of *ROHR* on the initial chamber pressure (reprinted with permission from G. T. Sato[25]).

theoretically,[24,25] and the influence of initial ambient temperature, pressure, oxygen concentration, injection pressure and nozzle geometry have been indicated. It is believed that the combustion of the diesel spray takes place in three stages. This three-stage combustion is indicated by the fact that the *ROHR* diagram has two distinct peaks. The first stage is the pre-mixed flame burning which is initiated by the appearance of the first visible flame in the mixing region of the spray. The flame, which is bluish in colour, propagates towards the tip of the spray and the *ROHR* reaches its first maximum as the flame reaches the spray tip. In the second stage the inner core of the spray burns until the end of the injection, this point being the second peak of the *ROHR*. The burning régime is one of diffusion burning and the flame is bright white, indicating the presence of a diffusion flame. Finally, the remainder of the spray, after the termination of the injection, burns as a diffusion flame. The propagation velocity of the flame in the first stage of the combustion has been determined to be in the range 50–60 m s^{-1}.

6. DROP SIZE AND VELOCITY DISTRIBUTION DIAGNOSTICS

The importance of photographic techniques in diesel spray diagnostics has been discussed earlier. Although being very informative and despite development of automated analysis, these techniques can be laborious, expensive and time-consuming. These disadvantages have given rise to the invention and application of a number of optical techniques employing lasers for the measurement of drop size and velocity distribution. These techniques have been reviewed extensively by several research workers[26–28] and in this section they are briefly outlined. However, before describing some of the available optical measurement techniques, it is important to make a distinction between two types of size distribution that are measurable, depending on the measurement technique employed.[29] The distinction is between temporal and spatial size distribution.

Spatial size distribution is obtained when the measurement is made such that the actual size distribution contained within the test volume is measured. Temporal size distribution is produced when allowance is made for different velocities of droplets passing through a small test volume. The temporal size distribution thus indicates the mass flux of droplets of different size. Single flash photography and single pulsed holography are examples which yield spatial distribution measurement, while laser–doppler anemometry (LDA) is an example of a temporal distribution measurement technique. The two types of distribution can be related to each other provided that the size–velocity distribution for the drops is known.

6.1. Holography

The major advantage of this technique is that a three-dimensional record of the fuel spray can be made. This record can then be either reconstructed and analysed in much the same way as a photograph or a second drop sizing technique can be employed to analyse the hologram by utilizing its diffraction pattern.[30]

A hologram is the result of interference of the scattered and unscattered light from a particle field when the incident light is monochromatic and coherent laser light. The illumination time of the spray depends on the drop velocity and diameters and must be such that the drop does not displace by more than a tenth of its diameter during the exposure. Therefore the employment of a Q-switched pulse laser is essential in the holographic measurement of the moving spray.

In general there are two possible configurations for a holographic set-up. For the in-line holographic technique the reference beam and the beam producing the diffraction pattern are the same. On the other hand, the off-axis holographic set-up has these two beams separated. The in-line technique is the more practical technique for spray measurement studies. Figure 13 shows the optical set-up for the production and reconstruction of a hologram.

Thompson[31] has discussed the merits of holographic particle sizing techniques and its application to drop size and velocity distribution measurement using double-pulse holography. The minimum distance of the holographic plate from the spray is an important factor and depends on the smallest measurable drop diameter in the spray and the wavelength of the illuminating light. It can be shown that this distance must be greater than d^2/λ. The lower limit of drop diameter measurable in a spray is believed to be approximately 5 μm. However, Bexon *et al.*[32] have used a holographic technique to measure the size distribution of aerosol particles.

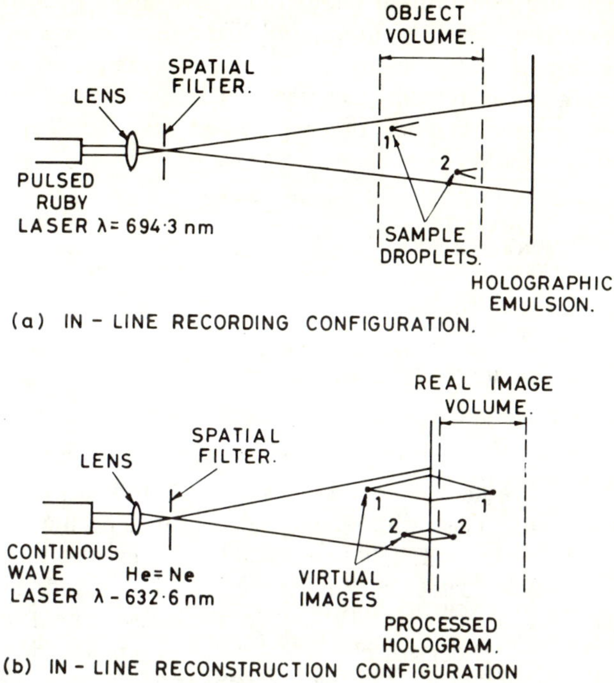

Fig. 13. Optical set-up for recording and reconstruction of in-line holography (reprinted with permission, *Progr. Engng Combust. Sci.*, 1977; Jones[28]).

6.2. Diffraction-based Techniques

A spherical particle scatters light producing a Fraunhofer diffraction pattern of rings when illuminated by a beam of monochromatic coherent light. The diffraction pattern is uniquely related to the diameter of the particle. Hence information on the light energy distribution of this diffraction pattern can yield the diameter of the particle. This technique can be applied to measure the size distribution of drops in a spray. Dobbins *et al.*[33] used a high-pressure mercury light as the illuminating beam and a single photomultiplier to collect the scattered light. The Sauter mean diameter of the polydispersed particle field was deduced from the measured light energy, and the particle concentration was calculated from the beam attenuation. Cornillault[34] developed an instrument for measurement of powders in

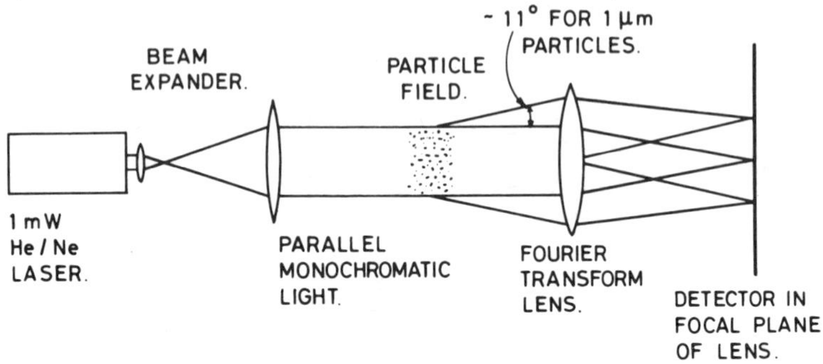

Fig. 14. General optical set-up for diffraction-based particle sizing techniques.

the range 2–500 μm. The technique uses a single photodetector in front of which is a rotating mask with rectangular apertures positioned radially on it, so that the full light energy distribution could be measured using a single detector. An instrument developed by Swithenbank *et al.*[35] has been extensively used in the measurement of size distribution of the dispersed phase in many two-phase flow problems. The most recent version of this instrument has a very fast data acquisition system which enables its application to the fast transient diesel sprays. The drop diameter range from 2 μm to several mm can be measured depending on the focal length of the collection lens, the lower limit is due to Fraunhofer diffraction theory considerations[36] and the upper limit depends on the laser beam diameter and other physical effects. A typical optical set-up for a diffraction-based drop sizing instrument is shown in Fig. 14.

One of the major problems associated with measuring both the size and velocity distribution of drops in diesel sprays is the high drop density in the region of interest. Photographic and holographic techniques cannot be used in these regions. Similarly, the light scattering techniques also suffer from this shortcoming. It had originally been indicated that for the light scattering instruments the attenuation of incident light must be less than 50%. This limit represents the onset of importance of multiple scattering, which renders the measured light energy distribution biased and unreliable. The effect of multiple scattering on the measured light energy distribution and the calculated size distribution has been studied experimentally and theoretically[37,38] and a series of correction equations has been derived which enables the measurement of drop size distribution in the region of spray where over 95% of the incident light is scattered. Although these studies have been very useful and reliable measurements of drop size distribution can be made in regions where other techniques cannot be used, it is still impossible to apply the technique to regions where the bulk of atomization takes place in the diesel sprays. A further shortcoming of diffraction-based techniques described is that the measured size distribution is the result of the diffraction from all the drops in the incident beam path. Therefore the size distribution is that of the 'line of sight' measurement. For most purposes this information is sufficient, but if more detailed size distribution within the spray is required a tomographical transformation of the measured light energy distribution at several radial positions can be carried out to provide a better spatial resolution.[39]

6.3. Laser–Doppler Anemometry (LDA)

Double-exposure photographs or holograms can be used to produce information on drop size and velocity distribution. A faster and more reliable technique is LDA. This technique can produce simultaneous measurements of size and velocity distributions of the drops in the spray. Unlike the measurement techniques described so far, this is essentially a single particle counter and this limits its applicability to relatively dilute sprays. Particle sizing interferometry, as this technique is sometimes referred to, has been described in the literature.[40,41] The basic concept is to produce a test volume at the intersection of two laser beams originating from the same source. A fringe pattern consisting of dark and bright fringes is formed within this test volume as a result of interference of the two beams with each other (see Chapter 15). As a drop passes through the test volume it scatters light, the amplitude of which depends on the drop diameter, and the frequency of which is indicative of drop velocity. The size range of the drops that can be measured by this technique is 3 μm to 5 mm depending on the fringe spacing. Drop velocities of the order of $1000\ \mathrm{m\ s^{-1}}$ can be measured using LDA. A typical LDA set-up is shown in Fig. 15. It is essential that only one drop should be present in the test volume at any one time, which makes the technique of little applicability to the measurement of diesel sprays. A problem with the

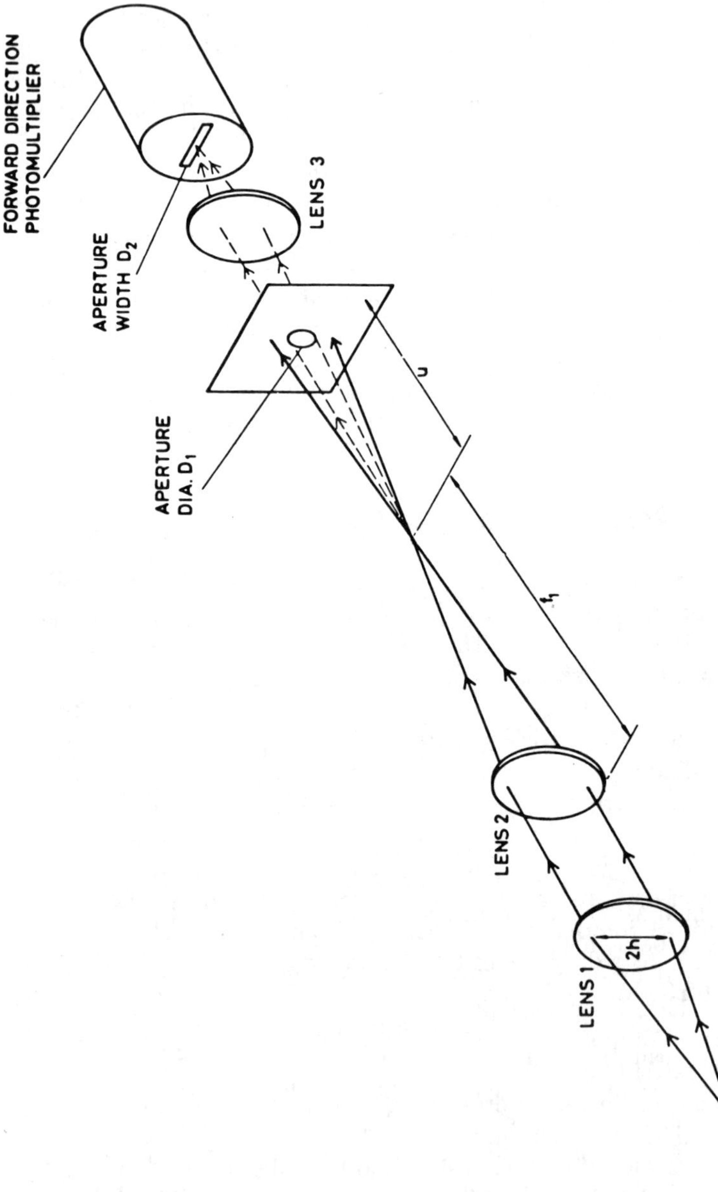

Fig. 15. Typical LDA optical set-up.

technique is the Gaussian light intensity distribution of the laser beams. The consequence of this is that a large drop traversing the edge of the sample volume may give the same scattered light amplitude as a small drop crossing through the centre of the test volume. This ambiguity has been solved in many ways, the most notable method being to tailor the laser beam to produce a top-hat light intensity distribution[42] and the use of two-colour LDA.[43]

7. SIZE AND VELOCITY DISTRIBUTION MEASUREMENT: RESULTS

The size and velocity distribution of the liquid drops, as well as other characteristics in the diesel spray, depends on a large number of factors, which can be divided into three categories: (1) the liquid fuel properties; (2) the injection characteristics; and (3) the ambient air properties.

The major fuel properties are density, viscosity and surface tension: the effect of each on the spray characteristics can be studied by the use of multi-fuel mixtures consisting of heavy diesel, light diesel, kerosene, gasoline and other fuels and liquids. These fuels can be mixed in various proportions to produce the desired physical properties. The most important injection properties are the injection pressure and nozzle hole geometry. The injection pressure of the liquid varies with time; therefore the injection pressure referred to is usually the opening pressure of the nozzle. The opening pressure of the nozzle is usually in the range 100–500 bar, but the maximum injection pressure can be as high as twice the opening pressure. The hole diameter is usually less than 0·5 mm and the length to diameter ratio is of the order of 4·0.

The main ambient air property is the density. The ambient pressure can be as high as 50 bar and the ambient temperature at the point of injection can be as high as 700 K, depending on the engine compression ratio and other design features. The effects of some of these parameters have been presented earlier. Some of the recent studies and results on drop size and velocity distribution are presented here.

Elkotb[10] carried out an extensive experimental study of diesel sprays and made measurements of drop size distribution for various operating conditions. The size measurement technique employed was one of sampling on a soot-coated glass slide and the subsequent photography of the slide. The Sauter mean diameter, which is of relevance in evaporation and combustion of the drop, and the linear diameter, which is a measure of the effectiveness of the atomization process, were measured from the photographs. The investigation was carried out for different operating conditions, and the sample cell was situated several centimetres downstream of the nozzle.

The effect of fuel viscosity and surface tension on diesel spray drop size distribution was studied by Tabata[44], who confirmed some of the findings of the work by Elkotb. The measurement technique used in this study was a laser diffraction technique. The spray was a continuous one and hence no information on drop size distribution variation with time was obtained, even though the injection pressure was varied from 30 to 900 bar.

Takeuchi *et al.*[45] used a photographic technique to measure the drop size distribution and spray tip penetration for the spray as a function of time. The start of the injection was detected using a laser beam to detect the needle lift of the injector. Ewan[30] used an in-line holographic technique to measure the drop size distribution in a diesel spray. The measurements were made at fairly large distances below the nozzle, which illustrated the difficulties in obtaining reliable results near the nozzle, where the measurements are of greatest interest and benefit as input data for mathematical models of diesel spray combustion. The variation of peak size with time and distance from the nozzle is shown in Fig. 16. The variation of the number percent of drops less than 15 μm with time and distance from the nozzle is presented in Fig. 17.

As a result of the experimental work carried out by various research workers, a large number of empirical equations have been derived expressing the Sauter mean diameter of the spray as a function of time, and the physical properties of liquid and ambient air, as well as other injection properties. One such equation based on dimensional analysis and experimental findings has been derived by Elkotb[10]:

$$D_{32}\quad (\mu\text{m}) = 6156\nu^{0\cdot385}\sigma^{0\cdot737}\rho_L^{0\cdot737}\rho_g^{0\cdot06}\Delta P^{-0\cdot54} \tag{10}$$

Although the overall injection velocity and spray tip velocity can be measured very easily using photographic or laser light attenuation techniques, there has been little work done on the measurement of the velocity of individual drops in the region of interest. The available techniques for the simultaneous measurements of drop diameter and velocity can only be applied to the dilute region of the spray where

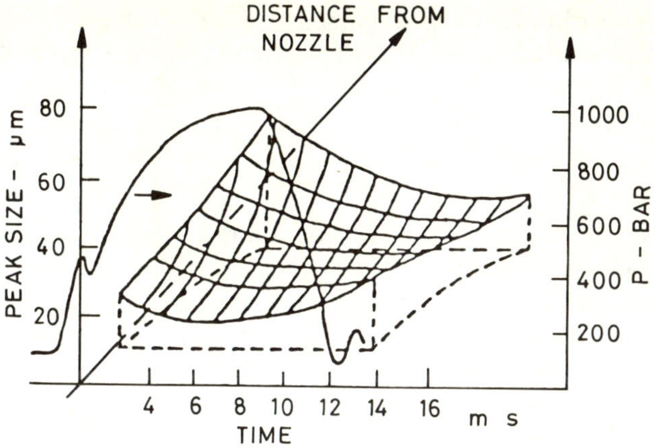

Fig. 16. Variation of peak size with time and distance from the nozzle (Ewan[30]).

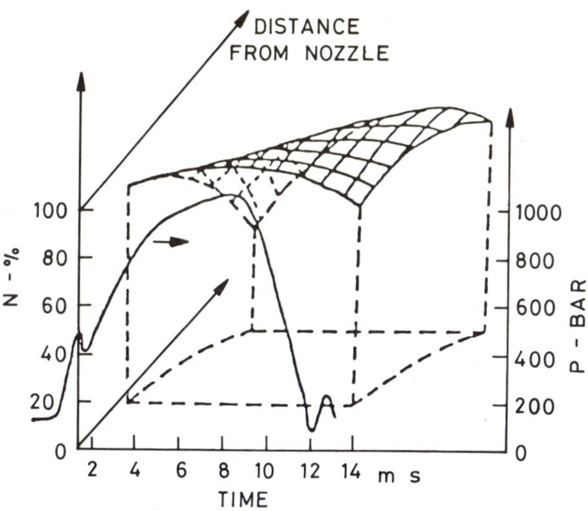

Fig. 17. Variation of the number percent of drops less than 15 μm with time and distance from the nozzle (Ewan[30]).

either very little mass of the spray is dispersed or it is so far away from the nozzle that the acquired data is of little use.

8. CONCLUSION

It has been shown that a great deal of work has already been done in measurement and characterization of diesel sprays. A large amount of data, correlations, tables and charts have been produced by various researchers as a result of their efforts to express various spray characteristics as a function of operating conditions. Recent years have seen the development and application of reliable and sophisticated measurement techniques to the measurement of drop size, velocity, concentration and other parameters of diesel sprays. However, there is still a need for new techniques for measurement of drop size and velocity distribution in the dense and near the nozzle regions of the spray where the bulk of the atomization takes place. Therefore, although much has been established about the behaviour and characteristics of diesel sprays in recent years, more work is required to fully elucidate the outstanding problems of diesel engine fuel injection processes.

REFERENCES

1. Lilly, L. R. C., *Diesel Engine Reference Book*. Butterworth, London, 1984.
2. Hillier, V. A. M. & Pittuck, F., *Fundamentals of Motor Vehicle Technology*. Hutchinson, London, 3rd edition, 1984.
3. Yule, A. J., Mo, S. L., Tham, S. Y. & Aval, S. M., Diesel spray structure. 3rd ICLASS, London, 1985.
4. McCreath, C. G., Roett, M. F. & Chigier, N. A., A technique for measurement of velocity and size of particles in flames. *J. Phys. E: Scient. Instrum.*, **5** (1972) 601–4.
5. Finlay, I. C. & Welsh, E., Photographic techniques for determining velocities of liquid droplets flowing in an air stream. *J. Photogr. Sci.*, **16** (1968) 71–8.
6. Yule, A. J., Chigier, N. A. & Cox, W. W., *Measurement of Particle Size in Sprays by Automated Analysis of Spark Photographs*. Heyden, London, 1978.
7. Pirouz-Panah, V. & Williams, T. J., A technique for an application of light pipes to transient fuel jet droplet photography. *J. Phys. E: Scient. Instrum.*, **13** (1980) 32–3.

8. Reitz, R. D., Atomization and other break-up regimes of a liquid jet. PhD thesis, Department of Aerospace and Mechanical Science, Princeton University, Princeton, N.J.
9. Rayleigh, W. S., On the instability of jets. *Proc. Lond. Math. Soc.*, **4**(1879) 4–13.
10. Elkotb, M. M., Fuel atomization for spray modelling. *Progr. Energy Combust. Sci.*, **8** (1982) 61–91.
11. Weber, C., On the breakdown of a fluid jet. *Math. Mech.*, **11** (1931) 136.
12. Ranz, W. E., On spray and spraying. *Bulletin 65,* Department of Engineering Research, Penn State University, 1956.
13. Castleman, R. A., Mechanism of atomization of liquids. *J. Res., U.S. Natl Bur. Stand.*, **6** (1931) 281.
14. DeJuhas, K. J., Dispersion of sprays in solid injection oil engines. *Trans. Am. Soc. Mech. Engrs,* **53** (1931) 65.
15. Bergwerk, W., Flow pattern in diesel nozzle spray holes. *Proc. Inst. Mech. Engrs,* **173** (1959) 655.
16. Rupe, J. H., On the dynamic characteristics of free liquid jets and a partial correlation with orifice geometry. Tech. Report No. 53, Joint Propulsion Laboratory, 1962.
17. Giffen, E. & Muraszaw, A., *The Atomization of Liquid Fuels.* John Wiley, New York, 1953.
18. Wakuri, Y., Fuji, M., Amitani, T. & Tsuneya, R., Studies on the penetration of fuel spray in a diesel engine. *Bull. Jpn. Soc. Mech. Engrs,* **3**(9) (1960) 123–30.
19. Fujimoto, H., Tanaba, H., Kumiyashi, H. & Sato, G. T., Investigation on the characteristics of diesel spray. *Bulletin Jpn. Soc. Mech. Engrs* **25**(200) (1982) 249–56.
20. Jong-Yul Ha, Hayashi, A. *et al.*, Investigation on the initial part and spray formation delay of diesel spray. SAE paper 830451, 1983.
21. Fujimoto, H., Tanabe, H., Kuniyashi, H. & Sato, G. T., Investigation on combustion in medium-speed marine diesel engines using model chamber. 15th CIMAC, 1983.
22. Jong-Yul Ha, Norimasa, I., *et al.*, Experimental investigation of the entrainment into diesel spray. SAE paper 841078, 1984.
23. Law, C. K. & Law, H. K., A d^2-law for multicomponent droplet vaporization and combustion. *AIAA J.*, **20**(4) (1981) 522–7.
24. Fujimoto, H. & Shimada, T., 'Investigation on combustion in diesel engines using a constant volume combustion chamber, *Bull. Jpn. Soc. Mech. Engrs,* **22**(174) (1979) 1818–25.
25. Sato, G. T., Tanabe, H., Sugihara, H. & Fujimoto, H., Studies on diesel spray and its combustion, 19th FISITA, Melbourne, paper 82027, 1982.
26. Hirleman, E. D., Non-intrusive laser particle diagnostics. AIAA 18th Thermophysics Conference, Montreal, June 1983.
27. Stevenson, W. H., Spray and particulate diagnostics in combustion systems: a review of optical methods. Central States Section Meeting of Combustion Institute, West Lafayette, Indiana, USA, 1978.
28. Jones, A. R., A review of drop size measurement—the application of techniques to dense fuel sprays. *Prog. Energy Combust. Sci.*, **3** (1977) 225–34.

29. Tishkoff, J. M., Spray characterization: practices and requirements. *Optical Engng,* **23**(5) (1984) 557–60.
30. Ewan, B. C. R., Houdard, T. & Bastenhof, D., The holographic measurement of diesel sprays. International Conference on Instrumentation in Aerospace Simulation Facilities, St Louis, France, 1983.
31. Thompson, B. J. & Dunn, P., Advances in far-field holography theory and applications. Society of Photo Optical Instrumentation Engineers, Vol 215, Recent Advances in Holography, 1980.
32. Bexon, R., Gibbs, J. & Bishop, G. D., Automatic assessment of aerosol holograms. *J. Aerosol Sci.,* **7** (1976) 397–407.
33. Dobbins, R. A., Crocco, L. & Glassman, I., Measurement of mean particle sizes of sprays from diffractively scattered light. *AIAA J.,* **1**(8) (1983) 1882–6.
34. Cornillault, J., Particle size analyser. *Appl. Optics,* **11**(2) (1972) 265–8.
35. Swithenbank, J., Beer, D. S. *et al.,* A laser diagnostic technique for the measurement of droplet and particle size distribution. 14th Aerospace Sciences Meeting, AIAA, Washington, DC, 1976.
36. Van de Hulst, H. C., *Light Scattering by Particles.* Dover, New York, 1981.
37. Felton, P. G., Hamidi, A. A. & Aigal, A. K., Measurement of drop size distribution in dense sprays by laser diffraction. 3rd ICLASS, London, 1985.
38. Dodge, L. G., Change of calibration of diffraction based particle sizers in dense sprays. *Optical Engng,* **23**(5) (1984) 626–30.
39. Yule, A. J., Ahseng, C. *et al.,* A laser tomographic investigation of liquid fuel sprays. 18th Symposium on Combustion, Canada, 1980.
40. Bachalo, W. D., Hass, C. F. & Hartwell, C. A., An instrument for spray droplet size and velocity measurement. ASME Winter Annual Meeting, 1979.
41. Robinson, D. M. & Chu, W. P., Diffraction analysis of doppler signal characteristics for a cross beam laser doppler velocimeter. *Appl. Optics,* **14**(9) (1975) 2177–83.
42. Ereaut, P. R., Ungut, A. *et al.,* Measurement of drop size and velocity in vaporizing sprays. 2nd ICLASS, Madison, Wisconsin, 1982.
43. Yeoman, M. L., Azzopardi, B. J. *et al.,* Optical development and application of a two colour LDA system for the simultaneous measurement of particle size and particle velocity. Internal Report AERE-R 10468, United Kingdom Atomic Energy Authority, Harwell, 1982.
44. Tabata, M., Aria, M. & Hiroyasu, H., Effect of fuel viscosity and surface tension on diesel spray drops. 3rd ICLASS, London, 1985.
45. Takeuchi, K., Murayarra, H. *et al.,* Droplet size distribution in diesel fuel spray. *Bull. Jpn. Soc. Mech. Engrs,* **26**(215) (1983) 797–804.

8

Turbulent Flows in Reciprocating Internal Combustion Engines

R. J. Tabaczynski

Ford Motor Co., Dearborn, Michigan, USA

1. General features of a turbulent flow

2. Turbulence in reciprocating engines
   2.1. Introduction
   2.2. Definition of turbulent flow parameters

3. Qualitative features of in-cylinder flows
   3.1. General flow patterns
   3.2. The turbulent flow field
   3.3. Turbulence scales: macroscales
   3.4. Turbulence scales: microscales

4. Conceptual modelling of engine turbulence
   4.1. Introduction
   4.2. The intake process
   4.3. The compression process
   4.4. Swirl and combustion chamber effects

5. Summary

1. GENERAL FEATURES OF A TURBULENT FLOW

In this section some of the fundamental assumptions underlying the description of a turbulent flow are presented.[1] First, it is useful to discuss the basic features of a turbulent flow. Turbulent flows are highly diffusive. This high diffusivity results in increased rates of momentum, heat and energy transfer. For example, turbulent flames propagate faster than laminar flames and turbulent heat transfer allows the heating of a room to occur in seconds rather than hours. Another feature of turbulent flows is that they always occur at high Reynolds numbers. This implies that turbulence originates from instabilities in the laminar flow. These instabilities originate from the interaction of viscous and inertial forces which are highly non-linear. Turbulent flows are always three-dimensional and rotational. They contain distributed regions of vorticity which undergo vortex stretching in order to maintain or increase the amount of vorticity. Without this vortex stretching mechanism, the turbulence would decay rapidly. This implies that turbulent flows are highly dissipative: that is, without a supply of energy to maintain the loss to viscosity, the turbulence will decay rapidly. Finally, turbulence is a characteristic of the flow. It is not a property of the fluid like viscosity or density but rather a feature of a particular flow field. It depends on the dimensions and velocity gradients of the particular flow condition. Students and researchers who concentrate on turbulence attempt to develop formulations that are valid for particular classes of turbulent flows. The internal combustion engine is one such class of flows and, although it has many features that are similar to other turbulent flows, various constants in the formulation will be different than those for shear flows or wakes etc.

One of the basic assumptions in turbulence theory is that the kinetic energy of an eddy $u'^2/2$ decays in the characteristic time l/u', where l is the characteristic size of the eddy and v' is the turbulence intensity namely the r.m.s. value of the velocity. This concept is truly one of the cornerstone concepts of turbulence theory and is used frequently in developing the dynamic equations governing the turbulent flow. The rate at which the kinetic energy dissipates is referred to as the dissipation ε and can be defined as

$$\varepsilon = Cu'^2/(l/u')$$

or

$$\varepsilon = Cu'^3/l$$

Another basic concept is that the viscosity (ν) of the fluid is only important at the small scales of turbulence. These small scales have several interesting features. They tend to have small time scales relative to the mean flow. This implies that whatever the time scale of the mean flow for dissipation the small scales can adjust quickly to dissipate the energy via viscous effects. Hence the smallest scales of turbulence depend only on the viscosity of the fluid and the dissipation of energy (i.e. the energy supplied from the large scales). With this information, one can form length, time and velocity scales as follows:

$$\eta = (\nu^3/\varepsilon)^{1/4}, \qquad \tau = (\nu/\varepsilon)^{1/2}, \qquad \upsilon = (\nu\varepsilon)^{1/4}$$

These scales are referred to as the Kolmogorov microscales. From examining these scales, one can see that the major difference between a turbulent flow of lower and higher Reynolds number is the size of the small scale structure. The higher the Reynolds number the smaller the small scales of turbulence.

Before leaving this section, a comment on the scales of turbulence is in order. The conceptual modeller of turbulent flows needs information on the scaling for characteristic velocities and scales. In general, the largest scales in a turbulent flow can be the size of the vessel: for example, the diameter of a pipe or the diameter of a blunt body in a flow. For turbulent jets in the atmosphere no natural scale is present and other assumptions have to be made. Hence scales can vary from the size of the vessel to the Kolmogorov scales which are dominated by viscous effects. Scaling the characteristic velocity is accomplished by taking a fraction of the velocity difference in the regions of high shear. Typically, the characteristic velocity is on the order of $0{\cdot}1(U_1 - U_2)$.

In the following sections of this chapter these features of a turbulent flow field will be discussed with respect to reciprocating engines. In addition, the simple scaling laws will be used to develop a simple model for engine turbulence.

There are very significant differences in the in-cylinder air pattern of flow and turbulence required in the diesel (DI) engine and the spark-ignition (SI) engine. In the former a high degree of swirl is required to bring the air into the fuel spray and to establish mixing in order to obtain a near-stoichiometric air:fuel ratio for good combustion. In the SI engine high turbulence is necessary, and swirl can usefully be used to produce this condition in order to give fast burning of the premixed petrol/air charge. However, the fluid dynamic principles are the same and both types of engine will be considered together, differentiation being made where necessary.

2. TURBULENCE IN RECIPROCATING ENGINES

2.1. Introduction

The mean and turbulent flow fields that are produced in a reciprocating engine have a major influence on the transport processes that occur in these devices. In the spark-ignition (SI) engine the speed at which the flame travels across the chamber, the degree of air–fuel stratification, the emissions formation, heat transfer, and lean-limit capability of an engine system all depend heavily on the fluid motions that exist in the combustion space. In the diesel engine designers have used masked valves, helical ports, bowls in pistons and pre-chambers in an attempt to bring these flows under control, in the development of a specific combustion process and to improve the smoke-limited performance. They have used flow visualization techniques, hot-wire anemometry and, more recently, computational fluid mechanics and laser-based experimental techniques to gain some understanding of the flow processes and the effects they have on the engine performance. In this chapter, the general features of the turbulent flow in engines will be discussed. The emphasis will be on understanding the general features of the flow, developing some conceptual scaling models to explain these general features and finally to discuss how changes in the flow process affect the combustion and mixing processes.

2.2. Definition of Turbulent Flow Parameters

A turbulent flow field is generally characterized by the measurement of a mean velocity U, the turbulence intensity u', a characteristic turbulent integral time scale L_t, a characteristic turbulent integral length scale L_x and the distribution of the turbulent kinetic energy with respect to frequency or energy spectrum E. Since the engine is a periodic device in which the flow may not be quasi-steady, the classical definitions of these parameters using conventional time-averaging techniques may not be applicable. In general, it is necessary to perform ensemble averaging over a large number of engine cycles in place of time averaging. This technique is used by most researchers in this field and the definitions put forth by Witze[2] and Fraser *et al.*[3] will be used in the following description. The instantaneous velocity at a given crank angle position ϕ in the cycle is defined as:

$$U(\phi) = \langle U(\phi) \rangle + u(\phi) \tag{1}$$

where $u(\phi)$ is the fluctuating velocity and $\langle U(\phi) \rangle$ is defined as:

$$\langle U(\phi) \rangle = 1/N \sum U(\phi + in\pi) \tag{2}$$

and n is the number of strokes in the cycle ($n = 4$ for conventional four-stroke engines and 2 for two-stroke engines). The turbulence intensity $u'(\phi)$ can also be defined in a similar manner and is given by:

$$u'(\phi) = \sqrt{(1/N \sum [U^2(\phi + in\pi) - \langle U(\phi)\rangle^2])} \tag{3}$$

The mean velocity and turbulence intensity, as defined by eqns (2) and (3), are the most frequently measured quantities in engine turbulence experiments. The difficulty in performing time scale and length scale measurements in an unsteady flow, such as exists in an engine cylinder, has limited the amount of experimental data on these quantities. Recently, Fraser *et al.*[3] measured integral length scales using the following definition:

$$L_x = \int R_x \, dx \tag{4}$$

where

$$R_x = 1/N \sum [u(x_0)u(x_1)/(u'(x_0)u'(x_1))] \tag{5}$$

This measurement requires the use of two probes, the spacing of which is systematically varied, and the correlation of large amounts of data. Due to the relative difficulty of such a technique, most investigators measure the integral time scale and attempt to infer a length scale from the time scale data and the Taylor hypothesis. The integral time scale is defined in a similar manner to the integral length scale, except for the correlation of the data, which is made at two times in the cycle, i.e.:

$$L_t = \int R_t \, dt \tag{6}$$

where

$$R_t = 1/N \sum [u(t_0)u(t_0 + t)/(u'(t_0)u'(t_0 + t))] \tag{7}$$

The integral length scale is then inferred through the Taylor hypothesis:

$$L_x = L_t\langle U\rangle \tag{8}$$

Equation (8) is only valid when a strong mean flow can be defined, and is generally used when the flow is quasi-steady. Another scale that is of importance is the Taylor microscale. This scale has been used by some researchers in the development of turbulent combustion models

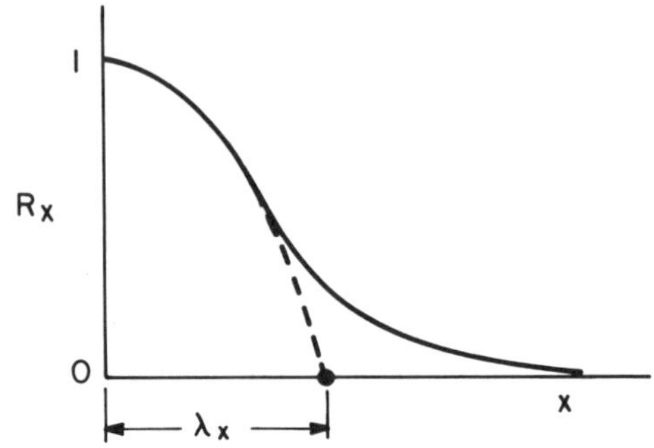

Fig. 1. Correlation coefficient R_x and estimate of λ_x (from Dent & Salama[5]).

to define the appropriate turbulent chemical reaction time (see Tabaczynski *et al.*).[4] It is also thought to be important in the diffusion processes which occur in diesel engines that lead to the production of particulate emissions. The Taylor microscale, λ, can be determined from the correlation curves R_x and R_t, as shown in Fig. 1, from Dent & Salama.[5] It is defined by the second derivative of the correlation coefficient evaluated at the origin x_0 or t_0, and can be formally expressed as:

$$\lambda_x^2 = -2/(\partial^2 R_x/\partial x^2), \qquad \lambda_t^2 = -2/(\partial^2 R_t/\partial t^2) \qquad (10, 11)$$

The microscale can be used as a measure of the fluctuating strain rate in a turbulent field via the expression u'/λ. This fluctuating strain rate can be used to aid in understanding the effect of strain on the combustion and mixing processes.

In order to help understand the processes which control the dissipation and production of turbulence in an engine, it is useful to determine the kinetic energy of turbulence u'^2 as a function of frequency at various times in the engine cycle. This distribution of turbulent kinetic energy with respect to frequency is called the power spectral density function, $E(k)$, where k is the wavenumber. $E(k)$ can be determined by obtaining the Fourier transform of the correlation coefficient R_t. Once $E(k)$ is obtained it can be used to assist in interpreting the data by comparing the experimental spectrum with power spectra that are a result of classical turbulence theories. The most common power spectra is that developed by Kolmogorov (Tennekes & Lumley[6]), who used the concepts of local isotropy of the small scales of turbulence (i.e. that the small eddies have the same velocities in all directions) and local equilibrium (i.e. that the small eddies respond instantaneously to changes in the large mean flows) to develop the expression:

$$E(k) = a\varepsilon^{2/3}k^{-5/3} \tag{12}$$

where a is a constant, ε is a measure of the dissipation of turbulent kinetic energy and k is the wavenumber. By comparing eqn (11) with data for turbulent kinetic energy near the time of combustion, or at any other time in the cycle, it is possible to determine if the turbulent flow is in equilibrium, which could assist in the development of models for the turbulent flow or in the general interpretation of turbulence data.

3. QUALITATIVE FEATURES OF IN-CYLINDER FLOWS

3.1. General Flow Patterns

One of the major features of reciprocating engines is the production of high shear flows during the induction process. These high shear flows are thought to be the major production source of the turbulent field that persists and is modified during the compression process. To test this hypothesis, in an early work, Clark performed an experiment in an SI engine in which the intake and exhaust strokes were eliminated, and observed a rapid decrease in the rate of flame propagation, indicating the elimination of the turbulent field. After a number of years, Semenov[7] repeated this experiment using a hot-wire anemometer to measure turbulent velocities in a motored engine and showed the rapid decay in velocity fluctuations as the intake and exhaust process were eliminated (Fig. 2). He thus concluded that the intake stroke was responsible for the generation of the initial turbulent field. It is the production and modification of this initial turbulent field during the induction and compression strokes that occupies the thoughts of the development engineers.

3.1.1. The Mean Flow Field

In order to modify systematically the turbulent flow processes, it is necessary first to understand qualitatively the roles of the various mean flows that are produced during the induction and compression processes in creating the final turbulent field at the time of combustion. In this section, the general features of the mean flow field will be

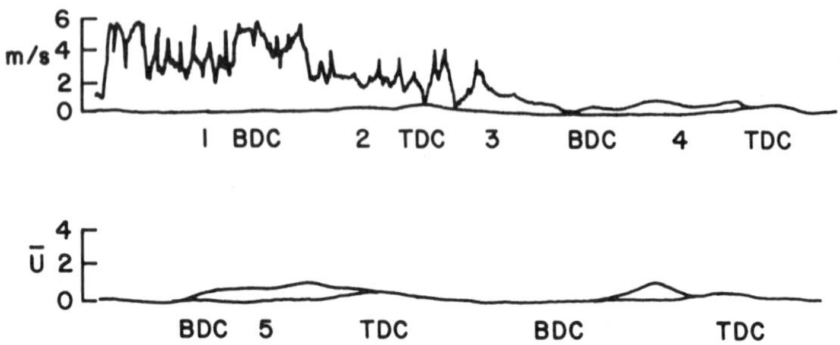

Fig. 2. Oscillogram trace of velocity during transition from normal cycles to a cycle in which the intake and exhaust are absent: 1, intake; 2, compression; 3, expansion; 4, second compression; 5, third compression (from Semenov[7]).

discussed as a precursor to the discussion of the interrelationship between the mean and turbulent flow fields. The major mean flow processes are shown schematically in Figs 3 and 4 for the induction and compression processes. During the induction process (Fig. 3), a jet flow is produced by the intake valve as the piston moves downward, drawing fresh mixture into the cylinder. Basically, three major flow patterns are created: a large recirculating region having a toroidal vortex structure, which fills the major portion of the cylinder volume

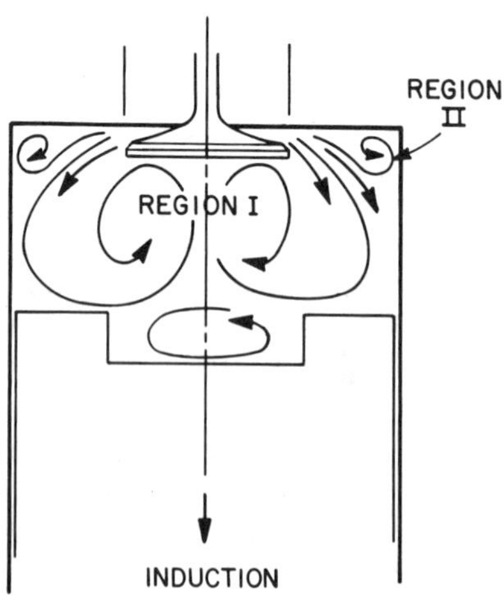

Fig. 3. Schematic of mean in-cylinder flows during intake.

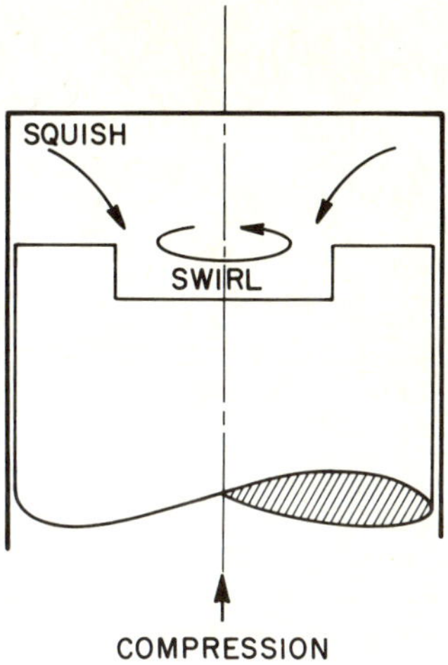

Fig. 4. Schematic of mean in-cylinder flows during compression.

(region I, which is the curling up of the intake jet due to the presence of the cylinder walls and piston); a smaller region of recirculating flow in the vicinity of the cylinder head/bore wall interface (region II, caused by the separation of the jet flow from the cylinder head); and a swirling flow in the form of a forced vortex around the cylinder axis, the strength of which is determined by port design characteristics. Soon after the intake valve closes, the toroidal vortex structure disappears, leaving the structure of the flow field to be determined by the level of swirl created during the induction process and its modification via chamber geometry during the compression process. In the compression process (Fig. 4), the swirl velocity decays due to the shear at the cylinder boundaries. It can be increased by changing the shape of the combustion space to a more compact form such as the bowl-in-piston geometry shown in Fig. 4, a typical geometry for direct injection diesel engines. In addition to the modification of the swirl velocity during compression, a jet flow, commonly called squish, can be created by the piston and cylinder head contours and the amount of clearance between the cylinder head and piston crown near top centre of the compression process (region II in Fig. 4).

From this brief discussion of the mean flow processes that can occur in an engine cylinder, it is apparent that the structure of the mean flow field is quite complex and that measurements at a point in such a flow field need careful interpretation, and should be supplemented by some type of visualization techniques. In the following discussion, the

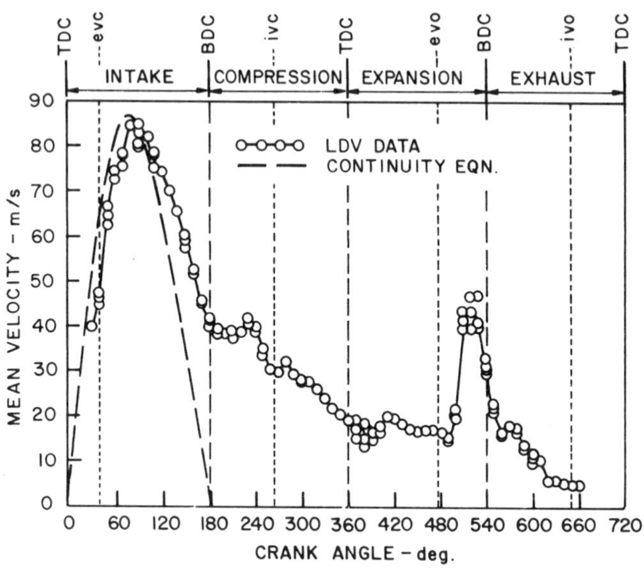

Fig. 5. LDA measurements of the mean velocity. The dashed line is the theoretical induction velocity computed from the one-dimensional continuity equation (from Witze[8]).

relative magnitudes and general behaviour of the mean velocity will be discussed for some typical, but idealized, geometries.

The mean velocity versus crank angle measured at one point in the chamber of a research engine is shown in Fig. 5 (from Witze[8]) for a typical non-swirling engine flow. The turbulent velocity corresponding to this mean velocity history will be discussed in the next section. During the intake stroke the mean velocity increases as the piston begins to draw fresh charge into the cylinder and reaches a maximum at approximately the same crank angle position as the maximum piston velocity. The mean velocity then decreases following, approximately, the instantaneous piston speed. However, the mean velocities do not approach zero near bottom dead centre but rather have a slow decay which continues into the compression stroke as the mean kinetic energy of the flow dissipates due to interactions with the cylinder boundaries and internal fluid shear. This example (Fig. 5) was for a geometry which did not intentionally produce swirl during the intake process, and for a simple disc-shaped combustion space. If a swirl-generating intake port is used in conjunction with a bowl-in-piston geometry, as shown in Fig. 6, a mean tangential velocity (swirl) can be produced, which will be enhanced as the piston moves upward on the compression stroke. The general behaviour of the swirling flow, during the intake and compression strokes, is controlled by the decay of the inducted angular momentum and the chamber shape. During the intake stroke, the swirl is produced by the action of the intake port and cylinder wall boundaries. Throughout the entire process, the total angular momentum is decaying due to the shearing action of the cylinder walls. However, the absolute value of the swirl velocity can be quite high due to the bowl-in-piston geometry and the conservation of angular momentum. As shown in Fig. 7, the swirl velocity begins to increase as the piston approaches top centre of the compression process, reaching a maximum at top dead centre. The maximum swirl velocity in the piston bowl can be quite high, as indicated in Fig. 7, reaching a maximum value of near 20 m s^{-1}. These high swirl velocities

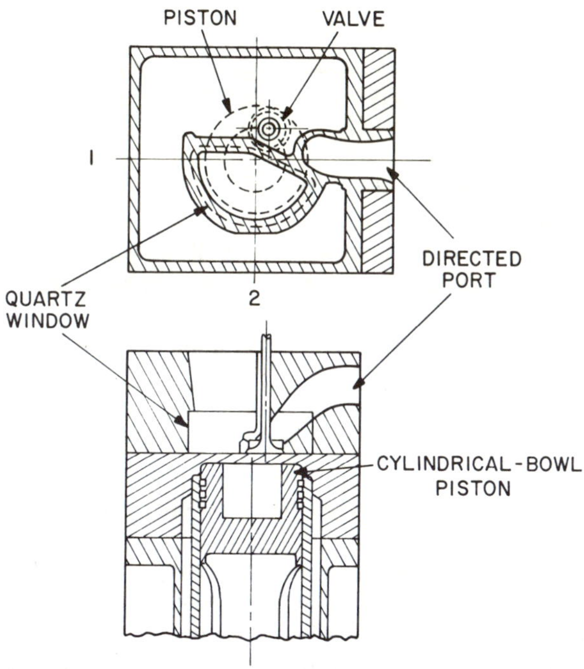

Fig. 6. Schematic illustration of the optical DISC engine, indicating measurement orientations 1 and 2 (from Fansler[14]).

Fig. 7. Measured and calculated swirl velocity near the edge of a piston bowl versus crank angle (from Kondoh *et al.*[15]).

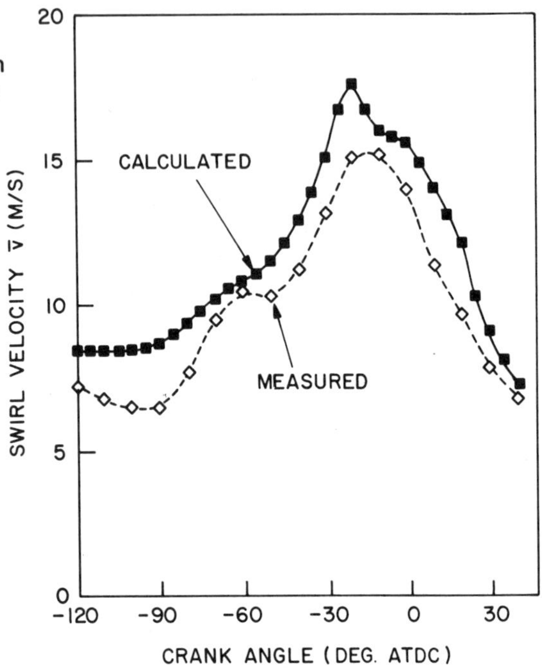

and the fluid motions produced by the shape of the piston bowl can significantly increase the amount of turbulence in the chamber at the time of combustion. One of the characteristics of the swirl velocity during the compression stroke is its near-solid-body behaviour. A typical swirl velocity profile, showing the solid body profile at TDC compression, is shown in Fig. 8, where the maximum swirl velocity occurs at the edge of the piston bowl.

The final mean flow that will be discussed is normally referred to as the squish velocity. One of the major roles of the squish velocity is its influence on the recirculating or secondary flows that are produced in the piston bowl. Squish is utilized in both spark-ignition and diesel engines. Shown by Arcoumanis *et al.*[9] in Fig. 9, are the types of secondary flows that result for swirling and non-swirling flows as the amount of squish is varied for a diesel combustion chamber. These flow pattern changes produce air motion which although not as high in velocity as the swirling flows, can significantly affect the fuel–air mixing process for direct injection diesel engines and can alter the turbulence level due to the differences in fluid shearing. However, for typical spark-ignition engines, with less than 40% of the piston area involved in the squish process, the effect of the squish shear velocity on the turbulence production is minimal. This is a result of the small amount of mass that is present in the squish regions when the velocities are the highest. Note that the highest squish velocities occur near 30° BTDC, as shown in Fig. 10, where the amount of mass in the squish region is small compared to that in the main portion of the combustion space.[10]

3.2. The Turbulent Flow Field

The following discussion relates to spark-ignition engines except where diesel engines are particularly specified.

The turbulent flow field in an engine cylinder is at best difficult to characterize. In general, it is spatially and temporally non-uniform and hence any single-point measurement as a function of time or crank angle can not adequately be used to describe the in-cylinder flow.

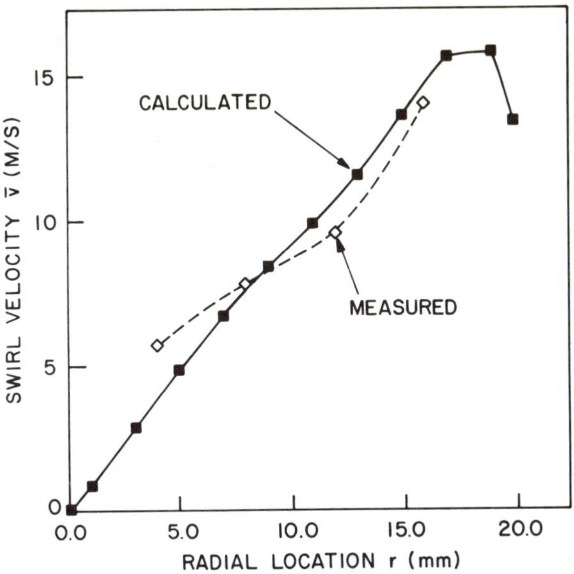

Fig. 8. Measured and calculated swirl velocity versus radial position at TDC of the compression process (from Kondoh *et al.*[15]).

Fig. 9. Schematic diagrams of air flow in piston bowls at TDC compression: (a) and (b) have zero swirl; (c) and (d) have swirl (from Arcoumanis *et al.*[9]).

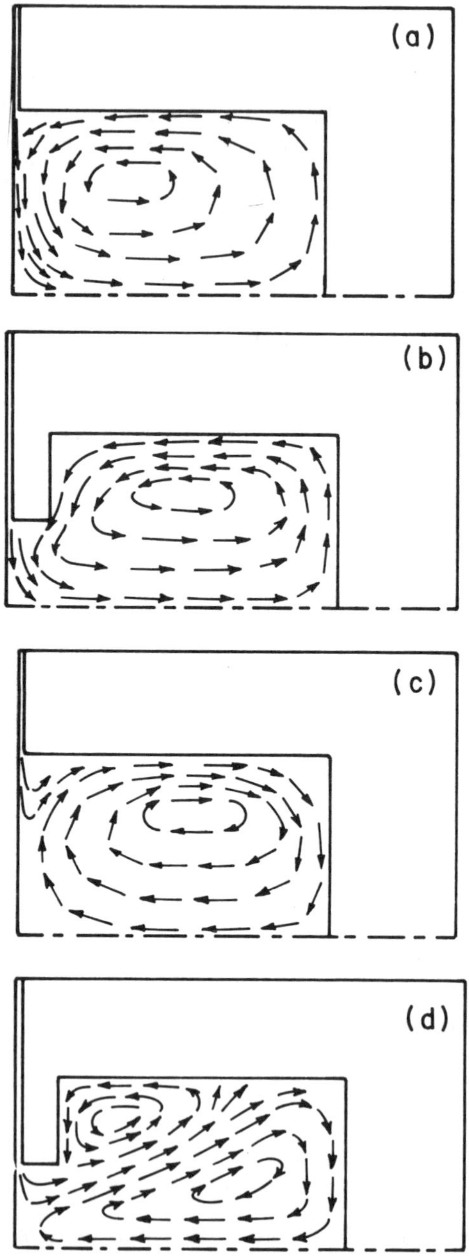

Nevertheless, it is useful and necessary to use such measurements to describe the gross behaviour of the flow. In this section, the general behaviour of the turbulent intensity, macro- and micro-length scales and energy spectrum will be discussed. Throughout the discussion, the relevance of the various quantities to engine combustion and performance parameters will be addressed.

The most common turbulence parameter that researchers have attempted to characterize in an engine is the turbulence intensity. This is a direct result of the role the turbulence intensity plays in the combustion process. Engine researchers recognized the fact that in order for the engine to operate successfully over a wide speed range the turbulence intensity must scale almost linearly with engine speed. This conclusion came from the observation that simple turbulent combustion theories stated that the turbulent flame speed, S_t, was proportional to the turbulence intensity:

$$S_t/S_L = au' + b \tag{12}$$

The supposition of a near-linear relationship between u' and engine speed has been verified by many researchers and a composite of their findings is shown in Fig. 11 (Bopp *et al.*[11]). Although a near-linear dependence is observed for a given location in the cylinder and a

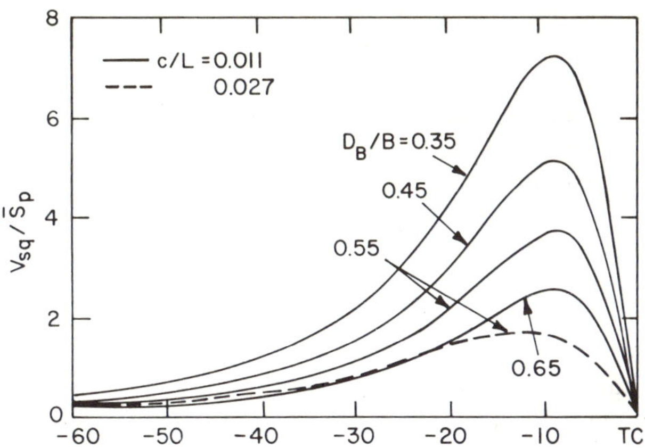

Fig. 10. Theoretical squish velocity divided by mean piston speed for bowl-in-piston chambers for different ratios of bowl diameter divided by cylinder bore, D_B/B, and clearance height divided by stroke, c/L. $B/L = 0{\cdot}914$, $V_B/V_d = 0{\cdot}056$, crank radius/connecting rod length = 3·76 (from Shimamoto & Akiyama[10]).

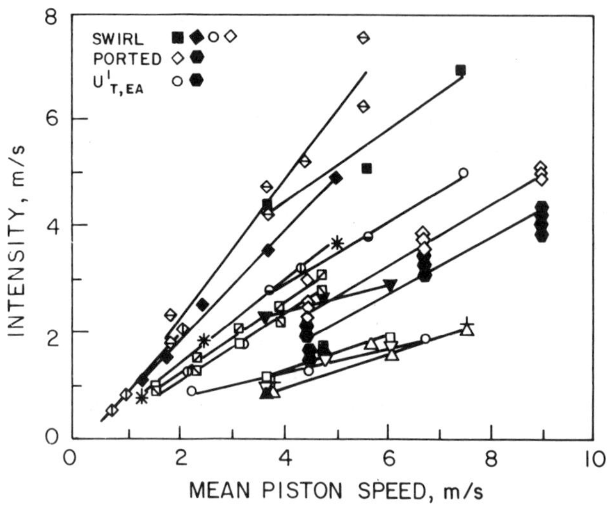

Fig. 11. Cycle-resolved turbulent intensity $U'_{T,EA}$ and ensemble averaged r.m.s. fluctuation velocities at TDC for a number of flow configurations and chamber geometries as a function of mean piston speed. Ported engines are for two-stroke configurations. Data includes engines with intake-generated swirl (from Bopp *et al.*[11]).

specific engine geometry, the slopes of u' versus speed are not universal. This observation of near-linear behaviour with non-constant slopes will be discussed in the section on conceptual models for engine turbulence. From Fig. 11 it is apparent that the absolute value of u' at any given speed is not universal. It is a function of engine design variables, and the trade-offs that the engine designer selected for speed of combustion versus engine power.

Since the value of u' at the time of combustion is not a universal value for a given engine speed, it is important to determine the history of u' throughout the intake process. By studying the history of u' versus crank angle, the engine researcher may be able to determine the relevant design variables which could alter u' at the time of combustion in the most efficient manner. One such measurement by Witze[8] is shown in Fig. 12, where u' is measured versus crank angle. The corresponding mean velocity versus crank angle characteristics were shown in Fig. 5. Several typical features of the turbulent flow can be described with these figures. Beginning with the intake process, both the mean and turbulent velocities increase as the piston moves downward, pulling charge into the cylinder and reaching a maximum near the time of maximum piston velocity. Both the mean and turbulent velocities decrease as the piston decelerates. However, the turbulent velocity decays faster than the mean, indicating a rapid dissipation of the initial turbulent intensity. Another noteworthy feature is the relative magnitudes of the mean and turbulent velocities throughout the cycle. This study indicates that u' is approximately 10% of the mean flow velocity. Such a ratio is typical of turbulent shear flows and indicates that engine turbulence is not unlike classical turbulent flows. After the intake valve closes, this set of measurements shows a constant decay of both the mean and turbulent velocities until the exhaust valve opens and the rush of gases out of the cylinder produces shear and additional turbulence. Other investigators (see Figs 13–16) have shown similar mean and turbulence velocity characteristics for the engine cycle.[7,12–14] One of the regions of interest is the compression process. During this portion of the cycle, it appears that the turbulence undergoes a very slow rate of decay, for engine configurations that do not have bowl-in-piston geometries. This

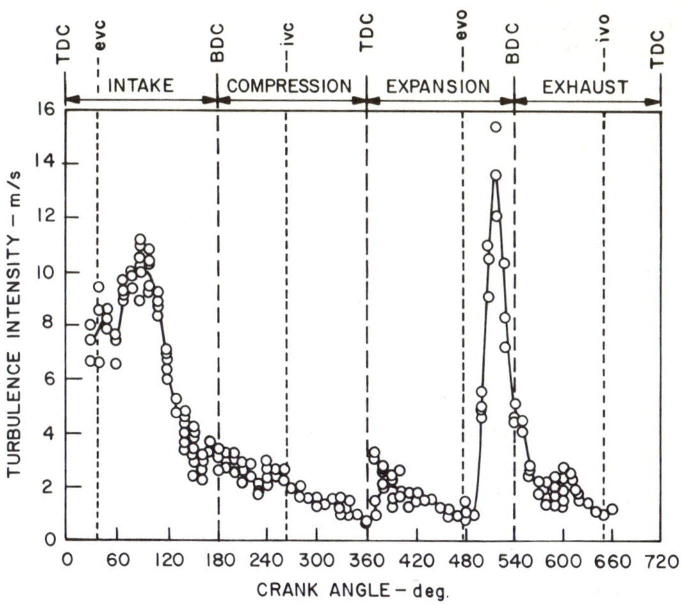

Fig. 12. LDA measurements of turbulence intensity versus crank angle for the entire engine cycle, non-firing engine (from Witze[18]).

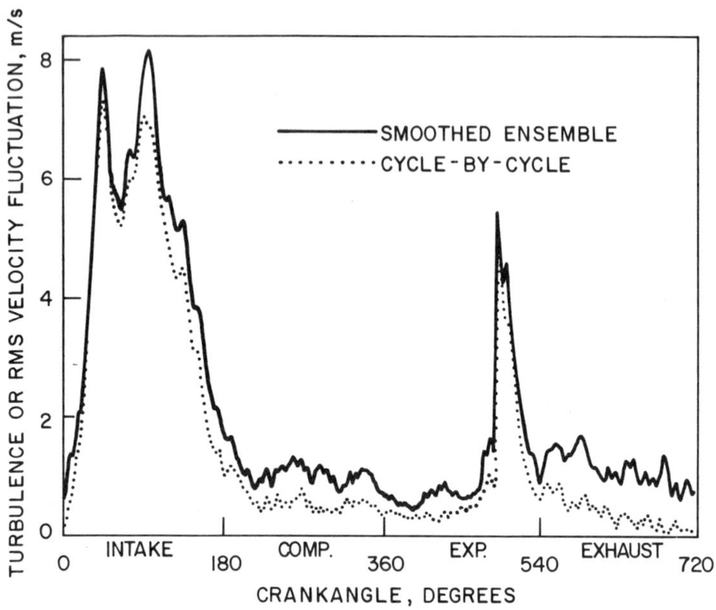

Fig. 13. Comparison of smoothed ensemble and cycle-by-cycle turbulence measurements using LDA (from Rask[12]).

implies that in many engine geometries the turbulent decay tends to be offset by the turbulent production which results from the decay of swirl and the shear produced by the compression process itself (see Figs 13–16, taken from selected references). Figure 16 is for the optical access engine described by Fig. 6, which has a bowl-in-piston geometry. For this configuration an increase in u' is observed which begins about 30° before TDC and reaches a peak value of approximately 2·5 times the value at 30° before TDC. A similar set of results have been predicted by researchers who model the turbulence in engines using computational fluid mechanics. One such result[15] is presented in Fig. 17, which shows the turbulent kinetic energy (basically $\frac{3}{2}u'^2$ for isotropic turbulence). This figure shows the effect of different geometries on the turbulent field. For the 'deep dish' or bowl-in-piston geometry similar to that of Fig. 6, the value of u' increases by approximately 2·5 times, showing agreement with the observed experimental data.

Before leaving this section on turbulence intensity, the reader must be warned that the values of u' reported by researchers doing experimental work are a result of the type of averaging being performed. In most experiments ensemble averaging is used. With this technique, it can be difficult to separate turbulence, defined as the r.m.s. of the mean velocity, from a fluctuating mean velocity. For example, in engines with non-swirling flows and very quiescent flow fields, ensemble averaging can lead to a high measured r.m.s. velocity. In fact, this r.m.s. velocity may be higher than the r.m.s. velocity in a swirling flow, which would imply that the swirling flow had a lower turbulence level. This could be an erroneous conclusion due to misinterpretation of the averaging process. Whenever possible, other techniques should be used to support the point measurement data, such as combustion rate data and turbulent time scale data obtained with hot-wire anemometers.

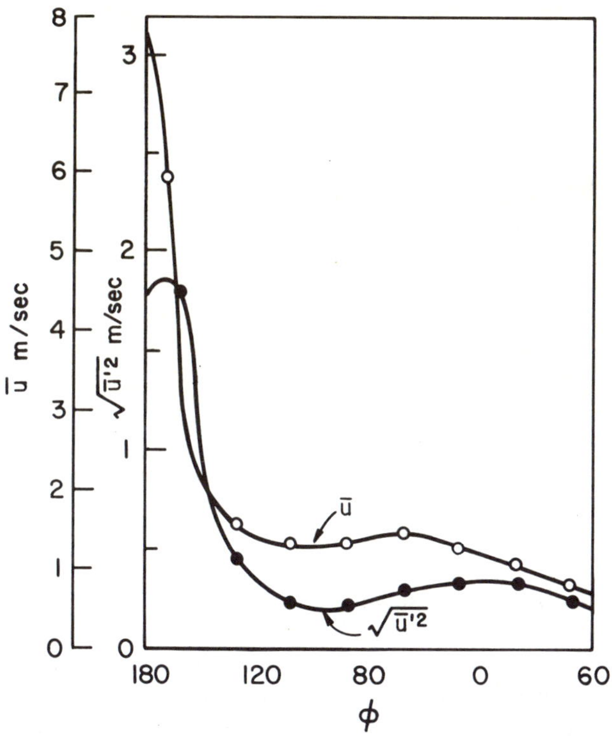

Fig. 14. Mean velocity and turbulence intensity versus crank angle for intake and compression using hot-wire anemometry (from Semenov[7]).

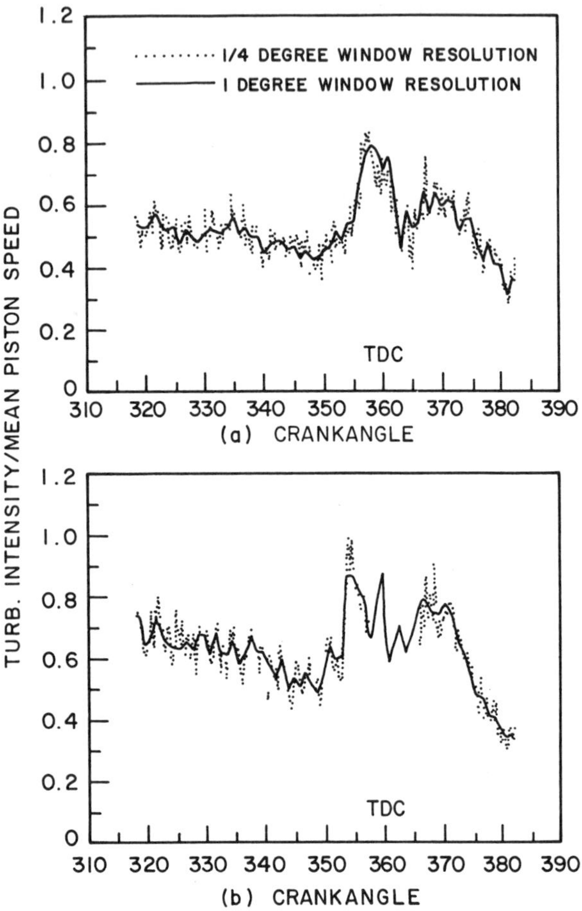

Fig. 15. LDA-measured turbulence intensities calculated using 1-degree and 1/4 degree window resolution applied to same data (from Hall *et al.*[13]).

3.3. Turbulence Scales: Macroscales

In order to understand the behaviour of the turbulent intensity during the engine cycle, it is necessary to understand what controls its rate of production and decay. The behaviour of the integral length scale and time scale during the entire induction–compression process controls the rate of decay of the turbulent flow through a quantity called the

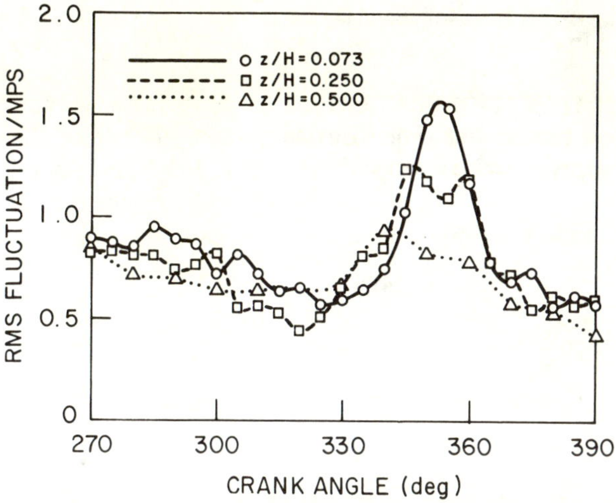

Fig. 16. Ensemble-averaged r.m.s. velocity fluctuation intensity at three depths. Measurement locations $r/R = 0{\cdot}8$, orientation 1, engine speed 600 rpm (from Fansler[14]).

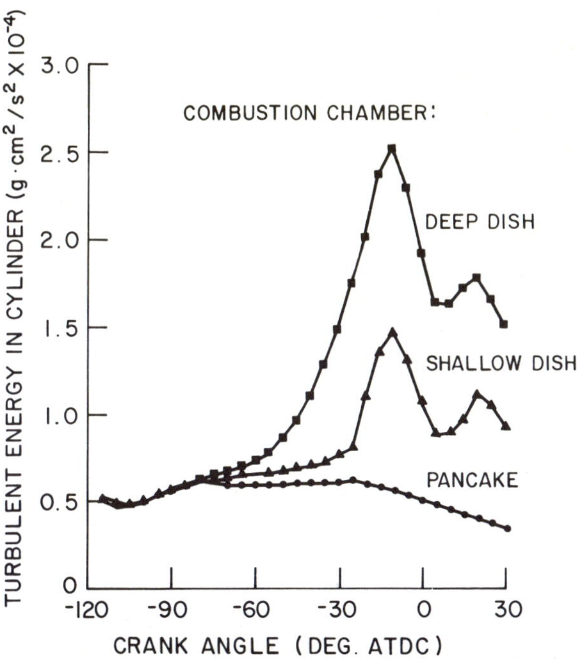

Fig. 17. Influence of combustion chamber geometry on spatially integrated turbulent kinetic energy versus time (from Kondoh *et al.*[15]).

dissipation (ε). In fact, one of the cornerstone concepts in turbulence theory is the statement that the time constant for the decay of the turbulent kinetic energy ($\frac{3}{2}u'^2$) is proportional to L_x/u' (the integral time scale), where L_x is the integral length scale. In mathematical terms, the rate of decay of turbulent kinetic energy can be written as:

$$\mathrm{d}(u'^2)/\mathrm{d}t = -A(u'^2)/L_t \tag{13}$$

or

$$\mathrm{d}(u'^2)/\mathrm{d}t = -A(u'^2)/L_x/u' = -A(u'^3)/L_x \tag{14}$$

and the dissipation rate of turbulent kinetic energy is given as:

$$\varepsilon = Au'^3/L_x \tag{15}$$

The measurement of either L_t or L_x in an engine is difficult because of the time-varying nature of the flow field and the lack of an adequate measurement technique. In order to perform this measurement, the time history of the turbulent flow must be obtained within a given cycle. These measurements can be envisioned conceptually, but are very difficult to obtain due to practical problems associated with obtaining a high speed density for laser-based techniques and the variable properties of the fluid for analyzing hot-wire signals. Further discussions on these techniques are beyond the scope of this chapter, and the reader is encouraged to study some of the recent works referenced in this discussion for more information.

Even though the task of measuring scales in an engine cylinder is formidable, several investigators have reported data on L_t and L_x. Most of the data has been obtained for simple combustion chambers having a disc or pancake shape near top centre. One of the striking features of the data[16] is shown in Fig. 18, in which L_t and L_x are plotted versus crank angle for the intake and compression stroke. The integral length scale is almost constant throughout the cycle, having a value of approximately 3 mm for the unshrouded valve and 6–7 mm for the shrouded valve. If such constancy is correct, the construction of a simplified conceptual model of the turbulent field becomes possible. The behaviour of L_t is also shown in Fig. 18, and it continues to increase with time. This trend would be expected if the amount of turbulent production during compression is small and the turbulent decay process resembled that of grid turbulence, which is probably correct for the simple geometries under study. For grid turbulence, $u'L_x$ is constant as the flow continues downstream of the grid and, since there is no production mechanism, u' decreases and L_x increases,

Fig. 18. Turbulent integral time and length scales during intake and compression (from Lancaster[16]).

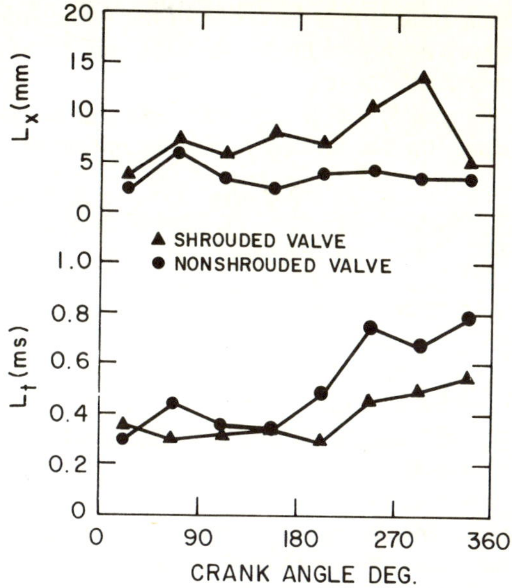

causing L_x/u' to increase at a faster rate than L_x. The process of obtaining the turbulent length scale is to first obtain the turbulent time scale and then use Taylor's hypothesis to deduce the length scale. This, as previously discussed, can be valid for engine flows where a definite mean flow exists. However, for many combustion chambers a near-zero mean is possible, and other techniques must be used. Fraser *et al.*[3] measured the turbulent length scale directly by systematically varying the distance between two probes and determining the correlation coefficient R_x, as discussed in the section on turbulence definitions. The integral length scale for the ported engine studied by Fraser *et al.* is shown in Fig. 19. The general feature of the data is that the integral length scale decreases as TDC is approached and then increases during expansion, ranging in value from 3 mm at 320° to 2 mm at TDC. This same data is plotted in Fig. 20, where it is non-dimensionalized by the instantaneous chamber height H, showing that after TDC the integral length scale appears to be proportional to the chamber height. At TDC the proportionality constant is approximately 0·2, which is in agreement with the data of Lancaster,[16] whose data ranged from 0·175 to 0·275 for the case of swirling flow. In a theoretical study of turbulence in engines, Morel & Mansour[17] calculated the behaviour of the integral length scale, assuming that the

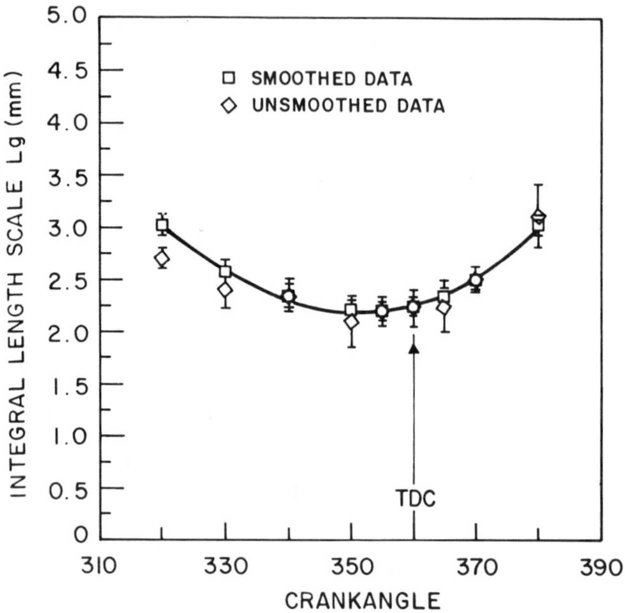

Fig. 19. Turbulent integral length scale versus crank angle for smoothed and unsmoothed data (from Fraser *et al.*[3]).

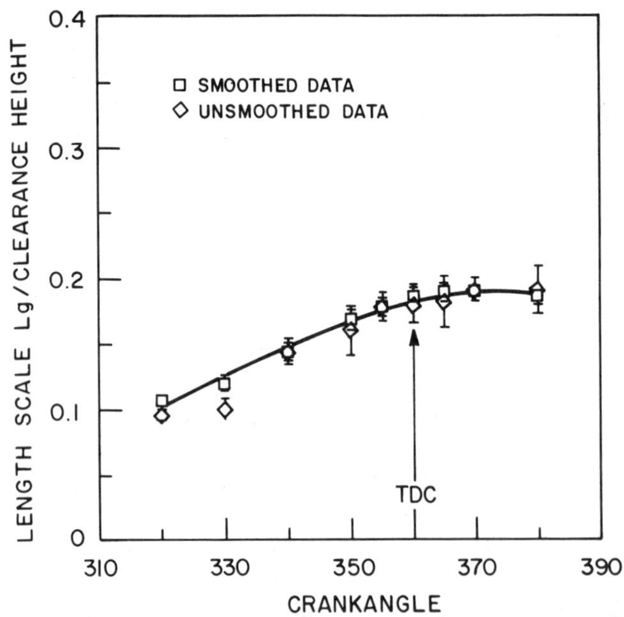

Fig. 20. Turbulent integral length scale/instantaneous clearance height versus crank angle (from Fraser *et al.*[3]).

initial scale was 0·1 of the stroke. Their values for L_x/H range from 0·28 to 0·5 for a non-swirling flow.

3.4. Turbulence Scales: Microscales

While the macroscales of turbulence are important for describing the decay rate of the turbulent kinetic energy, the microscales are significant in understanding the mass, momentum and heat transfer processes. Unfortunately, only indirect evidence is available on the role of the microscale in the determination of the transport processes that occur in engines. One of the major uses of the Taylor microlength scale is in engine combustion modelling. In particular, the model of Tabaczynski *et al.*[4] uses the Taylor microlength scale to define a characteristic turbulent reaction time τ_c, defined as:

$$\tau_c = \lambda_x / S_L \tag{16}$$

This implies that the unburned mixture undergoes laminar burning over the length scale λ_x. It is important to note that there is a distribution of macro- and microscales in a turbulent field, and that the use of one value for λ_x represents some mean value. There is also indirect evidence that the microlength scale affects the diffusion of air and fuel in diesel combustion. In a study of directed versus helical ports, researchers observed that, for identical swirl rates, the directed (higher inlet velocity) ports gave lower soot emissions than the helical ports (Monaghan & Pettifer[18]). It is conjectured that this is a result of the smaller microscales produced by the directed ports.

The most commonly measured microscale is the Taylor microtime scale. This measurement can be made with an uncalibrated hot-wire by either measuring the autocorrelation coefficient, R_t (Dent & Salama[5] and Lancaster[16]) or by performing peak-to-peak analysis of the uncalibrated hot-wire data (Haghgooie *et al.*[19]). The typical behaviour of the microtime scale with engine speed is shown in Figs 21 and 22. If the microtime scale is related to the integral time scale and u' through the classical relationships for isotropic equilibrium turbulence, then λ_t should vary as $(u')^{-1\cdot5}$ or as $\text{rpm}^{-1\cdot5}$, since u' is proportional to rpm. This behaviour is shown on Fig. 21. Typical values for the microtime scale range from 0·05 to 0·2 ms and are typically from 0·25 to 0·3 of the integral time scale, depending on the level of turbulent intensity generated during the intake process. In order to calculate diffusion

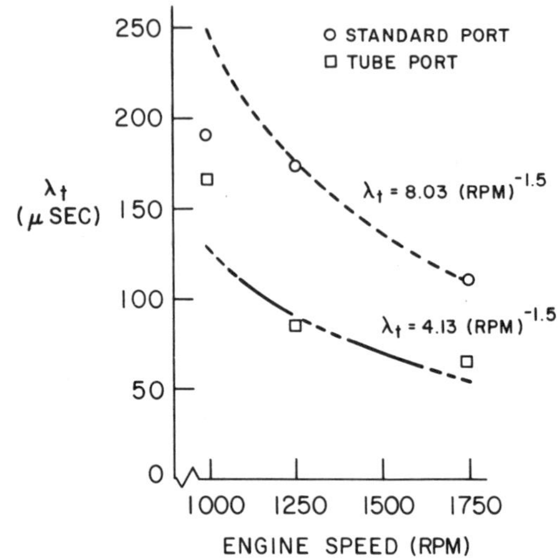

Fig. 21. Turbulent microtime scale for two inlet port configurations as a function of engine speed (from Haghgooie *et al.*[18]).

times or chemical reaction times for processes that occur over the small scales of turbulence, it is necessary to know the magnitude and behaviour of the microlength scale.

The Taylor microlength scale is a difficult quantity to measure directly. Generally, the microtime scale is measured and the microlength scale inferred through the Taylor hypothesis such that:

$$\lambda_x = \lambda_t \langle U \rangle \tag{17}$$

This approach is acceptable where a well defined mean flow exists, such as in engines having swirl ports: however, in low swirl engines

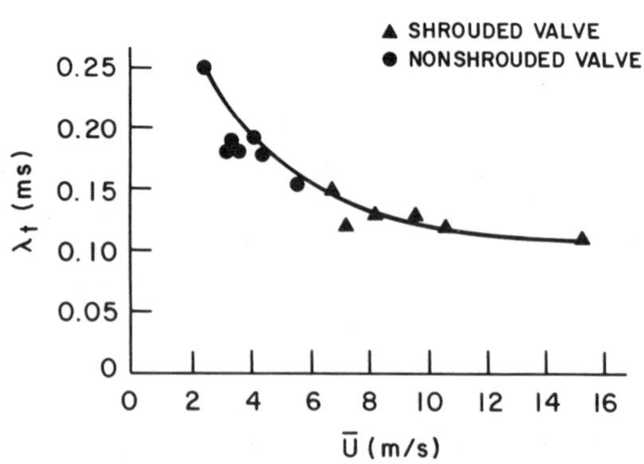

Fig. 22. Turbulent microtime scale versus mean velocity for the 45° period prior to TDC compression (from Lancaster[16]).

such a technique is not accurate because a zero mean flow can exist. In a firing engine, two alternative approaches can be used to measure the microlength scale, either directly using a micro-Schlieren or shadowgraph technique or indirectly using an ion probe and a measure of the flame-front velocity. A typical micro-shadowgraph of the flame structure in a homogeneous charge engine is shown in Fig. 23. Using such micro-shadowgraphs Smith[20] compared measured versus calculated values of the Taylor microlength scale versus engine speed and found good correlation between the photographs and the calculated value based on a measured u' and an assumed integral scale of 3·8 mm. The calculated Taylor microlength scale is shown in Fig. 24. An interesting feature of the data in Fig. 24 is that it obeys the classical scaling law for the microscale. Basically, the ratio of the micro- to integral length scale should vary inversely with the square root of the turbulent Reynolds number. Since the integral scale was constant and u' is proportional to rpm, then $\lambda_x = C(\text{rpm})^{-0\cdot5}$, which is the approximate behaviour of the data in Fig. 24. The microtime scale results from ion probe measurements were shown previously in Fig. 21. Using the flame speed, as the mean velocity, a microlength scale can be calculated using the Taylor hypothesis (Fig. 25). Such a measurement should be valid for swirling or non-swirling flows since the mean velocity is determined by the rate of flame propagation and not from the cold flow data. By comparing the results for the macro- and microlength scales, the ratio of macro- to microlength scales is of the order of 4·0. This ratio varies with engine speed and load, since the macrolength scale is relatively insensitive to engine operating condition and the microscale is dependent on the turbulent Reynolds number, which is both a function of engine speed and load.

3.4.1. Energy Spectra

The distribution of turbulent kinetic energy with respect to frequency is an important relationship to aid our understanding of the engine turbulent flow field. A large number of measurements do not exist. However, several researchers, dating back to the early work of Semenov,[7] have measured the energy spectrum with respect to time in the engine cycle. This measurement is quite important to researchers attempting to formulate models of the in-cylinder turbulent flow, since they can determine when various assumptions are valid. We will discuss these assumptions and their regions of validity in this section.

The energy spectra data that will now be discussed are plots of the

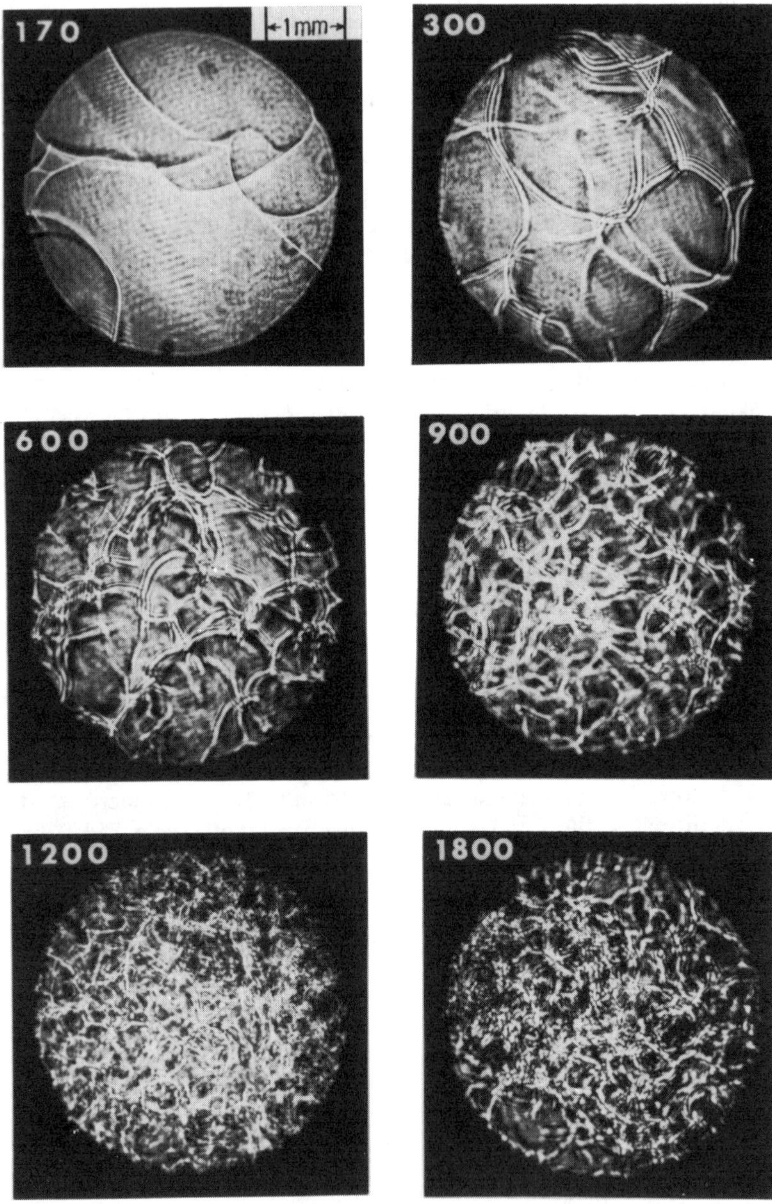

Fig. 23. Microshadowgraphs of flames propagating toward the viewer, showing increased wrinkling as the engine speed increases. Engine rpm is indicated in the upper left-hand corner of each frame (from Smith[20]).

distribution of turbulent kinetic energy with respect to frequency (rather than wavenumber, since scales are much harder to measure in an engine). One of the earliest measurements of this type was performed by Semenov[7] and is shown in Fig. 26. This figure clearly shows the shift of turbulent kinetic energy to lower and lower frequencies during the intake process; the turbulent kinetic energy at 60° after TDC intake having the highest frequencies and that at 180° after TDC having the lowest frequencies. This is to be expected since the high shear flows of the intake jet are diminishing and the rate of turbulent decay is highest when the frequencies are the highest. Also shown in Fig. 26 is the distribution of turbulent kinetic energy at TDC

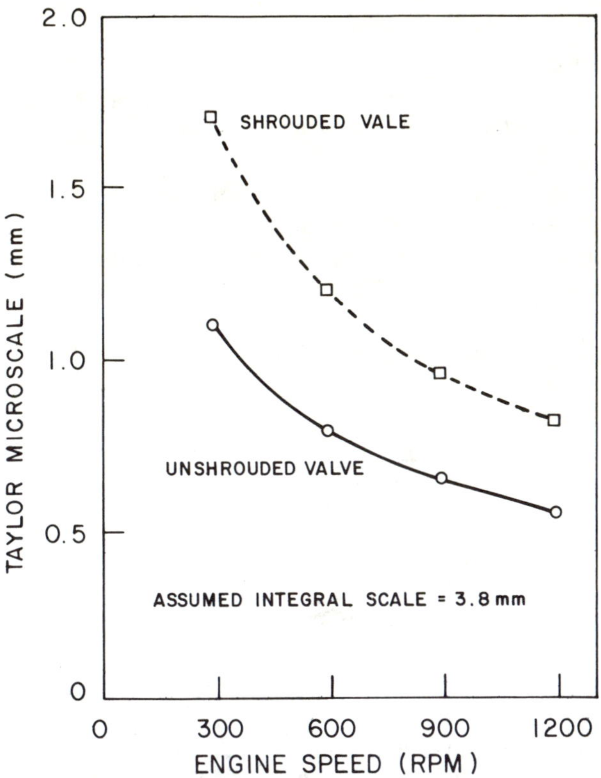

Fig. 24. Calculated Taylor microscale as a function of engine speed (from Smith[20]).

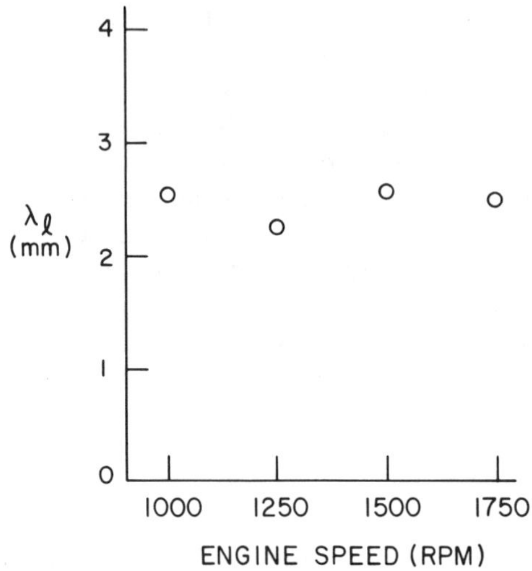

Fig. 25. Microlength scale, calculated from microtime scale data; versus engine speed for a standard inlet port (from Haghgooie *et al.*[19]).

compression. At TDC compression a significant fraction of the turbulent kinetic energy is at the lowest frequencies.

Semenov also varied the compression ratio and measured the energy spectrum at TDC compression (Fig. 27). The interesting feature of this figure is that the frequency of the turbulent kinetic energy shifts to higher frequencies as the compression ratio increases. If the turbulent

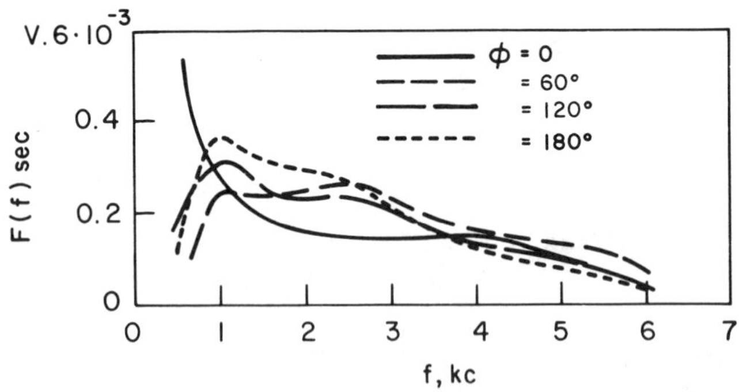

Fig. 26. Turbulent energy distribution functions $F(f)$ versus frequency, f. $\phi = 0°$ is for TDC compression. $\phi = 60°$, 120° and 180° curves are after TDC intake. Compression ratio = 6 and engine speed = 900 rpm (from Semenov[7]).

flow was governed by the shape of the vessel, i.e. the integral scale decreased as compression ratio increased and u' did not change substantially, then the frequency would increase. The set of curves given in Figs 26 and 27 tend to lend credence to the supposition that the turbulent scales of the flow are being governed by the chamber dimensions near TDC compression. A similar set of data was presented by Dent & Salama[5] and is shown in Fig. 28. This figure also shows the shift to lower frequencies as time progresses from intake to TDC of compression. In order to understand the turbulent decay process in engines in more detail, Tsuge *et al.*[21,22] studied the turbulent flow produced by a grid pulled through a cylinder. Their results are quite similar to Semenov's and Dent & Salama's. Tsuge *et al.* characterized three regimes of the flow: (i) the pre-relaxed region where the jet flows are dominant; (ii) the transition region; and (iii) a relaxed region where the shape of the vessel controls the scales of the turbulence and hence the decay process. It is clear that all three regions exist in engines. However, it appears that near TDC the flow does relax to the chamber shape. This is shown by the low-frequency content of the turbulent kinetic energy near TDC and the results of Fraser *et al.*,[3] previously shown in Fig. 20, showing that the turbulence scale is proportional to the chamber height in the vicinity of TDC.

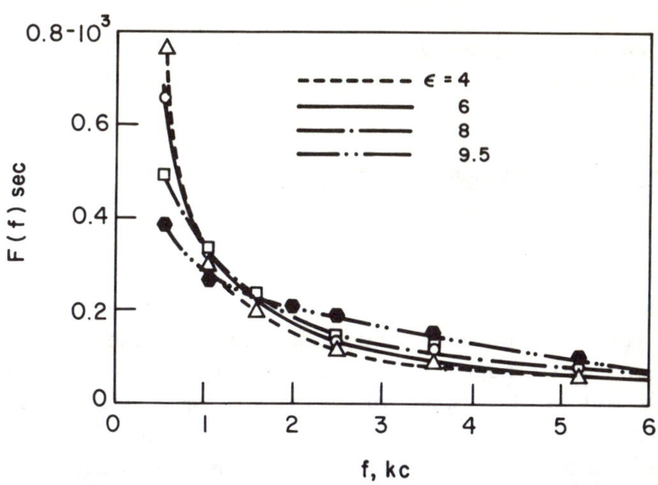

Fig. 27. Turbulent energy distribution $F(f)$ at TDC compression for various compression ratios (4, 6, 8 and 9·5); engine speed = 900 rpm (from Semenov[7]).

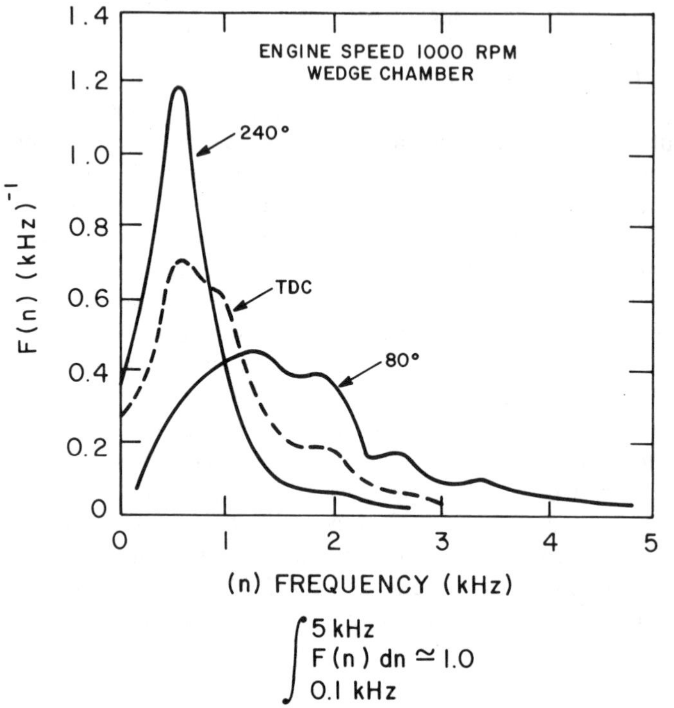

Fig. 28. Turbulent energy distribution $F(n)$ at three times in the engine cycle (from Dent & Salama[5]).

4. CONCEPTUAL MODELLING OF ENGINE TURBULENCE

4.1. Introduction

In order to develop an understanding of the basic features of engine turbulence, it is useful to develop some simple heuristic models for the turbulence decay and production processes. In this section, several simple models will be developed for the generation and decay of turbulence during the intake and compression strokes of the engine cycle. The basic thrust of these models is not accuracy of prediction, but rather for understanding the phenomena. The goal is an explanation of why the flow responds to various geometric and flow-related parameters.

4.2. The Intake Process

One of the simplest ways to view the intake event of an engine is to assume that the cylinder is empty at a time $t<0$ and is filled instantaneously with an initial level of turbulent kinetic energy at $t=0$. This turbulent kinetic energy then decays until the compression process begins. Hence, the turbulent kinetic energy undergoes a pure decay process and the governing equation can be written as:

$$\mathrm{d}K/\mathrm{d}t = -\varepsilon \tag{18}$$

where

$$\varepsilon = AK^{3/2}/L \tag{19}$$

and

$$K = K_1, \quad L = L_1 \qquad \text{at } t=0 \tag{20}$$

It is necessary to determine good estimates for K_1 and L_1, as well as for the constant A in eqn (19). From basic turbulence theory for shear flows, the constant A has a value of approximately 0·1 (see Morel & Mansour[17]). If the turbulent diffusivity, $K^{1/2}L$, is assumed to be constant, then the dissipation term can be written as

$$\varepsilon = 0{\cdot}1K^2/(K_1^{1/2}L_1) \tag{21}$$

In order to determine $K_1^{1/2}L_1$, it is necessary to model both the initial length scale and the turbulent kinetic energy.

The initial value of K should be related to the mean kinetic energy that is inducted into the cylinder during the intake stroke. Since it has been determined that the intake process is the main production mechanism for turbulence in engines, and since the turbulence intensity is generally some fraction of the intake shear flow, the initial turbulent kinetic energy can be estimated by first determining the total mean kinetic energy inducted and then assuming that a 'constant' fraction is transferred into turbulence. If U_p is defined as the mean piston speed, A_p as the area of the piston, S as the engine stroke and A_v as the effective area of the inlet valve, then the mean velocity past the valve, U_v, can be written as

$$U_v = A_p/A_v U_p \tag{22}$$

where

$$U_p = S(\text{rpm})/30 \tag{23}$$

Since A_p/A_v has a value near 10 for typical production engines and since the turbulence intensity is approximately equal to 0·1 times the shear velocity past the valve, the initial turbulence intensity is approximately equal to the mean piston speed, U_p. Hence, the initial kinetic energy can be written as

$$K_1 = \tfrac{3}{2}[0{\cdot}1A_p/A_v U_p]^2 \tag{24}$$

In order to solve eqn (18) it is necessary to model L during the intake process. During the intake process, the turbulent jet, which is created at the intake valve by the piston motion, is rolled up into a vortex that fills the space between the intake valve and the piston top. This jet flow process differs considerably from that of a turbulent jet into an open atmosphere. For the classical free turbulent jet, the width of the jet is independent of the Reynolds number (jet velocity) of the flow and the width of the jet is assumed to be proportional to the turbulent integral scale. In a closed vessel, such as an engine cylinder, the jet is 'forced' to fill the cylinder in some fashion and hence must roll-up on itself, resembling a series of 'shear layers' more than a single free jet. The remaining task is to calculate the effective width of the jets in order to estimate the integral scale. To calculate the width of the shear layers, it is necessary to model the size of the recirculation zone as a function of engine geometry. From flow visualization studies, one can observe that the vortex size depends on the instantaneous stroke for the case where the stroke is less than the bore, and on the bore diameter once the stroke exceeds the cylinder bore. Hence, the assumptions will be made that the recirculation zone cross-sectional area, A_R, can be modelled as

$$A_R = \pi S^2/4, \quad S < B; \qquad A_R = \pi B^2/4, \quad B < S \tag{25, 26}$$

The length of the jet, L_j, can be calculated by multiplying the average jet velocity by the length of time for the intake stroke. Using U_v for the jet velocity and noting that the time for the intake stroke is $t = 30/\text{rpm}$, the length of the jet can be expressed as

$$L_j = (A_p/A_v)U_p \times 30/\text{rpm} = (A_p/A_v)S \tag{27}$$

By equating the area of the jet, $L_j L_i$, with the area of the recirculation zone and noting that the integral scale is generally modelled as one-quarter of the width of the jet, the initial integral scale can be expressed as

$$L_i = (\pi/16)(A_v/A_p)S \qquad S < B \tag{28}$$

$$L_i = (\pi/16)(A_v/A_p)B^2/S \qquad B < S \tag{29}$$

Using eqns (24), (28) and (29), the initial turbulent diffusivity can now be estimated, and has the following value

$$K_i^{1/2}L_i = 0{\cdot}024S^2 \times \text{rpm}/30, \qquad S < B \tag{30}$$

$$K_i^{1/2}L_i = 0{\cdot}024B^2 \times \text{rpm}/30, \qquad B < S \tag{31}$$

Equation (18), which governs the decay of turbulent kinetic energy, can be written as

$$\mathrm{d}K/\mathrm{d}t = -4K^2/[\beta^2 S^2 \times \text{rpm}/30] \tag{32}$$

where $\beta = 1$ for $S < B$ and $\beta = B/S$ for $B < S$. Solving eqn (32) in terms of crank-angle degrees, where $t = \theta/(6\text{rpm})$, the expression for the turbulent kinetic energy is given by

$$K/K_i = 1/[1 + 0{\cdot}06(\theta/180)(A_p/A_v)^2\beta^{-2}] \tag{33}$$

Note that it will be assumed that wherever the term β occurs the correct value will be inserted, dependent upon bore : stroke ratio. If we assume that all the kinetic energy is placed into the cylinder at mid-stroke and we calculate the final value at BDC ($\theta = 90°$) in terms of the turbulent intensity u', then

$$u'_{\text{BDC}} = \{0{\cdot}1(A_p/A_v)S \times \text{rpm}/30\}/[1 + 0{\cdot}03(A_p/A_v)^2\beta^{-2}]^{1/2} \tag{34}$$

Example 1. To illustrate the behaviour of the turbulence at BDC for a typical engine and a simple modification which is the result of halving the intake lift, consider the following example. An engine having an 85 mm bore and 85 mm stroke operates at 1500 rpm. What is the initial turbulence intensity and the corresponding turbulence intensity at the end of the intake process for lifts of 10 mm and 5 mm respectively? Assume that the intake valve has a 42 mm diameter, 45° seat angle and a discharge coefficient of 0·8.

Step 1: calculate the effective *mean* valve area:

$$A_v = C_d \pi d_v \cos 45\, (2/\pi)L_v$$

where C_d is the discharge coefficient, d_v is the valve diameter, L_v is the valve lift and the first cos 45 yields the area normal to the valve seat. The $(2/\pi)L_v$ yields the mean valve lift, assuming a sinusoidal valve lift profile.

Case 1—10 mm valve lift:

$$A_v = 0{\cdot}8 \times 42 \times 0{\cdot}707 \times 2 \times 10$$
$$= 475{\cdot}1 \text{ mm}^2$$
$$A_p = 5674{\cdot}5 \text{ mm}^2$$
$$A_p/A_v = 11{\cdot}94$$

Calculate $U_p = S(\text{rpm})/30$;

$$U_p = 0{\cdot}085 \times 1500/30 = 4{\cdot}25 \text{ m s}^{-1}$$

Calculate U_v:

$$U_v = (A_p/A_v)U_p = 50{\cdot}76 \text{ m s}^{-1}$$

Calculate u_i':

$$u_i' = 0{\cdot}1U_v = 5{\cdot}08 \text{ m s}^{-1}$$

Then

$$u_f' = u_i'/(1 + 0{\cdot}03(A_p/A_v)^2)^{1/2}$$
$$= 5{\cdot}08/(1 + 4{\cdot}277)^{1/2} = 2{\cdot}21 \text{ m s}^{-1}.$$

Note that in 90 crank-angle degrees the turbulent intensity was decreased by more than half the initial value.

Case 2—$L_v = 5$ mm:

$$A_p/A_v = 21{\cdot}5$$

$$u_i' = 10{\cdot}16 \qquad \text{or simply twice the 10 mm lift case}$$

$$u_f' = 10{\cdot}16(1 + 0{\cdot}03(A_p/A_v)^2)^{1/2} = 10{\cdot}16/(1 + 17{\cdot}11)^{1/2}$$
$$= 2{\cdot}39 \text{ m s}^{-1}$$

Note that although twice the initial turbulence of the 10 mm valve lift case was introduced into the cylinder, only 8% more exists at the end of the intake stroke, assuming that the turbulent length scale is proportional to A_v/A_p as given by eqn (28). This example illustrates the non-linear decay of turbulence, showing that although twice as much u' is introduced due to the valve restriction, the final percentage difference at BDC for the two cases is almost negligible.

4.3. The Compression Process

During the compression process the turbulent kinetic energy undergoes both decay and production. The production of turbulent kinetic energy depends on the shear flows created during the compression process. These shear flows are produced by the combustion chamber shape and its interaction with the mean flow field. If there is no initial swirl and the chamber is of a simple disc geometry, the production of turbulent kinetic energy is governed only by the piston motion. Then the turbulent kinetic energy behaves according to the equation:

$$\frac{dK}{dt}=\frac{2K}{3\rho}\frac{d\rho}{dt}-\varepsilon \tag{35}$$

The first term in eqn (35) is a result of the strain produced by the piston motion which, through the continuity equation, is directly related to the change in density. The second term in eqn (35) is the dissipation of turbulent kinetic energy, as discussed previously. In order to solve eqn (35) analytically, it will be necessary to model the production term with an average value for $(1/\rho)\,d\rho/dt$.

The average value of $(1/\rho)\,d\rho/dt$ over the compression process is defined as

$$\langle(1/\rho)\,d\rho/dt\rangle=(1/T)\int(1/\rho)\,d\rho/dt \tag{36}$$

where T is the time for the compression process and is given by 30/rpm. After integration, eqn (36) becomes

$$\langle(1/\rho)\,d\rho/dt\rangle=\mathrm{rpm}/30\,\ln(\rho_f/\rho_i) \tag{37}$$

or since ρ_f/ρ_i is the engine compression ratio, cr,

$$\langle(1/\rho)\,d\rho/dt\rangle=\mathrm{rpm}/30\,\ln(cr) \tag{38}$$

Note that eqn (38) states that higher compression ratio engines should have more turbulent production. Substituting eqn (38) into eqn (35) and using the expression for dissipation as defined by eqns (30) and (31), the governing equation for the kinetic energy during compression becomes

$$dK/dt=(\mathrm{rpm}/45)K\ln(cr)-120K^2/(S^2\beta^2\times\mathrm{rpm}) \tag{39}$$

Equation (39) can be solved analytically as a function of time. Since the combustion event occurs near TDC, the expressions for K_{TDC} and u'_{TDC} are given, with reference to the values at BDC, as

$$K_{\mathrm{TDC}}/K_{\mathrm{BDC}}=\tfrac{1}{6}\ln(cr)/[\tfrac{1}{6}\ln(cr)\exp(-\tfrac{2}{3}\ln(cr)) + K_{\mathrm{BDC}}/U_p^2\beta^{-2}(1-\exp(-\tfrac{2}{3}\ln(cr)))] \tag{40}$$

and

$$u'_{\text{TDC}}/u'_{\text{BDC}} = [K_{\text{TDC}}/K_{\text{BDC}}]^{1/2} \tag{41}$$

where it is noted that $K = \frac{3}{2}u'^2$. To illustrate the effect of production and dissipation on the final value of u' near TDC, an example problem will be considered.

Example 2. Using the data and results of Example 1, calculate the final value of u' at TDC for both the 5 mm and 10 mm valve lift cases, assuming a compression ratio of 9.

10 mm valve lift:

$$u'_{\text{BDC}} = 2{\cdot}21 \text{ m s}^{-1}, \qquad U_p = 4{\cdot}25 \text{ m s}^{-1}$$

$$\tfrac{1}{6}\ln(cr) = 0{\cdot}366, \qquad \exp(-\tfrac{2}{3}\ln(cr)) = 0{\cdot}231$$

$$K_{\text{BDC}}/U_p^2 = \tfrac{3}{2}(2{\cdot}21)^2/(4{\cdot}25)^2 = 0{\cdot}4056$$

Substituting the above values into eqns (40) and (41) one obtains

$$u'_{\text{TDC}}/u'_{\text{BDC}} = [0{\cdot}366/\{(0{\cdot}366 \times 0{\cdot}231) + (0{\cdot}4056 \times 0{\cdot}769)\}]^{1/2}$$
$$= 0{\cdot}961$$

or

$$u'_{\text{TDC}} = 2{\cdot}12 \text{ m s}^{-1}$$

5 mm valve lift:

$$u'_{\text{BDC}} = 2{\cdot}39 \text{ m s}^{-1}$$

$$K_{\text{BDC}}/U_p^2 = \tfrac{3}{2}(2{\cdot}39)^2/(4{\cdot}25)^2 = 0{\cdot}4744$$

$$u'_{\text{TDC}}/u'_{\text{BDC}} = [0{\cdot}366/\{(0{\cdot}366 \times 0{\cdot}231) + (0{\cdot}4744 \times 0{\cdot}769)\}]^{1/2}$$
$$= 0{\cdot}902$$

or

$$u'_{\text{TDC}} = 2{\cdot}16 \text{ m s}^{-1}$$

Note that at the end of the compression process the turbulence levels for the two valve lift cases are essentially identical. This indicates that slightly more decay occurred for the low lift case than for the higher lift.

In addition, this example illustrates that the decay approximately equals the production during the compression process, as was previously discussed when the data for turbulence intensity was reviewed. One more observation can be made with respect to u' at TDC. Since u' at TDC is approximately equal to its value at BDC, and since u' at

BDC is proportional to the engine rpm from eqn (34), the linear relationship of u' with rpm shown in Fig. 11 can be explained. The variation of slope from engine to engine is more complex and depends heavily on the bore:stroke ratio, displacement per cylinder and the initial length scale of the turbulence. The initial length scale is a function of the inlet system and the flow field it creates during the intake process. Changes in the initial length scale can significantly affect the final turbulence levels observed at TDC. This can also cause the slope of u' versus rpm to change between engine configurations.

A final caution concerning the validity of the data presented in Fig. 11 is that the r.m.s. velocity presented in this data includes cyclic variations in the mean flow as well as turbulence. This fact can lead one to believe that turbulence in one geometry is higher or lower than that of another. However, cyclic variations may be the only cause of this discrepancy, which will lead to an erroneous conclusion concerning the rate of combustion.

4.4. Swirl and Combustion Chamber Effects

The introduction of swirl via modifications to the inlet port and chamber geometry and the further modification of the mean flow field by the combustion chamber shape will result in a modification of the turbulent field near the time of combustion initiation. Engine designers have historically proposed compact chambers with and without swirling inlet flows to speed up the combustion process. In this section a brief discussion will be presented on the effects of squish, chamber shape and swirl in the production process for turbulent kinetic energy. The discussion will be qualitative, and the reader is encouraged to read the references cited for details in the modelling of these processes.

One technique used to increase the rate of flame propagation in spark-ignition engines is to design a *compact* chamber. This generally implies that the chamber will have significant squish area as well as a more spherical volume in the vicinity of the spark plug. Combustion is observed to proceed at a faster rate than for a disc chamber in these compact chambers, due to at least two phenomena. First, the initial flame kernel is allowed to grow in a spherical fashion for a greater length of time. Second, the compactness of the chamber produces more turbulence due to the squish-generated flows and the increased fluid shear produced by increasing the initial level of swirl via conservation of angular momentum and by the interaction of the squish-generated flows with the swirling flows. The degree to which the turbulence is enhanced depends heavily on the amount of squish created in the chamber (compactness of the chamber) and the initial swirl level. These effects are difficult to model in an heuristic manner since the mean flow field character changes and is dependent upon the amount of squish and swirl. An example of the change in flow field that can occur due to the interaction of swirl and squish was shown in Fig. 9. From this figure we observe that, for moderate levels of squish and simple chamber geometries, the recirculation zone is approximately the same size with or without swirl, even though it has a different flow direction. For complex, 'undercut-bowl-in-piston', geometries the flow becomes considerably more complex. Hence, the heuristic type of model might have some chance of success for relatively low-squish compact chambers where it can be assumed that the flow patterns do not undergo drastic change. Borgnakke *et al.*[23] modelled the swirl and squish effects on the turbulent kinetic energy production. These terms would be added to eqn (35) for the rate of change of turbulent kinetic energy. The production term for the squish-generated flows is given by

$$p_{\mathrm{Sq}} = (0{\cdot}1 V_{\mathrm{Sq}})^2 \tag{42}$$

and the production term for swirl is given by

$$P_{\mathrm{Sw}} = \nu_t(\partial V_\theta/\partial r - V_\theta/r)^2 \tag{43}$$

where V_θ is the tangential velocity and ν_t is the turbulent diffusivity. The implications of eqn (43) are that the production of turbulent kinetic energy in swirling flows occurs due to the deviation of the flow from solid-body rotation. This is the only way fluid shear occurs. Therefore, if the chamber geometry causes non-solid-body rotation it has the opportunity for creating turbulence. Equation (42) is more straightforward; the turbulence is produced simply by the shear velocities caused by the squish flows.

A critical step in the modelling process is to postulate what happens to the integral scale when high degrees of squish or swirl are introduced into the chamber, because this parameter 'controls' the dissipation of the turbulence. The length scale is much more difficult to model than the mean flows, and the heuristic modeller often refers to more complex models or flow visualization experiments to gain insight into the process. Using the example of modelling swirl turbulence production, one can develop an appreciation of the role of the integral scale, L, in the dissipation process. In a recent paper, Davis *et al.*[24] developed an heuristic model for the contribution of swirl to the turbulent kinetic energy via the dissipation of the mean angular momentum. In this model the integral scale was assumed to be the same for swirling or non-swirling flows during the intake stroke. In Fig. 29 are shown the flow patterns for swirling and non-swirling flows computed from a multi-dimensional model (El-Tahry[25]). It is apparent that the recirculation zone is larger for the swirling flows, implying that the integral scale would be larger and hence less dissipation would occur throughout the process and higher turbulence levels would be

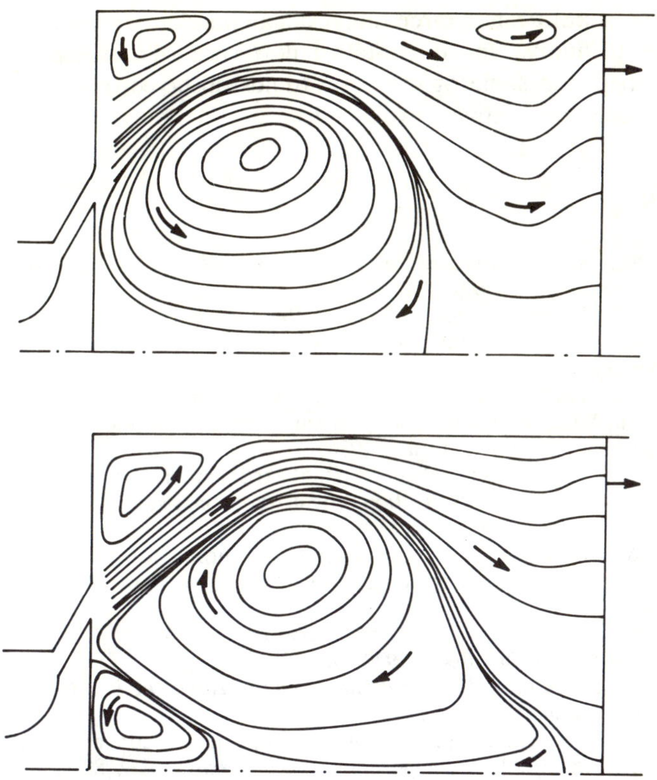

Fig. 29. Streamline contours of swirling and non-swirling flows (from El-Tahry[25]).

present with swirl. This is a simplification, of course: however, this type of insight and observation will lead us to a better understanding of turbulent flows in engines.

5. SUMMARY

This chapter has presented a rudimentary treatment of engine turbulence which is intended to give the engineer insight into the complex processes that occur in an engine cylinder. As the reader may have surmised, engine turbulence is a complex phenomenon which is currently being actively studied by researchers. Our understanding is still at an elementary level, and it will only be improved by the development of sophisticated three-dimensional simulations coupled with extensive qualitative and quantitative flow visualization experiments. It is hoped that some of the basic trends observed in engine operation have been qualitatively explained.

REFERENCES

1. Bradshaw, P., *An Introduction to Turbulence and its Measurement.* Pergamon Press, Oxford, 1971.
2. Witze, P. O., Hot wire measurements of the turbulent structure in a motored spark ignition engine. 14th AIAA Aerospace Sciences Conference, January 1976.
3. Fraser, R. A., Felton, P. G., Bracco, F. V. & Santavicca, D. A., Preliminary turbulence length scale measurements in a motored IC engine. SAE paper 860021, February 1986.
4. Tabaczynski, R. J., Ferguson, C. R. & Radhakrishnan, K., A turbulent entrainment model for spark-ignition engine combustion. SAE paper 770647, October 1977.
5. Dent, J. C. & Salama, N. S., The measurement of turbulence characteristics in an internal combustion engine cylinder. SAE paper 750886, October 1975.
6. Tennekes, H. & Lumley, J., *A First Course in Turbulence.* The MIT Press, Cambridge, Massachusetts, 1972, pp. 263–5.
7. Semenov, E. S., *Studies of Turbulent Gas Flow in Piston Engines.* NASA Technical Translation F-97, 1963.
8. Witze, P. O., A critical comparison of hot-wire anemometry and laser doppler velocimetry for I.C. engine applications. *Trans. SAE,* **89** (1980) 711–40 (paper 800132).
9. Arcoumanis, C., Bicen, A. F. & Whitelaw, J. H., Squish and squish–swirl interactions in motored model engines. *J. Fluid Engng,* **105** (1983) 105.

10. Shimamoto & Akiyama, A study of squish in open combustion chambers of a diesel engine. *Bull. Jpn. Soc. Mech. Engrs,* **13**(63) (1970) 1096–103.
11. Bopp, S., Valfidis, C. & Whitelaw, J. H., The effect of engine speed on the TDC flowfield in a motored reciprocating engine. SAE paper 860023.
12. Rask, R. B., Comparison of window, smooth-ensemble, and cycle-by-cycle data reduction techniques for laser doppler anemometer measurements of in-cylinder velocity. In *Fluid Mechanics of Combustion Systems,* eds T. Morel, R. P. Lohmann & J. Rackley, ASME, 22–23 June 1981.
13. Hall, M. J., Bracco, F. V. & Santavicca, D. A., Cycle-resolved velocity and turbulence measurements in an IC engine with combustion. SAE paper 860320, 24–28 February 1986.
14. Fansler, T. D., Laser velocimetry measurements of swirl and squish flows in an engine with a cylindrical piston bowl. SAE paper 850124, February 1985.
15. Kondoh, T., Fukumoto, A., Ohsawa, K. & Ohkubo, Y., An assessment of a multi-dimensional numerical method to predict the flow in internal combustion engines. SAE paper 850500, February 1985.
16. Lancaster, D. R., Effects of engine variables on turbulence in a spark-ignition engine. SAE paper 760159.
17. Morel, T. & Mansour, N. N., Modelling of turbulence in internal combustion engines. SAE paper 820042, 1982.
18. Monaghan, M. L. & Pettifer, H. F., Air motion and its effect on diesel performance and emissions. SAE paper 810255, February 1981.
19. Haghgooie, M., Kent, J. C. & Tabaczynski, R. J., Turbulent time-scale measurements in a spark-ignition engine using hot wire anemometry and fast response ion probes. In *Flows in Internal Combustion Engines,* ed. T. Uzkan. ASME Winter Annual Meeting, Phoenix, Arizona, 14–19 November 1982.
20. Smith, J. R., The influence of turbulence on flame structure in an engine. In *Flows in Internal Combustion Engines,* ed. T. Uzkan. ASME Winter Annual Meeting, Phoenix, Arizona, 14–19 November 1982.
21. Tsuge, M., Kido, H., Kato, K. & Nomiyama, Y., Decay of turbulence in a closed vessel. *Bull. Jpn. Soc. Mech. Engrs,* **17** (1974).
22. Tsuge, M., Kido, H., Kato, K. & Nomiyama, Y., Decay of turbulence in a closed vessel. *Bull. Jpn. Soc. Mech. Engrs,* **16** (1973).
23. Borgnakke, C., Davis, G. C. & Tabaczynski, R. J., Predictions of in-cylinder swirl velocity and turbulence intensity for an open chamber cup in piston engine. *Trans. SAE,* **90** (paper no. 810224).
24. Davis, G. C., Mikulec, A., Kent, J. C. & Tabaczynski, R. J., Modelling the effect of swirl on turbulence intensity and burn rate in S.I. engines and comparison with experiment. SAE paper 860325, 1986.
25. El-Tahry, S. H., A comparison of three turbulence models in engine-like geometries. *Proceedings of International Symposium on Diagnostics and Modelling of Combustion in Reciprocating Engines, COMODIA 85, Jpn. Soc. Mech. Engrs,* Tokyo, Japan, September 1985, pp. 203–13.

9

Combustion in Gasoline Engines

D. Bradley

Department of Mechanical Engineering, University of Leeds, Leeds, UK

1. Introduction
   1.1. Control of engine combustion
   1.2. Turbulent burning
   1.3. Laminar burning velocity
   1.4. Straining of laminar flames
   1.5. Flammability limits
2. Turbulent burning
   2.1. Steady state
   2.2. Unsteady state
   2.3. Correlation of experimental turbulent burning velocities
   2.4. Experimental turbulent explosion bombs
   2.5. Mass burning rate
3. Ignition
   3.1. The spark
   3.2. Flame straining and spark-assisted propagation
4. Combustion and engine performance
   4.1. Initial burning and cyclic variations
   4.2. Burning during compression ahead of the flame
   4.3. Burning at different engine speeds
   4.4. Engine performance
   4.5. Autoignition and knock

NOTATION

A Area of flame surface
C_p Specific heat at constant pressure
D Diffusion coefficient

$E(F)$	Defined by eqn (17)
f	Volume fraction of completely burnt gas and frequency
$\bar{F}$	Dimensionless frequency $= f\tau_a$
$\bar{F}_k$	Reciprocal of dimensionless time $= \tau_a/t_k$
i	Turbulence intensity $= u'/U$
I	Normalized CH emission intensity
k	Thermal conductivity
K	Karlovitz stretch factor
l_θ	Size of energy-containing eddies
L	Integral length scale of turbulence
Le	Lewis number
q	Volumetric heat release rate
$q_{\max}$	Maximum volumetric heat release rate
r	Flame front radius
R_L	Turbulent Reynolds number
s	R.m.s. strain rate
s'_k	Developing strain rate
S'_k	Dimensionless developing strain rate $= s'_k\tau_a$
$S(f)$	Power spectral density function
$\bar{S}(\bar{F})$	Dimensionless power spectral density function $= S(f)/u'^2\tau_a$
S_s	Flame speed
t	Time
$\bar{t}_c$	Dimensionless chemical lifetime
t'	Fixed time
t_k	Elapsed time
$\bar{t}_k$	Dimensionless elapsed time $= t_k/\tau_a$
$\bar{t}_s$	Dimensionless strain lifetime
T	Temperature
T_a	Activation temperature
u	Velocity
u'	R.m.s. turbulent velocity
u'_k	Effective r.m.s. turbulent velocity
u''	R.m.s. value of u'
u_l	Laminar burning velocity
u_{ls}	Laminar burning velocity of strained flame
u_t	Turbulent burning velocity
u_{tk}	Developing turbulent burning velocity
u_{tr}	Mass burning velocity
U	Mean velocity
x	Distance along x axis

α Strain rate
$\bar{\alpha}$ Dimensionless geometric strain rate $= \alpha\tau_a$
β Zeldovich number
δ_l Reaction zone thickness of laminar flame
ε Specific rate of dissipation of turbulent energy
ε_k Effective specific rate of dissipation of turbulent energy
θ Dimensionless temperature rise $= (T - T_u)/(T_b - T_u)$
θ_c Distance to centroid of area
λ Taylor microscale
Λ Burning velocity eigenvalue
ρ Density
$\bar{\rho}_b$ Mean density of gas behind flame front
τ Turbulent reaction time
τ_a Autocorrelation time
ν Kinematic viscosity

Subscripts
b Adiabatic burnt gas conditions
u Unburnt gas conditions

1. INTRODUCTION

1.1. Control of Engine Combustion

The rate of burning in the cylinder exercises considerable influence upon the power and efficiency of the engine. Combustion should occur close to top centre and be rapid enough to utilize to the fullest possible extent the maximum compression, and yet not be so rapid as to induce strong shock waves in the gas or intolerable impact loadings on bearings and other mechanical components. Inherently, a lean mixture gives a higher thermodynamic efficiency than does a rich one.

It is desirable to control the orderly propagation of the flame front and this must come through the control of the fluid flow, the rates of chemical reaction and the coupling between the flow and the chemistry. Means of achieving this include control of the mixture composition and, possibly, charge stratification within a combustion space. Control of the flow must involve suitable design of the inlet manifold and valves, as well as the piston crown and combustion volume, for all of these influence the flow pattern, swirl and generation of turbulence. Feedback from appropriate sensors can control both the mixture strength and, through the use of swirl control valves in the inlet ports the degree of turbulence.[1]

1.2. Turbulent Burning

The flow field and turbulence parameters can be computed from the mass and momentum conservation equations, in conjunction with an appropriate model of turbulence. The coupling of these to the chemistry is difficult. Ideally, a conservation equation is required for each chemical species and this should account for all possible reactions. Westbrook & Dryer[2] have listed 255 reaction rates relevant to hydrocarbon oxidation and, consequently, the computational task can be formidable, even allowing for the dominance of some reactions. Dixon-Lewis & Islam[3] used 48 reactions in their modelling of the laminar methane–air flame. There has been progress in the chemical kinetic modelling of laminar flames, particularly with hydrogen and methane as fuels, but the higher hydrocarbons still present some problems. Furthermore, the computational effort to solve coupled chemical reaction and flow equations in three-dimensional, time-dependent, turbulent flows within complex geometries is overwhelming.

Simplification is achieved through the use of experimentally measured, or empirically estimated, parameters that embody the chemical reaction rates. Two such parameters are the laminar burning velocity and the profile of volumetric heat release rate against temperature for a laminar flame. The relevance of these parameters to turbulent combustion of the same mixture arises from the rapid chemistry and molecular transport in a flame. At least initially, the introduction of turbulence into the flow merely wrinkles the laminar flame surface, to present a greater surface area and increase the overall burning rate. At higher levels of turbulence when the laminar flame becomes stretched, fragmented and eventually extinguished, the reality is more complex and not fully understood. However, the simplifying concept of a laminar flamelet[4] has been widely used. In a turbulent flow field of premixed gases it enables the heat release rate source term in the energy equation to be expressed in terms of the laminar flame profile of heat release rate versus temperature and a probability density function of the fluctuating temperature.[5,6] The fractional temperature rise provides a single reaction progress variable. While such a modelling approach is attractive for complex, recirculating turbulent flows in which there is no readily discernible flame front, in the case of the SI combustion process which is characterized by a propagating flame front, the concept of a turbulent burning velocity is more appropriate. This latter approach is explored in this chapter, but first it is necessary to review briefly some aspects of laminar burning.

1.3. Laminar Burning Velocity

This basic physico-chemical parameter is the velocity with which the cold premixed gases flow into the laminar flame. A chemical kinetic mathematical model can yield the profiles of all species concentrations and their reaction rates, as well as those of velocity and temperature, with the burning velocity, u_l, as a unique solution. The sum of the rates of enthalpy change associated with each reacting species yields the volumetric heat release rate. The computed temperature and heat release rate profiles through an atmospheric methane–air flame of equivalence ratio 0·84 are shown in Fig. 1.

Mathematical modelling of the laminar flames of the higher hydrocarbons is more difficult, but simplification of the chemical kinetics has made this possible for the alkanes up to octane, C_8H_{18}. The fuel molecule is attacked initially by O, H and OH radicals, generated in the oxy–hydrogen reactions, at known rates to form alkyl radicals. At high temperatures it is possible that both the decomposition of such radicals by alkene elimination and the subsequent alkene oxidation are not rate-limiting processes. Warnatz[7] has used these assumptions to compute premixed laminar flame structures and burning velocities. In reality, alkyl decomposition reactions are complex, but Warnatz replaced these by a simple reaction path leading to the formation of the CH_3 alkyl and the C_3H_6 alkene radicals. The oxidation of C_1 and C_2 hydrocarbons is modelled in some detail, along with the oxidation of H_2 and CO. With such an approach, the burning of the higher

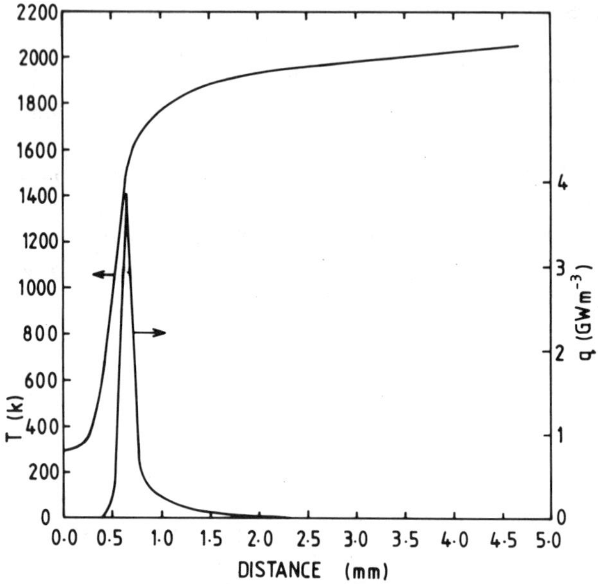

Fig. 1. Computed heat release rate profile through an atmospheric CH_4–air flame of equivalence ratio 0·84 (courtesy of G. Dixon-Lewis).

alkanes is based firmly on earlier work on the kinetics of CO, H_2 and CH_4 oxidation.

Key chain propagation and branching reactions are:

$$H + O_2 \rightarrow OH + O \tag{i}$$

$$OH + O \rightarrow H + O_2 \tag{ii}$$

$$O + H_2 \rightarrow OH + H \tag{iii}$$

$$OH + H \rightarrow O + H_2 \tag{iv}$$

$$OH + H_2 \rightarrow H_2O + H \tag{v}$$

Important three-body recombination reactions are:

$$H + H + M \rightarrow H_2 + M \tag{vi}$$

$$H + OH + M \rightarrow H_2O + M \tag{vii}$$

$$H + O_2 + M \rightarrow HO_2 + M \tag{viii}$$

The HO_2 is consumed in reactions with the reactive radicals O, H and OH. Methane, CH_4, is formed from CH_3, and CH_3 is formed from CH_4. Formaldehyde can arise from

$$CH_3 + O \rightarrow CH_2O + H \quad \text{(ix)}$$

and

$$C_2H_3 + O_2 \rightarrow CH_2O + CHO \quad \text{(x)}$$

This reacts to form CHO which in turn creates CO, which is oxidized subsequently by:

$$CO + OH \rightarrow CO_2 + H \quad \text{(xi)}$$

Values of burning velocities can be computed on the basis of such schemes, but because of modelling uncertainties, experimentally measured values are currently used in practice. Although burning velocities are of primary importance in combustion, experimental errors still arise in the variety of measurement techniques that are used for their determination[8] and there is an appreciable spread in the values from different workers and techniques. Most often, u_l has been measured under atmospheric conditions, and it has not always been possible to apply the most accurate techniques at the high temperatures and pressures relevant to engines.

In Fig. 2 are compared values of u_l obtained by different workers, with different techniques, for iso-octane–air mixtures of different equivalence ratios at atmospheric pressure and 333°K. Results have been reduced to this common datum with the expressions of Metghalchi & Keck[9] for the variations of u_l with temperature and pressure. These workers obtained values at elevated temperature and pressure in bomb explosions, from measurements of pressure rise. Their values are presented along unburned gas isentropes for different initial conditions.

The influence of the changing temperature and pressure is important in engine combustion. Adiabatic compression of the charge leads to an overall increase in u_l: a decrease on account of the increase in pressure, that is more than compensated by an increase due to the increase in temperature. Gülder's bomb measurements[10] were at a pressure of one atmosphere over a range of initial temperatures, during the pre-pressure period, while Hamid[11] employed the double-kernel technique[12] during this phase. Gibbs & Calcote[13] used a burner for their measurements.

The classical, one-dimensional, laminar flame theory of Zeldovich & Frank-Kamenetskii[14,15] relates the laminar burning velocity, u_l, to the

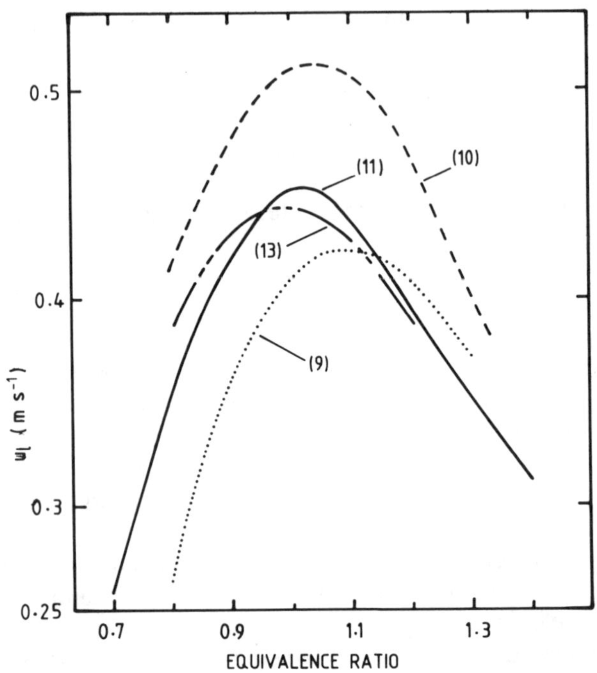

Fig. 2. Iso-octane–air laminar burning velocities at atmospheric pressure and 333 K: references numbered on curves.

profile of volumetric heat release rate, q, against temperature, T. This and subsequent expressions are of the form:

$$u_l = \left\{ \left(\int_{T_u}^{T_b} kq \, \mathrm{d}T \right) \Big/ [C_p^2 \rho_u^2 (T_b - T_u)^2 \Lambda] \right\}^{0 \cdot 5} \quad (1)$$

Suffix 'u' denotes the unburnt and suffix 'b' the adiabatically burnt condition, k is the thermal conductivity, C_p the specific heat at constant pressure, ρ_u the initial density, and Λ the burning velocity eigenvalue in the conservation equation for the flame. The eigenvalue assumes a value of 0·5 in the early theory of Zeldovich & Frank-Kamenetskii. Spalding[16] expressed Λ in terms of the centroid of the area of the heat release rate integral. For a variety of possible heat release rate – temperature profiles he found

$$\Lambda = 0 \cdot 5 - 0 \cdot 6604(1 - \theta_c) - 0 \cdot 4283(1 - \theta_c)^2 \quad (2)$$

where θ_c is the centroid, based on the dimensionless temperature rise, and $\theta = (T - T_u)/(T_b - T_u)$. The present author has found these two equations to give accurate values of burning velocity.

Other expressions have been derived for the eigenvalue, based on the maximum heat release rate, q_{max}, occurring close to T_b and on the expression for it being of Arrhenius form in T_b. The associated activation energy divided by the gas constant gives an activation temperature, T_a. The Zeldovich & Frank-Kamenetskii expression for burning velocity, with $\Lambda = 0{\cdot}5$, becomes[17]:

$$u_l = \{2k_b q_{max}/[\beta C_p^2 \rho_u^2 (T_b - T_u)]\}^{0{\cdot}5} \quad (3)$$

In this form, β is the Zeldovich number,[18] given by

$$\beta = T_a(T_b - T_u)/T_b^2 \quad (4)$$

A constant value of q, equal to q_{max}, and extending over a temperature interval of $(T_b - T_u)/\beta$ in the integral of eqn (1), with $\Lambda = 0{\cdot}5$, gives eqn (3). Thus β^{-1} defines an effective dimensionless temperature width of the reaction zone. Practical values of β range between 5 and 15. The value of T_a for a mixture can be found from the slope of a plot of $\ln u_l$ against T_b^{-1}.[19] Asymptotic analyses have yielded more accurate, two-term expansions for Λ and have shown the effects of flame straining.[18]

1.4. Straining of Laminar Flames

The burning velocity is the unidirectional velocity of the cold gases orthogonally into the flame front. The flame structure might be approximated to an initial preheat zone, followed by a reaction zone. In practice, the flow is often two-dimensional with a velocity component along the flame front that stretches, or strains, the flame. The balance of the mass, conductive and convective fluxes then is changed and, with it, the heat release rate – temperature profile and the burning velocity. The Lewis number, *Le*, expresses the ratio of molecular transport coefficients for energy and mass:

$$Le = k/\rho C_p D \quad (5)$$

The diffusion coefficient, D, is usually taken to be that of the deficient reactant in the original mixture.

The expression of Batchelor[20] for the rate of strain of a material surface,

$$\alpha = \frac{1}{A}\frac{dA}{dt} \tag{6}$$

has been used for flame straining,[21,22] with A as the area of the flame surface. It is analytically useful to express α in dimensionless form as the Karlovitz stretch factor, K, by dividing it by a chemical strain, or reciprocal chemical lifetime, u_l/δ_l, where δ_l is the reaction zone thickness of the unstrained laminar flame:

$$K = \alpha\delta_l/u_l \tag{7}$$

Often associated with this is the approximate expression for δ_l, in which v is the kinematic viscosity of the mixture,

$$\delta_l = v/u_l \tag{8}$$

Since the first analysis of Klimov,[21] a variety of approximate asymptotic treatments have shown the variations of strained flame laminar burning velocity, u_{ls}, with strain rate. Figure 3 is based upon the analytical expressions of Tromans,[23] but with the dimensionless laminar flame strain rate replaced by the turbulent strain rate, K, discussed in the next section. The form of the curves is the same with laminar strain, and the figure shows the variation of u_{ls}/u_l with dimensionless strain rate to depend upon Le. It also depends upon β. At a sufficiently high rate the flame can be extinguished, but such analyses can only act as a guide. More precise numerical approaches involve detailed chemical kinetics and, along with these, valuable experimental data[24] are accumulating to quantify the effects of straining.

1.5. Flammability Limits

Lower and upper fuel concentrations, below and above which no burning is possible are indicated in Fig. 2. These comprise the limits of flammability. Flame propagation depends upon the generation in the reaction zone of a relatively high concentration of active radicals, particularly O, H and OH. The fall in adiabatic temperature as a flammability limit is approached makes this more difficult and the flame becomes very sensitive to flame straining. Close to the lean limit, quite low strain rates cause extinction and, because of this,

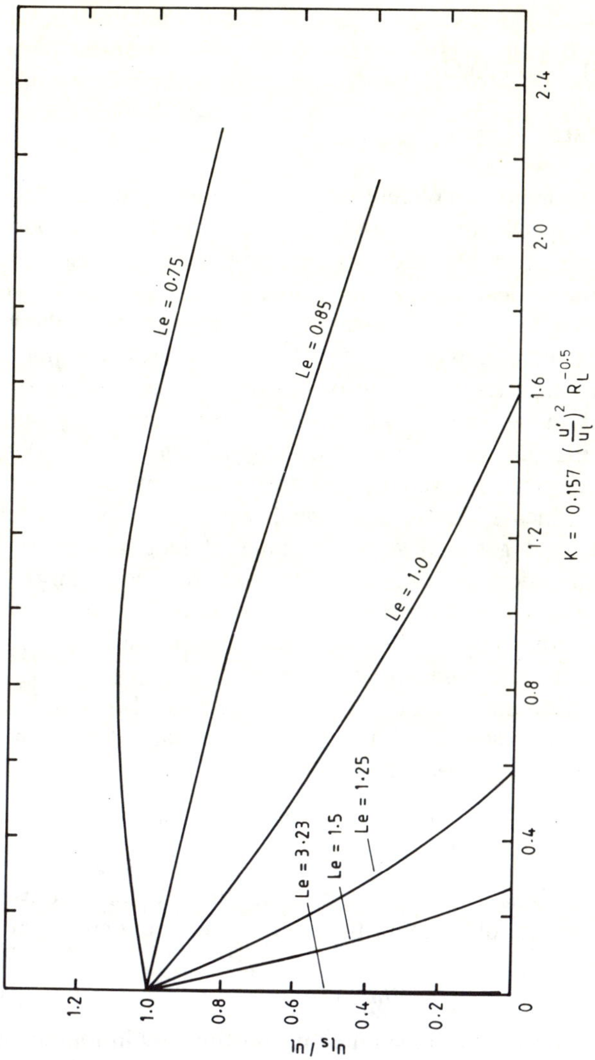

Fig. 3. Theoretical effect of strain rate upon laminar burning velocity.[19,23] $T_a/(T_b - T_u) = 8$.

below a limiting fuel concentration a flame cannot be stabilized. Measured fuel concentrations for flame extinction at low strain rates are in good agreement with the observed limits of flammability.[25]

2. TURBULENT BURNING

2.1. Steady State

At present, there is no completely satisfactory theory to quantify the turbulent burning velocity u_t, and, as with the laminar burning velocity, recourse must be made to experiment. There has been a multiplicity of theoretical approaches, ranging from the early concepts of flame wrinkling to more recent asymptotic analyses that allow for the effects of stretch and curvature.[26–28] The present author and co-workers have expressed the rate of burning by the product of the rate of decay of eddies and the amount of mixture chemically reacted in strained laminar flamelets during the eddy lifetimes.[19,29] A dimensionless eddy lifetime is given by the turbulent Reynolds number, $R_L = u'L/\nu$, in which u' is the root mean square (r.m.s.) turbulent velocity of the cold gas and L is the integral length scale of the turbulence. Dimensionless chemical lifetimes for both large and dissipative eddies involve the ratio u'/u_l, as well as R_L, and u_t/u_l can be expressed in terms of R_L and u'/u_l. These dimensionless groups are implicit in most of the theoretical approaches.

The influence of flame straining due to turbulence also is embodied within these groups, together with Le. The r.m.s. rate of strain in the cold gas, s', can be expressed[30] by

$$s' = [\overline{(\partial u/\partial x)^2}]^{0 \cdot 5} = [\varepsilon/15\nu]^{0 \cdot 5} = u'/\lambda \tag{9}$$

where u is the velocity in the x direction, ε is the specific rate of dissipation of turbulent energy and λ is the Taylor microscale. Equating ε to the rate of decay of the energy-containing eddies of size l_θ gives

$$\lambda^2/l_\theta \propto \nu/u' \tag{10}$$

With l_θ replaced by L, the constant of proportionality in the equation has been ascribed a value of 40·4 for isotropic turbulence and $R_L > 60$,[29] although the relationship may also depend upon R_L.[31] When α in eqn (7) is replaced by s' from eqn (9), and eqn (8) is invoked, the Karlovitz stretch factor becomes

$$K = 0{\cdot}157(u'/u_l)^2 R_L^{-0{\cdot}5} \tag{11}$$

The analytical expressions of Tromans,[23] with the dimensionless laminar flame strain replaced by K, give the ratio of the strained laminar burning velocity to the unstrained value, u_{ls}/u_l, as a function of K, as shown in Fig. 3. This relationship has been found for different values of Le and the ratio $T_a/(T_b - T_u)$, which is related to β.[19] With a value of $T_a/(T_b - T_u)$ of 8, the relationships are shown, for different values of Le, in Fig. 3. These demonstrate, at least qualitatively, how the turbulent strain rate might affect the laminar flamelets.[32] For lean gasoline–air mixtures the Lewis number is greater than one, and greater than it is for rich mixtures. As a result, reduction of the burning velocity by strain, and ultimately flame quenching, occur more readily in lean mixtures.

Flame quenching has been the subject of a separate experimental study[33] and tentative bounds suggested for such quenching are as follows:

$$u'/u_l \geqslant 0{\cdot}71 R_L^{0{\cdot}5} \qquad \text{for } R_L < 300 \tag{12}$$

and

$$u'/u_l \geqslant 3{\cdot}1 (R_L/Le^2)^{0{\cdot}25} \qquad \text{for } R_L > 300 \tag{13}$$

Conceptually, K can be regarded as a ratio of chemical to eddy lifetimes, with flame quenching when K is greater than order unity. In fact, with K given by eqn (11), eqn (13) is equivalent to

$$K\,Le > 1{\cdot}5 \tag{14}$$

2.2. Unsteady State

The situation is more complicated when a turbulent flame develops from a point source. Spark ignition initially creates a predominantly laminar flame kernel, for which most of the spectrum of turbulence is experienced as convection. In the case of isotropic turbulence with no mean flow, the small kernel is convected randomly in space with a convection velocity close to u'.[34] In an engine, if the spark gap is close to a solid surface, flame propagation may be impeded when the kernel is convected towards it, with the additional possibility of thermal quenching at the surface.

Fig. 4. Variations of power spectral density function with dimensionless frequency.

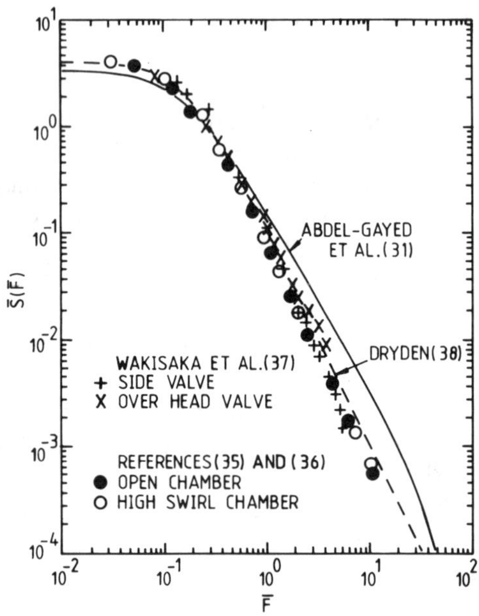

As the flame propagates, the reaction front is affected first by the higher frequencies of the turbulence spectrum and then, progressively, by the lower ones. The frequency band affecting flame propagation, initially by wrinkling the flame surface, might be assumed to extend from the highest frequency to a threshold frequency given by the reciprocal of the time elapsed from sparking. At any instant, the effective r.m.s. turbulent velocity for the propagation of the flame front, u'_k, is given by the integral of the power spectral density function against the frequency, from the highest frequency up to the reciprocal of the elapsed time.

Abdel-Gayed *et al.*[31] have derived the dimensionless power spectrum, shown by the full-line curve in Fig. 4, from laser–doppler measurements of isotropic turbulence in a fan-stirred bomb. The dimensionless power spectral density function, $\bar{S}(\bar{F})$, is $S(f)/u'^2\tau_a$, where $S(f)$ is the usual power spectral density function and τ_a is the autocorrelation time. The frequency, f, is normalized to give a dimensionless frequency $\bar{F} = f\tau_a$.

The circled experimental points in the figure are for two different engines; one with traditional values of turbulent velocities, and the other with high swirl and turbulence.[35,36] Values over ten cycles and within 100° crank angle about TDC are shown. Crossed experimental points are those of Wakisaka *et al.*[37] for two different engines. All of these engine results were obtained with hot wires. The broken-line curve represents the relationship proposed by Dryden[38] and is a good fit to these points. Differences from the full-line curve might be explained by anisotropy, measurement errors and neglected additional variables.

After an elapsed time of t_k, the effective r.m.s. velocity u'_k, is quantified by:

$$(u'_k)^2 = u'^2 \int_{\bar{F}_k}^{\infty} \bar{S}(\bar{F})\, d\bar{F} \tag{15}$$

where $\bar{F}_k$ is the reciprocal of the dimensionless time, $\bar{t}_k$, equal to t_k/τ_a. For a mean velocity U, Taylor's hypothesis gives $\bar{t}_k = t_k U/L$. The full-line curve of Fig. 4 yields the temporal development full-line curve of u'_k/u' with $\bar{t}_k$, shown in Fig. 5. This demonstrates how the effective r.m.s. turbulent velocity experienced by a growing flame kernel increases with time towards the fully developed value of u', which arises from the entire spectrum of turbulent energies. A similar development could be derived from the broken curve in Fig. 5.

High values of both mean, U, and turbulent, u', velocities give rise to a high value of u'_k. Because the velocity U is generally proportional to the engine speed, the time for the piston to move through a given crank angle is inversely proportional to U. The value of L is almost invariant with engine speed, and hence the dimensionless elapsed time from ignition, $\bar{t}_k$, for motion through a given crank angle must be the same for all speeds, and the abscissa scale on Fig. 5 must be always proportional to the elapsed crank angle.

The effective flame strain rate also has a spectral band dependence and a similar integration yields the strain rate experienced by the flame kernel. It can be shown[31] that the contribution to ε from the highest spectral frequency to F_k is given by

$$(\varepsilon_k/\nu)\tau_a^2 = (8\pi^2 u'^2/U^2) \int_{\bar{F}_k}^{\infty} \bar{F}^2 \bar{E}(\bar{F})\, d\bar{F} \tag{16}$$

where

$$\bar{E}(\bar{F}) = \frac{\bar{F}^3}{2} \frac{d}{d\bar{F}} \left[\frac{1}{\bar{F}} \frac{d\bar{S}(\bar{F})}{d\bar{F}} \right] \tag{17}$$

Equation (9) shows the strain rate to be proportional to $\varepsilon^{0.5}$, and the

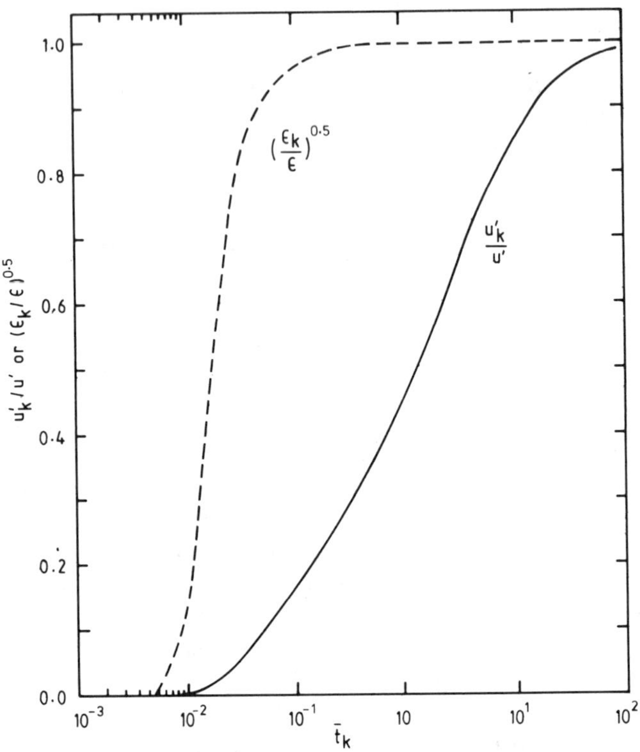

Fig. 5. Development of effective r.m.s. turbulent velocity and strain rate with dimensionless time.

broken curve in Fig. 5 shows the variation of $(\varepsilon_k/\varepsilon)^{0.5}$ with $\bar{t}_k$. Strain rate development is more rapid than is that of u'_k.

2.3. Correlation of Experimental Turbulent Burning Velocities

The above considerations formulate the relevant dimensionless groups to correlate experimental values of turbulent burning velocities. Abdel-Gayed *et al.*[31] have correlated 1650 available values from measurements on burners and stirred explosion bombs. For the latter, account was taken of the temporal development of the r.m.s. velocity and values of u'_k were found from Fig. 5. In all cases the strain rate experienced by the flame kernel was fully developed and values of K

for the purposes of the correlation were obtained from eqn (11). The general scatter of the results masked the detailed influence of Le, which nevertheless can be important in making lean mixtures more prone to flame quenching.[32]

In Fig. 6 are shown the values of u_t/u_l plotted against u_k'/u_l for $Le \leqslant 1{\cdot}3$, while in Fig. 7 are shown correlations for $Le > 1{\cdot}3$. There was no discernible influence of β. The full-line curves show values of K and the broken curves values of R_L, where $u' = u_k'$. The data cover many fuels and equivalence ratios. Individual experimental points are not shown: the appreciable scatter in them is attributable to both experimental errors and limitations in the implicit theory. There must be inherent limits to the accuracy of such correlations, with their neglect of the detailed chemical kinetics and use of a Lewis number that is based only on the deficient initial reactant. The regime of flame quenching at the high values of K is shown on both figures.

In addition to indicating flame quenching, the value of K indicates the nature of the flame front. Up to a value of about 0·15 the flame would appear to be a wrinkled continuous laminar flame sheet. As K increases the wrinkling becomes more severe, with the formation of peninsulas and islands of hot and cold gas and the break-up of the front. This is followed by instants of partial quenching, due to the inevitable distribution of strain rate values about a mean. Finally, when K is sufficiently high, the flame is quenched. These regimes also are indicated on Fig. 6.

With ignition from a point source the burning velocity increases to the fully developed value. The temporal development can be obtained from the values of u_t/u_l and u_k'/u_l in Figs 6 and 7, and the variation of u_k'/u_l with $\bar{t}_k$ in Fig. 5. These yield

$$u_{tk}/u_t = [1 - \exp(-0{\cdot}2(\bar{t}_k)^{0{\cdot}75})]^{0{\cdot}5} \tag{18}$$

where u_{tk} is the turbulent burning velocity at time t_k. This expression covers a wide range of strain rates and Lewis numbers.[31] The fully developed value of the turbulent burning velocity is found from Fig. 6 or 7 for a given value of R_L, at the point where $u_k' = u'$.

The correlations presented here are for turbulent burning in the absence of strong pressure fields. Their presence introduces new factors, such as flame-front instabilities. Vorticity is generated when the strong density gradients in a flame are orthogonal to pressure gradients.[39]

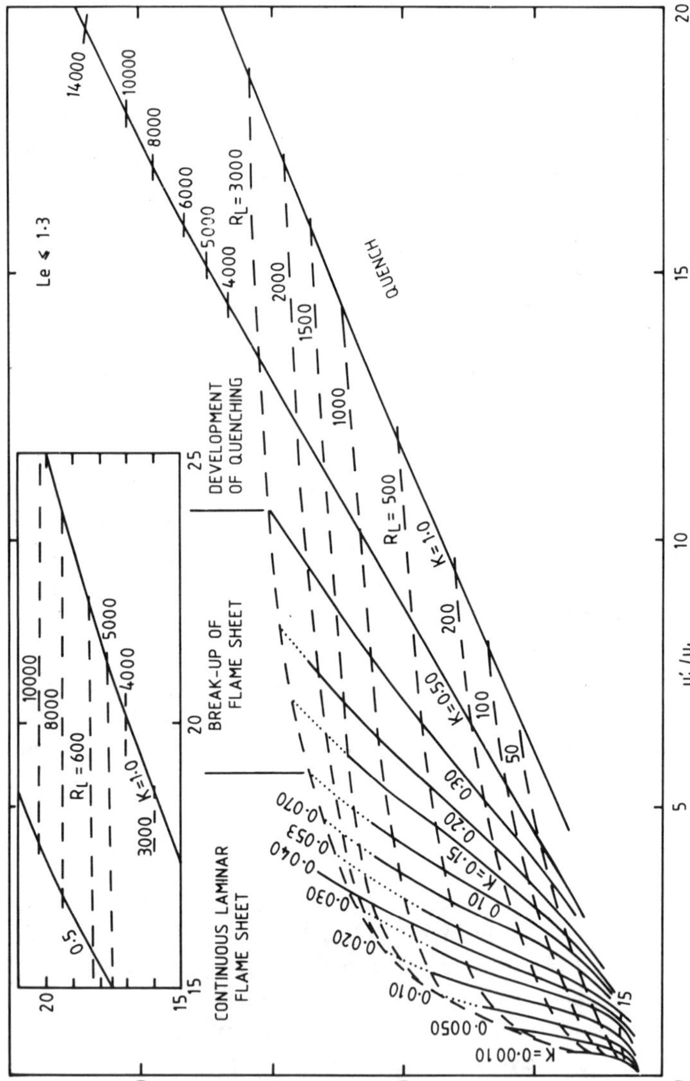

Fig. 6. Variations of u_t/u_l with u'_k/u_l at different values of the dimensionless strain rate, K, for $Le \leqslant 1{\cdot}3$. Broken curves show different values of R_L.[31]

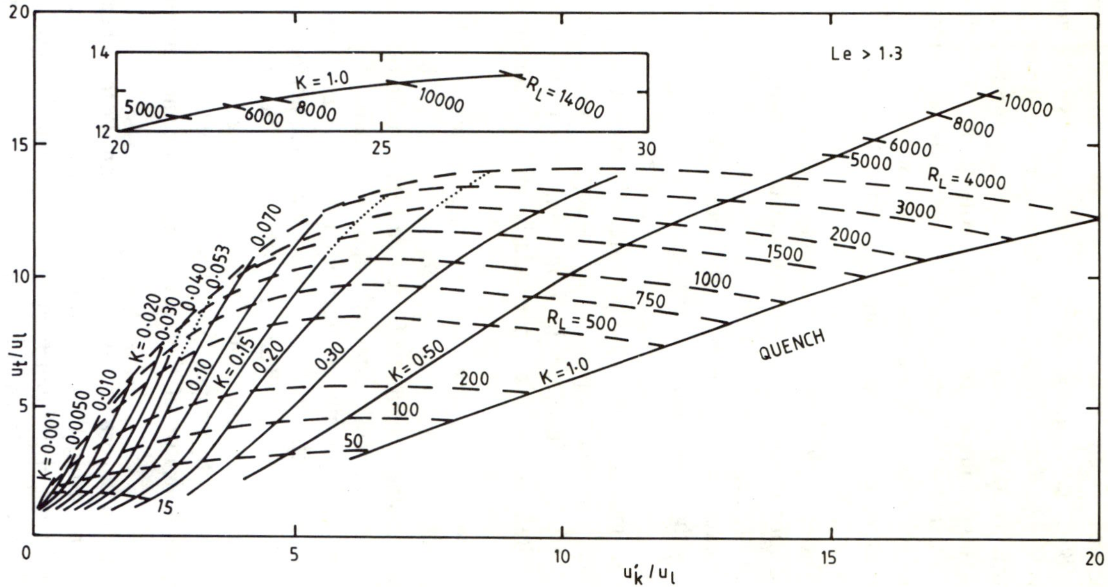

Fig. 7. Variations of u_t/u_l with u'_k/u_l at different values of the dimensionless strain rate, K, for $Le > 1{\cdot}3$. Broken curves show different values of R_L.[31]

2.4. Experimental Turbulent Explosion Bombs

Much of the burning velocity data at high values of R_L was obtained in fan-stirred bombs. Such bombs, first used by Schloesing and de Mondesir in 1864, have subsequently proved attractive for measurements of developing flame fronts, turbulent flame structure, burning velocities, flame speeds, pressure waves, rates of pressure rise and radiative emissions, under controllable conditions of turbulence. Bomb measurements have the advantages over those on burners that there are no problems of flame stabilization and of continuous gas flows. Engine measurements are inherently difficult, as is the control of the variables.

Semenov,[40] after earlier hot-wire anemometry in a motored engine, showed that if four identical fans are symmetrically disposed within a vessel and run at the same speed, the turbulence within the central region is isotropic. Measurements of the variations of pressure and flame radius with time enabled burning rates to be obtained.[41,42] Andrews *et al.*[43] used the double-kernel technique, as well as hot-wire anemometry, to measure turbulent burning velocities in a cylindrical bomb with four fans: subsequently, laser–doppler velocimetry was employed.[19,44] Multiple jet flows into and out of a bomb also have created controlled turbulence, particularly in experimental dust explosions.[45,46]

Other bomb researchers have created time-varying turbulence by the movement of a perforated plate. Ohigashi *et al.*[47] triggered such movement by releasing a magnetic force that retained the plate against the force of a spring, while Iinuma[48] allowed a pair of perforated plates to move outwards under electromagnetic forces. Checkel & Thomas[49] employed mechanical force to drive a single perforated plate across a bomb, and studied flame growth under the conditions of decaying turbulence. The emission intensity from the CH radical was measured, as it is directly proportional to the mass burning rate, and is a more sensitive indicator of that rate than is the measurement of pressure rise.

The Leeds group have also measured this intensity to indicate the mass burning rate, using the apparatus shown in Fig. 8.[50] The bomb comprised a cylinder of 305 mm diameter and 305 mm length, with a 150 mm diameter concentric window in each of the two end plates. These enabled the progress of the explosion flame to be recorded by helium–neon laser schlieren high-speed photography during the pre-pressure period. The four identical fans could be driven at any speed

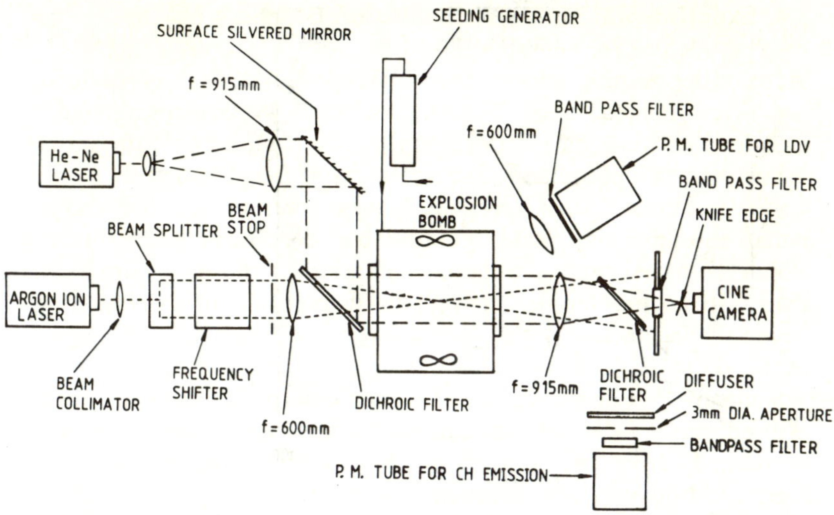

Fig. 8. Turbulent explosion bomb with facilities for schlieren photography, laser–doppler velocimetry and measurement of CH emission.

up to 11 000 rpm by separate 8 kW induction motors with independent speed control. The argon ion laser was for laser–doppler velocimetry. A dichroic filter at 45° to the axis of the vessel reflected 94% of the CH emission at 431·5 nm wavelength onto the photomultiplier (P.M.) tube, yet transmitted 90% of the helium–neon schlieren light.

2.5. Mass Burning Rate

The turbulent burning velocity, u_t, is the rate of propagation of the smoothed turbulent flame front into the unburnt gas. It is usually found from two-dimensional photographs of the three-dimensional phenomenon. Peninsulas and islands of unburnt gas can exist behind the flame front, particularly at the higher values of K. This implies that u_t is not necessarily a true measure of the mass burning rate. For a smoothed spherical flame front of radius r,

$$u_t 4\pi r^2 \rho_u = \frac{4\pi}{3} \frac{dr^3 \bar{\rho}_b}{dt} \tag{19}$$

where $\bar{\rho}_b$ is the mean density of the gas within the sphere of radius r. As the turbulent flame develops from the small spark kernel, the

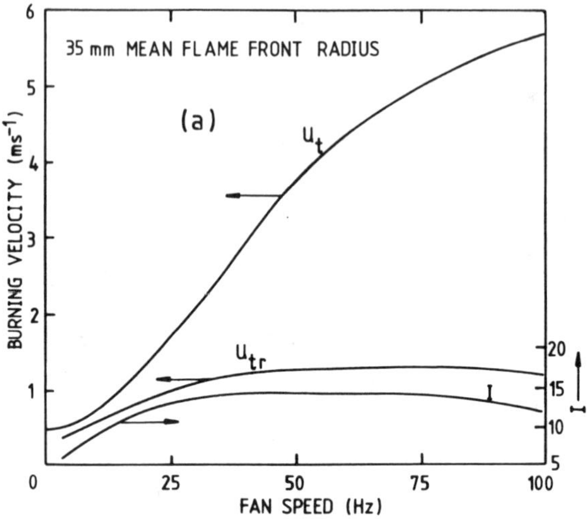

Fig. 9. Values of burning velocities and CH emission intensities at different fan speeds.

thickness of the turbulent flame brush increases and, because of the spherical geometry, an appreciable proportion of the gas behind the smoothed front may be incompletely burnt.

Let the density of the equilibrium, adiabatically burnt gases be ρ_b. Behind the smoothed front the density, $\bar{\rho}_b$, might be regarded as comprised of a volume fraction, f, of completely burnt gas and a remaining fraction, $1-f$, of unburnt gas. Hence

$$\bar{\rho}_b = f\rho_b + (1-f)\rho_u \tag{20}$$

Now define a mass burning velocity u_{tr} that, still associated with r, expresses the mass burning rate, as distinct from a flame-front propagation rate, such that

$$u_{tr} 4\pi r^2 \rho_u = \frac{4\pi}{3}\frac{\mathrm{d}r^3 f\rho_b}{\mathrm{d}t} \tag{21}$$

Equations (19)–(21) yield

$$u_{tr} = (S_s - u_t)(\rho_u/\rho_b - 1) \tag{22}$$

and u_{tr} can be found from experimental measurements of the flame speed, S_s ($=\mathrm{d}r/\mathrm{d}t$) and u_t.

In Fig. 9 are shown values of u_{tr} found in this way, at different fan speeds, using the apparatus shown in Fig. 8.[50] Values of u_t were found by the double-kernel technique and these are shown, together with those of the CH emission intensity, normalized by an appropriate measured value from a laminar explosion, I. The mixture was stoichiometric CH_4–air at a temperature of 55°C and atmospheric pressure. Values are for a mean flame-front radius of 35 mm. The similarity of the curves for u_{tr} and I suggests that both of these independently derived parameters measure the mass burning rate. The difference between u_t and u_{tr} increases with fan speed. At a fan speed of 4000 rpm the value of K is 0·44. A similar effect might be anticipated with an increase in engine speed and, indeed, Beretta *et al.*[51] have found appreciable incomplete combustion behind the mean flame front in engines. At an engine speed of 1044 rpm the volume fraction of completely burnt gas was only about 60%, somewhat lower than the bomb values. This proportion persisted into the region of increasing pressure.

Quantification of the burning rate behind the flame front is difficult. One approach is to adopt empirical burning laws; another is to quantify the surface areas of discrete laminar flamelets. In the former category, a widely used assumption is that the mass burning rate is equal to the total mass of unreacted gas behind the flame front, divided by a reaction time, τ. Equation (21) would yield

$$\mathrm{d}r^3 f\rho_b/\mathrm{d}t = r^3(1-f)\rho_u/\tau \tag{23}$$

where τ is a characteristic turbulent reaction time. It is readily shown that, of the unburnt mass behind the flame front at time t', the mass fraction of it that has burnt at a later time, t, is $1-\exp[-(t-t')/\tau]$. This expression was used by Blizzard & Keck[52] as a convenient analytic form.

The value of τ was taken by Tabaczynski *et al.*,[53] as well as subsequent workers, to be λ/u_l. A reaction time of half this value also appears in the two-eddy theory of burning[29] and rests upon a turbulent structure suggested by Tenekes.[54] In this, dissipative eddies are comprised of vortex tubes with a diameter of the order of the Kolmogorov microscale and a spacing of the order of the Taylor microscale.

Quantification on the basis of the surface areas of flamelets is at an early stage of development and is dependent upon the availability of experimental data. zur Loye & Bracco[55] have examined the flame structure that is revealed in an engine by a thin sheet of laser light, about 200 μm thick and of 10 ns duration, and measured flame areas. The scattered light from seeded particles of titanium oxide, formed by the reaction of titanium tetrachloride with the water vapour in the air, was collected by a digital camera.

3. IGNITION

3.1. The Spark

Combustion must be initiated with precise and controllable timing. Although it is possible to trigger chemical reactions radiatively by a burst of appropriate photons, spark ignition is favoured for its convenience. Electrical breakdown of the gas between spark electrodes occurs on a time scale of nanoseconds. The temperature of the dissociated and ionized gas can be as high as 60 000 K. After breakdown, the spark discharge occurs in a narrow channel with a high current density. After about 0·5 μs the discharge channel diameter is about 0·4 mm and the temperature 10 000 K.[56]

Successful ignition implies flame propagation throughout the mixture. This involves first establishing chemical reactions that lead to local ignition. However, this does not necessarily ensure flame propagation beyond a critical radius against the effects of flame straining. By means of high activation energy asymptotic analysis, Deshaies & Joulin[57] have shown that to do this a continuous power source is required. A source of hydrogen atoms or molecules also is beneficial.

Because the composition of the gases in the discharge channel is uncertain, it is difficult to model the chemical kinetics of ignition with precision. The creation of a high proportion of H atoms in excess of equilibrium by the discharge is more effective for ignition than is the dispersion of the same energy over equilibrated gases.[58] For the first stage, the concept of a critical volume within which the exothermic chemical reactions of flame propagation must become established is a useful simplification. The radius of such a volume is of the order of the flame thickness and the complexities of the chemical kinetics can be partially embodied in the use of the ignition delay time as a parameter. Such times are a function of initial composition, temperature and pressure.

It follows that a key problem in spark ignition is the relatively slow

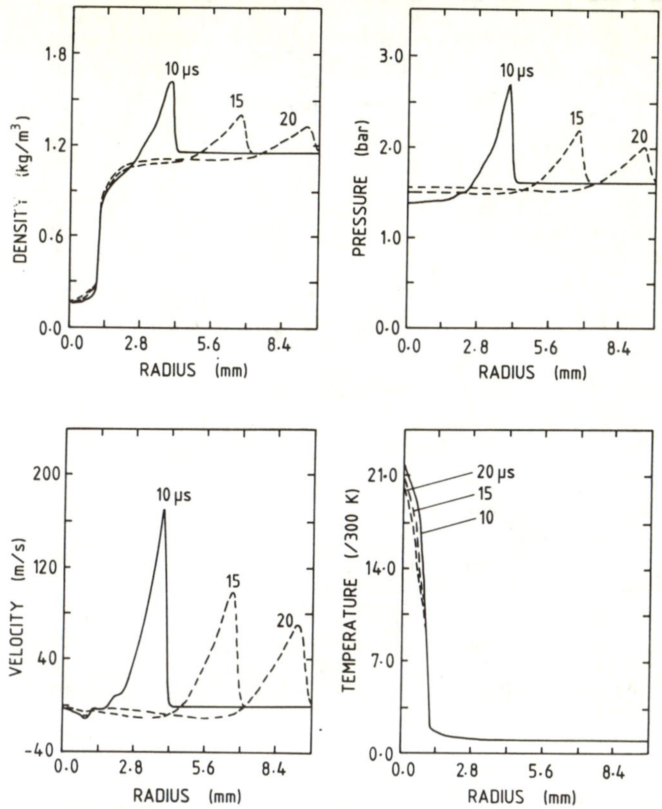

Fig. 10. Shock and thermal waves generated by a spark after 10, 15 and 20 μs.[59]

outward diffusion of the thermal energy and active atoms and radicals, that are initially highly concentrated within the narrow discharge channel. A high energy input in the early stages of the discharge aids this process, not only because of the greater heating, but also because of the strength of the shock wave that is generated by it. Results obtained from a mathematical model for a spark discharge energy of 10·1 mJ at a constant power of 13·1 W in atmospheric air[59] are shown in Fig. 10. The rapid propagation of the shock wave at time intervals of 10, 15 and 20 μs contrasts with the slow conductive spread of the thermal wave.

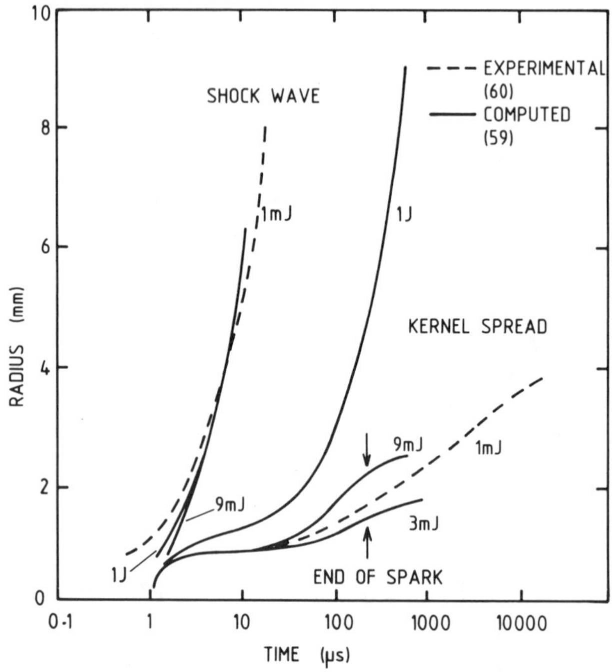

Fig. 11. Shock wave propagation and hot kernel spreading after spark initiation. Experimental data are for CH_4–air mixture; equivalence ratio = 0·67.

As the initial energy input is increased, the radially outward velocity associated with the shock wave enhances the thermal spreading. In Fig. 11 are shown computed radius–time relationships for the shock wave and hot kernel for energies of 3 mJ, 9 mJ and 1 J at constant electrical powers of 15, 45 and 200 W, respectively, together with some experimental measurements. Whatever the duration of the spark, the thermal spreading rate sharply decreased when the sparking ceased. Whether or not localized ignition occurs at a given radius can be estimated from such a model and ignition delay times.[34]

These computations show the better energy dispersion that follows a concentrated burst of electrical energy in the early stages of the discharge. This has been demonstrated experimentally by Ziegler *et al.*,[61] who compared short-duration high-current pulses in the breakdown mode with longer-duration glow discharges. Their photographs also showed a greater separation of the kernel from the electrodes with the former, due to the stronger pressure pulse.

Arc discharges might seem advantageous because they sustain high currents, but, unfortunately, electrode erosion can be excessive. 'Spark-aider' fuel additives, comprising alkali and alkaline-earth metal compounds, create deposits with a low electron work function on the electrode surface. These enhance electrical breakdown and enable a higher current density to be attained. After breakdown, there is strong gas heating in the cathode fall, with appreciable energy loss to the electrode surface.[62] Kalghatgi[63] has attributed improved engine performance with spark-aiders to a reduction in the cathode fall and consequent smaller energy loss to the electrodes. This gives more consistent flame development and higher flame speeds in the early stages.

3.2. Flame Straining and Spark-assisted Propagation

Although localized ignition may occur at a given radius, for an ensuing flame to propagate it must be able to overcome extinction by flame stretch. In the absence of strain from the turbulent flow field, the geometric strain rate, α, experienced by a spherical flame of radius r, is given by eqn (6) as

$$\alpha = \frac{2}{r}\frac{\mathrm{d}r}{\mathrm{d}t} \qquad (24)$$

From eqns (7) and (8) the associated Karlovitz stretch factor becomes

$$K = \frac{2}{r}\frac{\mathrm{d}r}{\mathrm{d}t}\frac{\nu}{u_l^2} \qquad (25)$$

Lewis & von Elbe[64] have shown how flame stretch might quench a flame and derived an expression for K, essentially the same as eqn (25). This they evaluated, over a wide range of fuel–oxygen mixtures, for a critical value of r equal to half the experimentally measured quenching distance. Most of these critical values of K lay between 0·5 and 1·0, to confirm the relevance of flame stretch in ignition.

Turbulence does not significantly increase the initial rate of spread of the spark kernel,[59] but the effects of flame straining can be paramount. At present, we cannot quantify precisely the strain rate acting on a laminar flamelet in a turbulent flow field, but we shall again assume that, at a dimensionless time $\bar{t}_k$ after sparking, it can be derived from eqn (16). In eqn (9), at this time, let it be denoted by s'_k to give a dimensionless turbulent strain rate S'_k when it is multiplied by τ_a. Thus

$$S'_k = \left[\frac{8\pi^2 i^2}{15}\int_{\bar{F}_k}^{\infty} \bar{F}^2 \bar{E}(\bar{F})\,\mathrm{d}\bar{F}\right]^{0\cdot 5} \tag{26}$$

where i is the turbulence intensity, u'/U.

The geometric strain rate, α, can similarly be made dimensionless to give $\bar{\alpha} = \tau_a \alpha$. If at time t_k the flame radius in eqn (24) is r_k and the flame speed S_s ($= \mathrm{d}r/\mathrm{d}t$) is assumed to be constant, then $r_k = S_s t_k$ and

$$\alpha = 2/t_k \tag{27}$$

At very small values of t_k the dimensionless strain rate is approximated by $\bar{\alpha}$, and at larger values of t_k by S'_k integrated over the complete spectrum of turbulence, to give $u'\tau_a/\lambda$. Between these limits it is assumed, for convenience, that the strain rates are additive. It is again conceptually useful to consider the reciprocal of the strain rate as an eddy, or strain, lifetime, and a dimensionless strain lifetime, $\bar{t}_s$, is defined by

$$\bar{t}_s = 1/(S'_k + \bar{\alpha}) \tag{28}$$

Values of $\bar{t}_s$, based upon the full-line curve turbulence spectrum of Fig. 4 and a value of i of 0·2, are plotted against $\bar{t}_k$ in Fig. 12.

A similarly dimensionless chemical lifetime, $\bar{t}_c$, is given by

$$\bar{t}_c = \delta_l / u_l \tau_a \tag{29}$$

From eqn (8) and because $\tau_a = L/U$ and $i = u'/U$, it is readily shown that

$$\bar{t}_c = \left(\frac{u'}{u_l}\right)^2 \frac{1}{iR_L} \tag{30}$$

In practice, the effects of straining are coupled with the complex initiating chemical reactions. However, the discussion is continued in terms of the simplified sequential pattern of an ignition delay, followed by the effect of straining upon the embryonic flame. This effect depends upon the relative magnitudes of $\bar{t}_s$ and $\bar{t}_c$. Values of $\bar{t}_c$ are represented in Fig. 12 by broken horizontal lines. Line 'a' represents a limit condition at which $\bar{t}_c$ is equal to the developed eddy lifetime, $\bar{t}_s$, represented by the reciprocal of the fully developed strain rate, equal

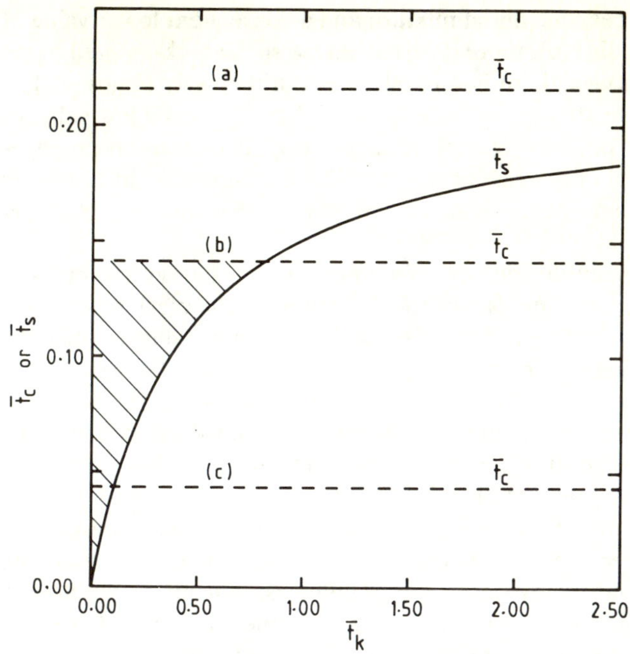

Fig. 12. Comparisons of dimensionless eddy lifetime, $\bar{t}_s$, and chemical lifetime, $\bar{t}_c$. Shaded area shows régime of spark-assisted propagation.

to $u'\tau_a/\lambda$. When the chemical lifetime is greater than this value, no flame propagation is possible. This corresponds to a value of K greater than unity. Strictly, the experimentally determined flame quenching condition of eqn (14) gives, for $Le = 1$, a limiting K value of 1·5, but here the reasoning is intended to be illustrative rather than precisely quantitative.

When $\bar{t}_c$ is reduced to the value shown by line 'b' flame propagation eventually can occur, provided that the flame survives transit through the shaded regime in which the chemical is greater than the eddy lifetime. In practice, this is achieved by the energy input from the spark. This must be sufficient to reduce the local chemical lifetime in this regime to less than $\bar{t}_s$. The spark energy may be predominantly thermal or, more effectively, in the form of active radicals, as in a plasma jet.[65]

With a more reactive initial mixture and a consequent lower value of K, as shown by line 'c', there is less need for such spark-assisted flame propagation, though there is a requirement in the early stages. A low value of $\bar{t}_c$ can be attained by both a high value of u_l and a low value of u'. The greater need for a minimal amount of spark-assisted propagation at the higher values of K partially explains the increases in minimum ignition energy with u'/u_l that were measured by Ballal & Lefebvre.[66]

Theoretical quantification of minimum ignition energy is not easy, because of the coupling of electrical discharge, chemical kinetic and flame straining effects. Indeed, Deshaies & Joulin[57] suggest that the concept of a critical power might be more useful. However, the essential requirements for ignition are reasonably clear: an initial high-energy pulse to enhance breakdown and shock-assisted thermal spread, followed by further power input to ensure localized ignition and spark-assisted propagation.

An increase in spark energy beyond the minimum necessary to survive the effects of flame straining can increase the flame speed above the normal steady state value. The measurements by De Soete[67] of initial flame speeds for different spark energies show flames to accelerate to the steady state flame speed from an initially low value at low energies and to decelerate to it from an initially high value at high energies. Locally, the spark-assisted flame might be regarded as electrically boosted. The greater the spark energy, the greater is the flame speed, as also is the CH emission intensity in the early stages.[59] It is interesting that the effects of the spark can persist for some milliseconds. Although it is difficult to increase burning velocities by the application of electrical fields, such an increase has been observed just after ignition, suggesting the persistence also of spark-originated ionization.[62]

Ignition problems are aggravated for lean mixtures. To achieve a sufficiently high burn rate, the low value of u_l is compensated by a designed increase in u'. This leads to high values of $\bar{t}_c$. Not only is localized ignition more difficult due to the greater laminar flame thickness, but Fig. 12 shows that greater spark assistance is necessary also for initial flame propagation. A value of $\bar{t}_c$ close to line 'a' would be prone to some flame quenching, because of the inevitable distribution of eddy lifetimes about the mean value. Design remedies might include a lower value of u' or, by charge stratification, a higher value of u_l around the ignition point.

4. COMBUSTION AND ENGINE PERFORMANCE

4.1. Initial Burning and Cyclic Variations

With point source ignition it takes, on average, about one third of the total combustion time to burn about 1% of the charge. As a result, variations in the initial flame speed have important effects upon pressure development and engine performance. These early variations are the fundamental cause of cyclical variations in engine performance.[68] There is scope for minimizing some of the random variations and for alleviating their effects by higher flame speeds.

For example, sparking is not a precisely repeatable phenomenon, particularly for mixtures close to a flammability limit. The variability of electrode surface finish contributes to this, as breakdown is associated with the strength of the localized electric fields. Smooth electrode surfaces of low electron work function should be sought to give consistent electrical breakdown and enhanced power input for spark-assisted propagation.

A high initial burn rate also might be achieved by torch ignition from a prechamber and by plasma jets.[69,70] Both devices are triggered outside the main combustion volume. The ejected flame gases or plasma provide an extended, wrinkled flame surface that enhances the initial burning rate. Stratification of the charge with a mixture of higher burning velocity around the ignition point, has the same effect.

The inherent variations in turbulence in the early stages create inevitable changes in burning rates. Al-Khishali *et al.*[44] measured the r.m.s. velocity within different time windows from a long sample of velocity fluctuations in the fan-stirred bomb shown in Fig. 8. For a given time window duration, the r.m.s. velocity was found for many different windows and the r.m.s. value, u'', found for this distribution. This increased as the duration of the window decreased. For a time window value of $\bar{t}_k$ equal to one, the ratio of u'' to u' was about 0·2. A similar variation in u' might be anticipated in the early stages of flame propagation and, with it, a consequential cyclical variation in the burning velocity, u_{tk}. Such a variation can be overridden, at least partially, by spark-assisted enhancement of the flame speed. An associated cyclical variation, in both magnitude and direction, is that of the convective velocity of the ignition kernel, referred to in Section 2.2. This effect can be minimized by locating the spark gap remote from other solid surfaces.

Inevitable variations in mixture composition also create cyclical variations in the burning rate. Additionally, it has been demonstrated computationally that such variations arise inevitably from the non-linearity of reaction rates. Calculations by Daily[71] of the exhaust gas temperatures of successive cycles show inherent cyclical variations. These are greatest with the least reactive mixtures and with the most recirculation of exhaust gas.

4.2. Burning during Compression Ahead of the Flame

During this ensuing period the part of the turbulence spectrum that is relevant to flame propagation continues to extend to lower frequencies. Because of the compression, the cold mixture turbulence is changing, the laminar burning velocity is increasing and the flame shape is distorting to that of the combustion space. Mixture turbulence might be calculated from the conservation of eddy mass and of angular momentum, which assumes the product of the length scale and the turbulent velocity is constant.

Hynes & Sheppard[72] have developed a zero-dimensional model of engine cycle combustion that uses the values of u' at ignition and u'_k given by Fig. 5, with values of u_l for iso-octane–air mixtures drawn from Ref. 9. From the values of u'_k/u_l and K, the value of u_t/u_l is found from either Fig. 6 or Fig. 7, to yield u_t. Combustion behind the smoothed flame front is modelled by the approach of Blizard & Keck,[52] described in Section 2.5, with $\tau = \lambda/u_l$. Appropriate engine turbulence parameters, dependent upon engine speed, are taken from empirical correlations and allowance is made for heat transfer.

This model yielded the relationship of u_t/u_l to u'_k/u_l, as well as the change of K during flame propagation, shown by the blackened dotted curve of Fig. 13. This is for iso-octane as fuel, with an equivalence ratio of 0·8 (air:fuel ratio of 18·5), a compression ratio of 8:1, an engine speed of 3400 rpm and a high-turbulence combustion chamber.[73]

As the flame grows, u'_k increases, not only because of the broadening of the effective turbulence spectrum for the flame, but also because of the compression of the charge. As a result, the turbulent burning velocity increases. However, the increase in u'_k also raises the strain rate and the propensity to quenching. Further flame growth produces a rise in temperature, leading to an increase in u_l. This, in turn, reduces

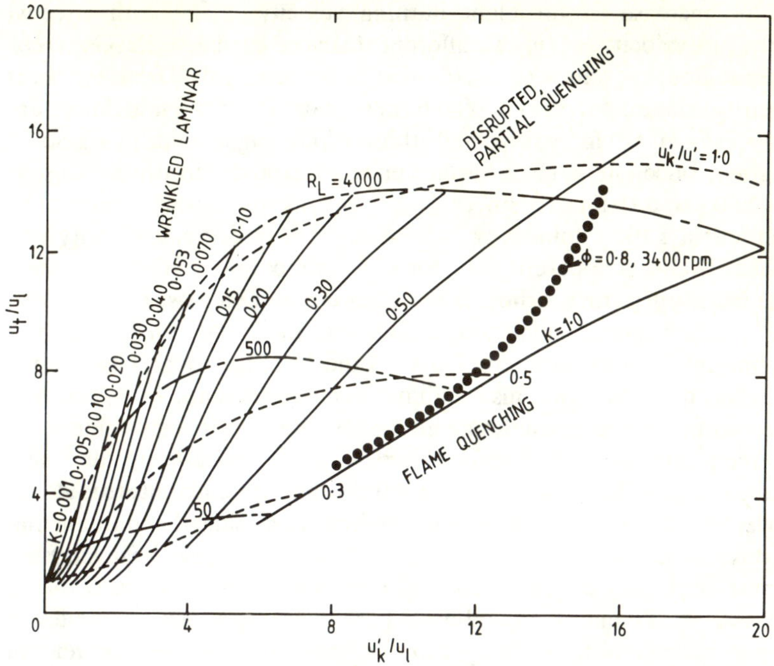

Fig. 13. Burning velocities in engines. Dotted curves show changes during a single explosion: dashed curves show effect of increasing engine speed.

the value of K, as shown in the figure, and quenching becomes less likely. Flame quenching is most likely in the early stages of flame development, when both the temperature and u_l are relatively low. This is particularly true for lean mixtures (low u_l) and high turbulence (high u'), exemplified by the present conditions, since these lead to high values of K. This might be countered by spark-assisted propagation. At an engine speed of 1000 rpm combustion follows a curve close to that for $K = 0{\cdot}1$, while at the same speed with a stoichiometric mixture the combustion would be within the wrinkled flame regime.

4.3. Burning at Different Engine Speeds

The effects of different engine speeds are shown by the three dashed curves in Fig. 13. These demonstrate the diminishing returns, in terms of the increase of turbulent burning velocity, u_t/u_l, with effective turbulent velocity, u'_k/u_l, at different stages of the flame development, represented by values of u'_k/u' of 0·3, 0·5 and 1·0. These loci have been calculated for values of 0·65 m s^{-1} for u_l, of 1·5 mm for L and $2{\cdot}6 \times 10^{\times 6}$ m^2 s^{-1} for ν, typical of those which might occur in a gasoline engine. Since, at a given equivalence ratio and throttle setting, u' increases approximately linearly with engine speed, these curves also demonstrate the diminishing rate of increase in burning velocity with speed. It also is apparent that, for a given mixture strength, the flame quench point is approached as the engine speed increases.

To achieve good engine combustion both the value of u_t and the rate of burning behind the flame front should increase roughly in proportion to engine speed. This facilitates the maintenance of as much of the combustion as possible at the highest effective compression ratio, whatever the speed. The loci for increasing speed shown by the dashed curves clearly demonstrate the problems that are encountered. At the lower values of u'_k/u_l the design goal of increasing the burning rate almost proportionally to the increase in engine speed is achieved. However, in the plateau region of these curves the regime is that of the disrupted flame sheet and partial gas phase quenching of the flame. As a consequence, not only can the thermal efficiency, power and driveability be expected to deteriorate with increasing speed, but the hydrocarbon and carbon monoxide exhaust emissions can be expected to increase. It would appear inadvisable to allow K to exceed a value of about 0·3. Accurate chemical kinetic modelling and quantitative predictions of such emissions are difficult, but the present approach indicates regimes in which they might become excessive, due to gas phase flame quenching.

It will be recalled that the mass burning rate is significantly less than that suggested by the value of u_t. With a value of the turbulent reaction time, $\tau = \lambda/u_l$, eqns (10) and (23) suggest a mass burning rate behind the flame front that, for a given temperature, pressure and length scale, varies as $(u_l(u')^{0{\cdot}5})$. Although u' increases with engine speed, the effective value of u_l might be reduced by flame straining.

4.4. Engine Performance

The engine model[73] demonstrates that the accurate modelling of the early stages of combustion is important, for the same physical reasons that these stages control the cyclical fluctuations. The model also shows the decline in performance associated with the onset of localized flame quenching. The thermodynamic advantages of lean burning can only be realized if the lower burning velocity is compensated for by enhanced turbulence, to ensure that the bulk of the heat release occurs close to top centre. This might be difficult to achieve throughout the full range of engine speeds, without encountering at least partial flame quench through excessive turbulence.

The generation of a high value of u' creates increased pumping losses, with a reduction in volumetric efficiency and an associated detrimental effect on the overall engine performance. This performance can be controlled, not only through the spark timing, but also through sensitized control of the mixture strength and turbulence. To achieve high engine torques, pumping losses should be minimized and burning rates should be high. These aims are best attained with a low value of u' and a near-stoichiometric mixture. The thermal efficiency would not be high and a catalytic converter might be necessary to reduce nitric oxide and carbon monoxide emissions. To achieve high cruising efficiency, when the power is reduced, the associated throttling of the air might beneficially be linked to the creation of cylinder swirl and turbulence. This would enable a leaner mixture to be burnt, with greater efficiency, and the throttling would not be as detrimental as at maximum power.

Desired performance regimes in the lean combustion control system, reported by Matsushita *et al.*[1], are shown in Fig. 14. These are in the form of contours of air:fuel ratio, on a graph of intake manifold pressure against engine speed. The system comprises a mixture control sensor in the exhaust gases, together with a means of measuring intake manifold pressure and engine speed, to give the air flow rate. From this information, a microprocessor controls the desired performance contours. Control of the air:fuel ratio is through the amount of fuel injected, and control of the turbulence is through the opening or closing of an inlet swirl control valve, located in a bypass of the intake port. The control valve is activated by a diaphragm and is closed when high swirl is required. Closure increases the turbulence, but reduces the volumetric efficiency.

Operational regimes are marked in Fig. 14. At full load, the swirl control valve is open and the air:fuel ratio is about 13. When cruising, the valve is closed, the intake manifold pressure decreases and the air:fuel ratio is about 21. As the engine speed increases, the mixture

Fig. 14. Performance contours of air : fuel ratio.

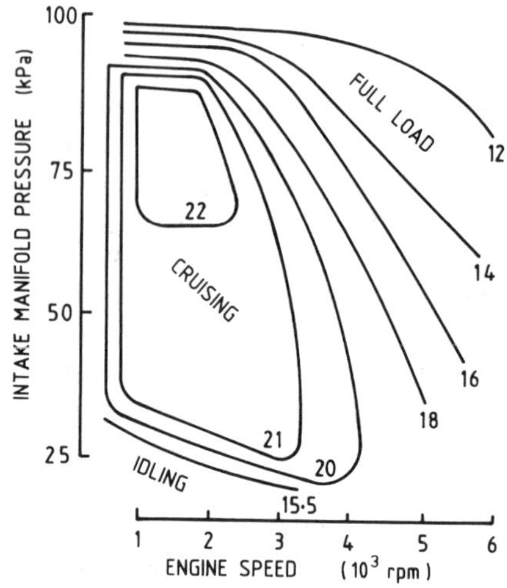

becomes richer. Idling is characterized by strong throttling and poor mixing, with the necessity to enrich the mixture for satisfactory running.

4.5. Autoignition and Knock

Orderly combustion in the deflagration wave can be disturbed by autoignition of the end gas, ahead of the flame front. Autoignition can not only cause knock, but also run-on, which occurs when the ignition system is switched off. High-speed photography at NACA some time ago showed that autoignition does not always give rise to knock, nor does knock necessarily originate in the end gas.[74] Photographs confirmed the presence of unburnt gas behind the visible flame front, as reported in Section 2.5, and also showed that autoignition does not necessarily result in complete reaction.

An increase in the turbulence, provided that it does not lead to quenching, increases the propagation rate and allows less time for the end gas chemical reactions of autoignition to develop. To avoid knock, the combustion chamber flame transit time should be less than the induction time. Values of induction time have been presented as the sum of the durations of two sequential events. The first is associated with chain reactions that culminate in cool flames and the second with the more rapid, final release of the chemical enthalpy. Empirical expressions have been presented for both of these times, in terms of mixture composition, temperature and pressure.[64] Both times decrease with pressure. The first also decreases with temperature, with which the second increases.

However, the complexity of the chemistry demands a somewhat more detailed approach. Autoignition is due to an exothermic degenerately branched chain reaction, the rate of which depends upon the prior accumulation of the branching agent and on the reaction history, not solely on the instantaneous state. The detailed chemistry seems impossibly complex, not least because a gasoline might be comprised of about 150 compounds.

Kirsch & Quinn[75,76] have studied the two-stage processes in a rapid compression machine and developed a generalized chemical kinetic approach that, while avoiding detail, still retains the essential chemical features. For autoignition to occur, the formation of the branching agent should not be favoured by high temperature, but it should be enhanced by the accumulation of intermediate reactive products. The dissociation of this agent is the main source of radicals in the exothermic degenerately branched chain that gives the main heat release. A model of engine combustion can yield the changing conditions of the unburnt gas which, when used in such a chemical model of autoignition, can predict the possible subsequent onset of knock.

The simplified high-temperature chemical kinetic scheme discussed in Section 1.3 has insufficient kinetic detail of intermediate hydrocarbon reactions for application to preflame reactions at lower temperatures. Attempts are currently being made to remedy this for knock prediction, and Axelsson *et al.*[77] have introduced site-specific H atom abstraction reactions and octyl radical decomposition into more detailed schemes of *n*-octane and iso-octane oxidation.

Of course, to predict autoignition is not the same as to predict the details of the knocking process. This is a complex task, still requiring experimental enlightenment. In Fig. 15 are shown high-speed laser light shadowgraphs, obtained by Nakagawa *et al.*,[78] of flame propagation and autoignition under heavy knocking conditions, together with two associated pressure records from pressure transducers on opposite walls of the combustion chamber. The exposure time was $0{\cdot}8\ \mu s$, with a film speed of 30 000 frames per second. To locate the high density gradients associated with autoignition, film images were photo-digitized by a 700×500 array of pixels. Dark regions indicate high

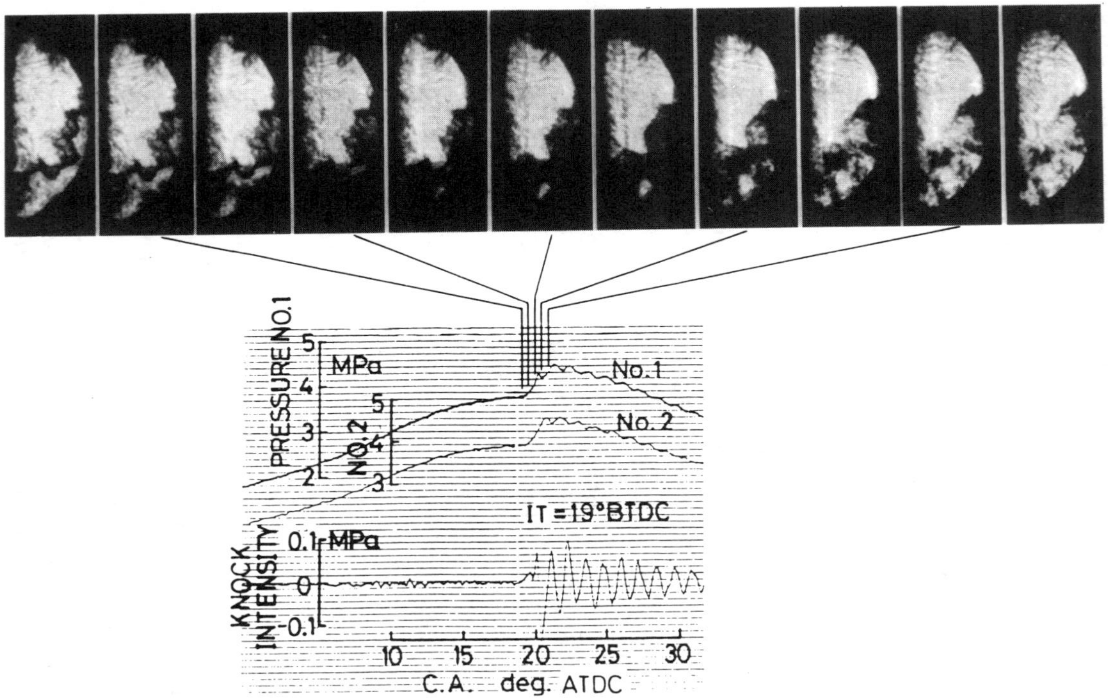

Fig. 15. Shadowgraphs of main flame and autoignition in an engine cylinder with heavy knocking. Pressure records are for two transducers mounted opposite each other (reprinted with permission of the Society of Automotive Engineers, 1984[78]).

temperature gradients. The second frame shows regions of autoignition in the unburnt gas at the cylinder wall. The expansion rate of the autoignited gas is more rapid than that of the main propagating flame and extends over the whole unburnt gas in three to four frames, within one degree of crank angle (CA). During this time the main flame front hardly moves.

The autoignited gas subsequently becomes very bright in a further three to four frames. Pressure rise is coincident with pronounced autoignition and there are subsequent pressure oscillations. With less severe autoignition no such oscillations are observed. Such photographs cannot yield three-dimensional information, but from the two-dimensional projections, it appeared that out of 24 observed instances of autoignition, ten occurred in the vicinity of the cylinder wall, six in the middle of unburnt gas and one in front of the flame. The remainder were at combinations of these locations.

Sudden, localized energy release at autoignition can create strong pressure waves that move outwards and reflect at the solid boundary of the combustion volume. If there are several centres, the pressure waves and their reflections interact to give, in some instances, still higher pressure pulse amplitudes. The way in which pressure waves propagate from a single, isolated centre of rapid energy release, in this case an electrical one, is shown in Fig. 10. This also shows the high gas velocities that are generated. It is these that can halt the main flame propagation as viewed on the film. The interaction of waves from ignition centres, randomly dispersed in time and space, and combined with wall reflections, is complex in the extreme.

For the simplified conditions of Fig. 10, up to 10 μs the volumetric energy release rate is about $10\,\mathrm{GW\,m^{-3}}$, of the same order as the volumetric chemical energy release rate of the steady state gaseous flame of Fig. 1. The sudden transient chemical energy release of autoignition might be even larger and an associated generation of strong pressure waves would be expected. These waves might, in turn, trigger reaction in the already highly reactive unburnt mixture, wherever it is located. For example, from induction time data for a stoichiometric propane–air mixture, it has been calculated[34] that the shock wave from an energetic spark might ignite such a mixture up to a distance of about 10 mm. Such ignition would generate further pressure waves to reinforce earlier ones. A system of relatively ordered combustion collapses into a chaotic one.

In terms of practical engine operation, it is desirable to devise feedback control, such that the engine might run in a regime just outside the threshold of knock damage. This can be achieved by means of a knock sensor that signals a control box, which actuates the retardation of the spark.[79]

Other experimental techniques have involved the use of electrostatic probes[80] to detect, through enhanced ionic currents, the onset of knock. This has been combined with the photographic recording of reaction fronts. Rapid compression machines have proved useful and high-speed photographs in one such[81] havc shown knock to start with the autoignition of a narrow region of end gas. This expanded with supersonic velocity, to form a sharp-fronted shock wave which was reflected at the cylinder wall. Violent pressure waves were generated.

REFERENCES

1. Matsushita, S., Inoue, T., Nakanishi, K., Kato, K. & Kobayashi, N., *Development of the Toyota lean combustion system.* SAE paper 850044, 1985.
2. Westbrook, C. K. & Dryer, F. L., Chemical kinetic modelling of hydrocarbon combustion. *Prog. Energy Combust. Sci.* **10** (1984) 1.
3. Dixon-Lewis, G. & Islam, S. M., Flame modelling and burning velocity measurement. *Nineteenth Symposium (International) on Combustion.* The Combustion Institute, Pittsburgh, 1982, p. 283.
4. Bray, K. N. C., The interaction between turbulence and combustion. *Seventeenth Symposium (International) on Combustion.* The Combustion Institute, Pittsburgh, 1979, p. 223.
5. Bradley, D., Gaskell, P. H., Lau, A. K. C., Missaghi, M. & Chin, S. B., Mathematical modelling of turbulent combustion. *Proc. Inst. Mech. Engrs,* International Conference on Computers in Engine Technology. Institution of Mechanical Engineers, London, 1987, p. 315.
6. Bradley, D., Kwa, L. K., Lau, A. K. C., Missaghi, M. & Chin, S. B., Laminar flamelet modelling of recirculating premixed methane and propane-air combustion. *Combust. Flame,* **71** (1988) 109.
7. Warnatz, J., Chemistry of high temperature combustion of alkanes up to octane. *Twentieth Symposium (International) on Combustion.* The Combustion Institute, Pittsburgh, 1985, p. 845.
8. Andrews, G. E. & Bradley, D., Determination of burning velocities—a critical review. *Combust. Flame,* **18** (1972) 133.
9. Metghalchi, M. & Keck, J. C., Burning velocities of mixtures of air with methanol, isooctane, and indolene at high pressure and temperature. *Combust. Flame,* **48** (1982) 191.
10. Gülder, O. L., Burning velocities of ethanol–isooctane blends. *Combust. Flame,* **56** (1984) 261.

11. Hamid, M. N., Fundamentals of turbulent combustion related to gasoline engines. Ph.D. Thesis, University of Leeds, 1986.
12. Andrews, G. E. & Bradley, D., Determination of burning velocity by double ignition in a closed vessel. *Combust. Flame,* **20** (1973) 77.
13. Gibbs, G. J. & Calcote, H. F., Effect of molecular structure on burning velocity. *J. Chem. Engng. Data,* **4** (1959) 226.
14. Zeldovich, Y. B. & Frank-Kamenetskii, D. A., K teorii ravnomernogo rasprostraneniya plameni. *Dokl. AN SSSR,* **19** (1938) 693.
15. Zeldovich, Y. B. & Frank-Kamenetskii, D. A., Teoriya teplovogo rasprostraneniya plameni. *Zh. Fiz. Khim.,* **12** (1938) 100.
16. Spalding, D. B., I. Predicting the laminar flame speed in gases with temperature-explicit reaction rates. II. One-dimensional laminar flame theory for temperature-explicit reaction rate. *Combust. Flame,* **1** (1957) 287, 296.
17. Zeldovich, Y. B., Barenblatt, G. I., Librovich, V. B. & Makhviladze, G. M., *The Mathematical Theory of Combustion and Explosions.* Consultants Bureau, New York, 1985.
18. Williams, F. A., *Combustion Theory,* 2nd edn. Benjamin/Cummings, Menlo Park, California, 1985.
19. Abdel-Gayed, R. G., Al-Khishali, K. J. & Bradley, D., Turbulent burning velocities and flame straining in explosions. *Proc. R. Soc. Lond.,* **A391** (1984) 393.
20. Batchelor, G. K., The effect of homogeneous turbulence on material lines and surfaces. *Proc. R. Soc. Lond.,* **A213** (1952) 349.
21. Klimov, A. M., Laminarnoe plamya v turbulentnom potoke. *Zh. Prikl. Mekh. Tekhn. Fiz.,* **3** (1963) 49.
22. Williams, F. A., A review of some theoretical considerations of turbulent flame structure. *AGARD Conference Proceedings* (164), 1975, p. II: 1–1.
23. Tromans, P. S., The interaction between strain fields and flames—a possible source of combustion variations in spark ignition engines. *ASME Symposium on Fluid Mechanics of Combustion Systems,* Boulder, Colorado, 1981, p. 201.
24. Law, C. K., Zhu, D. L. & Yu, G., Propagation and extinction of stretched premixed flames. *Twenty-first Symposium (International) on Combustion.* The Combustion Institute, Pittsburgh, 1988, p. 1419.
25. Puri, I. K. & Seshadri, K., The extinction of counterflow premixed flames burning diluted methane–air, and diluted propane–air mixtures. *Combust. Sci. Technol.,* **53** (1987) 55.
26. Andrews, G. E., Bradley, D. & Lwakabamba, S. B., Turbulence and turbulent flame propagation: a critical appraisal. *Combust. Flame,* **24** (1975) 285.
27. Libby, P. A. & Williams, F. A., Turbulent flows involving chemical reactions. *Ann. Rev. Fluid. Mech.,* **8** (1976) 351.
28. Bray, K. N. C., Turbulent flows with premixed reactants. *Turbulent Reacting Flows,* ed. P. A. Libby & F. A. Williams, Topics in Applied Physics Vol. 44. Springer-Verlag, 1980, p. 115.
29. Abdel-Gayed, R. G. & Bradley, D., A two-eddy theory of premixed turbulent flame propagation. *Phil. Trans. R. Soc. Lond.,* **301** (1981) 1.

30. Taylor, G. I., Statistical theory of turbulence. *Proc. R. Soc. Lond.*, **A151** (1935) 421.
31. Abdel-Gayed, R. G., Bradley, D. & Lawes, M., Turbulent burning velocities: a general correlation in terms of straining rates. *Proc. R. Soc. Lond.*, **A414** (1987) 389.
32. Abdel-Gayed, R. G., Bradley, D., Hamid, M. N. & Lawes, M., Lewis number effects on turbulent burning velocity. *Twentieth Symposium (International) on Combustion.* The Combustion Institute, Pittsburgh, 1985, p. 505.
33. Abdel-Gayed, R. G. & Bradley, D., Criteria for turbulent propagation limits of premixed flames. *Combust. Flame,* **62** (1985) 61.
34. Akindele, O. O., Bradley, D., Mak, P. W. & McMahon, M., Spark ignition of turbulent gases. *Combust. Flame,* **47** (1982) 129.
35. Benjamin, S. F., BL lean burn combustion research. Magerbetrieb Beirn Ottomotor—lean burn engines. VDI Berichte 578, Wolfsburg, 1985.
36. Lawes, M., Effects of turbulence on combustion in engines. Ph.D thesis, University of Leeds, 1987.
37. Wakisaka, T., Hamamoto, Y. T. & Kinoshita, S., Turbulence characteristics in internal combustion engines. *Bull. Jpn. Soc. Mech. Engrs.*, **26** (1983) 254.
38. Dryden, H. L., Turbulence investigations at the National Bureau of Standards. *Fifth International Congress for Applied Mechanics,* 1938, p. 362.
39. Oran, E. S. & Boris, J. P., Detailed modelling of combustion systems. *Prog. Energy Combust. Sci.*, **7** (1981) 1.
40. Semenov, E. S., Measurement of turbulence characteristics in a closed volume with artificial turbulence. *Combust. Explos. Shock Waves,* **1** (1966) 57.
41. Karpov, V. P., Semenov, E. S. & Sokolik, A. S., Turbulent combustion in an enclosed space. *Proc. Acad. Sci. USSR, Phys. Chem. Sect.*, **128** (1959) 871.
42. Sokolik, A. S., Karpov, V. P. & Semenov, E. S., Turbulent combustion of gases. *Combust. Explos. Shock Waves,* **3** (1967) 36.
43. Andrews, G. E., Bradley, D. & Lwakabamba, S. B., Measurement of turbulent burning velocity for large turbulent Reynolds numbers. *Fifteenth Symposium (International) on Combustion.* The Combustion Institute, Pittsburgh, 1975, p. 655.
44. Al-Khishali, K. J., Boston, P. M., Bradley, D., Lawes, M. & Pegg, M. J., The influence of fluctuations in turbulence upon fluctuations in turbulent burning velocity. *Combustion in Engineering,* Vol. 1. Institution of Mechanical Engineers, London, 1983, p. 175.
45. Kauffmann, C. W., Srinath, S. R., Tezok, F. I., Arisoy, A. & Leisch, S. O., Fundamental investigations pertaining to the three different modes of dust combustion. *First International Colloquium on Explosibility of Industrial Dusts,* Part 2. Polish Academy of Sciences, Warszawa, 1985, p. 23.
46. Bond, F., Fresko, M., Knystautas, R. & Lee, J. H. S., Influence of

turbulence on dust explosions. *First International Colloquium on Explosibility of Industrial Dusts,* Part 2. Polish Academy of Sciences, Warszawa, 1985, p. 37.
47. Ohigashi, S., Hamamoto, Y. & Kizima, A., Effects of turbulence on flame propagation in closed vessel. *Bull. Jpn. Soc. Mech. Engrs.,* **14** (1971) 849.
48. Iinuma, K., A study of turbulent flame propagation in closed vessels. Department of Aeronautics, Faculty of Engineering, University of Tokyo, 1977.
49. Checkel, M. D. & Thomas, A., Turbulent explosions in closed vessels. *Combustion in Engineering,* Vol. 1. Institution of Mechanical Engineers, London, 1983, p. 181.
50. Abdel-Gayed, R. G., Bradley, D., Lawes, M. & Lung, F. K.-K., Premixed turbulent burning during the early stages of an explosion. *Twenty-first Symposium (International) on Combustion.* The Combustion Institute, Pittsburgh, 1988, p. 497.
51. Beretta, G. P., Rashidi, M. & Keck, J. C., Turbulent flame propagation and combustion in spark ignition engines. *Combust. Flame,* **52** (1983) 217.
52. Blizzard, N. C. & Keck, J. C., *Experimental and theoretical investigation of turbulent burning model.* SAE paper 740191, 1974.
53. Tabaczynski, R. J. T., Trinker, F. H. & Shannon, B. A. S., Further refinement and validation of turbulent flame propagation model for spark-ignition engines. *Combust. Flame,* **39** (1980) 111.
54. Tenekes, H., Simple model for the small-scale structure of turbulence. *Phys. Fluids,* **11** (1968) 669.
55. zur Loye, A. O. & Bracco, F. V., *Two dimensional visualization of premixed charge flame structure in an I. C. engine.* SAE paper 870454, 1987.
56. Maly, R. & Vogel, M., Initiation and propagation of flame fronts in lean CH_4-air mixtures. *Seventeenth Symposium (International) on Combustion.* The Combustion Institute, Pittsburgh, 1979, p. 821.
57. Deshaies, D. & Joulin, G., On the initiation of a spherical flame kernel. *Combust. Sci. Technol.,* **37** (1984) 99.
58. Dixon-Lewis, G. & Shepherd, I. G., Some aspects of ignition by localised sources, and of cylindrical and spherical flames. *Fifteenth Symposium (International) on Combustion.* The Combustion Institute, Pittsburgh, 1975, p. 1483.
59. Bradley, D. & Lung, F. K.-K., Spark ignition and the early stages of turbulent flame propagation. *Combust. Flame,* **69** (1987) 71.
60. Ziegler, G. F. W., Wagner, E. P. & Maly, R. R., Ignition of lean methane–air mixtures by high pressure glow and arc discharges. *Twentieth Symposium (International) on Combustion.* The Combustion Institute, Pittsburgh, 1985, p. 1817.
61. Ziegler, G. F. W., Maly, R. R. & Wagner, E. P., Effect of ignition system design on flammability requirements in ultra-lean turbulent mixtures. *Combustion in Engineering,* Vol. 1. Institution of Mechanical Engineers, London, 1983, p. 81.
62. Bradley, D., The effects of electric fields on combustion processes.

Advanced Combustion Methods, ed. F. J. Weinberg. Academic Press, London, 1986, p. 331.

63. Kalghatgi, G. T., Improvements in early flame development in a spark ignition engine brought about by 'spark-aider' fuel additives. *Combust. Sci. Technol.,* **52** (1987) 427.
64. Lewis, B. & von Elbe, G., *Combustion, Flames and Explosions of Gases,* 3rd edn. Academic Press, New York, 1987.
65. Weinberg, F. J., Electrical discharge-augmented flames and plasma jets in combustion. *Advanced Combustion Methods,* ed. F. J. Weinberg. Academic Press, London, 1986, p. 277.
66. Ballal, D. R. & Lefebvre, A. H., Ignition and flame quenching in flowing gaseous mixtures. *Proc. R. Soc. Lond.,* **A357** (1977) 163.
67. De Soete, G. C., Propagation behaviour of spark ignited flames in early stages. *Combustion in Engineering,* Vol. 1. Institution of Mechanical Engineers, London, 1983, p. 93.
68. Young, M. B., *Cyclic dispersion in the homogeneous-charge spark-ignition engine: A literature survey.* SAE paper 810020, 1981.
69. Boston, P. M., Bradley, D., Lung, F. K.-K., Vince, I. M. & Weinberg, F. J., Flame initiation in lean, quiescent and turbulent mixtures with various igniters. *Twentieth Symposium (International) on Combustion.* The Combustion Institute, Pittsburgh, 1985, p. 141.
70. Edwards, C. F., Stewart, H. E. & Oppenheim, A. K., *A photographic study of plasma ignition systems.* SAE paper 850077, 1985.
71. Daily, J. W., *Cycle-to-cycle variation: a chaotic process?* SAE paper 870165, 1987.
72. Hynes, J. & Sheppard, C. G. W., SI engine cycle simulation using the Leeds turbulent burning velocity correlations, submitted for publication, 1989.
73. Bradley, D., Hynes, J., Lawes, M. & Sheppard, C. G. W., Limitations to turbulence-enhanced burning rates in lean burn engines. *International Conference on Combustion in Engines.* Institution of Mechanical Engineers, London, 1988, p. 17.
74. Miller, C. D., Knock as shown by photographs taken at 40 000 and 200 000 frames per sec. *SAE Q. Trans.,* **1** (1) (1947) 99.
75. Kirsch, L. J. & Quinn, C. P., A fundamentally based model of knock in the gasoline engine. *Sixteenth Symposium (International) on Combustion.* The Combustion Institute, Pittsburgh, 1977, p. 233.
76. Kirsch, L. J. & Quinn, C. P., Progress towards a comprehensive model of hydrocarbon autoignition. *J. Chim. Phys.,* **82** (1985) 459.
77. Axelsson, E. I., Brezinsky, K., Dryer, F. L., Pitz, W. J. & Westbrook, C. K., Chemical kinetic modeling of the oxidation of large alkane fuels: *n*-octane and iso-octane. *Twenty-first Symposium (International) on Combustion.* The Combustion Institute, Pittsburgh, 1988, p. 783.
78. Nakagawa, Y., Takagi, Y., Itoh, T. & Iijima, T., *Laser shadowgraphic analysis of knocking in S.I. engine.* SAE paper 845001, 1984.
79. Kiyokane Kaji *et al., Development of a knock sensor.* SAE paper 861375, 1986.

80. Sheppard, C. G. W. & Ibrahim, E.-S. A. A., S.I. engine ion probe diagnostics. In: *Instrumentation for Combustion and Flow in Engines*, ed. D. F. G. Durao, J. H. Whitelaw & P. O. Witze, Kluwer Academic Publishers, Dordrecht, 1989, p. 355.
81. Hayashi, T., Taki, M., Kojima, S. & Kondo, T., *Photographic observation of knock with a rapid compression and expansion machine.* SAE paper 841336, 1984.

10

Combustion in the Diesel Engine

S. Matsuoka

Science University of Tokyo, Tokyo, Japan

1. Introduction
2. Behaviour of spray
   - 2.1. Spray characteristics in evaporation
   - 2.2. The density and momentum distribution in the fuel spray
   - 2.3. The influences of swirl on the behaviour of the spray
3. Behaviour of the free-spray flame and its combustion mechanism in a rapid compression machine (RCM)
   - 3.1. The initial combustion stage
   - 3.2. Air entrainment into the spray flame
   - 3.3. Self re-entrainment of burnt-gas
   - 3.4. The influence of spray turbulence on the mixture formation of a (free) spray
   - 3.5. Wall impingement and the stratified structure inside the flame
   - 3.6. Main combustion stage
4. Combustion in DI diesel engines (small size about 100 mm bore) with piston cavity combustion chamber
   - 4.1. Engine performance characteristics
   - 4.2. Instrumentation accuracy
   - 4.3. Behaviour of in-cylinder swirl and cloud of flame mass
   - 4.4. Combustion characteristics index and *ROAE* in-cylinder
   - 4.5. Spatial and temporal distribution of temperature, equivalence ratio, soot and NO of the flame in a DI engine
   - 4.6. Interference of swirl and squish flow in the cavity combustion chamber and their optimization
   - 4.7. Internal structure of a diesel spray flame
5. Conceptual classification of diesel combustion system from the viewpoint of the optimization of the air utilization and burnt-gas disturbance
6. Concept of an optimization strategy from the viewpoint of low fuel consumption, low particulate and low NO emission

1. INTRODUCTION

The complicated mechanism of combustion in the diesel engine and the fact that sufficiently precise measurements of the chemical and physical phenomena are beset with difficulties mean that many phases of the combustion process still defy satisfactory analysis.

The troubled diesel engine designer still awaits a solution of problems that will give him reliable criteria for the design of a combustion chamber for a new engine that will give good economy with low smoke and noise together with low exhaust gas pollutants, namely oxides of nitrogen and particulates, as required by the emission regulations.

The systematic research work[1–7,9–26,29,30,33–40] presented in this chapter made by the author, his research staff and students at the Tokyo Institute of Technology (1970–84) is an endeavour to elucidate some of these problems. To simplify such phenomena a large single-stroke rapid compression machine (RCM) was designed for a large part of the research investigation as this allows much better access to instrumentation and *image analyzed* high-speed ciné photography. However, to ensure that true engine conditions are being reproduced, correlation with actual engine investigations were incorporated into the project.

2. BEHAVIOUR OF SPRAY

2.1. Spray Characteristics in Evaporation

The evaporation process of the diesel fuel spray was obtained from experiments and prediction,[1–3] from the viewpoint of thermodynamical heat balance and the integration of a cloud of single droplets. Kamimoto and the author *et al.* predicted that the main dominating factors in the evaporation rate are the surrounding temperature, droplet diameter and spray volume. Immediate evaporation of the droplets takes place in less than 0·5 ms after completion of injection.

2.2. The Density and Momentum Distribution in the Fuel Spray

In practice in a spray, it is clear that there exists a liquid phase at the core of the spray centre. This will be seen by comparing the focused

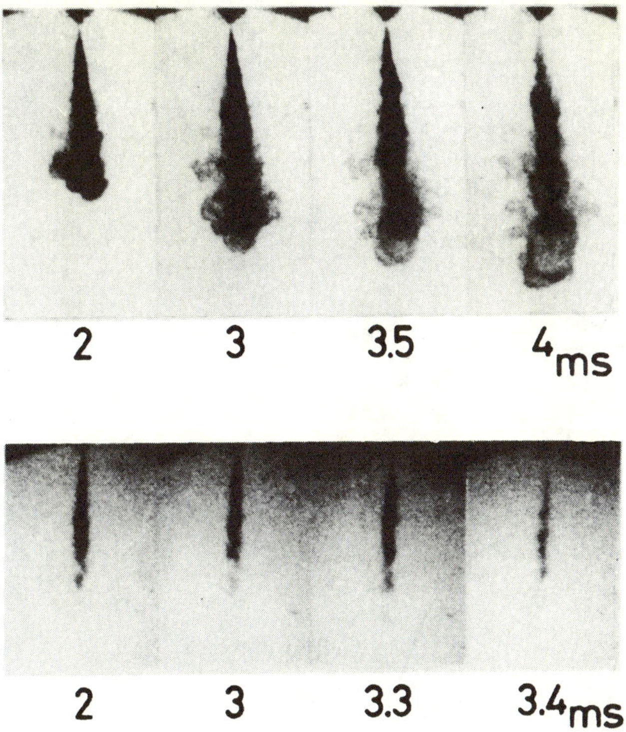

Fig. 1. Evaporating diesel spray in quiescent chamber taken by the focused shadow (top) and the back-illumination (bottom) photo techniques. Surrounding condition: gas = N_2; pressure = 3·4 MPa; density = 16·9 kg/m^3; temperature = 670 °K, injection condition; dia. of orifice = 0·2 mm; injection pressure = 25·5 MPa, injection duration = 3·3 ms; fuel = *n*-tridecane.[5,9]

shadow photography shown in Fig. 1 (top) with back illuminating photography shown[4,5,9] in Fig. 1 (bottom). It is shown, however, in Fig. 2 that in the vicinity of the nozzle hole, that is within 40 mm, there is a high fuel concentration towards the centre.[6] This region possesses therefore a strong momentum.

2.3. The Influences of Swirl on the Behaviour of the Spray

In the swirling combustion chamber only the peripheral portion of the spray, with a weak momentum, is carried away by the swirl, and the

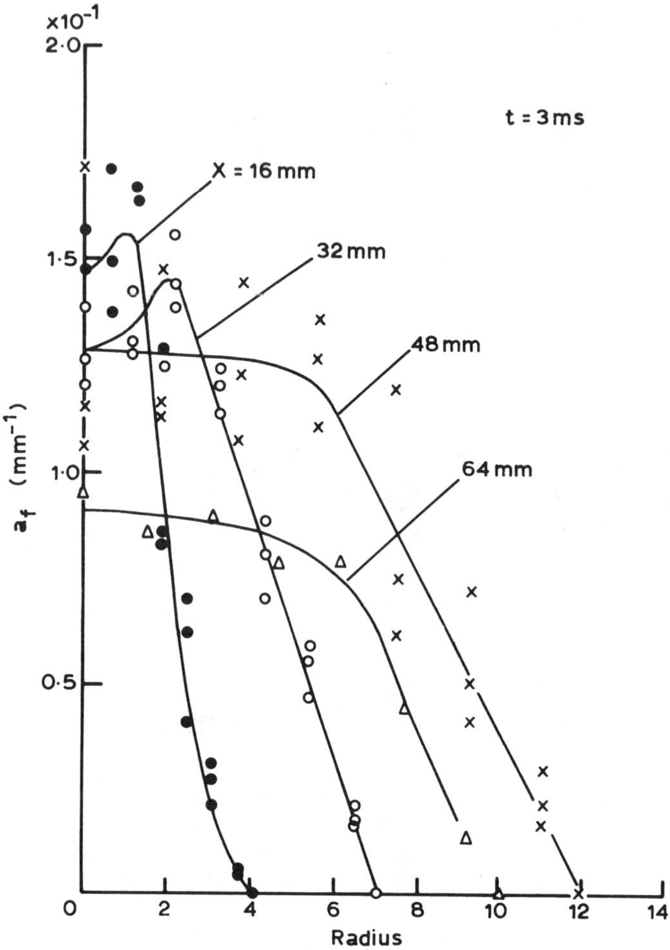

Fig. 2. Distribution of a_f (droplet space density m^2/m^3) in radial direction at spray B (see Fig. 5).[6]

main part of the spray with the high momentum is only slightly affected or destroyed by the weaker gas flow momentum. This is verified as follows.

The comparison of measured and predicted spray development[7] is shown in Fig. 3. On the right hand side is shown our modified 2-D model, and the behaviour of the centre column of the main jet portion is described as being the same as the 2-D model by Sinnamon *et al.*[8] at the centre of figures.

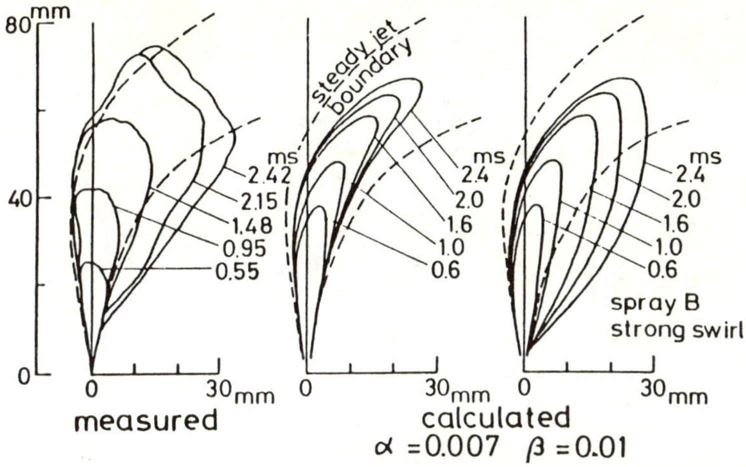

Fig. 3. Comparisons of development of spray B (see Fig. 5) between predictions and measurements. Left: solid line shows measured boundary. Middle: solid line shows boundary calculated by simple 2-D model.[8] Right: solid line shows boundary calculated by modified 2-D model.[7] (α, β, entrainment parameters along axial and normal directions).[8]

The trailing boundary at the downstream side can be determined from an envelope formed by trajectories of tracer particles and its travel distance can be expressed by time multiplied by the local swirl velocity.

The predicted results as shown on the right hand side in Fig. 3 agree well with the experimental results on the left hand side. This suggests to us, in a medium size diesel engine (c. 100 mm diameter), that the mixing of the jet spray in the cavity is not promoted by the swirl before wall impingement but is promoted after the spray momentum is diffused along the impinged wall by the sweeping action of the swirl.

3. BEHAVIOUR OF THE FREE-SPRAY FLAME AND ITS COMBUSTION MECHANISM IN A RAPID COMPRESSION MACHINE (RCM)

3.1. The Initial Combustion Stage

The initial spike on the rate of heat release (ROHR) curve, as is well known, is the initial combustion stage as the result of the combustion

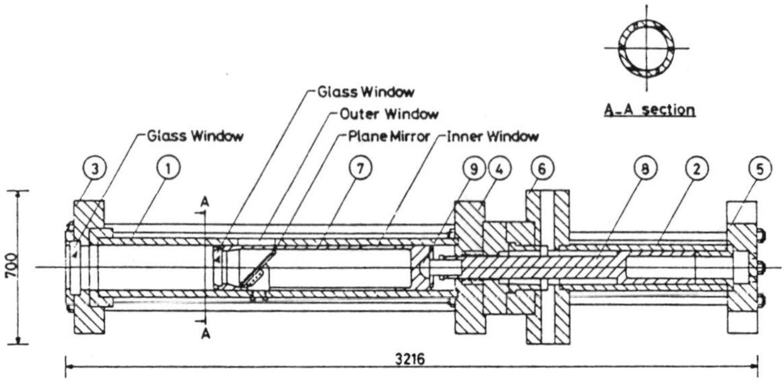

① Combustion Cylinder
② Driving Cylinder
③ Head Block
④ Intermediate Block
⑤ Tail Block
⑥ Controlling Block
⑦ Compression Piston
⑧ Driving Piston
⑨ Ball Joint

Fig. 4. Schematic diagram of large-scale rapid compression machine of Tokyo Institute of Technology (combustion chamber 200 ϕmm × 40 mm).[9,10]

of the peripheral lean mixture of the spray, as mentioned in Section 2.3.

3.2. Air Entrainment into the Spray Flame

The spray volume rather relates to the evaporation process, but this is influenced by the air entrainment and fuel evaporation in the spray. The author and co-researchers designed[5,9,11] a large-size (200 mm diameter) rapid compression machine (RCM) as shown in Fig. 4, and Kobayashi and the author *et al.* observed a very large number of single-shot free-sprays flames as shown in Fig. 5.[10,11] They measured the change in the volume of spray flame from these photographs and estimated the rate of air entrainment (*ROAE*) by the method of thermodynamical calculation referred to in the literature.[9–11]

The diesel spray, with its strong momentum accompanying the initial explosive combustion, enables it to entrain a considerable amount of the surrounding air in the vicinity of the spray. The measured relationship between air entrainment quantity *vs* time is shown in Fig. 6. In this figure, the free-spray flames S, M and L obtained in the RCM (but with different ignition delays) are the same

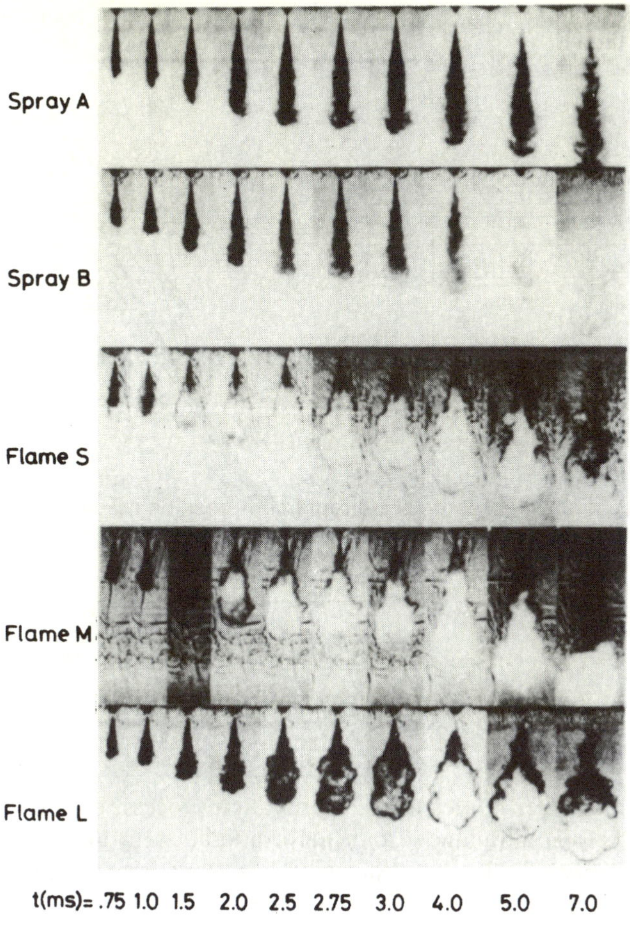

Spray A	Spray B	Flame S	Flame M	Flame L
in N_2	in N_2	in Air	in Air	in Air
1·49 MPa	3·22 MPa	3·16 MPa	3·21 MPa	3·22 MPa
293 K	627 K	710 K	667 K	642 K
17·1 kg/m^3	17·3 kg/m^3	15·5 kg/m^3	16·8 kg/m^3	17·5 kg/m^3

Fig. 5. Photograph of non-evaporated, evaporated and three fired diesel sprays under the same injection condition by means of RCM and focused shadow photography.[10,11]

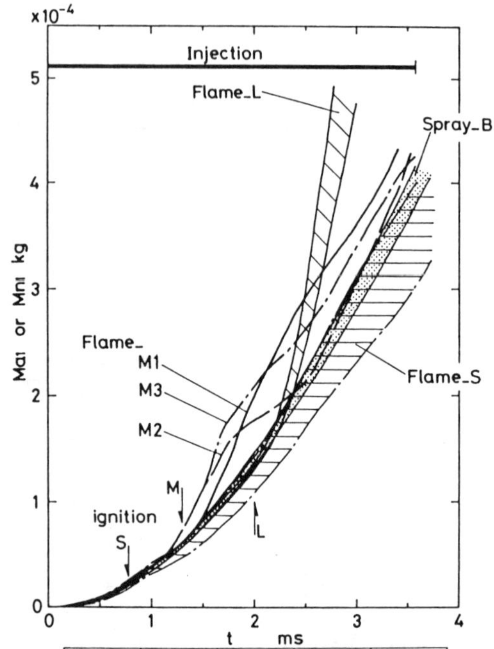

Series	gas	P MPa	T_2 K	ρ_2 kg/m^3
Spray	N_2	3.22	627	17.3
Flame_L	Air	3.22	642	17.5
Flame_M	Air	3.21	667	16.8
Flame_S	Air	3.16	710	15.5

Fig. 6. Measurement of air entrainment vs time in diesel sprays and flames in RCM.[9–11]

sprays as the non-burning spray B. Although they have the same injection characteristics, flames M and L with longer ignition delays than flame S show an explosive increase in the initial combustion rate, which causes the air entrainment mass to become larger than in spray B and flame S.

The relationship between the *ROAE* and injection characteristics, etc., shown here is formulated in Ref. 30, from the data of Fig. 6.

3.3. Self Re-entrainment of Burnt-gas

During the initial combustion period of the diesel spray, because the spray continues to feed into the flame of its own making, it is easy to imagine that its own burnt-gas is re-entrained (*REOBG*) into the spray in the same way as the new air is entrained as shown in Section 3.2. This supposition was described by the author in Refs. 15–19, and confirmed by various data obtained as follows:

(a) Verification by high speed photography: the photographs in Fig. 5, illustrate the self-re-entrainment of the burnt-gas quite well.

(b) verification by a comparison between the *ROHR* curves obtained in the RCM and a DI engine: typical data of the *ROHR* curves of the free-spray flames I, II, III in the RCM[10,11,30] and in the DI engine obtained[20,21] by varying the ignition delay period are shown in Figs. 7(a) and (b). For the purpose of comparing the *ROHR* curves of each main combustion stage, we may take the cases at medium (II and −15°)† and short (I and −5°) ignition delays for initial combustion in the RCM in Fig. 7(a), and for the DI in Fig. 7(b).

RCM (Fig. 7(a)):

$$(ROHR)\text{max II} < (ROHR)\text{max I} \tag{1}$$

but at −15° CA and −5° CA in the DI engine (Fig. 7(b));

$$(ROHR)\text{max} -15° \geqslant (ROHR)\text{max} - 5° \tag{2}$$

(some of the data in the DI engine gives a clearer difference of (*ROHR*)max with varied injection timing.)

These show that the relationship between initial combustion and (*ROHR*)max in the two combustion chambers (Figs 7(a) and (b)) are somewhat different. A brief interpretation will be made as follows. The reason for the difference is presumed to be a result of the differences in the internal structure of the flame due to the presence or absence of wall impingement. In the period when the initial combustion increases, both in the RCM and in the DI engine, it is presumed that the re-entrainment of burnt-gas will increase as the entrainment of air increases.

In the case of free-spray in the RCM, the burnt-gas will be entrained into the fuel spray zone rather earlier than air due to a degree of stratification. From this, it may be presumed that the burnt-gas will trap to entrain the air into the spray zone and then mixing occurs, although not completely. Therefore, in this free-spray case, the shorter the ignition delay the smaller the burnt-gas entrainment, and the larger the air entrainment, then the higher the (*ROHR*)max becomes.

† Crank position in degrees before top dead centre (BTDC) is given as −ATDC (minus after top dead centre).

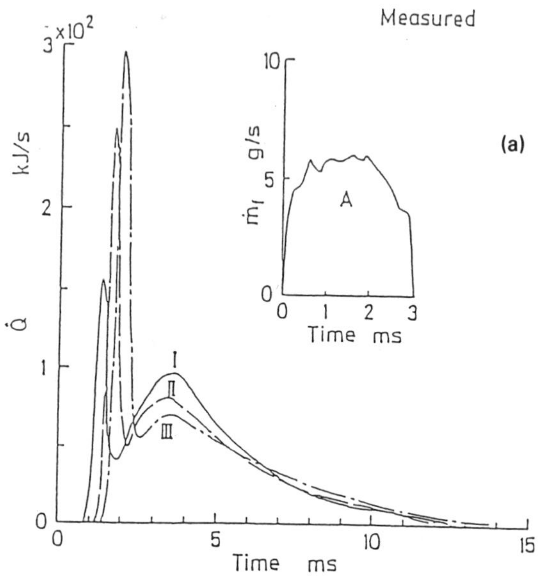

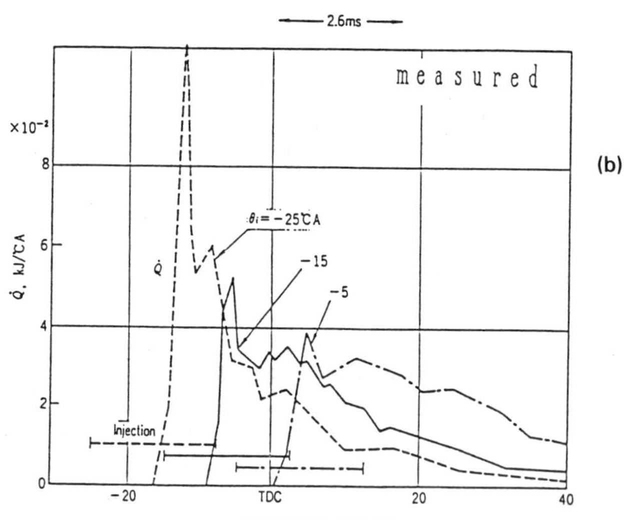

Fig. 7.(a) $\dot{Q}$ = *ROHR* when ignition delay timing is varied at RCM (0·8 ms, 1·2 ms, 2·1 ms; Fig. 5).[10,11,30] (b) $\dot{Q}$ = *ROHR* when injection timing is varied in DI engine (SR = 0, shallow dish, 1250 rpm, $\lambda = 2{\cdot}0$; Fig. 20).[20,21]

In the case of the DI engine in Fig. 7(b), however, because there exists wall impingement mixing of the abovementioned stratified zone is promoted. Therefore, in the DI engine case, the longer the ignition delay the larger the air entrainment and this will dominate the (*ROHR*)max rather than the effect of *REOBG*. However, there would be an optimum ignition delay timing to excite the highest (*ROHR*)max. These are the factors which differentiate Figs. 7(a) and (b).

(*ROHR*)max also depends on the mass of fuel remaining after initial combustion is finished. In the DI engine, the retard injection timing will also be a further reason for the decrease in the (*ROHR*)max. These three factors together explain the phenomena mentioned above. A more detailed explanation will be given in part (f) and Sections 3.3 and 3.5.

(c) Verification by the measurement of temperature and soot concentration within the free-spray flame is as follows: the values of the flame temperature distribution and soot concentration distribution are measured over a cross-section at right angles to the axis of one of the free-sprays in an RCM.[4,17,22–24] Figure 8 shows that in the free-spray flame in the RCM, the temperature in the central section becomes quite high (2200 °K) while the soot concentration becomes

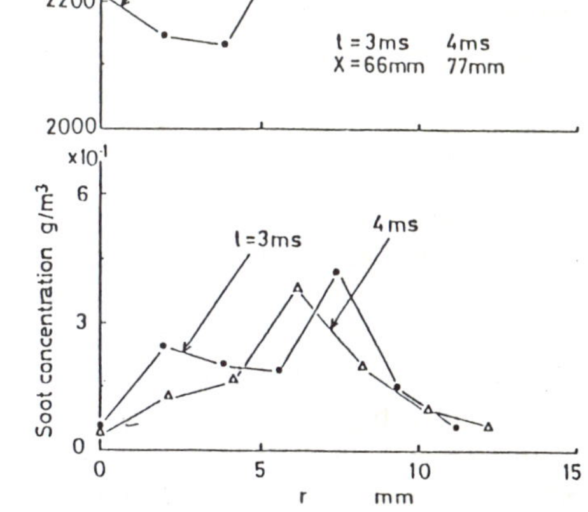

Fig. 8. Flame temperature and soot concentration distribution in the radial direction with RCM flame, measured by the two-colour method and calculated using the onion peeling model.[4,17,19,24]

very low. Since the central core portion of diesel spray has a high fuel concentration and strong momentum, the existence of a non-burning low-temperature spray may be expected. However, the burnt-gas formed by the initial combustion which contains a small amount of oxygen is presumed to enwrap and invade the following injected fuel spray, causing the temperature at the centre of the spray-flame to become high and the soot concentration to become low.

(d) Verification by measured values of the soot oxidation rate in a DI engine, compared with the method of graphite rod and shock wave tube etc. is as follows: Figure 9 is an Arrhenius plot of a comparison between our data from Figs. 30–32, and the soot oxidation rate obtained from other non-diesel experiments as mentioned above and referred to in Ref. 25. This shows that in a diesel engine with a high partial pressure of oxygen within a combustion chamber, the soot oxidation rate is lower compared with other examples, indicating that in the latter the oxygen partial pressure is low.

From these comparisons, it is presumed that the reason for the drop in the soot oxidation rate in our engine data is mainly caused by the drop in the partial pressure of oxygen in the centre of the spray by the re-entrainment of burnt-gas (*REOBG*).

(e) Verification by the high speed laser shadowgraph photography of the soot formation process in the RCM is as follows: according to a study of Kamimoto *et al.* at the Tokyo Institute of Technology,[26–28] the soot in free-spray in a RCM is formed in the fuel-lean region near the flame tip rather than in a fuel-rich region in the vicinity of the nozzle (Fig. 10). The interpretation of this process in Ref. 28 is that the evaporated fuel requires time to be pyrolyzed before it arrives at the tip of spray flame.

The author suggests that the new fuel spray travels through the burning high temperature (with scarce oxygen) zone, which was formed by the re-entrainment of burnt-gas of the initial combustion which enwrapped the spray, towards the flame tip and as it is fuel-rich it is pyrolyzed. After this, the spray will meet new air entrained from outside of the envelope of burnt-gas at the leading portion of the spray flame, and will burn accompanied by the formation of soot. If more turbulence remains at this time, then more air is entrained and more soot is oxidized and vanishes (Figs. 10, 11(a) and (b)).

(f) Verification by the model calculation using a new three zone model is as follows: The author and co-workers attempted a calculation to obtain the value of *ROHR*, using a new three zone model of

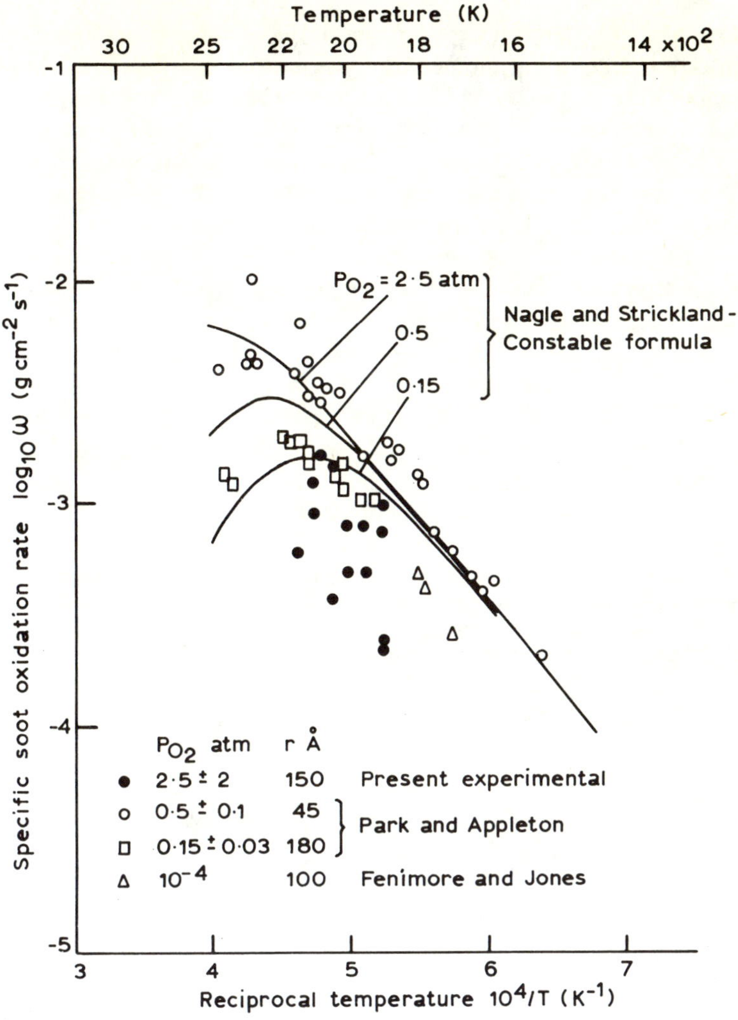

Fig. 9. Arrhenius plot of specific oxidation rate showing a comparison with our own and previous data.[25] Our data are calculated from Figs 30–32.

which the schema is shown in Fig. 12 from Ref. 29. In our three zone model, the injected fuel is mixed not only with the entrained air as shown in Refs 30 and 31 but also with the re-entrained burnt-gas (*REOBG*) which enwrapped the zone 1 of fuel spray mixture.

In the case where the ignition delay is long, naturally the amount of

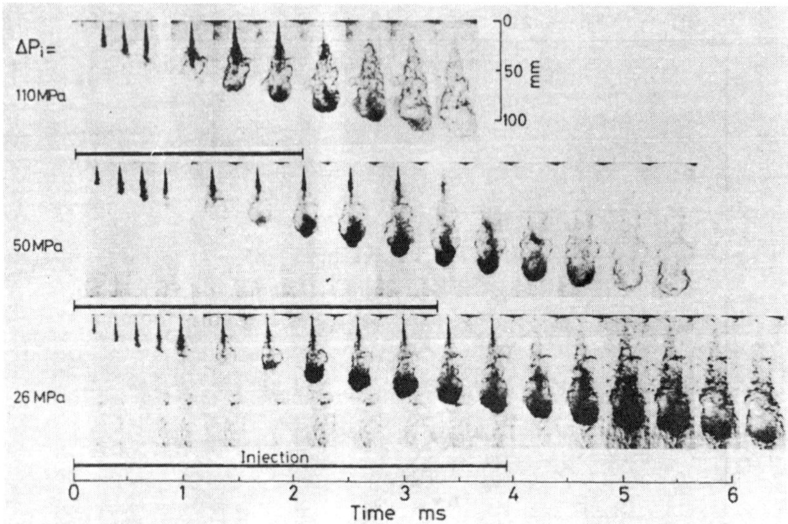

Fig. 10. High-speed focused shadow photography using Argon laser at free-spray in RCM in three different injection pressures. Black portion indicates mostly the shadow of particulates. Density difference appears as a slight shadow and it shows the envelope of flame.[28]

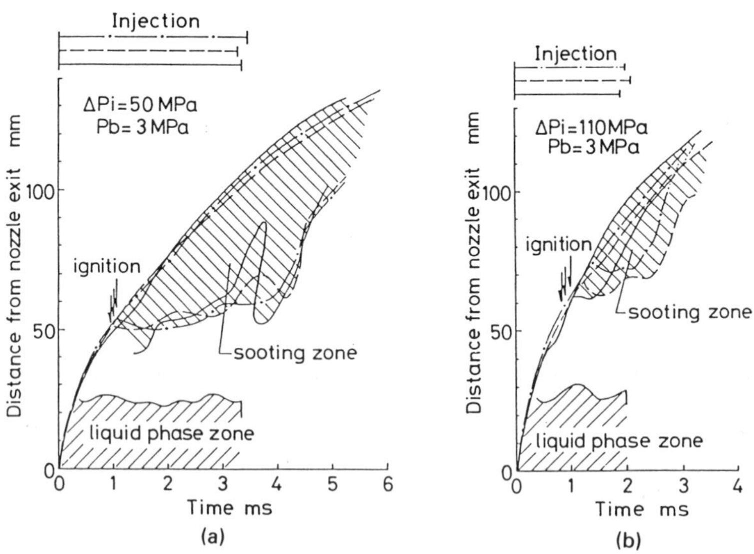

Fig. 11. Illustration of the concept of soot formation in a diesel flame.[28] (a) $\Delta P = 50$ MPa; (b) $\Delta P = 110$ MPa.

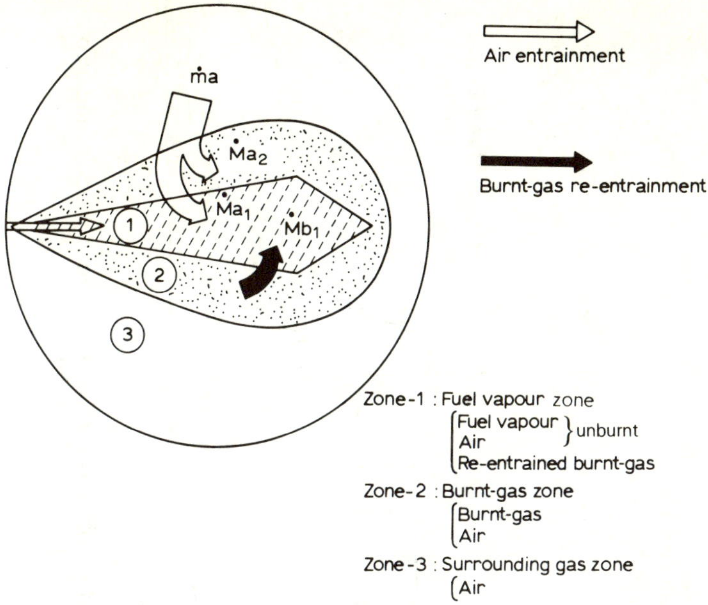

Fig. 12. Schema of three zone diesel flame model.[29]

the initial combustion becomes large. In this three zone model, in such a case, the fuel spray mixture of zone 1 is enwrapped with the thicker overcoat of the burnt-gas compared with the case of short ignition delay. Therefore *REOBG* into zone 1 is increased and it accompanies the decrease of the mass flow rate of the mixed amount of fuel in zone 1, though the mixing rate is the same. This causes a decrease in the accelerated rate of heat release (*AccROHR*) accompanied by the pyrolysis formation, which will be the sooting as mentioned in part (e) of this section. Our calculated result is shown on the right-hand side of Fig. 13, compared with the results of the two zone model (centre) of Refs 30 and 31 and the measured result (on the left-hand side of Fig. 7(a)). Our results agree well with those measured in Fig. 7(a).

From the several kinds of data and verifications shown above, the existence of the phenomena of self re-entrainment of burnt-gas (*REOBG*) into the fuel spray is virtually confirmed. It can be schematized as a Christmas tree type stratified-spray flow as shown in Fig. 14 with more details shown in Fig. 19 (the concept of stratification is described in Section 3.5).

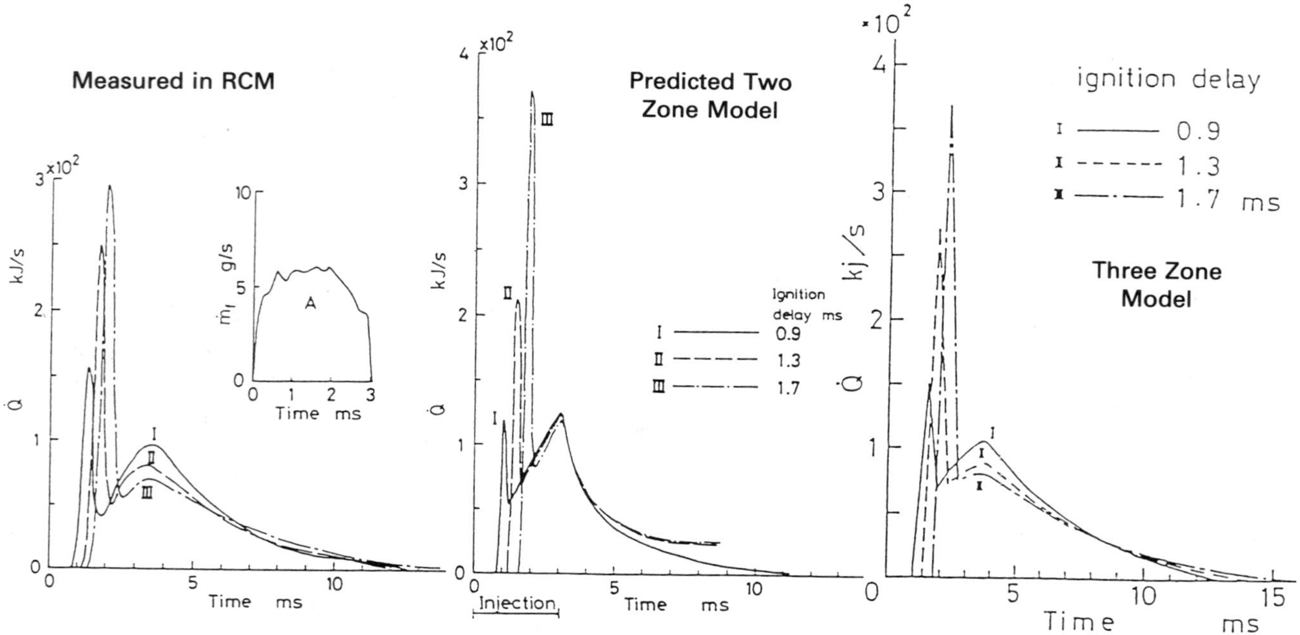

Fig. 13. Comparison of predicted *ROHR* by three (right-hand side)[29] and two (middle)[30] zone models, and also with a measured one in RCM (left-hand side, as Fig. 7(a)) with varied ignition delay timing.

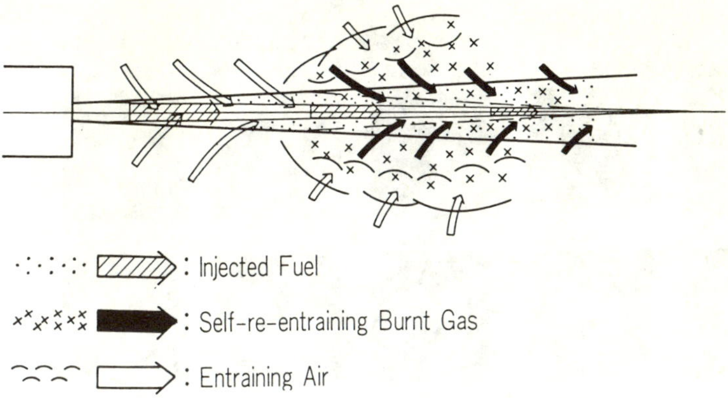

Fig. 14. Schema of Christmas tree type stratified spray flame.[19]

3.4. The Influence of Spray Turbulence on the Mixture Formation of a (Free) Spray

According to the study of RCM of the Tokyo Institute of Technology, as already shown in Section 3.3(e), Kamimoto *et al.*[28] clarified that the high pressure injection method contributes considerably to the decrease of soot formation inside the spray. Figures 10 and 11 show that the higher the injection pressure the lower the sooting is. The flow rate of the mixed amount of fuel mass is known to be proportional to the amount of entrained air, turbidity[32] but diluted by the entrained burnt-gas amount etc.[29]

According to the data of Fig. 6, and the calculation of *ROAE*,[29,30] the total amount of entrained air in the spray is greater than the stoichiometric amount (calculated results are omitted here). From this, the author presumes that the process of low soot formation in a high injection pressure condition (Figs 10 and 11) is mostly dependent on the high mixing rate characteristics resulting from the fast diffusion and mixing of the entrained air into the central core of spray.

However, when the injection rate is not sufficiently high, as frequently in the conventional DI engine, the entrained burnt-gas and low mixing condition inside the spray flame form the high-temperature, fuel rich condition at the central core of the spray forming in turn the fog of soot at the centre. The low soot formation is needed for strong turbulence inside the spray flame. Figure 19 and Ref. 29 indicate this process.

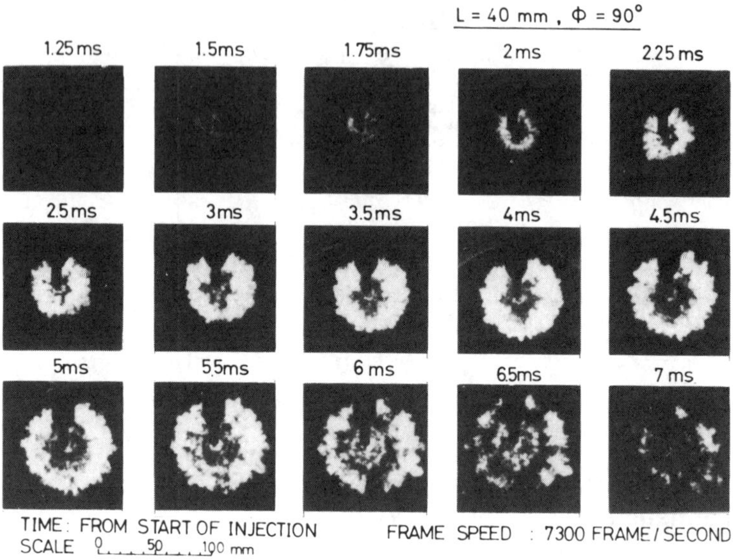

Fig. 15. High-speed photographs when spray flame impinges at right angles on to the wall in RCM ($L = 40$ mm, $\phi = 90°$).[18,19]

3.5. Wall Impingement and the Stratified Structure Inside the Flame

Photographs of a spray flame impingement on the wall surface in the RCM[18,19,33] are shown in Fig. 15 (right-angle impingement) and in Fig. 16 (slanting-angle impingement). A comparison of the true *ROHR* (heat losses are not included) for various cases of wall impinging flames with a free-spray flame is shown in Fig. 17. The (*ROHR*)max in all the wall impinging cases is larger than that of a free-spray flame. The calculations are made from measurements of pressure and heat flux.[33]

The heat flux data[33] in the case of right-angle impingement ($L = 40$ mm and $\phi = 90°$) suggest that the flame temperature is increased because the turbulence has increased at the tip of the flame. Also, the flame temperature in this area is about 100°K higher compared with a free-spray flame.[23,24] The reason for this is the enhanced promotion of mixing by impingement. The importance of turbulence promotion is shown not only from the injection itself, as already mentioned in Section 3.4, but also from that made by the wall impinging action of the spray jet. However, it is shown in Fig. 16, in the case of slanting impingement, that the flame spread is greatly diverted to the progression side only and develops into a triangular pattern rolled up scroll. The total spray flame volume for the case of slanting impingement is mostly occupied by this triangular portion which is smaller than the total volume for the case for right-angle impingement which forms the 'doughnut' pattern shown in Fig. 15.

In spite of this observation, the (*ROHR*)max for the slanting impingement is almost the same as in the case of right-angled impingement shown in Fig. 17. This indicates that the *ROHR* flux per unit volume and the local combustibility in the rolled-up portions of the flame obtained by slanting impingement is considerably enhanced, although the turbidity in the flame must be decreased compared with the case of thc right-angled impingement.

This suggests that in a case of slanting impingement, the increased slipping speed and surface area of the spray spread promote the fuel evaporation and mixing and also generate a large scale vortex in the shape of a rolled-up carpet, which causes the exchange of the three composites

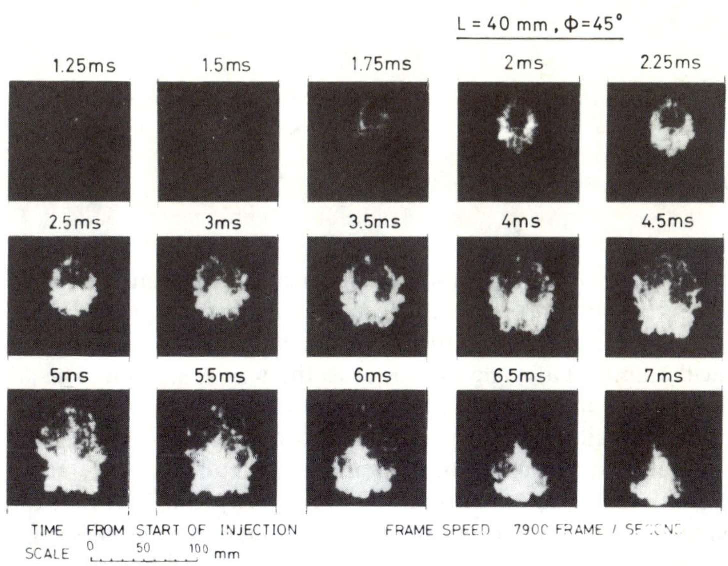

Fig. 16. As Fig. 15 but with slanting-angle impingement ($L = 40$ mm, $\phi = 45°$).[18,19]

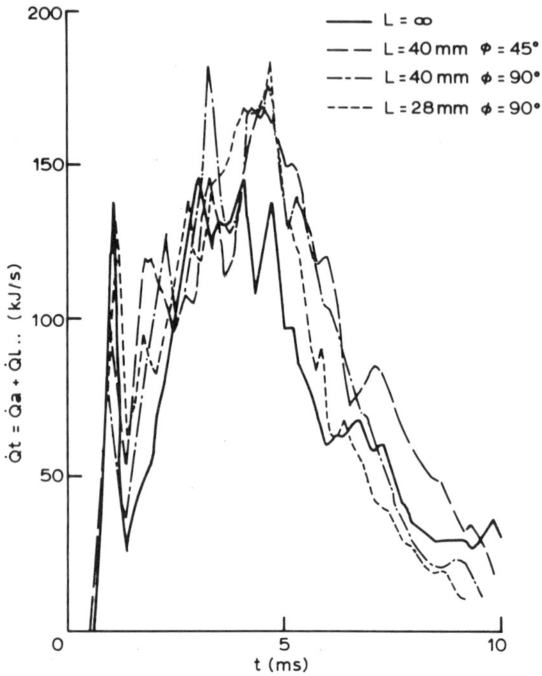

Fig. 17. Values of *ROHR* when impingement of RCM flame is varied. $\dot{Q}t$ = total heat release rate; $\dot{Q}a$ = apparent heat release rate; $\dot{Q}L$ = heat loss rate at impinging wall side. $L = \infty$ means in the case of free-spray.[17,18,33]

of air, fuel and burnt-gas within the spray by stirring and mixing to a near-perfect mixing process, which in turn increase *ROHR*. If this is the case, it will be desirable to design the curvature of the cavity wall so that the impinging angle of the jet to the wall is smooth. This idea may be realized in practice[42] (see Section 5.2).

In summary, if this interpretation is correct, it suggests that the diesel spray flame consists of a stratified structure of fuel, air and burnt-gas. This is an important base to verify that the burnt-gas entrainment phenomenon is of similar significance to that of air entrainment. Figure 18 shows the schema of the difference in the pattern attained with wall impingement at a right-angle compared with that of a slanting angle.

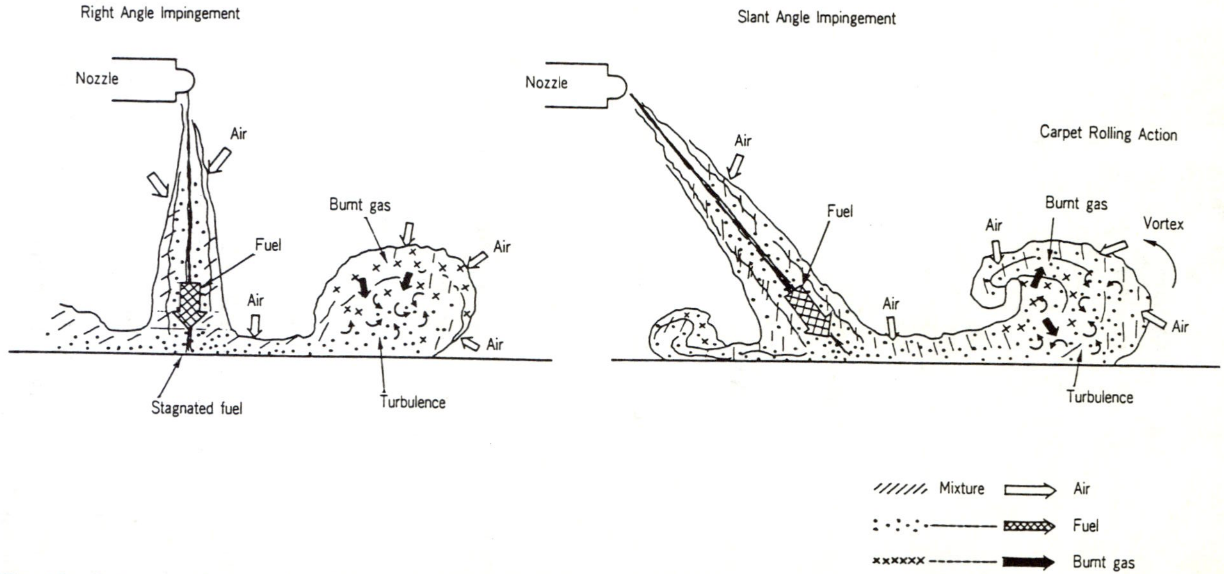

Fig. 18. Schema of spray impingement patterns along the wall.

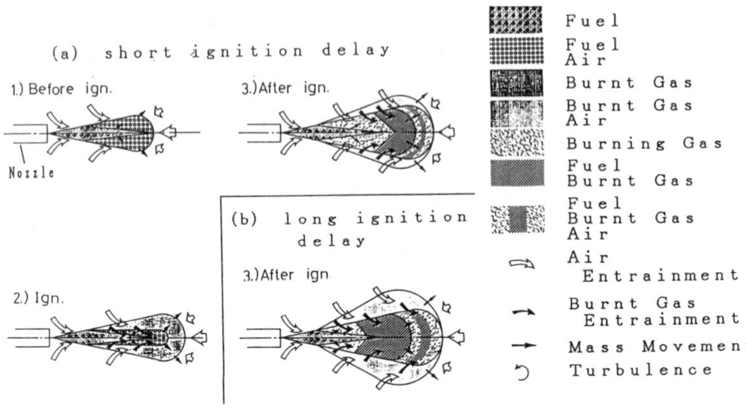

Fig. 19. Schema of *REOBG* phenomenon in RCM free-spray flame.[29]

3.6. Main Combustion Stage

It is already a known fact that the first half of the *ROHR* of the main combustion period (i.e. accelerated rate of heat release; *AccROHR*) is controlled by the mixing rate excited by the turbulence of the injection spray. But if the interpretation of this phase, as already mentioned in the preceding Sections 3.3 and 3.5, is correct, *AccROHR* will also become dominated by the amount of *REOBG*. The greater the amount of *REOBG,* the slower the *AccROHR* becomes and the more the soot is formed, as mentioned in Sections 3.3(e)(f) and 3.4.

Schematic diagrams in Fig. 19 illustrate this process for a free-spray in the RCM. In the case of a long ignition delay, the fresh air surrounding the set will be trapped and entrained into the central area where there will be an air-deficient high temperature condition (see Fig. 19(b)). One can see the mist of sooting in this illustration as already mentioned.

For the case of the wall-impinging spray flame in the DI engine shown in Fig. 7(b), the schematic diagram in Fig. 18 illustrates this process. The amount of each of the three composites of fuel, air and burnt-gas with their wall impinging action generates a rather complicated fluid condition. Too much entrainment air with too much burnt-gas will cause rather irregular combustion. It sometimes causes a steep rise in the *ROHR* curves in the first half period, but occasionally (and simultaneously) prolongs the tail of the end-burning, as shown in

Fig. 7(b). Therefore, in the case of wall impingement, there is an optimum ignition delay timing to achieve the best combustion, as mentioned in Section 3.3(b).

For the latter half period of the *ROHR* of the main combustion, the influence of turbidity decreases and re-entrainment of burnt-gas increases. The mixing action becomes slow and the chemical action spreads its domination.

4. COMBUSTION IN DI DIESEL ENGINES (SMALL SIZE ABOUT 100 MM BORE) WITH PISTON CAVITY COMBUSTION CHAMBER

4.1. Engine Performance Characteristics

Figure 20 shows the performance and exhaust characteristics of the test DI engine when several parameters are varied as shown.[20,21]

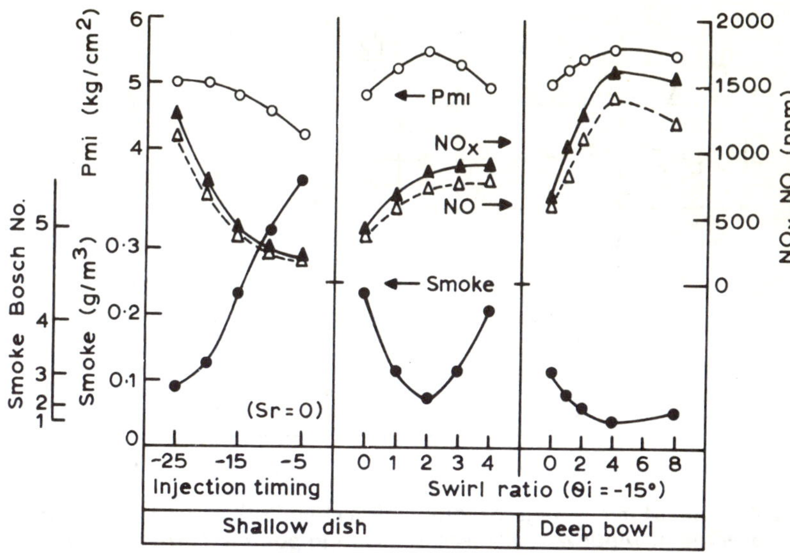

Fig. 20. Engine test conditions and bulk exhaust measurements at an engine speed of 1250 rpm and overall equivalence ratio of $\phi_m = 0{\cdot}50$. (Fuel amount is kept constant at all conditions).[20,21]

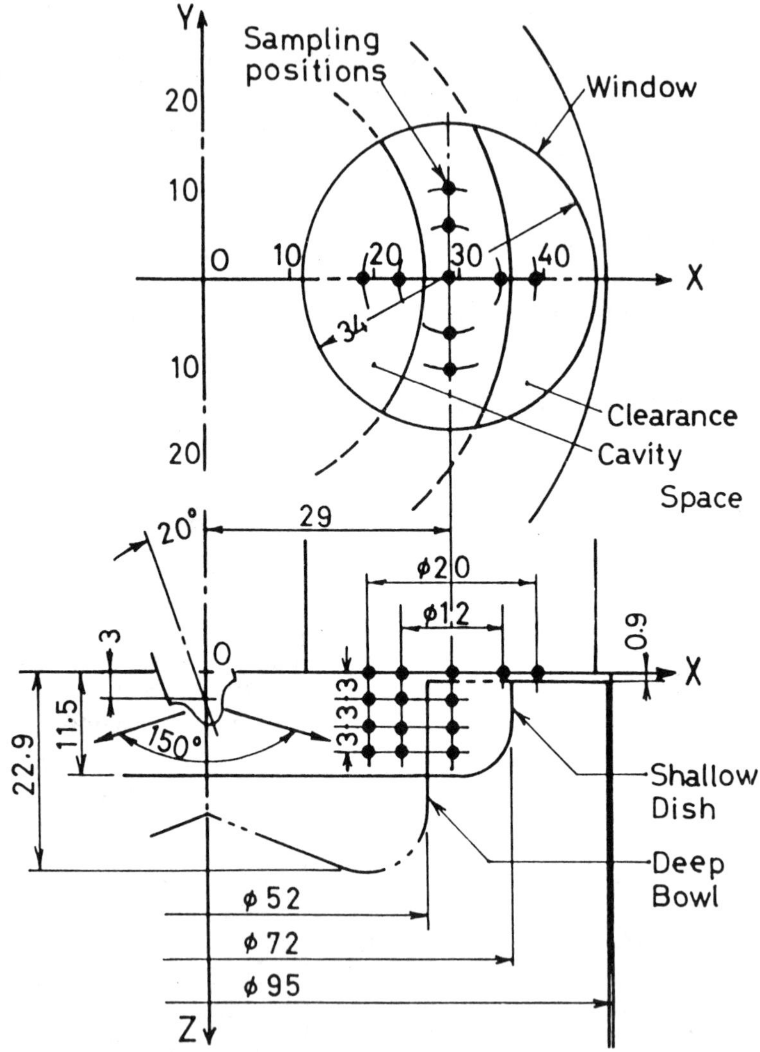

Fig. 21. Test DI diesel engine (combustion chamber), geometry, position of observation window by camera, measuring point of flame temperature and gas sampling.[20,21]

Test engine specifications are as follows:

Bore, ϕ95 mm;
Stroke, 110 mm;
Nozzle diameter, 0·20 mm × 4;
Spray angle, 150°;
Cetane Number 57;
Compression ratio, 14·6;
Valve opening pressure, 17 MPa;
Fuel, diesel oil JIS No. 2;
Specific Density, 0·849(15°C).

For these tests the standard operating conditions are fixed as follows:

Engine speed, 1250 rpm;
Without swirl in shallow dish cavity combustion chamber;
Injection timing, −15°ATDC;
Air excess ratio, 2·07–2·00 ($\phi_m = 0{\cdot}483$–$0{\cdot}5$).

The geometry of the combustion chamber is indicated in Fig. 21.

Under the standard operating conditions, the most suitable swirl ratio (SR) from the viewpoint of soot and SFC, was 2 for the shallow dish type and 4 for the deep bowl type as shown in Fig. 20.

4.2. Instrumentation Accuracy

Details of measuring techniques used in various researches are shown in Refs 13, 14, 20, 21, 34–39. Measurement of local flame temperature was made by a two-colour method as shown in detail in Refs. 34–36. In Fig. 22, Matsui and the author *et al.* presented the comparative data of local soot concentration in the flame of the above DI engine by means of the two-colour spectroscopic method[34–37] and the direct gas sampling method.[16,20,21] Both sets of data results gave good agreement.

Figure 23 presents a comparison of the equivalence ratio ϕ_{fa} in the flame, by means of a measurement obtained by the calculation of the data from the rate of air entrainment (*ROAE*) and the rate of injection (*ROI*), with ϕ_{fg} by the direct gas sampling method, which was measured and calculated using 12 chemical species including soot concentration.[20,21]

As previously mentioned *ROAE* is measured by the method discussed in Section 3.2. But in the DI engine case, the measurement of the volume change of the spray flame shown in Section 3.2 in RCM is impossible. Here, it was measured by the representative flame temperature T_f, instead of by volume, by means of a two-colour method, scanning across the flame sections. Both sets of data

Fig. 22. Comparison of the local soot concentrations in flame, obtained by the two-colour method and the gas sampling method.[16,37]

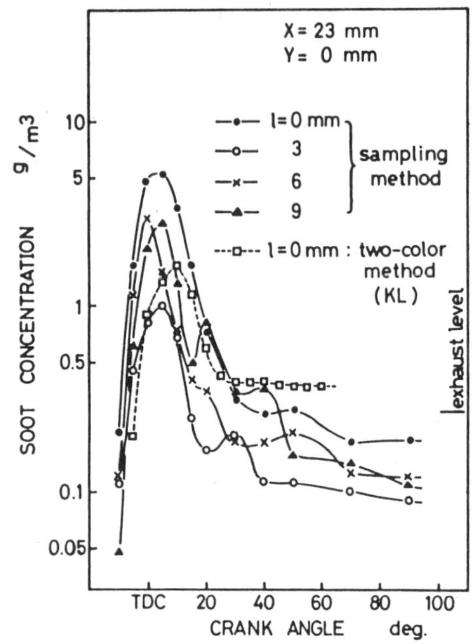

(abovementioned and here) show good agreement under the quite different principles of instrumentation.

The data shown in Figs 24 and 25 and Figs 29–32 are all measured by the previously discussed instrumentation. These are the combined results of the phenomena of chemical reaction, convection and dilution. In the case of SR = 0, where the gas state in the vicinity of the TDC ± 20° is in a rather quiescent condition, the data obtained are influenced predominantly by chemical reaction, but it must be noted that 20° ATDC convection and dilution exert a progressive influence. In such cases, careful interpretation is needed. Measurement of swirl ratio is observed from the ciné photographs directly.

4.3. Behaviour of In-cylinder Swirl and Cloud of Flame Mass

In Figs 26(a) and (b), Ahn and the author *et al.* present maps of time resolved isothermal lines made from the image analyzed measurements[4,22–24] by applying the two-colour method to the high-speed ciné-photographs shown in Figs 27 and 28 (Fig. 27 SR = 2, Fig.

28 SR = 4) in the combustion chamber of a shallow dish type DI engine mentioned above. The following are noteworthy points in these figures:

(a) The highest flame temperatures at 10° ATDC are (2234 °K SR = 0, 2510 °K SR = 2, 2540 °K SR = 4). This shows that the higher the swirl ratio, the higher the temperature becomes.

(b) When there is no swirl, the temperature is high at the flame tip, while where there is a swirl the flame temperature is highest on the downstream tip side. NO concentration is the highest in this area.[20]

(c) The white bright flame portion in the photographs of Figs 27 and 28, seems to indicate a good combustion condition, and the flame seems to be well diffused. However the emissivity of these portions of the flames was rather low and because the mass is small they contain only a small amount of energy; therefore these contributions were omitted here. The maps indicate that the flame mass does not seem to be greatly diffused, distorted or diluted and appears to be moving along with the swirl. The destruction of these masses is very important and

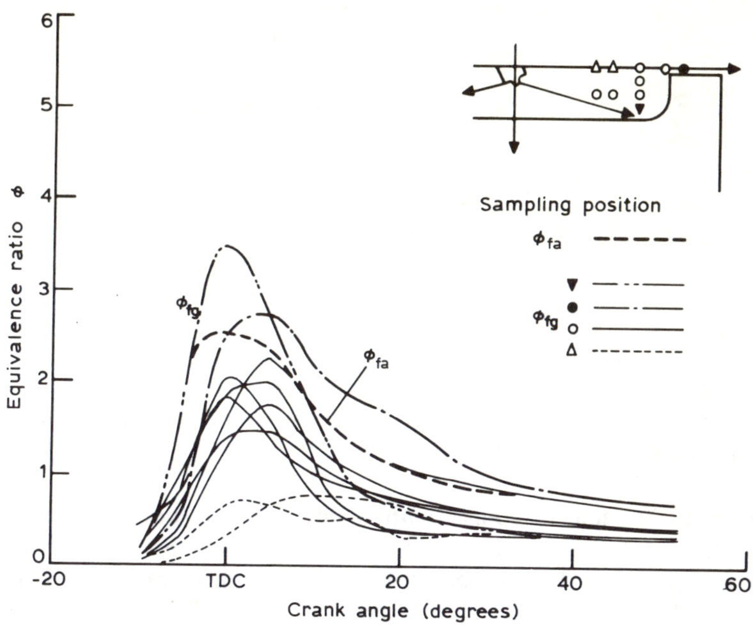

Fig. 23. Comparison of local ϕ_{fg} in flame, by the direct gas sampling method and ϕ_{fa} by *ROAE* and *ROI*.[16,20]

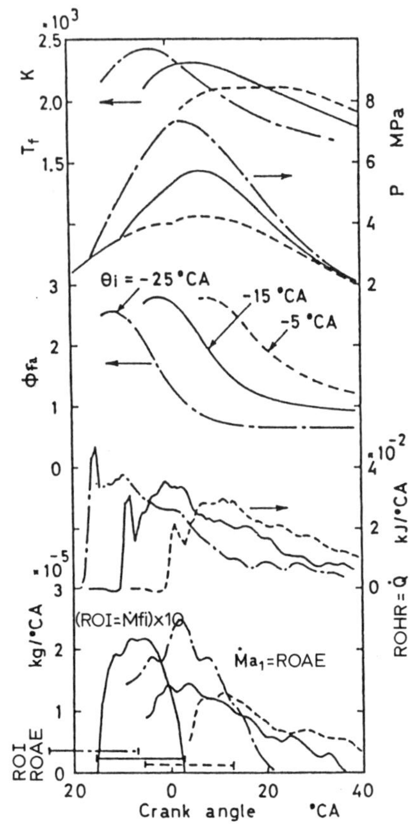

Fig. 24. Influence of injection timing θ_i to the indexes (all measured) of combustion characteristics[39]: T_f, temperature of flame (two-colour method); ϕ_{fa}, equivalence ratio (in the flame, see Fig. 23); $\dot{Q}$, *ROHR*; $\dot{M}a$, *ROAE* (rate of air entrainment); $\dot{M}f_1$, *ROI*.

will increase *ROHR*. It must be noted, however, that the ciné photograph catches only the flame surface state.

(d) When the specific point within the flame is observed it seems that the flame mass is revolving around itself.

The data obtained by the gas sampling method shown in Figs 31 and 32 indicate that the combustibility at the inside cavity in both the shallow and deep bowls becomes better as the swirl speed is increased.[20,21]

The data in Fig. 26(a) and (b) may be interpreted to mean that the diesel spray flame after impinging upon the cavity inner wall, becomes

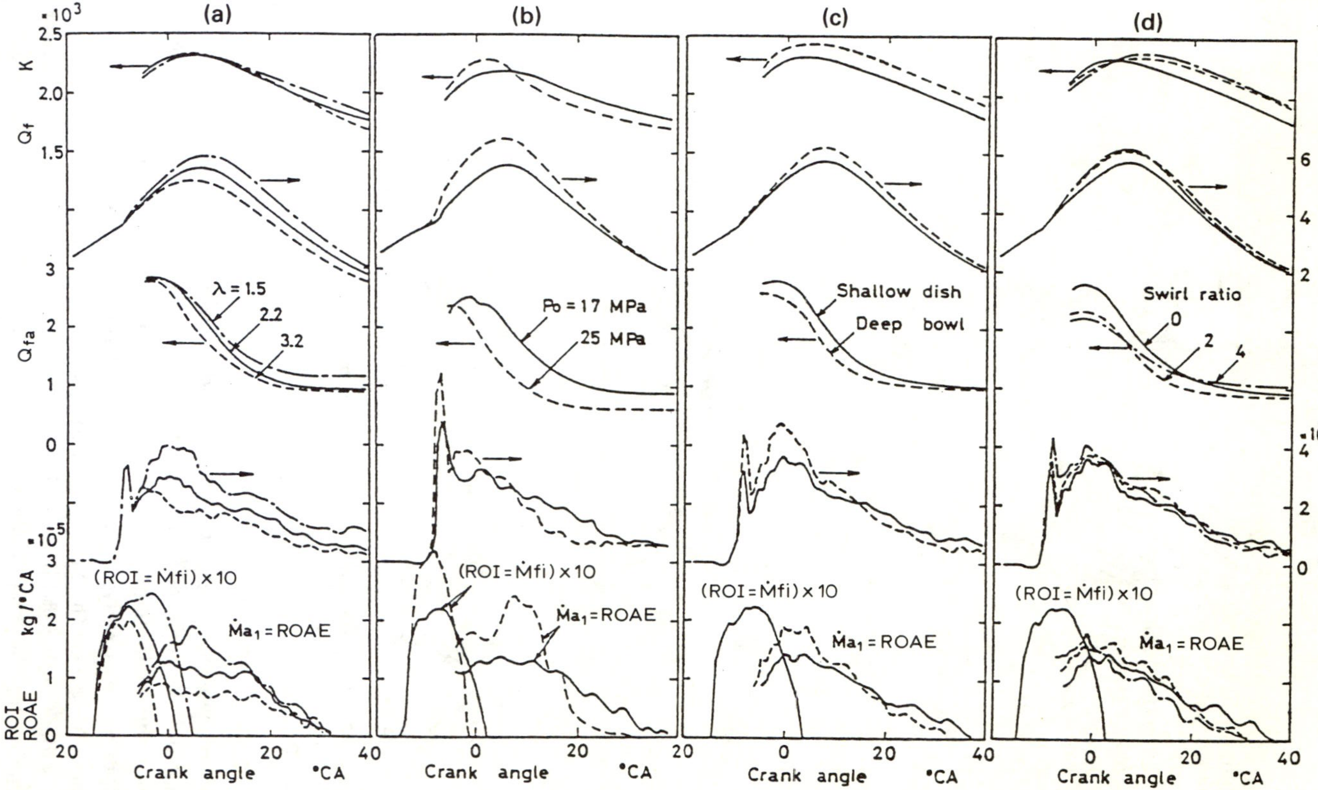

Fig. 25. Influence of: (a) air excess ratio λ; (b) injection valve opening pressure, p_0; (c) combustion chamber shape; (d) swirl ratio *vs* combustion characteristics indexes (see Fig. 24).[39]

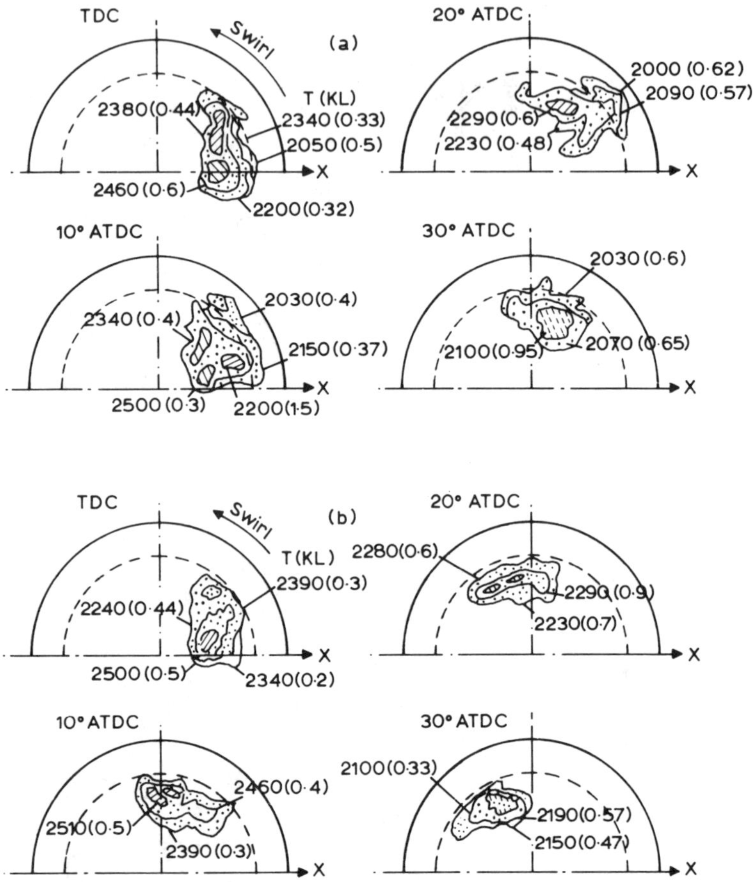

Fig. 26. Temperature and soot concentration distribution map of a DI engine when the swirl ratio is (a) 2 and (b) 4.[4,23] ($\theta_{inj} = -15°$ ATDC, 1250 rpm, $\lambda = 2{\cdot}0$, shallow dish).

a revolving mass. The peripheral portion with a small momentum will not only be torn off (Section 2.3), but also become the source of NO, and the main flame portion with a strong angular momentum will continue to revolve against the flow of the swirl and is needed at the time to release the heat. The strong shear of these revolving masses in a short time period is essential to achieve better combustion in a diesel engine, and will be discussed in Section 5.

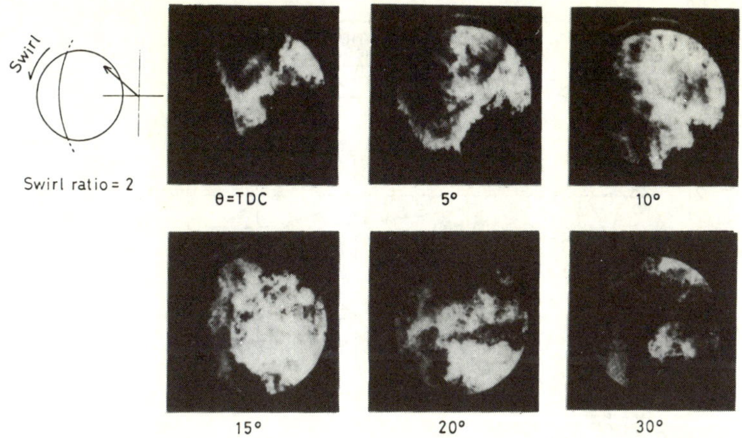

Fig. 27. High-speed photographs of engine flame ($\theta_{inj} = -15°$ ATDC, 1250 rpm, SR = 2, shallow dish).[20,21]

4.4. Combustion Characteristics Index and *ROAE* In-cylinder

Figures 24 and 25 show combustion characteristics *vs* crank angle as the injection timing, etc. is varied under standard conditions. In these figures $Ma_1 = ROAE$ is a unique index. Figures 24 and 25 indicate how these combustion characteristics are affected when five combustion

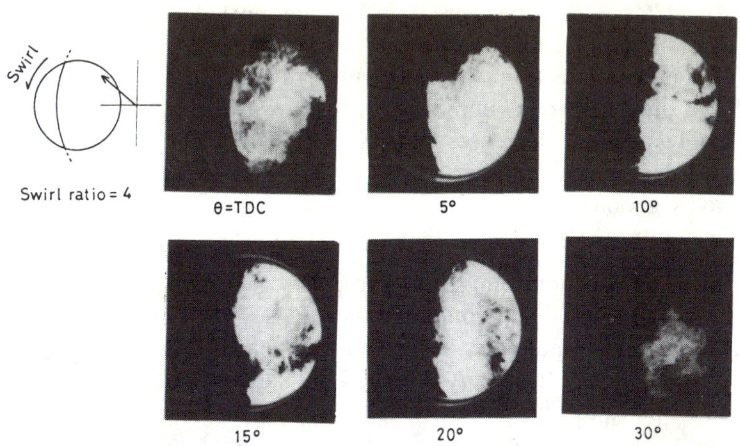

Fig. 28. As Fig. 27, but when swirl ratio is varied to 4 (shallow dish).[20,21]

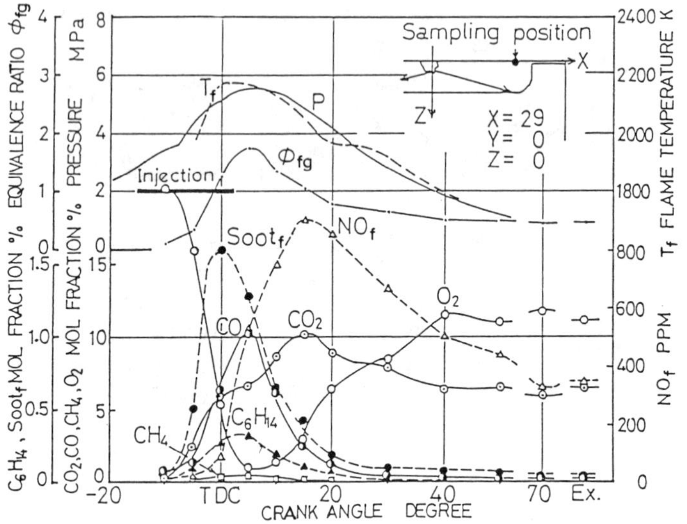

Fig. 29. An example of a time history map of combustion products, P, T_f and ϕ_{fg} in flame *vs* CA in the case of the shallow dish ($\theta_{inj} = -15°$ ATDC, 1250 rpm, $\lambda = 1{\cdot}8$, SR = 0)[20,21]

parameters are varied under standard conditions; namely

(a) When the fuel amount is increased (the excess air ratio λ becomes small) with $(ROI = \dot{M}f_1) \times 10$, the $ROAE = \dot{M}a_1$ increases, but $ROAE/ROI \doteqdot d(\text{surface area})/d(\text{fuel amount})$ decreases so that ϕ_{fa} becomes large.

(b) When the injection valve opening pressure P_0 is increased, *ROAE* is high; after it impinges on the wall, *ROAE* increases considerably and ϕ_{fa} decreases rapidly. The effect of high injection pressure is discussed in Section 5.2.

(c) Cavity geometry.

(d) Swirl ratio (SR).

The effect of injection timing, SR and cavity geometry are described precisely in Section 4.5 and compared with Figs. 30–32 of NO_f, $Soot_f$, T_f and ϕ_{fg} at a local point inside the flame, measured by the method shown in Section 4.2. In Figs 7(b) and 34 are shown the *ROHR* curves which correspond to Figs 24 and 30–32.

Generally, as mentioned, the *ROAE* has a great influence on ϕ_f (equivalence ratio in the flame) and T_f (flame temperature) and in turn on *ROHR*. Measurements of flame temperature, T_f clarified the considerable differences when compared with the thermodynamic mean temperature, *T*, of the gas derived from cylinder pressure.

4.5. Spatial and Temporal Distribution of Temperature, Equivalence Ratio, Soot and NO of the Flame in a DI Engine

Aoyagi and the author *et al.* presented the data for the spatial and temporal distribution of flame temperature, T_f by means of the two-colour method,[34–36] equivalence ratio ϕ_{fg} soot$_f$, and NO_f by means of the gas sampling method,[20,21] in the flame of a DI diesel engine, measured by these methods under the conditions shown in Sections 4.1 and 4.2. Figure 29 gives an example.

Summing up a large number of these data, Figs 30–32 show the

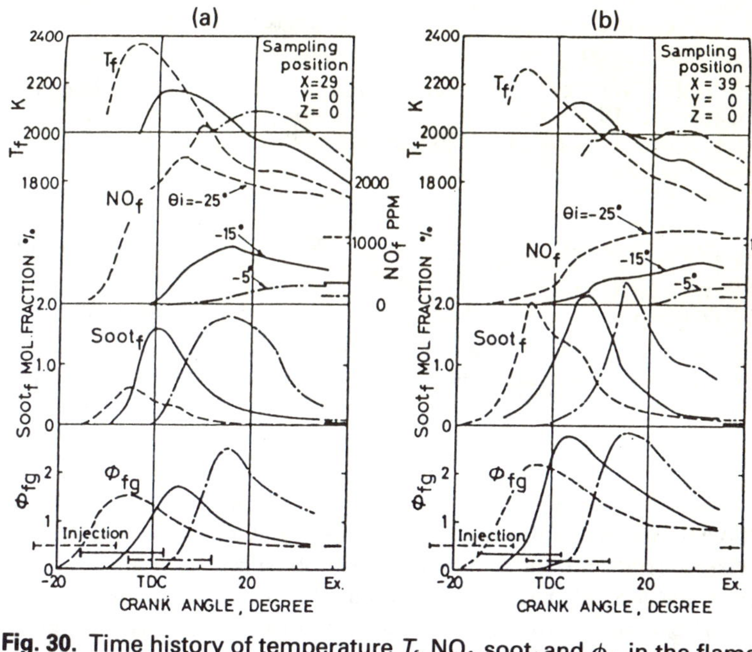

Fig. 30. Time history of temperature T_f, NO_f, soot$_f$ and ϕ_{fg} in the flame of the shallow dish DI engine when injection timing is varied: (a) cavity portion; (b) clearance portion (1250 rpm, $\lambda = 1{\cdot}8$ SR $= 0$, shallow dish).[20,21]

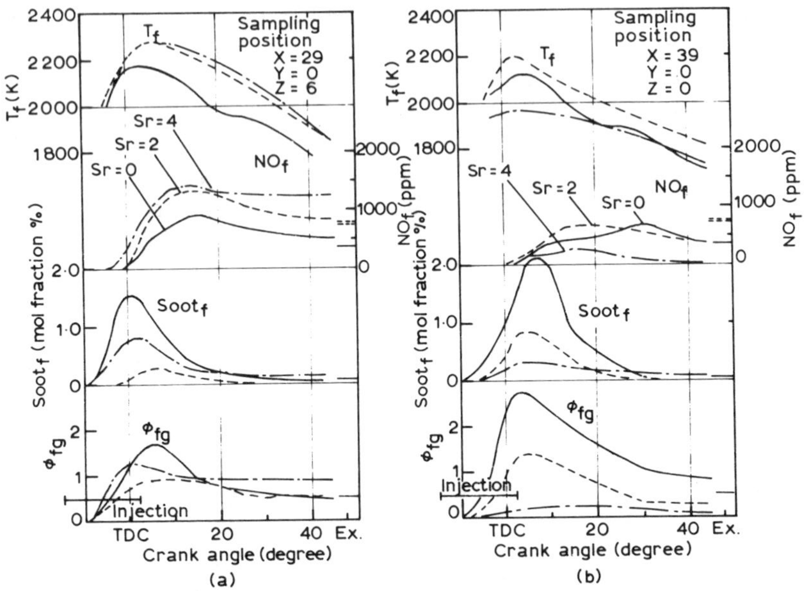

Fig. 31. Time history of temperature T_f, NO_f, soot$_f$ and ϕ_{fg} in the flame of the shallow dish DI engine when the swirl ratio is varied: (a) cavity portion; (b) clearance portion (1250 rpm, $\lambda = 1{\cdot}8$, $\theta_{inj} = -15°$ ATDC, shallow dish) (see Figs 27 and 28).[20,21]

relationships between T_f, ϕ_{fg} soot$_f$ and NO_f *vs* crank angle, when parameters such as injection timing, SR and combustion cavity geometry are varied. The engine performance of the indicated mean effective pressure (P_{mi}) and emission characteristics of smoke (g/m^3) and NO_x and NO (ppm) are shown in Fig. 20, while the fuel amount is kept constant in all conditions.

The effects of injection timing on *ROHR* in Fig. 7(b), to engine performance in Fig. 20, on *ROAE* in Fig. 24 and on T_f, ϕ_{fg}, soot$_f$ and NO_f in Fig. 30 are shown respectively. In these figures, when the injection timing is advanced to −25° ATDC, the increase of *ROAE* is remarkable (Fig. 24) and seems to be simply related to the rise of T_f and NO_f and a drop of ϕ_{fg} and soot$_f$ (Fig. 30). In this case, an increase of *ROAE* is accompanied by an increase of initial combustion and will inevitably accompany an increase of *REOBG* as mentioned in Sections 3.3–3.6. On the other hand, in the case of a turbo-charged DI diesel engine, it always accompanies weak initial combustion although SFC and smoke are good under the same excess air ratio (λ) condition compared to a naturally aspirated (NA) engine. The major reason is said to be that the air is sufficient; careful engineers, however, also know that it may depend on the mixing being promoted by the increase of in-cylinder gas momentum.

The example of the turbo-charged engine indicates that the performance gain obtained by this advance of injection timing is mainly obtained by the promotion of mixing, accompanying the increase of entrained total gases into the spray flame as mentioned in Section 3.5, not only on account of the *ROAE* increase.

In Fig. 31, the swirl ratio SR was varied from 0 to 4 in the shallow dish combustion chamber. As already mentioned the optimum SR is 2, and when SR becomes 4, the flame temperature in the shallow dish case is at its highest level, this is because of the strong mixing caused by the swirl in the cavity, and the equivalence ratio ϕ_{fg} becomes higher than in the case of SR = 2. This is caused by the flame pinching inside

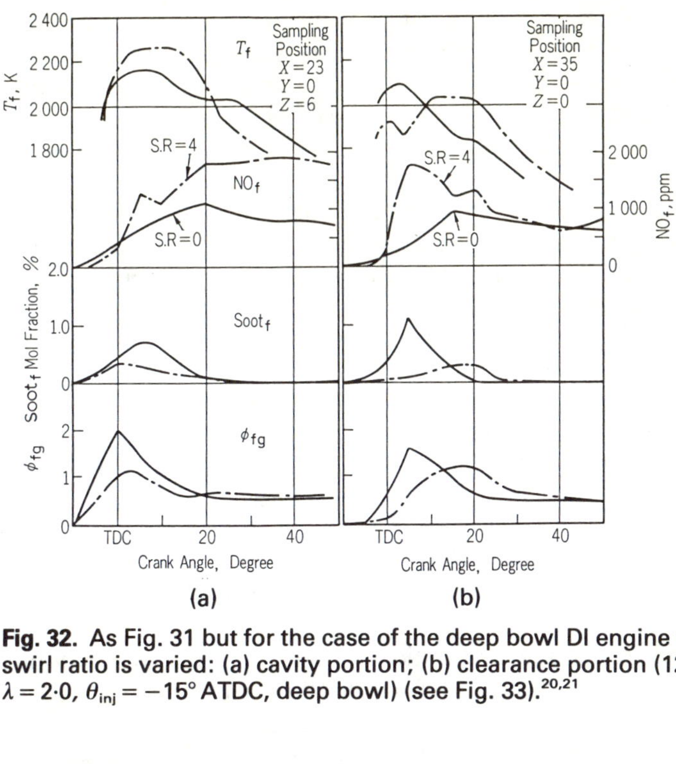

Fig. 32. As Fig. 31 but for the case of the deep bowl DI engine when the swirl ratio is varied: (a) cavity portion; (b) clearance portion (1250 rpm, $\lambda = 2{\cdot}0$, $\theta_{inj} = -15°$ ATDC, deep bowl) (see Fig. 33).[20,21]

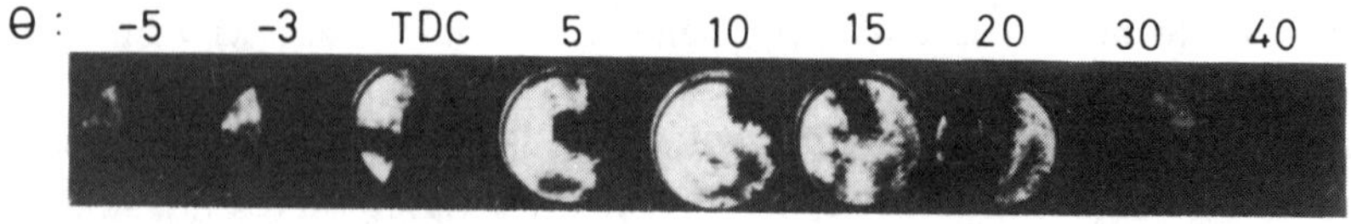

Fig. 33. High-speed photographs of engine flame (1250 rpm, $\theta_{inj} = -15°$ ATDC, SR = 4, deep bowl).[20,21]

the cavity through the trapping phenomenon of the squish flow which results in a higher soot concentration, as shown in Fig. 31. Therefore, in the case of SR = 4, the combustion in the cavity is good, as shown in Figs. 31 and 34, but the total engine and emission performance, as shown in Fig. 20, is worse; namely the output P_{mi} is decreased, and NO and soot emissions both become higher. These phenomena also occur in the clearance portion of the same combustion chamber in the case of SR = 4. The data in Fig. 31(b) and photographs in Fig. 28 show that the flame remains within the cavity and does not reach the clearance portion where the air utilization become worse.

In contrast to this, Figs 32(a) and (b) show the deep bowl case when the swirl ratio is changed from 0 to 4. In the piston cavity, at SR = 4, and at a low equivalence ratio, high temperature with low soot but high NO emission is realized. This is caused by the spread of air utilization by the introduction of a strong squish flow with strong swirl. In the clearance portion shown in Fig. 32(b), the overflow of the burnt-gas (high temperature portion) from the cavity is a little delayed, and at 20° ATDC the flame temperature and equivalence ratio increase but they soon decrease again by convection and dilution. The high-speed ciné-photographs of Fig. 33 also support this fact. In this area, the temperature is low, but high NO in the piston cavity overflows and NO emission becomes higher.

Soot decreases due to the high utilization of air at high temperature. From the engine performance diagram in Fig. 20, where the SR increases from 4 to 8, NO emission becomes a little higher, but high output with a lower SFC and low soot are realized.

In the case of the deep bowl, absolute values of output P_{mi}, SFC and soot become better than those for the shallow dish at each optimum condition. This shows that the air utilization improved with high swirl in the deep bowl cavity.

The overall trends already mentioned generally agree well with the *ROHR* curves shown in Fig. 34, indicated in the case of the shallow dish when SR is varied from 0–4 and for the deep bowl with SR = 4, but the difference in the emission characteristics which appear in the case of SR 2–4 in the shallow dish case are somewhat difficult to judge from these curves. It will be noted that the *ROHR* curves are not as sensitive to the difference in soot formation, especially that generated in the clearance zone.

In summary, from the above examples with the combustion chamber in the piston, it is seen to be desirable to mix the spray flame with strong air swirl, but when the swirl is too strong, as in the case of shallow dish, the flame mass remains in the cavity even when the squish flow is present. In the case of the deep bowl, it is presumed that squish flow becomes stronger so that the flame mass, even if it rotates and mixes with the strong swirl, nevertheless floods to the outside of the cavity, helped by the reverse squish. This phenomenon is explained in the following section.

These results suggest that there exists an optimum swirl speed relationship to the squish action to mix and utilize the inner and outer air of the cavity bowl.

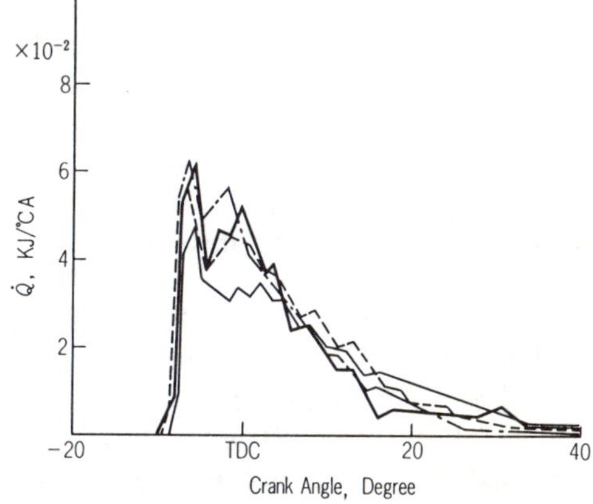

Fig. 34. Values of *ROHR* when swirl ratio is varied in shallow dish and deep bowl (1250 rpm, $\lambda = 2{\cdot}0$, $\theta_{inj} = -15°$ ATDC).[20,21]

4.6. Interference of Swirl and Squish Flow in the Cavity Combustion Chamber and Their Optimization

According to the results of the numerical calculations by multi-dimensional modelling made in Ref. 40 regarding the compounded flow pattern of swirl and squish flow during the inlet and outlet processes of the squish flow from the combustion cavity, it is made clear that the swirl within the cavity and the squish flow interfere with each other.

One of the examples of a complicated flow pattern is indicated in Fig. 35. On the left-hand side of Fig. 35 is shown the pattern when the clearance distance is fixed at ≒2 mm, and the swirl flow is at SR = 1. As the piston moves from −20° ATDC to +20° ATDC, positive squish flow and reverse squish flow will form a flow pattern, as shown in the figure. The positive squish dominates the whole swirl in the cavity, as shown in Fig. 35 (left-hand side), and the swirl will overflow from the cavity through the action of reverse squish.

As the swirl ratio becomes stronger from 1·0 to 5·0 at TDC, the squish flow pattern changes a lot, as shown in Fig. 35 (right-hand side). In the case of a high swirl ratio, the squish flow will only influence the peripheral portion of the swirl, and the swirl will remain in the cavity. This is because the centrifugal force generates a pressure field barrier to prevent the entrainment of the squish flow into the combustion bowl and this varies according to the swirl intensity; consequently it changes the shape and intensity of the secondary vortices. These data explain phenomenologically the over-swirl state in a small DI diesel engine quite well.

The flow patterns shown in Fig. 35 involve the qualitative relationship between the air swirl and squish flow but the revolving mass of spray and combustion are not taken into account. However, from the composite flow pattern shown, it is suggested that the size of the revolving mass of the spray flame, its rotational direction and angular momentum as it interacts with the swirl and squish flow interfere with each other and affect the mixing process.

Regarding the interpretation of optimization of the swirl speed, it is necessary to consider the cooling effect. But the data in the case of the shallow dish in Fig. 20, indicate that when the SR is increased the smoke becomes worse. This tendency is difficult to explain by the cooling effect. Also, it is necessary to consider that when SR is increased in this kind of engine, the spray tip will overlap with its

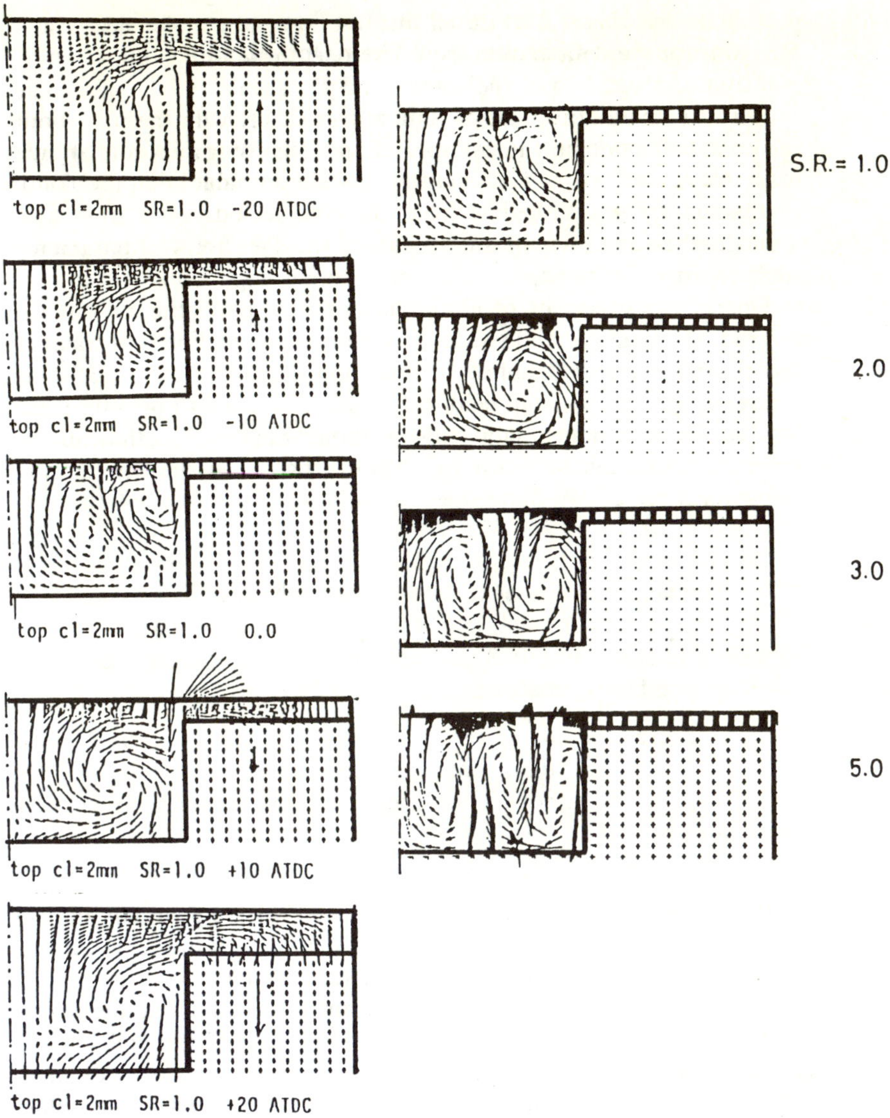

Fig. 35. Interfered flow in cylinder between swirl and squish flow, by the computation of a multi-dimensional model (axisymmetric two-dimensional model). Left: effect of piston movement from −20° ATDC to +20° ATDC when SR = 1·0 (top clearance = 2 mm); Right: effect of swirl ratio varied from SR = 1·0 to 5·0 when piston is at TDC.[40]

neighbour and this will cause an increase in smoke and a decrease in P_{mi}. But, for the data in the case of the deep bowl in Fig. 20, when SR is increased from 4 to 8, the smoke and P_{mi} are not made any worse. At SR 8 the spray tip overlaps almost completely with the neighbouring sprays. Therefore the interpretation of the optimization of swirl ratio in Sections 4.4 and 4.5 may be more reasonable than the theory mentioned above but the latter cannot be neglected.

In these cases, not only the discussion of the effects of parameters such as injection timing or the wall slanting angle (this will also be shown in Section 5, pattern No. 3, using the *REOBG* theory) but also the effects of the protuberance of a squish-tip combustion chamber, the number of nozzle holes, or wall temperatures of materials like ceramics, etc., it will be difficult for these parameters to be interpreted by the optimization relationship without the introduction of the burnt-gas re-entrainment theory. More progress of the optimization theory is expected.

4.7. Internal Structure of a Diesel Spray Flame

From the data of Figs 30–32 and other literature already mentioned, the following has been abstracted.

(a) The flame temperature T_f is sometimes higher than 2100 °K, even when ϕ_f in the flame is larger than 1·8.
(b) $Soot_f$ (carbonaceous substances containing a high amount of H) is formed even above 2000 °K.
(c) NO_f is formed even when ϕ_f is large.

The states and phenomena in (a), (b) and (c) are generally difficult to explain from the standpoint of a chemical reaction in the premixture flame. From these data the following arguments become possible. In a diesel spray flame there is such an extremely stochastic and turbulent diffusion condition that within the cloud masses of the spray flame, there are a large number of reacting micro-clouds consisting of small- and large-scale turbulence of various intensities. These turbulent flows mix with the fuel, air and burnt gases, and consequently the equivalence ratio and oxygen partial pressure undergo changes at every moment over a wide range. Thus the following picture of combustion may be constructed: there exists a number of reacting micro-clouds which, even when over-rich from the average level because they contain a certain range of lean ϕ micro-clouds, may reach a high temperature such as in (a) and may form NO such as in (c). In the reverse case, over-rich ϕ micro-clouds among the average lean flame, may form soot such as in (b) because of an air-deficient high-temperature condition.

5. CONCEPTUAL CLASSIFICATION OF DIESEL COMBUSTION SYSTEM FROM THE VIEWPOINT OF THE OPTIMIZATION OF THE AIR UTILIZATION AND THE BURNT-GAS DISTURBANCE

In the preceding sections, the author has demonstrated that the diesel combustion process is dominated not only by air utilization but also by the inevitable burnt-gas disturbance. Indeed, the author would like to emphasize that an approach for good air utilization combustion is not always compatible with the burnt-gas treatment. For example, as shown in Section 3.5, in the case of right-angle impingement of the fuel spray, as it generates stronger turbulence at the wall then air utilization or the degree of mixing with fuel will be higher compared to slanting wall impingement, but the rate of heat release per unit volume is presumed to be lower. This is because, as has been shown, the burnt-gas situation is not as good in the case of the former as with the latter.

Here, the author would like to discuss the following four classified concepts of the combustion system which can be discerned or optimized from the abovementioned two opposing flow patterns:

(1) *Pattern No. 1* belongs to the spark ignition direct injection stratified charge type engine (e.g. Texaco and Ford Proco, Chapter 5) and may be termed SI-DISC engine. Here the atomized fuel and air mix intimately in a stratified mixture cloud, which will burn with a normal pre-mixture propagation pattern. Therefore there is no problem of interference with the burnt-gas, which remains behind the burning zone. However, it is not easy to obtain this stratified cloud of good mixture under all conditions of speed and load.

Ohnishi's direct injection impinging diffusion stratified charge (OSKA) combustion system[41] seems to be another way of conceiving this concept, as shown in Fig. 36(a). In this concept the fuel spray evaporates by impinging on a protuberance located in the piston cavity, thereby assisting in the formation of the symmetrical disc cloud

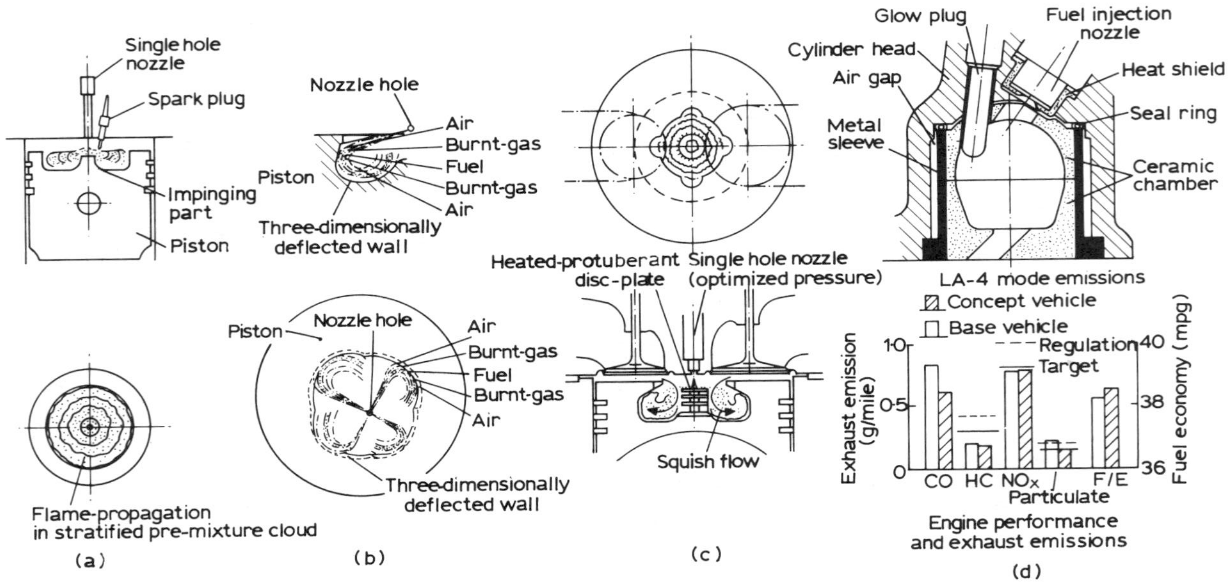

Fig. 36. Classification of diesel combustion system from the viewpoint of the air utilization and burnt-gas disturbance. (a) Example of Pattern No. 1: Ohnishi's OSKA SI-DISC combustion system;[41] (b) example of Pattern No. 2: Matsuoka's CR-M CI-DI combustion system;[42] (c) example of Pattern No. 3: Ohnishi's OSKA-D CI-DI combustion system;[44] (d) example of Pattern No. 4: MAZDA's ceramic CI-IDI combustion system.[45]

of an ignitable mixture, almost regardless of the conditions of speed and load of the engine, and achieving more than 40% brake thermal efficiency with methanol and 36% with gasoline fuel.

(2) *Pattern No. 2* is appropriate to the conventional wall-impinging DI engines, where we have a diffusion flame surrounding the fuel injection zone: because the combustion chamber is a compact closed volume, it is unavoidable that burnt-gas will mix with clean air. As the combustion progresses the mixing of burnt-gas with clean air continues and it is desirable that this burnt-gas is diffused as quickly and homogeneously as possible with the clean air. Most diesel engine designers know that the slanting wall impingement is effective for starting, etc., but that a fully developed theory has not been established.

Matsuoka's carpet-rolling combustion system (CR-M),[42] in which the intentionally designed three-dimensional slanting spray impinges on the smooth curved wall of the combustion chamber, is a typical illustration of this concept, which is based on Section 3.5 (see Fig. 36(b)). The MAZDA Titan DI diesel engine[42] succeeds with this CR-M combustion system, under an optimized manner of gas-flow incorporating the slanting-angle wall-impinged spray with its beneficial characteristics, the fuel consumption is kept constant while the soot and NO emission clear the regulations of the Japanese government (1987) and noise emissions are considerably minimized, as shown in Figs 37–39.[42]

To achieve this Pattern No. 2, where the goal is fast and homogeneous mixing, a possible means is the formation of very strong turbulence inside the piston cavity chamber, for example by high pressure fuel injection. As shown in Figs 10 and 11, as far as the free-spray in RCM is concerned, the high pressure injection is quite effective in solving the abovementioned two opposing problems. However, for the application to a small size engine the data is not yet available.

(3) *Pattern No. 3*: in this pattern the concept is the wall evaporation fast combustion system. Air is introduced into the fuel spray as soon as it has evaporated and reached the thermal-decomposition state in order to achieve the fast combustion, but also simultaneously to separate and carry the burnt-gas to the downstream side of the burning zone by the action of the momentum of the spray and gas-flow. It is interesting to interpret the MAN-M combustion system[43] from the viewpoint of this burnt-gas treatment.

Fig. 37. CR-M combustion system: effect of slanting angle (α against cylinder axial direction) to SFC and smoke.[42]

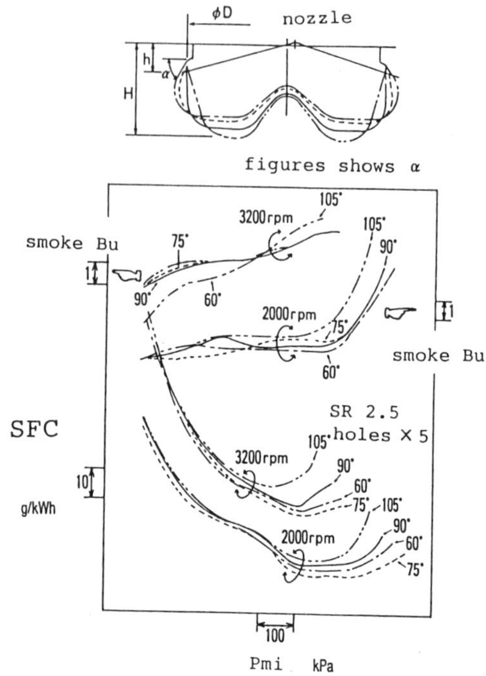

Ohnishi's OSKA-D combustion system (same type of OSKA-SI mentioned above in Section 5(1), but a compression ignition system[44]) seems to belong to this concept. The schema of OSKA-D is shown in Fig. 36(c). To accomplish this combustion process, it is desirable to use a rather low pressure-injection designed to give a thin liquid-fuel film disc diffusing along the radial direction, having impinged and evaporated on the heated protuberant disc-plate located in the piston cavity. The squish jet-flow from the clearance portion of the piston is introduced along the vertical direction against the mixture-disc-plate-flow and fast combustion will be achieved as soon as it reaches the thermal decomposition state. The burnt-gas which was carried to the downstream side contains little unburnt fuel in it, and Ohnishi's data[44] claims that it contains 1/10 of the smoke density compared to the conventional DI diesel engine at the same 3/4 load output.

(4) *Pattern No. 4* is appropriate to the IDI engine, where it emits rather low smoke and NO_x emissions but SFC is not equal to the

DI-engine. When one sees this type of engine from the two aforementioned viewpoints, one notices that the treatment of burnt-gas is quite different from the DI engine. The fuel is mixed almost perfectly with its own burnt-gas under an air-deficient high-temperature closed condition. As soon as this mixture is injected into the main chamber through a connecting orifice, fast combustion is promoted with less smoke and NO emission. (These components will be minimized if the temperature is well controlled as mentioned in Section 6.)

According to the classification of Section 5, the IDI diesel engine developed by MAZDA with a ceramic divided chamber[45] (Fig. 36(d))

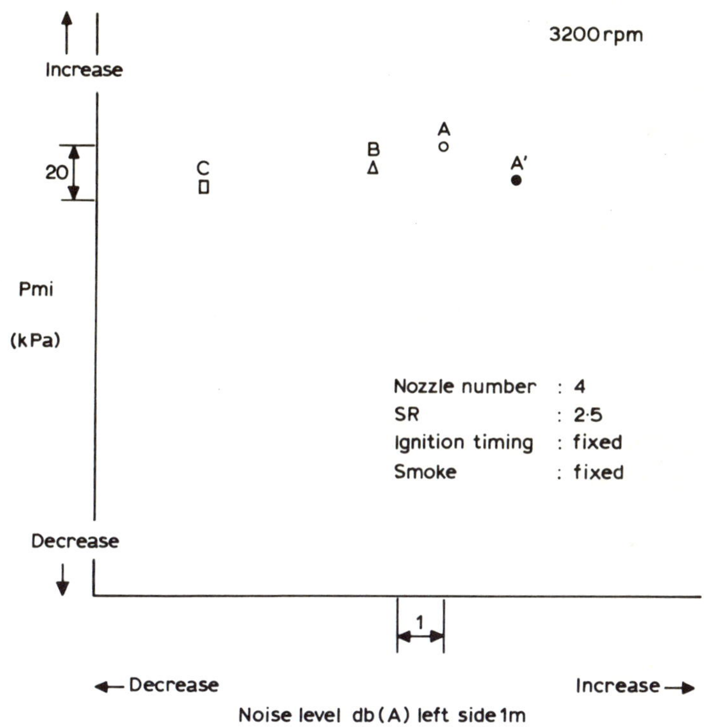

Fig. 38. CR-M combustion system: comparison of the effect of combustion chambers shape and circumferential slanting angle to P_{mi}.[42] (A: conventional round type; A′: as A, with protuberant turbulence generators; B: clover shaped with circumferentional slanting angle; C: as B, with more slanting angle) (all have $\alpha = 75°$ against cylinder axial direction; see Fig. 37).

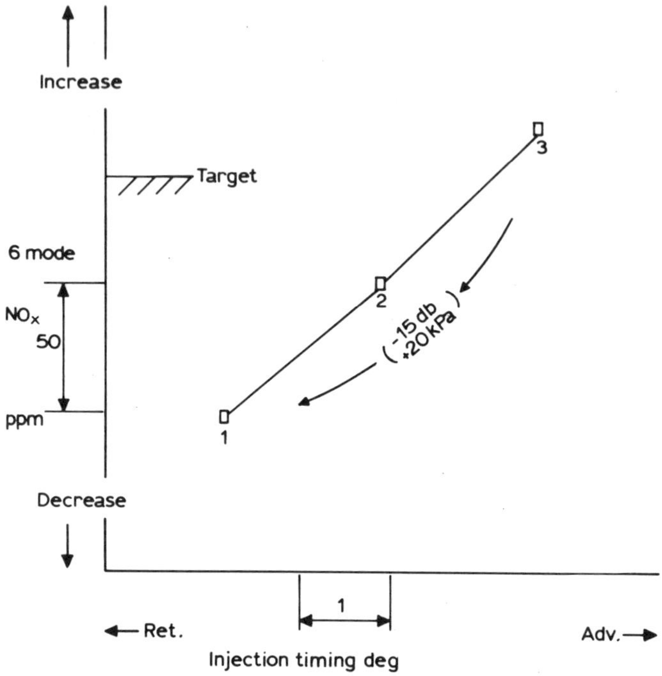

Fig. 39. CR-M combustion system: effect of injection timing to NO emission (C-type in Fig. 38).[42]

seems to belong to this No. 4 pattern. The combustion in the divided chamber stage of this engine is under high temperature with a ceramic wall, and is in a more air-deficient condition due to the application of normal EGR. The result indicates, as shown in Fig. 36(d), low smoke and nearly the same NO_x emission with a low SFC value as the conventional IDI engine.

The author would like to propose the application of the variable throat method of IFP[46] with direct EGR to the divided chamber through the 3rd valve on this ceramic IDI engine to reduce not only smoke but SFC and NO emission.

Tentatively, the author would like to propose that Patterns No. 1 and No. 3 are combined together to form an OSKA hybrid-combustion system, where under light load conditions, the combustion system is switched to the No. 3 diffusion combustion system and under the heavy load condition is switched to the No. 1 pre-mixture combustion system. This new hybrid combustion concept is not only based on the concept in Ref. 47 by the author himself (namely, under a light load condition, the diffusion combustion system has smaller pumping loss as it has no throttle valve but can burn faster under a very lean (as total) mixture ratio, and under a heavy load condition, the pre-mixed combustion system has a high output, is not smoky and can burn faster), but is also based on the new concept to conquer the opposing problems of *ROAE* and *REOBG* phenomena under each abovementioned operating condition. The result is that the hybrid combustion system will have the lowest SFC, NO, and smoke performance characteristics at all running conditions.

6. CONCEPT OF AN OPTIMIZATION STRATEGY FROM THE VIEWPOINT OF LOW FUEL CONSUMPTION, LOW PARTICULATE AND LOW NO EMISSION

In the concept of combustion promotion mentioned previously, there is generally trade-off relationships between output and soot emission, and soot and NO emissions.

Regarding the former trade-off problem of output and soot emission, fast combustion, including the problem of Section 5, is a typical solution. Fast combustion in a diffusion flame means a fast mixing rate inside the spray. This might be achieved, for example, by high pressure injection or by the impinging spray against the wall or opposed injection, etc.

Regarding the latter trade-off problem of soot and NO emissions, in general, fast combustion is accompanied by high NO emission. In Fig. 40, the limiting temperature range of NO and soot inside the cylinder, as shown already in Figs 30–32 and from Ref. 20, are summed up. According to these latter figures, the soot concentration in the flame drops, due to the oxidation, with crank angle position as it moves from TDC to nearly 20° ATDC. The temperature also goes down and soot oxidation nearly stops; as a result exhaust smoke in the tail pipe keeps a constant level when the equivalence ratio ϕ_{is} is less than 0·8 (ϕ_{is} is the value when ϕ_f in the flame becomes nearly constant) as shown in Figs 30–32.

These soot oxidation stop points arranged in Fig. 40, show that on average all points are included in the temperature range 1875 °K ± 50 °K. When the soot oxidation stops the exhaust smoke level relates

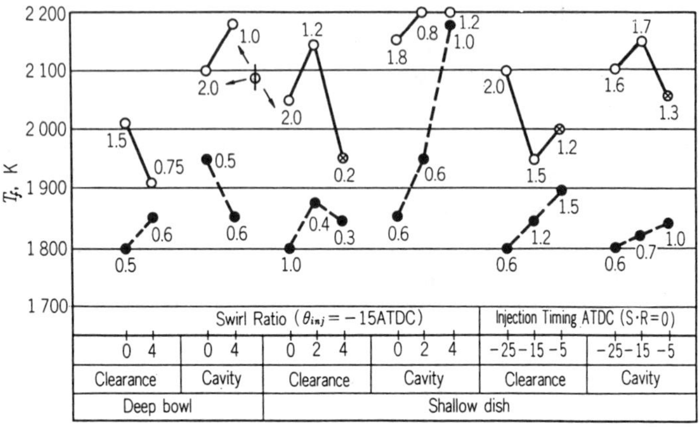

Fig. 40. Temperature range of upper limit for NO and lower limit for soot in diesel flame (1250 rpm, λ = 1·85–2·0) (●), soot$_f$ oxidation stop point; (○), NO$_f$ over 500 ppm point; (⊗), very low emission. (see Figs 30–32).

nearly with $\phi_{is} = 0{\cdot}8$ (Fig. 41)[20]. Figure 9 shows the Arrhenius plot of specific oxidation rate *vs* flame temperature as noted in Section 3.3. As is shown in this figure, the measurement of the oxidation rate was only possible over 1850 °K and below this point oxidation froze and the rate became very low.

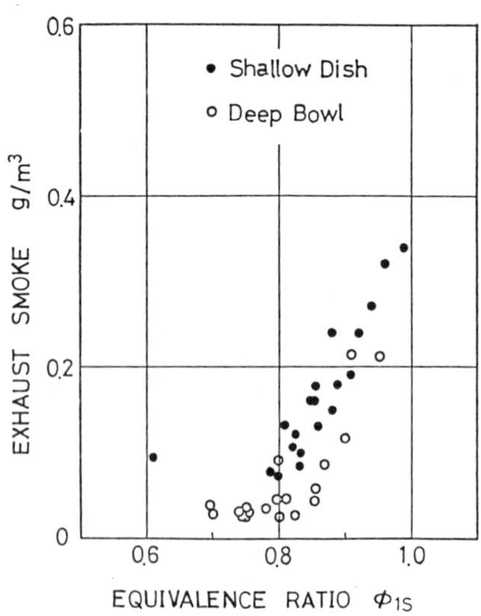

Fig. 41. Relationship between smoke in exhaust-pipe and ϕ_{is} (ϕ_{is} is a value when ϕ_f in flame becomes nearly constant) (see Figs 30–32).

Similarly, regarding NO formation, in Fig. 40 the temperature by which NO increases over the 500 ppm point is lined up on average in the range 2100 ± 100 °K. 500 ppm of NO in an exhaust pipe corresponds to the critical flame temperature of NO formation; that is about 2100–2200 °K.[20] These data generally agree well with the theoretical viewpoint and experiments reported in the past. Therefore it can be recognized that when the DI diesel engine is operated between 1800 and 2200 °K flame temperature, NO and soot or particulate emissions are all kept down.

To lower the flame temperature, retarding the injection timing and the application of EGR are most effective, the resulting consequence being an increase in SFC. In these cases, the combustion may be speeded up using the aforementioned combustion patterns in Section 5, the increase of SFC will then be minimized and NO emission effectively decreased.

As an outcome of the researches described above, the author would like to propose the following concept, where it would seem reasonable to optimize NO and soot in the engine by detecting the flame temperature and its duration, for example, by means of a radiation intensity sensor with a feedback system to control the parameters which influence the flame temperature.

The related data in Figs 30–32 are limited to the engine operating condition of 1250 rpm and $\lambda \fallingdotseq 1{\cdot}8$, but the flame temperature in diesel diffusion combustion is not influenced much by speed and load as noted in the literature.[35,36] The author has described here the basic concept of the strategy of diesel combustion, but more data are still required to elucidate the complex interaction of the fuel spray, air and burnt-gas.

REFERENCES

1. Kamimoto, T., Research on the evaporation and combustion of fuel spray in a DI diesel engine. Dr Engn. thesis, Tokyo Institute of Technology, 1975.
2. Kamimoto, T., Matsuoka, S., Sugiyama, H. & Aoyagi, H., Evaporation process of fuel spray in diesel engine. *Trans. Jpn. Soc. Mech. Engn.*, **40** (November 1974) 339.
3. Kamimoto, T. & Matsuoka, S., Prediction of spray evaporation in reciprocating engine. Paper 770413, *Trans. SAE* **86** (1977).
4. Ahn, S. K., Research on the diesel spray and flame by means of image analysis method. PhD thesis, Tokyo Institute of Technology, 1981.

5. Kobayashi, H., Kamimoto, T. & Matsuoka, S., Research on diesel combustion by means of big size rapid compression machine (mechanism, performance characteristics and experimental example of a trial machine). *Proc. 1st Jpn. Soc. Auto. Engn.–Jpn. Soc. Mech. Engn.*, Joint Symposium on I.C.E., August 1979.
6. Kamimoto, T., Ahn, S. K., Chang, Y. J., Kobayashi, H. & Matsuoka, S., Measurement of droplet diameter and fuel concentration in a non-evaporating diesel spray by means of image analysis of shadow photographs. Paper 840276, *Trans. SAE*, **93** (1984), *Proc. 3rd Jpn. Soc. Auto. Engn.–Jpn. Soc. Mech. Engn*, Joint Symposium on I.C.E., July 1982.
7. Kobayashi, H., Kamimoto, T., Matsuoka, S. & Yagita, M., Prediction of transient sprays in swirling flow via a modified 2-D jet model. Paper 860332, *Trans. SAE* **95,** 1986.
8. Sinnamon, J. F., Lancaster, D. R. & Stiener, J. C., An experimental and analytical study of engine fuel spray trajectories. SAE paper 800135, 1980.
9. Kamimoto, T., Kobayashi, H. & Matsuoka, S., A big size rapid compression machine for fundamental studies of diesel combustion. Paper 811004, *Trans. SAE* **90,** 1981.
10. Kobayashi, H., Kamimoto, T. & Matsuoka, S., Photographic and thermodynamic study of diesel combustion in a rapid compression machine. Paper 810259, *Trans. SAE,* **90,** 1981.
11. Kobayashi, H., Kamimoto, T. & Matsuoka, S., Trial made of RCM for the fundamental studies of combustion and heat transfer of diesel engine. *Trans. Jpn. Soc. Mech. Engn.*, **48** (January 1982) 426.
12. Kobayashi, H., Research on diesel combustion by means of rapid compression machine. PhD thesis. Tokyo Institute of Technology, 1982.
13. Kamimoto, T., Matsuoka, S., Matsui, Y. & Aoyagi, Y., The measurement of flame temperature and the thermodynamical analysis of combustion processes in a DI diesel engine. *Proc. Inst. Mech. Engn.* (Conference of Diesel Combustion in Engine.), C-96/75, July 1975; *Proc. Jpn. Soc. Mech. Engn.*, April 1975, 750–5.
14. Kamimoto, T., Matsuoka, S. & Aoyagi, T., Research on DI diesel combustion (thermodynamical analytical method by means of measured flame temperature). *Trans. Jpn. Soc. Mech. Engn.* **43** (September 1977) 373.
15. Matsuoka, S. & Kamimoto, T., An Image on Diesel Spray Flame. *Proc. 3rd Jpn. Soc. Auto. Engn.–Jpn. Soc. Mech. Engn.* Joint Symposium on I.C.E., July 1982.
16. Matsuoka, S., Kamimoto, T. & Matsui, Y., An introduction to the research on combustion mechanism and formation-extinction process of NO_x and soot particulate in DI engine. *18th Proc. Féderation international dés sociétés, d'ingénieurs des techniques de l'automobile, FISITA, Hamburg,* VDI-Be, 1980, p. 370.
17. Matsuoka, S., Creation of image on diesel spray and flame by means of RCM. Paper 830447, *Trans. SAE,* **92,** 1983.
18. Matsuoka, S., Kamimoto, T. & Kobayashi, H., Photographic and image analysis studies of diesel engine (interpretation and conceptual image). *20th Proc. Féderation international dés sociétés, d'ingénieurs des techniques de l'automobile, FISITA, Vienna.* Paper 845009, 1984.

19. Matsuoka, S., Clarification of black box in diesel combustion. *Trans. Jpn. Soc. Mech. Engn.* (Perspective review), **51,** (461-3-11) (January 1985).
20. Aoyagi, Y., A gas sampling study on the formation processes of soot and NO in a DI diesel engine. PhD thesis, Tokyo Institute of Technology, 1980.
21. Aoyagi, Y., Kamimoto, T., Matsui, Y. & Matsuoka, S., A gas sampling study on the formation processes of soot and NO in a DI diesel engine. Paper 800254, *Trans. SAE,* **89** (1980); 1st report: *Trans. Jpn. Soc. Mech. Engn.,* **46** (403) (March 1980) 540–549; 2nd report: *Trans. Jpn. Soc. Mech. Engn.* **47** (414) (Feb. 1981) 392–401.
22. Ahn, S. K., Matsui, Y., Kamimoto, T. & Matsuoka, S., Measurement of flame temperature distribution in a diesel engine by means of image analysis of high speed photographs. *Proc. 13th High Speed Photography & Photonics,* Tokyo, 1978.
23. Ahn, S. K., Kamimoto, T., Matsui, Y. & Matsuoka, S., Measurement of flame temperature in a DI diesel engine by means of image analysis of nega-color photographs. *Proc. 1st Jpn. Soc. Auto. Engn.–Jpn. Mech. Engn.* Joint Symposium on I.C.E., August 1979; Paper 810183. *Trans. SAE* **90** (1981); *Trans. Jpn. Soc. Mech. Engn.,* **47** (May 1981) 417.
24. Ahn, S. K., Kobayashi, H., Kamimoto, T. & Matsuoka, S., Research on the diesel spray flame by means of image analysis using high speed ciné camera. *Proc. 1st Jpn. Soc. Auto. Engn.–Jpn. Soc. Mech. Engn.* Joint Symposium on I.C.E., January 1981; *Jpn. Soc. Mech. Engn.* paper 810–12 (October 1981).
25. Kamimoto, T. & Matsuoka, S., Soot oxidation rate in the flame in a DI diesel engine. *Proc. Inst. Mech. Engn.,* C-103/82 (1982).
26. Chang, Y. J., Research on the diesel spray and flame with an RCM. PhD thesis, Tokyo Institute of Technology, 1985.
27. Chang, Y. J., Kobayashi, H., Matsuzawa, K. & Kamimoto, T., A photographic study of soot formation and combustion in a diesel flame with an RCM. *Proc. of International Symposium on Diagnostics and Modeling of Combustion in Reciprocating Engines. COMODIA 85,* Tokyo, Japan, September 1985.
28. Kamimoto, T., Yokota, H. & Kobayashi, H., Effect of high pressure injection on soot formation process in RCM to simulate diesel flame. Paper 871610, *Trans. SAE,* **96** (1987).
29. Matsuoka, S. & Yoshizaki, T., Model verification of burned-gas re-entrainment phenomenon and the soot formation mechanism in diesel combustion (free spray flame in RCM). Paper 890440, *Trans. SAE,* **98** (1989).
30. Kobayashi, H., Kamimoto, T. & Matsuoka, S., Prediction of the rate of heat release of an axisymmetric diesel flame in RCM. SAE paper 840519, 1984.
31. Kamimoto, T., Chang, Y. J. & Kobayashi, H., Rate of heat release and prediction of a diesel flame in RCM. SAE paper 841076, 1984.
32. Khan, I. M., Greeves, G. & Wang, C. H. T., Factors affecting smoke and gaseous emission from direct injection engines and a method of calculation. SAE paper 730139.

33. Kamimoto, T., Takahashi, H., Kobayashi, H. & Matsuoka, S., Convective heat transfer of an impinging diesel flame in a RCM. SAE paper 821035, 1982; *Trans. Jpn. Soc. Mech. Engn.*, **49-B** (April 1983) 440.
34. Kamimoto, T., Matsuoka, S. & Sakai, K., Research on the measurement of flame temperature in diesel engine. *Jpn. Soc. Mech. Engn.* Paper 720–14 (August 1972).
35. Matsui, Y., Research on the optical measurement of flame temperature and soot concentration in DI diesel engine. PhD thesis, Tokyo Institute of Technology, 1979.
36. Matsui, Y., Kamimoto, T. & Matsuoka, S., A study on the time and space resolved measurement of flame temperature and soot concentration in a DI diesel engine by the two-color method. Paper 790491, *Trans. SAE,* **88** (1979); *Trans. Jpn. Soc. Mech. Engn.*, **44** (January 1978) 377.
37. Matsui, Y., Kamimoto, T. & Matsuoka, S., Formation and oxidation processes of soot particulates in a DI diesel engine—an experimental study via the two color method. SAE paper 820464, 1982.
38. Kamimoto, T., Aoyagi, T. & Matsuoka, S., Research on the combustion of a DI diesel engine (analysis of diffusion combustion by means of measured flame temperature). *Jpn. Soc. Mech. Engn.* Paper 740–16 (Nov. 1974).
39. Kamimoto, T., Aoyagi, Y., Matsui, Y. & Matsuoka, S., The effects of some engine variables on measured rate of air entrainment (*ROAE*) and heat release (*ROHR*) in a DI diesel engine. SAE paper 800253, *Trans. SAE,* **89** (1980); *Trans. Jpn. Soc. Mech. Engn.*, **47** (January 1981) 413.
40. Matsuoka, S., Kamimoto, T., Urushihara, T., Mochimaru, Y. & Morita, H., LDA measurement and a theoretical analysis of the in-cylinder air motion in a DI diesel engine. Paper 850106, *Trans. SAE,* **94** (1985).
41. Ohnishi, S. & Katoh, S., New mixture formation of DISC-SI engine (OSKA). SAE paper 871689, 1987.
42. Kitamura, T. & Matsuoka, S., A development of carpet-roll combustion system of slanting wall impingement in DI diesel engine. *Proc. Jpn. Soc. Auto. Engn.* 891. 891020. (May 1989).
43. Meurer, J. S., Evaluation of reaction kinetics eliminates diesel knock—the M-combustion system of MAN. *Trans. SAE,* **64** (1956).
44. Ohnishi, S. & Katoh, S., A development of direct injection impinging diffusion diesel combustion system (OSKA-D). Private open forum at Kanazawa Bunka-Hall, June 1989.
45. Sakurai, S. & Matsuoka, T., Development of low particulate engine with ceramic swirl chamber. SAE paper 861407, 1986.
46. Brisson, B., Ecomard, A. & Eyzat, P., A new diesel combustion chamber—the variable throat chamber. SAE paper 730167, 1973.
47. Matsuoka, S. & Tasaka, H., Theoretical approach to elucidate the better fuel consumption of SI and CI engines. *19th Proc. Féderation International dés Sociétés, d'ingénieurs dés Techniques de l'automobile FISITA,* Melbourne. SAE paper 82015, 1982.

11

Computer Simulation of Fluid Flow and Combustion in Reciprocating Engines

D. B. Spalding

CHAM Ltd, London, UK

1. Introduction
   1.1. What is meant by computer simulation of IC engine combustion?
   1.2. Benefits and limitations of computer simulation
   1.3. Historical development of the subject
2. Fundamentals of computational fluid dynamics
   2.1. Formulation of the mathematical problem
   2.2. Numerical methods of solution
   2.3. The quality of the solutions of the equations
3. Computational representation of combustion
   3.1. Distributed-reaction representation
   3.2. The two-fluid model of turbulent combustion
   3.3. Further developments
4. Concluding remarks
   4.1. The current status of numerical simulation of flow and combustion in reciprocating engines
   4.2. How computer simulation can assist engine designers today
   4.3. The future

1. INTRODUCTION

1.1. What is Meant by Computer Simulation of IC Engine Combustion?

In the present context, computer simulation of IC engine combustion is the use of a digital computer to predict, from the fundamental laws of physics and chemistry, the *distributions* of velocity, pressure, temperature and other gas properties throughout the space above the piston, and throughout the cycle.

The word 'distributions' has been emphasized so as to distinguish the present topic from those simulations of in-cylinder processes of a purely thermodynamic character, which postulate that the gas is in the same state at all points at a given time; and also from those which include chemical-reaction-rate information in the analysis, while still postulating local uniformity.

The latter analyses may be termed 'zero-dimensional'. The present chapter, by contrast, is concerned with one-, two- and three-dimensional simulations, the emphasis being on the latter because all real engine-cylinder geometries *are* three-dimensional.

Computer simulations in this sense do take account of thermodynamics and chemical kinetics; but, instead of *presuming* instantaneous uniformity of conditions, they *calculate* the non-uniformity. This involves attention to fluid-dynamic phenomena, including turbulence; and it necessitates defining within the combustion space a 'grid' of cells, within each of which the pressure, temperature and other properties are to be computed.

For physically accurate simulation, these cells must be very numerous; for example, in excess of ten thousand. Since the amount of arithmetic to be executed increases in proportion (at least) to the number of cells, accurate simulations require large and fast computers. Fortunately, the cost of these is diminishing and their availability is increasing.

1.2. Benefits and Limitations of Computer Simulation

It is much cheaper to run a computer than to build a new engine; and the results are more quickly available. This is why engine manufacturers are increasingly turning to computers for guidance as to whether a proposed design change will or will not be beneficial. There is another reason: computer simulations can sometimes provide information which it is too difficult to obtain experimentally; in so doing they assist designers to gain a deeper understanding.

Nevertheless, *exclusive* reliance on computer simulations is not, and perhaps never will be, justifiable; for even the best simulations leave some aspects of the real situation out of account; and some of them rest on very uncertain foundations.

Computer simulations of engine behaviour are most trustworthy when:

(a) the geometrical shapes (of the combustion chamber, valves and associated ducts, etc.) are properly represented;
(b) the initial and boundary conditions are correctly introduced;
(c) the physical and chemical-kinetic properties of the materials (fuel, air, combustion products) are known, and included; and
(d) the equations expressing the relevant laws of physics, in their application to the particular circumstances, are accurately solved.

Unfortunately, fulfilling requirements (a), (b) and (d) costs money; often therefore they are incompletely met. As to requirement (c), the current state of science is such that many of the relevant properties of the materials, particularly when they are in that semi-chaotic state of matter known as turbulence, cannot be precisely prescribed.

When to employ a computer simulation, and how much confidence should be placed in it, is thus a matter requiring the exercise of judgement, based on experience and knowledge. The main purpose of the present chapter is to explain the factors which have to be taken into account when that judgement is exercised.

1.3. Historical Development of the Subject

Computer simulation of practical combustion processes, in the sense in which it is to be understood here, appears to have begun around 20 years ago, one of its first embodiments being in a computer program developed by W. M. Pun and the present writer.[1] That program was for two-dimensional steady flow; but its adaptation to the transient phenomena of diesel-engine combustion was quickly proposed.[2]

Before that proposal could be put into effect, a newer and better method of formulating and solving the fluid-flow equations was invented,[3] and embodied in a computer program;[4] and it was this which formed the basis of what may well have been the first (two-dimensional) computer simulator of a reciprocating-engine flow.[5]

The first three-dimensional combustion-chamber calculation known to the present writer was a steady-flow one;[6] and the first three-dimensional reciprocating-engine-flow simulation appears to be that of Markatos & Mukerjee[7] in 1972.

Since that time, several computer programs capable of performing such calculations have been developed;[8–12] and engine designers and researchers are actively using them, and developing them further.

No attempt will be made in this chapter to recount the recent history of the subject, which is in rapid flux. Instead, attention will be concentrated on those features which are fundamental, and therefore change slowly or not at all.

2. FUNDAMENTALS OF COMPUTATIONAL FLUID DYNAMICS

2.1. Formulation of the Mathematical Problem

2.1.1. Basic Laws of Physics

The flow of fluids is governed by a set of 'balance' laws, all of which can be expressed verbally as:

> 'The rate of increase of ϕ with time *plus* the net outflow of ϕ from the space in question *equals* the net rate at which ϕ is being produced there.'

What is ϕ? It is a symbol used to represent any of the following:

- the mass of material in the specified volume;
- the mass of a specific component (e.g. oxygen gas) of that material;
- the thermal-energy content of the total mass or of a specific component;
- the turbulence-energy content of the material;
- the momentum of the material in a specified direction;
- or some other property of interest.

What is meant by 'outflow', in the above basic law, is one or all of:

- bulk motion;
- molecular-scale random motion (diffusion); and
- larger-scale near-random motion (turbulent mixing).

The 'specified volume' is a three-dimensional space, of finite size, often referred to as a 'computational cell'. The whole of the combustion chamber (if that is the subject of study) is imagined to be subdivided into a large number of such cells, usually in an orderly manner. The simplest subdivision of a cylindrical space is effected by

Fig. 1. Subdivision of a cylindrical space into cells.

defining a series of planes running through the symmetry axis, another series of planes which are normal to that axis, and a set of concentric cylinders. The volumes defined by these three sets of surfaces are the computational cells. This subdivision is illustrated in Fig. 1.

2.1.2. Differential Equations

Even though all computer simulations are based on the consideration of cells having finite volumes, it is common to express the basic laws of physics by way of differential equations, that is to say in terms of infinitesimally small volumes. Therefore, in order to make the concepts understandable to mathematicians, the laws are expressed as follows:

$$\frac{\partial}{\partial t}(\rho\alpha\phi) + \mathrm{div}(\rho\vec{\alpha}v\phi + \vec{J}_\phi) = S_\phi \tag{1}$$

Here $\partial/\partial t$ denotes the 'rate of increase with time' and 'div' denotes the 'rate of outflow from unit volume. The symbol ρ stands for density, i.e. mass per unit volume; α stands for the proportion of the volume which is occupied by the material (sometimes called 'phase') in question; $\vec{v}$ represents the bulk velocity of that material, in magnitude and direction; $\vec{J}_\phi$ represents the transport of the entity ϕ by random molecular or turbulent motion; and S_ϕ represents the source of entity ϕ per unit volume.

Often $\vec{J}_\phi$ is taken to be proportional to the negative gradient of ϕ, or of $\alpha\phi$, thus:

$$\vec{J}_\phi = -\Gamma_\phi \alpha \,\mathrm{grad}\, \phi \tag{2}$$

or

$$\vec{J}_\phi = \Gamma_\phi \,\mathrm{grad}\,(\alpha\phi) \tag{3}$$

where Γ_ϕ is an 'exchange coefficient' for entity ϕ. However, such a simple formulation is an approximation to which Nature often declines to conform.

The differential equation (1) is written in the so-called co-ordinate-free form, which means that no decision has been made about the shapes of the infinitesimal volumes to which it is applied.

Alternative forms express the same physical laws by reference to specific co-ordinate systems such as Cartesian, cylindrical polar, curvilinear orthogonal, general curvilinear etc.; but there is no need to go into such matters here.

However the differential equations are expressed, they are numerous, non-linear and strongly coupled with each other. Therefore, it is impossible to obtain their solution (i.e. the distributions of pressure, temperature, velocity, etc.) by classical methods of analysis; instead, *numerical methods* must be used. Fortunately, adequate methods of this kind have been invented; and digital computers can be programmed to put the methods into operation.

Numerical methods always operate on finite intervals of space and time. In a sense, therefore, they do not solve the *differential* equations at all. Only when the number of computational cells in the grid is infinitely great, which of course never happens, can the solution of the differential equations be arrived at.

2.1.3. Auxiliary Relations

The differential equations (or their numerical counterparts) do not completely specify the flow-simulation calculation which is to be performed. Information of two further types is required, namely (a) auxiliary relations, and (b) initial and boundary conditions. The former are discussed here.

Auxiliary relations are (usually) algebraic equations enabling one quantity of relevance to the calculation to be computed from one or more others. For example, the density of an ideal gas may be computed from its pressure and temperature; and the density of a mixture of such gases may be obtained from those quantities *and* the set of mass fractions of individual components which defines its composition.

The above examples come from the subject of thermodynamics; others involved in engine combustion belong to the subjects of chemical kinetics (reaction rates), transport properties (viscosity, diffusion coefficients, thermal conductivity), heat transfer (emissivities, absorptivities, heat-transfer coefficients at the cylinder walls), turbulence (rates of generation and dissipation of energy), and multiphase flow (gas–droplet interactions, size-distribution laws). Clearly the creator of a computer code for simulating flow and combustion processes in IC engines requires a compendious knowledge of physics and chemistry; and the user of such codes should at least understand the general nature of the auxiliary relations which they embody.

2.1.4. Initial and Boundary Conditions

The differential equations and auxiliary relations define the substances in question and the laws which they obey; but it is the initial and boundary conditions which determine the processes which they undergo.

The *initial conditions* determine how things stand at the start. For example, they might comprise a complete statement of the distributions of pressure, temperature, velocity components, turbulence properties and gas composition at bottom dead centre. With these defined, the calculation can begin by 'marching through time', as the phrase is, i.e. deducing conditions at the next instant of time from those prevailing at the previous instant.

Here it should be immediately admitted that the required initial-condition information is seldom available; instead, what is known is that the conditions in the cylinder at the beginning of one stroke are those which were created at the end of the previous stroke. The way out of the difficulty is therefore to make an arbitrary start, and to continue the calculations for a sufficient number of engine cycles for the nature of that start to have lost its influence.

There is, however, no possibility of dispensing with prescriptions of *boundary conditions*; for these are the mathematical representations of such physical factors as the opening and closing of valves, the introduction of energy at a spark or of liquid fuel through an injector, and the loss of heat through the cylinder wall.

Even here, however, the simulation-code user may have to be content with incomplete information, as when, for example, he knows how much gasoline–air mixture enters through the inlet-valve port but is uninformed about its distribution. He can then presume that the distribution is uniform; or he can experiment with various non-uniformities so as to establish the seriousness or otherwise of his ignorance; or he can extend the domain of flow simulation upstream into the inlet manifold.

Initial conditions are expressed by specifications of the values of the dependent variables, at *all* cells in the numerical grid, at the *start* of the computation. Boundary conditions may be expressed by specifications of the values of the variables at *some* of the cells at *later times*; but they may also be expressed by specifications of '*sources*' (i.e. rates of inflow) of mass, momentum and energy into some of the cells.

2.2. Numerical Methods of Solution

2.2.1. Discretization of Space and Time

It has already been stated that the space in the engine cylinder is subdivided, for the purposes of analysis, into volume elements; these are known as either 'cells' or 'control volumes'. In Fig. 2 a single cell is illustrated, and also the notation which is commonly employed to distinguish it from its neighbours. *P* is the 'typical point' within the cell; and the corresponding points in the six neighbouring cells are *N* (for 'north'), *S* (for 'south'), *E* (for 'east'), *W* (for 'west'), *H* (for 'high') and *L* (for 'low').

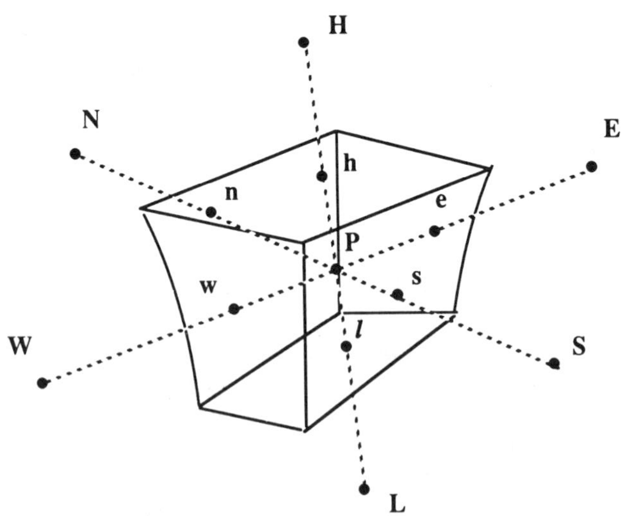

Fig. 2. A computational cell.

The straight lines joining P to its neighbours intersect the faces of the cell boundaries at the points n, s, e, w, h and l.

The time dimension is similarly 'discretized', in the sense that conditions at the points P, N, S, etc. are computed only for a finite number of instants of time.

What happens between the selected geometrical locations and time instants? That question is answered by presumption: interpolation formulae are employed which embody the reasonable expectation that variations of fluid properties are smooth, and that maxima and minima are (often) absent. Some commonly used interpolation formulae will be mentioned in subsection 2.2.2.4.

2.2.2. *Finite-volume Equations*

2.2.2.1. *Explicit, implicit and other forms.* The equations which a computer-simulating code actually solves are the result of applying the balance law enunciated in Section 2.1.1 to each cell, to each time interval, and to each dependent variable ϕ. The equations are usually formulated thus:

$$\phi_P = f(\phi_N, \phi_S, \phi_E, \phi_W, \phi_H, \phi_L, \phi_P) \tag{4}$$

where the subscripts to ϕ indicate the locations to which the values of dependent variable ϕ pertain, and $f(\ldots)$ denotes 'some function of ...'. The equation in this form allows the value of ϕ at point P to be deduced from the values of ϕ at the neighbouring points.

The appearance of ϕ_P on the right-hand side of eqn (4) may have been noted. How can P be a neighbour of itself? The answer to this question is that 'neighbourhood' can be temporal as well as spatial: the ϕ_P on the right-hand side must be the ϕ at point P at an *immediately earlier time instant.* Obviously a better notation is needed; for example, the earlier-time ϕ_P can be distinguished as ϕ_P^-, and the later-time ϕ_P as ϕ_P^+.

This distinction can be made also in respect of all the other neighbour variables. Thus, the following equation would connect the late-time ϕ_P with early-time neighbours:

$$\phi_P^+ = f(\phi_N^-, \phi_S^-, \phi_E^-, \phi_W^-, \phi_H^-, \phi_L^-, \phi_P^-) \tag{5}$$

This is known as the 'fully explicit' form of the finite-volume equation, because it expresses ϕ_P^+ explicitly in terms of known (i.e. early-time) quantities.

Alternatively, the 'fully implicit' form may be used, namely:

$$\phi_P^+ = f(\phi_N^+, \phi_S^+, \phi_E^+, \phi_W^+, \phi_H^+, \phi_L^+, \phi_P^-) \tag{6}$$

wherein the only early-time value on the right-hand side is ϕ_P^-. In this form, ϕ_N^+, ϕ_S^+, etc. are not known; for they are new-time values. The whole set of equations must therefore be solved simultaneously.

Many other forms may be envisaged, and some are sometimes used. Thus, if the superscript $\sim$ is used to denote 'arithmetic mean of late- and early-time values', i.e.

$$\phi^{\sim} = \tfrac{1}{2}(\phi^+ + \phi^-) \tag{7}$$

the equation which results is:

$$\phi_P = f(\phi_N^{\sim}, \phi_S^{\sim}, \phi_E^{\sim}, \phi_W^{\sim}, \phi_H^{\sim}, \phi_L^{\sim}, \phi_P^-) \tag{8}$$

This is known as the Crank–Nicholson form; and it is recommended and used by some practitioners of computer simulation.

2.2.2.2. *Freedom of choice.* The newcomer to numerical mathematics may be perplexed by the above indications that the practitioner is free to choose one form rather than another. 'Does it not matter?', he may well ask; 'And is not one formulation more accurate than the others?'

There is indeed some freedom of choice; and the reason is that *all formulations are wrong,* to some extent and in some circumstances, because of the finite sizes of the control volumes and time intervals; but *all must agree* as the sizes are indefinitely reduced.

Given that the freedom exists, the question of which to choose on a particular occasion can be answered according to the weight which the chooser gives to such desiderata as: simplicity of concept; numerical accuracy for given sizes of cells and time steps; speed of computation; freedom to choose any time step without encountering numerical instabilities; and so on. Many books have been written on such topics; but their contents cannot be summarized here.

2.2.2.3. *The 'linear-seeming' form.* From now on, the superscript-free form of eqn (4) will be used, in order that the reader's mind should not be influenced towards one or other of the possibilities just indicated. Attention will therefore be devoted to other aspects of the finite-volume equations.

Usually, these equations are expressed in the form of an equation of linear appearance, thus:

$$a_P\phi_P = a_N\phi_N + a_S\phi_S + a_E\phi_E + a_W\phi_W + a_H\phi_H + a_L\phi_L + b_P\phi_P + c \quad (9)$$

wherein the a's and b_P, are coefficients having the dimensions of mass per unit time, and c is a constant with the dimensions of Φ per unit time. Here Φ stands for that entity of which ϕ is the amount per unit mass; thus Φ is momentum when ϕ is velocity (i.e. momentum per unit mass).

The a's on the right-hand side of the equation express the influences of bulk motion (convection), and of molecular or turbulent random motion (diffusion) across the faces of the cell. The term b_P on the right represents the influence of the time variation, and equals the early-time density–volume product divided by the time interval. The a_P on the left-hand side expresses further influences of bulk and random motions, and of the variations in time. Finally, c represents the source rate of Φ per unit volume, and can have many forms and significances. Thus it represents pressure-gradient and body-force contributions to the balance of momentum when ϕ has the significance of velocity; it represents the chemical-source terms when ϕ is a chemical-species concentration; and it can include a variety of energy sources when ϕ stands for temperature, internal energy and enthalpy. Often, c is expressed as a linear function of ϕ_P, i.e. through:

$$c = c_0 + c_1\phi_P \quad (10)$$

If the ϕ_P in question is evaluated for the same time level as that on the left-hand side, eqn (9) can then usefully be rewritten as:

$$(a_P - c_1)\phi_P = a_N\phi_N + a_S\phi_S + a_E\phi_E + a_W\phi_W + a_H\phi_H + a_L\phi_L + b_P\phi_P + c_0 \quad (11)$$

The word 'linear-seeming' appears in the heading of the present subsection. The reason is that, although the above equations have the *appearance* of linear equations, they commonly lack its reality; for the quantities a, b and c are themselves often functions of the dependent variables ϕ, and are therefore not known in advance. For example, the density ρ enters the expressions for a's and b's; and ρ frequently depends upon pressure, enthalpy or concentration, each of which may be a ϕ. The coefficients are therefore themselves functions of the variables to be solved for. It is for this reason that iterative methods of solution are preferred: guesses have to be made about the unknown quantities; then these are gradually updated, as new information becomes available.

2.2.2.4. *Interpolation formulae.* The term $a_N\phi_N$, in eqns (9) and (11), expresses the contribution of the value of ϕ_N to the convective and diffusive fluxes of the entity Φ across the cell face in which point n is located. Now these fluxes are properly expressed as;

$$[A\alpha(\rho\langle v\rangle - \Gamma\langle \text{grad}\,\phi\rangle)]_n$$

where A represents the cell-face area, α represents the volume fraction of the phase in question, ρ represents the fluid density, $\langle v\rangle$ represents the inwardly directed normal component of fluid velocity, Γ represents the exchange coefficient of entity Φ, $\langle \text{grad}\,\phi\rangle$ represents the normal component of the gradient of ϕ, and the subscript n implies that evaluation at the interface point n is required.

A difficulty immediately appears: we may know fluid properties at the grid nodes P and N; but how are they to be computed for point n? The question applies to each of the quantities α, ρ, $\langle v\rangle$, ϕ, Γ and $\langle \text{grad}\,\phi\rangle$; and great care has to be taken in answering it.

The difficulty can be resolved by the employment of *interpolation formulae,* a topic which was first referred to at the end of Section 2.2.1. One such formula is the arithmetic-mean one; for example, ϕ_n might be deduced from:

$$\phi_n = \tfrac{1}{2}(\phi_P + \phi_N) \tag{12}$$

or, if account is taken of the fact that n does not, in general, lie halfway between points P and N, the following formula might be used:

$$\phi_n = \frac{\phi_P\{nN\} + \phi_N\{Pn\}}{\{nN\} + \{Pn\}} \tag{13}$$

where $\{nN\}$ equals the distance between n and N, and so on.

Both the above are *linear*-interpolation formulae; but quadratic and higher-order formulae are also sometimes used, the constants in which depend upon, for example, the magnitude and direction of the local velocity, and on whether sources are present. The deviser of an interpolation formula is always seeking to compensate for his actual ignorance of how the fluid properties vary between grid nodes by reasonable guesses, based upon general experience and locally available knowledge. If perfect guesses could be made, very few grid points would be needed; but this can happen only occasionally and by chance.

Interpolation in time is just as necessary in respect of the coefficients as it is in respect of the ϕ's. The latter topic was already broached in subsection (a) above, where the possibility of using the superscripts $+$, $-$ and $\sim$ was discussed. It will readily be recognized that such superscripts are also required for α, ρ, Γ, etc.; for these quantities also, in general, vary with time. Indeed, even the area A may vary, in an internal-combustion engine simulation; for it is common to cause the whole grid system to contract and expand, concertina fashion, as the piston rises and falls.

Although just as necessary as interpolation in space, interpolation in time presents somewhat fewer options. The reason is that the 'velocity' through the time dimension is always the same (i.e. 1 second per second); and there is therefore no possibility of diffusion in the time dimension.

The subject of interpolation formulae is too large for treatment here; but three topics arising in it will now be explained in order that the newcomer to the subject, who is sure to find them referred to, can understand their significance. The topics are 'central differences', 'upwind differences' and 'the staggered grid'.

The 'central-difference' interpolation procedure is easy to explain: it is represented by eqn (12). In words, it runs: *presume that the cell-face values of any variable are the arithmetic means of the values at the two grid points closest to the cell face.*

The 'upwind-difference' interpolation procedure is a little more subtle. It can be expressed (for the variable ϕ) as:

$$\phi_n = \phi_N \text{ if bulk motion is from } N \text{ to } P$$

and

$$\phi_n = \phi_P \text{ if bulk motion is from } P \text{ to } N \qquad (14)$$

Its physical significance is this: when the wind comes from the north, the conditions in the north cell are transmitted by bulk motion into its neighbour P; and otherwise they are not.

Finally, the 'staggered-grid' concept may be regarded as, in part, a means of rendering interpolation for velocity unnecessary; for it entails storing velocities at points such as n, s, e and w, where they are needed for the evaluation of bulk-motion coefficients. As it turns out, interpolation for velocities is not totally eliminated thereby, but merely displaced to another part of the calculation; but the practice does have many conveniences, and is widely used.

2.2.2.5. *Initial and boundary conditions.* The finite-volume equations express for large segments of space what the differential equations express about infinitesimally small ones; but they also can be employed to express the initial and boundary conditions.

The initial conditions fit easily into their conceptual framework: they are simply the values of all dependent variables ϕ, for all the points in the grid, for the earliest time-level appearing in the calculation.

The boundary conditions are best expressed by way of contributions to the source term, c. Thus, if the value of ϕ at point p is to be fixed to ϕ_P^*, the values of c_0 and c_1 in eqn (14) may be set to $10^{20}\phi_P^*$ and -10^{20} respectively. Then eqn (11) reduces to:

$$10^{20}\phi_P = 10^{20}\phi_P^* + \text{negligible terms}$$

i.e. to:

$$\phi_P = \phi_P^* \tag{15}$$

as required.

Alternatively, if the boundary condition is of the flux-per-unit area kind, c_0 should be set equal to F times A, where F is the flux in question, and c_2 should be set equal to zero.

In general, boundary conditions are neither of the fixed-value nor fixed-flux kind; then c_0 and c_1 have less-extreme value. They may also, on occasion, be functions of the ϕ's at or near the point.

2.2.2.6. *The mass-conservation equations.* Before leaving the topic of the finite-volume equations, it is necessary to pay special attention to the mass-conservation equation. This may be derived from eqn (9) by setting ϕ's to unity (because it is *mere* mass, not a property of mass, that is in question).

If a single-phase flow is in question, which will be presumed so as to simplify the discussion, the source term c can be set to zero, as can also the diffusive contributions to the a's. The result is a simplified equation which can be written:

$$\begin{aligned}(V\rho/\delta t)_P^+ = {} & (A\rho\langle v\rangle)_n + (A\rho\langle v\rangle)_s \\ & + (A\rho\langle v\rangle)_e + (A\rho\langle v\rangle)_w \\ & + (A\rho\langle v\rangle)_h + (A\rho\langle v\rangle)_l \\ & + (V\rho/\delta t)_P^-\end{aligned} \tag{16}$$

Here the + superscript has been supplied for the transient term in the left-hand side and the − superscript to that on the right, because there is no other choice; but superscripts have been omitted from the other terms because options remain open.

As it stands, this equation differs from eqn (9) in having no equivalent to ϕ. This can be changed, if the density ρ is required to take over the role, with the result:

$$a_P\rho_P^+ = a_N\rho_N + a_s\rho_s + a_E\rho_E + a_W\rho_W + a_H\rho_H + a_L\rho_L + b_P\rho_P^- \quad (17)$$

where the a's and b are appropriately defined by way of interpolation assumptions.

This form is indeed used by some CFD practitioners, especially those who concern themselves with transient compressible flows, in which the $a_P\rho_P^+$ and $b_P\rho_P^-$ terms predominate; but it is less often used when the other terms are more important, as happens to be the case in reciprocating-engine flows.

Often therefore, eqn (16) is employed as a means of obtaining not density but pressure. How this is done will be explained in subsection 2.2.3.2 below.

2.2.3. Methods of Solving the Equations

2.2.3.1. *The equations for a single variable.* Let it be supposed that the finite-volume equations of Section 2.2.2 are truly linear, in that all the a's, b's and c's in them are known; and let attention be concentrated on how the values can be found for a single ϕ (for example, the temperature, T) at a single time step, and for all points of the grid. The task is thus to find values of T which satisfy equations of the type:

$$(a_P - c_1)T_P = a_N T_N + a_S T_S + a_E T_E + a_W T_W + a_H T_H + a_L T_L + b_P T_P + c_0 \quad (18)$$

How such equations are best solved depends upon the time superscript.

Thus, if all the T's on the right are T^-'s, i.e. early-time values, and only T_P on the left has a + (late-time) superscript, simple evaluation of the right-hand sides is all that is needed. Unfortunately, this method can be used only when the time-derivation term b_P is larger than $(a_P - c_1)/2$ for all points in the grid; for, when this condition is not satisfied, the solution becomes 'numerically unstable'. This means that the computed values oscillate, with increasing amplitude, as time progresses.

When all the T's on the right except T_P have + (late-time) superscripts, determination of the solution is more laborious. However, various methods are available, details of which may be found in specialist works. For example, the Gauss–Seidel method involves computing T_P^+ for each point of the grid in turn, using always the last-calculated value of the values of T_S^+, T_W^+ and T_L^+, in a 'journey' through the grid from south, west and low to north, east and high. Then, when the north-east-high corner of the grid has been attained, the adjustment process starts again; and it continues until all the equations are sufficiently well satisfied. This process is known as iteration; and the state of equation-satisfaction arrived at is known as convergence.

Names of other popular methods are as follows: Jacobi, A.D.I. (alternating-directions implicit), successive over-relaxation, strongly implicit and matrix-inversion. All have advantages and disadvantages, advocates and detractors. All arrive at the same solution, if continued for long enough.

No attempt will be made here to summarize the known facts or to make recommendations; however, a warning will be given. Many of the published comparative studies have concerned systems of equations in which the coefficients are all equal, or of the same order of magnitude. Yet, the equations which arise in practical flow-simulation calculations often possess coefficients with enormous (i.e. several-orders-of-magnitude) variations from place to place; and it is from this fact that much of the difficulty of solution stems. It is therefore wise always to enquire, when the merits of a new method are being presented, as to whether the method has been proved to retain its advantages in varying-coefficient circumstances.

Of the methods mentioned above, the matrix-inversion method differs from the others in an important respect: no iteration is involved. This may appear to be advantageous; yet it has its own disadvantage, namely that it requires a large amount of computer storage for the holding of results of intermediate calculations.

Partly for this reason, the matrix-inversion procedure is little used in computational fluid dynamics; the other reason is that, even if it were to be used for the solution of sets of linear-seeming equations, iteration would still be needed when they are not truly linear; for, as soon as one solution for the ϕ's has been found, the coefficients of the equations have to be updated. Economy is therefore sometimes found to reside in methods in which the updating of the coefficients occurs within the iterative solving procedure.

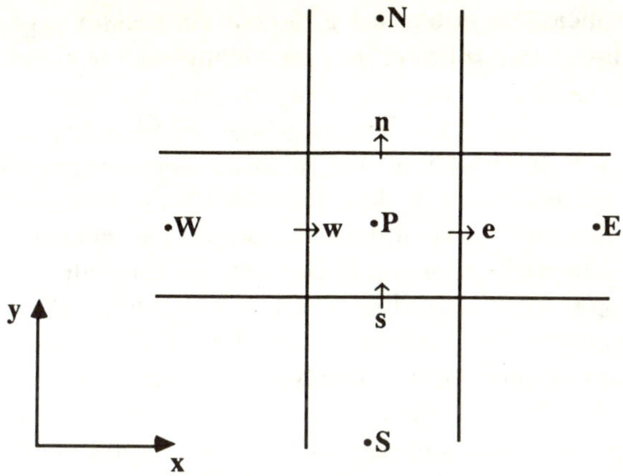

Fig. 3. Single cell cluster in two-dimensional staggered grid.

2.2.3.2. *Coupled variables.* The literature of numerical analysis is predominantely concerned with the solution of a set of equations for a single variable. Yet the problems which arise in computational fluid dynamics concern *coupled* variables. Specifically, special attention has to be given to the coupling between the velocities and the pressure.

In Fig. 3 is illustrated a single cluster of cells in a two-dimensional grid, of the 'staggered' variety, according to which the pressures are calculated for points P, N, S, E and W, the x-direction velocities for points e and w, and the y-direction velocities for points n and s. The coupling of the pressure at P with the velocities u_e, u_w, v_n and v_s will now be explained.

The value of u_e, for example, depends upon the value of the local pressure gradient in the positive-x direction, $(p_E - p_P)/(x_E - x_p)$. Similarly, u_w is influenced by $(p_W - p_P)$, v_n by $(p_P - p_N)$ and v_s by $(p_S - p_P)$. So, if the pressure-field values are known, the values of all the u's and v's can probably be computed by single-variable-solver techniques.

But what set of equations can be used for obtaining the pressures? The answer is: the mass-conservation equation, in one of the forms described in subsection 2.2.2.6; for, if arbitrary pressure fields are chosen, there is no reason to suppose that the momentum-equation-satisfying velocity fields will then obey the mass-conservation equation. A set of pressure values must be sought which leads to velocities which *do* satisfy this equation, for all cells within the field.

The most common means of determining the correct pressure is to derive equations for the pressure *correction*, p', having the form:

$$a_P p'_P = a_N p'_N + a_S p'_S + a_E p'_E + a_W p'_W + a_H p'_H + a_L p'_L + c \quad (19)$$

i.e. the same form as is possessed by the equation for the general property ϕ. Here p' is that pressure which must be added to p in order that, next time the momentum equations are solved, the associated mass-conservation errors shall be smaller. But how should values of the coefficients a_p etc., and of the additive constant c, be derived?

The provenance of c is easy: it can be taken as the imbalance of the mass-conservation equation (16) which results from the insertion into its terms of the velocities obtained by solving the momentum equations. Then, if all the c's are zero (i.e. there are no errors), all the pressure corrections will be zero also, whatever the values of the a's; and if c is positive, p'_P will increase in proportion, as is to be expected physically.

This last thought suggests where the a's should come from: they should reflect the physical tendency, when p'_P rises about its neighbours p'_N, p'_S etc., for fluid to flow out of the cell, or at least flow in less fast. They should thus reflect the influence of pressure differences between grid points on velocities at cell faces.

2.2.3.3. ***Some particular methods.*** The interrelationships between the momentum and continuity equations have been expressed in equation form by numerous authors. Among the first were Harlow & Amsden,[13] who devised the MAC (marker and cell) method, and Chorin.[14] Both papers were published in 1968.

The present writer's contribution began with the SIVA method[15] published in 1973, and continued with SIMPLE (Ref. 3) and SIMPLEST.[16] Numerous variants of these methods can be found in the literature, of which probably the most important is that published by Rhie & Chow,[17] who showed how a certain difficulty which had prompted Harlow & Amsden to propose the 'staggered-grid' arrangement, and other authors, including the present one, to follow their proposal, could be circumvented without its use.

When more than one phase is present, as when liquid-fuel droplets are injected into a diesel-engine cylinder, the problem of coupling between equation sets becomes more severe; for there are then (at least) two sets of velocities to compute for each computational cell; and the gas and liquid phases interact by way of heat, mass and momentum transfer, and by sharing the space between them. Solution procedures for such multi-phase processes have been devised and published by Amsden & Harlow,[18] the present author[19] and several later writers. They all make use, in one form or another, of the pressure-correction principle.

For further information, the interested reader is referred to specialist publications, for example, the book by Patankar,[20] and to a recently published handbook.[21]

2.3. The Quality of the Solutions of the Equations

2.3.1. Convergence

It has been mentioned above that all methods of solving the equations of computational fluid dynamics proceed by iteration, i.e. by systematic successive adjustments of the computed variables, made in directions tending to reduce the imbalances in the equations to zero. Figure 4 illustrates what happens, by plots of a selected variable at a selected 'monitor' cell, M, and of a selected 'residual' E (i.e. sum of absolute imbalances for all cells for the selected variable, $\Sigma\,|i|$, against iteration number N. As N increases, M tends to a finite asymptotic value, and R tends to zero. This is what is required. It is called 'convergence'.

The opposite of convergence, namely 'divergence', is illustrated in Fig. 5, where the values of M exhibit oscillations of increasing amplitude and the residuals increase rather than decrease.

It should be understood that whether divergence or convergence

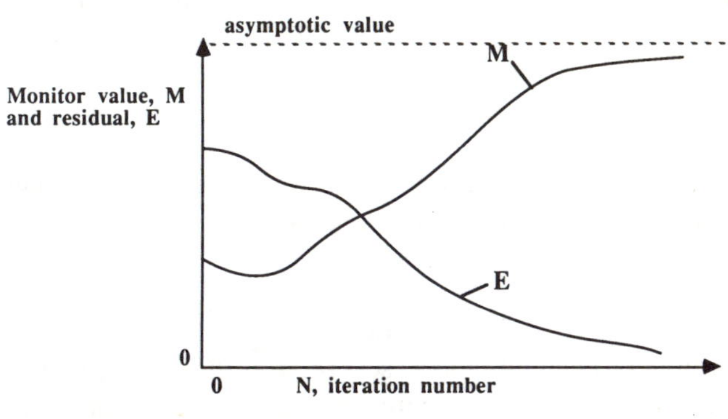

Fig. 4. Normal convergence behaviour.

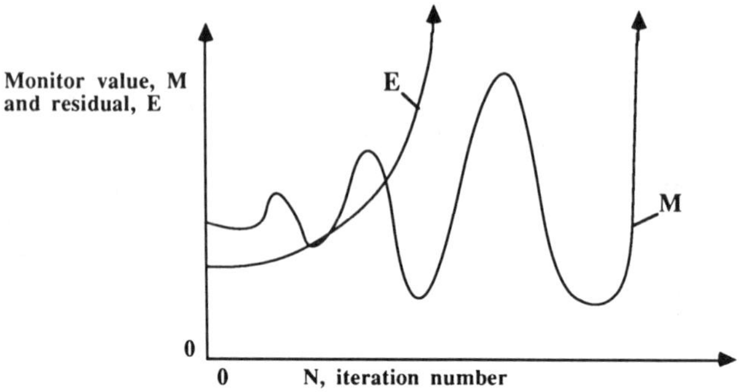

Fig. 5. Divergence.

occurs depends on the solution *method,* not on the equations which are being solved. This point can be illustrated by way of the example shown in Fig. 6, where two curves are shown on a graph of y versus x. Each curve represents an equation connecting y with x; and the fact that the two equations intersect shows that they do indeed have a solution. The dotted lines on the graph represent two solution procedures, a and b, of which the first is convergent and the second divergent.

The vertical broken lines represent solution-method moves of the kind: 'find the y value from the given x-value'; and the horizontal broken lines represent moves of the kind: 'find the x-value from the given y-value'. Clearly the a-sequence, which obtains y from curve 2 and then x from curve 1, is convergent (i.e. leads to the intersection between 1 and 2); whereas the b-sequence, which obtains y from 1 and

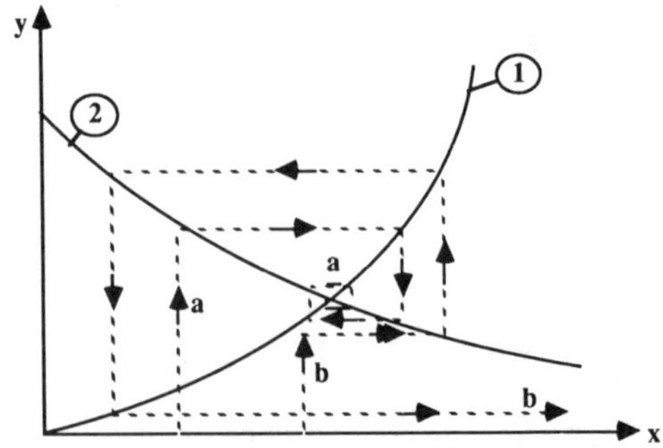

Fig. 6. Convergent (a) and divergent (b) solution procedures.

x from 2, is divergent. This proves the point: convergence is a property of the method of solution.

2.3.2. Economy

It is not enough that a solution method should converge: it must do so rapidly, so that the residuals have reduced to an acceptable level after a number of iterations that is sufficiently small for the computer times and costs to remain affordable.

There are large differences in the rates of convergence of alternative methods applied to the same problem; and a given method may also provide differing convergence rates according to the settings of such adjustable parameters as 'under-relaxation' factors. The influence of the latter can be seen from Fig. 7, which shows the same two curves as appeared on the previous figure. In this case the method used, designated *c*, employs a 50% under-relaxation factor; this means that horizontal broken lines proceed only halfway to their intersection with the next curve. Clearly, convergence proceeds via a different path from that followed in method a.

What would happen if a different under-relaxation factor had been employed; or if under-relaxation had been applied to the vertical moves instead of (or as well as) to the horizontal ones? It is not hard to discern. If *exactly* the right under-relaxation factor could be selected, the solution would be arrived at in a single move. The relevance of this observation to the present discussion is that great economy is, in principle, capable of being effected by correct selections of solution method and of parameter settings; moreover, a method which allows the optimal settings to be learned during the

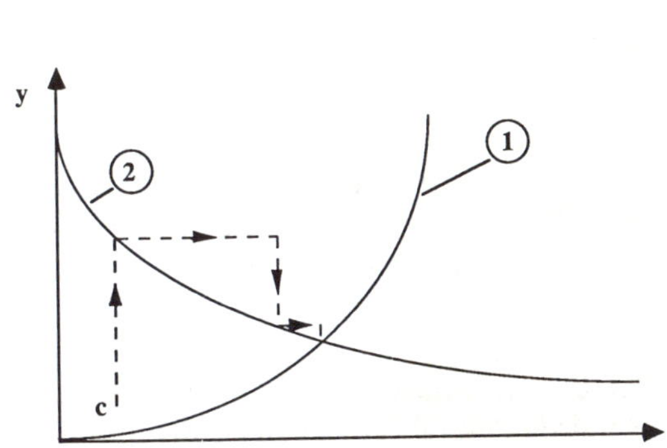

Fig. 7. The convergence behaviour of solution method c, which employs 50% under-relaxation.

course of the solution is especially advantageous. Because of the great complexity and variety of CFD problems, few computer programs are currently equipped with automatic devices for selecting relaxation parameters; but this situation can be expected to change as competition between software vendors increases.

2.3.3. Numerical Accuracy

One way to effect economy is simply to stop the iteration process while the residuals, though reducing, are still large. Then economy has been gained, but the accuracy of the solution is limited. Since, for engineering applications, three-significant-figure accuracy is often all that is required, it seldom is worthwhile to allow iteration processes to continue until the round-off error of the computer prevents further reduction of residuals.

There are other economies to be made, at the expense of numerical accuracy, namely by using a coarse rather than a fine grid, and by using few rather than many time steps for a transient simulation. The economy results because fewer numerical operations have to be performed the fewer are the points in the four-dimensional space which have to be attended to; the deterioration of numerical accuracy results from (what is called) discretization error, i.e. the inability of a few points joined by straight lines (or other interpolators) to fit a complex curve.

The practitioner of computer simulation, especially when he is concerned with the complex, transient, three-dimensional phenomena which occur in an engine cylinder, is constantly in this dilemma: he needs fine grids and numerous time steps in order to resolve adequately the important details of the phenomena which he is simulating; and yet he has not a fast enough computer, or cannot afford to use it, for the fine-grid small-time-step runs which he needs to perform.

Fortunately, the power of computers is increasing rapidly, and their costs are diminishing; although the dilemma will always exist, its severity decreases year by year.

2.3.4. Physical Realism

The ability to attain greater numerical accuracy at smaller expense, which is what the computer-hardware manufacturers both promise and provide, will not bring to an end the computer-simulator's problems. It will enable him to solve his equations precisely and quickly; but whether Nature will conform to those solutions is a quite different question.

This is the question of physical realism, i.e. of whether the physical (and chemical) behaviour of the fluids in the combustion chamber is adequately described by the equations which are being solved; and it is a question which is far from being answerable affirmatively at the present time.

If the flow in a combustion chamber were wholly laminar, if the fuel were (say) a mixture of hydrogen and carbon monoxide, and if no droplets or smoke particles were present, the affirmative answer *could* be given; for the validity of the equations is not in doubt. However, the presence of turbulence, of more complex chemical reactions, and of multi-phase phenomena, make physical realism a hope for the future rather than an immediate prospect.

It is important to adopt and maintain a properly balanced view of current capabilities and limitations. The shortcomings of current models of turbulence, for example, are not so severe as to render computations based upon them unworthy of belief; yet this belief must be tinged with caution and constantly checked against such experimental data as are available.

2.3.5. Display

Even numerically accurate and physically realistic computer simulations are of little value unless they can be understood; and the tables of numbers which are most readily produced by a numerical solution procedure are very hard to understand.

Fortunately, computer-graphics hardware and software have more than kept pace with the abilities of researchers to formulate, and of analysts to solve, the equations of fluid flow. It is therefore now easy to present the results of a flow simulation on a VDU screen by way of contour diagrams, vector plots, streamlines and perspective views. The user of a suitable graphics program is free to view the object of his choice from various viewpoints, to magnify or diminish selected parts of the display, and to add text and pictures of his own. Indeed, the danger is now that he will be so beguiled by the richness of colour and profusion of computed detail that he may accord to the underlying model more credence than it deserves.

2.3.6. *Accessibility*

A mathematical model of physical phenomena is for some people a work of art which it suffices to contemplate: actual use of the model is, for them, unnecessary. For the imagined readers of this book, however, more is needed. It is therefore necessary to indicate that the equations and methods of solution described above can be and have been incorporated into computer programs, which can be run on a wide range of computers, from supercomputers at the one extreme to personal computers at the other.

The computer programs likely to interest the present readers are of two kinds: special-purpose ones such as KIVA,[8] which have been built specifically for simulating internal-combustion-engine phenomena, and general-purpose ones such as PHOENICS,[9] which are capable of simulating such phenomena, but can do many other things as well.

It is reasonable to expect that the special-purpose programs will be better supplied with engine-specific features, because their creators will have had time and incentive to concentrate their attention on such matters. The general-purpose ones, on the other hand, should be more efficient and user-friendly because they will have been more extensively used, have enjoyed the input of greater resources, and have been subjected to a wider variety of tests. Further remarks on this topic will be found in Section 4.2.

3. COMPUTATIONAL REPRESENTATION OF COMBUSTION

3.1. Distributed-reaction Representation

3.1.1. *General Ideas*

3.1.1.1. *Chemical kinetics.* Relationships such as eqn (9), which are the central focus of computational fluid dynamics, express aspects of both physics and chemistry. In what follows, it is the chemistry which will receive attention: ϕ will have the significance of a composition variable; and of especial importance will be the source term c.

For example, ϕ might stand for m_{O_2}, the mass of molecular oxygen in unit mass of the local mixture, a quantity which equals 0·232 for pure air but which falls to zero in products of combustion. Then c would stand for the mass rate of production of molecular oxygen in the computational cell; and it would be a negative quantity, having a magnitude which depended upon the local composition, temperature and pressure of the gaseous mixture.

The main difficulty about the computer simulation of combustion resides in the ascription of values to the quantity *c*. This difficulty has two components.

The first component is that, in order to describe correctly the influence of gas composition, one must take account of the concentrations of very large numbers of chemical substances, for example those having the symbolic representations: O_2, O, H, H_2, OH, H_2O, CO, CO_2, N_2, NO, CH_4, CH_3, etc., to name but a few. The determination of such compositional influences is the concern of the science of chemical kinetics, which is highly advanced; indeed it is *too* advanced for the CFD specialist, who seldom has space in his computer for all the information that the chemical kineticist can supply.

The second component of the difficulty results from the fluctuating nature of composition and temperature in a turbulent flow. This will be explained in Section 3.1.4 below.

3.1.1.2. *The simple chemically reacting system.* The purpose of the present chapter is to convey general ideas about combustion modelling to persons to whom the subject is new. Fortunately these can be conveyed, without detailed explanations of the full complexities either of chemical kinetics or of turbulence, by reference to an idealization known as the 'Simple Chemically Reacting System'. This, abbreviated to SCRS, will now be described. A fuller account can be found in Ref. 22.

The SCRS implies that the fuel-combustion process involves the direct combination of fuel and oxygen, in fixed proportions, to form products; and it can be represented thus:

$$1 \text{ kg fuel} + s \text{ kg oxygen} \rightarrow (1+s) \text{ kg products}$$

This implies that intermediate products are absent.

If the fuel is a hydrocarbon, i.e. is composed of the elements carbon and hydrogen alone (which is very close to the truth for most internal-combustion-engine fuels), the products will consist of carbon dioxide and water vapour, usually written: CO_2 and H_2O. Typically there are about twice as many hydrogen atoms as carbon atoms in the fuel; so the reaction can be written:

$$(CH_2)_n + (n + n/2)O_2 \rightarrow nCO_2 + nH_2O$$

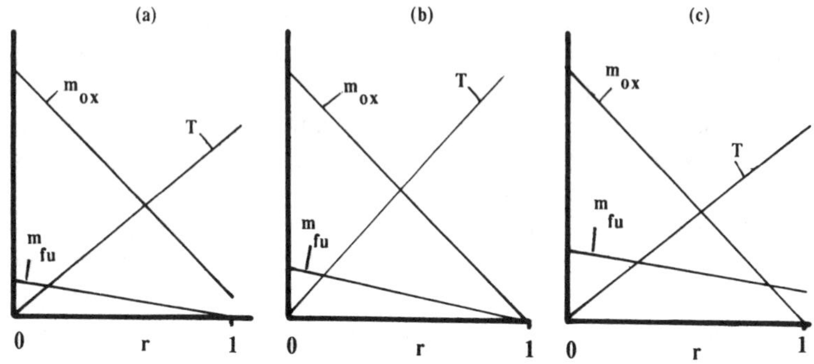

Fig. 8. Relations for adiabatic reactions.

Since the atomic weights of carbon, hydrogen and oxygen are approximately 12, 1 and 16 respectively, it is easy to work out that the value of the 'stoichiometry factor', s, in the fuel–oxygen reaction must be about:

$$s = \{(n + n/2) \times 16 \times 2\}/\{(12 + 2) \times n\} = 3{\cdot}4$$

i.e. 3·4. So 3·4 kg of oxygen is needed to burn 1 kg of fuel; and, since the mass fraction of oxygen in air is 0·232, the needed quantity of air is 3·4/0·232, i.e. 14·66 kg.

3.1.1.3. *Reactedness and temperature rise.* Combustion reactions are associated with the transformation of 'locked-up' chemical energy into thermal energy, meaning that, unless the products are cooled artificially, their temperature becomes appreciably higher than that of the reactants from which they derive; rises of 1500 K and more are common.

As a consequence, the course of a chemical reaction can be represented by the graphs shown in Fig. 8, in which (a), (b) and (c) represent respectively:

(a) a 'lean' mixture, having too little fuel to burn all the oxygen;
(b) a 'stoichiometric' mixture, having just the right (i.e. 1:s) proportions of fuel to oxygen; and
(c) a 'rich' mixture, in which there is too little oxygen to burn all the fuel.

The abscissa quantity, r, is the 'reactedness', sometimes called a 'reaction-progress variable'. It has the value zero for the fuel–air mixture in its unburned state, and the value unity in its fully burned state. The quantities m_{fu}, m_{ox} and T vary linearly with r, according to the SCRS; they stand for mass fraction of fuel, for mass fraction of oxygen and for temperature, respectively.

3.1.1.4. *Reaction rate.* Combustion reactions proceed more rapidly, other things being equal, the higher is the temperature. Since such reactions *produce* temperature rises, the occurrence of explosions (i.e. of rapidly accelerating reactions) is easily understood.

However, reactions cannot proceed beyond the point at which all the reactants have been consumed; and they inevitably slow down as this point is approached. Consequently, the typical form of the variation of reaction rate R (measured, say, in kg of fuel consumed per cubic metre per second) with reactedness r is as shown in Fig. 9.

R is not quite zero in the zero-reactedness condition; but it is often small enough to be neglected. Thus, a gasoline and air mixture at room temperature (fortunately) reacts at a rate which is totally insignificant. If it is heated, however, for example by sudden compression, the reaction rate becomes appreciable.

The shapes of the curves shown in Fig. 9 are of great significance; for most of the characteristics of combustion phenomena, namely ignition, explosion and extinction, can be explained by reference to them.

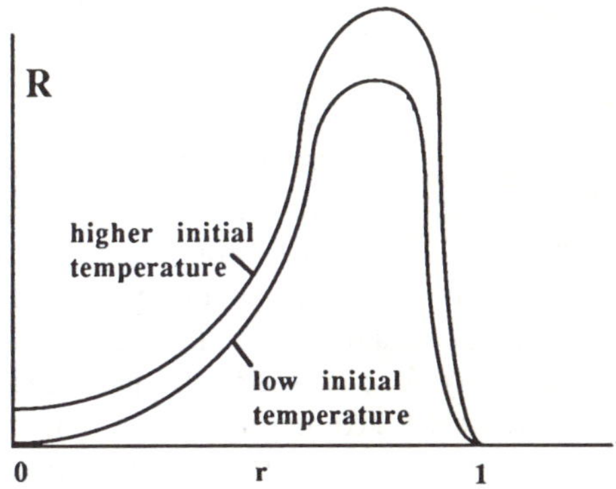

Fig. 9. Reaction rate versus reactedness for an adiabatic fuel–air mixture.

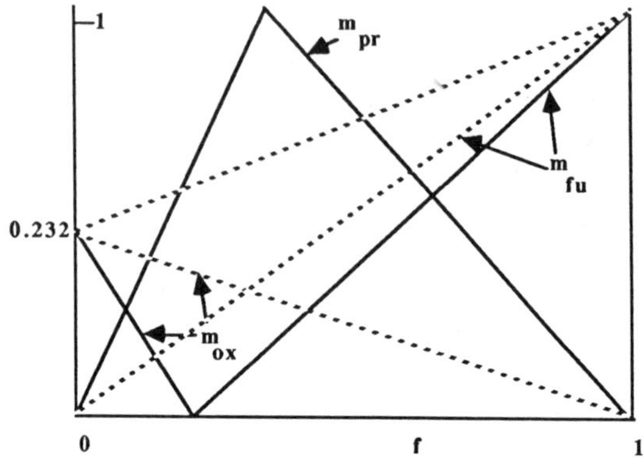

Fig. 10. Concentration relations for fully reacted (full lines) and for unreacted (broken lines) gases.

3.1.1.5. *Mixture fraction.* The fuel–air mixture which enters a reciprocating engine is rarely quite uniform, even in a gasoline engine; and in a diesel engine non-uniformities of composition are essential consequences of the way in which the fuel is injected. It is therefore useful to have a quantitative means of characterizing the non-uniformity.

The one which is often chosen in computer-simulation work is the so-called 'mixture fraction', f, defined as the mass fraction in the local mixture of material derived from the fuel. This definition implies that f remains constant, in a body of fuel–air mixture that is isolated from its surroundings, as reactedness proceeds from zero to unity.

The relation of f to other concentration variables is shown in Fig. 10, where m_{pr} stands for the mass fraction of products. The full lines represent conditions for fully reacted ($r = 1$) gas, while the broken lines represent conditions for totally unreacted ($r = 0$) gas. For the latter, of course, the m_{pr} line coincides with the baseline. Between the broken and the full lines lie conditions of intermediate reactedness.

Finally, the SCRS assumptions entail that, for an adiabatically reacting fuel–air mixture, the reaction rate R can be represented as a function of the two parameters r (reactedness) and f (mixture fraction), and that this function has the general properties shown on the contour diagrams in Fig. 11. The maximum reaction rate is exhibited by the stoichiometric mixture (because it is the hottest) at a reactedness of around 0·8. This shift towards the high-reactedness extreme is a consequence of the steep rise of reaction rate with

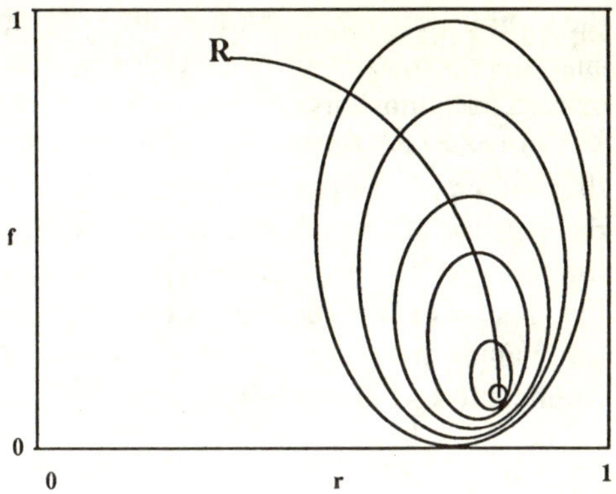

Fig. 11. Variation of *R* with *r* and *f*.

increase in temperature, involving typically a doubling of the rate for every ten degrees Celsius.

3.1.2. The Steadily Propagating Laminar Flame

The Simple Chemically Reacting System is an idealization; so, although conditions in a gasoline engine are far from steady, why not continue the idealization process a further stage, imagining as an object for contemplation and numerical simulation a *steadily* propagating plane flame? This possibility has been explored, by many researchers; and it is instructive to consider here the main features of their findings.

What occurs is illustrated in Fig. 12: a plane flame travels towards

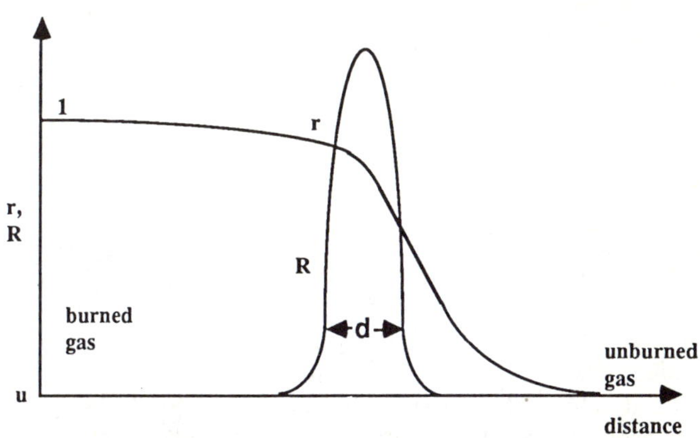

Fig. 12. Profiles of *r* and *R* in a steadily propagating pre-mixed flame.

the unburned gas (supposedly of uniform mixture fraction f), characterized by uniform profiles of reactedness, of reaction rate and of related quantities such as density and velocity.

The thickness of the flame (δ, say) is governed by the consideration that the amount of reaction per unit surface area, i.e. $\rho_u S_u$ where ρ_u is the density of the unburned gas and S_u is the speed of propagation of the flame towards it, must be proportional to *both* the thickness times the space-average reaction rate, $R_{\text{ave}}\delta$, *and* to the rate of diffusion into the flame, measured by Γ/δ, where Γ is the product of diffusion coefficient and density. Thus:

$$\rho_u S_u \propto R_{\text{ave}}\delta \propto \Gamma/\delta \tag{20}$$

Two important results follow, namely:

$$\delta \propto (\Gamma/R_{\text{ave}})^{1/2} \tag{21}$$

and

$$S_u \propto (\Gamma R_{\text{ave}})^{1/2}\rho_u \tag{22}$$

Actual values of exchange coefficient Γ ($\equiv$ diffusion coefficient times density) and R_{ave} (space-averaged reaction rate) are such that δ proves to be of the order of 1 mm at atmosphere pressure and even less at the pressures prevailing in reciprocating engines.

This means that, if the propagation speed is to be accurately calculated by a computer code based upon the principles described in Section 2, the finite volumes into which space is divided must be extremely small, for example 0·01 mm. Therefore, if the grid size is fixed throughout, and the flame has to be tracked through 10 cm of travel, there must be 10^4 subdivisions in the relevant direction!

Numerical analysts have been ingenious enough to devise moving-grid techniques[23] which avoid this difficulty, when calculating the propagation speed is the only objective of the computation. However, when the flame simulation is only one aspect of a three-dimensional transient-flow simulation for an engine combustion chamber, their techniques are impracticable.

The grid-size problem will be discussed in more detail below. Before leaving the topic of the steadily propagating plane flame, however, it will be remarked that this idealization, coupled with that of the Simple Chemically Reacting System, has just provided important insight into what proves to be a fundamental obstacle to the computer simulation of flame phenomena, namely the *disparity of scales* between flames and combustion chambers. Even very extreme simplifications, such as these two idealizations are, can be extremely valuable. It is a fact that the possessors of powerful computers, and users of computer programs which can handle great and realistic detail, should never allow themselves to forget.

3.1.3. The Grid-fineness Problem

Computers of currently available size can handle fluid-dynamics-simulating grids containing tens of thousands of points with fair ease, hundreds of thousands on special occasions, and a million or more almost never. Suppose then that a three-dimensional grid were to have 27 000 points; then, if these were equally distributed in the three directions, the grid could consist of $30 \times 30 \times 30$ computational cells.

To fit an engine cylinder, a polar co-ordinate grid might be used, as illustrated by Fig. 1; so the cells might consist of the volumes formed by the intersections of 30 horizontal planes, 30 concentric cylinders, and 30 vertical planes passing through the common axis of the cylinders.

Because events of importance take place at all points within the combustion space, there is no compelling reason to make the grid non-uniform. Therefore, if the cylinder diameter were 12 cm, say, the cells of the computational grid would be $12/(2 \times 30)$, i.e. 0·2 cm, apart in the radial direction. In the axial directions the cells would be similarly spaced, when the piston was at bottom dead centre. In the circumferential direction the intercell distances would be larger near the periphery and smaller near the axis.

The significance of these observations is that the *cell dimensions are appreciably greater than the flame thicknesses,* which were indicated in Section 3.1.2 to be 0·1 cm or less; and this would remain true if all the grid dimensions were halved (increasing the number of nodes to 214 000) or quartered (increasing it to 1 912 000). The consequence of *this* is that it is quite impracticable for the distribution of chemical reaction within the flame to be represented by the numerical methods described above.

Proposals have been made to surmount this obstacle by employing highly non-uniform grids, the idea being to pack numerous very small cells into the reaction-zone region, so that numerical accuracy can be achieved. However, the practical realization of this notion involves much complexity of programming; and it is highly likely that any improvement of accuracy in respect of flame-propagation speed would be accompanied by serious *loss* of accuracy in respect of fluid dynamics. This is not the way to proceed.

Mindful of this difficulty, some authors (e.g. Ref. 24) have ingeniously proposed that the flame could be artificially thickened, in the numerical computation, by multiplying the presumed transport properties by a large factor and dividing the presumed reaction-rate constants by the same quantity. The idea behind this proposal is that the speed of propagation will be unchanged, because only the ΓR_{ave} *product* affects the propagation speed, as eqn (22) revealed, whereas it is Γ / R_{ave} which influences the thickness.

However, this idea is little used, mainly because there are more serious obstacles in the way of representing the rate of distributed reaction. They derive from the presence of turbulence, and of two-phase effects. These will now be discussed, in order.

3.1.4. The Turbulent Fluctuation Problem

3.1.4.1. *The nature of the problem.* In a turbulent flow there exist, at any instant in time, variations of velocity, temperature and concentration which are significant over extremely small distances, e.g. less than 0·1 mm; and the fluid properties at a point vary significantly in times which are also extremely small, e.g. less than 0·001 seconds. Therefore, if an accurate representation of a turbulent flow is to be made, very fine grids are required, coupled with very small time steps.

Numerical simulations in which these conditions are fulfilled are known as 'direct simulations'. They are made by researchers investigating fundamental problems of turbulence research; but they involve very great computational expense, even for flows of the greatest simplicity; and they are totally impracticable, at the present time, for internal combustion engine purposes.

Fortunately, methods have been devised whereby some aspects of turbulent-flow phenomena in engineering equipment can be rather well simulated. These involve the use of 'turbulence models', i.e. systems of additional differential and auxiliary equations, to be added to those governing the distributions of mass, momentum and energy, which purport to describe how *statistical* properties of turbulence such as time-average energy and dissipation rate are distributed; and from these properties *effective* values of viscosity and other exchange coefficients are deduced for introduction into the equations for mass, momentum and energy themselves. From these equations, time-averaged values of pressure, velocity, temperature and concentration

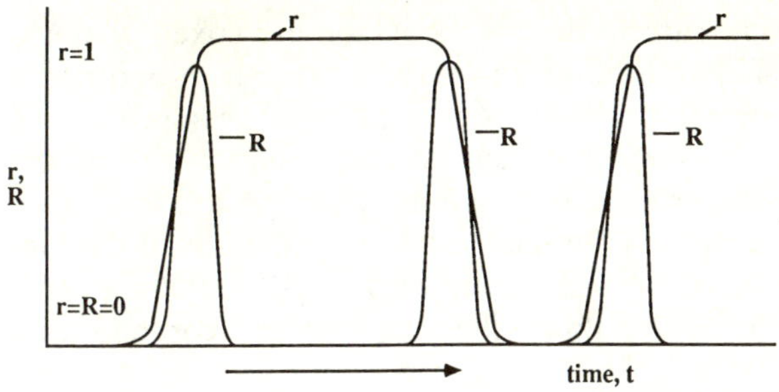

Fig. 13. Sketch of the way in which reactedness *r* might vary with time, and of its consequences in respect of the variations of reaction rate *R*.

may be computed. A simple introduction to the subject is provided by Ref. 25.

Despite the fact that the turbulence-model approach became practicable only in the late 1960s, there is now a large literature on the subject; and, although some deficiencies still remain, the existence of turbulence models has made it possible to predict numerous flows of engineering importance with fair accuracy.

Unfortunately, the said deficiencies are especially severe when it is the rates of chemical reaction which have to be predicted; and the reason is that knowledge of *time-averaged* values of temperature and concentration does not suffice to permit time-averaged *reaction rates* to be computed. Figure 13 helps to explain why.

The diagram shows the variation with time of reactedness *r*, at a single point, which might arise when fragments of fully burned ($r = 1$) gas are thrust by turbulent exchange into a mass of gas which is not yet burned ($r = 0$). The *r–t* curve thus consists of a series of horizontals, at two levels, linked by rather steep joining curves.

The *R–t* curve expresses the consequences in terms of reaction rate of the *r–t* curve. Why it has the 'spiky' form indicated is explicable by consideration of Fig. 9, and what gave rise to it: *R* is zero when $r = 0$ or 1; and *R* has finite values only for a narrow range of intermediate *r*'s.

The following consequence is immediately obvious: the time-averaged reaction rate at the point is directly proportional to the thickness of the *flanks* of the hot-gas intrusions; but the time-averaged reactedness is mainly determined by the relative thicknesses of the hot- and cold-gas intrusions themselves. The average *R* depends little on the average *r*, but greatly on the shape of the *r–t* curve.

A similar conclusion can be drawn about non-pre-mixed combustion. If (to simplify matters) the reaction rate is taken as proportional to the product of concentrations, $m_{fu} \times m_{O_2}$, the time-averaged rate is proportional to the time-averaged value of the product, *not* to the product of the time-averaged concentrations. Thus the time-averaged concentrations of fuel and oxygen might both be finite; but, if the fuel and oxygen are present at the same point *only at different times* (as is easily possible in a fluctuating flow), the time-averaged concentration product is always zero.

3.1.4.2. *Some proposed solutions to the problem*

(i) The use of the laminar flame speed, S_u. Since chemical reaction takes place on the flanks of the intrusions of hot gas into cold, or of fuel-rich gas into oxygen-rich, the amount of interface area per unit volume is likely to have a strong influence on the flow. This quantity has the dimensions of a reciprocal length, which latter might be regarded as the fragment size, sometimes called the length-scale of the turbulence. Let it be given the symbol l.

If the flame is a pre-mixed one, it *might* be the case that a flame propagated into a cold-gas fragment from a hot-gas one at the same speed as if it were a plane flame in an infinite medium. Then the mass rate of burning of gas mixture per unit volume might be computed from:

$$\langle R \rangle = \rho_u S_u / l \tag{23}$$

where $\langle R \rangle$ is the time-averaged reaction rate and ρ_u and S_u are the quantities encountered in Section 3.1.2 above.

Since values of S_u are known for many fuel–air mixtures as a function of initial composition and temperature, and since l can be estimated from turbulence models concerned with hydrodynamics, for example the $k-\varepsilon$ turbulence model,[26,27] eqn (23) is a formula which can be used for estimating reaction rates in turbulent pre-mixed flames. However, it requires modifying by some factor which ensures that the rate falls to zero when the time-averaged reactedness $\langle r \rangle$ equals unity (because there is nothing left to burn) or zero (because there are no hot-gas fragments from which flame can spread.

The following formula would meet these requirements:

$$\langle R \rangle = \{1 - (2\langle r \rangle - 1)^2\} \rho_u S_u / l \tag{24}$$

but any even-number exponent of $(2\langle r \rangle - 1)$ would do.

The present author is not aware of any use having been made of just this formula; but such use would not be unreasonable.

(ii) The use of the fluctuation velocity, u'. Experimental observations of some turbulent flames, for example those which spread from flame-holders mounted in streams of fast-moving pre-mixed gas, reveal that the time-averaged chemical reaction rate is remarkably insensitive to the chemical composition, and to the initial temperature, of the gas mixture. Since S_u *is* sensitive to both these factors, it appears that eqn (24) cannot fit all the data.

What does appear to influence $\langle R \rangle$ is the flow velocity, so much so that doubling this velocity very nearly doubles the reaction rate. Since it is hard to see how the velocity can be *directly* influential, one must look for an indirect influence, which can be found in the *fluctuation* velocity, i.e. the root-mean-square value of the instantaneous departure from the time-averaged velocity. This quantity is often given the symbol u'.

It follows that a formula which could reasonably be employed for predicting time-averaged reaction rates in turbulent flames is:

$$\langle R \rangle = \{1 - (2\langle r \rangle - 1)^2\} C\rho u'/l \qquad (25)$$

which is like eqn (24) but with $C\rho u'$ substituted for $\rho_u S_u$, where C is a constant to be determined from experiment.

(iii) The 'eddy-break-up' model. Thoughts of the above kind led the present author, long ago,[28] to propose that the reaction rate should be proportional to the local time-averaged velocity gradient; and indeed much success has been achieved with the use of the formula:

$$\langle R \rangle = C(1 - \langle r \rangle)\rho\,|\text{grad}\langle u \rangle| \qquad (26)$$

where $\langle u \rangle$ is the time-averaged flow velocity, and C is a constant of the order of 0·4.

The connection between eqns (25) and (26) is best seen if it is recalled that Prandtl,[29] when deriving his 'mixing-length theory' for turbulent exchange, argued that u', which he needed as a measure of the random motion, could be approximated by the product of l and $|\text{grad}\, u|$; thus u'/l and $|\text{grad}\, u|$ are reasonably taken as proportional to one another.

The $(1 - \langle r \rangle)$ factor in eqn (26) has the effect of making $\langle R \rangle$ zero when $\langle r \rangle$ equals 1; but it allows $\langle R \rangle$ to be finite when $\langle r \rangle$ equals zero, which is a defect. The more elaborate function of $\langle r \rangle$ in eqns (24) and (25) is therefore qualitatively preferable.

Equation (26) is not especially useful for simulation of reciprocating-engine combustion; for it is hard to identify a velocity gradient. However, a quantity having the same dimensions is Ω, the number of crank revolutions per second; and it is reasonable to use this because all |grad u|'s, or (u'/l)'s for that matter, are likely to be proportional to it. A plausible formula for the reaction rate in a reciprocating engine is thus:

$$\langle R\rangle = \{1-(2\langle r\rangle - 1)^2\}C\rho\Omega \tag{27}$$

No doubt the 'constant' C would depend on position in the cylinder and time in the cycle.

(iv) Combination models. Formula (27) would imply that the combustion in the engine would exactly keep pace with an increase in engine speed, which is not quite true; and it has no place for an influence of mixture ratio. Since it is known that mixture ratio *does* have an influence, it is reasonable to seek to re-introduce the laminar flame speed, S_u, so as to express this effect.

How to introduce it is entirely speculative; but the following formula has the right qualitative behaviour:

$$\langle R\rangle = \{1-(2\langle r\rangle - 1)^2\}\rho/(C_1/\Omega + C_2 l/S_u) \tag{28}$$

Clearly, when S_u is very small, the term containing it dominates the denominator, and makes the reaction rate very small. When S_u is large, on the other hand, S_u has no influence on $\langle R\rangle$ whatever.

Equation (28) is probably the best of the simple formulae for combustion of pre-mixed gases in reciprocating engines.

3.1.4.3. *Discussion.* Although more elaborate concepts and formulae have been proposed, both by the present author[30–32] and others,[33,34] the most important fact for the newcomer to grasp is that all combustion-rate formulae for turbulent flows remain speculative.

This situation is not likely to change rapidly, for both turbulence and chemical kinetics are highly complex phenomena; and the non-linearities represented by the (still highly simplified) R–r curve of Fig. 9 amplify the effects of small deviations from regularity.

Nevertheless, progress is being made. Some ideas regarded as promising by the present author will be presented in Section 3.2 below.

3.1.5. The Two-phase Problem

3.1.5.1. *The nature of the problem.* Whereas in gasoline engines it can be presumed that all the fuel is in the vapour phase by the time that the spark initiates combustion, the situation is quite different in diesel engines; for in these the fuel is injected near top dead centre in the form of droplets, which have still not fully vaporized by the time that the combustion is well advanced.

In some designs of engine, the liquid droplets pass right through the gas space and impinge on the cylinder wall, forming a thin film there; vaporization and combustion of this fuel take place subsequently.

Both the droplets themselves, and the film of liquid on the wall, are of small dimensions in comparison with the allowable size of computational grid. The scale-disparity problem connected with the simulation of thin flames, and of turbulent eddies, therefore arises again here: it is impracticable to simulate the details of two-phase flow even on computers of the largest conceivable kind.

Nevertheless, simulations of some kind *must* be made; so methods have had to be devised. Some of these will now be outlined, in general terms.

3.1.5.2. *Some approaches to solving the problem*

(i) Presuming that vaporization is instantaneous. Some problems are best coped with by being ignored; and indeed, for some combustion systems burning liquid fuels, for example those in furnaces and gas-turbine combustion chambers, the vaporization time of the droplets may be so short compared with their flight time to the oxygen-containing region that ignoring the two-phase aspects of the phenomenon may entail little unrealism.

However, to treat a diesel-engine fuel spray as a fuel-gas jet would not represent the true situation well; for example, it would completely negate the possibility of simulating the impingement of liquid on the wall. This practice is therefore not to be recommended, except when only a very rough indication is needed of how the diesel-engine combustion process will proceed. Ignition, for example, could definitely not be simulated well.

(ii) Calculating the trajectories of individual droplets. Although computational fluid dynamics is not yet sufficiently advanced for droplet sizes in diesel sprays to be predicted simply from knowledge of fuel properties and injector geometry, sufficient experimental data exist concerning the so-called atomization process for estimates of the droplet sizes to be made; and their injection velocities can be deduced from the injection pressure.

It is therefore possible to imagine that each injected droplet travels through the gas space along a curved path, and at a varying velocity, determined by the forces exerted upon it by the air in the combustion chamber. Since formulae exist for the vaporization rate of droplets, as a function of relative velocity, the size-diminution rate of the droplet can be simultaneously calculated, so that it is possible to predict where the droplet finally disappears, i.e. where vaporization is complete, unless of course the droplet hits a solid surface first.

The influence of the air on the droplet is matched by an influence of the droplet, indeed of *all* the droplets, on the air; so the complete set of interactions must be taken simultaneously into account. This can be done, however, without significant difficulty, by way of iteration.

(iii) The full two-phase treatment. Computational fluid dynamics has for some years been capable of simulating flow phenomena in which fluids of two distinct phases move simultaneously, but at different velocities and with varying volume fractions. The driving force for much of the research was concern about possible loss-of-coolant accidents in nuclear power plants, in which steam–water mixtures were thought likely to flow through and out of the coolant system; but the techniques so developed are of general applicability. They can be used for the prediction of diesel-engine injection and combustion.

The full two-phase treatment of fuel-spray and gas motion involves recognizing that the two fluids (liquid fuel and gaseous air) intermingle on a scale that is fine in comparison with the size of the computational grid, and that the extent to which they share space can be expressed (albeit imperfectly) by way of a 'void-fraction', α, which varies from zero for cells filled with liquid to unity for cells filled with gas. α varies from cell to cell, and from instant to instant; and computation of its distribution in space and time is one of the objects of the computation.

The two phases are presumed also to have different average velocities, and indeed different average temperatures, in each cell at any instant; and these velocity and temperature distributions are also to be computed. This is done by solving the set of differential equations similar to those described in Section 2.1.2. Algorithms effecting this solution have been published, e.g. by Amsden & Harlow,[18] and by the present author;[19] and they have been incorporated into at least one general-purpose computer program.[9]

Of course, the auxiliary relations which represent the processes of interchange of heat, momentum and mass between the phases must be separately supplied; and these entail a considerable amount of uncertainty, not least because the two-phase method provides no information on how the whole population of droplets departs from its mass-mean in respect of size, velocity and temperature. A *multi-phase* model can in principle provide information of this kind; but that involves much more computation, and is rarely used.

The two-phase treatment can, it should be mentioned, simulate the impingement of the liquid phase on the wall because it readily allows for differences of velocity between the liquid and the gas at (nominally) the same location; and it reflects the fact that pressure gradients which suffice to make the gas flow tangentially to the wall are insufficient to deflect the flow of the denser liquid to the same extent.

It cannot yet be said that experience of using the full two-phase treatment for the simulation of fuel-spray behaviour in diesel-engine combustion has demonstrated its actual (as distinct from potential) superiority to the individual-droplet treatment; but only a few groups in the world have been using it so far, so that judgement should be suspended.

3.1.5.3. *Discussion.* Despite a very large multinational research effort, given urgency by apprehensions regarding the safety of nuclear power plants, two-phase-flow research has not yet advanced to the point at which even steam–water flow phenomena can be predicted with reliability. The main reason is that the degree of interspersion of one phase in the other, measured by the size distributions of droplets (when the liquid is the sparser fluid) and bubbles (when it is the vapour phase which is more sparse) depends upon physical phenomena (disruption, collision and coalescence) which are hard to express via simple formulae.

Steam and water are simple substances, engaging for example in no chemical reaction and involving no differences of composition; diesel fuel and air, by contrast, are capable of many more interactions. It is therefore not surprising that ability to predict the behaviour of fuel sprays is still far from adequate.

However, what is difficult to perform to a high standard of precision is not necessarily impossible to perform to a lower standard. So it is in this case: engine designers *are* beginning to use computer simulations of two-phase flow to assist them in their design work; and the fact that increasing numbers are doing so is an indication that they are finding their endeavours to be worthwhile.

3.2. The Two-fluid Model of Turbulent Combustion

3.2.1. General Ideas

It has been explained that turbulence–chemistry interactions on the one hand, and two-phase phenomena on the other, both make the realistic simulation of engine combustion by means of computers rather difficult. Nevertheless, there is a certain sense in which having both difficulties to contend with simultaneously has led to the development of ideas and techniques which may ultimately provide solutions. That, at any rate, is the hope that inspires researchers who concern themselves with the *two-fluid model of turbulence*. The purpose of this section is to present the grounds for that hope.

It has long been recognized that turbulent flames, as Fig. 13 has already illustrated, are 'spotty'; that is, that hot- and cold-gas fragments intermingle within a flame in the manner of an archipelago in the sea or a lake system on land. This concept was first expressed, so far as the present writer knows, by Shchelkin.[35] Wohlenberg[36] independently proposed something similar, with emphasis on the differences in the fuel: air ratio between the two interspersed gases. Howe & Shipman[37] developed a mathematical theory for a uniform fuel: air ratio on the presumption that the interspersed gas fragments travel through the flame at equal velocities.

Recently, attention has been focused on the fact that the fragments may have significantly *different* velocities, i.e. they 'slip' relative to one another. Bray & Libby,[38] who credit Kuznetzov[39] with the first publication, explained how this arises from the different accelerations experienced by lighter and heavier gas fragments when subjected to the same pressure gradient. Their co-workers Moss[40] and Shepherd[41] have provided experimental confirmation of this 'slip' between the two phases. Phillips[42] has recently published some calculations for flame propagation that incorporate this effect.

Making use of the mathematical apparatus alluded to in Section 3.1.5, namely that developed primarily for the analysis of two-phase flows,[19] the author has proposed a general theory of turbulent flow, both with and without reaction, that is based on the two-fluid (i.e. 'lakes and islands') idea. This theory allows the relative motion of the gas fragments to be computed, as well as their individual temperatures, fuel: air ratios, and states of completeness of reaction.

The 'two-fluid' model of turbulence requires for its completion the determination of laws of fluid–fluid interaction in respect to heat, mass and momentum transfer, and fragment size changes. At present, these laws are known only approximately and comparisons with experimental data are needed for their better determination. Research of this kind is now in progress.

The present author outlined his approach to a two-fluid model of turbulent combustion in engines at the 1983 Conference of the Institution of Mechanical Engineers.[43] Other references to the subsequent work of the author and his colleagues include: Spalding,[44] Spalding & Wu[45] Ilegbusi & Spalding,[46] and Malin & Spalding.[47,48] The interested reader is asked to turn to these for a more detailed account than can be presented here, where the following summary must suffice:

(1) The basic concept is that turbulent gases in an engine consist of 'fragments' of hot gas intermingling with 'fragments' of cold gas. These two fluids rub up against each other (because ordinarily they are travelling at different velocities) and engage in heat, mass and momentum transfer.
(2) The relative motion of the two fluids is caused by their different densities, whereby lighter gas can respond more rapidly to pressure gradients than can heavier gas.
(3) The rates of interfluid transfer of mass, momentum and energy are proportional, at high Reynolds numbers, to the local relative velocity; and this is how increases of piston speed proportionately increase the rate of evolution of the combustion process.
(4) Although rates of chemical transformation are not directly influenced by the turbulence enhancement, they are affected indirectly; for, until the intensity of combustion becomes very great indeed, it is the rate of transfer of material across the boundaries between hot and cold fragments which governs the rate at which fuel and air can mix before burning; or, if they are already mixed, the rate at which they can be heated sufficiently to ignite.
(5) The system of equations that must be solved, when the implications of the two-fluid assumptions are worked out, consists mainly of *two* sets of Navier–Stokes and scalar-conservation equations, coupled together by algebraic relations representing the interfluid transfers of mass, momentum and energy. Numerical methods for solving these coupled equations have been developed, and are embodied in widely available computer codes, as was explained in Section 3.1.5 above.
(6) The said algebraic relations express some empirical knowledge,

extended by the theory of dimensional analysis, and by extrapolation and speculation. Their conception, formulation, investigation and settlement constitute the major activity of two-fluid-turbulence-model research.

(7) Of particular importance, and current uncertainty, are the equations governing fragment size; for the latter quantity has a major influence on interfluid transfer rates.

The remainder of this chapter concerns an embodiment of the two-fluid ideas in the PHOENICS computer code,[9] and the application of this code to the simulation of pre-mixed combustion in idealized reciprocating-engine circumstances. It is in part a shortened version of a paper presented in the Second International PHOENICS Users Conference,[49] to which the reader is referred for further information.

Section 3.2.2 concerns a one-dimensional idealization, so chosen as to bring into prominence what the two-fluid model alone can do, namely predict the turbulence which can arise in the absence of mean-flow velocity gradients.

Section 3.2.3 concerns a two-dimensional idealization of the same phenomenon, with the addition of swirl, 'squish' and other initiators of the motion known as the Rayleigh–Taylor instability.

These studies are presented so as to provoke thought about how turbulence, combustion, density fluctuations and pressure gradients may interact in reciprocating-engine circumstances. Their implications are discussed in Section 3.2.4, together with some indications of how the two-fluid model is being further developed.

3.2.2. A One-Dimensional Representation of Turbulent Combustion in a Gasoline Engine

3.2.2.1. *The problem considered.* A single compression stroke of a piston–cylinder combination is considered, followed by a single expansion stroke. The valves are taken as closed throughout, and the walls as impervious to heat. The gas in the cylinder is taken to be at rest, at the start of the compression process, and to be without turbulence of any kind.

Some combustion is imagined to have occurred at the start; indeed the gas in one-tenth of the cylinder volume farthest from the piston is taken as being fully burned. Both the hot gas and the unburned mixture which occupy the remaining nine-tenths of the volume are regarded as compressible; and the ratio of the densities of the two gases is taken as a constant, namely 0·3. This departure from strict thermodynamic correctness is made for convenience; and it is not such as to invalidate the conclusions which will be drawn.

Of course, this representation of the combustion process is idealized in the extreme, indeed almost to the point of absurdity; for it is as though the spark advance is 180°, and the flame which it has induced is supposed to have somehow spread itself radially right across the head of the cylinder without propagating significantly in an axial direction.

These departures from reality have not been introduced because two- or three-dimensional models cannot be handled by the computer program. They *can* be; indeed two-dimensional simulations are shown in Section 3.2.3, and three-dimensional simulations have also been made (although not reported here). It is to emphasize the following point that the one-dimensional idealization has been chosen.

This flow situation has been envisaged because it is one for which conventional turbulence models will predict no combustion whatever (if the slow laminar flame propagation is neglected). The two-fluid model predicts rather intense combustion, such as would (the author believes) undoubtedly be experienced in practice. How this comes about is as follows:

(1) During the first half of the compression stroke, the gases are accelerating. The interface between the hot and cold gases then remains perfectly plane and stable.
(2) During the second half, the piston decelerates, and the gases of course do so also. The changed sign of the (pressure gradient) × (density gradient) product now makes the interface unstable, so that fragments of dense unburned gas penetrate the hot-gas layer, and fragments of hot gas move towards the piston face.
(3) The relative motion of the hot- and cold-gas fragments sets in train transfers of heat, mass and momentum between the two fluids. Specifically, hot-gas fragments engulf the surrounding cold unburned mixture, and set it alight. The interpenetration of the two fluids thus precipitates further combustion.
(4) The deceleration (i.e. decrease of positive-direction velocity) of the piston continues during the half-stroke after top dead centre: the pressure gradient in the gas is still such as to thrust the hot-gas fragments preferentially towards the piston and away from its original position adjacent to the cylinder head.

Combustion therefore continues vigorously; the proportion of hot gas to cold correspondingly increases.

(5) In the final half of the expansion stroke, the pressure gradient resumes its original sign. Now the tendency is for hot-gas fragments to be thrust back towards the cylinder head, and for a cold-gas layer to accumulate against the face of the piston. By now, however, little cold gas actually remains.

To perform the experiment outlined would of course be very difficult; for the initially plane interface between hot and cold gas would be hard to contrive; and the effects of walls could not be totally eliminated. However, computer programs can simulate easily many situations that defy experimental contrivance. This will now be illustrated.

3.2.2.2. The method of solution. Problems of the type in question can be solved by means of the PHOENICS computer code;[9] for this is equipped not only for representing the balance equations of heat, mass and momentum of two-phase systems, but also with simple means of introducing initial-value, boundary-condition and fluid-property data, and with means of solving the equation system economically.

The mathematical theory of two-phase flow, and the structure and methodology of PHOENICS, have been described in print in many places (e.g. Ref. 50); and the input-files relating to the calculations to be described below are provided in Ref. 9. It should therefore suffice to state here that the relevant equations were solved by means of PHOENICS, in its two-phase mode, and to pass on to a presentation and a discussion of some of the results of the calculation.

3.2.2.3. *Typical results*

The reference calculation. In Fig. 14 are shown plots of the computed volume fraction, R_2, of hot-gas fragments, versus axial distance, at crank angles 90°, 180° (top dead centre), 270° and 360° (bottom dead centre again) for a reference calculation. It will be seen that, by 90°, there has been very little departure of the R_2 profile from its original rectangular form. By 180°, by contrast, interpenetration of hot and cold gases has begun. By 270° it is far advanced, and a hot-gas layer has been drawn right up to the face of the piston; moreover, a significant amount of combustion has occurred, causing the average value of R_2 to be greatly increased. Finally, when the crank angle has

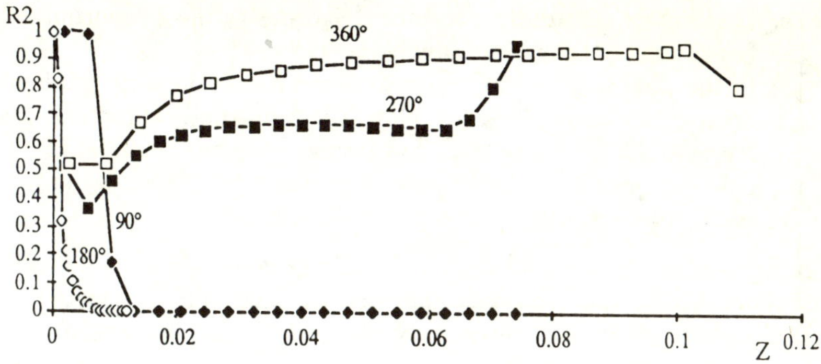

Fig. 14. Variation of hot-gas volume fraction (plotted vertically as R_2) against axial distance Z (in cm), below the upper wall of the combustion chamber, for four different crank angles.

attained 360°, although the deceleration of the piston has caused cooler gas to begin again to congregate there, there is no reversal of the tendency towards completeness of combustion.

The corresponding pressure versus crank angle curve is shown in Fig. 15. The curve would have been symmetrical about the 180° vertical, had combustion been absent; and its greater values on the right than on the left are the confirmation that the evidence of combustion shown in Fig. 14 is matched by a corresponding increase in pressure.

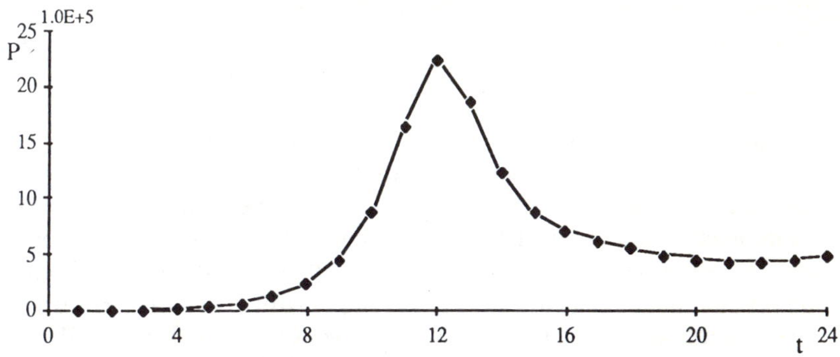

Fig. 15. Variation of cylinder pressure P versus time t, on a scale for which $t = 0$ and $t = 24$ represent bottom dead centre and $t = 12$ represents top dead centre.

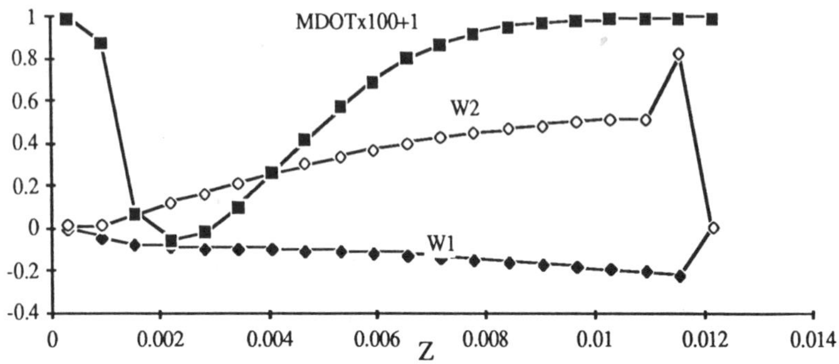

Fig. 16. Variation of combustion rate MDOT, and axial velocities of cold and hot gas, W1 and W2 respectively, versus axial distance z at $t = 12$.

Finally, in Fig. 16 are plotted the profiles of gas velocity and interfluid mass-transfer rate, valid for 180° crank angle. The figure reveals that the differences between the velocities are very large, with the hot gas moving relatively to the cold gas in the direction of the piston. The mass-transfer rate, to which the combustion rate is proportional, has its greatest (negative) value near the region in which (as Fig. 14 reveals) the product $\alpha(1-\alpha)$ is largest.

3.2.2.4. *Discussion.* The results are entirely in accordance with the qualitative expectations expressed in Section 3.2.2.1.

What is worth emphasizing is that no conventional turbulence model, such as the k–ε one, could have predicted the phenomenon at all; for they are all single-fluid models, lacking the possibility of simulating the 'sifting' of one fluid through another under the combined influences of density fluctuations and pressure gradients.

Quantitative realism is of course not yet to be sought from such calculations, because of the idealized character of the assumptions concerning thermodynamic properties, interfluid transport and dimensionality. Thus, no real reciprocating-engine process is one-dimensional; and even two-dimensional flows of this kind exist only in research laboratories possessing especially designed equipment. Nevertheless, it is useful to examine what new phenomena two-dimensionality entails; this is therefore the subject of the next section.

3.2.3. A Two-dimensional Representation of Turbulent Combustion in a Gasoline Engine

3.2.3.1. *The problem considered.* In the previous section, gas properties were allowed to vary with axial distance and time alone. Now, variations with radial distance will also be permitted. Such variations will be provoked by three different factors in the problem specification; namely, swirl, a radial variation in the initial distribution of hot-gas concentration, and a bowl in the piston.

What can be expected is that, by reason of the so-called Rayleigh–Taylor instability, the pressure gradient prevailing between 90° and 270° crank angle is such as to cause any slight departures from one-dimensionality to be greatly enlarged. Consequently, large-scale two-dimensional motions will be superimposed on the intermingling process of the hot and cold fluids: as a consequence, the overall rate of combustion will be affected.

Such large effects of small initial inhomogeneities have been demonstrated both experimentally and computationally by Andrews,[51] and no doubt by others, in experiments which, though they were conducted with fluids of small compressibility, reveal very well the Rayleigh–Taylor instability phenomenon.

3.2.3.2. *Method of solution.* The PHOENICS computer program which produced the results shown above can also perform two-dimensional flow simulations (and 3-D ones as well, although this capability will not be illustrated here); moreover, axial symmetry is one of its available geometry options; and the equation systems for the radial and swirl velocity components can be easily switched on. This program was therefore used for the calculations which are now to be presented.

3.2.3.3. *Typical results*

The effect of swirl. In Fig. 17 is shown, by way of the curve marked A, the variation of cylinder pressure versus crank angle for precisely the same conditions as those pertaining to Fig. 18, except for the fact that the gases present in the cylinder initially are swirling in solid-body rotation about the cylinder axis.

Comparison of the two diagrams shows that the swirling gases exhibit increased pressures, from shortly before 180° crank angle to

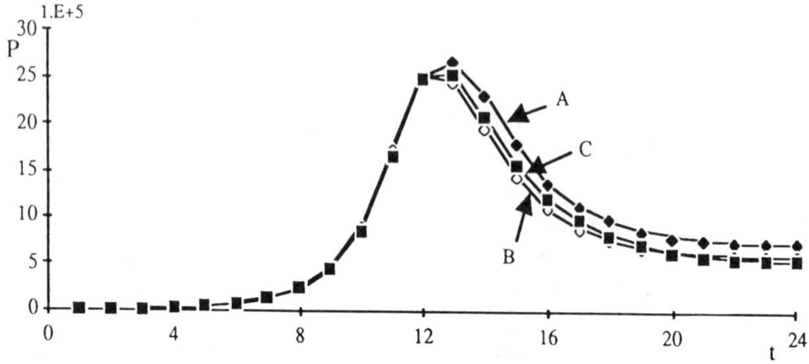

Fig. 17. Variation of cylinder pressure with time for: A, two-dimensionality caused by an imposed initial swirl about the axis; B, two-dimensionality caused by a reduced size of initial hot-gas region; C, two-dimensionality caused by the presence of a bowl in the piston.

the end of the stroke. This can be interpreted as resulting from a more rapid and complete combustion process.

The PHOENICS computation yields detailed information about the development of the process; and this information can be displayed graphically in various ways. Because of limitations on space, a single diagram must suffice here, namely Fig. 18. This shows a contour diagram, plotted by way of a line printer, of the volume fraction of hot gas at top dead centre.

The diagram is not to scale; and it is drawn with the piston surface at the top and the cylinder head at the bottom. The left-hand boundary corresponds to the axis of symmetry and the right-hand boundary to the cylinder wall. The printed numbers in the body of the diagram indicate bands of volume-fraction value, scaled between the maximum and the minimum values in the field, which are here 0·999E-4 and 0·812E-4.

Inspection of the diagram reveals a row of 8's close to the piston face, which shows that hot-gas fragments have reached there; and the 2's and 6's on the left of the diagram, contrasted with the 0's on the right, are an obvious indication that hot gas has migrated to the inner region of the cylinder, as a consequence of the radial pressure gradient induced by the swirl.

At the instant in question, PHOENICS reports that the average mass fraction of hot gas is 0·2767, whereas it was only 0·0791 at 90°

2-fluid TMs	27	Computed results.
& recip. engs.	----	Distributions of hot-gas volume fraction
1987	43	at 180 degrees crank angle.

```
 PATCH(MAP        , CONTUR,  1,    1,   1,    10,   1,   20,   1,   24)
PLOT(MAP        , R2 ,   0.000E  + 00,  1.000  E + 01  )
CONTOUR PLOT OF R2
     X-PLANE; Y HORIZONTAL, Z VERTICAL
 CELL AVE = 2.882E-01 MASS AVE = 2.767E-01
 MIN VALU = 8.121E-04 MAX VALU = 9.990E-01
 PHOENICS RECTANGULAR CONTOUR PLOT
 N, S, E, W . . . DENOTE: NORTH, SOUTH, EAST . .
Axis                      Piston surface                          Wall
 0 HHHHHHHHHHHHHHHHHHHHHHHHHHHHHHHHHHHHHHHHHHHHHHH
 1 S8888              8888888888  88888 66 4 2  0N
 2 S     666          88 6 4 22   000000000000000N
 3 S66666   44     222  0000000000000000000000000N
 4 S                            00000000000000000N
 5 S                             0000000000000000N
 6 S                              000000000000000N
 7 S                               00000000000000N
 8 S2222222                         0000000000000N
 9 S22222222222                      000000000000N
10 S2222222222222                    000000000000N
11 S222222222222222                   00000000000N
12 S22222222222222222222              00000000000N
13 S22222222222222222222222222         0000000000N
14 S22222222222222222222222222222       000000000N
15 S2222222222222222222222222222222     000000000N
16 S              222222222222222222     00000000N
17 S                              222     0000000N
18 S6666666666666666          44444 22     000000N
19 S                               8888 66 44 2 0N
20 S                              0 00000000 8  6N
20  LLLLLLLLLLLLLLLLLLLLLLLLLLLLLLLLLLLLLLLLLLLLL
Axis                      Cylinder head                           Wall
```

By this time, convexity of the hot-cold interface caused by swirl has been augmented by the deceleration in the piston. A Rayleigh–Taylor instability is developing.

Fig. 18. Line-printer-plotted contour diagram showing values of hot-gas volume fraction at top dead centre.

crank angle, and rises to 0·9056 at 270° crank angle. Evidently, at top dead centre, the combustion process has just got into its stride.

The effect of an initial departure from one-dimensionality. A corresponding calculation has been performed in which the hot-gas 'slab' which exists near the cylinder head at the start of the stroke extends only halfway from the axis to the cylinder wall. The pressure versus crank angle variation is shown as curve B in Fig. 17.

Evidently the pressure rise is smaller than in the earlier case, even though, because there is a greater proportion of unburned gas initially, the potential for pressure rise is greater.

Inspection of the corresponding volume-fraction printouts, for which there is no space here, reveals the reason: the hot gas migrates towards the cylinder axis earlier in the cycle than before; and the centrifugal-force field tends to keep it there, reducing the relative motion between adjacent hot- and cold-gas fragments which (in the present model) is the necessary precursor to combustion.

The effect of the presence of a bowl in the piston. A further calculation has been performed in which the piston surface is not plane; instead a 'bowl' is present, centrally placed, with a radius equal to half that of the cylinder itself. Such features are more common in diesel than in gasoline engines; but it remains interesting to see how the resulting 'squish' action affects the combustion process.

All that there is space to discuss here is the curve marked C in Fig. 17, which shows the pressure versus crank angle variation. Evidently the presence of the bowl has had a deleterious effect on the combustion. Inspection of the contour diagram of hot-gas volume fraction reveals that this is again the consequence of the formation of a hot-gas core which does not penetrate to the surrounding cold-gas envelope.

3.2.3.4. *Discussion.* The two-fluid model has been shown to be capable of responding to influences of swirl, of initial distributions of gas condition, and of combustion-space geometry. The predictions to which it has led appear to be qualitatively plausible.

Many more computations have been performed than it has been possible to present here; and they have been subjected to grid-refinement and time-step tests so as to ensure that the conclusions drawn are not invalidated by the presence of excessive numerical error. The main conclusion remains the same: even the simple model described above shows some encouraging signs of simulating turbulent combustion with qualitative realism.

The question now is: What must be done to approach *quantitative* realism? This will now be discussed.

3.3. Further Developments

3.3.1. Developments of the Two-fluid Model

The mass-transfer option used in the calculations described was the simplest available in PHOENICS, but there are others; and one making the mass-transfer rate proportional to $(1-\alpha)\alpha(0{\cdot}5-\alpha)$ is to be preferred. The reason is that this formulation recognizes that entrainment can proceed in either direction, and that the more sparse fluid is likely to entrain the less sparse. It has been used by Wu,[52] who has applied the two-fluid model of turbulent combustion to flame spread in ducts.

Introduction of the above option is easy, but it is best coupled with at least two other refinements. The first is the switching on of a finite-reaction-rate feature, which takes account of the facts that cold gas is *not* instantly burned as soon as it is entrained into a hot-gas fragment, and that hot gas entrained by cold-gas fragments is not quenched instantly, or even at all.

The second refinement would be a representation of the gas-density calculation which did more justice to the laws of thermodynamics than do the assumptions described in Section 3.2.2 above. There is no difficulty about this; indeed all the refinements in question here would have been switched on for the study reported here, had it not been desired to focus full attention on the influence of density-difference alone, without diversion of attention to how the density difference varied with other factors.

Finally, there exist other factors which influence the relative motion of interspersed fragments besides pressure gradient. The latter was brought into prominence above because it is commonly neglected; but a complete two-fluid model must allow for the role of mean-flow shear on the relative motion. The aforementioned paper by Wu shows how this can be done.

3.3.2. Improved Representation of the Engine and the Phenomena Occurring Within the Cylinder

Many phenomena were left deliberately out of account in the study described above; but, when greater realism is required, it will be appropriate to introduce them. They include:

- heat transfer to the walls of the combustion space;
- the complex initial air motion which results from inflow through valves;
- the spark-ignition process, represented as a localized short-duration source of heat;
- the non-uniformities of initial-gas composition, resulting from incompleteness of scavenging, or from direct fuel injection; and
- the process of vaporization of injected liquid fuels.

Furthermore, the true three-dimensional shapes of real gasoline- and diesel-engine combustors must be introduced before the two-fluid turbulence model can be used by engine developers. Once again, there is no essential difficulty; but the computational expense is certain to increase greatly, unless radical improvements in hardware and/or software are made.

4. CONCLUDING REMARKS

4.1. The Current Status of Numerical Simulation of Flow and Combustion in Reciprocating Engines

At the time of writing (September 1988), computer simulation of fluid flow in reciprocating engines has certainly proved to be of value to engine designers. For example, Blair[53] has demonstrated how PHOENICS can be used for distinguishing better from worse scavenging arrangements in two-stroke engines, and a video film produced by the Toyota company[54] shows how the same program is assisting designers to understand and favourably influence flow patterns in four-stroke engines.

No doubt similar practical advantages have attended the employment of other simulation programs for reciprocating engines, such as KIVA.[8]

It is, however, not yet possible to make the same claim with regard to computer simulation of the combustion and two-phase phenomena which take place in the engine cylinder. The reasons are those which have been discussed above; they are partly numerical, resulting from the restricted grid size and computing speed attainable by currently available computers; and they are partly physical, associated with the incompleteness of knowledge regarding turbulence and combustion.

This situation is not likely to change rapidly. Even the largest supercomputers which exist or are being planned will be unable to provide the fineness of grid which will render unnecessary the use of such idealizations as grossly simplify chemical kinetics, restrict information about turbulence to its statistical properties, and represent multi-phase phenomena by way of the interpenetrating-continua concept.

This by no means implies, however, that computer simulation of combustion in gasoline and diesel engines is without practical utility even today. If his expectations are suitably adjusted to what is attainable, the engine designer can already learn from computer models how to distinguish better concepts from worse ones; and he can achieve an insight into the processes occurring in his engines which will stimulate his inventiveness, and guide him in the interpretation of his experimental observations.

In the following, and final, section, some suggestions will be given as to how designers can get best value from the available software tools.

4.2. How Computer Simulation can Assist Engine Designers Today

Engine designers do not emerge from universities fully equipped with the knowledge which they require. Instead they must spend many further years in acquiring knowledge of how engines have been designed in the past, how their designs have influenced performance, what practical constraints of money, materials and manufacturing have limited the designs which could be experimented with, and what are the limitations set by legislation and user acceptance.

Since they also have to become proficient in testing, in general design techniques, and in the skills of operation within a team of variously endowed specialists, it is not to be expected that they should have found time to study computational fluid dynamics or combustion theory in great depth. It is therefore small wonder that computer simulation of engine-combustion phenomena is not easily embarked upon by the designer, even when he is persuaded of its potential value to him.

The following discussion has therefore been written so as to provide the sufficiently enthusiastic designer with advice on how to get started. It has been structured in a question-and-answer manner.

Question 1: Where to Acquire the Computer Program?

It is hardly necessary to explain why the designer must acquire a computer program created by others: the creation of computer programs is specialist work, comparable in difficulty to the designing of engines; only specialists can do it.

A large automobile company might have sufficient resources to enable it to hire a team of computer specialists who would create a special program for its own exclusive use. However, the specialists are in short supply; team-creation is a lengthy business; and the outcome of the endeavour is far from certain. In practice, therefore, this route is rarely taken.

Nevertheless, since computer programs differ so greatly in their character and capabilities, it is important for engine specialists desirous of acquiring one to have sufficient knowledge to become an 'enlightened shopper'. At least he should consider the following questions.

Question 2: Multiple-use or Special-use Programs?

There exist, as has already been stated, special-purpose computer simulators for reciprocating engines, such as KIVA, and general-purpose programs, such as PHOENICS, which can simulate reciprocating-engine phenomena and many others also. Which kind of computer code is to be preferred?

In favour of the special-purpose code will be its directness of approach, and its presentation in a language and context that the engine designer will find congenial to him. It may be smaller, and more economical to run, than the general-purpose program, because it is not burdened with features that are not needed for engine simulations. Moreover, because the same program may be in the hands of several collaborating design and research groups, it may furnish for its users a convenient means of communicating problems, solutions and interesting findings.

For the general-purpose program it can be said that, *its* input language being the same for *all* fluid-flow simulations, its use facilitates communication between practitioners of rather different specialities, for example: engine-cylinder designers; researchers concerned with inlet manifolds and exhaust pipes; lubrication modellers; vehicle aerodynamicists; and others having nothing to do with power production or automotive engineering. Moreover, although the program will inevitably contain much that is not needed for engine simulation, there are countervailing advantages. Because many users will have tested it in many ways, it is likely to be relatively 'bug-free'; and its efficiency as a 'number-cruncher' may well be of the highest, because resources will have been available to fund the necessary development effort.

Best for the engine designer would be a combination: he should acquire a program which will *look like* a special-purpose one, and will be widely used by other internal combustion engine specialists; but the program should in essence be a general-purpose one, so that the abovementioned advantages can be gained in addition.

Question 3: What Kind of Input Procedures?

Computer programs written by experts *for* experts often allow their users to insert the problem-defining data in speedy and compact ways. Those who use the programs every day can quickly learn them; and frequency of use keeps the memory fresh.

The engine-designer, by contrast, is likely to intersperse short periods of program use with longer ones of physical testing, discussions with management, interactions with the production department, and so on. When he returns to his computer modelling, therefore, his memory of how to proceed may have been wiped almost clean.

Fortunately, stimulated by the development of 'user-friendly' programs for the personal computers which are used in offices and homes, the creators of computer programs for engineering use have learned the necessity of making their programs easy to use. Either they provide 'menu-type' data-input 'front ends' to their flow-simulation programs; or they devise 'icon-and-mouse' procedures which enable calculations to be set up by swift hand and finger motions. Whichever technique is used (and they are often combined), they both express the recognition that burdens on the program-user's memory are to be of only the slightest; and, should he make any error (as he certainly will), the data-input program should immediately bring this to his attention, and provide the necessary advice.

The informed seeker after computer programs for simulating engine flow and combustion will therefore ask the would-be vendor searching questions about what user-assisting means are provided for setting up the problem.

Question 4: What Means are Required for the Display and Interpretation of Output?

Introducing the problem-specifying data to the program is only the first of the difficulties which the program-creator must help the non-specialist user to overcome. An even greater one is how to interpret and profit from the results of the calculations which the program thereafter produces.

These results can be expressed in various ways, of which the easiest to provide, namely tables of numbers, is the least easy to use. What the engine designer needs is, firstly, an intelligently composed *visual display,* and secondly a *condensation* of numerical output into a few key quantities.

The visual displays are of two kinds: the computer-generated equivalents of what the test-engineer is accustomed to looking at, such as 'indicator diagrams' (i.e. plots of cylinder pressure versus crank angle), and temperature–time traces from thermocouples in walls and ducts; and three-dimensional flow visualizations, enabling the user to 'see', and comprehend, what is (according to the computer program) happening within the chamber.

The condensations comprise such quantities as brake-mean-effective pressure, specific fuel consumption and heat rejection rates to the cooling water, to the piston and to the lubricating oil. These quantities, which are what the traditional engine designer most wants to know, are in truth rather easy to deduce from the numerical results which the computer program generates; but it is not for the engine designer to concern himself with their deduction. The program should do the job for him.

Once again, the discerning seeker after flow-simulation software will inquire closely into what will be provided in respect of data-digestion techniques; for the computer program does nothing for him at all if he cannot understand its output, in the time that he can spare for its study.

4.3. The Future

Twenty years ago, computer simulation of flow and combustion processes in engines was dreamed of, albeit naïvely; but belief in its attainability was confined to a handful of individuals. Now it is an actuality; and opinions about its potential benefit are beginning to be founded upon firm experience rather than on speculation.

The practical benefits in terms of quantifiable improvements in the performance of engines are still hard to document; but the speed of advance, both in simulation techniques and in the associated hardware for computation and display, is very great.

Twenty years from now, this author firmly believes, computer simulation of his engines will be something that no top-class designer will dream of doing without.

REFERENCES

1. Pun, W. M. & Spalding, D. B., A procedure for predicting the velocity and temperature distributions in a confined, steady, turbulent, gaseous, diffusion flame. In *XVIIU International Astronautical Congress*. Pergamon Press/PWN Polish Scientific Publishers, 1968, pp. 3–21.
2. Spalding, D. B., Predicting the performance of diesel engine combustion chambers. Closing lecture at IMechE Symposium on Diesel-Engine Combustion, London. *Proc. Instn Mech. Engrs* **184** (Part 3J) (1969/70).
3. Patankar, S. V. & Spalding, D. B., A calculation procedure for heat, mass and momentum transfer in parabolic flows. *Int. J. Heat Mass Transfer,* **15** (1972) 1787–806.
4. Patankar, S. V. & Spalding, D. B., The EASI program. Report, CHAM Ltd, 1974.
5. Watkins, A. P., Calculation of flow and heat transfer in the combustion chamber of a reciprocating engine. MSc thesis, Dept of Mech. Engng, Imperial College, University of London.
6. Zuber, I., *Ein mathematisches Modell des Brennraums*. Monographs and Memoranda no. 12. Staatliche Forschungs Institut fuer Maschinenbau, Bechovice, Czechoslovakia, 1972.
7. Markatos, N. C. & Mukerjee, T., 3-Dimensional computer analysis of flow and combustion in automotive internal combustion engines. In *Mathematics and Computers in Simulation*; *Trans. IMACS,* **23**(4) (1981).
8. Amsden, A. A., Ramshaw, J. D., O'Rourke, P. J. & Dubowicz, J. K., KIVA: A computer program for two- and three-dimensional fluid flows with chemical reactions and fuel sprays. Report LA-10245-MS, Los Alamos National Laboratory, 1985.
9. Spalding, D. B., A general-purpose computer program for multi-dimensional one- and two-phase flow. Lehigh IMACS Conference, July 1981; *Mathematics and Computers in Simulation,* **XIII** (1981) 267–76.
10. Gosman, A. D., Tsui, Y. Y. & Watkins, A. P., Calculation of unsteady three-dimensional flow in a model motored reciprocating engine and comparison with experiment. In *Proceedings of 5th International Symposium on Turbulent Shear Flows,* Cornell University, New York, 1985.
11. Ramos, J. I., Lectures on mathematical models of diesel engines. In *Advanced Course on Computer Simulation of Fluid Flow, Heat and Mass*

Transfer and Combustion in Reciprocating Engines, Dubrovnik, Yugoslavia, Hemisphere Publishing, New York, 1987.

12. Bracco, F. V., Modelling of two-phase, two-dimensional unsteady combustion for internal combustion engines. In *Stratified-charge Engines.* Instn Mech. Engrs Conference publications 1976–11, pp. 167–187.
13. Harlow, F. H. & Amsden, A. A., Numerical calculation of almost incompressible flow. *J. Comput. Phys.,* **3** (1968) 1.
14. Chorin, A. J., Numerical solution of the Navier–Stokes equations. *Maths Comput.,* **22** (No. 104) (1968) 745–62.
15. Caretto, L. S., Gosman, A. D., Patankar, S. V. & Spalding, D. B., Two calculation procedures for steady, three-dimensional flows with recirculation. In *Proceedings of 3rd International Conference on Numerical Methods in Fluid Mechanics,* Vol. II, 1973, 60–68.
16. Spalding, D. B., Mathematical modelling of fluid mechanics, heat transfer and mass transfer processes; a lecture course. Report HTS/80/1, Dept of Mech. Engng, Imperial College, 1980.
17. Rhie, C. M. & Chow, W. L., Numerical study of the turbulent flow past an airfoil with trailing edge separation. *AIAA J.,* **21**(11) (1983) 1525–32.
18. Harlow, F. H. & Amsden, A. A., Numerical calculation of multi-phase fluid flow. *J. Comput. Phys.,* **17**(1) (1975) 19–52.
19. Spalding, D. B., The calculation of free-convection phenomena in gas–liquid mixtures. ICHMT Seminar, Dubrovnik. Also in *Turbulent Buoyant Convection,* ed. N. Afgan & D. B. Spalding, Hemisphere, Washington, DC, 1977, pp. 569–86.
20. Patankar, S. V., *Numerical Heat Transfer and Fluid Flow.* Hemisphere, Washington, DC, 1980.
21. Minkowycz, W. J., Sparrow, E. M., Schneider, G. E. & Pletcher, R. H., *Handbook of Numerical Heat Transfer.* John Wiley, New York, 1988.
22. Spalding, D. B., *Combustion and Mass Transfer.* Pergamon Press, Oxford, 1979.
23. Spalding, D. B., Stephenson, P. L. & Taylor, R. G., A calculation procedure for the prediction of laminar flame speeds. *Combustion & Flame,* **17** (1971) 55–64.
24. Bracco, E. V. & O'Rourke, P. J., *Prog. Energy Combust. Sci.* **7** (1981) 103.
25. Launder, B. E. & Spalding, D. B., *Mathematical Models of Turbulence,* Academic Press, London and New York, 1972.
26. Harlow, F. H. & Nakayama, P. I., Transport of turbulence energy decay rate, LA-3854, Los Alamos Sci. Lab., University of California, 1968.
27. Launder, B. E. & Spalding, D. B., The numerical computation of turbulent flows. *Computer Methods in Applied Mechanics and Engineering,* **3** (1974) 269–89.
28. Spalding, D. B., Mixing and chemical reaction in steady confined turbulent flames. *13th Symposium (International) on Combustion,* The Combustion Institute, Pittsburgh, PA, 1971, p. 649.
29. Prandtl, L., Bericht uber Untersuchungen zur ausgebildeten Turbulenz. *Z. angew Math. Mech.* (*ZAMM*), **5**(2) (1925) 136–9.

30. Spalding, D. B., Theories of turbulent reacting flows. *AIAA,* Paper 79-0213 (1979).
31. Spalding, D. B., Chemical reactions in turbulent fluids. *Levich 60th Birthday Conference on PhysicoChemical Hydrodynamics,* Advance Publications, Vol 1, 1978, p. 521.
32. Ma, A. S. C., Sun, R. L. T. & Spalding, D. B., Application of ESCIMO to the turbulent hydrogen–air diffusion flame. *13th Symposium (International) on Combustion,* The Combustion Institute, Pittsburgh, PA, 1982, pp. 393–402.
33. Borghi, R. & Moreau, P., Turbulent combustion in a pre-mixed flow. *Acta Astronautica,* **4** (1977) 321–41.
34. Dopazo, C. & O'Brien, E. E., Statistical treatment of non-isothermal chemical reactions in turbulence. *Combustion Science and Technology,* **13** (1976) 99–122.
35. Shchelkin, K., *Soviet Physics—Technical Physics.* Vol. 13, p. 520.
36. Wohlenberg, W. J., Minimum depth of flame front for stable combustion . . . in a gaseous system at constant pressure. *4th Symposium (International) on Combustion,* Williams and Wilkins, Baltimore, 1953, p. 796.
37. Howe, N. M. & Shipman, C. W., A tentative model for rates of combustion in confined turbulent flames. *10th Symposium (International) on Combustion,* The Combustion Institute, Pittsburgh, PA, 1965, p. 1139.
38. Bray, K. N. C. & Libby, P. A., Countergradient diffusion in premixed turbulent flames. *AIAA J.,* **19** (1981) 205.
39. Kuznetsov, V. R., *Fluid Dynamics* (USSR), **14** (1979) 328.
40. Moss, J. B., Simultaneous measurements of concentration and velocity in an open pre-mixed flame. *Combust. Sci. Technol.,* **22** (1980) 115–29.
41. Shepherd, I. G. & Moss, J. B., Measurements of conditioned velocities in a turbulent pre-mixed flame. AIAA paper 81-0181, 1981.
42. Phillips, H., Towards a two-fluid model for flame acceleration in explosions. H & SE Explosion and Flame Laboratory, Buxton, England, 1983. Submitted to the 9th International Colloquium on Dynamics of Explosions and Reactive Systems.
43. Spalding, D. B., Chemical reaction in turbulent fluids. *J. Phys. Chem. Hydrodyn.,* **4**(4) (1983) 323–36.
44. Spalding, D. B., Towards a two-fluid model of turbulent combustion in gases with special reference to the spark-ignition engine. In *Conference on Combustion in Engineering,* Vol. 1, Instn Mech. Engrs, London, 1983, pp. 135–42.
45. Spalding, D. B. & Wu, J. Z., Numerical studies of propagating flames exhibiting the Landau and Rayleigh–Taylor instabilities. *J. Phys. Chem. Hydrodyn.,* **7**(5/6) (1986) 353–84.
46. Ilegbusi, J. O. & Spalding, D. B., Application of a two-fluid model of turbulence to turbulent flows in conduits and free shear-layers. *J. Phys. Chem. Hydrodyn.* **9**(1/2) (1987) 161–81.
47. Malin, M. R. & Spalding, D. B., Flow and heat transfer in two-dimensional turbulent wall jets and plumes. *J. Phys. Chem. Hydrodyn.,* **9**(1/2) (1987) 237–74.

48. Malin, M. R. & Spalding, D. B., A two-fluid model of turbulence and its application to heated plane jets and wakes. *PCH J.*, **5** (1984) 339–61.
49. Spalding, D. B., Computer simulation of turbulent combustion in reciprocating engines. In *Proceedings of 2nd International PHOENICS User Conference,* CHAM Ltd, London, 1989.
50. Spalding, D. B., The PHOENICS equations, CHAM/TR/99 Doc Rev 02, CHAM Ltd, 1989.
51. Andrews, M. J., Turbulent mixing by Rayleigh–Taylor instability. PhD thesis, Imperial College, University of London, CFDU Report CFD/86/10, 1986.
52. Wu, J. Z., The application of the two-fluid model of turbulence to ducted flames. In *Proceedings of PHOENICS User Conference,* 1987.
53. Sweeney, M. E. G., Swann, G. B. G., Kenny, R. G. & Blair, G. P., Computational fluid dynamics applied to two-stroke-engine scavenging. SAE paper 851519, 1985.
54. Yamada, T., Inoue, T., Yoshimatsu, A., Hiramatsu, T. & Konishi, M., In-cylinder gas motion of multivalve engine—three-dimensional numerical simulation. SAE paper 860465, 1986.

12

The Theory of Wave Action Approaches Applied to Reciprocating Engines

D. E. Winterbone
UMIST, Manchester, UK

1. Introduction
2. Equations of flow in one-dimensional form
   2.1. Linear waves
   2.2. Homentropic finite waves
   2.3. Non-homentropic finite waves
3. Methods of calculating compressible flow
   3.1. Linear waves
   3.2. The method of characteristics
   3.3. Finite difference techniques
   3.4. Conclusions
4. Pipe boundary conditions
   4.1. Open end boundary
   4.2. Closed end boundary
   4.3. Partially open end (or nozzle) boundary
   4.4. Change of cross-section
   4.5. Entropy discontinuity in pipe
   4.6. Valve and port boundaries
   4.7. Multi-pipe junctions
   4.8. Carburettor and fuel injection equipment
   4.9. Compressor
   4.10. Turbine

1. INTRODUCTION

It has long been realized that the design of the inlet and exhaust manifolds has a large effect on the performance of reciprocating engines. The unsteady nature of the induction and exhaust processes means that the effect of the manifold on charging and discharging is extremely dependent upon the engine speed. This is because the *impedance* (or *admittance*) of the manifold is a function of the frequency of the pulses entering it. The outcome of this is that it is possible to *tune* the manifolds to give a particular engine power output characteristic as a function of speed. In the case of a racing engine the manifolds will be designed to produce high power outputs at high speeds; this will produce the maximum specific power (power/weight) from the engine, but probably at the expense of flexibility. This is not an insurmountable problem in a racing car where the driver is skilled at obtaining the maximum performance from his vehicle by the use of the tachometer and the gearbox. However, the average driver does not want such a temperamental machine and road-going engines are often tuned to give a much more forgiving engine characteristic. Most modern car engines are designed to give a high torque at low engine speed; this means that as the engine slows down the torque rises, obviating the need to change gear. A concomitant advantage of tuning the engine to give this power characteristic is the ability to use a higher overall gear ratio, which results in a lower engine speed for the same vehicle speed; this improves the fuel economy of the vehicle. Probably more improvement in vehicle fuel economy has been achieved from this approach than any other engine design change. (The lean-burn technology now being introduced will allow improvements in combustion to make their contribution to engine economy.)

The design of manifolds can be done in various ways, although for naturally aspirated (or unsupercharged) engines the only suitable method is one which considers the passage of waves in the pipes. The reason for this is the very large effect the unsteady flow processes have on the volumetric efficiency of the engine. Such methods are referred to as *wave action* calculations. It is possible to use simpler techniques with turbocharged engines because the components of the turbocharger (the compressor and turbine) act as filters to the waves, and 'force' the gas into the cylinders; such engines are less susceptible to wave effects. These techniques are usually given the generic title of the *filling and emptying method*. This is because they consider the manifolds and the cylinders to be a series of interconnected volumes which are filled or emptied by the passage of gas through their boundaries; viz. the valves or the turbomachine. A detailed description of the techniques used for analysing the processes occurring in internal combustion engines is given in Horlock & Winterbone[1] and an excellent qualitative description of factors affecting air flow in engines is given by Annand & Roe.[2]

A complete simulation of an inlet or exhaust manifold would require a three-dimensional model of the equations of fluid mechanics, including some model of the turbulence level in the fluid. It is currently beyond the power of present-day computers to cope with such a *multi-dimensional* problem. Many of the basic building blocks exist but the cost of putting together the data for the analysis of a manifold with many pipes coming together at a large number of junctions is beyond the budgets of most research groups. It is possible that, as computer power rises and running costs decrease, this type of problem will become soluble. Interesting papers have already appeared in which parts of manifolds have been treated either as two-dimensional or three-dimensional pipes. Dimitriades, Leschziner, Winterbone, Alexander & Sierens[3] report an analysis of a complex junction using a three-dimensional code. A similar study of the flow through the intake valve has been undertaken by Seppen[4] using a program entitled SIFLEX, based on the generalized three-dimensional turbulence program (PHOENICS) written by Spalding (see Rosten & Spalding[5]). To be able to tackle the problem Seppen has had to assume steady, incompressible flow; the latter assumption would seem to be unrealistic for flow through a valve, but assumptions of constant density are often required in such calculations. Gosman & Ahmed[6] are also examining the detailed flow through inlet valves to gain a better understanding of the actual flow pattern entering the engine cylinder. However, although these studies show that various components of the internal combustion engine manifold can be treated in such detail, the analysis is limited to quasi-steady flow. Chapman[7] published a paper in which he considered the unsteady flow of gas in a two-dimensional manifold, and Tosa, Shimoda & Oikawa[8] have reported a similar calculation in which a two-dimensional analysis was applied to the pipework junction regions, while one-dimensional analysis was used elsewhere. Hence, there is progress towards multi-dimensional calculations of flow in engine manifolds. However, the methods being used in industry to design intake and exhaust manifolds are, on the whole, *one-dimensional formulations* of the fluid dynamic equations of compressible flow. Such a simplification leads to substantial reductions in the complexity of the computer programs, with a concomitant reduction in running cost, but at the expense of some empiricism and loss of detailed definition of the flow field. This chapter will concentrate on these simplified equations because they can be used to provide extensive and useful results today.

Waves occur in the manifolds of engines because of the periodic nature of the induction and exhaust processes. These processes act upon the compressible fluid in the manifolds, viz the air or exhaust gas, and pressure waves are transmitted through it. The pressure waves travel at the local velocity of sound in all directions *relative* to the fluid. If the solution is limited to one space dimension then this means that the waves travel in two possible directions; notionally they can travel to the right or the left. There is a large body of literature on waves in fluids and this is well summarized by Lighthill.[9] This reference, which approaches the problem from a mathematical viewpoint, shows that small-amplitude waves behave in a mathematically linear manner whereas large-amplitude waves are non-linear. Linear behaviour means that the effect of two waves interacting is simply the sum of the effects when each acts individually. This is true of sound waves and might be applied to the wave action in the intake manifold but is not really applicable to the pressure waves that occur in the exhaust manifold of internal combustion engines. The waves in engines are generally *non-linear* and require more complex methods of calculation. This point will be returned to later.

The basic methods for evaluating compressible flow in inlet and exhaust systems of internal combustion engines will be developed in this chapter. The boundary conditions necessary to simulate the whole range of engines, from naturally aspirated to turbocharged, will also be derived here. The application of wave action techniques to engines will be left until Chapter 13, where naturally aspirated and turbocharged engines will be discussed. When considering naturally aspirated engines particular attention will be given to tuned intake systems and methods for evaluating the tuned frequencies. In addition, the use of wave action techniques in exhaust systems will be introduced, and an approach for evaluating the efficiency of manifolds will be described. The approach will be limited to the unsteady flow of air alone in the inlet manifold; this means that in the case of the petrol engine the fuel is neglected. With the current trend towards multipoint fuel injection in these engines, especially with tuned manifolds, this is not such a shortcoming but care has to be taken when designing manifolds for engines with carburettors or single-point (throttle body) fuel injection. Low, Baruah & Winterbone[10] did some idealized calculations on the passage of fuel droplets in pulsating flow. (This is discussed briefly in Chapter 6 [Nightingale]). While such a calculation provides some interesting data on droplet velocities and evaporation it disregards one of the major problems of the petrol engine, viz. that a large proportion of the fuel deposits itself on the manifold wall due to gravity or centrifugal forces. For this reason it is essential to concentrate simply on the flow of air or a homogeneous fuel–air mixture. (It should be noted that a major cause of charge maldistribution in spark-ignition engines is the maldistribution of the fuel.)

2. EQUATIONS OF FLOW IN ONE-DIMENSIONAL FORM

Four equations are necessary to define the flow of a single-phase compressible fluid. These equations, the continuity, momentum, energy and state equations, are shown in differential form as eqns (1)–(4). A derivation of these, from basic principles, is given in Benson[11]:

$$\frac{\partial \rho}{\partial t}+\rho\frac{\partial u}{\partial x}+u\frac{\partial \rho}{\partial x}+\frac{\rho u}{F}\frac{\mathrm{d}F}{\mathrm{d}x}=0 \tag{1}$$

$$\frac{\partial u}{\partial t}+u\frac{\partial u}{\partial x}+\frac{1}{\rho}\frac{\partial p}{\partial x}+G=0 \tag{2}$$

where

$$G=f\frac{u^2}{2}\frac{u}{|u|}\frac{4}{D}$$

$$\left(\frac{\partial p}{\partial t}+u\frac{\partial p}{\partial x}\right)-a^2\left(\frac{\partial \rho}{\partial t}+u\frac{\partial \rho}{\partial x}\right)-(\kappa-1)\rho(q+uG)=0 \tag{3}$$

$$p/\rho=RT \tag{4}$$

These equations constitute a set of quasi-linear hyperbolic partial differential equations, and a feature of their solution is that it contains propagation of information by waves. This means that if a disturbance is introduced into the fluid it does not simply diffuse through the fluid, in the manner of heat conduction in a metal bar, but it is carried through the fluid by a wave, like a ripple on a pond. In the flow of a compressible gas in the pipework of an internal combustion engine the disturbance takes the form of pressure.

2.1. Linear Waves

In some cases the magnitude of the pressure disturbance is small relative to the local pressure level and this enables the equations to be simplified; these simplified equations are referred to as the *linearized wave equations*. Typical examples of 'linear waves' are sound waves. A feature of these waves is that they obey the principle of *superposition*; viz. if two waves cross the combined amplitude is simply the sum of the two individual amplitudes. The equations of linear waves are evaluated by regarding as small all departures of the properties of the fluid from a state of rest and uniform density; the products of these quantities are neglected. In the case of the momentum equation (eqn (2)) this means that the convective term $u\, du/dx$ is neglected.

The linearized forms of the momentum and continuity equations can be combined, in the case where the entropy level of the fluid remains constant, to give:

$$\frac{\partial^2 p_e}{\partial t^2} = a^2 \frac{\partial^2 p_e}{\partial x^2} \tag{5}$$

where p_e is the excess pressure above the undisturbed pressure level.

Equation (5) is applicable to waves of *small* amplitude and has the great advantage of having an analytical solution. Unfortunately, it is not really suitable for fully evaluating the effects that occur in the inlet and exhaust pipes of IC engines where the waves are usually large in amplitude, but it can be a useful method for evaluating inlet manifold designs. Lighthill[9] describes many of the techniques which have been developed for treating these small waves, including simple propagation in a constant area pipe, the effect of branches, cavities, and open and closed ends. He also extends his analysis to frictional effects. An interesting product of linear wave theory is the combination of a pipe and a reservoir; this is normally known as the *Helmholtz resonator* and the theory associated with this has been applied to inlet manifold design of diesel engines by Cser.[12] Soedel and co-workers[13] describe the application of linear wave theory to the design of manifold systems for reciprocating compressors. It is possible that this technique could be extended to the design of tuned inlet manifold systems for IC engines. These ideas will be returned to later in the text.

If the linearized, isentropic equations are being used then the problem becomes directly analogous to the transmission of electricity in a power-line and there is a body of knowledge available for dealing with this. An interesting text on such methods is that by Metzger & Vabre.[14] It is possible that synthesis design techniques can be developed from this approach to give specific tuning characteristics to induction manifolds.

2.2. Homentropic Finite Waves

In the previous section waves of very small amplitude were discussed. In this section any restriction on the size of the wave will be removed, but it will still be assumed that the compression or expansion processes caused by the wave passage are isentropic. In fact, the larger assumption will be made that the whole field of solution is at the same entropy level; this is termed *homentropic flow*. Homentropic flow is a more general condition than simply assuming that all processes in the flow are isentropic because it means that *all points* in the flow régime are at the same entropy. Such an assumption introduces the useful simplification that the energy equation (eqn (3)) can be replaced by a simple isentropic relationship. A derivation of the equations of homentropic flow is given in Ref. 11. The final form of the homentropic equations, also neglecting changes of area of cross-section F, is given below:

$$\frac{1}{\rho}\frac{\partial \rho}{\partial t}+\frac{u}{\rho}\frac{\partial \rho}{\partial x}+\frac{\partial u}{\partial x}=0 \tag{6}$$

$$\frac{1}{\rho}\frac{\partial p}{\partial x}+\frac{\partial u}{\partial t}+u\frac{\partial u}{\partial x}=0 \tag{7}$$

$$a^2=(\partial p/\partial \rho)_s \tag{8}$$

$$p/p_{\text{ref}}=(a/a_A)^{2\kappa/(\kappa-1)} \tag{9}$$

Homentropic flow can only be achieved under ideal conditions of frictionless, adiabatic flow of a constant composition fluid. Both friction and heat transfer change the entropy levels of the fluid, and variable composition (and hence gas properties) also means that fluid at the same temperature and pressure will have different entropy levels.

It is possible to transform eqns (6)–(9) from partial differential equations into total differential equations. The transformation is due to Riemann,[15] and is called the *method of characteristics*. This method recognizes that information propagates through the fluid in a clearly defined manner; namely at the speed of sound relative to the local fluid velocity. The waves travel both in the direction of fluid flow and in the opposite direction, and these carry with them the information required to evaluate local conditions in the fluid at any point in space and time. The derivation of the equations in characteristics form is

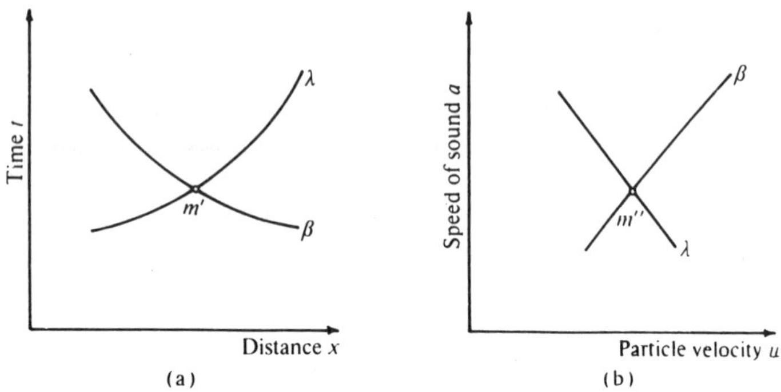

Fig. 1. Method of characteristics diagrams: (a) position diagram; (b) state diagram.

given in detail by Benson[11] and will not be repeated here; the resulting equations are simply quoted here as:

$$\mathrm{d}x/\mathrm{d}t = u \pm a, \qquad \mathrm{d}a/\mathrm{d}u = \mp(\kappa - 1)/2 \tag{10}$$

The significance of these equations can be depicted on two companion diagrams; the *position* and *state* diagrams, which are shown in Fig. 1. The *right-moving characteristic* is usually called the λ-characteristic and the *left-moving* one the β-characteristic. These are given by:

$$\lambda = A + \frac{\kappa - 1}{2} U \tag{11}$$

and

$$\beta = A - \frac{\kappa - 1}{2} U \tag{12}$$

where $A = a/a_{\mathrm{ref}}$ and $U = u/a_{\mathrm{ref}}$. Hence

and

$$\left.\begin{aligned} A &= (p/p_{\mathrm{ref}})^{(\kappa-1)/2\kappa} = (\lambda + \beta)/2 \\ U &= u/a_{\mathrm{ref}} = (\lambda - \beta)/(\kappa - 1) \end{aligned}\right\} \tag{13}$$

A and U are non-dimensionalized velocities and a_{ref} is the speed of sound at a convenient point in the system. In homentropic flow the choice of a_{ref} is important because it defines the entropy level for the whole flow calculation. The value of a_{ref} is not so important in non-homentropic flow.

At this point it is worthwhile considering what the difference is between a homentropic, finite wave and a linear wave. Basically, in a pattern of linear waves the effect of interactions is simply the sum of the effects of the individual waves acting separately; with finite amplitude waves this is not the case because in these the exponentations and products that occur in the equations cannot be neglected. What this means in physical terms is that the amplitude of the pressure has an effect on the way the wave propagates. A simple example of this is the propagation of a pressure wave down a pipe. If the wave has a finite amplitude then the higher-pressure region will travel at a higher velocity than the low-pressure part. This means that the high-pressure parts of the wave tend to overtake the low-pressure ones and the wave steepens; the limiting case is for the wave, which was initially rounded, to become a steep-fronted shock wave; an infinitesimal, linear wave will maintain its initial shape. Hence, although homentropic flow has been assumed, the non-linear analysis is capable of demonstrating the behaviour measured with real waves.

Initially the *characteristic equations* were solved by graphical methods; these solutions were first applied to engines by Jenny[16] and de Haller.[17] An example of such a graphical solution for a single-cylinder engine with one exhaust pipe is shown in Fig. 2 (from Benson[11]). It can be seen that even for this case, in which quiescent flow was assumed in the exhaust pipe prior to the exhaust valve opening, the solution is quite complex; for a multi-cylinder engine with many pipes and junctions the graphical solution becomes almost unmanageable! Benson[11] does give some examples of such calculations.

The advent of the digital computer solved the problem of tedious drawing techniques but it required a reformulation of the method used to solve the equations. A discussion of how compressible unsteady flow can be calculated on a computer is given in Section 3.

2.3. Non-homentropic Finite Waves

It is also possible to solve the equations of unsteady, compressible flow without restricting the conditions to those of constant entropy; such a solution will be termed *non-homentropic*. It is useful to show the *method of characteristics* equations for such flow because they highlight the important features. The *characteristics* equations are derived in detail in Benson[11] and will simply be presented here. The substantive difference between the solutions for homentropic and non-homentropic flow is that the latter régime requires an extra equation

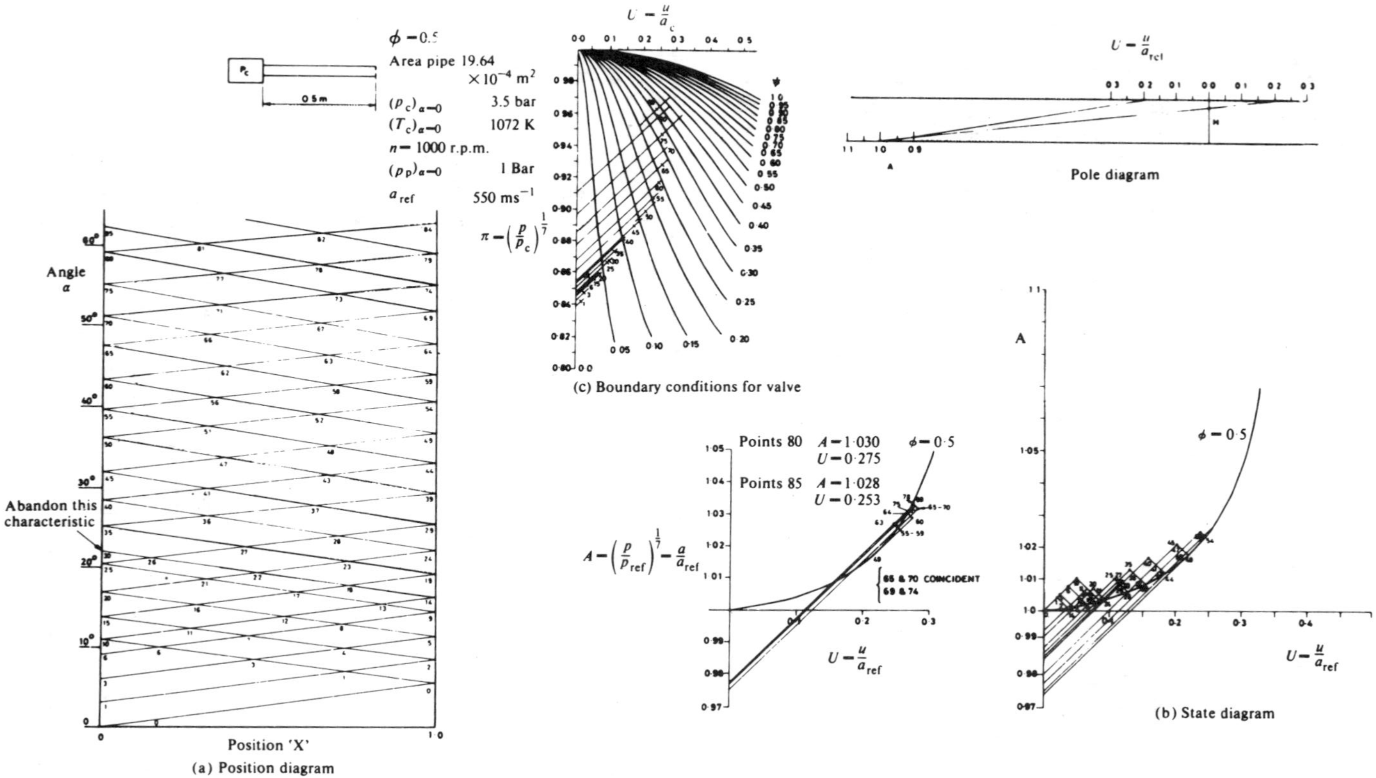

Fig. 2. Method of characteristics applied to single-cylinder engine with single exhaust pipe: (a) position diagram; (b) state diagram; (c) boundary conditions for valve; (d) boundary conditions at nozzle.

to depict the passage of 'entropy' in the fluid. This results in the introduction of an additional characteristic, called a *path line,* which travels at the local velocity of the fluid. The significance of this set of equations is that pressure travels along *waves,* which propagate in both directions at the local speed of sound *relative to the fluid,* whereas temperature, composition, 'entropy' and other *particle-related properties,* travel at the *local fluid velocity*. Obviously, in non-homentropic flow, it is possible to account for friction, heat transfer and other factors.

The equations of non-homentropic flow are given below.

2.3.1. Wave Equations (for a λ Characteristic)

Direction condition:

$$dX/dZ = U + A \tag{14}$$

Compatibility condition:

$$\lambda = A + \frac{\kappa - 1}{2} U \tag{15}$$

The change in Riemann variable, dλ, is

$$d\lambda = dA + \frac{\kappa - 1}{2} dU$$

$$= -\frac{(\kappa - 1)}{2} \frac{AU}{F} \frac{dF}{dX} dZ + \frac{A}{A_A} dA_A \tag{16}$$

$$-\frac{(\kappa - 1)}{2} \frac{2fx_{\mathrm{ref}}}{D} U^2 \frac{U}{|U|} \left\{1 - (\kappa - 1)\frac{U}{A}\right\} dZ$$

$$+\frac{(\kappa - 1)^2}{2} \frac{qx_{\mathrm{ref}}}{a_{\mathrm{ref}}^3} \frac{1}{A} dZ \tag{17}$$

In these equations the non-dimensionalized time, Z, equals $a_{\mathrm{ref}}t/L$, and the non-dimensionalized entropy level, A_A, equals a_A/a_{ref}, where a_A is the speed of sound at the reference pressure, p_{ref}, due to an isentropic change of state from the pressure, p.

2.3.2. Path Line Equations

Direction condition:

$$dX/dZ = U \tag{18}$$

Compatibility equation:

$$dA_A = \left(\frac{\kappa - 1}{2}\right) \frac{A_A}{A^2} \left(\frac{q x_{\text{ref}}}{a_{\text{ref}}^3} + \frac{2 f x_{\text{ref}}}{D} |U|^3\right) dZ \tag{19}$$

The relationship between pressure and speed of sound is given by

$$p/p_{\text{ref}} = (a/a_A)^{2\kappa/(\kappa-1)} = (A/A_A)^{2\kappa/(\kappa-1)} \tag{20}$$

Methods of solving the equations of unsteady flow will be described in Section 3.

3. METHODS OF CALCULATING COMPRESSIBLE FLOW

The equations developed above do not, in general, have analytical solutions. The only group which can be solved analytically is the linearized one. For this reason it is worthwhile considering such solutions, especially in the inlet manifold where the wave amplitude is usually small. Linear solutions can take two forms: the simpler is the prediction of the tuned frequencies of the pipework arrangement, the more complex is to solve the linearized wave equations on a step-by-step basis. Both of these approaches have been adopted, and are described below.

The most accurate method for evaluating the flow in intake and exhaust pipes is one which will solve the non-homentropic equations of unsteady gas motion. Such techniques will also be described below. In addition, a simpler method of solving the homentropic flow equations will also be introduced: this has been widely used in both intake and exhaust pipes.

3.1. Linear Waves

Engleman[18] proposed a lumped parameter linearized approach that was extended by Vorum[19,20] who used it in a rather empirical manner to develop a design synthesis technique. It will be shown later that such a method, while it will give good results in some circumstances, is not completely reliable. Chapman, Novak & Stein[21] introduce a more complex lumped parameter approach. Ohata & Ishida[22] and Matsumoto & Ohata[23] have also developed linearized analytical solutions for engine inlet manifolds.

The latter methods are based on the analytical solution of eqn (5) which can take the form

$$p = k\rho_0 a_0^2(A_1 \exp(j(\omega t - kx)) - B_1 \exp(j(\omega t + kx)) \qquad (21)$$

and

$$q = j\omega F(A_1 \exp(j(\omega t - kx) + B_1 \exp(j(\omega t + kx)) \qquad (22)$$

where q is the volume flow rate of air at x and t, k is the wavenumber $(=\omega/a_0)$ and $j=\sqrt{-1}$.

These equations may be manipulated in various ways to give different types of solution. The simplest approach is to solve only for the natural frequency of the system. This can be done by equating the pressure to infinity when the mass, or volume, flow is not equal to zero. A more satisfactory use of these equations is to solve them using a step-by-step method of some form. In fact the equations are those of a delay-line and can be solved using such an approach (see Metzger & Vabre[14]). Examples of the use of this method are given in Section 4.1.1 of Chapter 13.

3.2. The Method of Characteristics

3.2.1. Homentropic Flow

This solution method was proposed by Riemann in 1885, well before the discovery of the digital computer. The beauty of approach is that it recognizes the inherent property of the equations to generate waves, and it capitalizes on these to convert from partial to ordinary differential equations. Undoubtedly the *method of characteristics* gives an insight into unsteady compressible flow that is not available from any other approach. An example of a simple graphical solution using this method is shown in Fig. 2; it can be seen that the solution is tedious, and it becomes even more so if allowance is made for junctions and changes of entropy level. It would be possible to computerize the graphical approach and evaluate all the wave crossing points using simple co-ordinate geometry. Jones & Brown[24] describe such a solution. There are drawbacks in this because it becomes necessary to store a very large amount of wave data on an irregular grid; such a program can be very cumbersome and would probably be time consuming. The numerical analyst is much happier if he can reduce the problem to one using a rectangular grid structure; the techniques described will all be based on such a grid.

The equations for the grid-based technique were originally proposed by Hartree[25] and they are also described by Shapiro.[26] The first application of these methods, referred to as the mesh methods, to internal combustion engines appears to have been by Benson, Garg & Woollatt.[27] It is probably true to say that up to the present time the *mesh method of characteristics* is the most widely used approach for solving unsteady flow in reciprocating engines, and there are many reports in the literature of the use of this technique (e.g., see Refs 28–34).

To understand how a grid-based method can be developed it is necessary to consider the position diagram in the graphical method of characteristics. Three nodes in the *position diagram* ($s-1$, s, $s+1$) at two intervals of time (r, $r+1$) are shown in Fig. 3. In homentropic flow, a characteristic leaving point ($s-1$, r) will travel through the X–Z diagram with a fixed slope given by $\mathrm{d}X/\mathrm{d}Z = b\lambda - a\beta$, where a and b are constants dependent on the ratio of specific heats of the gas, κ [$a = (3-\kappa)/(2(\kappa-1))$ and $b = (\kappa+1)/(2(\kappa-1))$]. The full derivation of these equations is contained in Benson,[11] to which the reader is referred for further details. It can be shown that for a stable solution it is necessary that the maximum time step is within the Courant, Fredrichs & Lewy[35] *stability criterion.* This states that the size of the time step, on a fixed X grid, is limited to the minimum value of $\mathrm{d}Z/\mathrm{d}X = 1/(A + |U|)$ for all the grid points. The significance of this in terms of Fig. 3 is that point R ($r+1$, s) is always below (or in the limit, coincident with) the point at which the characteristic crosses the grid line. This means that points P (r, $s \pm 1$) and S (r, s) are within the *domain of dependence* of R, and information from these points is available at R when the next calculation will be performed there. In mathematical terms it means that the conditions at R are always evaluated by *interpolating* the values on the characteristics. Figure 3, which is based on Fig. 6.13 in Ref. 11, introduces the concept of interpolating in the time domain: there is no reason why the situation cannot be changed so that the interpolation takes place in the distance domain. In fact, Benson[11] uses a time interpolation for the homentropic technique and a distance interpolation for the non-homentropic one; the first technique is probably easier to visualize, whereas the second has some advantages when allowing for changes in properties as the wave propagates.

The result of this method is that the value of λ or β can be calculated at point R if the values of these parameters are known at

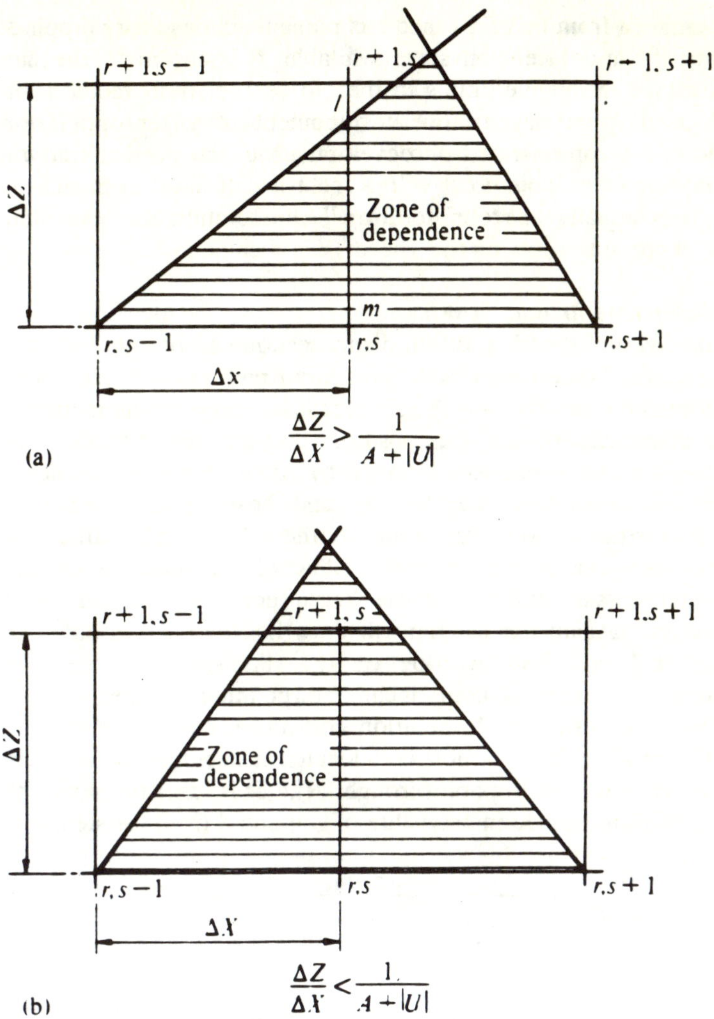

Fig. 3. Grid proportions for stable solution. Grid points: $R=(r+1, s)$; $S=(r, s)$; $P=(r, s\pm 1)$.

the relevant points on the previous time step. The resulting equation, for homentropic flow, is

$$\lambda_R = \lambda_S + \frac{\Delta Z}{\Delta X}(b\lambda_P - a\beta_P)(\lambda_P - \lambda_S) \tag{23}$$

It can be seen from eqn (23) that it is possible to solve for all points in the pipe if the adjacent ones are available. It is also apparent that this is the case for all the points in the pipe except the two ends, or *boundaries*. A large part of the development of any program for unsteady flow is concentrated on developing the *boundary conditions* for the flow régime. This point will be returned to later in Section 4 because it is equally appropriate for all the solution methods that might be proposed.

3.2.2. Non-homentropic Flow

The equations developed in the previous section can be quite readily extended to non-homentropic flow. The major philosophical difference between this type of flow and the simpler homentropic type is that in this the characteristics on the position diagram are not lines of constant λ. As the waves pass through the flow field they are acted upon by irreversibilities (e.g. friction and heat transfer), or pass through fluid stratifications, which modify the value of the characteristic. Equations (17) and (19) showed the effect of such interactions, and now it is necessary only to relate these equations to the finite difference grid. Again this section will draw heavily on Benson[11] and only the final equations will be quoted. As stated before, the non-homentropic method uses linear interpolation in the distance domain and this results in the solution grid having the form shown in Fig. 4. The first requirement now is to locate the *departure point of the characteristic* which will pass through R'. This can be done by assuming that the change in the value of λ is small over the step, and then

$$\frac{\delta X}{\Delta X}=\frac{b\lambda_{I_l}-a\lambda_{II_l}}{\dfrac{\Delta X}{\Delta Z}+b(\lambda_{I_l}-\lambda_{I_{l-1}})-a(\lambda_{II_l}-\lambda_{II_{l-1}})} \tag{24}$$

(It should be noted that λ and β have been replaced by the generalized Riemann variables λ_I and λ_{II}; these are essentially the same as λ and β but enable general equations to be derived.)

Having located this point the value of λ_{IP} can be calculated from eqn (25):

$$\lambda_{I_P}=\lambda_{I_l}-\frac{\delta X}{\Delta X}(\lambda_{I_l}-\lambda_{I_{l-1}}) \tag{25}$$

However, in this non-homentropic case it is also necessary to evaluate the change in λ that takes place during the passage of the

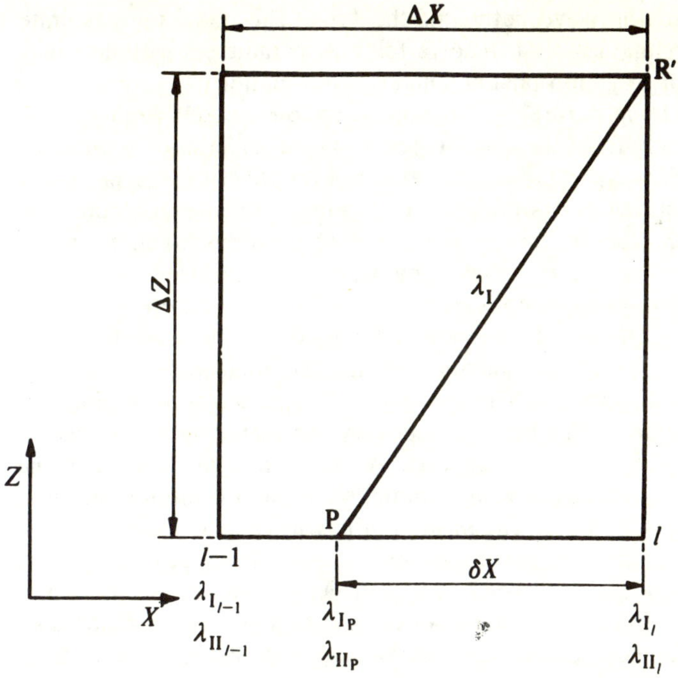

Fig. 4. Solution for λ_l in subsonic flow.

wave from P to R'. In its travel the wave can be acted upon by four influences; change of cross-sectional area (dF/dX), friction (f), heat transfer (q), and stratification which causes changes of entropy level (A_A). The first of these is not really a non-homentropic effect because it is not an irreversibility; it is just convenient to consider it here. It is possible to solve the equations for homentropic flow with area change. All of these phenomena cause the value of λ to be altered as the wave passes through the field; this can result in stability problems in the calculation, particularly in the case of area change.

3.3. Finite Difference Techniques

In reality the mesh method of characteristics, described above, is a finite difference technique. It differs from the more usual ones because it calculates the values of Riemann variables at various points taking cognizance of the wave nature of the flow. The more normal finite difference techniques tend to solve for flow parameters without really imposing a propagation phenomenon on the solution.

The hyperbolic nature of the flow equations makes them behave quite differently from the other types of partial differential equations, viz. elliptical and parabolic. This also has an effect on the necessary boundary conditions. An excellent summary of the problems associated with stability and convergence of solutions when trying to solve fluid flow problems is given by Morton,[36] and a review of finite difference methods for non-linear hyperbolic partial differential equations is given by Sod.[37] These papers discuss the problems of fluid flow simulation in general mathematical terms; the engineer is often more pragmatic in his approach to modelling his particular flow situation and will use the techniques developed by the mathematicians, testing them in his own sphere of interest. A major area of concern in the calculation of unsteady flow in a compressible fluid is the formation of shock waves in the fluid. The *method of characteristics* requires careful formulation to deal with such problems, and the approach used to introduce finite shock waves in the graphical technique is outlined in Benson[11] (Chapter 5). The numerical solution of the characteristics equations has not been developed to deal with shocks, but in most internal combustion engine problems these do not build up; an exception is in the exhaust system of two-stroke engines (see Wallace & Boxer[38]).

3.3.1. Lax–Wendroff Technique

There is a group of solutions that is referred to as the Lax–Wendroff methods; the most commonly used variants in engine calculations are the *one-step* and *two-step methods*[39] and both of these are second order accurate based on the Taylor series. The two-step method will be described here, and the equations derived.

Continuity equation:

$$\frac{\partial \rho}{\partial t} + \frac{\partial(\rho u)}{\partial x} = -\rho u \frac{\mathrm{d}(\ln F)}{\mathrm{d}x} \tag{26}$$

Impulse equation:

$$\frac{\partial}{\partial t}(\rho u) + \frac{\partial}{\partial x}(\rho u^2 + p) = -\rho u^2 \frac{\mathrm{d}(\ln F)}{\mathrm{d}x} - g\rho \tag{27}$$

Energy equation:

$$\frac{\partial}{\partial t}\left[\left(\frac{u^2}{2}+e\right)\rho F\,\mathrm{d}x\right]+\frac{\partial}{\partial x}\left[F\rho u\left(\frac{u^2}{2}+e+\frac{p}{\rho}\right)\right]\mathrm{d}x=q\rho F\,\mathrm{d}x \qquad (28)$$

For a perfect gas,

$$e=\frac{RT}{\kappa-1}=\frac{p}{\rho}\frac{1}{\kappa-1}$$

It useful to introduce the following symbolic vector notation:

$$\bar{\mathbf{W}}(x,t)=\bar{\mathbf{W}}=\begin{bmatrix}\rho\\ \rho u\\ \rho\dfrac{u^2}{2}+\dfrac{p}{\kappa-1}\end{bmatrix}$$

and function $\bar{\mathbf{F}}$, where:

$$\bar{\mathbf{F}}(\bar{\mathbf{W}})=\begin{bmatrix}\rho u\\ \rho u^2+p\\ u\left(\rho\dfrac{u^2}{2}+\dfrac{\kappa}{\kappa-1}p\right)\end{bmatrix}$$

and

$$\bar{\mathbf{C}}(\bar{\mathbf{W}})=\begin{bmatrix}\rho u\\ \rho u^2\\ u\left(\rho\dfrac{u^2}{2}+\dfrac{\kappa}{\kappa-1}p\right)\end{bmatrix}\frac{\mathrm{d}\ln F}{\mathrm{d}x}+\begin{bmatrix}0\\ g\rho\\ -q\rho\end{bmatrix}$$

The system of equations (26)–(28) then becomes a first order hyperbolic partial differential equation in the vector form:

$$\frac{\partial\bar{\mathbf{W}}}{\partial t}+\frac{\partial\bar{\mathbf{F}}}{\partial x}+\bar{\mathbf{C}}=0 \qquad (29)$$

in which $\bar{\mathbf{W}}$ is the desired vector of the solution. Equation (29) may be converted to a difference equation by applying the Taylor's series of function $\bar{\mathbf{W}}(x, t)$ for point (i, n) in direction t (Fig. 5) and considering the first two terms of the expansion. If a central difference approach is used for $\partial\bar{\mathbf{F}}/\partial x$, together with the mean value of vector $\bar{\mathbf{W}}$ at point (i, n) then the desired vector of solution at mesh point $(i, n+1)$ can be obtained in two steps.

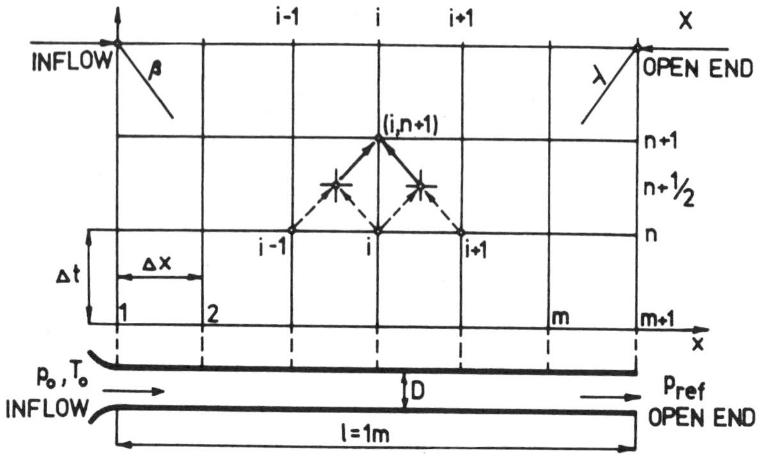

Fig. 5. Grid structure for two-step Lax–Wendroff technique: de Haller problem.

In the first step:

$$\bar{\mathbf{W}}_{i+1/2}^{n+1/2} = \tfrac{1}{2}(\bar{\mathbf{W}}_{i+1}^{n} + \bar{\mathbf{W}}_{i}^{n}) - \frac{\Delta t}{2\Delta x}(\bar{\mathbf{F}}_{i+1}^{n} - \bar{\mathbf{F}}_{i}^{n}) - \frac{\Delta t}{4}(\bar{\mathbf{C}}_{i+1}^{n} + \bar{\mathbf{C}}_{i}^{n}) \quad (30)$$

and in the second one:

$$\bar{\mathbf{W}}_{i}^{n+1} = \bar{\mathbf{W}}_{i}^{n} - \frac{\Delta t}{\Delta x}(\bar{\mathbf{F}}_{i+1/2}^{n+1/2} - \bar{\mathbf{F}}_{i-1/2}^{n+1/2}) - \frac{\Delta t}{2}(\bar{\mathbf{C}}_{i+1}^{n} + \bar{\mathbf{C}}_{i-1/2}^{n}) \quad (31)$$

Solution of eqns (30) and (31) results in the evaluation of a state vector (p, ρ, u) for the second to the penultimate points in space at time level $(n+1)$. The new state is evaluated explicitly from the initial values; the states at the ends of the pipe are found from boundary conditions.

Morton[36] explains that one of the advantages of the two-step method is that it reduces the 'damping factor' which is implicit in the first order solutions. Basically all these schemes introduce an effective viscosity term, and the lower the order of the scheme the greater the effect.

Variants of Lax–Wendroff schemes have been used widely to simulate the wave action occurring in engine manifolds. One of the difficulties of these, and any other true finite difference methods, is that they can only be applied to the nodes inside the pipe; another technique is required to get to the boundaries. The best method of incorporating the boundary conditions in such a solution would appear to be the *method of characteristics* since this contains the means of transmitting information between two nodes.

Poloni, Winterbone & Nichols[40] give a comparison between the two-step Lax–Wendroff method and method of characteristics for a simple 'blowdown' problem. They show that the Lax–Wendroff method can be both faster and more accurate in terms of mass flow than the method of characteristics. However, if the equations defined above are applied to a situation in which there is a temperature (or entropy) discontinuity then certain instabilities can be introduced. This aspect of the solution method is reported by Poloni, Winterbone & Nichols[41] who show a direct comparison of Lax–Wendroff and method of characteristics techniques for a simple discharge process with a temperature discontinuity and an engine exhaust process. In the former case, the variation in pressure and velocity in a pipe connected to a reservoir have been compared with steady state results evaluated analytically. The following conclusions were reached. First, the Lax–Wendroff technique conserves mass better than method of characteristics because it is a conservative set of equations; however, both methods achieve better accuracy than 0·2% for prediction of mass flow. The Lax–Wendroff method was also found to be faster than method of characteristics by a factor of approximately 2 (in the case of temperature shock this is 1·4). Both methods gave good accuracy in predicting the variation of pressure with time for the exhaust system, as shown in Fig. 6.

Bulaty[42,43] has used Lax–Wendroff methods to simulate the flow in a turbocharged diesel engine. He recognizes that such techniques have the tendency to predict unrealistic overshoots at discontinuities and attempts to overcome this by using a *one-step method with flux correction*. He shows that such an approach is able to cope with quite complex exhaust arrangements and achieve good predictions. It is suggested that the Lax–Wendroff technique can be substantially faster than method of characteristics, especially in cases with large property (pressure and temperature) discontinuities.

Azuma, Yura & Tokunaga[44] describe the application of a two-step Lax–Wendroff technique to simulate medium- and low-speed marine diesel engines. It should be noted that these authors had previously[28] used the *method of characteristics* to solve a similar problem. The authors state that they had encountered difficulties in obtaining

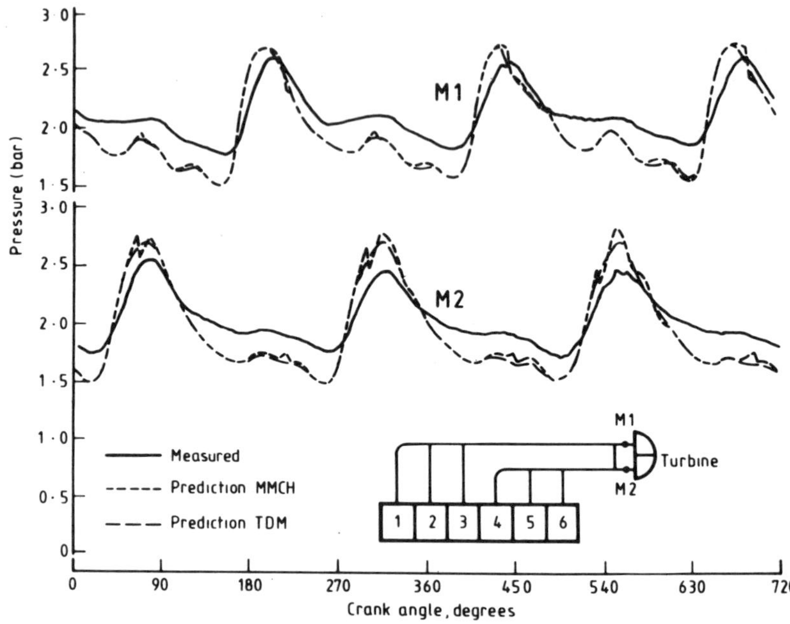

Fig. 6. Comparison of pressure predictions by method of characteristics and Lax–Wendroff technique.

accurate air-flow calculations while using the *method of characteristics*, even though the pressure variation in the manifolds was calculated to acceptable accuracy. The results presented in Ref. 44 do not seem to show an absolute superiority for the Lax–Wendroff technique over the method of characteristics; in fact in two out of three simulations reported the values of mass flow calculated by characteristics were just as good as those obtained by the other method. In the third example there was quite a large discrepancy but, in the author's experience, this could easily be due to poor definition of meshes or path lines. Azuma *et al.* use the *method of characteristics* to go from the pipe to the boundary.

3.3.2. Other Methods Used for Solving Compressible Flow in Engines

The methods described above are all derived in a rigorous mathematical manner from the equations of motion and consideration of mathematical expansions of the various equations. There is quite a large number of other methods proposed in the literature which are either more idiosyncratic or less rigorous than those introduced above. It is felt that the *method of characteristics* and Lax–Wendroff techniques offer better starting points than those described below, simply because they are well defined and there is a good body of experience associated with them; an advantage claimed for most of the newer methods is speed of solution.

3.3.2.1. *Runge–Kutta solution.* Lakshminarayanan, Janakiraman, Gajendra Babu & Murthy[45] describe a method in which they solve the partial differential equations of fluid motion (eqns (1)–(4)) using a fourth order Runge–Kutta technique, which is usually reserved for the solution of first order ordinary differential equations. The equations of flow are discretized into finite difference equations using a central-difference formulation, and linear extrapolation is used to introduce the boundaries. The predictions obtained on an engine calculation were in good agreement with both measured pressures and velocities; it is also claimed that the program was faster than using the *method of characteristics*. An advantage claimed for this method is that it is able to use a time step equal to 2·83 ($\sqrt{8}$) times the usual stability step; this should result in increased speed of computation.

3.3.2.2. *Random-choice method (RCM).* Iwamoto & Deckker[46] have applied the random-choice method to the calculation of unsteady flow in pipes; this is also referred to by Sod.[37] The advantage of this method over many of the other finite-difference techniques is that it does not require the introduction of an artificial viscosity term to limit overshoots at discontinuities. Iwamoto & Deckker compare their results with rig tests that approximate an engine case, and claim that the technique will better track shocks and other discontinuities (e.g. entropy) than alternative approaches.

3.3.2.3. *General finite difference techniques.* Walter & Chapman[47] report the use of a mixture of upwind and centred differencing to formulate the finite difference equations. They evaluate the flux terms using the upwind difference approach and calculate a 'time-centred' pressure term. They solve the resulting equations with an algorithm called FRAM, which is designed to limit the amount of numerical diffusion, while conserving the large gradients in field variables.

Meisner & Sorenson[48] have used McCormack's method to integrate the finite difference equations. This was specially developed for integration of the flow equations where there are waves in two directions. It is a two-step method, using forward differencing on the first, predictor, step and backward differencing on the second, corrector, step. The results shown were based on frictionless flow and, while obtaining good agreement for the basic frequency of the waves in the manifold, they exhibit substantial differences in pressure amplitude.

3.3.2.4. *Other methods.* There are various other simpler methods of setting up an equation for the compressible flow of gas in engine manifolds. These are usually derived intuitively from lumped parameter representations of the gas in the pipes and do not seem to offer any advantages over the more satisfactory approach of developing the finite difference equations from the original differential ones. Such approaches will not be considered here.

3.4. Conclusions

There are a number of methods for evaluating wave motion in the manifolds of engines. These all have their own protagonists who champion their merits. The technique that has probably been used most widely is the *method of characteristics*. This can be used in both the homentropic or non-homentropic forms; the latter is the most realistic for engines in which there is combustion. This is because there will be severe entropy gradients between the inlet and exhaust flows; there can also be large entropy differences in an individual manifold if there is scavenge flow, or backflow. The major disadvantages with this method, when applied as the mesh (finite difference) method, are that it is a first order technique because of the interpolation used, it does not have an explicit mass flow balance equation, and it has difficulties catering for shock waves. Careful choice of the mesh size and numbers of path lines help to overcome the first two problems and, fortunately, shock waves are not common in engine pipework, except in the exhaust of highly rated two-strokes. Most of the examples given later in chapter 13 will be based on the method of characteristics.

Lax–Wendroff methods have also been widely used, and some examples of calculations using this technique will be given. The results will be compared with those from the *method of characteristics*. This approach is also capable of dealing with entropy changes, although it does not preserve the discontinuities as well as pathlines can achieve. The Lax–Wendroff method can be first or second order, but chapter 13 will only consider the latter; it has good mass conservation properties.

4. PIPE BOUNDARY CONDITIONS

Section 3 concentrated on methods of solving the equations of wave motion in a pipe. It was stated there that an important aspect of the calculation was the boundary conditions. Every pipe has two ends associated with it, and hence has two boundaries. There are various types of boundary and these must be simulated in a satisfactory manner: the best way to accurately incorporate the boundaries in all wave action techniques is by the *method of characteristics*. This is because the only way to get to the end of the pipe in a logical manner is to go along a wave: hence the following section can be applied to all calculations, not only method of characteristics. Benson[11] describes how to tackle the various boundaries encountered in engines, and this section will only summarize the simpler ones. Typical boundaries for which methods exist are: open end, closed end, inflow boundary, valve boundaries, junctions (both with and without pressure losses), pulse converters, orifices or valves in pipes, carburettors, turbines and compressors. These all have one feature in common: they are all treated as if the flow is *quasi-steady*. This basically means that the characteristic of the boundary is not a function of the frequency of the incident waveform; such an assumption is more appropriate to a simple boundary such as an open end than one with a relatively large volume, such as a turbine.

4.1. Open End Boundary

The main feature of a simple open end boundary is that the pressure at the pipe exit is equal to that outside the pipe. This can be easily written into the method of characteristics equations as:

$$\lambda_{out} = 2 - \lambda_{in} \tag{32}$$

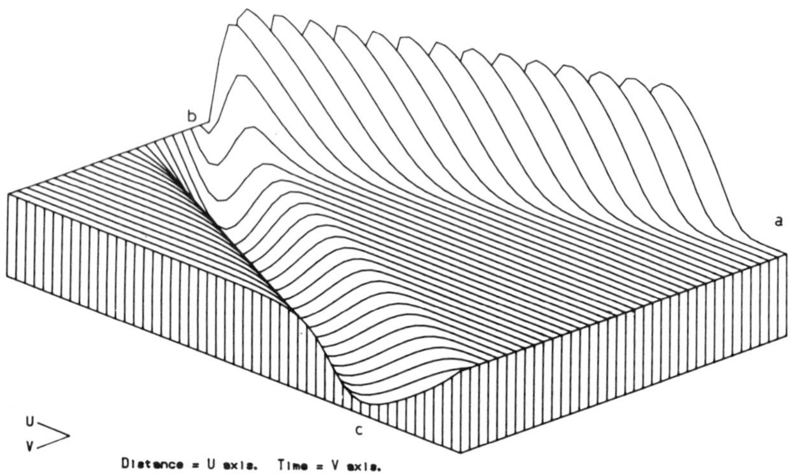

Fig. 7. Pressure wave interaction with open end.

The effect of such a boundary is to invert an incident wave: thus an incident compression wave is reflected as a rarefraction wave, and vice versa. An example of this is shown in the isometric distance–time–pressure diagram in Fig. 7 (taken from Winterbone & Nichols[49]). The wave travels along the constant cross-sectional area pipe until it reaches the open end where it is reflected back (the physical arrangement is similar to Fig. 8). In practice the open end boundary is not as clearly defined as implied by this approach. It is found that the establishment of three-dimensional flow structures at this boundary tends to increase the 'effective length' of the pipe by around one pipe diameter.

4.2. Closed End Boundary

The important feature of a closed end boundary is that it does not transmit flow: hence the velocity at the end is zero. This means that an incident compression wave is reflected as an equal wave. This can be defined as

$$\lambda_{out} = \lambda_{in} \tag{33}$$

Figure 8 is a similar diagram to Fig. 7, but for a closed end. It can be seen that the pressure at the closed end actually rises higher than the amplitude of the wave because of the momentum change involved in reversing its direction. There tend to be various closed ends in manifolds, e.g. at the extremities of inlet manifold plenums and closed valves.

4.3. Partially Open End (or Nozzle) Boundary

Many boundaries are neither open ends nor closed ones, but fall some way between these extremes. Obviously the way in which waves are transmitted by, or reflected from, such a boundary will be a function of the area ratio of the hole to the pipe: the nozzle area ratio will be

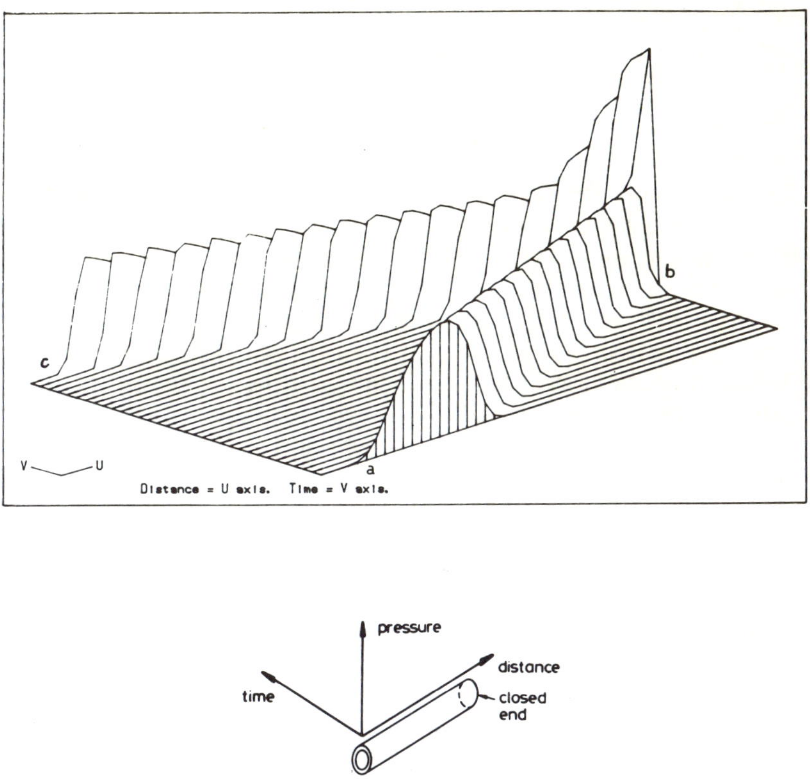

Fig. 8. Pressure wave interaction with closed end.

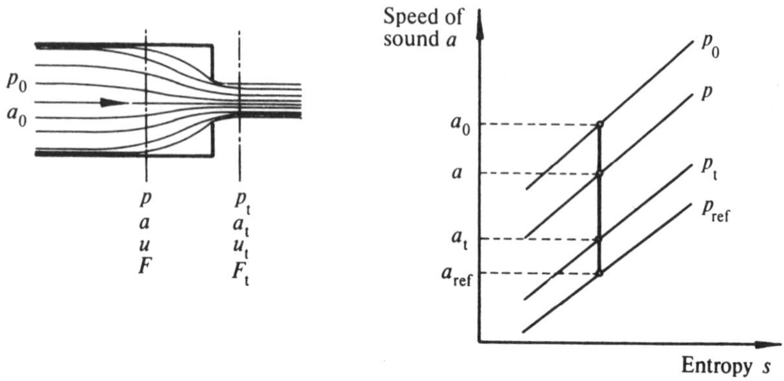

Fig. 9. Flow through a partially open end (or nozzle) boundary.

denoted by

$$\varphi = F_t / F_p \tag{34}$$

In Fig. 9 are shown the main parameters defining flow through a partially open end, both on a physical and a speed of sound–entropy (a–s) diagram. The flow can be either sonic or subsonic, and in the latter case the pressure at the throat of the flow is assumed to be the downstream static pressure (no allowance is made for pressure recovery). The effective area ratio of the nozzle should be used and account must be taken of the coefficient of discharge, C_d, in evaluating φ.

The continuity and energy equations can be applied to the nozzle arrangement and a relationship between the non-dimensional speed of sound, A, the non-dimensional particle velocity in the pipe, U, and the area ratio, φ, can be derived. This relationship is shown in a graphical form in Fig. 10. To use this information in a numerical procedure it is necessary to relate λ_{out} and λ_{in}; this can be achieved by manipulating the equations defining the boundary. Benson[11] describes the derivation of these equations, which are summarized below.

For subsonic flow:

$$\lambda_{out} = 2A - \lambda_{in} \tag{35}$$

where A can be evaluated from the equation

$$(A^{4/(\kappa-1)} - \varphi^2)(\lambda_{in} - A)^2 - \frac{\kappa - 1}{2}(A^2 - 1)\varphi^2 = 0 \tag{36}$$

while, for sonic flow,

$$\lambda_{\text{out}} = K_s \lambda_{\text{in}} \tag{37}$$

where

$$K_s = \left\{1 - \frac{\kappa - 1}{2}\varphi\left(\frac{1}{A_{\text{cr}}}\right)^{(\kappa+1)/(\kappa-1)}\right\} \Big/ \left\{1 + \frac{\kappa - 1}{2}\varphi\left(\frac{1}{A_{\text{cr}}}\right)^{(\kappa+1)/(\kappa-1)}\right\} \tag{38}$$

and A_{cr} can be evaluated from

$$\varphi^2 - \left\{\frac{\kappa + 1}{\kappa - 1} - \left(\frac{2}{\kappa - 1}\right)(A_{\text{cr}})^2\right\} A_{\text{cr}}^{4/(\kappa-1)} = 0 \tag{39}$$

Partially open end boundaries are encountered in a number of areas of the engine, but the most important application is in obtaining an approximate representation of the turbine in a turbocharged engine. At low pressure ratios an axial flow turbine behaves like an orifice, although a more realistic model for a 50% reaction turbine (in which the enthalpy drop is split between the nozzles and the rotor) is to use two orifices in series. However, the partially open end is a reasonable approximation of the device if no accurate characteristics are available. It is also possible to use this boundary at other regions in the pipework where certain restrictions occur.

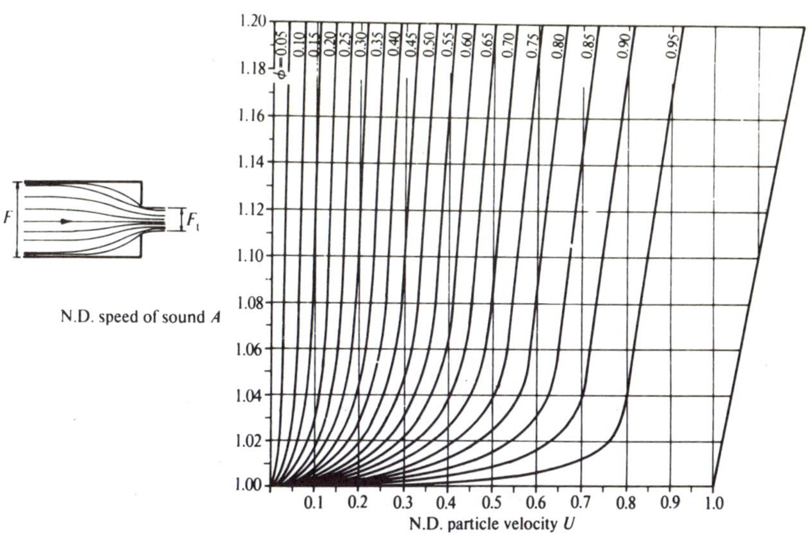

Fig. 10. Boundary chart for a partially open end (or nozzle); $\kappa = 1{\cdot}4$.

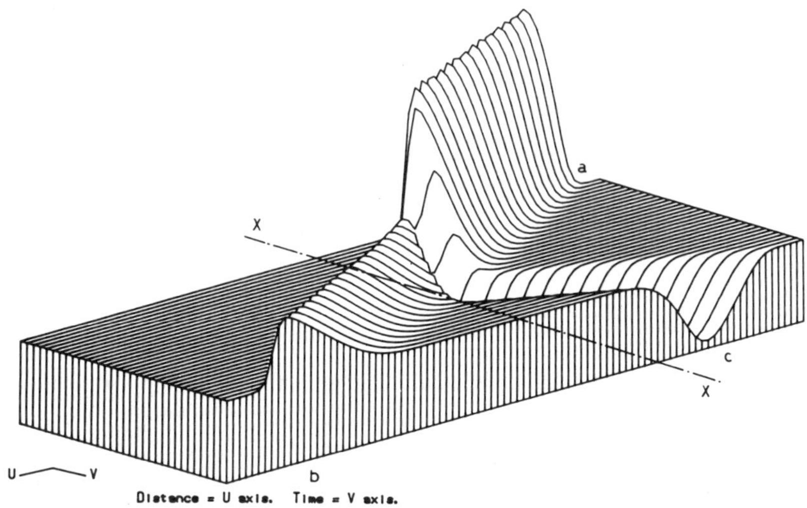

Fig. 11. Pressure wave interaction with sudden enlargement in pipe.

4.4. Change of Cross-section

This section will consider a change of cross-section without pressure loss: this is a simple two-pipe junction. Benson[11] derives the general equations for a multi-pipe junction (which appears as a change of section to the gas flow in the engine) but these equations will be simplified to deal with just two pipes. The Riemann variables of the incident and reflected waves are then related by

$$\lambda_{\text{out}.q} = K_T - \lambda_{\text{in}.q} \qquad \text{for } q = 1, 2 \tag{40}$$

where

$$K_T = \sum_{q=1}^{q=2} K_q \lambda_{\text{in}.q} = \frac{2}{F_T}(F_1\lambda_{\text{in}.1} + F_2\lambda_{\text{in}.2}) \tag{41}$$

and $F_T = F_1 + F_2$.

It should be noted that a corrected value of λ_{in} should be used in non-homentropic flow: this includes an entropy correction to take account of the possible difference in entropy levels between the pipes on either side of the junction. The concept of corrected λ values will not be covered in this text and the reader is referred to Benson.[11] The way in which a simple wave interacts with an area change junction is shown in Fig. 11. In the case of a sudden enlargement the incident pressure wave is partially transmitted by the junction but a rarefaction wave is reflected back into the original pipe. A similar effect occurs at a multi-pipe junction, and such a phenomenon can be used to assist the scavenging process in the engine cylinder: this is the essence of tuning the manifolds.

4.5. Entropy Discontinuity in Pipe

Entropy discontinuities occur in the gas in the pipes associated with engines when an interface between hot and cold gas exists. These are not really boundaries but they do exhibit behaviour similar to them. At the interface between gases at two temperatures the pressure wave experiences a large change in entropy, and this makes itself apparent in the $d\lambda$ term of the characteristic equations. The result is that the wave does not simply pass through the interface but is partly reflected and partly transmitted, just as light is when passing from air into glass. In Fig. 12 is shown a simple wave hitting a temperature (and hence entropy) discontinuity: there are transmitted and reflected waves similar to the effect of an area change, but in this case the discontinuity is set in motion by the wave!

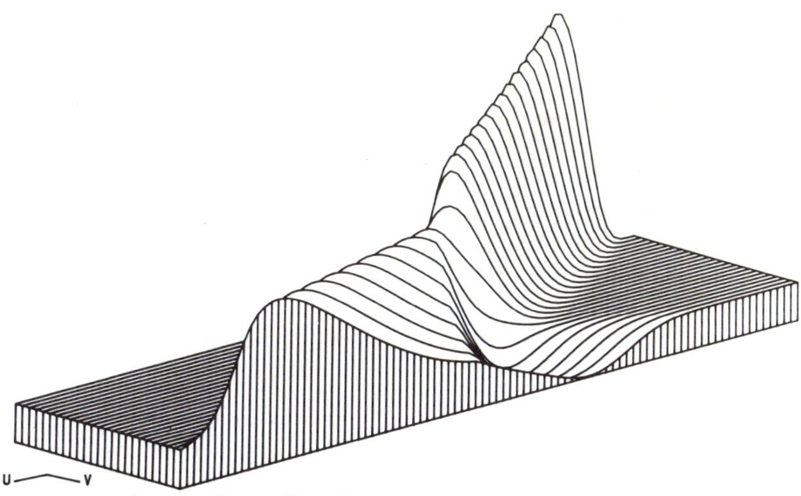

Fig. 12. Pressure wave interaction with temperature discontinuity, 300–900 K.

4.6. Valve and Port Boundaries

Valves and ports play an important role in the design of IC engines, and are characterized by the period they are open, together with the area available for flow. In this text the term valve will be used for poppet valves, as used in four-stroke engine intake and exhaust systems, while ports will be taken to refer to piston-controlled holes in the liners of two-stroke engines. A qualitative discussion of the flow through these devices is given in Annand & Roe[2] and the reader is referred to that text. Work, using laser–doppler anemometry, is currently being performed by Vafidis & Whitelaw[50] and Tindal, Cheung & Yianneskis.[51]

While valves and ports are necessary simply to transfer gas into and out of the cylinder, they also serve the purpose of generating air motion inside the cylinder itself. This results in a compromise in valve design between maximizing the flow area of the assembly and achieving the desired quantity of air motion in the cylinder. For example, in older engines shrouded inlet valves were fitted to generate sufficient swirl to aid combustion; while this was necessary to enable the engine to operate efficiently over a broad range of speed and load, the shroud obstructed the flow of air into the cylinder. In modern high-speed direct injection diesel engines helical ports are used to induce high levels of swirl into the gas in the cylinder to enhance the rate of combustion. The ports on two-stroke engines serve a similar dual role: they transfer the gases into and out of the cylinder but in addition they impart air motion; in this case, to increase the efficiency of scavenge. These factors make valves and ports complex fluid-dynamic arrangements, and it is only now that the multi-dimensional calculation techniques are being developed to produce design procedures; previously design was performed largely intuitively with empirical verification.

It has long been recognized that the flow pattern through a valve changes with the lift (see Fig. 13): on the inlet valve, at low lifts (Fig.

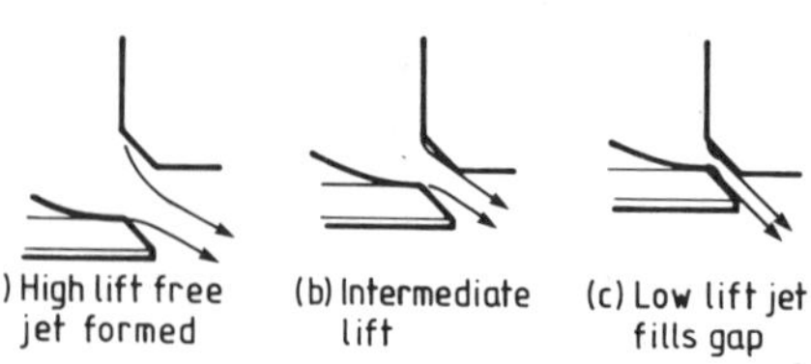

Fig. 13. Flow patterns through a valve.

13c) the flow remains attached to the valve and valve seat, while at high lifts (Fig. 13a) a free jet is formed. The effect of these variations in flow pattern is to alter the coefficient of discharge, C_d, of the valve. Careful choice of the valve parameters (e.g. valve diameter and lift, and valve seat angle), together with detailed design of the valve and port enable the effective area (C_dF_v) of the assembly to be optimized. Similar design variations apply to ports, where care has to be taken to achieve acceptable C_d values. The flow pattern also changes with exhaust valves and this will affect the C_d variation which occurs: although the effect of the exhaust valve C_d is not as marked on air flow as that for the inlet valve, it will affect the pumping work required to move the air through the engine, and reduce the available energy reaching the turbocharger if one is fitted. Typical C_d values vary from around 0·8 to 0·9 at low lifts down to 0·5 at high lifts, for both inlet and exhaust valves. The value of C_d is also a function of the pressure ratio across the valve, decreasing slightly with increasing pressure ratio. At high lifts the value of C_d is not a function of Reynolds number (Re), but at low lifts it tends to decrease with increasing Re because the viscous effects become more important.

The different flow régimes for an exhaust valve are shown in Fig. 14. It can be seen that with both subsonic and sonic flow there is a substantial increase in entropy across the valve, even though the enthalpy remains substantially constant. This model is known as the constant-pressure model because it is assumed that the flow from the valve throat to the exhaust pipe takes place without any pressure recovery: the static pressure at the valve throat equals that in the pipe. The constant-pressure model is the one most commonly used for simulating flow through exhaust valves. (Other models of flow through valves and ports are discussed in Benson.[11]) Application of the continuity and energy equations to the flow across the valve enable the boundary to be written in a form suitable for use in simulation models. It is necessary to define the value of λ_{out} as a function of the various valve parameters, and the value of λ_{in}. Benson[11] derives the valve boundary equations in detail, and only the results will be stated here.

It is necessary to introduce certain non-dimensional parameters to define the valve area, and flow regimes. These are:

$$\psi = F_t/F_p \tag{42}$$

where F_t is the effective area of the valve, including the coefficient of

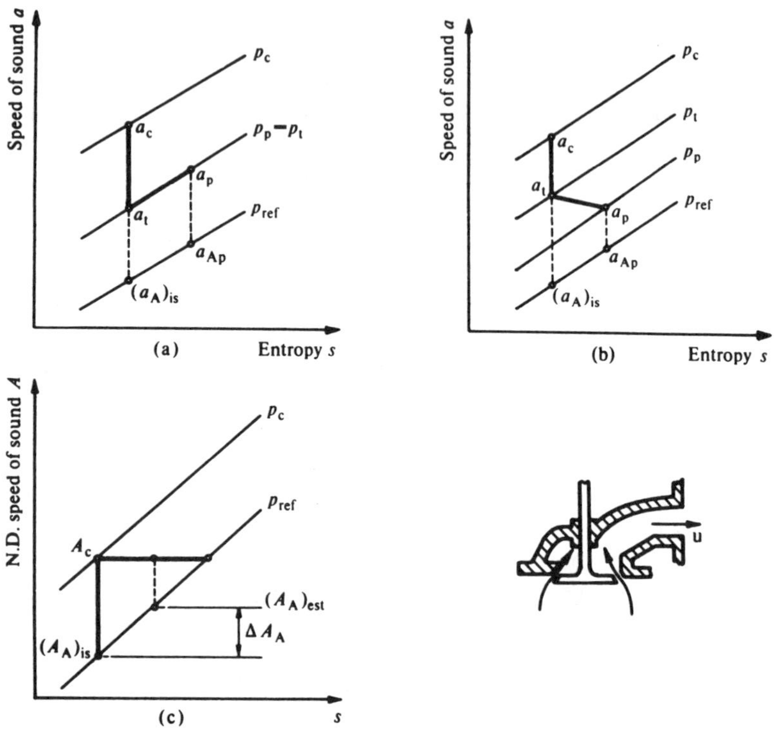

Fig. 14. Outflow through a valve: non-homentropic flow.

discharge, C_d; F_p is the pipe cross-sectional area;

$$\pi = \text{non-dimensional pressure ratio} = (p_p/p_c)^{(\kappa-1)/2\kappa} \tag{43}$$

$$U = \text{non-dimensional velocity} = u/a_c \tag{44}$$

and

$$a_{A_c}/a_{A_p} = \text{`entropy' change across the valve} \tag{45}$$

Manipulation of the equations, together with the use of these parameters, enables the valve boundary chart to be constructed, as shown in Fig. 15(a). This chart is ideal for use with graphical methods but it is necessary to write the relationships embodied in using such a chart in algebraic form for use in a numerical technique. Again, the results from Benson[11] will simply be quoted here. There are four basic relationships required to define the flow regimes: subsonic flow through the valve throat; sonic boundary at the valve throat; sonic flow through the valve throat; and sonic flow in the pipe. These relationships are defined on the boundary chart, as shown in Fig. 16. Manipulation of the equations gives

$$\lambda_{\text{out}} = \left(\frac{3-\kappa}{\kappa+1}\right)\lambda_{\text{in}} + \frac{2}{(\kappa+1)}\{(\kappa^2-1)A_c^2 + 2(1-\kappa)\lambda_{\text{in}}^2\}^{1/2} \qquad (46)$$

and then, for subsonic flow,

$$p_p/p_c = \left[\frac{1}{2C}\{\psi(\psi^2+4C) - \psi^2\}^{1/2}\right]^{\kappa/(\kappa-1)} \qquad (47)$$

where

$$C = \frac{\kappa-1}{2}U^2\Big/\left(1-\frac{\kappa-1}{2}U^2\right)^2 \qquad (48)$$

and, for sonic flow,

$$p_p/p_c = \psi\left(\frac{2}{\kappa+1}\right)^{(\kappa+1)/2(\kappa-1)}\left(1-\frac{\kappa-1}{2}U^2\right)\frac{1}{U} \qquad (49)$$

An iterative procedure must be used to perform these calculations because the actual values of λ_{in} and λ_{out} used are corrected for the entropy change across the valve.

If there is flow from a pipe to the cylinder (normal flow in the case of an inlet valve, and reverse flow in the exhaust valve) then a modified partially open end boundary is used (this was discussed in Section 4.3). It is adapted to the valve in the manner shown in Fig. 15(b), where the symbols have the same meaning as above. In this case φ is used to define the valve opening, and this has the same meaning as ψ in the previous valve equations: it simply relates to the inflow boundary. The method of solution is the same as in Section 4.3.

The derivation of these boundary equations has assumed that the valves act as quasi-steady devices, i.e. the rate of change of the properties with distance is very much greater than that with time and there are no accumulation effects in the control volume defining the valve. This assumption has been investigated by Blair & Blair,[52] who assessed the quasi-steady flow assumption (for intake valves) on an unsteady flow rig. Their initial tests of the valves were to see the dependence of the coefficient of discharge, C_d, on pressure ratio and Reynolds number: they found, as did Annand and others, that C_d was a function of pressure ratio for outflow from the cylinder through either valve (i.e. normal flow for the exhaust valve, and reverse flow for the inlet valve),. but it seemed to be unaffected by pressure ratio for inflow through the inlet valve. The actual pressure level did not affect the values of C_d measured. Having evaluated the quasi-steady data they proceeded to test this under unsteady flow conditions on a rig. They found that up to the speeds normally encountered on engines, the quasi-steady assumption is quite adequate: it is weakest when the flow goes from the cylinder into the pipework.

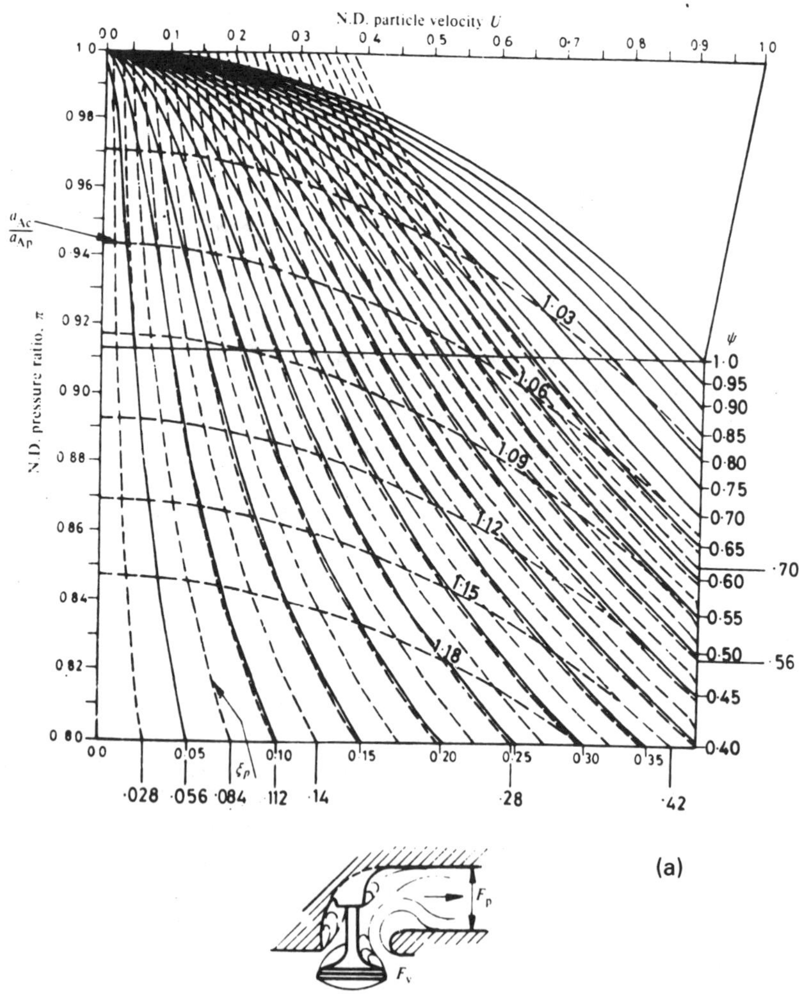

Fig. 15. Valve flow boundary: (a) outflow from cylinder.

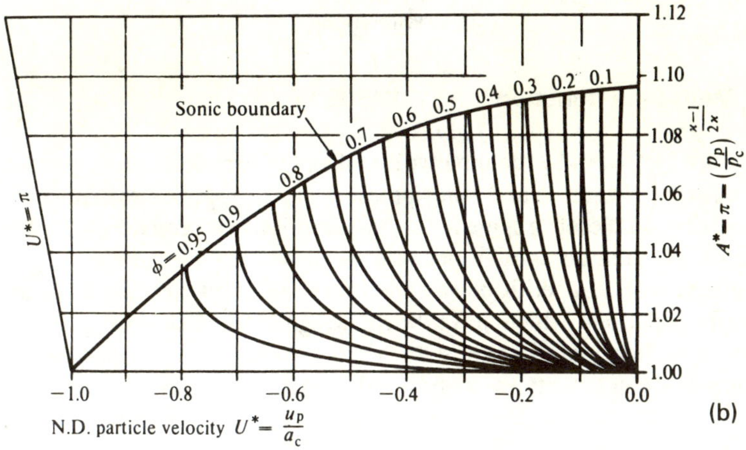

Fig. 15—*contd.* (b) inflow to cylinder.

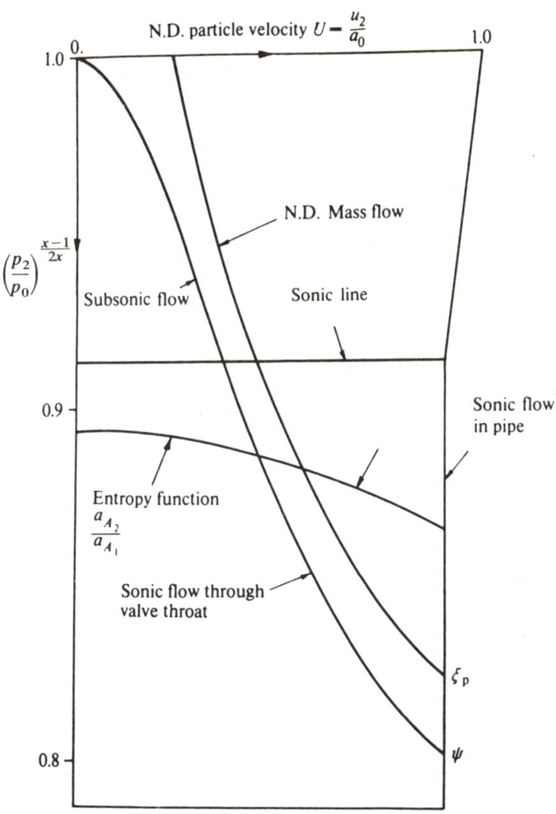

Fig. 16. Flow regimes encountered in valve boundary for outflow from a cylinder.

4.7. Multi-Pipe Junction

There are two ways for dealing with multi-pipe junctions: the constant- (or equal) pressure, and the pressure loss approaches. In most turbocharged diesel engines the assumption of constant-pressure junctions is often acceptable because the flow velocities are quite low: the situation is changed if pulse converters are fitted because these are basically junctions which have been designed to give high pressure losses in preferential directions. In petrol (or gasoline) engines the gas flow rates are substantially higher and then it seems necessary to take account of these flow losses, where they can have an effect on the volumetric efficiency of the engine.

4.7.1. Constant- (or Equal-) Pressure Junctions

The characteristics of such junctions are fully defined by the geometric areas and hence there is no necessity to perform flow measurements on the junctions. This is a great advantage because flow tests are both time-consuming and expensive. The equations for such boundaries are derived in Benson[11] as a generalized version of the two-pipe junction equations given in eqns (40) and (41) and will not be repeated here.

A great advantage of the constant-pressure junction theory is that there is no limit to the number of pipes that may be joined at a point. Hence it is a very powerful way of obtaining a first design for a multi-pipe junction. The approach has been used for a five-way junction without any difficulty, and should be capable of treating even more pipes at a junction.

4.7.2. Pressure Loss Junctions

In all flows in junctions there is a pressure loss; sometimes that loss is very small and may be neglected, while at other times it can affect the results calculated. Deciding whether there will be a large effect is difficult in marginal cases, but as a guide pressure losses should be included in high-speed engines where the gas velocities will be high, and in engines fitted with pulse converters. The designer is hampered by a lack of generalized data and it is often necessary to flow test individual junctions (sometimes cutting the manifold into sections to allow the problem to be discretized).

There is a dearth of information on pressure loss values, particularly with compressible fluids. Some data has been obtained by Benson, Woollatt & Woods[53] and this is summarized in Benson.[11] The data

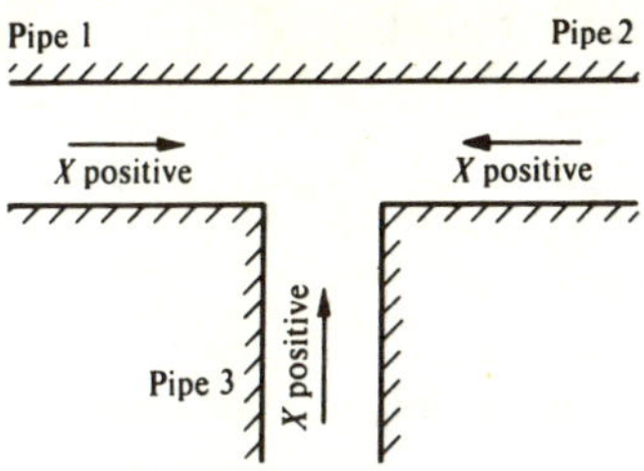

Diagram to illustrate sign convention and pipe numbering for junction

Basic flow type	Flow type	u_1	u_2	u_3
A	I	+	−	−
	II	−	+	−
B	III	−	−	+
C	IV	−	+	+
	V	+	−	+
D	VI	+	+	−

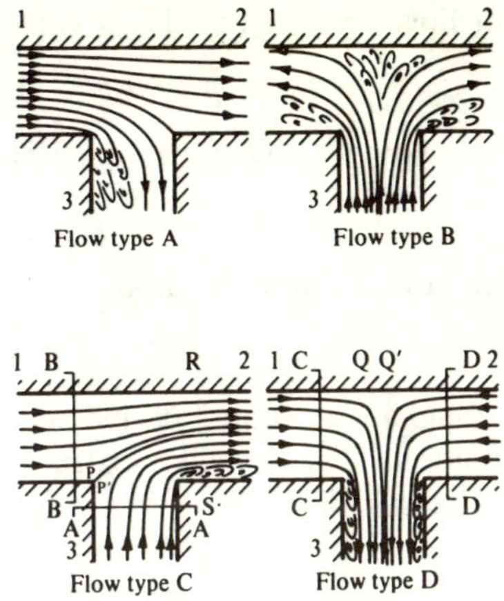

Fig. 17. Flow patterns at equal area T-junction.

reported is limited to equal-area 'T' junctions and the authors attempt to represent the results by a 'quasi-momentum' model. The equal area 'T' junction is a special case of the generalized junction because it is symmetrical: very few manifolds are as simple as this. An advantage of this symmetry is that the number of different flow fields is reduced from six to four, and hence the number of loss coefficients is eight. The most comprehensive data sheets for loss coefficients are those by Miller,[54] and some other values are given by Bingham & Blair.[29]

The possible flow directions for a 'T' junction are shown in Fig. 17 and the quasi-momentum equations are given below. For flow type A,

$$p_1 - p_2 = C_1(\rho_2 u_2^2 - \rho_1 u_1^2) \tag{50}$$

$$p_1 - p_3 = C_2 \rho_3 u_3^2 \tag{51}$$

For basic flow type B (flow type III)

$$p_1 - p_2 = C_3(\rho_2 u_2^2 - \rho_1 u_1^2) \tag{52}$$

This type is symmetrical. It is found experimentally that $p_3 - p_1$ is a function of $\rho_1 u_1^2$ only; thus it follows, by symmetry, that

$$p_3 - p_1 = C_3 \rho_1 u_1^2, \qquad p_3 - p_2 = C_3 \rho_2 u_2^2 \tag{53,54}$$

Basic flow type C (flow type IV, V) gives

$$p_1 - p_2 = C_4(\rho_2 u_2^2 - \rho_1 u_1^2), \qquad p_1 - p_3 = C_5(\rho_2 u_2^2 - \rho_1 u_1^2) \tag{55,56}$$

For flow type V (Fig. 17) it is found experimentally that

$$p_1 = p_3, \qquad \text{i.e.} \qquad p_1 - p_3 = 0.$$

Basic flow type D (flow type VI) gives

$$p_1 - p_3 = C_6 \rho_3 u_3^2 \tag{57}$$

The flow is symmetrical and is a function of $\rho_3 u_3^2$ only. Similarly,

$$p_2 - p_3 = C_6 \rho_3 u_3^2 \tag{58}$$

Bingham & Blair[29] derive slightly different equations for these flow situations. These include extra coefficients in the momentum equations as shown below:

$$p_1 - p_2 = C_1 \rho_2 u_2^2 - C_2 \rho_1 u_1^2 \tag{59}$$

$$p_1 - p_3 = C_3 \rho_3 u_3^2 - C_4 \rho_1 u_1^2, \qquad \text{etc.} \tag{60}$$

They also showed, from experimental values, that the value of the loss coefficient was related to the branch angle, θ_d, by the linear relationship

$$C = 1{\cdot}6 - \frac{1{\cdot}6}{167}\,\theta_d \tag{61}$$

This approach of using pressure loss junctions has been quite widely applied to engine simulations by many workers, including Azuma *et al.*,[28,44] Benson[11] and Bingham & Blair.[29]

4.7.3. Pulse Converter Junctions

Pulse converters are specially designed pressure loss junctions in which an attempt is made to allow preferential flow in certain directions. Pulse converters are used on medium- and low-speed turbocharged diesel engines, and have the effect of smoothing the pressure pulsations upstream of the turbine: they also isolate the cylinders from interference through the exhaust manifold. Medium- and low-speed four-stroke diesel engines are designed with large valve overlaps (i.e. the time between inlet valve opening and exhaust valve closing) to improve scavenging of the cylinder and to decrease the exhaust temperature to improve turbine durability. Two-stroke engines require a large valve overlap to enable the scavenge process to be completed in the time available. This means that these engines are very susceptible to interference from other cylinders, which will disrupt the gas flow through them resulting in poor scavenge and backflow into the inlet, which can cause fouling.

The ideal pulse converter is a 'gas dynamic rectifier' which permits flow in one direction with a low pressure loss but impedes flow in the opposite direction. This effect can be achieved in various ways: Azuma *et al.*[28,44] simply fitted an orifice plate into the exhaust pipe of a two-stroke marine engine to stop interference, and this approach has been adopted independently by Wichmann.[55] Undoubtedly this can have a beneficial effect on the performance of the individual cylinders, even though it raises the exhaust back-pressure during the blowdown period. Birmann[56] and Zehnder[57] proposed much more elegant devices which are based on ejectors, in which gas flow in one manifold will provide suction to draw the gas out of the other manifold. The arrangement suggested by Zehnder is a symmetrical design which is used to connect together two pulse exhaust manifolds, thus isolating each from the effects of the other. Benson & Alexander[58] also considered this type of pulse converter in a form suitable for an automotive diesel engine. The Birmann pulse converter is an asymmetrical design and can be fitted at the exit from each cylinder. This approach has been adopted by SEMT Pielstick for their engine range[59] and has also been used by GEC–Ruston Diesels.[60] An analysis of such manifolds has been performed by Chan, Winterbone, Alexander & Nichols[61] and models have been developed for these.

Pulse converters as such are not appropriate to naturally aspirated engines because there is no necessity to retain the energy of the exhaust gas. However, many exhaust pipe designs attempt to include the ejector characteristics associated with the pulse converter, as can be seen in the 'banana bunch' manifolds fitted to higher performance cars. In turbocharged petrol engines there is not the need to design manifolds for high transmission efficiency at high speed because there is surplus turbine power which must be removed by the waste-gate: however, improved design might result in a higher torque at low speed with a concomitant improvement in driveability.

Mathematical models for symmetrical pulse converters are included in Benson.[11] These relate the pressure loss coefficient to the common branch Mach number and the ratio of mass flows between the various pipes. It has since been found that the more satisfactory parameter is the square of the Mach number, and this can be supported by considering the generalized equations of compressible flow, as propounded by Shapiro.[26] A similar equation can be developed from the incompressible approach of Hager,[62] and this is summarized in Ref. 61. The use of Mach number, with its implication of compressible

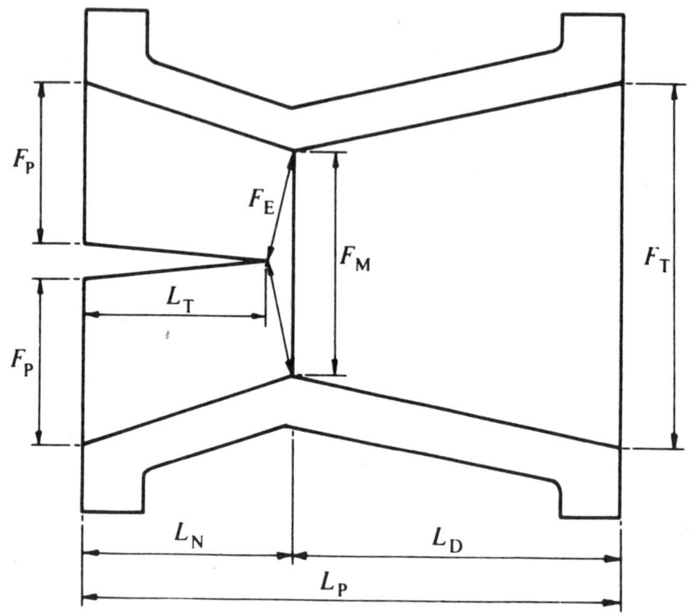

Fig. 18. Symmetrical pulse converter for automotive use.

flow, may seem to be unnecessary when considering engine manifolds: however, in steady flow tests it is often found that there is choking of the flow at the *throat of the junction* even when the Mach number in the pipe is quite low.

Benson[11] and Horlock & Winterbone[1] describe pulse converter geometries in some detail. It is convenient to define pulse converters in terms of area ratios, and typical parameters (based on Fig. 18) are:

$$\text{ejector nozzle area ratio,} \qquad \psi_N = F_E/F_P \tag{62}$$

$$\text{diffuser area ratio,} \qquad \psi_D = F_M/F_T \tag{63}$$

These define the throttling and pressure recovery action of the device respectively. In practice, the ejector nozzle area ratio is usually the value quoted because this has the major effect on the rectifier action of the pulse converter. Typical values of ψ_N vary between 0·5 and 1·0; the latter value defines a junction with no basic throttling which relies on the directionality of the pipes for its rectifying action.

In Fig. 19, from Benson & Alexander[58], is shown the variation of

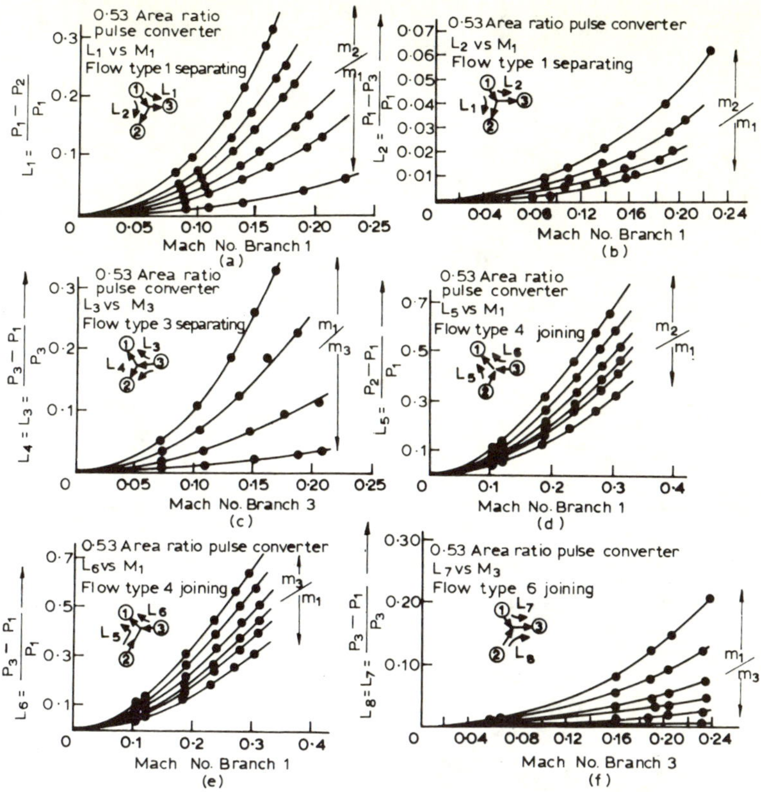

Fig. 19. Loss coefficients for symmetrical pulse converter junction (Sulzer type).

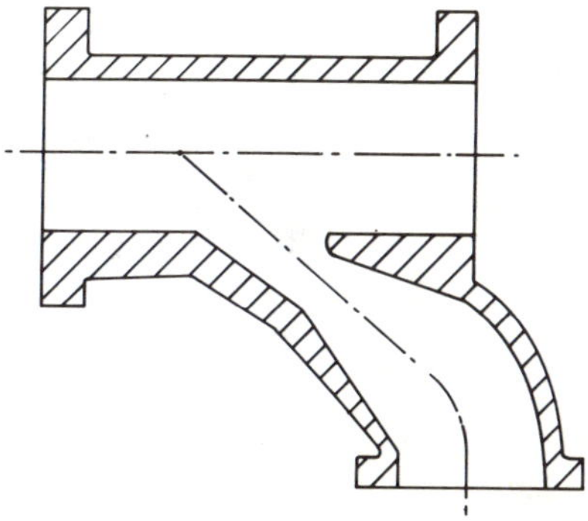

Fig. 20. Asymmetrical pulse converter junction.

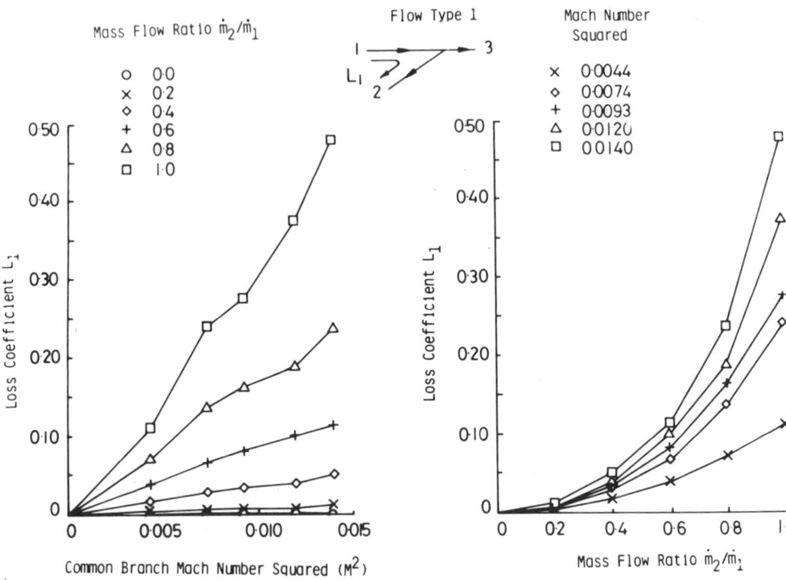

Fig. 21. Loss coefficients for asymmetrical pulse converter junction.

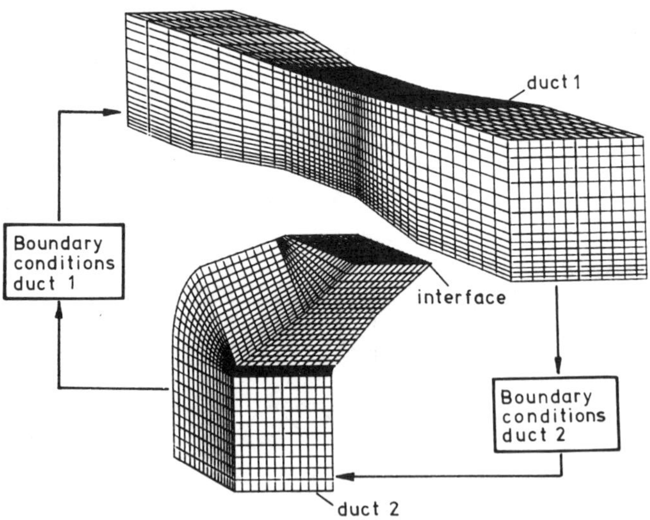

Fig. 22. Grid for calculating flow at a three-dimensional pulse converter junction.

loss coefficient for a junction ($\psi_N = 0{\cdot}53$) of the form shown in Fig. 18. It can be seen that the loss coefficient is substantially larger in the direction of flow around the bend than when it is travelling through the pipework towards the turbine. Such a loss coefficient will tend to rectify the flows towards the turbine. A typical asymmetric pulse converter is shown in Fig. 20 and the loss coefficients ($\psi_N = 0{\cdot}4$) for this are given in Fig. 21. Some multi-dimensional calculations have been performed on these pulse converter junctions. Dimitriades *et al.*[3,63] report such a calculation based on the grid shown in Fig. 22; typical of the flow results predicted are those in Fig. 23, and a

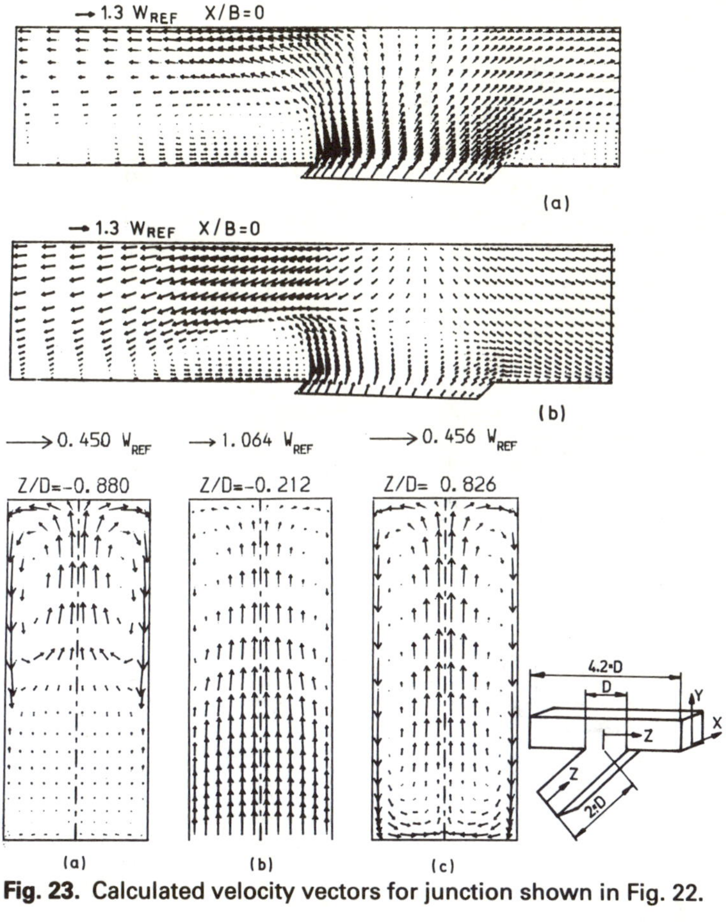

Fig. 23. Calculated velocity vectors for junction shown in Fig. 22.

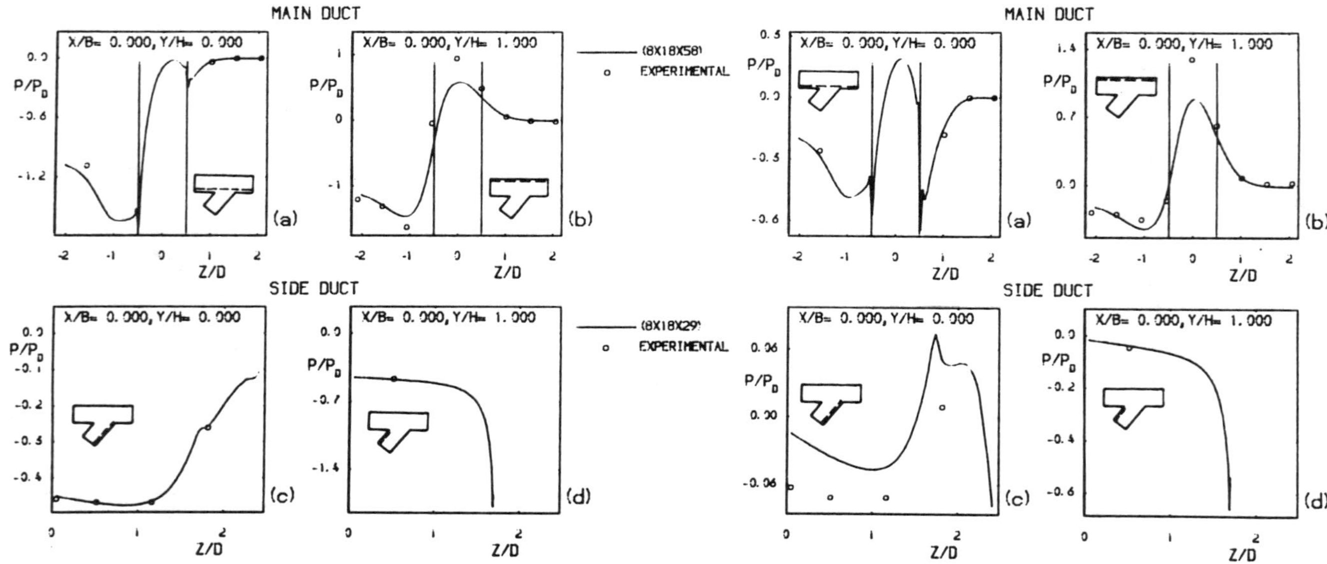

Fig. 24. Comparison of calculated and measured pressure distributions in three-dimensional junction.

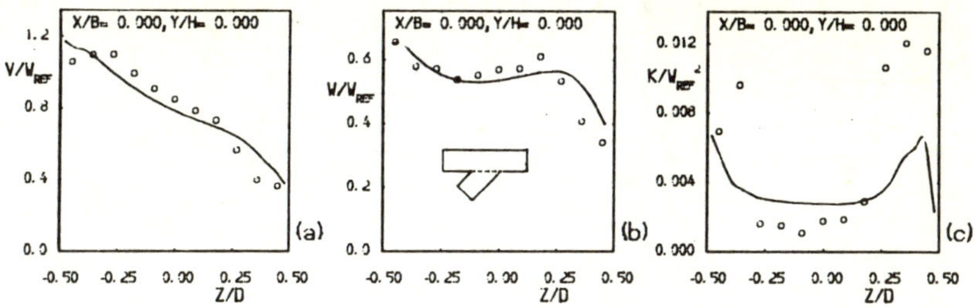

Fig. 24. —*Contd.*

comparison of the measured and calculated pressure losses is given in Fig. 24.

4.8. Carburettor and Fuel Injection Equipment

The purpose of a carburettor or fuel injection equipment in a petrol engine is to provide the cylinders with the correct air:fuel ratio to achieve satisfactory combustion: that air:fuel ratio will be defined by the particular requirements of the operating condition and might be weak or rich. Since the petrol engine is quantitatively controlled (i.e. the quantity of charge entering the cylinder varies with operating point, whereas load control in the diesel engine is achieved by varying the quality of the charge) it is necessary to have some method of altering the flow of charge. This role is performed by the throttle, which enables the power output of the engine to be reduced below the maximum simply by restricting the air flow. In a carburettor the fuel flow is metered by an orifice to be approximately proportional to the air flow. A fuel injection system performs the same purpose but the air and fuel are related by an electronically controlled metering system. The throttle valve in both systems acts as a classical thermodynamic device in which the flow is approximately adiabatic (particularly if the fuel addition is neglected), and is hence at constant stagnation enthalpy but with a pressure drop. In both cases the fuel will be neglected in analysing the flows, simply because the fuel does not behave 'one-dimensionally' and becomes centrifuged out of the air stream in certain areas of the manifold, while being re-evaporated later before induction into the cylinder (see Chapter 6). Fortunately, the energy involved in the evaporation of the fuel is not large enough

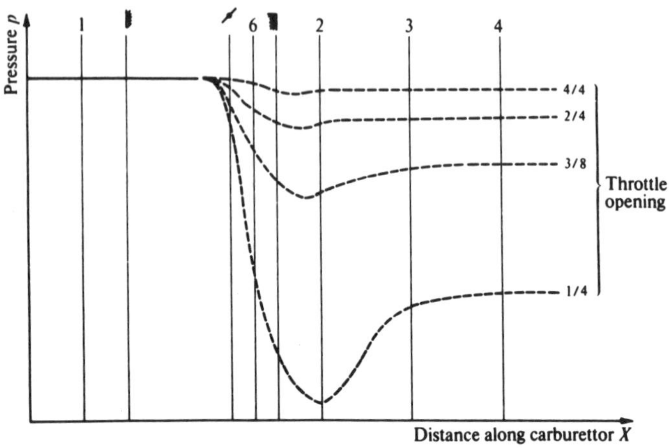

Fig. 25. Pressure variation along a carburettor and duct at various throttle settings.

to have a major effect on the gas dynamics of the air, and it can be neglected at this stage (see Low *et al.*[10]).

Again, Benson[11] describes a model based on work by Benson, Baruah & Sierens[64] for a simple carburettor. This is based on the assumptions outlined above and enables the Riemann variables on either side of the carburettor to be related to the throttle opening by means of an empirically evaluated resistance coefficient, K, defined as

$$K = \frac{f_c K}{2} \quad \text{and} \quad f_c = \frac{\Delta p}{\frac{1}{2}\rho u^2} \tag{64,65}$$

In Fig. 25 is shown the variation of pressure through the carburettor for four different throttle angles, and in Fig. 26 is shown the variation of resistance coefficient with Mach number and throttle opening evaluated from the data in Fig. 25. The resistance coefficients have been used in the calculation of petrol engine performance, and typical predictions are shown in Fig. 27. Recently, Blair & Blair[52] have re-examined this boundary and proposed some improvements. First, they recommend that the resistance coefficient be evaluated using the pressure drop measured after the flow has fully developed rather than the maximum value proposed by Benson: this probably removes the need for the dynamic factor used in Ref. 64. They also made some minor modifications to the entropy calculation which apparently improve the stability.

4.9. Compressor

Compressor boundaries are required to be able to undertake full simulations of turbocharged engines. The initial design for a turbocharged engine will often be undertaken by assuming the level of boost and the appropriate inlet temperature, either with or without an intercooler. To obtain the final match of the turbocharger it is necessary to assess the performance with the compressor characteristics included in the model. Typical compressor characteristics are shown in Chapter 14: these are used for matching purposes in the form of empirical maps. Benson[11] explains how the basic pressure ratio and efficiency versus mass flow parameter characteristics are converted into curves relating λ_{out} to λ_{in} of the form shown in Fig. 28. These are based on a constant inlet pressure to the compressor: boundaries with variable inlet pressure are available, although they are not used extensively. It is assumed that the steady state characteristics of the compressor may be used under unsteady flow conditions without modification, i.e. a quasi-steady approach is adopted. It is possible to make allowance for the size of the compressor volute by adding an equivalent length to the pipework (see Benson[11]).

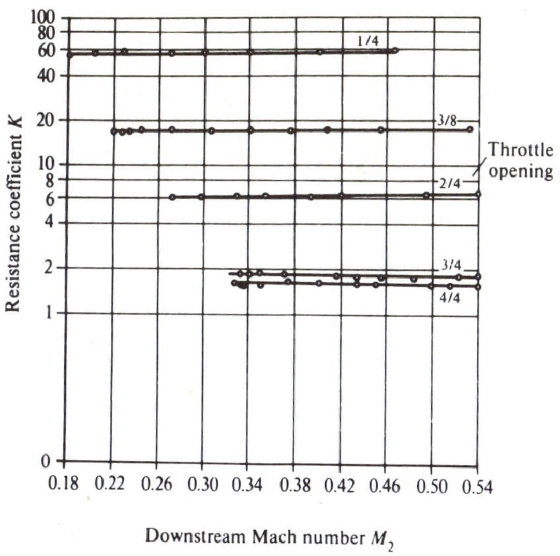

Fig. 26. Variation of resistance coefficient with downstream Mach number for a carburettor.

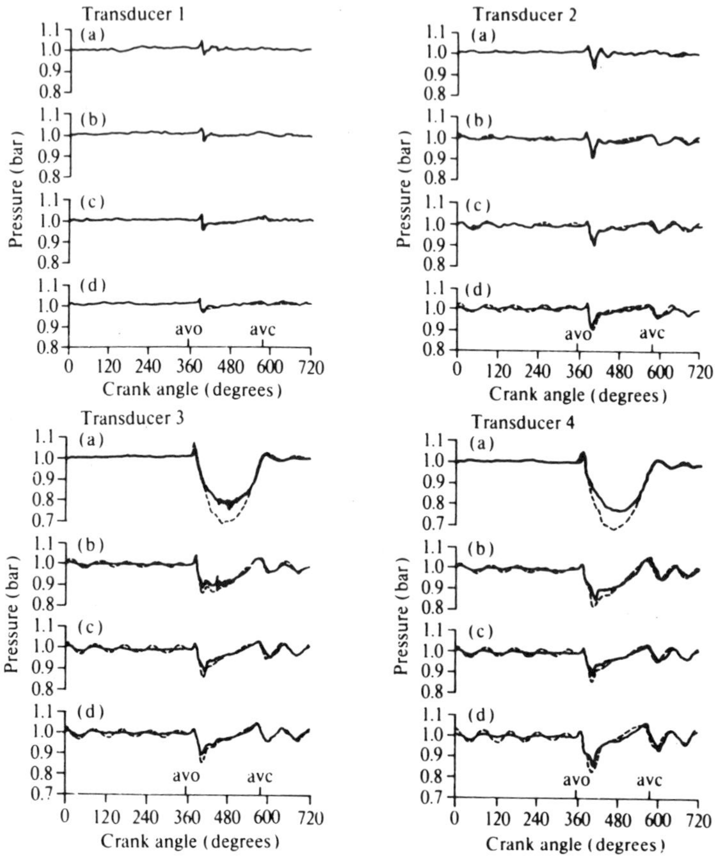

Fig. 27. Comparison of theoretical and experimental pressure variation for non-steady flow through a carburettor (engine speed = 1500 rpm): (a) 1/4 throttle; (b) 1/2 throttle; (c) 3/4 throttle; (d) full throttle. Transducers: 1, pipe inlet; 2, upstream of carburettor; 3, downstream of carburettor; 4, 25 cm downstream of carburettor. ———, Measured; - - - - - -, calculated.

4.10. Turbine

Turbine boundaries are also required for the simulation of turbocharged engines. As stated earlier, many initial design studies are performed with the turbine modelled as a nozzle. This approach is a rapid way of assessing the effect of changes of manifold design on the engine performance, although as the exhaust manifold pressure increases the nozzle boundary deviates further from the turbine characteristics. In Fig. 29 are shown the boundary curves that are obtained if the turbine is modelled as a simple nozzle, or two orifices in series: the latter is a better simulation of a typical turbine with 50% reaction since the enthalpy drop takes place in two stages. It can be seen that the simple nozzle ($\Lambda = 0$) chokes at a pressure ratio of 1·89, whereas the two-orifice model ($\Lambda = 0{\cdot}5$) does not choke until almost 3:1. The predicted curves are compared with a measured characteristic and this shows that the nozzle model overestimates the flow at low

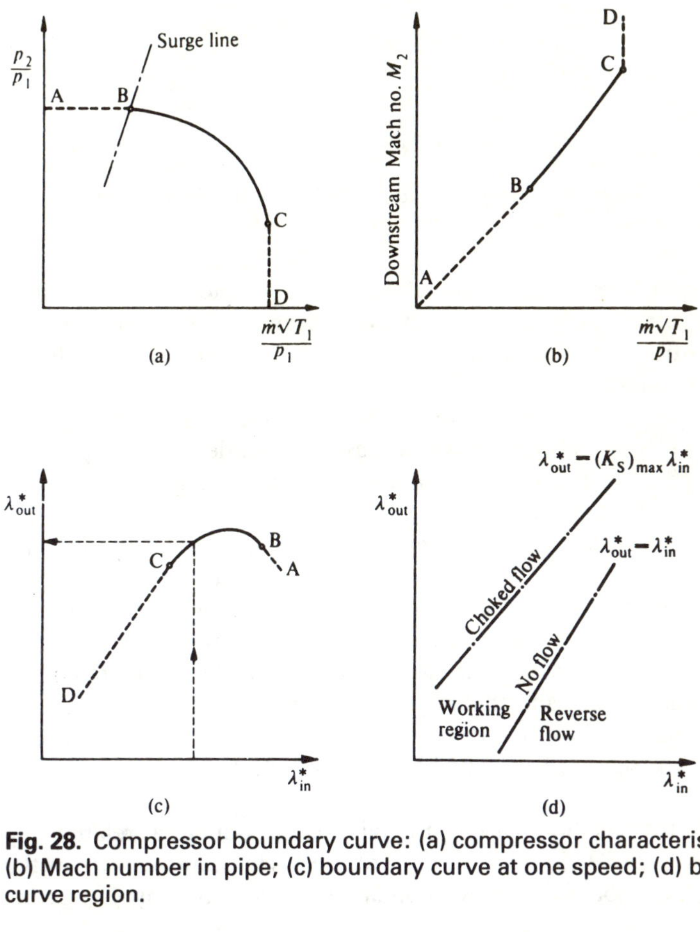

Fig. 28. Compressor boundary curve: (a) compressor characteristic; (b) Mach number in pipe; (c) boundary curve at one speed; (d) boundary curve region.

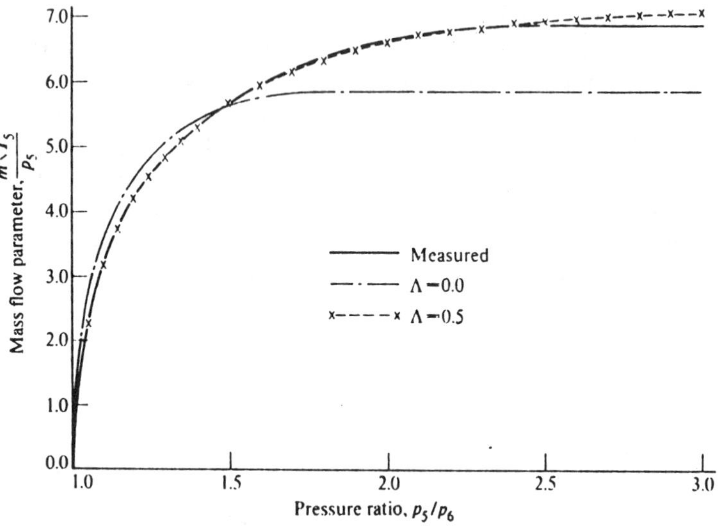

Fig. 29. Comparison between turbine modelled using simple nozzle ($\Lambda = 0$) and one modelled using two nozzles in series ($\Lambda = 0{\cdot}5$).

pressure ratios while underestimating it at high ones; the two-orifice model is substantially better. However, examination of the turbine characteristics given in Chapter 14 shows that rotational speed has a substantial effect on the mass flow characteristics of radial inflow turbochargers, due to the high centrifugal pressure gradient opposing the flow. Hence any approximation must be regarded with scepticism, since the effect of the actual turbine characteristics might substantially alter the conclusions reached.

The models used for representing turbines are quasi-steady ones based on the empirically measured characteristics. It is usually assumed that the pressure waves do not travel through the turbine; in other words, the back pressure downstream of the turbine is constant. The quasi-steady assumption also means that the volume of the turbine is neglected, and this might not be appropriate in the case of the radial turbine in which the casing has appreciable volume. This is referred to in Section 5.2.1 of Chapter 13 and Section 6.4.1 of Chapter 14.

As in previous boundaries, the aim is to obtain a relationship between λ_{out} and λ_{in}, taking account of the basic characteristics of the boundary. Benson[11] develops boundaries for turbines by converting the mass flow versus pressure ratio map from this field into one relating the two Riemann variables. The reader is referred to Ref. 11 for the detail on how the characteristics are processed.

REFERENCES

1. Horlock, J. H. & Winterbone, D. E., *The Thermodynamics and Gas Dynamics of Internal Combustion Engines,* Vol. II. Oxford University Press, 1986.
2. Annand, W. J. D. & Roe, G., *Gas Flow in the Internal Combustion Engine.* G. T. Foulis, Yeovil, 1974.
3. Dimitriades, K. P., Leschziner, M. A., Winterbone, D. E., Alexander, G. I. & Sierens, R., Computation of three-dimensional flow in manifold-type junctions. In *International Symposium on Flows in Internal Combustion Engines—III.* ASME, Miami Beach, 1985, pp. 57–62.
4. Seppen, J. J., Intake and exhaust processes in combustion engines, development of SIFLEX. In *Numerical Simulation of Fluid Flow and Heat/mass Transfer Processes,* ed. N. C. Markatos, D. G. Tatchell, M. Cross, & N. Rhodes, Springer-Verlag, 1986, pp. 73–84.
5. Rosten, H. L. & Spalding, D. B., PHOENICS-84 and beyond, ibid., pp. 3–29.
6. Gosman, A. D. & Ahmed, A. M. Y., Measurement and multi-dimensional prediction of flow in an axisymmetric port/valve assembly. SAE paper 870592, 1987.
7. Chapman, M., Two dimensional numerical simulation of inlet manifold flow in a four cylinder internal combustion engine. SAE paper 790244, Detroit, 1979.
8. Tosa, Y., Shimoda, K. & Oikawa, H., Calculation of two-dimensional unsteady flows in inlet pipe system and its application for V-8 resonant intake systems. In *International Symposium on Flows in Internal Combustion Engines—III.* ASME, Miami Beach, 1985, pp. 63–70.
9. Lighthill, M. J., *Waves in Fluids.* Cambridge University Press, 1978.
10. Low, S. C., Baruah, P. C. & Winterbone, D. E., One-dimensional air and droplets flow model which includes evaporation, heat transfer, and drag. Instn Mech. Engrs Conf., C65/81, Oxford, July 1981, pp. 37–46.
11. Benson, R. S., *The Thermodynamics and Gas Dynamics of Internal Combustion Engines,* Vol. I, ed. J. H. Horlock & D. E. Winterbone. Oxford University Press, 1982.
12. Cser, G., Some results of combined charging applications. Proc. Instn Mech. Engrs, Conf. on Turbocharging and Turbochargers, C64/78, London, 1978, pp. 141–6,
13. Soedel, W. & Dhar, M., *Gas Pulsations in Compressor Manifolds.* Short course text, Ray W. Herrick Laboratories, School of Mechanical Engineering, Purdue University, West Lafayette, Indiana.

14. Metzger, G. & Vabre, J. P., *Transmission Lines with Pulse Excitation.* Academic Press, 1969.
15. Riemann, B., Uber die fortpflanzung ebener luftwellen von endlicher schwingungsweife. *Bott. Abh.,* **8** (Math) (1885).
16. Jenny, E., Unidimensional transient flow with consideration of friction, heat transfer and change of section. *Brown Boveri Review,* **37** (11) (1950) 447.
17. de Haller, R., The application of a graphic method to some dynamic problems in gases. *Sulzer Tech. Rev.,* **1** (1945) 6.
18. Engleman, H. W., Design of a tuned intake manifold. ASME paper 73-WA/DGP-2, 1973.
19. Vorum, P. C., Short pipe manifold for four stroke engines. ASME paper 76 WA/DGP-4, 1976.
20. Vorum, P. C., Short pipe manifold design for four-stroke engines—2. *J. Engng Power, Trans. ASME,* **102** (4) (October 1980) 836–41.
21. Chapman, M., Novak, J. M., & Stein, R. A., A non-linear acoustic model of inlet and exhaust flow in multi-cylinder internal combustion engines. ASME paper 83-WA/DSG-14, 1983.
22. Ohata, A. & Ishida, Y., Dynamic inlet pressure and volumetric efficiency of four-cycle four cylinder engine. SAE paper 820407, Detroit, 1982.
23. Matsumoto, I. & Ohata, A., Variable induction systems to improve volumetric efficiency at low and/or medium engine speeds. SAE paper 860100, Detroit, February 1986.
24. Jones, A. D. & Brown, G. L., Determination of two-stroke engine exhaust noise by the method of characteristics. Acoustical Paper Awards Program, Nelson Industries Inc., Wisconsin, June 1981.
25. Hartree, D. R., Some practical methods of using characteristics in the calculation of non-steady compressible flow. Harvard University Report LA-HU-1, 1953. Also in Lister, M., The numerical solution of hyperbolic partial differential equations by the method of characteristics. In *Mathematical Methods for Digital Computers,* ed. Ralston, A. & Wilf, H. S. Wiley, Chichester, 1960, Chapter 15.
26. Shapiro, A. H., *The Dynamics and Thermodynamics of Compressible Fluid Flow,* Vol. II. The Ronald Press, New York, 1954.
27. Benson, R. S., Garg, R. D. & Woollatt, D., A numerical solution of unsteady flow problems. *Int. J. Mech. Sci.,* **6** (1964) 117–44.
28. Azuma, T., Tokunaga, Y. & Yura, T. Y., Characteristics of exhaust gas pulsation of constant pressure turbocharged diesel engines. *ASME J. Engng Power,* **102** (October 1980) 827–35.
29. Bingham, J. F. & Blair, G. P., An improved branched pipe model for multi-cylinder automotive engine calculations. *Proc. Instn Mech. Engrs,* **199,** No. D1 (1985) 65–77.
30. Gajendra Babu, M. K. & Murthy, B., Simulation and evaluation of a four-stroke single cylinder spark-ignition engine. SAE paper 750687, *Trans. SAE,* **84** (1975) 1631.
31. Sierens, R., van Hove, W. & Snauwert, P., Comparison of measured and calculated gas velocities in the inlet channel of a single cylinder reciprocating engine. *Proc Instn Mech. Engrs,* **198A,** No. 1 (1984).

32. Watson, N. & Janota, M. R., Non-steady flow in an exhaust system with a pulse converter junction. Instn Mech. Engrs. Conf. Internal Flows, Paper 27, Salford, 1971, pp. D17–D28.
33. Shimamoto, Y., Oka, M. & Tanaka, Y., A research on inertia charging effect of intake system in multi-cylinder engines. *Bull. Jpn. Soc. Mech. Engrs* **21,** No. 153 (March 1978) 502–10.
34. Desantes, J. M., Boada, F. & Corberan, J. M., Exhaust pipe design method for the optimization of the scavenging process. SAE paper 850083, SAE Congress, Detroit, February/March 1985.
35. Courant, R., Fredrichs, K. & Lewy, H., Uber die partiellen differenzengleichungen de mathematischen Physik. *Math. Ann.,* **100** (1928) 32–74.
36. Morton, K. W., Stability and convergence in fluid flow problems. *Proc. R. Soc. Lond. A,* **323** (1971) 237–53.
37. Sod, G. A., A survey of several finite difference methods for systems of non-linear hyperbolic conservation laws. *J. Comput. Phys.,* **27** (1978) 1–31.
38. Wallace, F. J. & Boxer, G. C., Wave action in diffusers for exhaust pipe systems with special reference to the scavenging of two-stroke engines. *Proc. Instn Mech. Engrs,* **170** (1956) 1131–56.
39. Lax, P. D. & Wendroff, B., Systems of conservation laws. *Communs Pure Appl. Math.,* **15** (1960) 217–37.
40. Poloni, M., Winterbone, D. E. & Nichols, J. R., Comparison of unsteady flow calculations in a pipe by the method of characteristics and the two-step differential Lax-Wendroff method. *Int. J. Mech. Sci.,* **29** (5) (1987) 367–78.
41. Poloni, M., Winterbone, D. E. & Nichols, J. R., Calculation of pressure and temperature discontinuity in a pipe by the method of characteristics and the two-step differential Lax–Wendroff method. ASME International Symposium on Flows in Internal Combustion Engines, FED, Vol. 62. Winter Annual Meeting, Boston, 13–18 Dec., 1987.
42. Bulaty, T. D. & Niessner, H., Calculation of 1-D unsteady flows in pipe systems of I.C. engines. *Trans. ASME, J. Fluids Engng,* **107** (1985) 407–12.
43. Neissner, H. & Bulaty, T. D., Improving conservation of mass, momentum and energy by appropriate flux correction at boundaries. *Z. Angew. Math. Mech.,* **66** (1986) 250–51.
44. Azuma, T., Yura, T. & Tokunaga, Y., Some aspects of constant pressure turbocharged marine diesel engines of medium and low speed. *Trans. ASME, J. Engng Power,* **105**(3) (1983).
45. Lakshiminarayanan, P. A., Janakiraman, P. A., Gajendra-Babu, M. K. & Murthy, B. S., Prediction of gas exchange processes in a single cylinder internal combustion engine. SAE paper 790359, Detroit, 26 February–2 March 1979.
46. Iwamoto, J. & Deckker, B. E. L., Application of random-choice method to the calculation of unsteady flow in pipes. SAE paper 851561, 1985. International Off-Highway & Powerplant Congress and Exposition, MECCA, Milwaukee, Wisconsin, 9–12 September, 1985.

47. Walter, G. A. & Chapman, M., Numerical simulation of the exhaust flow from a single cylinder of a two cycle engine. SAE paper 790243, Detroit, 1979.
48. Meisner, S. & Sorenson, S. C., Computer simulation of intake and exhaust manifold flow and heat transfer. SAE paper 860242, International Congress, Detroit, February 1986.
49. Winterbone, D. E. & Nichols, J. R., A graphic description of the interaction of waves with various boundaries. *Int. J. Mech. Sci.*, **27** (7/8) (1985) 543–50.
50. Vafidis, C. & Whitelaw, J. H., Intake valve and in-cylinder flow development in a reciprocating model engine. *Proc. Instn Mech. Engrs,* **200** (pt C), paper no. 41/86, 1986.
51. Tindal, M. J., Cheung, R. S. & Yianneskis, M., Velocity characteristics of steady flows through engine inlet ports and cylinders. SAE paper 880383, Detroit, 1988.
52. Blair, A. J. & Blair, G. P., Gas flow modelling of valves and manifolds in car engines. Paper C11/87, Instn Mech. Engrs Conf., Cambridge (UK), March 1986, pp. 131–44.
53. Benson, R. S., Woollatt, D. & Woods, W. A., Unsteady flow in simple branch systems. Instn Mech. Engrs, Thermo and Fluid Mech Convention, Cambridge, Paper 17, 1964, pp. 1–28.
54. Miller, D. S., Internal flow systems, BHRA Fluid Engineering, 1978.
55. Wichmann, Motorfabrikk A/S, Norway. Private communication.
56. Birmann, R., Exhaust energy converting means for internal combustion engines. US Patent 2406656, 1946.
57. Zehnder, G., Pulse converters on two-stroke diesel engines. *Brown Boveri Review,* **55** (1968) 414–9.
58. Benson, R. S. & Alexander, G. I., The application of pulse converters to automotive four-stroke cycle engines. SAE paper 770034, Detroit, 1977.
59. Curtil, R. & Magnet, J., Exhaust pipe system for high pressure charging. Instn Mech. Engrs Conf. Turbocharging and Turbochargers, C65/78, 1978, pp. 147–58.
60. Alexander, G. I., Nichols, J. R., Winterbone, D. E. & Sinha, S. K., Optimization of the design of an exhaust manifold system for improved turbocharger performance. Instn Mech. Engrs Conf. on Turbocharging and Turbochargers, C48/82, 1982, 175–83.
61. Chan, C. L., Winterbone, D. E., Nichols, J. E. & Alexander, G. I., A detailed study of compact exhaust manifolds applied to automotive diesel engines. C113/86, Instn Mech. Engrs, Conference on Turbocharging and Turbochargers, 1986, pp. 269–81.
62. Hager, W. H., An approximate treatment of flow in branches and bends. *Proc. Instn Mech. Engrs,* **198C** (4) (1984) 63–9.
63. Leschziner, M. A. & Dimitriades, K. P., Numerical simulation of three-dimensional turbulent flow in exhaust manifold junctions. Instn Mech. Engrs Conf. on Computers in Engine Technology, C09/87, Cambridge, 1987, pp. 183–90.
64. Benson, R. S., Baruah, P. C. & Sierens, R., Steady and non-steady flow in a simple carburettor. *Proc. Instn Mech. Engrs,* **18** (1974) 537–48.

13

The Application of Wave Action Techniques to Reciprocating Engines

D. E. Winterbone
UMIST, Manchester, UK

1. Introduction
2. Tuning of engine intake and exhaust systems: empirical approaches
   2.1. Intake manifold tuning
   2.2. Tuned inlet manifolds using resonators
   2.3. Volumetric efficiency
   2.4. Exhaust manifold tuning
3. Pipework arrangements for two-stroke engines
   3.1. Scavenging of two-stroke engines
   3.2. Exhaust system design for naturally aspirated two-stroke engines
   3.3. Supercharged engines
4. Calculation of wave action in manifold systems of naturally aspirated engines
   4.1. Inlet manifold calculations
   4.2. Exhaust manifold calculations
5. Calculation of wave action in manifold systems of supercharged or turbocharged engines
   5.1. Transmission efficiency of exhaust manifolds
   5.2. Diesel engines
6. The pressure exchanger
   6.1. Principle of operation
   6.2. Operating characteristics of pressure exchangers

1. INTRODUCTION

The concept of waves in the flow of a compressible fluid was introduced in Chapter 12. It was shown that the wave equations can be solved by a number of methods, and that techniques exist for small waves, as well as homentropic and non-homentropic flow. This chapter will describe the way in which wave action affects the performance of reciprocating engines, and how it can be used to improve that performance. The reader is referred to Chapter 12 for a detailed description of wave action methods.

The chapter commences with a section on semi-empirical methods of harnessing wave action effects to improve the breathing of an engine—commonly known as *induction tuning*. These approaches will be related to the full equations of flow. Then the fundamental structure of two-stroke motorcycle engine exhaust systems will be discussed as an addition to the effects described in Chapter 3. Following this, the application of the full wave equations to a wide range of inlet and exhaust manifolds will be described. Naturally aspirated and turbocharged engines will be discussed, together with their different exhaust system arrangements.

2. TUNING OF ENGINE INTAKE AND EXHAUST SYSTEMS: EMPIRICAL APPROACHES

2.1. Intake Manifold Tuning

Intake manifold tuning is sometimes referred to as 'induction ramming' due to the use of the dynamics of the fluid to ram the air into the cylinder. Studies of intake ramming go back many years: Broome[1] states that he has found references to this effect as far back as 1904, but a major body of the work occurred in the 1940s. It is possible to subdivide intake manifold tuning into two effects; those due to *waves* and those due to *inertia*. The first is caused by the pressure waves in the intake manifold and is similar to 'organ pipe' tuning. The second is caused by the momentum induced in the charge as the cylinder is filled; this can be transformed into a pressure rise in the cylinder. If the full wave equations are being solved then it is not necessary to separate these effects but the first effect is due to the wave characteristics and the second is due to the path lines.

Broome[1] defines two parameters for assessing manifolds:

$$\theta_t = 12\frac{NL}{a}, \qquad \text{where } N \text{ is in rpm} \tag{1}$$

and

$$\varphi_t = \theta_t^2(V_c/V_p) = 360(u/a)\theta_t \tag{2}$$

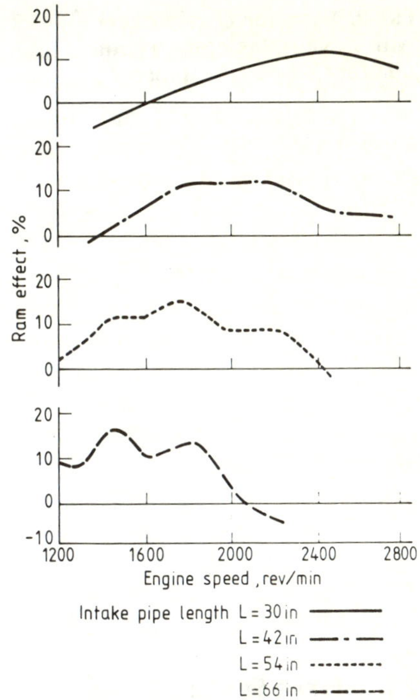

Fig. 1. Variation of ram effect with engine speed for four different length inlet pipes.

θ_t is the parameter which defines the wave action ram, and is equal to half the period of waves in the intake pipe expressed in degrees of crankshaft rotation. φ_t is the parameter for considering inertial ram. Examples of the use of these parameters are given in Ref. 1 and will be summarized here: basically the parameters are most appropriate to single-cylinder engines, although they can be applied to multi-cylinder arrangements if there is not too much destructive interference. Initially results for single-cylinder layouts will be considered. Broome[1] reports results obtained at Ricardo, and also by Williams[2]; both show the increase in volumetric efficiency with intake pipe length over a range of engine speeds—the Ricardo results are reproduced in Fig. 1. It can be seen that the volumetric efficiency is a function of engine speed for each of the lengths of intake pipe, and also that the shorter the pipe the higher the speed at which the manifold 'tunes'. It is also apparent that, while the short pipe shows a single resonance in the speed range considered, the longer pipes exhibit a number of resonances: some means of estimating the position of these resonances would be useful.

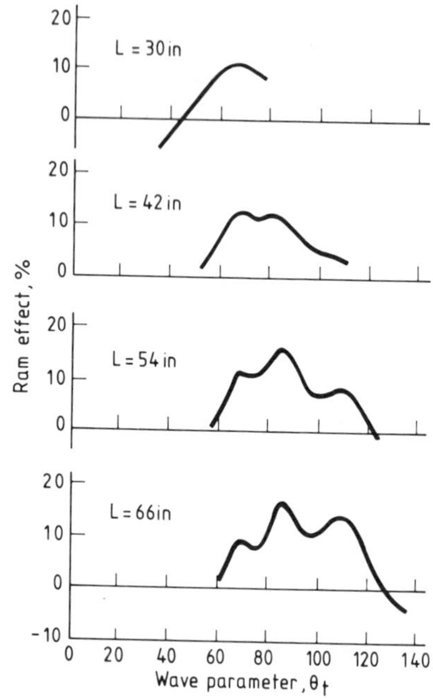

Fig. 2. Variation of ram effect with wave parameter for four different length inlet pipes.

2.1.1. Wave Effects

One way to formalize these effects in terms of a non-dimensional parameter of the inlet manifold is to replot the diagram as a function of the wave parameter, θ_t; this is shown in Fig. 2. The maximum value of the volumetric efficiency is achieved at values of θ_t approximately equal to 70, 90 and 115. Broome shows that θ_t can be evaluated from the time it takes for a rarefraction pulse to travel from the inlet valve to the entrance of the intake system and back to the valve again as a pressure pulse ($\delta t \simeq 2L/a$): this is the natural period of acoustic vibration of a pipe with an open and closed end. Equating δt to the crankangle rotation gives $\theta_t = 12NL/a$. If θ_t is less than the period the valve is open then the reflected pulse can influence the conditions in the cylinder. In fact, when the intake valve is open the intake pipe and cylinder act as a Helmholtz resonator (see Section 2.2), but here it is assumed that the presence of the valve tends to make the cylinder end behave as closed end boundary. The ram effect achieved from wave phenomena is a function of the period of propagation of the pulse and its relationship to the induction period; it is also affected by the standing waves that exist in the intake manifold from previous induction strokes. A condition for favourable resonance exists when the sense of the standing wave is such that it increases rather than decreases the following depression pulse. Only when $\theta_t \simeq \theta_d \simeq 720/2n$ is the standing wave a good fit to the new depression pulse (θ_d is the duration of the depression pulse caused by the intake valve and n is the wavenumber). It is also necessary to introduce a phase interval α to cover the effective shift of the new standing wave, but it is difficult to define this. Another, simple alternative approach can be used to give

$$(2n-1)\theta_t + \theta_d \simeq 720 \tag{3}$$

In practice θ_d is usually between 110° and 150° but is dependent on the level of ram, which will then elevate the cylinder pressure. The values of n which give the best ram are 3, 4 and 5; these result in $\theta_t \simeq 120$, 85 and 67°. The first two values give the highest ram but the optimum one depends on the actual valve timing because these values of θ_t are rather less than half the valve opening periods. These findings are supported by the experimental results given in Fig. 2, for an engine, with an effective induction period of about 210° which favoured $\theta_t = 85°$.

These theories work well with single-cylinder engines but there are limitations when applying them to multi-cylinder machines. The junctions in the manifold of a multi-cylinder engine make the tuning phenomenon much more complex. Whenever a wave interacts with a junction it will be partially transmitted and partially reflected (see Section 4.3 of Chapter 12); this means that waves are 'generated' at many places in the manifold. However, careful design of multi-cylinder manifolds can introduce beneficial effects and these can be used to broaden the engine torque–speed characteristic. It is possible to achieve some induction ramming when adjacent cylinders have siamesed intake manifolds, particularly if the intervals between successive induction pulses are equal. In a four-cylinder engine the central two cylinders (2 and 3) may be siamesed with good effect and θ_t should be either 70–83° or 42–50°. If the induction periods of siamesed ports overlap then the leading cylinder is always penalized relative to the trailing one.

2.1.2. Momentum Effects

Momentum effects tend to be greatest in short pipes because the effect relies on 'plug' flow being achieved. Application of the momentum equation to the gas in the pipe gives

$$F\delta p = \rho V_p \frac{\delta u}{\delta t} \tag{4}$$

In the intake pipe the density, ρ, remains roughly constant and variations can be neglected in the analysis. Also the velocity of the gas at the commencement of the induction stroke (due to residual waves) will be small and hence the maximum velocity achieved is approximately equal to the change in velocity, δu.

The change in velocity must also be closely related to the mean piston speed and swept volume, which can in turn be related to the engine bore; thus

$$\delta u \propto (d^2/d_p^2)SN \tag{5}$$

Also, the time interval available for the charging process is a function of the valve lift diagram and the engine speed ($\delta t \propto f(\theta_{\text{in}})/N$), which gives

$$\delta p \propto \frac{V_p N}{f(\theta_{\text{in}})d_p^2}\left(\frac{d}{d_p}\right)^2 SN \tag{6}$$

This can be simplified to

$$\delta p \propto \frac{L^2N^2}{f(\theta_{\text{in}})}\left(\frac{V_c}{V_p}\right) \tag{7}$$

for uniform cross-section pipes.

It would appear that a large value of δp would be most desirable, but obviously the time for accelerating the flow is approximately equal to that for decelerating it and so some compromise has to be reached. The useful outcome of the above analysis is that a dimensionless parameter φ_t can be defined:

$$\varphi_t \propto \frac{L^2N^2}{a^2}\left(\frac{V_c}{V_p}\right) \tag{8}$$

$$\propto 360(u/a)\theta_t \tag{9}$$

the constant 360 is used simply to retain similarity with Broome's results.

It is possible to relate φ_t to the parameters defining the natural frequency of a Helmholtz resonator by rearranging eqn (8) as

$$\varphi_t \propto \frac{L^2N^2}{a^2}\left(\frac{V_c}{f_pL}\right) \tag{10}$$

Fig. 3. Variation of ram effect with inertial parameter for four different length inlet pipes.

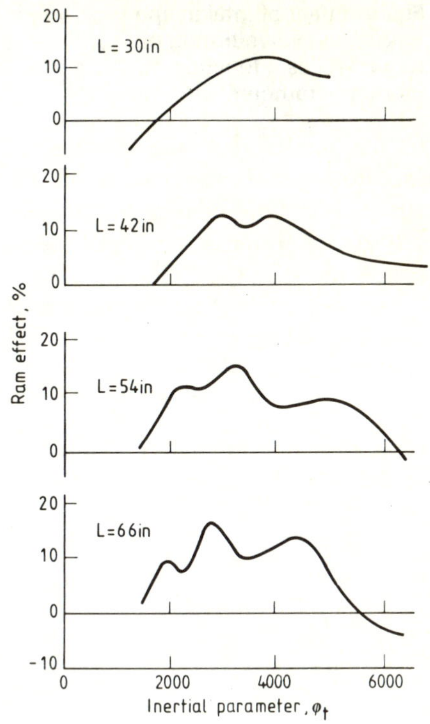

to give

$$N \propto a\{\varphi_t(f_p/LV_c)\}^{1/2} \tag{11}$$

It is shown in Section 2.2 that the resonant speed of a Helmholtz resonator is

$$N_{res} = \frac{15a}{\pi}\left(\frac{f}{LV}\right)^{1/2} \tag{12}$$

In Fig. 3 the data from Fig. 1 is shown plotted against φ_t as the abscissa. It can be seen that the characteristics of the engine are described much more compactly in terms of this parameter. In Fig. 4 are shown results obtained by Taylor, Livengood & Tsai,[3] in which the closing time of the inlet valve has also been introduced as a further variable; this effectively introduces $f(\theta_{in})$. Table 1 from Broome[1], lists the values of φ_t that may be used for optimizing ram effects in engines in which interference is not important.

Broome also shows that late closure of the inlet valve tends to increase the ram effect that can be achieved, but presumably caution

Fig. 4. Effect of intake pipe length and valve timing on ram effect as a function of inertial parameter.

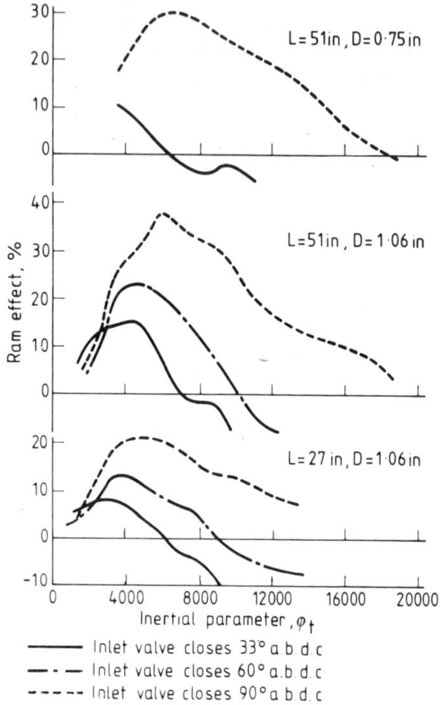

must be used in assessing these benefits because they are a comparison between the cylinder both with and without a manifold. It would have been more useful to have shown the change in trapped mass for the various valve timings.

While these momentum effects can be analysed on single pipes the situation is more complex when a multi-cylinder engine is being considered. Basically, if the induction periods do not interfere then the single-pipe theory is appropriate. On four-stroke engines when groups

Table 1. Intake Manifold Tuning Parameters

Effective valve, closing point, abdc	Optimum range of inertial ram parameter, φ_t	Mean gas velocity, $\bar{u}$ (m/s)
30	2500–3500	15–60
60	3500–5000	15–60
90	5000–7000	15–60

of two and three cylinders are connected then high degrees of ram can be achieved: sometimes a common manifold will give good breathing even if the induction periods overlap.

Broome[1] gives an example of the use of φ_t to assess the ram effect on a V-8, four-stroke engine. The crankshaft was a cruciform type and the Vee angle was 90°. This results in an uneven firing order on each bank and the consequence is poor filling of certain cylinders. A schematic diagram of the engine, indicating the firing order for each bank, is shown in Fig. 5. Also contained on that diagram are the values of volumetric efficiency measured for each cylinder: the change in efficiency is related to the value that was obtained with short stub pipes. It can be seen that the variation of volumetric efficiency is

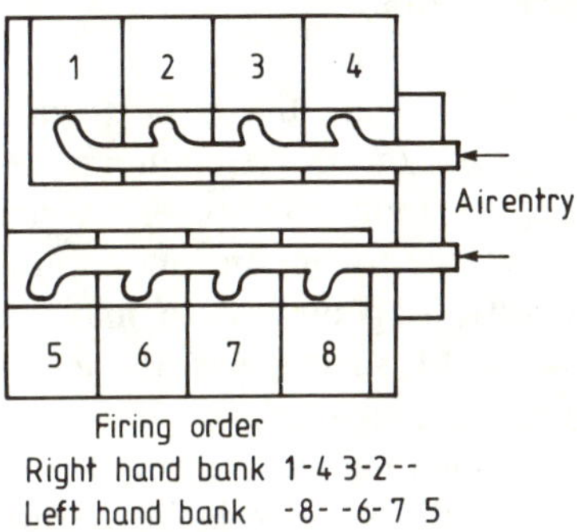

Measured Volumetric Efficiency of Individual Cylinders of a V-8 Diesel Engine with Standard Manifolds

	Right-hand bank				Left-hand bank			
Cylinder number	**1**	**2**	**3**	**4**	**5**	**6**	**7**	**8**
φt	4200	3500	*	2200	4200	3600	2800	*
Volumetric efficiency ηv per cent	86	86·5	77	84	86·5	85	82	77
Per cent change in ηv over stub pipe condition	+5·5	+4·5	−6·5	+3	+6	+4·5	+2·5	−4·5

* Values not significant owing to excessive overlap of following induction period. Effective closing point of inlet valves 53 deg a.b.d.c.

Fig. 5. Correlation between inertial parameter and volumetric efficiency for a V-8 engine.

almost 10%, which is equivalent to a difference of about 12% in the mass trapped in the cylinders. There is a weak correlation between the volumetric efficiency and φ_t for the individual cylinders.

All the previous analysis has been based on intake manifolds constructed from straight pipes. It has been found that there is a loss of ram effect if the flow pattern within the manifold is disturbed. Both bends and junctions in manifolds of multi-cylinder engines affect the general flow of the gas and will tend to detune the system and reduce the improvements in volumetric efficiency that can be obtained. It is usually found beneficial to place bends close to the inlet valve, since here they have less effect on the induction flow pattern.

2.2. Tuned Inlet Manifolds Using Resonators

There is a particularly useful method of changing the basic torque characteristics of an engine which is based on the Helmholtz resonator. It was shown previously that the individual pipes in the manifold will tune like organ pipes. However, to obtain low natural frequencies the pipe lengths are relatively long and there may be installation problems when attempting to fit such pipes to actual engine installations. It is possible to obtain a similar effect by using a resonator, which comprises a pipe and a resonating cavity. The theory of resonators is developed in detail in Lighthill,[4] where it is shown that the resonating cavity can act as an equivalent length of pipe: hence this approach is a method for reducing the size (or more specifically the length) of tuned manifold required to achieve a particular resonant speed. The resonator works by exchanging energy between the gas in the pipe and that contained in the vessel; it results in a system which has a clearly defined natural frequency.

The Helmholtz resonator can be analysed by applying the linearized wave equation (eqn (5) of Chapter 12) to the arrangement shown in Fig. 6(a). It is assumed that the pressure in the resonator is uniform, which implies that the wave travel time in the resonator is short compared to that in the pipes of the system, and then the resonator volume can be defined by its ability to absorb fluid; this is the capacitance of the volume,

$$C = \mathrm{d}v/\mathrm{d}p \tag{13}$$

Lighthill[4] shows that the natural frequency of the resonator is given

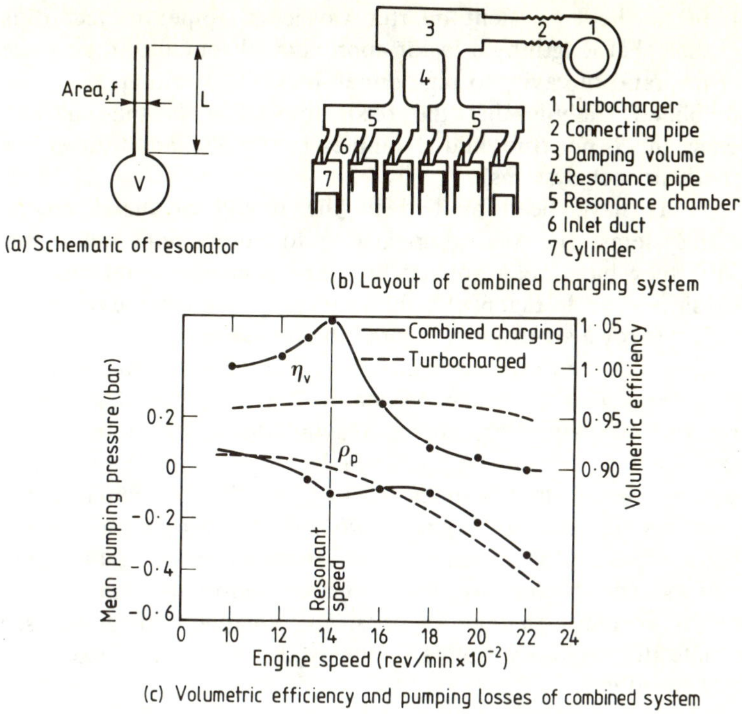

Fig. 6. Helmholtz resonators as intake tuning devices: (a) basic resonator system; (b) engine layout; (c) effect on volumetric efficiency and pumping losses.

by

$$\omega_0 = a(f/LV)^{1/2} \quad \text{rad/s} \tag{14}$$

$$= a/2\pi(f/LV)^{1/2} \quad \text{Hz} \tag{15}$$

The same result can be obtained by considering plug flow (i.e. the whole column of gas moving in phase) in the pipe with the gas in the plenum volume undergoing *isentropic* compression and expansion.

It can also be shown (see Lighthill[4]) that the resonating volume is equivalent to a pipe of length, l_0, where

$$l_0 = (a/\omega)\tan^{-1}(C\omega/Y) \tag{16}$$

and C and Y are capacitances and admittances, respectively, of the system.

This means that a resonator system is usefully compact when

compared to the equivalent quarter-wavelength pipe: in fact it is smaller than that length by a factor approximately equal to the square root of the ratio of cavity to pipe dimensions. (It would appear from this simplified analysis that the resonator will only tune at one frequency: this is not true, and a complex family of tuned frequencies also exists for such systems).

Resonators have been applied to the tuning of diesel engine induction systems. The original application to diesel engine tuning was by Cser[5]; they have been applied by many companies and research workers since then. In Fig. 6(b) is shown the way in which the resonator is constructed for a six-cylinder engine; the tuning pipes (labelled 4) in conjunction with the resonance chambers govern the natural frequency of the system. This particular resonator, for a six-cylinder engine, requires two resonance chambers to avoid destructive interference between induction periods. A four-cylinder engine could be designed with a single resonance chamber. In Fig. 6(c) it is shown how the resonator system (referred to as *combined charging*) improves the volumetric efficiency of the engine at its resonant speed of 1400 rpm. The performance is poorer off the resonance and there is an increase in pumping work at high engine speeds. The equation that can be used to evaluate the necessary size of tuning system to obtain a particular resonant speed is

$$N_{res} = \frac{15a}{\pi}\left(\frac{f}{LV}\right)^{1/2} \tag{17}$$

This has been evaluated by equating the natural frequency of the resonator to the firing frequency of the engine; the equation is developed for a four-stroke engine and contains the assumption that the interval between cylinders firing is even. Watson[6] has shown that it is substantially less attractive to apply tuned systems of this type when the firing interval is not even. Brands[7] suggests that the volume used for the resonator should be that of the plenum plus the intake ports, and additionally half the cylinder volume. The logic of including the last term is that the intake valves are open during the induction period, and this has the effect of allowing the volume of the cylinder to have an influence on the volume 'seen' by the fluctuating pressure. Vorum[8,9] also uses the mean cylinder volume in his equations.

It is apparent from eqn (17) that it is possible to trade off different parameters in the induction system and yet achieve the same natural frequency for the system. As an illustration of this, the parameters for

Table 2. Engine Parameters for Tuned Manifold

Cylinder swept volume, V_s (litres)	1·1
Inlet port diameter (mm)	37·5
Number of cylinders, n_c	4
Air inlet temperature (°C)	50
Tuned speed, N_{res} (rpm)	1500
Speed of sound of intake air, a (m/s)	360

a four cylinder, four-stroke diesel engine are given in Table 2, and it is required to design a tuned manifold which resonates at 1500 rpm.

The resonator volume, V, is defined by

$$V = \tfrac{1}{2}V_s + V_{port} + V_{plen} \tag{18}$$

where V_{plen} is the volume of the plenum chamber. In this case the plenum volume will be assumed to contain the port volume. In Fig. 7 is shown the way in which the plenum chamber volume must vary as the intake pipe length is changed if the resonant frequency is to remain constant; the diagram also shows the diameter of a cylindrical plenum chamber if the length available is 0·4 m. It can be seen that a broad range of sizes can be considered when designing a tuned system; basically, the product of the pipe length and volume must be constant for all combinations (hence the curve given in Fig. 7 is a rectangular hyperbola). Chen & Gu[10] state that the optimum diameter of the tuned inlet pipe should be between 0·6 times and equal to the

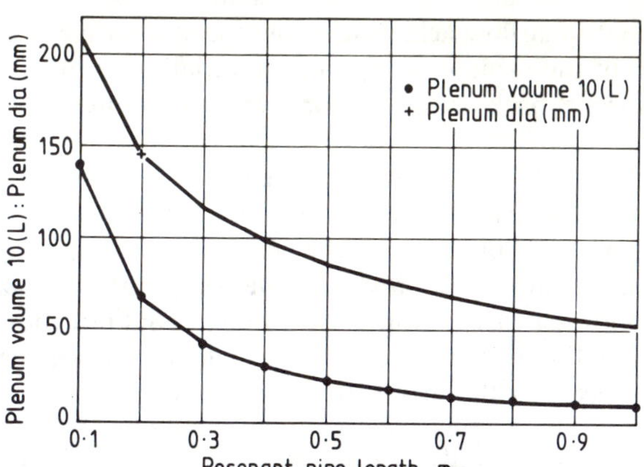

Fig. 7. Variation of resonator system parameters to achieve a particular tuned speed.

diameter of the inlet port. They also state that decreasing the cross-sectional area increases the tuning intensity; however, if the pipe cross-section is too small then frictional effects will become large and this will have a detrimental effect on the volumetric efficiency of the engine. Small-volume plenum chambers result in higher pressure fluctuations but also require longer pipes to achieve the same resonant frequency. If the volume of the chamber becomes too small then it will act like a pipe and wave action will occur in it, this will defeat the attempt to construct a Helmholtz resonator. Watson[6] used a ratio of resonant volume to cylinder volume of about six; this would suggest a pipe length of around 200–300 mm for the engine considered in Fig. 7, and would probably be acceptable for this particular engine and application. In such a system the plenum chamber would have a length of 400 mm and a diameter of 120 mm, which should not cause a packaging problem. If larger-diameter tuning pipes were used then the size of the tuned manifold would increase substantially. For example, if the diameter of the tuning pipes was 50 mm then, with pipes 300 mm long, the plenum chamber would have a volume of 8 litres and an internal diameter of 160 mm; such a manifold might be difficult to accommodate in some cases.

Knecht & Signer[11] report an ingenious design of combined charging system, where the pipework and plenum volumes are incorporated into the basic manifold. A sectioned photograph of the manifold is shown in Fig. 8(a), and this indicates how careful 'packaging' of the components can result in an acceptable total system. In Fig. 8(b) is shown the improvement in airflow achieved at low speed (relative to a turbocharged version of the same engine): such a variation in engine airflow will enable torque back-up to be improved with consequent benefits to driveability.

Vorum[8,9] proposes a method for designing tuned intake manifolds for multi-cylinder engines which is based on two resonators in series, but does not derive a method for assessing the actual values of η_v that will be achieved. In addition, he modified the single Helmholtz resonator equation following earlier work by Engelman, and incorporated (through the evaluation of the mean cylinder volume) the engine compression ratio, r, such that the tuned speed becomes

$$N_{\text{res}} = 642a\left\{\frac{f}{LV_s}\left(\frac{r-1}{r+1}\right)\right\}^{1/2} \tag{19a}$$

However, Vorum also found it necessary to include a further factor, k,

Fig. 8. Embodiment of combined charging system in a manifold: (a) physical arrangement; (b) improvement in performance.

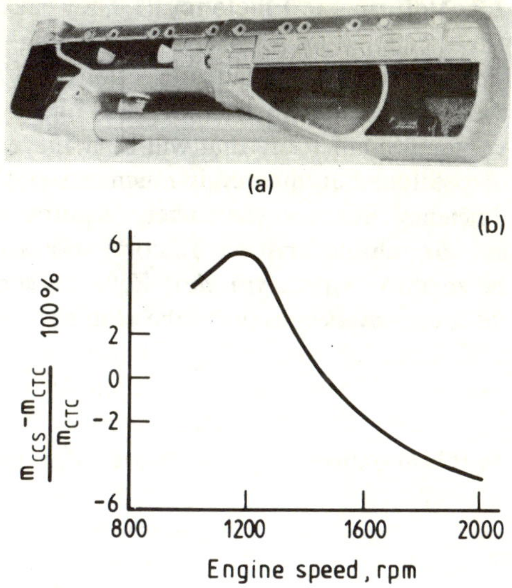

to allow for the variation of cam design, etc., giving

$$N_{res} = \frac{1350}{k} a \left\{ \frac{f}{LV_s} \left(\frac{r-1}{r+1} \right) \right\}^{1/2} \tag{19b}$$

Unfortunately the value of k does not appear to be well defined, and later additional factors are required for the flow velocities and other parameters. Nikpour[12] found that variation of k in the range suggested by Vorum resulted in a 55% change in the length of the primary pipe. It is apparent that, even with complex empirical coefficients, it is difficult to obtain a semi-empirical method for evaluating the tuning of multi-cylinder engines. It would appear that the only safe way of designing, or at least analysing, tuned manifolds is to use step-by-step simulation; the latter approach has the weakness that the technique cannot help the designer to get initial sizes for the manifold. Judicious use of the semi-empirical factors helps remove some of the uncertainty in obtaining initial designs for later analysis.

In conclusion, there are a number of methods available for evaluating the tuned frequencies of pipework systems. These are quite accurate in respect of the manifold frequency, but the tuning cannot be directly translated into a gain in engine performance due to the inherent dynamics of the intake valve and cylinder. This aspect of tuning will now be considered.

2.3. Volumetric Efficiency

The methods described in Sections 2.1 and 2.2 enable calculation of the frequency at which tuning will occur: they do not closely define the volumetric efficiency that will be achieved from the tuning. Evaluation of the tuned frequency is much easier than calculation of volumetric efficiency because the latter requires knowledge of the valve and cylinder characteristics. Taylor[13] shows how volumetric efficiency can be related to a *Mach index, Z,* when short pipes are used (this is also discussed by Ferguson[14]); the result is

$$\eta_v = 0{\cdot}58\left(\frac{\alpha_{ic} - \alpha_{io}}{\pi}\right)\frac{1}{Z} \tag{20}$$

In this equation, $(\alpha_{ic} - \alpha_{io})$ is the duration of inlet valve opening;

$$Z = \text{Mach index} = \frac{F_c}{\bar{F}_i}\frac{\bar{V}_p}{a} \tag{21}$$

$$\bar{F}_i = \frac{1}{\alpha_{ic} - \alpha_{io}}\int_{\alpha_{io}}^{\alpha_{ic}} C_{d_i} F_i \, \mathrm{d}\alpha \tag{22}$$

where C_{d_i} is the coefficient of discharge, F_i is the inlet pipe cross-sectional area, F_c is the cylinder bore area, $\bar{V}_p$ is the mean piston speed.

In Fig. 9 it is shown how closely this hyperbolic relationship follows measured values of volumetric efficiency when the flow is choked at the inlet valve. (The value 0·75 occurs because $(\alpha_{ic} - \alpha_{io})/\pi = 1{\cdot}3$ for the engine considered by Taylor.) The Mach index is not a true velocity but it characterizes the average gas speed that there would

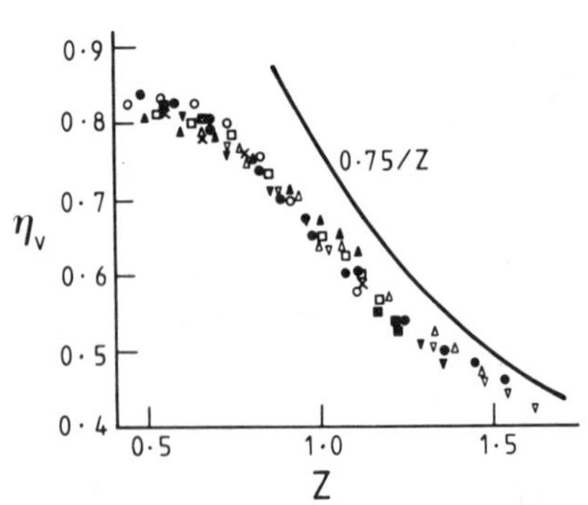

Fig. 9. Variation of volumetric efficiency with Mach index.

have to be at the valve to fill the swept volume with gas at the inlet manifold conditions. This approach is also discussed by Tabaczynski[15] and Fukutani & Watanabe.[16] Tabaczynski suggests that the Mach index, Z, can be used to indicate the improvement in volumetric efficiency that can be achieved as the parameters in eqn (20) are changed *for a given intake system and valve events*. He states that the major problem with the classical Z factor analysis is in determining the new volumetric efficiency against engine speed relationship when geometry changes (excluding tuning) are performed. Such geometry changes include changes in bore, stroke, and inlet valve size and timing. He does, however, show that the concept might be useful when attempting to assess the effect on engine performance of the steady flow pressure drops measured in various inlet manifolds.

Fukutani & Watanabe[16] propose replacing the Mach index, Z, by the *mean inlet Mach number, M_{im},* which gives more importance to the period that the inlet valve is open. As they state, the valve timings of engines vary widely depending on the purpose for which the engine will be used: the inlet valve period might be between 210° and 340° crank angle. The mean inlet Mach number is defined as

$$M_{im} = v_{im}/a \tag{23}$$

where

$$v_{im} = V_c \eta_v / 100 F_t \tag{24}$$

and

$$F_t = \int_{t_{io}}^{t_{ic}} C_{d_i} F_i \, \mathrm{d}t = \frac{1}{6N} \int_{\alpha_{io}}^{\alpha_{ic}} C_{d_i} F_i \, \mathrm{d}\alpha = \frac{\alpha_{ic} - \alpha_{io}}{6N} \bar{F}_i$$

where N is the engine speed (rpm).

These equations result in the following relationship between M_{im} and Z:

$$M_{im} = \frac{180 \eta_v}{100(\alpha_{ic} - \alpha_{io})} Z \tag{25}$$

Fukutani and Watanabe performed a number of rig tests to prove this hypothesis. The rig consisted of a variable speed electric motor which drove the three different 'engines' that were used; these were closely coupled to a large-volume inlet plenum, and had an open exhaust. On two of the engines the timing was maintained constant but the valve area integral was altered by means of different cams; on the third engine a variable timing mechanism was fitted to enable the point of inlet valve closure to be varied. In Fig. 10 it is shown that the

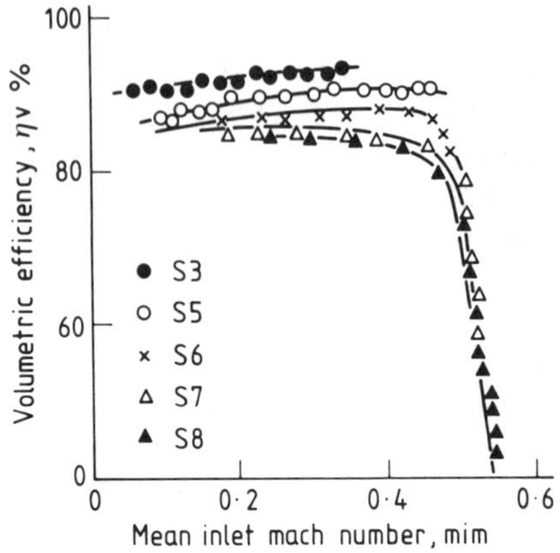

Fig. 10. Variation of volumetric efficiency with mean inlet Mach number.

volumetric efficiency of the engine decreases precipitously as M_{im} reaches 0·5; this is in contrast to the hyperbolic change of efficiency in terms of Z. The various lines in Fig. 10 (labelled S3 to S8) were obtained for different valve time area integrals over a range $S3 = 2{\cdot}3 \times 10^{-2}$ to $S8 = 0{\cdot}67 \times 10^{-2}\,\text{deg.m}^2$; this shows that the approach makes it unnecessary to test with each cam profile. In Fig. 11 is shown

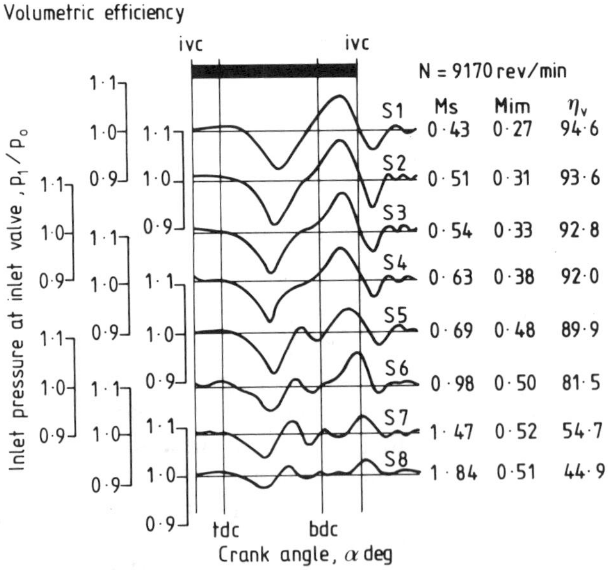

Fig. 11. Pressure pulsations upstream of inlet valve for various mean inlet Mach numbers.

the variation of pressure just upstream of the inlet valve for a range of values of M_{im}. It can be seen that at low values of M_{im} the trace shows the characteristic decrease in pressure as gas is drawn into the cylinder; as the value of M_{im} approaches 0·5 the pressure pulsations in the inlet tract become visible because the choked flow at the valve is causing pressure waves to be reflected from this boundary, rather than being absorbed by the cylinder. The effect of the point of inlet valve closure is shown in Fig. 12. The various sets of results labelled R1 to R3 again indicate the valve area integral, but in this case this is a variable quantity: R1 is between 6·20 and 8·26, R2 between 3·82 and 4·91, and R3 between 1·56 and $2{\cdot}02 \times 10^{-2}$ deg.m^2. It can be seen that the volumetric efficiency is higher at low engine speeds (i.e. low values

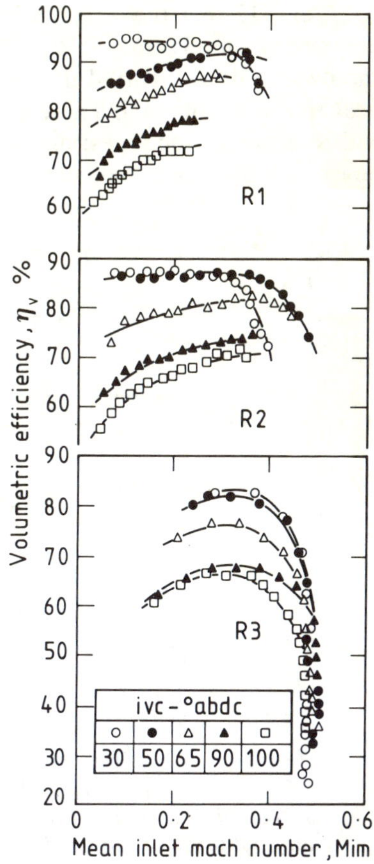

Fig. 12. Effect of inlet valve closure angle on volumetric efficiency.

of M_{im}) when the valve closes early, and the peak moves to higher speeds as the point of valve closure becomes later. It is also noticeable that values of M_{im} approaching 0·5 result in a rapid fall in volumetric efficiency irrespective of the actual point of closure of the inlet valve.

In summary, when designing engine inlet valves and ports it is necessary to ensure that there is sufficient valve time area integral to limit the value of *Mach index, Z,* or *mean inlet Mach number,* M_{im}. While the limiting value of M_{im} is more clearly defined than that of Z, there is the problem that M_{im} can only be calculated if the value of volumetric efficiency, η_v, is known. An iterative approach could be adopted to determine this. It also has to be recognized that this method is applicable only to the bare inlet system; that is, without any manifolding on either the inlet or exhaust. It would appear from Figs. 9, 10 and 12 that the volumetric efficiency of engines cannot exceed between about 85 and 95%, depending on the data considered (it should also be recognized that Figs 10 and 12 relate to a rig in which heat transfer is absent). These values are really only limits for engines with short induction pipes: tuning of the manifold enables this value to be exceeded. Hence, the actual value of volumetric efficiency that can

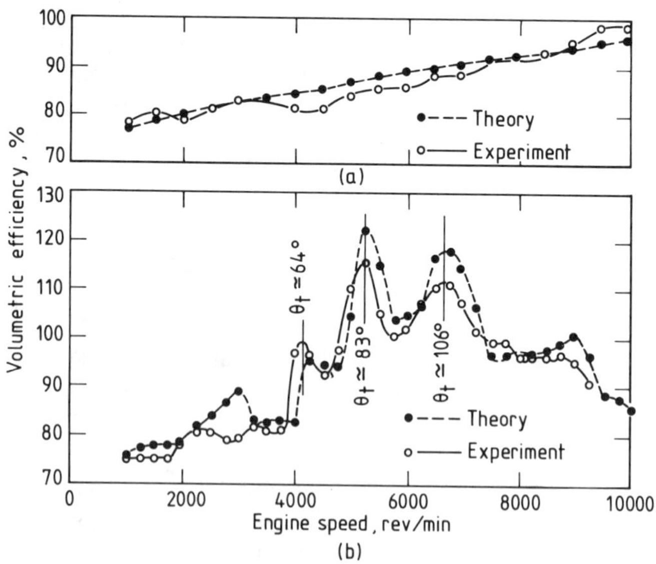

Fig. 13. Variation of volumetric efficiency with engine speed for two inlet pipe lengths: (a) short pipe (76 mm); (b) long pipe (457 mm).

be achieved is a combination of the effects of the inlet manifold parameters and the valve parameters.

Prosser[17] (Fig. 13) shows how the volumetric efficiency of a four-stroke engine varies with speed for a short intake pipe (76 mm) and a long one (457 mm): he also compares the measurements with calculations performed using method of characteristics. The agreement achieved is good and both closely obey the rules suggested by Broome[1] for pipe tuning (the values of θ_t at which the peaks occur with the longer pipe are 64°, 83° and 106°). At low speed the actual levels of η_v are calculated well but at high speed the prediction over-estimates the efficiency, presumably because friction was omitted in the simulation. Both Figs 13(a) and (b) show that the volumetric efficiency can be higher than 85%; the benefits of having a long pipe are quite marked and the best efficiency of 120% is achieved at about 5000 rpm, whereas the peak efficiency is not achieved by 10 000 rpm with the short pipe.

2.4. Exhaust Manifold Tuning

Exhaust manifold tuning has been used for many years to increase the power output of engines. Careful design of the exhaust pipe is essential for two-stroke engines, since these require high air flows to obtain good scavenge. The design of manifolds for such engines is described in Section 3.2 and also in Chapter 3. Exhaust pipe tuning is not so important for road-going four-stroke engines but it can be used, in conjunction with the inlet manifold, to modify the characteristics to give the required torque–speed curve. Designing exhaust pipes to resonate at specific frequencies is more difficult than designing a resonant inlet manifold because the pressure levels are substantially higher, there are variable entropy (temperature) levels in the gas, the exhaust temperature varies with load and speed, and it is also possible to get shock waves. The length of the exhaust pipe means that plug flow will not exist and hence the tuning is governed more by the wave motion. The parameter of importance is θ_t ($=12NL/a$), which will now be a function of speed and load.

There are a number of papers which consider the more empirical aspects of exhaust pipe design, but it is obvious that this part of the engine is not as amenable to simple theories as the induction side. Adams[18] considers a few of the available methods: two were based on the wave method described above. A further technique was surprising

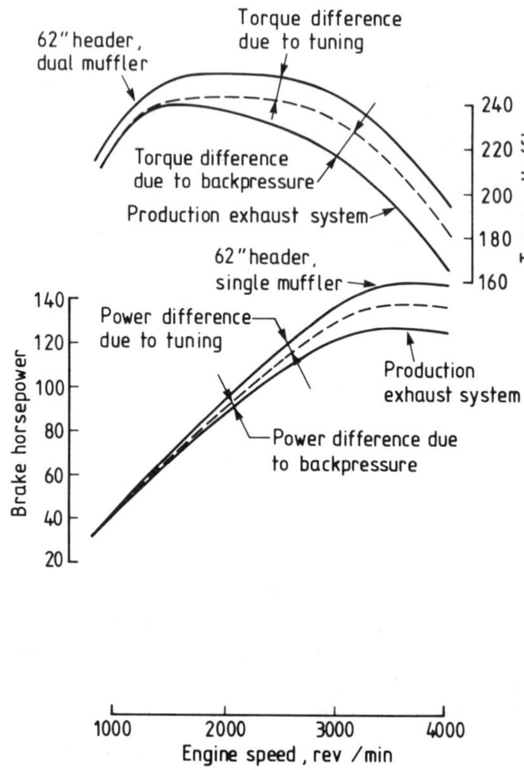

Fig. 14. Design of exhaust pipe for V-8 four-stroke engine.

because it did not suggest that the tuned length of the exhaust system would be dependent on the engine speed. The improvement that Adams achieved by redesigning an exhaust manifold is shown in Fig. 14. It can be seen that this comes from two sources: the effect due to tuning and that due to a reduction in exhaust system back pressure.

Vorum[8,9] includes a design technique for exhaust manifolds, but this seems to be extremely empirical.

The subject of exhaust manifold design for turbocharged engines will be returned to in Section 5 when the effect of different design assumptions will be assessed. The basic guidelines that can be obtained from the literature are to evaluate the exhaust pipe diameters to limit the mean gas velocity to less than 100 m/s at full speed and load, and to reduce the back pressure as much as possible consistent with noise and emissions restrictions; this will influence the devices fitted to the exhaust pipe.

3. PIPEWORK ARRANGEMENTS FOR TWO-STROKE ENGINES

3.1. Scavenging of Two-stroke Engines

The effect of air flow on the scavenge efficiency of two-stroke engines is described by Blair (Chapter 3). As early as 1914, Hopkinson[19] had shown that the purity of charge trapped in the cylinder of a two-stroke engine was a very strong function of the air flow through the ports. He had realized that the gas exchange process in such engines is very much one of mixing and dilution, rather than the positive exhaust and induction strokes of the four-stroke engine. Of course, the mixing concept is too simple and the actual exchange process is a very complex fluid-dynamic interaction of the gases in the cylinder; these effects are described in Chapters 3 and 4. Basically, the scavenge process is a combination of displacement of the exhaust gases by the incoming air, mixing of pure air and exhaust products, and some short-circuiting of the air directly to the exhaust ports. This means that the scavenge efficiency can be defined by a set of semi-empirical equations.

Benson & Whitehouse[20] suggest a scavenge equation which allows for short-circuiting, displacement and mixing scavenge. If the proportion of inflowing air which short circuits is y, then

$$(1-y)\lambda \leqslant x, \qquad \eta_s = (1-y)\lambda \tag{26a}$$

$$(1-y)\lambda > x, \qquad \eta_s = 1-(1-x)\exp(x-(1-y)\lambda) \tag{26b}$$

where x is the fraction of the cylinder volume which is initially subject to scavenge by perfect displacement, and λ is the scavenge ratio.

The scavenge ratio is a measure of the total flow of air through the inlet ports as a ratio of the amount of air required to fill the cylinder; viz.

$$\lambda = \frac{\text{mass of air flowing through inlet port}}{\text{mass of air which would fill cylinder at } p_a,\ T_a} \tag{27}$$

where p_a and T_a are the pressure and temperature in the intake pipe.

The scavenge efficiency (η_s) is defined as

$$\eta_s = \frac{\text{mass of air trapped}}{\text{mass of air which would fill cylinder at } p_a,\ T_a} \tag{28}$$

It is apparent from eqns (26a) and (26b) that the larger the value of scavenge ratio (λ) the larger will be the scavenge efficiency (η_s). The magnitude of η_s determines the quantity of fuel that can be burned in the cylinder, and hence the power output of the engine. Thus a major requirement in the design of two-stroke engines is to obtain a high air flow rate: this has to be achieved by careful design of the intake and exhaust systems because there is no pumping action available in the cycle. All naturally aspirated engines require some form of 'scavenge pump' to ensure that the air flows from the inlet port to the exhaust. This is achieved by means of the crankcase in the simple engines used for motor cycles, outboard motors etc. In larger engines, for example, the Detroit Diesel two-stroke uniflow scavenge engine and even larger marine engines, the air manifold is pressurized by some form of scavenge pump or blower. Different types of pump are used, ranging from positive displacement Roots blowers to centrifugal devices. Detroit Diesel[21] has also investigated an hydraulic auxiliary drive to provide sufficient power to assist the turbine of its turbocharged engine: this means that, at low load, the compressor is driven by a small oil-powered Pelton wheel, while at full load the turbocharger is self-sustaining.

3.2. Exhaust System Design for Naturally Aspirated Two-stroke Engines

In Fig. 1 of Chapter 3 is shown a typical exhaust system with expansion chamber design for a two-stroke petrol engine used in a racing motorcycle. It is built up from a number of different sections: there is a short section of constant cross-sectional area; two diffuser sections; a larger-diameter constant cross-section pipe; a converging section and a small cross-section tailpipe. What is the effect of all these various pipes, and why are they necessary? A qualitative description of the operation of such an exhaust and intake system was given in Horlock & Winterbone[22] and is repeated here for completeness. In Fig. 15(a) is depicted a typical two-stroke engine exhaust system, similar to that shown in Fig. 1 of Chapter 3 except that the first taper is

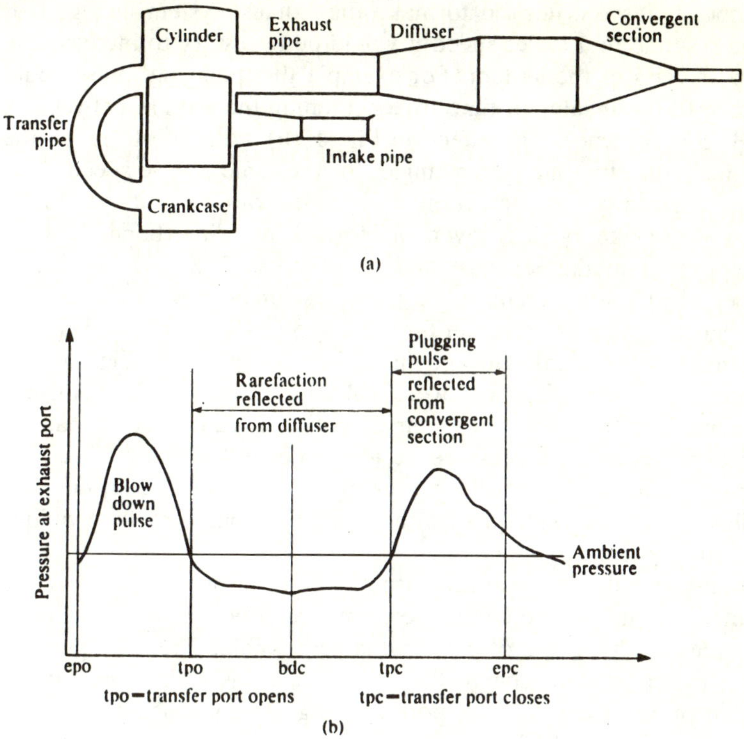

Fig. 15. Idealized two-stroke exhaust system: (a) schematic representation of engine; (b) exhaust pressure variation.

at only one angle. The idealized waveform is shown in Fig. 15(b), which is similar to that shown in Fig. 2 of Chapter 3. When the piston opens the exhaust port a pressure wave is transmitted down the exhaust pipe: this is partially reflected by the diffuser as a rarefaction wave and partially transmitted (see Fig. 11 of Chapter 12) but in this case the interaction is not so sharp since the diffuser is really a series of small sudden enlargements, and hence 'smears' the interaction. The rarefaction wave, on reaching the exhaust port, will assist scavenging of the cylinder by drawing gas through the transfer ports. The transmitted wave will eventually reach the converging section of the exhaust pipe and will be partially reflected as a pressure wave and partially transmitted to atmosphere. The lengths of the various sections of the exhaust system define the speed at which the engine will tune. If it were desired to make the exhaust system in Fig. 1 of Chapter 3 tune to a lower speed it would be necessary to increase the length of parts of the system. For example the 'plugging' pulse could be made to occur later in time by lengthening the section between A and B from 5 inches. The effect on Fig. 15(b) would be to move the 'plugging' pulse to a later crankangle, for the same engine speed.

A description of the interaction between the various processes in the simple two-stroke cycle is given in Horlock & Winterbone,[22] where predictions of intake, exhaust and transfer gas flows are shown. A number of pressure – crank angle diagrams for both the cylinder and the exhaust system are given in Figs. 3 and 5 of Chapter 3, and similar diagrams for the crankcase and inlet duct are given in Figs 6–8 of Chapter 3. These diagrams were calculated using a homentropic formulation of the method of characteristics. It can be seen that the predictions of exhaust pressure at low speed (4000 rpm) (Fig. 3 of Chapter 3) are substantially better than those at 8000 rpm. This possibly reflects an inadequacy of applying the homentropic technique to exhaust systems.

The intake side of a naturally aspirated two-stroke engine can also be tuned to improve the engine performance although there are space limitations on the length of pipe that is permissible. The main purpose of the inlet pipe is to charge the crankcase to as high a pressure as possible prior to the transfer port opening. The arrangement of the engine (see Fig. 10 of Chapter 3) means that this charge can be forced back out of the crankcase into the inlet duct during the pressurization process. A substantial improvement in power output can be achieved by fitting some form of valve between the inlet tract and the crankcase. High-performance engines have been fitted with disc valves (which rotate with the crankshaft) to phase the inlet event, or reed valves (which act as non-return devices); both of these have the benefit of restricting backflow from the crankcase.

3.3. Supercharged Engines

Crankcase scavenging is a suitable means of cleaning the exhaust products from the cylinders of small engines and is particularly useful for single-cylinder arrangements. On larger engines, particularly multi-cylinder ones, such an approach becomes less attractive and most large engines have separate scavenge pumps; these may be

Fig. 16. Two-stroke truck diesel engine, with scavenge blower and turbocharger.

switched off when the turbocharger is self-sustaining, giving increased efficiency. A typical arrangement is the Detroit Diesel engine, which is fitted with a Roots scavenge blower and a turbocharger (Fig. 16). The scavenge blower is essential for low speeds and loads when the turbocharger cannot supply a high enough boost pressure to exceed the exhaust back pressure: at higher speeds the turbocharger will become self-sustaining. In this case the Roots blower is driven directly by the crankshaft of the engine; for large marine diesel engines the scavenge blower might be driven by an electric motor. Descriptions of the various arrangements for different types of two-stroke engine are given in Lilly.[23]

4. CALCULATION OF WAVE ACTION IN MANIFOLD SYSTEMS OF NATURALLY ASPIRATED ENGINES

The basic techniques for calculating wave action have been described briefly in the previous sections. Now it is time to apply these methods to actual engines. The basic methods are the same for all the different engine types encountered, but it is convenient to examine the results according to engine type: any subdivision used for considering the literature is arbitrary and will produce idiosyncrasies in the development of the material. It has been decided here, for convenience, to consider naturally aspirated and supercharged engines separately. Pressure-charged engines are dealt with in Section 5. In the case of naturally aspirated engines, it has been decided to treat inlet and exhaust systems separately. This is an arbitrary subdivision which can produce some problems of interpretation because inlet and exhaust systems can interact quite severely, especially where the valve overlap (the period when both the inlet and exhaust valves are simultaneously open) is large. The result is reference back and forth in the text: while it would have been preferable to avoid this other problems of style would have been introduced. The development of the subject is such that detailed discussion of some results is precluded: a full description of many of the boundary models can be found in Benson,[24] and Horlock & Winterbone[22] give detailed discussion of many areas which are treated more discursively here.

Most of the applications reported have been evaluated using the method of characteristics, simply because this has been the most widely adopted approach to date; similar results would be obtained using any other suitable solution technique. Poloni, Winterbone & Nichols[25] and Azuma, Yura & Tokunaga[26] have shown that if method of characteristics and Lax–Wendroff techniques are used with the same boundary model they give almost identical results in respect of pressure variation, at least.

It has been shown that the performance of a naturally aspirated engine can be greatly influenced by the design of the inlet manifold. The basic problem in engine design is getting the right amount of air (and in the case of the carburetted petrol engine, fuel) into the engine to achieve efficient combustion. The air flow through an engine is a function of many parameters, but is limited by the ability of the engine to breathe freely. Equations (21) and (23) define inlet valve 'Mach numbers' which influence the volumetric efficiency of the engine. The denominator of both these expressions includes an inlet valve area term, and hence the inlet Mach number can be reduced by increasing the time area integral of the inlet valve. Fukutani & Watanabe[16] show how variation of the inlet valve area increases the volumetric efficiency: they also show the effect of the timing of inlet valve closure on efficiency (see Figs 10 and 12). There is a limit to how much the effective valve area of a two-valve cylinder head can be increased, simply because of geometric restrictions: lifting the valves higher than about a quarter of the valve diameter usually causes choking in the inlet port (see Annand & Roe[27]). It is possible to obtain some increase in the area available for valves by introducing some inclination in the cylinder head surface: examples of this approach are the hemispherical, wedge and pent-roof combustion chambers used in petrol engines.

Another method of obtaining better utilization of the area available on the cylinder head surface, whether plane or inclined, is to use more than two valves. Four- and six-valve cylinder heads have been used in diesel engines for some time. Four-valve heads have been common in high-performance and racing car engines for many years (Benz used such a design in 1910 for a speed trial engine, and Daimler used one in the 1914 French Grand Prix). The Rolls-Royce Merlin engine used in aircraft in World War II was fitted with a four-valve head, and also supercharged. At the less exotic end of the market, Triumph fitted a 16-valve four-cylinder engine to the Dolomite car in 1973.

However, it is only recently that four-valve cylinder heads have been embraced as a common solution to the manufacturers' quest for engines better able to meet the consumers' requirements. In addition to giving better breathing, the four-valve cylinder head will also enable emissions legislation to be met through better combustion of lean mixtures. Abthoff, Bruggemann, Huttebraucker & Zeilinger[28] describe such an engine and state that the air flow measured on a steady flow rig was 25% higher than the two-valve version at the same pressure drop (93 mbar). Besides improving the basic airflow, combustion is better because the flame travel times are reduced by mounting the spark plug centrally (see Fig. 17): this will have a beneficial effect on the resistance of the engine to detonation, enabling a higher compression ratio to be used. A further advantage is the reduced valve mass which enables higher rates of valve lift to be achieved; this can result

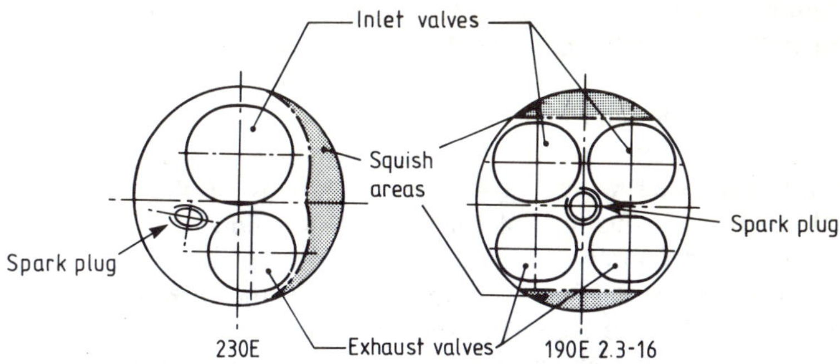

Fig. 17. Comparison of two- and four-valve cylinder arrangements.

in higher engine speeds or greater valve area integrals. All these factors tend to give improvements in the fuel consumption and rating of the engine. Abthoff *et al.* quote a volumetric efficiency of greater than 100% over a speed range from 3500 to 5800 rpm, and a maximum BMEP of 12·8 bar (185·6 lbf/in^2): these performance figures have been achieved on a 'top of the range' vehicle by careful attention to the design of the inlet and exhaust manifolds, and the use of fuel injection. Good performance can also be achieved on more modest vehicles by the application of similar techniques.

The Daimler–Benz engine described above is a relatively large one (2·3 litre) with an intrinsically high torque at low engine speeds. Smaller engines do not have this benefit and basically their design is a compromise between obtaining a high torque at low engine speed, to give good driveability, or high power from a high-speed engine. If the manifold is a fixed geometry one then it is not possible to obtain both of these features. Matsumoto & Ohata[29] state that while four-valve engines will give 25–30% more torque than their two-valve counterparts at high speeds this is achieved at the expense of reduced torque at medium engine speeds. Engines are now being developed with 'variable geometry' induction manifolds which attempt to combine both good high- and low-speed performance; these will be described below.

4.1. Inlet Manifold Calculations

Section 2.1 described empirical methods of intake manifold tuning. Tabaczynski[15] reviewed the techniques available and concluded that 'the simple models are of limited use since they can only give the tuning RPM [sic] and cannot yield actual volumetric efficiencies .... The one-dimensional models give good agreement with the data and can be used as a design tool'.

There are a number of one-dimensional models available, ranging from the linearized wave equations to solutions of the non-linear partial differential equations of compressible gas flow. The linearized equations are definitely not suitable for flow in the exhaust pipe because the wave strength is too large; however, they can be useful in the inlet manifold. The basic requirement of design is to synthesize an artifact, or system, which will meet a given requirement within a cost restraint. When applying this philosophy to the design of an inlet manifold the difficulty that is immediately encountered is the wide range of parameters available. There are some ground rules relating to basic flow velocities and these help in deciding the diameters of the basic pipework. After this there is not a lot of help available to enable the designer to achieve the air flow characteristics of the manifold. Hence the rather crude, and somewhat unreliable, methods proposed by Engleman[30] and Vorum[8,9] received a good deal of attention from manufacturers. Others have attempted to develop more comprehensive linear methods.

4.1.1. Linearized Techniques

Ohata & Ishida[31] show that the volumetric efficiency of an engine is most strongly influenced by the inlet manifold pressure around BDC just before inlet valve closure: pressure pulses elsewhere in the inlet period have little effect. The volumetric efficiency is then given by

$$\eta_v = \eta_{vo} + Cp_m \tag{29}$$

where η_{vo} is the volumetric efficiency if the pressure at the inlet valve is atmospheric, C is a ramming coefficient, and p_m is the mean pressure in the significant period (basically about 50° before inlet valve closure).

They then developed the equations for flow in a manifold system of the form shown in Fig. 18. This effectively consists of four (primary) pipes connected to the four engine cylinders, a plenum chamber, and a further (secondary) pipe breathing from the atmosphere. This is a

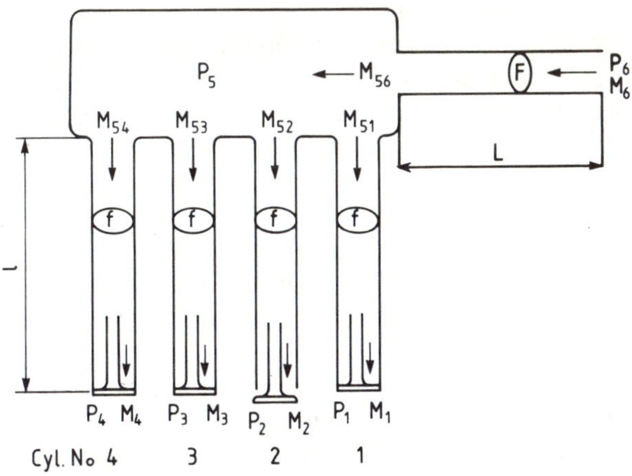

Fig. 18. Inlet manifold layout, defining parameters.

simple design of tuned manifold for a four-cylinder engine. It can be shown that the tuned frequencies for such a manifold are given by

$$\cos\frac{\omega l}{a}=0 \tag{30}$$

$$\frac{F}{f}\cot\frac{\omega L}{a}=\frac{\omega V}{fa}+4\tan\frac{\omega l}{a} \tag{31}$$

where F is the cross-sectional area of the secondary pipe, L is the length of the secondary pipe, f is the cross-sectional area of the primary pipe, l is the length of the primary pipe, a is the speed of sound, V is the volume of the plenum chamber, and ω is the angular frequency. If the number of cylinders is not four then the term $4\tan(\omega l/a)$ becomes $n_c\tan(\omega l/a)$.

Equation (30) defines the tuned frequencies of the individual primary inlet pipes, while eqn (31) defines the tuning of the whole manifold. Equation (30) is exactly equivalent to eqn (1) suggested by Broome,[1] and leads to tuned frequencies of $\omega=\pi a/2l$, $3\pi a/2l$, $5\pi a/2l, \ldots$; these frequencies give the values of θ_t (listed in Section 2.1). Similarly, the lowest frequency of eqn (31) is equivalent to a Helmholtz resonator: this can be shown by making $l=0$ (this results in a pipework arrangement like that shown in Fig. 6(a)) and taking the first term of the expansion of $\cot(\omega L/a)$, i.e. making the assumption $\omega L/a \ll 1$. For the pipework investigated by Ohata and Ishida the lowest frequencies evaluated from eqns (30) and (31) were 156 and 57 Hz respectively (the measured values were 157 and 56 Hz), showing that the plenum chamber and secondary pipe lower the natural frequency significantly. It is also possible to derive an expression for the difference in pressure between the plenum chamber and the inlet valve as

$$p_2-p_5=ja\left\{\frac{\sec\dfrac{\omega l}{a}-1}{A\cos\dfrac{\omega l}{a}}-\tan\frac{\omega l}{a}\right\}M_2 \tag{32}$$

where

$$A=\frac{\omega V}{fa}-\frac{F}{f}\cot\frac{\omega L}{a}+4\tan\frac{\omega l}{a}$$

and M_2 is the mass flow rate in pipe 2 of Fig. 18.

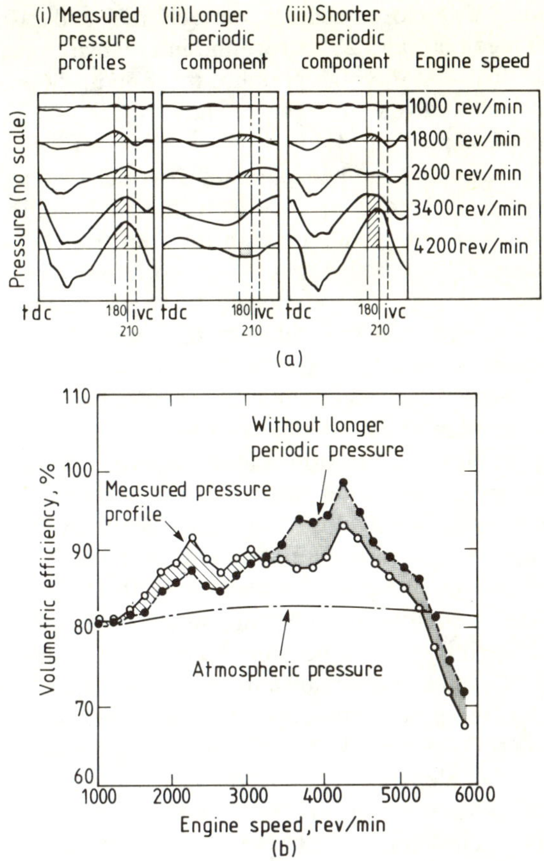

Fig. 19. Pressure variation in inlet manifold of Fig. 18 (at station 2): (a) measured profile—longer- and shorter-period components; (b) effect of tuning components on volumetric efficiency.

The first term of this expression contains the longer-periodic component of the pressure pulsation at the inlet valve associated with the overall system while the second one relates to the simple tuned primary pipe, i.e. the shorter-periodic component of pressure at the valve. These separated effects are compared to measured pressure traces in Fig. 19(a), and their effect on volumetric efficiency is given in Fig. 19(b). It can readily be seen that the measured curve is the sum of the two individual curves, and that the longer-period component assists the volumetric efficiency at speeds below 3000 rpm, while the shorter period component is beneficial around 4000 rpm. It is also apparent that to obtain the best charging of the cylinder that the maximum pressure upstream of the valve should be achieved not at inlet valve closure but at between 180° and 210° after TDC (which is between 50° and 20° before the inlet valve closes). This simply reflects the fact that the high pressure must be in the manifold at a time when the inlet valve is open sufficiently for the charging to benefit from the pressure pulse.

A complex lumped parameter model has been used by Chapman, Novak & Stein.[32] They solved the resulting equations using a Runge–Kutta technique with the step size controlled to $\delta t < 8\pi/\lambda_{max}$, where λ_{max} is the maximum value of the solution matrix: this is a much larger step size than allowed by the Courant–Fredrichs–Lewy criterion. The results calculated by this method are in good agreement with measured values.

Worth[33] has analysed inlet manifold traces obtained with a wide range of inlet manifolds, described below and shown in Fig. 29. Some of the detailed pressure traces are given later in Figs 31 and 32, where the measured values are compared with pressures calculated using a non-linear approach. An additional analysis performed by Worth, which is relevant to linear methods, was to obtain the frequency components of the pressure traces. In Figs 20 and 21 is shown the Fourier analysis of the pressure traces given in Figs 31 and 32 respectively. It can be seen from Fig. 20, which was obtained using a simple inlet pipe of 306 mm to each cylinder, that there are two basic resonances at all speeds. The lower frequency resonance is a broad spectrum and covers the first to fifth harmonic respectively: its form is largely independent of speed and it is the analysis of the suction trace. The other resonance is at a high harmonic number at low engine speed (11th harmonic at 2500 rpm) but moves to lower harmonics at higher speed, being the 6th harmonic at 5000 rpm. This resonance is associated with the natural frequency of the inlet pipe. In fact the tuning frequency for this pipe is around 250 Hz compared to the value of 299 Hz calculated using eqn (30), and neglecting end effects on the pipe. The analysis in Fig. 21 was based on a manifold similar to that of Ohata and Ishida,[31] and it can be seen that the resonant frequencies are now lower than previously observed. The frequency associated with the primary pipe is reduced because in this case the pipe length is 460 mm compared to the 306 mm long primary pipe used before. The characteristic low-frequency induction spectrum is still observed, as before, but the other pipe resonances are at around 190 Hz and 150 Hz. The values calculated using eqns (30) and (31) for this manifold were 198 Hz and 113 Hz respectively. Consideration of the

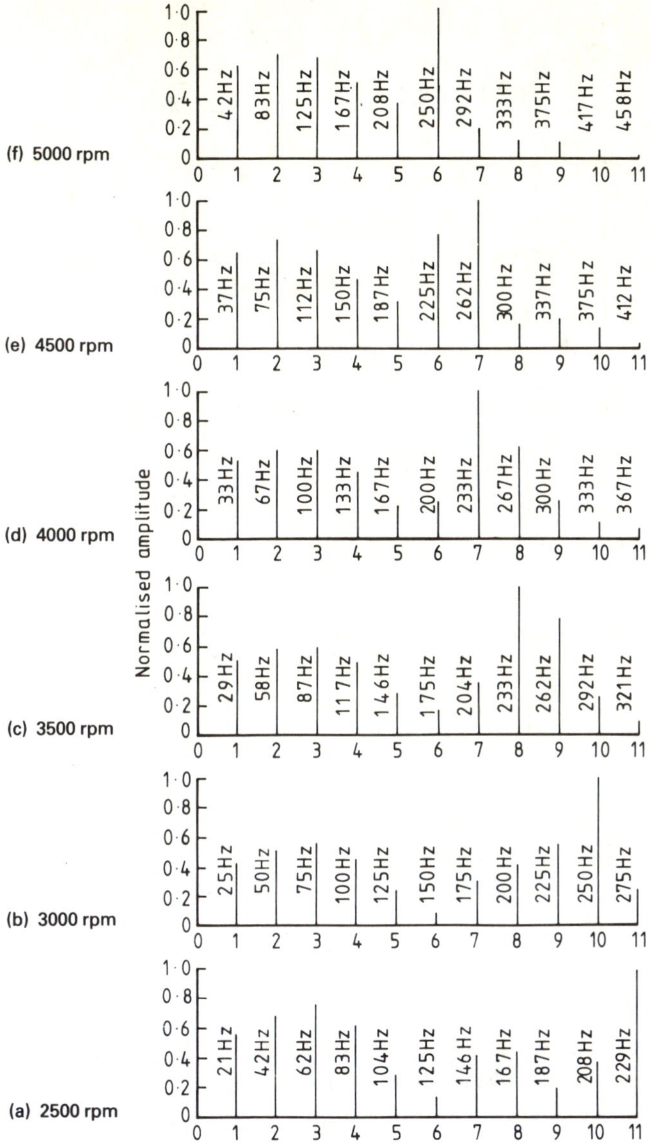

Fig. 20. Fourier components of pressure in intake pipes: 306 mm primary pipes (see Fig. 31).

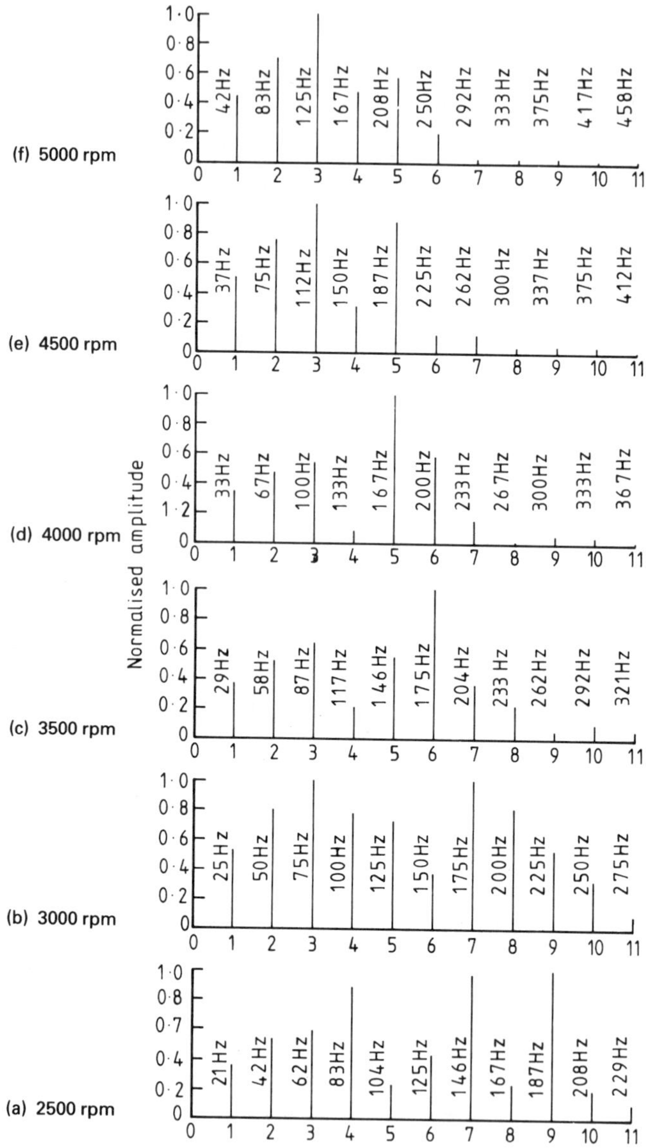

Fig. 21. Fourier components of pressure in intake manifold: 460 mm primary pipes plus plenum chamber (see Fig. 32).

overall engine results showed that the engine tuned at around 4500 rpm in both cases, and this coincides with the speed at which the pipe frequency is close to the third or fourth harmonic of engine speed. Such a conclusion is broadly in agreement with Ohata and Ishida, because at this frequency the pipe develops a strong positive pulse about 180° after the inlet valve opens, which is about 40° before it closes.

These results may also be compared to the approach of Broome[1] who proposed eqns (1) and (3), which relate the period of the pressure waves to crank angle (see Section 2.1). He suggests that the engine will tune at the speeds associated with θ_t equivalent to 120°, 85°, 67°, 55°, 46°, etc.: the best ram is achieved with the first two values but some effect will be observed at higher frequencies. These are equivalent to engine speeds of 11 960, 8472, 6678, 5482, and 4584 rpm respectively. Hence, this simple primary pipe is too short to achieve good performance with the lower harmonic numbers in the speed range of the engine, except for the fifth harmonic which tunes at about 4500 rpm: this was also observed from the engine performance. The longest primary pipe used by Worth, 460 mm, resulted in tuning speeds of 7956, 5635, 4442, 3646 and 3050 rpm: such speeds would be more appropriate to this engine, but are still high for improving driveability.

It is also possible to evaluate the tuning effect by solving the linear wave equations. A computer program based on a delay line model of

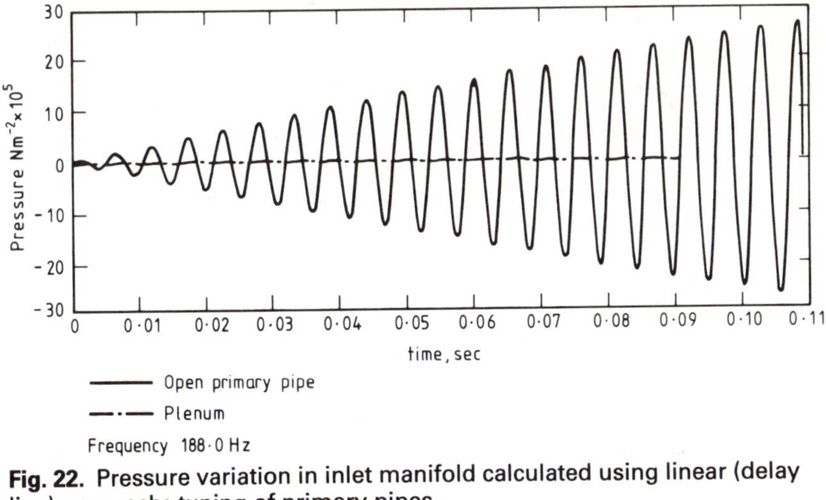

Fig. 22. Pressure variation in inlet manifold calculated using linear (delay line) approach: tuning of primary pipes.

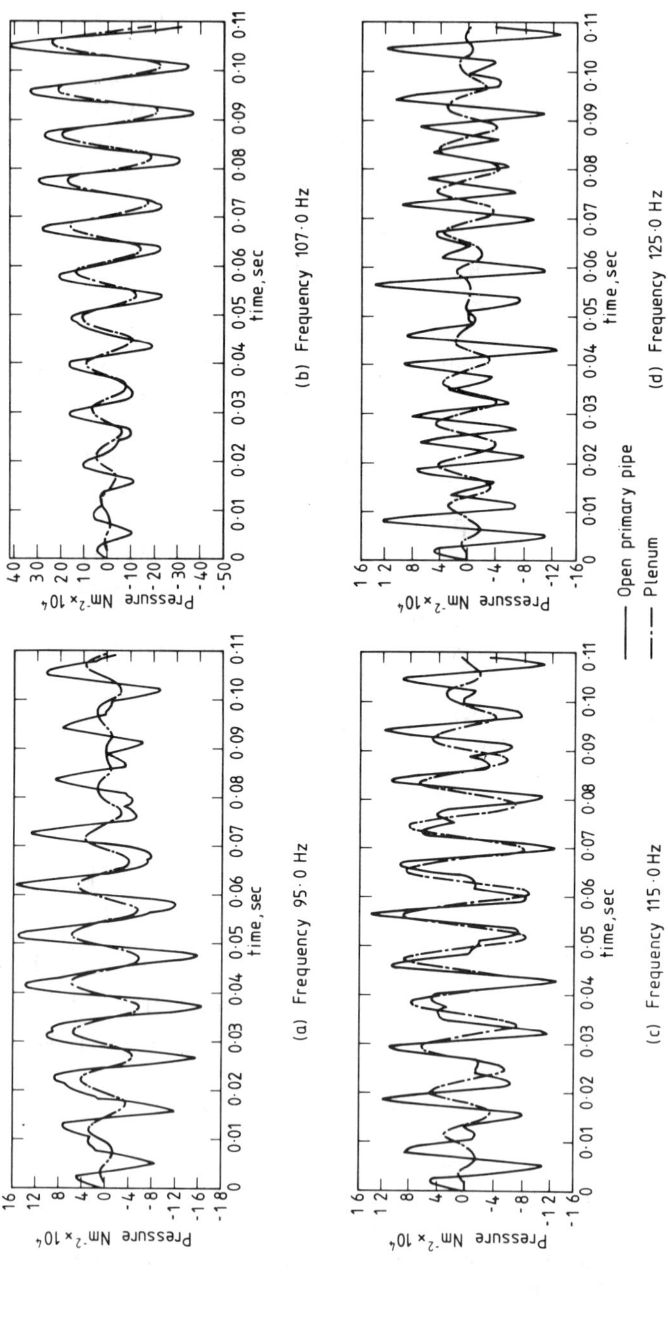

Fig. 23. Pressure variation in inlet manifold calculated using linear (delay line) approach: tuning of total system.

the form proposed by Boucher & Kitsios[34] (which was itself developed from the approach of Auslander[35]) has been written to simulate engine manifolds. In Figs 22 and 23 is shown the pressure variation in the inlet pipe to cylinder 2 of a manifold similar to that in Fig. 18 when subjected to a sinusoidal mass flow variation at the open end of that runner, and also the pressure variation in the plenum. Figure 22 was calculated at a frequency of 188 Hz (that obtained from eqn (30) for this particular manifold was 185 Hz—the difference is because the delay line approach is based on an integral number of delay lengths and these are not identical to the geometrical lengths) and it can be seen that the pressure variation all occurs in the primary pipe: the pressure variation in the plenum chamber is extremely small. Figure 23 was calculated with lower input frequencies of 95, 107 and 115 Hz, and the resonance can be clearly seen at about 107 Hz: the frequency calculated from eqn (31) was 107 Hz. Hence, while eqns (30) and (31) enable the resonant frequencies of manifolds to be obtained, a simple step-by-step model enables more detailed pressure variations to be examined. It is shown in Fig. 22 that the tuning is due simply to a longitudinal wave action in the primary pipes, whereas Fig. 23 indicates a resonance of the type found in the tuning of a Helmholtz resonator. In the latter case the pressures in the primary pipes and the plenum chamber are in phase, and air is 'pumped' into and out of the plenum chamber. Further advantages of the step-by-step approach are: (i) the primary pipes do not all have to be equal in length—although it is obviously beneficial for engine performance if they are; (ii) complex manifolds can be constructed without complete reformulation of the equations; (iii) interference between cylinders can be taken into account; (iv) the effect of pressure traces on volumetric efficiency can be included. A simple linear computer program of this type is much faster than one solving the non-linear equations. It should be an order of magnitude faster than a method of characteristics program, and Matsumoto & Ohata[29] quote a linear method 200 times faster than the non-linear ones. Hence, such an approach is ideally suited to an initial investigation of manifold arrangements.

Perhaps the most interesting advance in this area is the work by Matsumoto & Ohata,[29] who have applied a linearized method which has the ability to calculate volumetric efficiency. As they state, the choice of design for a variable geometry manifold is extremely time-consuming because of the many possible configurations that require investigation. While the non-linear wave equation methods are suitable analysis tools, they are of less use in this area. The requirement is for a technique which can analyse a variety of networks of pipe arrangements with sufficient accuracy to define those worthy of further investigation: then the non-linear techniques can be applied. Matsumoto & Ohata apply linearized equations in the frequency domain to the calculation of pressure waves in the inlet manifold: they show that the linearized equations will accurately represent the pressure variation in a naturally aspirated engine manifold up to 6000 rpm. They subdivide the induction process for multi-cylinder engines into two parts: the self-induction effect and the interference effect. The former is the variation in volumetric efficiency of a cylinder operating in isolation from the other cylinders, while the latter is the difference caused when the other cylinders are active. They show that the self-induction effect is the larger and then concentrate on this. A set of equations, similar to those of Soedel[36] and quoted in eqns (21) and (22) of Chapter 12, is developed for the usual pipe boundaries, including open and closed ends, and constant (or equal) pressure junctions: the valve is considered as an open end, and the cylinder volume change excites the pressure waves in the manifold through this. This equation set is solved for the pressure variation and mass flow variation throughout the manifold. The pressure variation is given by

$$P_1(s) = G_1(s)V_c(s) \tag{33}$$

where $G_1(s)$ is the transfer function matrix of the inlet manifold, and s is the complex operator, $s = j\omega$.

To use the technique all the engine parameters must be transformed to the frequency domain, which means that the cylinder volume, V_c, must also be considered in that sense. Typical results obtained are compared with the experimental values in Fig. 24; the agreement is extremely good considering the assumptions made, and the magnitude and phase are close throughout the whole speed range from 1000 to 7000 rpm. The authors then use the value of cylinder pressure at inlet valve closure to evaluate the volumetric efficiency from

$$\eta_v = (p_c/p_a)^{1/\kappa}(V_c/V_s)\}_{\alpha_{ivc}} \tag{34}$$

where p_a is the ambient pressure.

This equation is based on the assumption that the density of the gas in the cylinder is isentropically related to its pressure. In Fig. 25 the measured and calculated values of volumetric efficiency are compared over the speed range: the calculation technique is only able to show variations of this from the baseline value, in this case that at 1000 rpm.

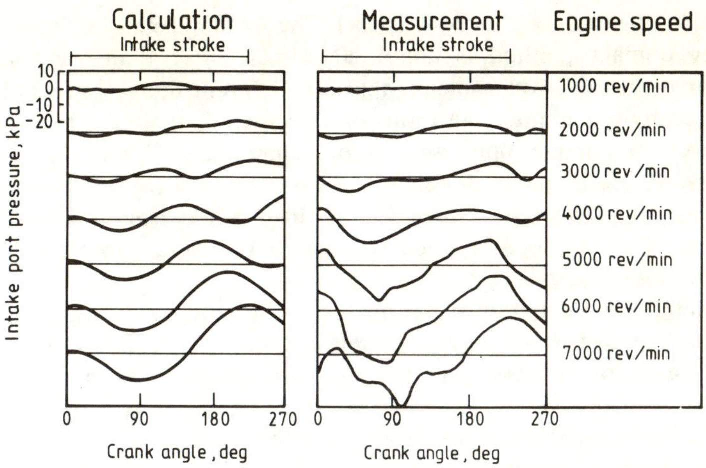

Fig. 24. Intake manifold pressure variations: calculation performed in frequency domain.

Again the agreement is good. The measured values show the variation of η_v with both single- and multi-cylinder valve operation. It can be seen that there is an interference effect between the cylinders, although in this case it is small. Investigation of the effects of interference between cylinders shows that it is possible to connect all the cylinders of a four-cylinder engine to a single manifold without deleterious results, whereas interference ruins the low-speed breathing of a six-cylinder engine. (It is necessary to split the manifold of a

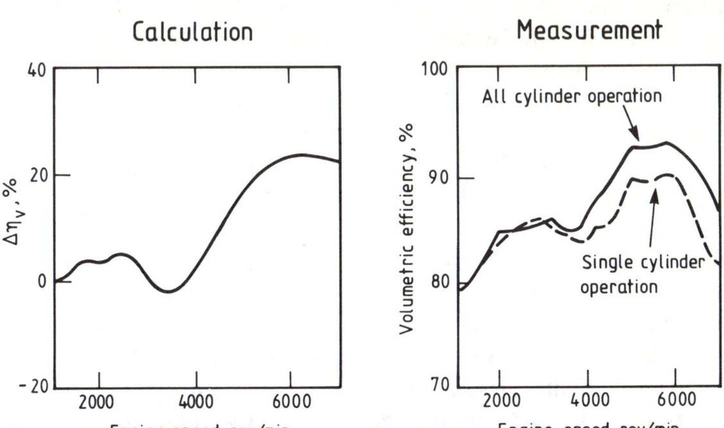

Fig. 25. Variation of volumetric efficiency calculated by frequency domain technique.

six-cylinder engine; see Section 4.1.3.) The authors state that the frequency domain method is about 200 times faster than the full non-linear technique, which means that 200 different manifolds could be analysed by this method (in terms of computer time) in the time it would take to analyse one using a full simulation. Some further developments are required to enable the method to cope with the interference effect, and this means that the method cannot always show detailed differences between engines of the same family, e.g. six-cylinder and four-cylinder versions.

Two features are apparent from this section. Firstly, linear methods give a good indication of tuned frequencies and can be extended to give pressure traces of acceptable accuracy. Secondly, they are extremely useful in the early stages of a design because it is easy to examine a broad range of manifold shapes with a view to achieving the desired performance from the engine, particularly in terms of volumetric efficiency and torque. Further development of step-by-step and frequency domain linear methods might extend the use of the technique for intake manifold design. This area has particular significance now because of the use of variable geometry intake manifolds (see Section 4.3).

4.1.2. Non-linear Techniques

Most of the finite difference techniques used to evaluate flow in internal combustion engines are non-linear. This heading includes all the approaches based on the method of characteristics and the typical finite difference methods derived from Taylor's theorem (i.e. Lax–Wendroff, etc.). These have been discussed in Section 3 of Chapter 12. Here the emphasis will be on validation of the methods. Basically, the methods are capable of predicting the variation of pressure and velocity in the pipework system: it would be ideal to validate the methods by comparison with measurements of these parameters. Most of the experimental data available is for pressure, simply because this is much easier to measure. However, some measurements of instantaneous mass flow have been made and these will be discussed.

Horlock & Winterbone[22] give an extensive discussion of the validation of a computer program based on the method of characteristics, and further validation of this particular program will not be given here. Bingham & Blair[37] describe the use of a homentropic method of characteristics approach to predict gas flow in the inlet and exhaust systems of a four-stroke petrol engine. It should be noted that the inclusion of pressure losses in a homentropic approach is inconsistent, but the inaccuracy in the inlet manifold will probably be small. In Figs 26 and 27 are shown some results obtained at 2000 and 4900 rpm respectively, at two points in the inlet manifold and two in the exhaust. At 2000 rpm the accuracy of prediction in the inlet manifold is good with both the constant-pressure and pressure loss junction models (the model is briefly described in Section 4.5.2 of Chapter 12); the same is not true at 4900 rpm, where there are quite significant deviations from the experimental values. The predictions with pressure loss junctions are significantly better than those with constant-pressure junctions. Bingham & Blair also show results at quarter throttle and the predictions are closer to measurement in this case, with the pressure loss junction model again being an improvement on the constant-pressure junction one. Benson[24] shows that the predictions with pressure loss junctions do not greatly improve the accuracy of calculation in the case of turbocharged diesel engine inlet manifolds. Hence it is more important to consider pressure losses in the junctions of high-speed engines, simply because the gas velocities encountered therein result in significantly greater losses.

Bingham[38] describes the validation of a computer program based on the homentropic method of characteristics. He considered five engines in total and calculated the performance of these over a range of loads. The results obtained for the pressure and velocity variations in the intake manifold are good, the agreement with the exhaust manifold pressures is also acceptable. The overall prediction of engine performance is impressive, although this will be strongly dependent on the 'combustion model' chosen. In Fig. 28 is shown useful data on the loss coefficients (defined as $K = \Delta p / \rho u^2$) of various manifold sections. Bingham also states that, to minimize losses, bends should have a radius of not less than twice the pipe diameter, and pipes entering plenum chambers should have a blending radius (bell–mouth) greater than 0·15 times the pipe diameter. He recommends that a useful design rule is to size the pipes in the inlet manifold to give mean velocities of 60–75 m/s at the maximum torque speed (based on assumed volumetric efficiency of 100%), with the total volume of the manifold less than half the engine swept volume. This is presumably a simple rule which ensures that the Mach index or mean inlet Mach number is small enough to give high volumetric efficiencies. Bingham gives a number of results which show good agreement between measured and calculated values. It is difficult to assess how a

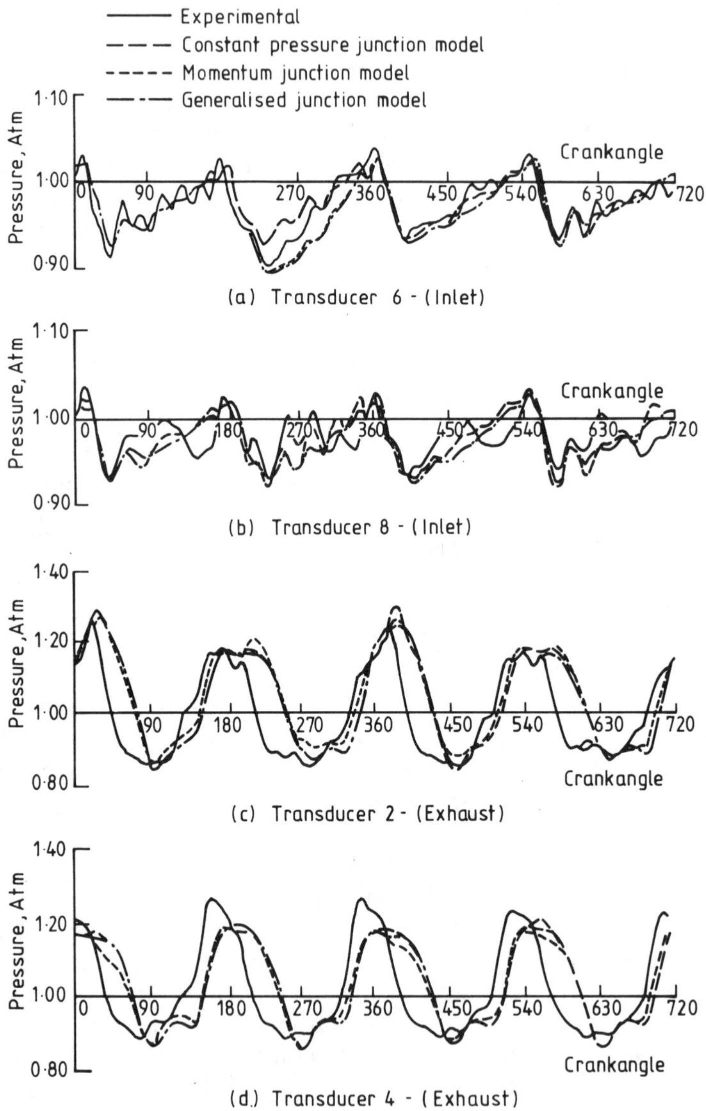

Fig. 26. Pressure variation in intake and exhaust manifold of a four-stroke petrol engine at 2000 rpm.

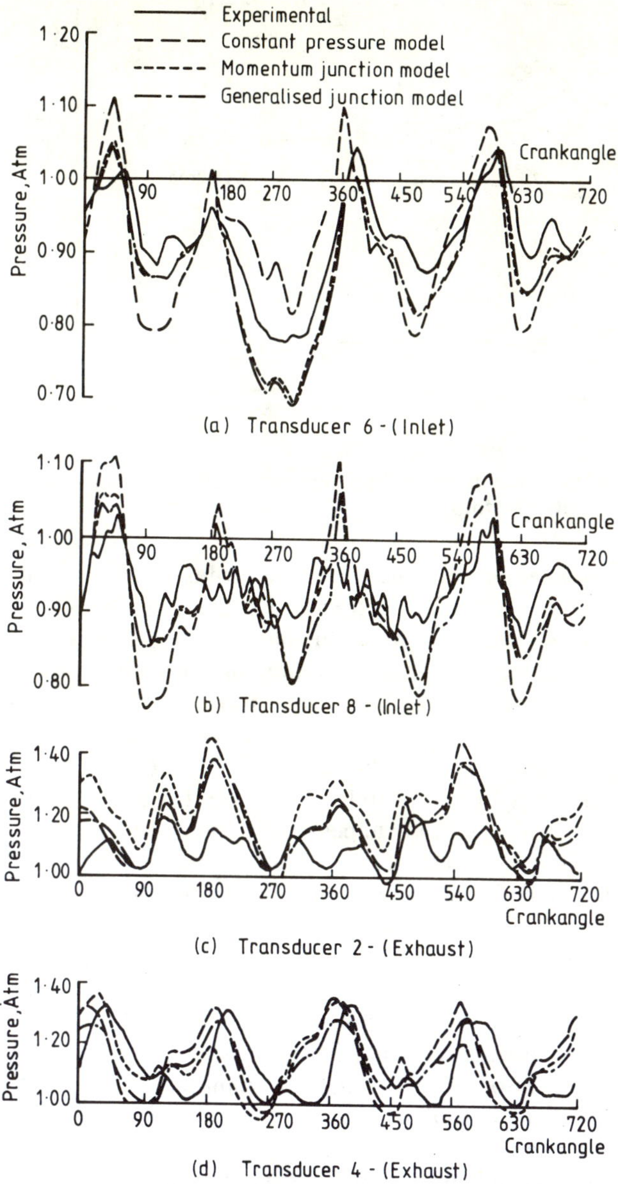

Fig. 27. Pressure variation in intake and exhaust manifold of a four-stroke petrol engine at 4900 rpm.

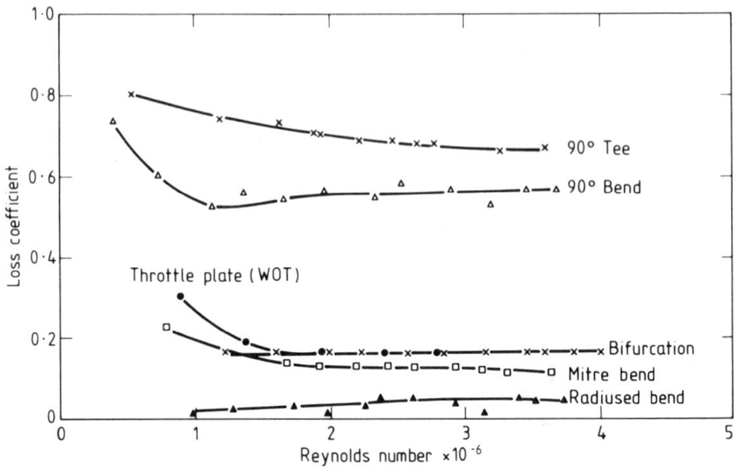

Fig. 28. Variation of loss coefficient with Reynolds number for various components of an inlet manifold.

homentropic program can achieve such accurate results over the whole engine operating range (especially considering the inaccuracies in pressure prediction in Fig. 27), and Bingham does not explicitly describe how the exhaust pipe reference conditions are defined: these are an essential component of the homentropic technique (see Benson[24]).

Recent work by Worth[33] in the author's laboratories at UMIST has concentrated on methods of tuning inlet manifolds for small petrol engines. A wide range of manifold shapes has been constructed from basic elements to enable the methods to be tested on a realistic set of possible manifolds. In Fig. 29 is shown an inlet manifold which has been constructed of individual modular components to make a particular manifold shape. It could be constructed from the three different primary pipe lengths available, to be connected to the three different sized plenum chambers (which, in turn, had different carburettor mounting points: central, top and end). Additionally, secondary pipes may also be fitted to the manifold. The whole structure of the manifold is similar to that proposed by Engleman[30] and Vorum[8,9] and, when the secondary pipes are fitted, can result in a system with two basic degrees of freedom, i.e. two tuning frequencies. A full set of results was obtained over most of the operating speed range of the engine, from 2500 to 5000 rpm at full throttle, for a total of 12 manifold arrangements. The different manifolds were permutations of open-ended primary pipes; small, medium and large plenum chambers, fitted with short, medium and long primary pipes. In addition, the carburettor body was mounted in the centre, on the top, or on the end of the plenum to test the effect of throttle mounting position on performance. A series of tests was also performed with secondary pipes of different lengths. To avoid variation in combustion parameters affecting the results, the fuel supply was from a multi-point fuel injection system. The various system parameters are shown in Table 3 and the engine arrangement is shown schematically in Fig. 30. The system shown in Fig. 30 is in the form required for simulating the engine on a full wave action computer program.

The primary aim of this project was the development of methods for designing inlet manifolds. The first part of the project, reported here, investigated how well an existing computer program, based on the non-homentropic method of characteristics, will predict a wide range of manifold arrangements. In Fig. 31 is shown the pressure variation

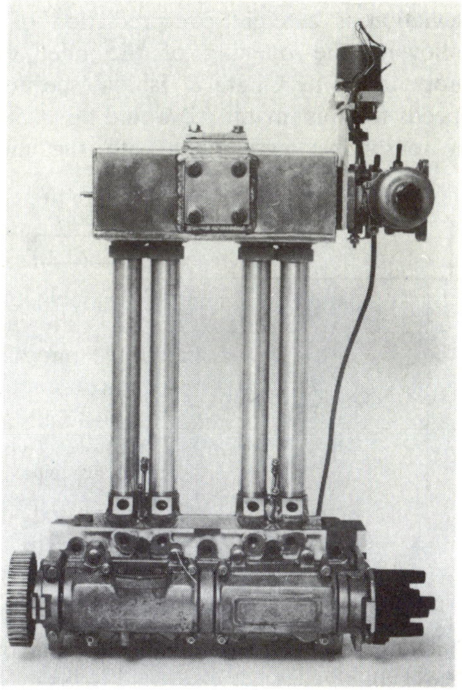

Fig. 29. Fabricated intake manifold for four-cylinder engine.

Table 3. Components of the Inlet Manifold

Engine type	Austin Rover 'S' series
Swept volume (cm^3)	1600
Bore (mm)	76·2
Stroke (mm)	87·6
Compression ratio (nominal)	9·6 : 1
Maximum rated speed (rpm)	
Primary pipes: short (mm)	230 } $\varphi = 30$ mm
medium (mm)	306 } $\varphi = 30$ mm
long (mm)	460 } $\varphi = 30$ mm
Plenum chamber volume: small (cm^3)	769
medium (cm^3)	1487
large (cm^3)	2948
Secondary pipes: short (mm)	126 } $\varphi = 45$ mm
long (mm)	1040 } $\varphi = 45$ mm

just upstream of the inlet valve for the engine operating with four medium length pipes breathing directly from the atmosphere, i.e. with open ends. It can be seen that the program accurately calculates the pressure pattern over the whole speed range considered: the only deviation is a slight overprediction of the depression immediately following the opening of the inlet valve. Based on the theory propounded by Ohata & Ishida (see eqn (30)), the expected tuning speeds for this 'manifold' would be 10 500, 7000 and 4200 rpm; this is in reasonable agreement with the measured results which show

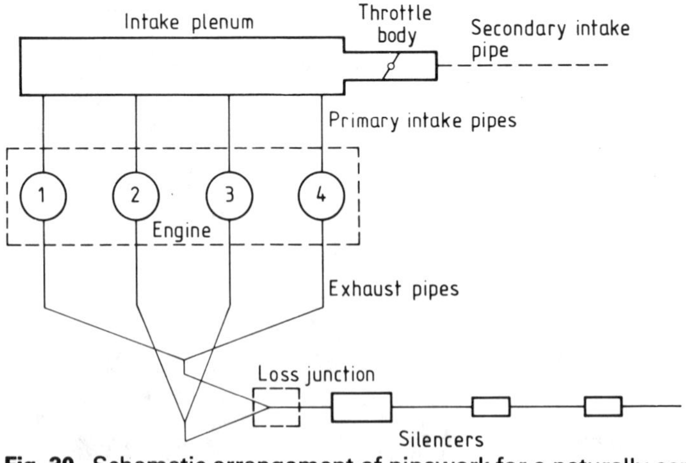

Fig. 30. Schematic arrangement of pipework for a naturally aspirated petrol engine.

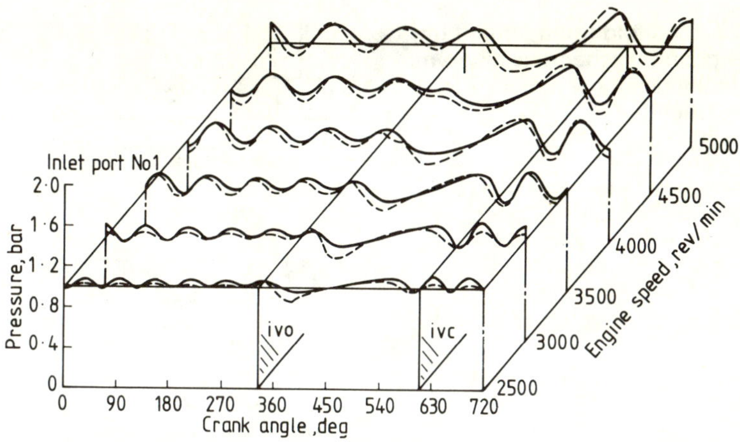

Fig. 31. Pressure variation in intake with medium length (306 mm) open ended pipes.

optimum wave superposition at between 4000 and 4500 rpm. The results obtained from the plenum chamber arrangements indicate that the largest effect is found at high engine speeds, probably because the primary pipes are never long enough to bring the tuned frequency well down the speed range. This is the dilemma of the vehicle designer: how can a system that will tune at low speed fit under the bonnet of the vehicle? A comparison of measured and predicted pressure traces for the large plenum with long pipes is shown in Fig. 32: the

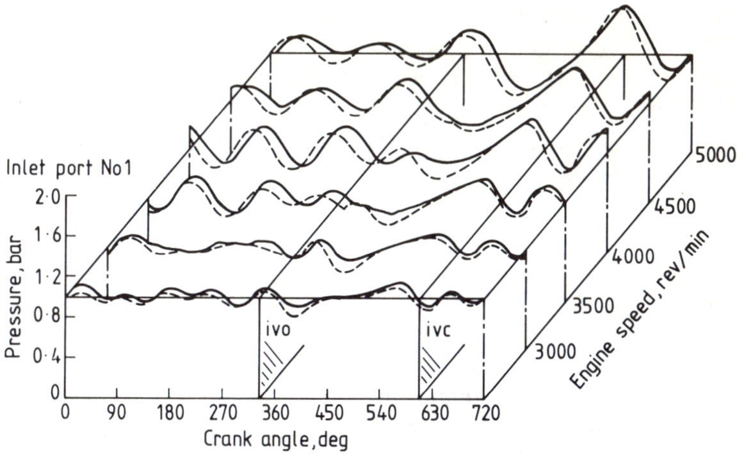

Fig. 32. Pressure variation in intake manifold with large plenum chamber (2948 cm^3) and long primary pipes (460 mm).

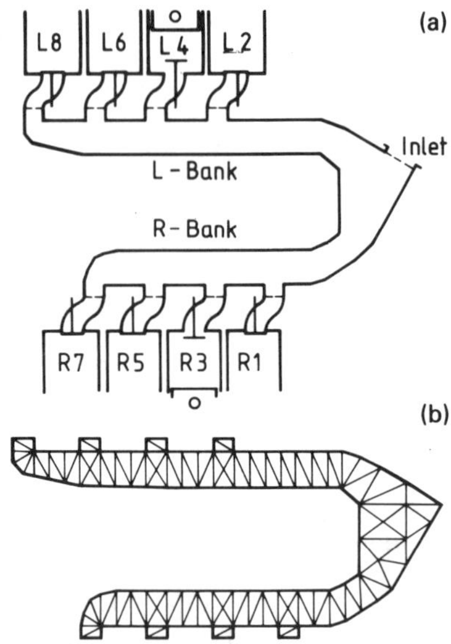

Fig. 33. Manifold arrangement for V-8 engine: (a) schematic diagram; (b) two-dimensional mesh arrangement.

agreement is good throughout the speed range. These results were chosen because they exhibit larger pressure variations than the other arrangements. It should be noted that constant-pressure junction theory was used in all these predictions, although it is known that the pressure losses occurring at the boundary between the primary pipes and the plenum can give high losses. Poor blending of the junction can ruin the performance of a system that is acceptable in theoretical terms; this was also found by Bingham & Blair (see above).

Takeyama, Ishizawa, Yoshikawa & Takagi[39] describe the use of the two-step Lax–Wendroff technique to predict the flow in the pipework of an automotive engine. The results are in good agreement with the measurements, which supports the applicability of this method for engine simulation. A novel feature which the authors have included in their approach is the modelling of junctions by the 'fluid in cell' (FLIC) approach. This enables a degree of two-dimensionality to be introduced into this region of the calculation, reducing the need for empirical loss coefficients. They also introduce the boundary conditions at pipe junctions in a form amenable to extending the Lax–Wendroff technique rather than using the method of characteristics. A more extreme version of this approach is reported by Tosa, Shimoda & Oikawa.[40] They used a two-dimensional mesh to evaluate the flow in the intake manifold of a V-8 engine, based on the mesh shown in Fig. 33, and evaluated the flow vectors shown in Fig. 34. The calculation time was reduced in the simulation by applying the two-dimensional mesh only in the regions where the geometric arrangement was liable to give such flow effects: the remaining areas were dealt with using a one-dimensional method. They also show the strength of the FLIC method for evaluating pressure losses in junctions, with an ability to differentiate between various flow angles.

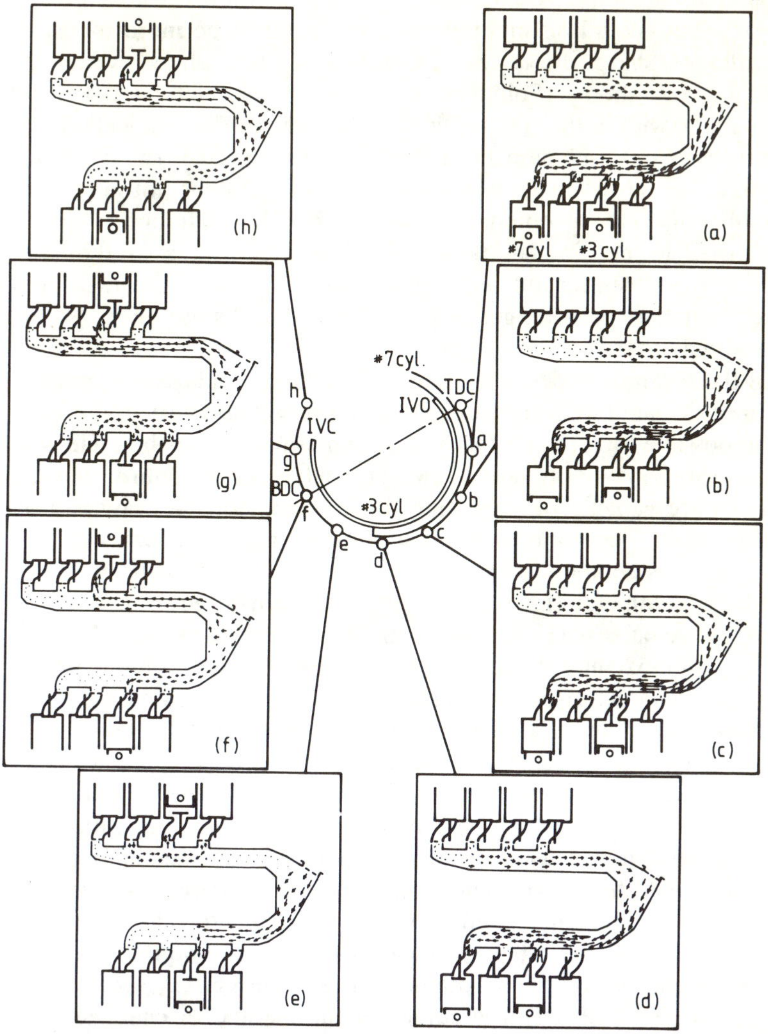

Fig. 34. Flow vectors calculated by two-dimensional technique (see Fig. 33).

Chapman, Novak & Stein[41] further develop the finite difference technique introduced by Walter & Chapman[42] when they used an algorithm called FRAM to solve the equations of flow. They claim that this enables a higher order difference scheme to be used when the flow is 'smooth' and introduces 'a minimal amount of dissipation only in those regions where a higher order algorithm would produce non-physical oscillations'. Comparisons of the FRAM scheme with a two-step Lax–Wendroff method show the latter to be more robust in the area of sharp discontinuities, retaining their shape while not exhibiting overshoots (Poloni *et al.*[43] found similar effects but overcame them within the Lax–Wendroff method). They modelled the junctions as small plenum chambers and introduced a small viscous loss term in the momentum equation: the model does not appear to contain sufficient geometric data to enable different junction geometries to be simulated. The calculated pressures in the inlet manifold and plenum are all in good agreement with the measured values over the speed range: surprisingly, no results are given for the exhaust manifold.

All the techniques described above have been validated by comparing the calculated and measured instantaneous pressure variations. It is also possible to correlate the predicted mass flow over the cycle with the measured mass flow rate. However, this does not confirm that the methods are capable of predicting the instantaneous velocity values. Sierens *et al.*[44] describe experiments on a single-cylinder engine in which the instantaneous gas velocity was measured using a laser–doppler anemometer (LDA). In Fig. 35 are shown the measured and predicted velocities 110 mm downstream of the carburettor of the engine, and 520 mm from the inlet valve. The non-homentropic method of characteristics was used for the calculation and the valve boundary was defined as a resistance coefficient (see Section 4.7 of Chapter 12). The agreement in the high-velocity region is good except as the velocity decreases around inlet valve closure; this is thought to be because of the spurious experimental readings obtained at this time when the seeding density is low. The particular LDA used did not have a frequency shifting device and hence could not differentiate between flow directions; this also resulted in the negative flow calculated just before 720° crank angle being indicated as a positive value. The level of agreement between the predictions and measurements is good enough to give faith in the accuracy of the method of characteristics calculation technique. It should be recognized that this method is

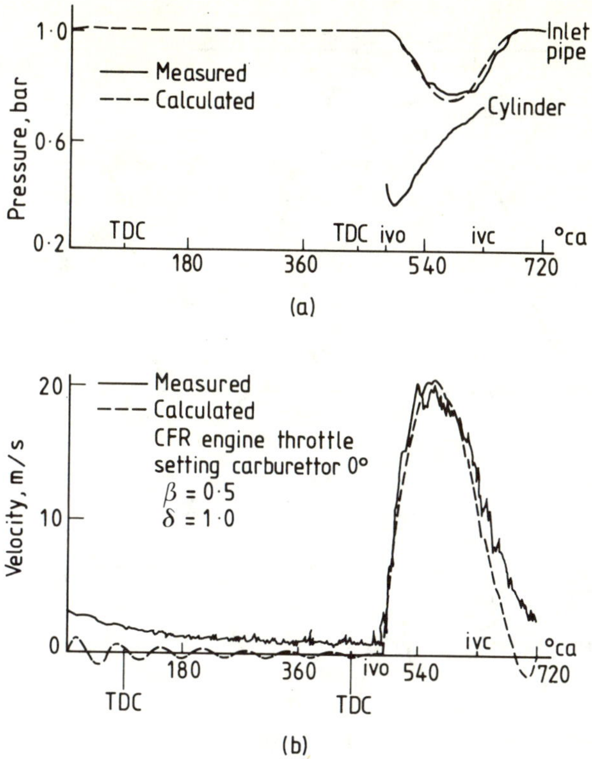

Fig. 35. Measured and calculated pressure and velocity in the intake pipe of a single-cylinder engine. Measurements by laser–doppler anemometry.

particularly susceptible to errors when evaluating velocity because the latter is based on the difference between the Riemann variables (see eqn (13) of Chapter 12).

Lakshminarayanan, Janakiraman, Babu & Murthy[45] have also attempted to measure the velocities in the intake and exhaust pipes of an engine, but they used an ultrasonic flowmeter. This consisted of a source of ultrasound (a modified spark plug) and two ultrasound receivers. By using two receivers it is possible to solve for the gas velocity without knowing accurately the temperature of the gas. It is also possible to evaluate the gas temperature from the readings obtained. They compared the ultrasound meter with a spark discharge device to assess its accuracy as a velocity measuring instrument, and a

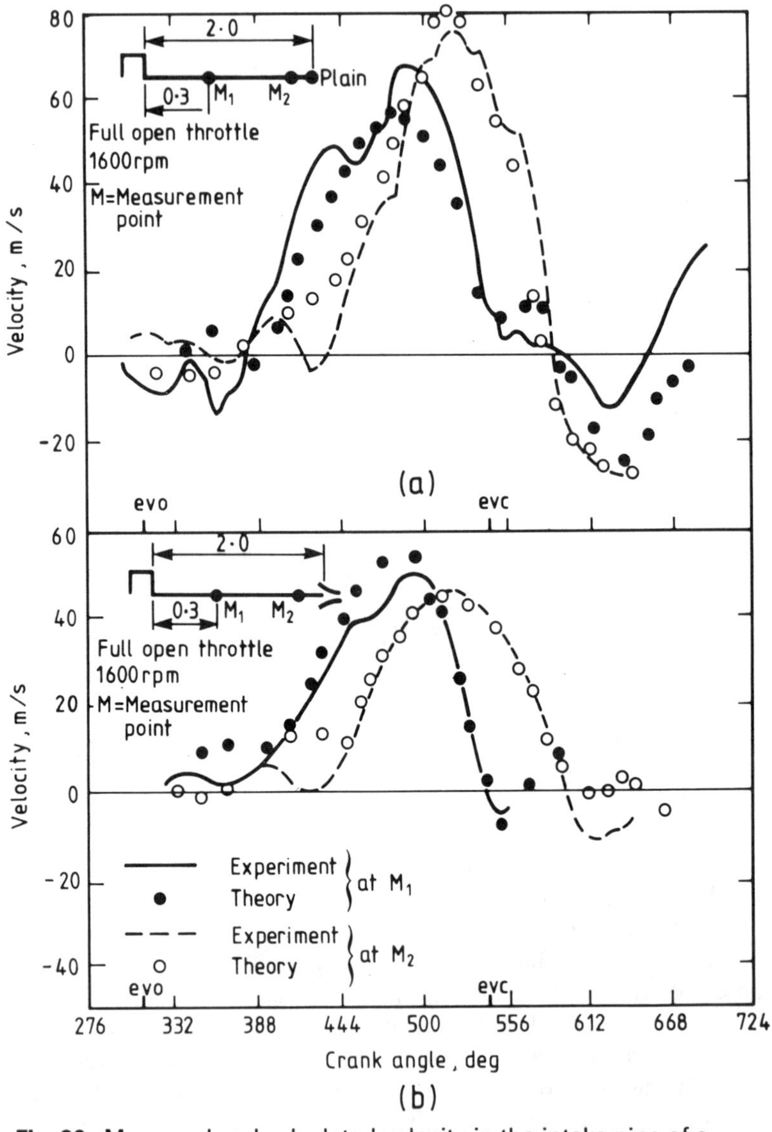

Fig. 36. Measured and calculated velocity in the intake pipe of a single-cylinder engine. Measurements by ultrasonic method.

resistance thermometer to see its ability for measuring temperature. In both cases the ultrasonic instrument was shown to follow the gas flow parameters as accurately as the other conventional devices. In Fig. 36 it is shown that the agreement between measurement and prediction of gas velocities, by a Runge–Kutta method (see Section 3.3.2.1 of Chapter 12), is again acceptable.

Hence it can be concluded that the non-linear methods of calculating unsteady gas flows in the intake systems of internal combustion engines will give a good assessment of the pressure variation in the manifold over a broad range of operating conditions. This section has concentrated on full throttle operating conditions, but part-throttle results are reported in Benson,[24] Bingham & Blair,[37] Sierens *et al.*,[44] Lakshminarayananan *et al.*[45,46] and by others. It would appear that part-throttle does not cause any particular difficulties until the intake vacuum becomes extremely low, when the carburettor boundary requires careful formulation to achieve acceptable results.

4.1.3. *Variable Geometry Inlet Manifolds*

In Fig. 37 is shown a typical manifold that can be designed using techniques of the type described above. It is a variable geometry manifold fitted to a 1998 cc V-6 petrol engine (bore 78 mm, stroke 69·7 mm). The engine has four valves per cylinder (intake valve diameter 30 mm, exhaust valve 26·5 mm) in a pent-roof combustion

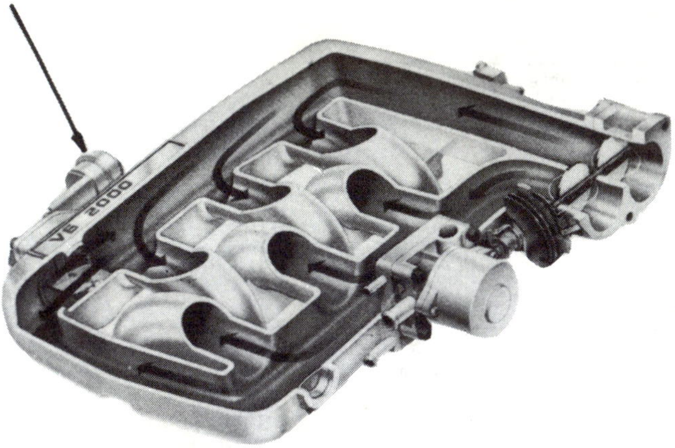

Fig. 37. Nissan Induction Control System (NICS) variable geometry intake manifold.

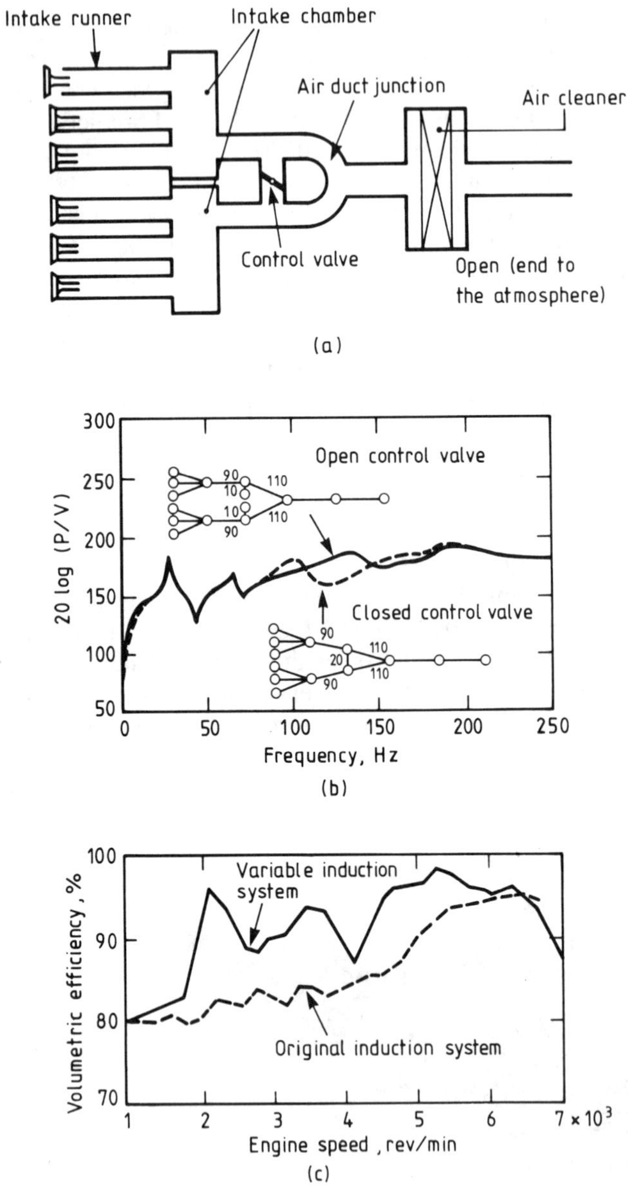

Fig. 38. Schematic diagram of variable geometry intake manifold by Toyota.

chamber with a centrally located spark plug. The valve timing is also variable by changing the camshaft phasing by 20° crank angle. The intake timing is advanced at low to medium speeds when the load is high; otherwise, it is maintained in the retarded condition. The variable geometry manifold, called the Nissan Induction Control System (NICS), is actuated by the valve indicated in Fig. 37. This is closed below 3400 rpm to increase the effective length of the tracts, thus tuning at a lower frequency associated with the total manifold system. Between 3400 and 5400 rpm the valve opens the connecting passage and makes breathing easier. It also uses the destructive interference between the six cylinders to move the 'node' of the tuning system to the small plenum chambers close to the cylinders, achieving tuning in the primary pipes alone. Above 5400 rpm the valve is closed again to take advantage of the ram effect that can be achieved. A slightly different embodiment of the same effect is described by Takeyama *et al.*,[39] when the control valve is only operated at one point of the engine range.

Matsumoto & Ohata[29] show a similar approach on a six-cylinder engine. The manifold and control valve arrangement are shown in Fig. 38. This design operates as a variable length manifold. When the control valve is closed the effective length of the pipes goes from the cylinders to the junction of the secondary pipes. Opening the control valve moves the node of the pressure pulsation to the interconnecting pipe, thus effectively shortening the system and increasing the tuned frequency. It can be seen that the variable induction tuned manifold system increases the volumetric efficiency of the engine substantially over the whole operating range of the engine: a simple manifold would tend to give improvements in one region to the detriment of performance at another. This can be seen from Figs 6 and 8, where a fixed geometry Helmholtz resonator was used. Note that the tuned system for a six-cylinder engine needs to consist of two subsystems, each containing three cylinders.

Tuned manifolds have also been applied to turbocharged diesel engines, and an example is given by Sato, Suenaga, Noda & Maeda.[47] In Fig. 39 is shown the Hino EP100-II engine; the long tuned inlet manifold is plainly visible, going from the charge-cooler in the radiator to the inlet manifold. This photograph depicts how a very long tuned pipe can be incorporated in a vehicle installation. In Fig. 40 it is shown how the manifold tuned length is varied to achieve the desired characteristics. The arrangement contains three control valves which

Fig. 39. Hino EP100-MkII engine with tuned inlet system.

are operated at various engine speeds to modify the engine output. The basic philosophy of this manifold arrangement is a Helmholtz resonator with *variable area* tuning pipes. How this is used to achieve a broad range of tuning will now be described. It can be seen from eqn (17) that the natural frequency of a Helmholtz resonator is defined by the parameter $(f/LV)^{1/2}$. Hence the point of resonance can be modified by changing the pipe area (f), the pipe length (L), or the plenum volume (V). An increase in the pipe area increases the natural frequency, whereas an increase in either the length or volume terms decreases the frequency. The Hino engine is tuned by two different

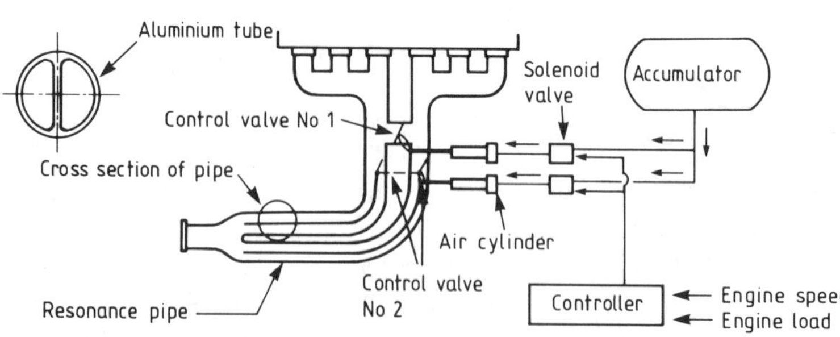

Fig. 40. Schematic of Hino manifold showing control valves.

mechanisms: first, by varying the pipe area (f) which is achieved by means of divided inlet pipes and, second, by varying the effective pipe length: this is shown in Fig. 41. To obtain the lowest frequency control valves 1 and 2 are closed, which gives the minimum flow area in the tuning pipes; this results in a tuned speed of about 700 rpm. In the next tuning region, control valves 2 are opened and the pipe area is effectively doubled. This increases the tuned speed by about 40% (i.e. $\sqrt{2}$) to around 1000 rpm. The tuned frequency of the manifold system can be increased still further by opening valve 1, which effectively shortens the pipes connected to the plenum chambers and gives a tuned frequency of around 1300 rpm. However, when valve 1 is open

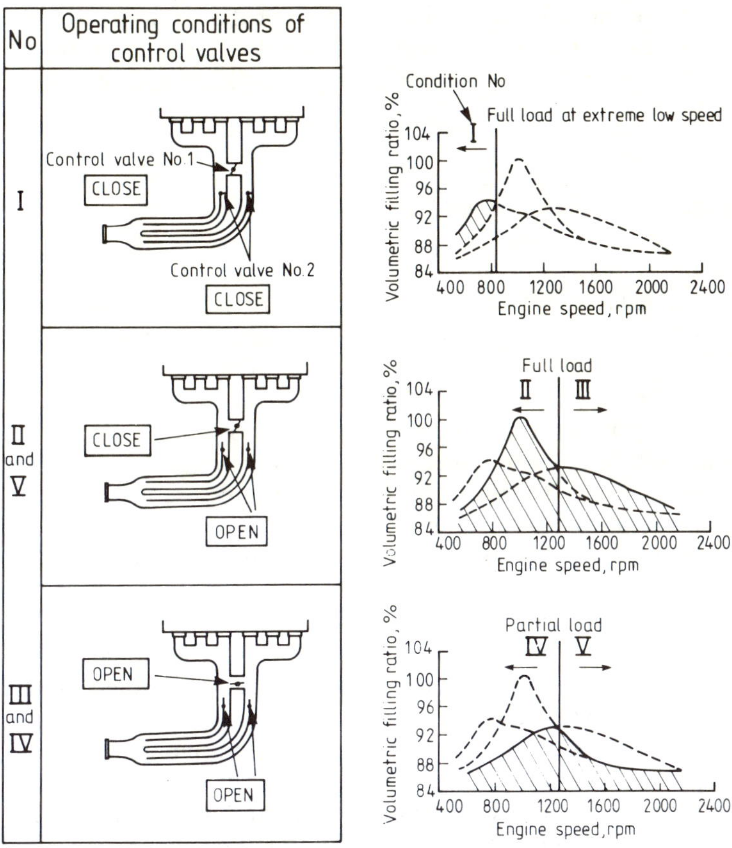

Fig. 41. Improvements in engine performance with tuned manifold.

the tuning is not as pronounced because of interference between the induction pulses. This is similar to the approach used above by Matsumoto & Ohata. The strategy adopted at part-load is not the same as at full load, and this is shown in the bottom diagram of volumetric efficiency against speed.

It can be seen that variable geometry manifolds enable the torque–speed characteristics to be broadened and this results in a more tractable engine. There are two benefits from this: (i) higher gear ratios can be used for the engine, giving better fuel consumptions; and (ii) fewer gear changes are required to achieve good driveability of the vehicle. The use of tuned manifolds based on long secondary pipes and plenum chambers, rather than using long primary pipes, will be important in automobile applications because of the trend in reducing bonnet heights. Variable geometry manifolds enable benefits in volumetric efficiency to be achieved over a broad range of engine speeds.

4.2. Exhaust Manifold Calculations

Many of the comments about inlet manifold simulation apply equally well to the exhaust manifold, but here there are two additional difficulties. First, the temperature is variable over a wide range during any one cycle, possibly from around 400 K up to more than 1000 K in the case of a petrol engine with a high scavenge ratio (i.e. a large valve overlap which enables a high throughflow of fresh charge). Secondly, the pipes are substantially longer, with a greater concomitant risk of shock waves forming. There is also the fact that pressure losses will be higher because of the increased flow velocities. When all these factors are considered it can be accepted that modelling the exhaust system will present more challenges than the inlet one. Fortunately, if the valve overlap is small, the accuracy of modelling the exhaust pipe will not have a large effect on the actual air mass flow through the engine since this is dominated by the ability of the engine to breathe through the inlet system: this also means that the effect on combustion might not be over-riding, since this is dependent on the aspirated air mass. However, while engines can operate with poorly designed exhaust systems there are benefits to be obtained by careful attention to this area of the power plant. In fact, on highly rated engines an efficient exhaust system can provide substantial performance gains.

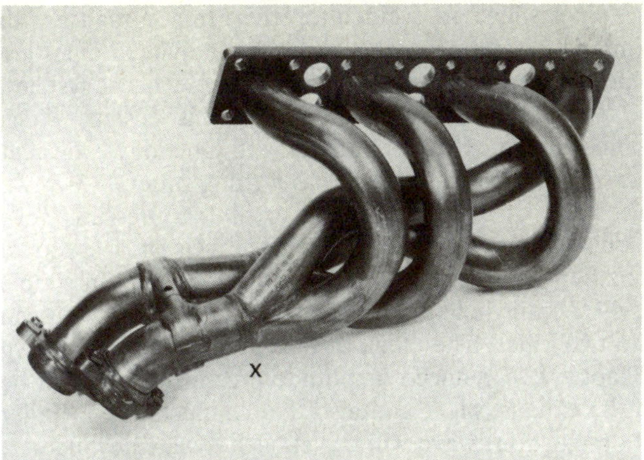

Fig. 42. Exhaust system on Daimler–Benz 2·3-16.

While there is a large amount of literature on the analysis of inlet manifolds, the same is not true for exhaust systems. Many of the papers available relate to single-cylinder engines, often with exhaust pipes more suitable for the laboratory than the road, and a large number concentrate on two-stroke engines (these are considered in detail in Chapter 3). To reduce the breadth of discussion, this section will review recent developments for multi-cylinder engines; additional references include Blair & Goulburn,[48] Takizawa, Uno, Oue & Yura,[49] Chapman *et al.*,[41] and Benson & Baruah.[50] In addition, Benson, Baruah and others[51,52] have produced papers on multi-cylinder engines fitted with exhaust gas recirculation (EGR) and catalytic converters, while Low & Baruah[53] developed a computer-aided design calculation for spark-ignition engines based on such techniques.

A typical exhaust pipe for a high-performance engine (the Daimler–Benz 2·3–16 introduced in Ref. 28) might be like that shown in Fig. 42: the lengths of the pipes from the cylinders to the first pipe junctions (X) are maintained the same for each of the cylinders. The connections after these junctions depend on whether the exhaust system remains split, or is joined to use one silencer system. It is common to join all cylinders into one exhaust pipe on in-line engines. Abthoff *et al.*[28] state that 'the exhaust system consists of high temperature steel pipes which are welded together. In accordance with the length determined during torque matching the individual pipes are united in two pipes just in front of connection to the exhaust system'. Annand & Roe[27] explain the reasoning behind this design, which relates to the manner in which waves interact with junctions. Consider the exhaust process in the engine (which has a firing order 1–3–4–2): when the exhaust valve of cylinder 1 opens it sends a pressure wave (the blowdown pulse) into the manifold. This travels at a speed of $u + a$ towards junction X, where three pipes join together. In simple terms the wave will see this junction as a change in area because it can now transmit into two pipes. The effect of this is similar to that shown in Fig. 11 (of Chapter 12): reduced amplitude pressure waves will be transmitted into the exhaust pipe, and also that connected to cylinder 4, but a rarefaction (suction) wave will travel back towards cylinder 1. If the length of the pipes is appropriate this rarefaction pulse can both draw residual gases out of, and pull fresh charge into, the cylinder. This tuning will tend to happen only over a limited range of speeds and hence care must be taken to ensure that the beneficial effects at one speed are not completely negated by detrimental effects at other speeds. The pipework is arranged to connect cylinders 1 and 4 because the exhaust periods of these cylinders do not overlap, and hence the pressure wave that went from the junction towards cylinder 4 will not affect its breathing (unless there is some residual undamped wave action at the time the valve opens). It should be noted that the junctions are very carefully designed to direct the exhaust gas towards the tailpipe of the vehicle and these will act like pulse converters, causing a much smaller amplitude wave to travel towards cylinder 4 than would be expected from a simple T-junction. Most engines have simpler exhaust pipes than the one shown, and often the pipes from all four cylinders will be joined in a single five-way junction.

The large variations in pressure and temperature make linear techniques unsuitable for exhaust pipe flows, and so this section will concentrate on the application of non-linear techniques. The simplest non-linear technique is the homentropic one. Benson (Figs 7.24 and 7.25 of Ref. 24) shows that it is essential in this method to choose the appropriate reference temperature, and hence entropy level, to achieve the correct results. It is difficult to assess the correct value in advance of the calculation, although an iterative approach can be adopted. This might give satisfactory results in engines where the gases in the exhaust system are of fixed composition, but if the scavenge ratio is large then the method will not be accurate across the hot- and cold-gas interface.

A typical recent example of the use of a homentropic technique to calculate engine performance is that by Bingham & Blair[37]; the results in the inlet manifold were considered above, and the agreement between measurement and prediction was good. The exhaust pressure traces in Figs 26 and 27 show a much poorer agreement between prediction and measurement. At 2000 rpm the agreement is reasonable in respect of amplitude but there is a phase error of about 30° crank angle: at 4900 rpm the agreement is extremely poor in respect of wave shape. It is apparent that there is a major shortcoming in the calculation technique, either in the pipe or at the boundaries, and the inclusion of pressure losses at the junctions does not overcome this. It is possible that the use of the homentropic method of characteristics has caused the wave calculations to become unrealistic in the hot exhaust gas flows, and this might explain the large deviations in wave shape. It must be remembered that the resultant wave pattern is composed of the interaction of a large number of waves; if these are wrongly calculated and go out of phase then the addition of the waves will change dramatically. For example, in Fig. 27(c) the calculations all predict a peak pressure at 180° crank angle, whereas the measurements show a local minimum close to that region: this could be because the phase of a wave has been displaced, causing it to be added to the previous wave shape rather than subtracted from it. Hence, it is possible that the wave travel times to various pipe junctions are incorrect, or even that the reflections from those junctions are wrong. Bingham[38] used the same computer program for engine analysis and showed good general agreement between gross engine performance measurements and predictions. As stated in the introduction to this section, the exhaust manifold pressure variation on an engine of this type, with a relatively small valve overlap, will not have a dramatic effect on the predicted airflow.

Recent work at UMIST[33] has investigated various exhaust manifold effects in an attempt to obtain satisfactory models of engine exhaust systems. The computer program is based on the non-homentropic method of characteristics, and includes friction, heat transfer, area change and models for pressure losses at various junctions. A schematic of the exhaust system is given in Fig. 30; it is very similar in layout to that in Fig. 42, but the components of this mass-produced vehicle engine will not be made to the same standard as those on the much more expensive Daimler–Benz. This could result in greater flow losses in the less aerodynamic junctions. Cylinders 1 and 4 are brought together, as are 2 and 3, prior to being joined at a further junction to give a single exhaust system. In modelling this arrangement it has been found that constant-pressure junctions result in the wrong pulse pattern, and to overcome the inaccuracies of the prediction a loss junction was included, as shown. The silencer boxes were also modelled as plenum chambers with pressure losses at entry and exit. Figure 43(a) and (b) compare predictions by two models with the measured values, at an engine speed of 3500 rpm and full throttle. In Fig. 43(a) it was assumed that the main junction was a constant-pressure one (Section 4.5.1 of Chapter 12), while for Fig. 43(b) pressure losses were included based on empirically obtained flow data (Section 4.5.2 of Chapter 12). It is shown in Fig. 43(a) that the use of the constant-pressure model results in an underestimation of the blowdown pulse (at about 180° crank angle), but there is overestimation of other pulses in the system. For example, the predicted pulse at around 400° is overestimated. This is the wave from cylinder 3 coming back through the pipework, and the overprediction indicates that the pressure losses in the junctions are too small; the pulse at 60° is a similar one from cylinder 2. It is also necessary to consider the pulse at 540° which is underpredicted; this is the blowdown pulse from cylinder 4 finding its way into this exhaust pipe (see the description above). It might be thought that the introduction of pressure losses will exacerbate this underprediction, but this will be seen not to be the case. In Fig. 43(b) are shown the improved predictions that can be achieved with a pressure loss model. The waves from cylinders 2 and 3 have been truncated at the junctions, but the blowdown pulse from cylinder 4 has been increased. Hence it is possible to obtain accurate predictions of the exhaust manifold pressures by attention to the detailed modelling of the system. Comparison of the exhaust pressure traces from the whole series of tests shows that the effect of the inlet manifold arrangement on the exhaust system parameters is marginal. In Fig. 44 are shown predictions, using the pressure loss model, of exhaust pressure for the whole speed range of the engine. The data provided for such a simulation would be the pipework specification, including pressure losses and carburettor model, and fuelling and combustion data. A full description of the requirements is given in Horlock & Winterbone.[22] The simulation accurately models the basic wave structure in the extensive exhaust system. The results at

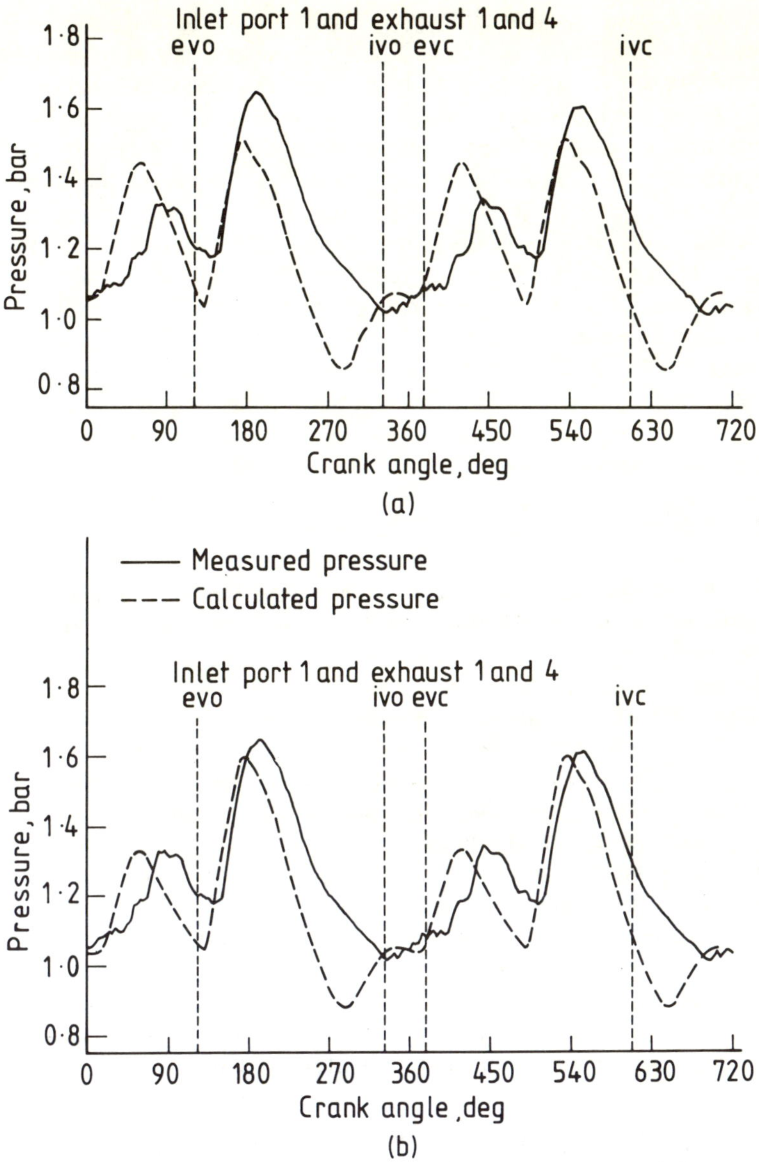

Fig. 43. Pressure variation in a petrol engine exhaust manifold at 3500 rpm: (a) neglecting pressure losses at junctions; (b) including pressure losses at junctions.

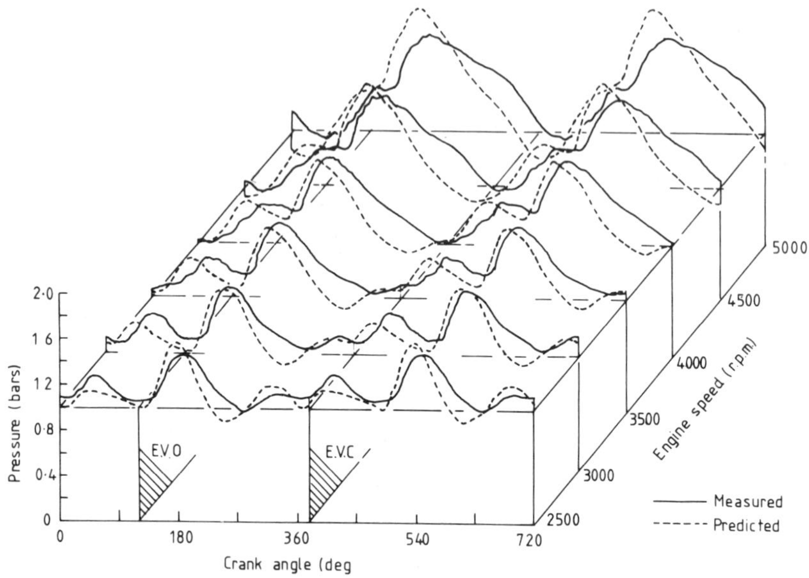

Fig. 44. Pressure variation in the exhaust manifold of a petrol engine operating at full throttle over the speed range 2500–5000 rpm: pressure loss junction.

5000 rpm show the largest deviation from measurement, and this probably reflects the increasing effect of pressure losses as the gas velocities in the pipe increase. [Comparison of these results with those in Fig. 73 (for a turbocharged petrol engine) indicates that the simulation of naturally aspirated engines can create bigger challenges than their pressure-charged counterparts. This is simply because of the larger influence of the total exhaust system in this case. The effect of losses at pipework junctions has been discussed above, and a further area that requires attention when modelling naturally aspirated engines is the silencer system. In this case the silencer has been modelled as a plenum of volume equal to that of the silencer. Discharge coefficients were included to allow for entry and exit losses.]

5. CALCULATION OF WAVE ACTION IN MANIFOLD SYSTEMS OF SUPERCHARGED OR TURBOCHARGED ENGINES

Wave action techniques can be used to analyse both supercharged and turbocharged engines, although there is no published information available on the former. There are two approaches to undertaking this task. The simpler is to consider the gas to be at a prescribed supercharge pressure at the entry to the inlet manifold and to represent the turbocharger (if fitted) by an equivalent area orifice. The more realistic approach is to consider the compressor and turbine as quasi-steady devices represented by their steady flow characteristics, as described in Section 4.8 of Chapter 12. There is a wealth of literature in this area, and this will be reviewed in this section.

First, it should be stated that it is actually easier to simulate pressure-charged engines than naturally aspirated ones because the flow directions are clearly defined by the pressure gradients—at least at high speed and load for turbocharged engines—and the main exhaust pipe (between the engine and turbocharger) is much shorter than the total system of a naturally aspirated engine. The only problem is that models are needed for the pressure-charging components: these are already available in quasi-steady form, but it is possible that non-quasi-steady simulations will be required to achieve ultimate accuracy due to the 'size' of the boundary. Turbochargers are discussed in Chapter 4, and the reader is referred to that chapter for further details on design requirements, including matching. Benson & Whitehouse[20] describe the basic requirements of the matching process, including the concepts of exhaust system efficiency; this will be returned to later. The basic aim in a turbocharger is to get sufficient energy to the turbine to drive the compressor at a condition that will give the desired boost ratio. In the case of a petrol engine that is reasonably easy because the exhaust gas enthalpy is high as a result of the near-stoichiometric combustion; for this reason petrol engine turbochargers are fitted with waste-gates to limit the boost achieved. The same is not true for a diesel engine, and truck diesel engines operate over a wide part of their range with the mean exhaust pressure higher than that in the inlet manifold. This requires careful design of the manifolds to ensure that the maximum possible useful energy reaches the turbocharger.

The differences between the inherent characteristics of reciprocating and rotational machines means that the air flow requirements, and pressure ratios of the devices do not change in a complementary manner. Hence, the matching of an engine and turbocharger over its whole speed range is a compromise; a fixed geometry rotational machine cannot satisfy the engine requirements at both ends of the operating envelope. For this reason both engine and turbocharger

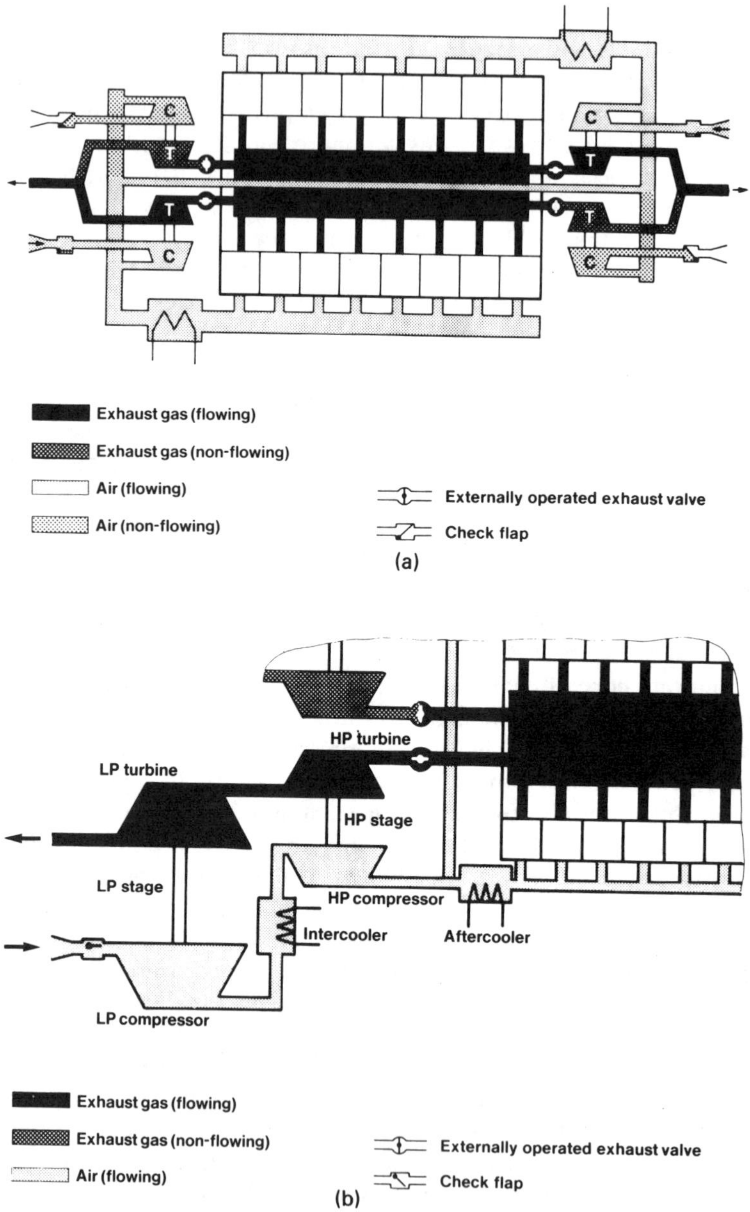

Fig. 45. Sequential turbocharging: (a) schematic arrangement of single-stage system; (b) schematic arrangement of two-stage system.

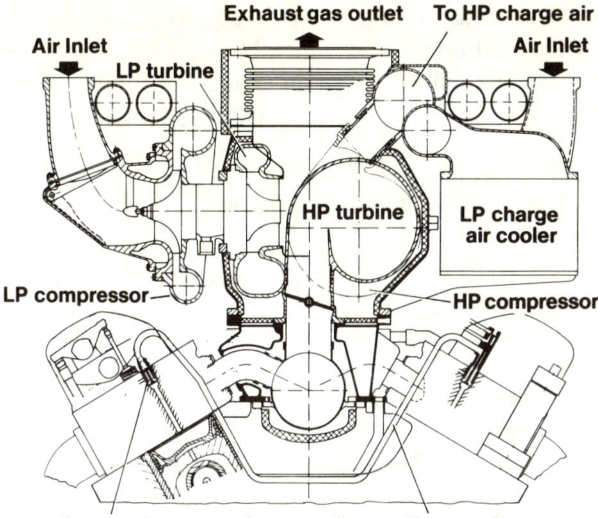

(c)

Fig. 45—*contd.* (c) turbocharger module for two-stage system, showing control valve.

manufacturers have attempted to improve the overall performance of engines by introducing some form of variability in the effective size of the turbocharger turbine. This can be achieved by a number of methods. The simplest way is to fit a waste-gate to the turbine: the turbine is then sized to give adequate boost at low speed (and load), and excessive boost pressure is avoided by by-passing exhaust gas around the turbine at high speed and load. An additional advantage of this approach is that the turbine transient response is improved because of its low moment of inertia. An extension of this concept is sequential turbocharging, in which a number of turbines are connected to the exhaust system through a series of valves (see Fig. 45[54]): the 'effective area' of the turbocharging device can be changed by connecting individual fixed area turbines in parallel. Another approach is to use a variable geometry turbocharger, in which the effective area of the turbine is changed by means of nozzles or some other arrangement. These turbocharger arrangements will not be discussed in detail here, except where they are relevant to calculation methods. All of them can be simulated using the available boundaries to define the appropriate pipework system. The basic aim of each of these approaches for turbocharging is to give the best utilization of the energy in the exhaust gas. This can be examined by an extension of the wave action techniques. The 'efficiency' of an exhaust manifold can be assessed by examining the degradation of useful energy along the pipe.

5.1. Transmission Efficiency of Exhaust Manifolds

Benson & Horlock[55] introduced an exhaust availability factor in an analysis in 1962, which related the energy of the gas entering the exhaust pipe to that entering the turbine; in other words, a transmission efficiency for the exhaust pipe. This is also referred to in Ref. 20, but no values or methods of evaluating this are given. Winterbone, Alexander and Nichols, with others, have produced a series of papers which examine this problem (Refs 56–58). They introduced this approach in an attempt to examine the effects of pulse converters on the propagation of energy in the exhaust pipe, because here it is necessary to balance the losses in certain parts of the manifold with gains found elsewhere. Consideration quickly shows that flow energy, i.e. enthalpy (H), is not an appropriate parameter to use. In the case of an adiabatic system (i.e. one with no heat transfer) the enthalpy entering the pipe is equal to that leaving it; even if heat transfer is considered this does not fully explain the degradation of energy that occurs. However, the usual methods of assessing available energy in thermodynamic systems (e.g. exergy, or availability[59]) are also not suitable unless a 'bottoming cycle' is being considered. The most appropriate parameter for estimating the degradation of energy in the exhaust pipe of an engine is the *turbine work function,* W_{is}. This is the isentropic work that can be obtained from the gas at any point in the exhaust system if it is expanded down to a datum condition, usually atmospheric pressure. Winterbone *et al.*[57,58] show the difference between the use of exergy and turbine work function. In Fig. 46 is shown the difference between enthalpy, exergy and turbine work function. The flow of exhaust gas through a valve can be considered to be isenthalpic (i.e. adiabatic), in which case there is no loss of enthalpy. However, the *quality* of that energy has decreased because there has been an increase in the entropy level; it is this concept of quality that both exergy and turbine work function bring to the analysis. The turbine work function available downstream of the valve is less than that in the cylinder, and the valve has introduced an

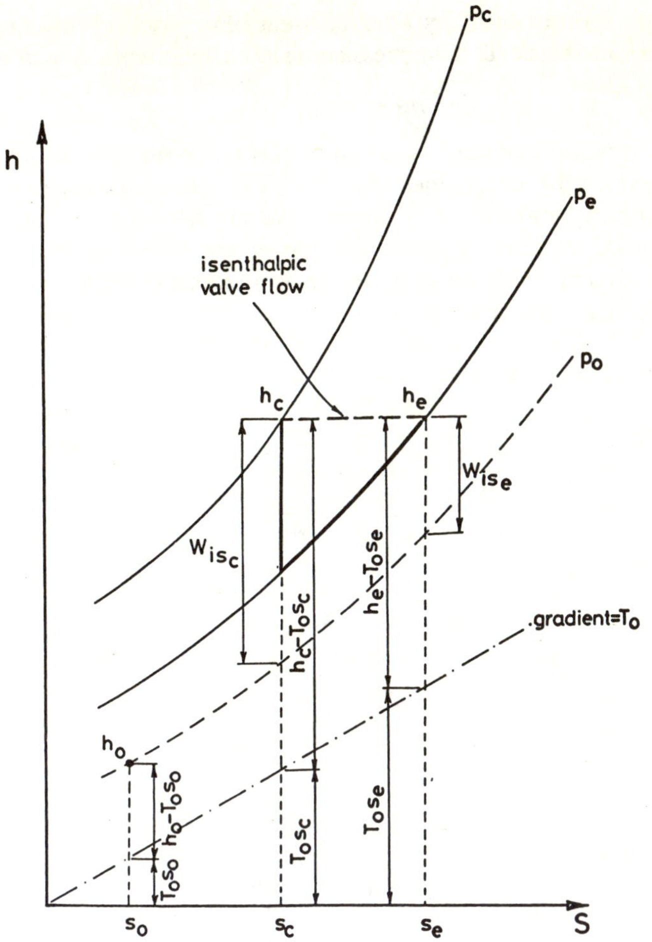

Fig. 46. *H–S* diagram showing energy and turbine work function variation across a valve.

inefficiency into the transmission process. Since a turbocharger turbine can be considered to be an adiabatic device then the maximum work which may be obtained from it is that which can be achieved during an isentropic process from the initial state to the dead state pressure (in this case ambient pressure). The specific work obtained from an isentropic process, w_{is}, is defined as

$$w_{is} = h_0 - h_{0_{is}} \tag{35}$$

here $h_{0_{is}}$ is the specific enthalpy after an isentropic process from the actual pressure to the dead state pressure. The turbine work function is simply

$$W_{is} = \dot{m}(h_0 - h_{0_{is}}) \tag{36}$$

The turbine work function is, as with exergy, a function of the entropy level since this determines the value of $h_{0_{is}}$. It is convenient to write eqn (36) in terms of the non-dimensional variables available from the method of characteristics solution of the non-homentropic unsteady flow equations. If the exhaust gas is assumed to be a perfect gas, the value of $h_{0_{is}}$ is given by:

$$h_{0_{is}} = \frac{a_{0_{is}}^2}{\kappa - 1} = \frac{A_{0_{is}}^2}{\kappa - 1} a_{\text{ref}}^2 \tag{37}$$

Now, for an isentropic process,

$$\frac{A^2}{A_{0_{is}}^2} = \left[\frac{p}{p_0}\right]^{(\kappa-1)/\kappa} = \left[\frac{p}{p_{\text{ref}}} \cdot \frac{p_{\text{ref}}}{p_0}\right]^{(\kappa-1)/\kappa}$$

therefore

$$A_{0_{is}}^2 = A^2\left[\frac{p_{\text{ref}}}{p}\right]^{(\kappa-1)/\kappa}\left[\frac{p_0}{p_{\text{ref}}}\right]^{(\kappa-1)/\kappa}$$

or

$$A_{0_{is}}^2 = A_A^2\left[\frac{p_0}{p_{\text{ref}}}\right]^{(\kappa-1)/\kappa} \tag{38}$$

From eqns (37) and (38) and the definition of h_0,

$$h_0 = \left\{\frac{A^2}{\kappa - 1} + \frac{U^2}{2}\right\} a_{\text{ref}}^2 \tag{39}$$

eqn (36) becomes

$$\dot{W}_{is} = \dot{m}\left\{\frac{A^2}{k-1} + \frac{U^2}{2} - \frac{A_A^2}{\kappa - 1}\left[\frac{p_0}{p_{\text{ref}}}\right]^{(\kappa-1)/\kappa}\right\} a_{\text{ref}}^2 \tag{40}$$

Equation (40) shows that the turbine work function is a function of the path line and wave characteristics.

Some examples of the results that can be achieved by such a technique will now be given. First, let us consider the loss of W_{is} across a valve. In Fig. 47(a) the variation of W_{is} with crank angle is depicted both in the cylinder and just downstream of the valve in the exhaust port. This particular diagram was evaluated with a pulse-converter junction situated close to the cylinder (i.e. a modular pulse converter, or a compact manifold—see Fig. 48, Section 4.7.3 of Chapter 12 and Section 5.2.3 of this chapter). The integrated value of W_{is} reduces as the gas passes through the valve, with most of the loss occurring

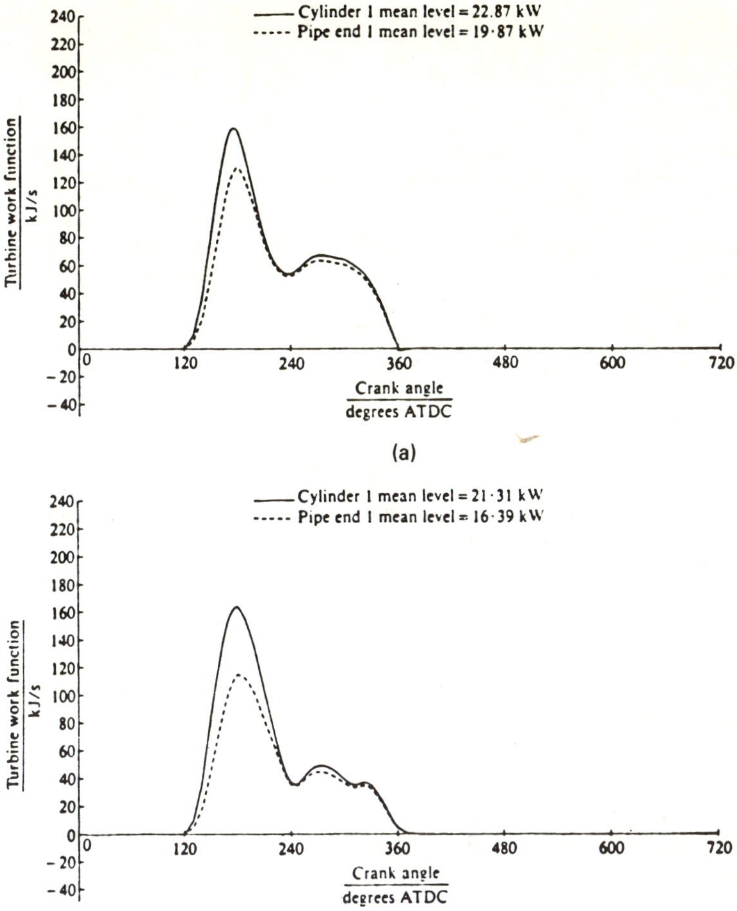

Fig. 47. Variation of turbine work function across the valve of a medium-speed diesel engine: (a) fitted with pulse converter junctions in compact manifold arrangement; (b) fitted with constant-pressure junctions in compact manifold arrangement.

during the blowdown period when the pressure differential across the valve is largest. If a transmission efficiency, π_T is defined as

$$\pi_T = (\dot{W}_{is})_{\text{downstream}}/(\dot{W}_{is})_{\text{upstream}} \tag{41}$$

then, in this case, it is 86·9%. In Fig. 47(b) is shown the variation of turbine work function for the same engine arrangement, but in this case the junctions in the manifold are considered to be loss free, i.e.

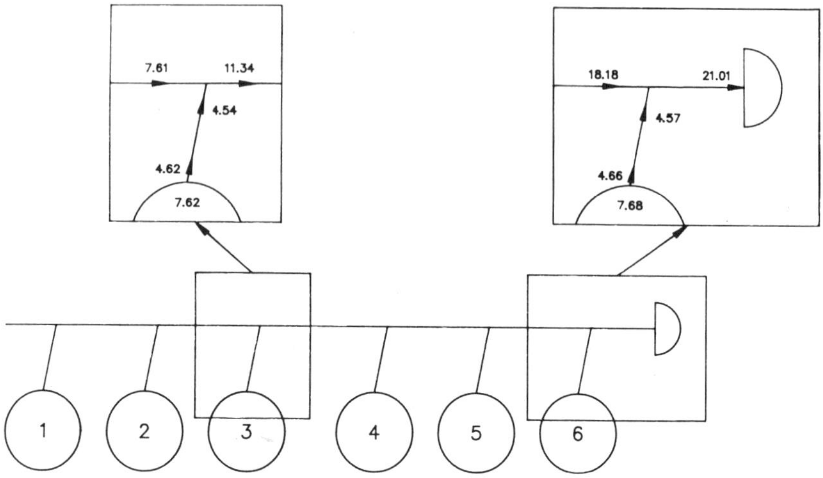

Fig. 48. Transmission of turbine work function down compact manifold of automotive-size diesel engine.

constant-pressure, junctions. It can be seen that there is a much larger loss of turbine work function in this case, with the transmission efficiency reducing to 76·9%; most of the loss occurs in the blowdown period in this case as well, and it is larger because the pressure does not build up in the exhaust port behind the valve. In fact, 21% more turbine work function enters the exhaust pipe with the pulse-converter junction. However, from here to the turbine the situation is different because the losses in the pulse-converter junctions reduce the level of turbine work function, whereas the constant pressure junctions do not. In Fig. 48 (for a different engine) it is shown how the level of turbine work function is degraded in a pulse-converted manifold, and the trade-off between efficiency of turbine work function transmission across the valve and its subsequent transmission down the manifold is demonstrated in Fig. 49.

The concept of transmission efficiency enables the characteristics of the different turbocharging systems to be understood. There are three basic exhaust manifold arrangements for turbocharged engines; pulse, constant-pressure and pulse-converted manifolds. These are each discussed in more detail later, but it is convenient to consider their individual merits at this stage. Pulse manifolds are commonly used on high- and medium-speed engines. They are designed on the Büchi principle and attempt to minimize the interference between individual cylinders. The pulse manifold has a smaller volume than other manifolds, and exhibits higher-pressure pulsations which reduce the period of unresisted expansion across the valve. These pressure pulses also transmit energy very efficiently from the cylinder to the turbine, where the problems then begin. It is necessary to use a multiple entry turbocharger to separate the pulses and this results in a lower efficiency unit: turbines also operate more efficiently under steady flow conditions, and thus again the pulsations lower the average efficiency of the turbine. In constant-pressure turbocharging a large-volume manifold is used and the period of unresisted expansion across the valve becomes very large. Oldfield & Watson[60] quote a transmission

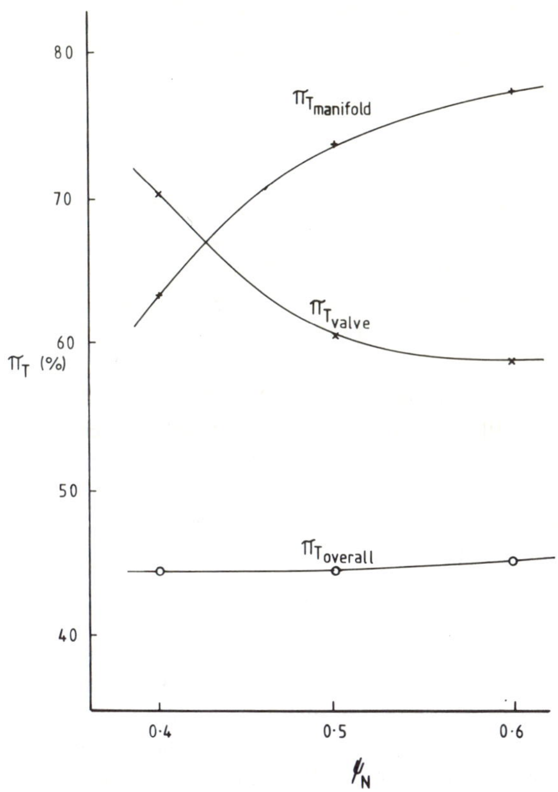

Fig. 49. Variation of transmission efficiency with pulse converter area ratio.

efficiency of about 50% for a valve in what was essentially a simulation of constant-pressure turbocharging. This is substantially lower than the efficiencies quoted above for pulse-converted and pulse manifolds, at 87% and 77% respectively. However, having entered the constant-pressure manifold the gas is then transmitted to the turbine very efficiently at low velocity, and the low-pressure fluctuations enable a high-efficiency single entry turbine to expand the gas in the most efficient manner. The pulse converter attempts to combine the advantages of both these systems. The narrow-bore exhaust pipes retain the integrity of the pulses, while the rectifying effects of the junctions enable a single entry turbine to be used. Here the efficiency of transmission across the valve is high, the transmission efficiency down the pipe is poor, but the turbine efficiency should be higher. Balancing these various options is the task of the engine designer.

5.2. Diesel Engines

5.2.1. Pulse Manifolding

The concept of pulse manifolding was patented by Büchi[61] in 1925. Büchi recognized that to obtain efficient turbocharging at the low pressure ratios of that time it was essential to transmit the energy from the cylinder to the turbine as pressure (and temperature) pulsations. This could be achieved by a small-bore, small-volume manifold; however, such a manifold led to interference between the cylinders if they all fed into one pipe. The characteristic of the pulse manifold is that the cylinders are grouped so that two cylinders connected to the same manifold do not have their exhaust valves open simultaneously. This approach has already been encountered in Fig. 42 with a manifold on a naturally aspirated engine. Such a criterion means that cylinders are connected in groups of two or three; Fig. 50 shows typical groupings of cylinders for four-, six- and eight-cylinder four-stroke engines. The most efficient grouping is the three-cylinder one because the quiescent period between pressure pulses is kept to a minimum consistent with acceptable interference between cylinders. In Fig. 51 is shown a typical pressure trace for one branch of a pulse manifold on a six-cylinder automotive engine operating at high load. The pulse pattern is quite distinct and the minimum pressure is well above the ambient level. The turbocharger is being driven all the time. The situation changes dramatically with an engine operating at low load. It

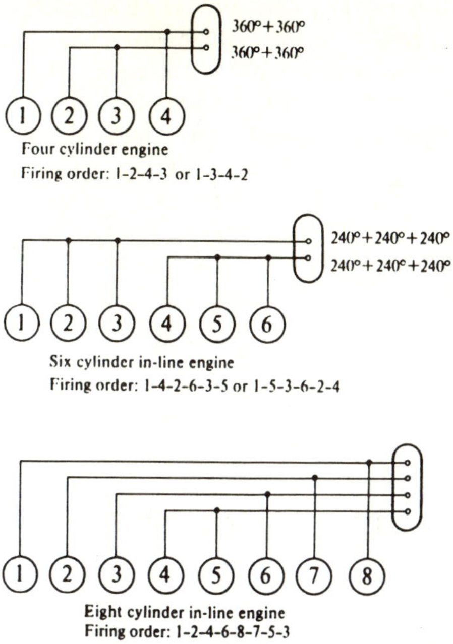

Fig. 50. Typical cylinder groupings for pulse exhaust manifolds.

is shown in Fig. 52 that under these conditions there are long periods when the gas enters the turbocharger at very low pressure and the efficiency is zero: in fact, under such conditions it is likely that the turbine acts as a compressor and absorbs work. This diagram shows how low the efficiency of the turbine goes under low-load operating conditions.

The simulation of pulse manifolds is reasonably straightforward, except that the turbine data will normally have been measured for both entries with equal upstream conditions. The usual approach is to assume that each entry receives its proportion of the total turbine flow, and that a twin-entry turbine (for example) acts like two single-entry turbines of half the capacity. Such an approach is a simplification of the actual operation. First, although the manifolds are separated right up to the turbine, there is interaction between the two entries around flow paths in the turbine. It is necessary to include an allowance for this in the model, and Winterbone *et al.*[58] incorporate a small pipe connecting the two manifolds: the size of the pipe was equivalent to

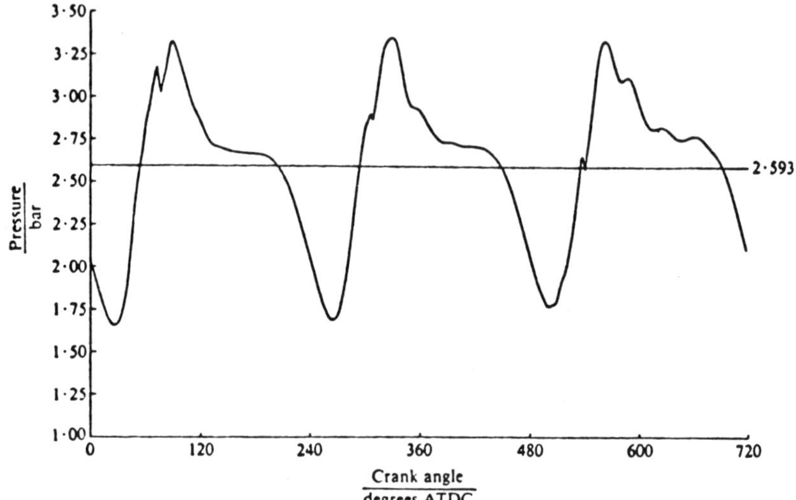

(a) Pressure

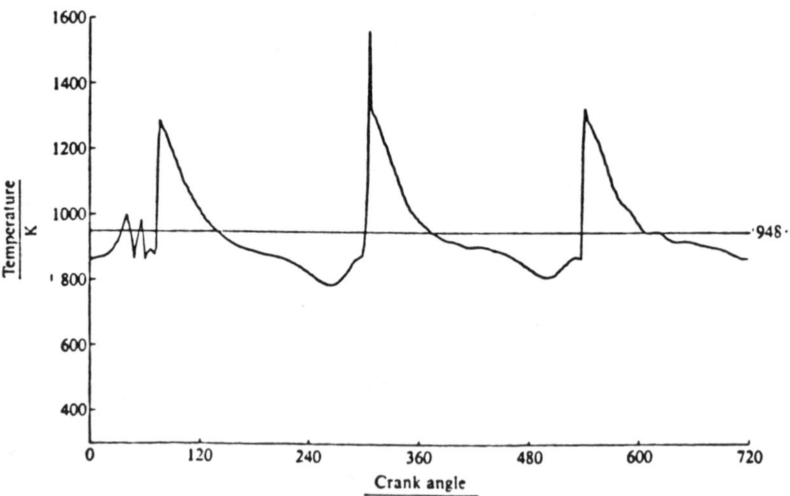

Fig. 51. Pressure and temperature in the exhaust manifold of a six-cylinder automotive engine at high load.

two mesh lengths and its diameter was one tenth of the main pipe area (Payri[62] also found that such a pipe was necessary.)

It has been known for some time that the flow through each entry of a twin-entry radial flow turbine is not symmetrical (see Benson[24] and Benson & Scrimshaw[63]), and this has recently been substantiated by Dale and Watson.[64] They measured the characteristics of a turbocharger under steady and unsteady conditions and showed that:

(a) the two entries of the turbine do not have identical characteristics;

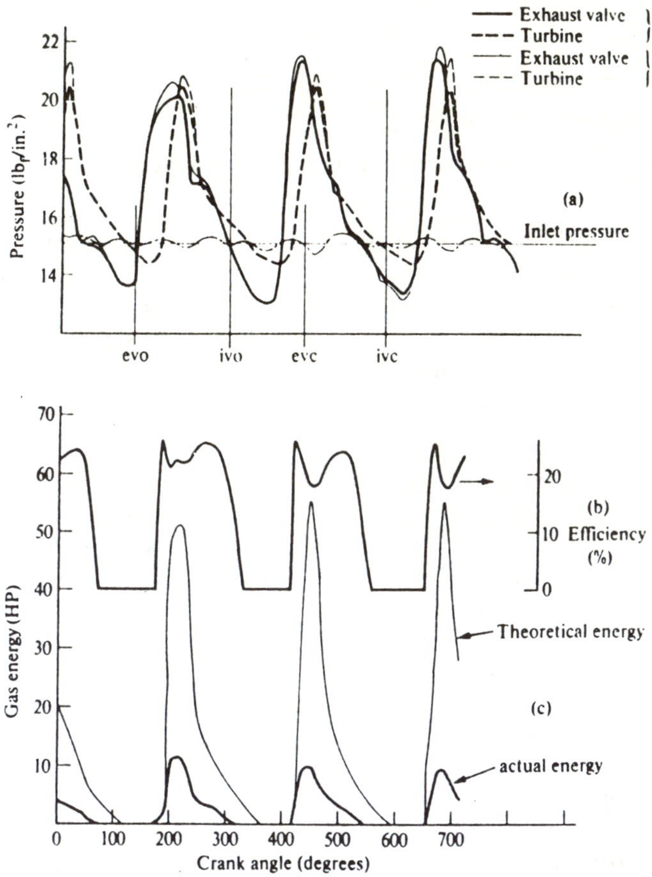

Fig. 52. Pressure variation and power at turbocharger for a medium-speed engine at low load.

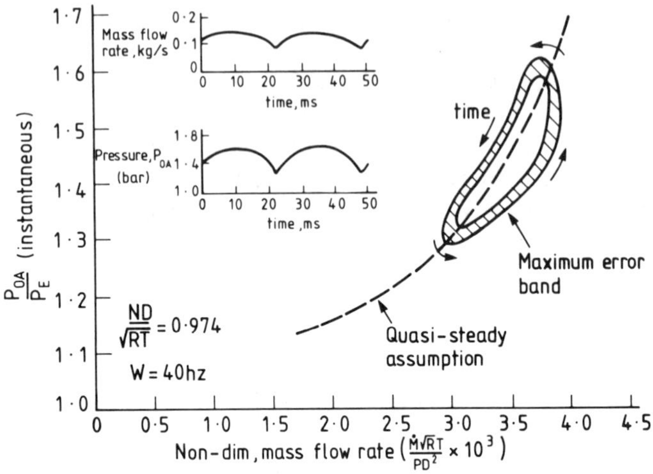

Fig. 53. Turbine operating locus under unsteady flow conditions.

(b) the turbine does not operate on its quasi-steady characteristic under unsteady flow conditions;

(c) while the turbine deviates from quasi-steady flow conditions, the actual error introduced by assuming such flow was relatively small for the turbine examined.

In Fig. 53 is shown the operating locus of pressure ratio and mass flow rate superimposed on the turbine swallowing curve for a turbocharger turbine, as measured on a pulse generator flow rig. The dotted line is the steady flow characteristic for the given speed. It can be seen that the locus deviates from the steady flow line, somewhat like a hysteresis curve in electromagnetism. This indicates that the turbine is not acting in a quasi-steady manner, and close examination of the pressure ratio and mass flow diagrams shows that the mass flow rates peaks slightly before the pressure ratio. The ability of the turbine to deviate from quasi-steady flow indicates that some form of mass and energy storage is taking place within the boundary. This is not unreasonable when the size of the turbine is considered, particularly that of the scroll casing. A more appropriate turbine model might include some allowance for this effect. Fortunately, Dale & Watson[64] state that the error introduced by the use of steady flow characteristics is acceptably small, largely because the time spent far away from the steady flow line is short; hence the time-average and cycle-average of the various parameters are close. However, they state that the 'mean efficiency with unsteady flow lies a few points below the steady flow line'; that is, the quasi-steady model assumption causes the turbine power to be overestimated.

5.2.2. Constant-pressure Manifolding

As stated above, the aim of the constant-pressure exhaust manifold is to present the turbocharger turbine with the most appropriate form of energy, namely steady flow. This is achieved in a large-volume exhaust system which can smooth out the individual firing pulses. Because of the high level of losses across the valves associated with such systems they are only appropriate for highly turbocharged engines. They are also most suited to engines running under steady conditions, or those in which the operating conditions vary slowly. Constant-pressure exhaust systems are most favoured in engines used for marine engines and those used in power generation.

Wallace (in Horlock & Winterbone[22]) discusses both the simulation and use of constant-pressure manifolding. He refers to results obtained by Janota, Hallam, Brock & Dexter[65] for an engine operated with a variety of exhaust pipes; small-bore, long manifolds were compared with equal volume manifolds in which the length and diameter were equal. It was found that if the exhaust manifold is of small volume and if the length is short the filling-and-emptying simulation technique will accurately predict the pressure and temperature variation at the turbine end of the pipe. As the aspect ratio of the exhaust pipe changes towards long, small-bore arrangements the filling-and-emptying simulation method becomes less appropriate. This is because the wave travel time is significant in long, small-bore manifolds but a large-volume manifold acts as an integrator, and the filling-and-emptying method simulates it in this way. (Azuma *et al.*[66] state that although filling-and-emptying techniques will give a good prediction of the gross parameters they are not capable of evaluating the detailed flow effects). Wallace also examines an engine operating over a range of boost pressure ratios from 1·5 to 2·5, and concludes that at high boost ratios the constant-pressure system, because of the more favourable turbine operating conditions, is more suitable than pulse manifolding.

Azuma *et al.*[26,66] describe the simulation of constant-pressure manifolds for marine diesel engines. The papers are based on method of characteristics and Lax–Wendroff techniques respectively, and

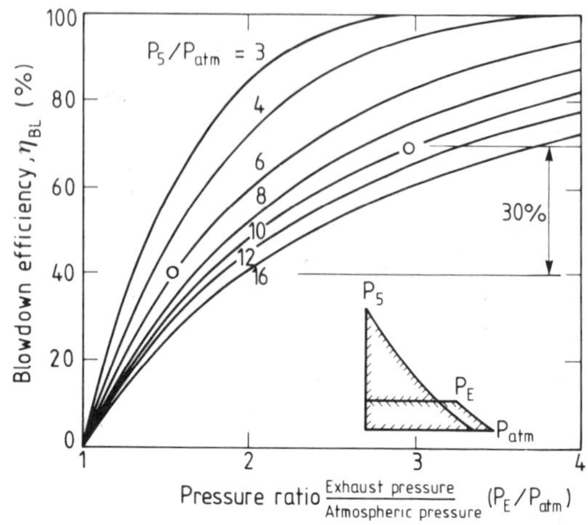

Fig. 54. Blowdown efficiency for pulse charging and constant pressure manifolds.

report a comprehensive investigation of constant-pressure manifolds as applied to large low-speed marine engines. The authors explain the benefits of this system and show (Fig. 54) that, as the boost pressure (and hence exhaust pressure) increases, the efficiency of the blowdown process with a constant-pressure exhaust system increases until it is comparable with that of the pulse charging system (p_5 is the release pressure in the cylinder at exhaust valve opening, and p_E is the pressure in the exhaust manifold). Hence at low boost levels, say $p_E/p_{\text{atm}} = 1{\cdot}5$ and $p_5/p_{\text{atm}} = 6$, the efficiency of transmitting the blowdown energy is about 40%, whereas $p_E/p_{\text{atm}} = 4$ and the same release conditions gives an efficiency of 93%. This concept of transmission efficiency has been introduced above, and will be returned to below. Azuma *et al.* emphasize that although the pressure variation is small it can have important effects on the engine performance. The trapped pressure can vary substantially between cylinders and, particularly in the case of two-stroke engines, there can be appreciable blowback into the inlet system, resulting in fouling of the ports. They also confirm that although the manifold is very large the filling-and-emptying method does not give a good detailed definition of the pressure variation in the exhaust system and the subsequent flows.

The results quoted by Azuma *et al.* show some of the interesting effects observed in constant-pressure systems. In Figs 55 and 56 are shown the pressure variations in two-stroke and four-stroke engines respectively. The engines are defined in Table 4. Careful examination of the pressure traces shows that the number of pressure peaks in the two-stroke engine diagram is 10, whereas there are only seven cylinders; for the four-stroke engine the numbers are eight and nine respectively. Hence the pressure variation does not simply reflect the number of cylinders connected to the manifold, as would be evaluated by the filling-and-emptying method. In fact, the frequency of pressure variation is that of a pipe closed at both ends ($a/2L$).

The effect of putting an orifice plate in the exhaust system (as shown in the schematic) is shown in Fig. 56. This brings about a large reduction in pressure pulsation, with a benefit in the breathing of the cylinders. Azuma *et al.* state that the fuel consumption was not adversely affected by the orifices, and was probably slightly improved.

Azuma *et al.*[26] compare the method of characteristics with the

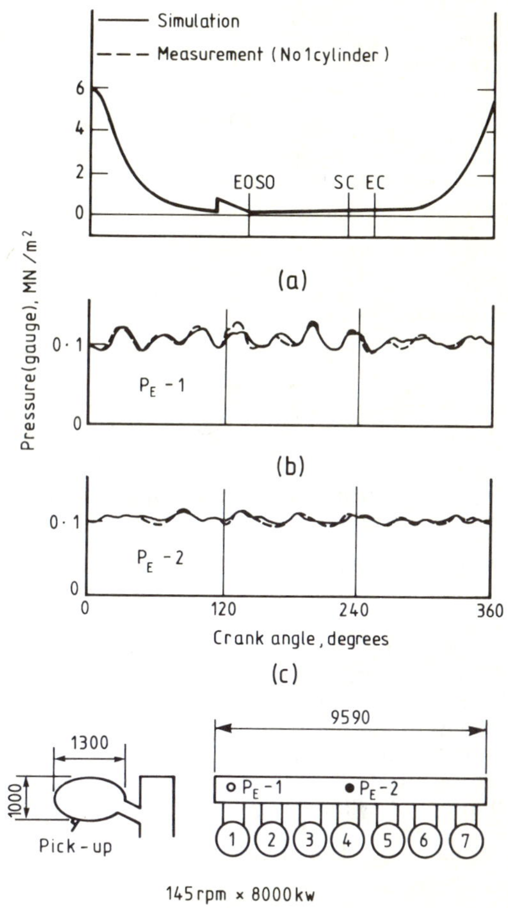

Fig. 55. Pressure variation in constant-pressure exhaust manifold: two-stroke engine.

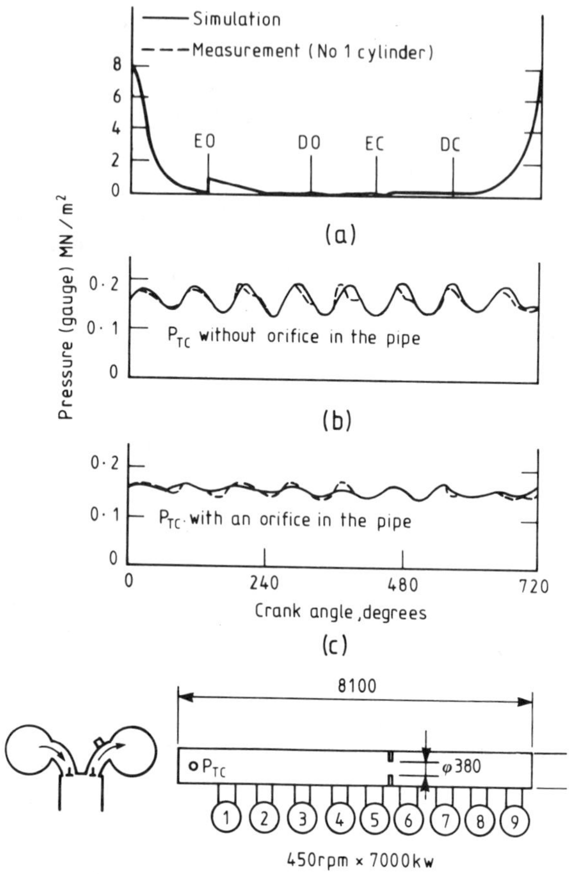

Fig. 56. Pressure variation in constant-pressure exhaust manifold: four-stroke engine.

Table 4. Data for Medium-speed Diesel Engines

Engine type	Two-stroke	Four-stroke
Speed (rpm)	145	450
Boost pressure (bar gauge)	1·17	1·90
Inlet temperature (K)	313	313
Release pressure (bar)	6·5	10·5
Cylinder bore (mm)	700	520
Stroke (mm)	1200	550
Number of cylinders	7	9
Firing order	1725436	135798642

Lax–Wendroff technique. They report extremely similar predicted variations of pressure but show that the mass conservation properties of the Lax–Wendroff method are better. This is in line with the conclusions of Poloni *et al.*[25] but the mass conservation quoted here suggests that the application of method of characteristics was not as accurate as can be achieved: this usually indicates a shortcoming in defining the path lines, and the author's experience is that the mass should balance within about 2%.

An interesting result quoted by Azuma *et al.* is the effect of the size of orifices in the exhaust system. For a pipe with $L/D \simeq 10$, a nozzle at the end of the pipe (simulating a turbine) of $\varphi > 0{\cdot}5$ can be regarded as an open end which reflects a negative wave, whereas a nozzle with $\varphi = 0{\cdot}25$ produces negligible reflection, with the result that the exhaust pipe is of infinite effective length. Lower values of φ approximate more closely to a closed end. If an orifice is placed in the centre of the pipe its ability to reduce the magnitude of waves is also dependent on its value of φ; to achieve any appreciable reduction in wave amplitude it is necessary for the orifice to have $\varphi < 0{\cdot}25$.

A further example of the use of constant-pressure exhaust manifolds comes from an investigation undertaken by the author for Wartsila–Wichmann, the only company remaining which manufactures large loop-scavenge engines. In Fig. 57 is shown the cross-section of that engine, while Fig. 58 is a schematic of an eight-cylinder version of the engine for calculation purposes: the engine data are given in Table 5. The loads indicated in this table follow the ‘propeller law’, with the engine power output proportional to speed cubed. This is typical of such applications and is also similar to the characteristic of an hydraulic dynamometer.

During tests Wichmann had found that the performance of the engine was not up to the level that would have been expected from the six-cylinder in-line version. This was thought to be brought about by interference caused by connecting cylinders on opposite banks, e.g. cylinders 1 and 5 which are phased 50° to each other because of the Vee angle of the engine. The firing order shows that the exhaust pulses of these interfere quite severely. To try to overcome this problem orifice plates had been fitted in the exhaust manifolds. These improved the engine performance but the reason for this was not obvious, even after measuring the exhaust manifold pressure traces. A simulation of the engine at the various speeds, both with the standard manifold and after fitting orifice plates just downstream of each cylinder, showed

Fig. 57. Wartsila–Wichmann WX28V two-stroke loop-scavenge diesel engine: bore 280 mm; stroke 360 mm; speed 600 rpm.

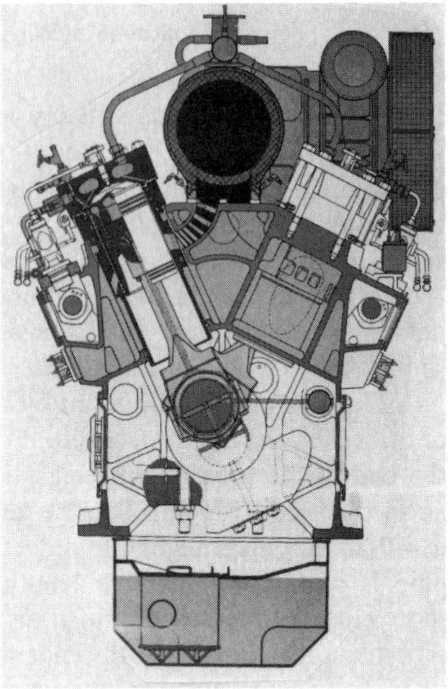

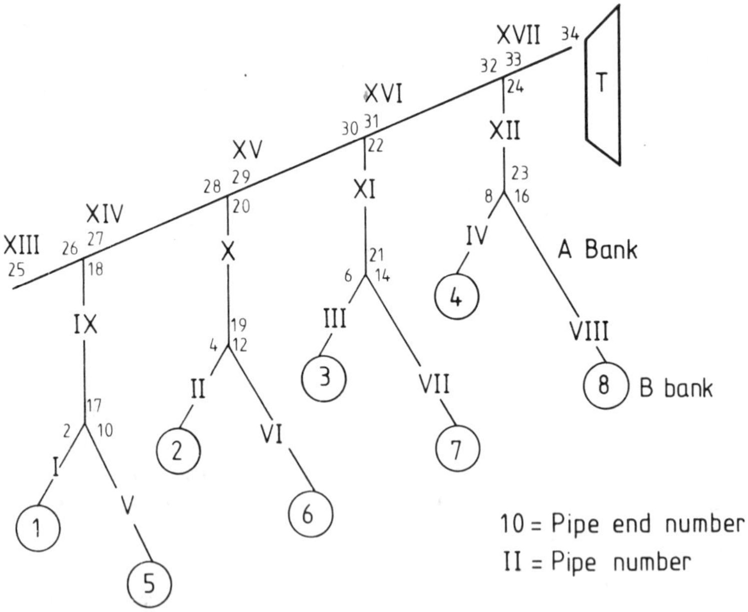

Fig. 58. Schematic of Wartsila–Wichmann engine showing pipe numbers.

Table 5. Basic Data for Wartsila–Wichmann Two-stroke Diesel Engine

Engine type	Two-stroke
Speed (rpm)	506–600
Load (% full)	60–100
Boost pressure (bar)	2·19–3·16
Inlet temperature (K)	318–319
Cylinder bore (mm)	280
Stroke (mm)	360
Number of cylinders	8
Firing order	15372648

how the improved performance was brought about. In Fig. 59 are shown the exhaust manifold pressures just downstream of the exhaust ports; Fig. 59(a) is for the standard manifold, while Fig. 59(b) is with the orifice plates fitted. It can be seen that the simulation predicts the measured changes, and hence the other trends can be used with some confidence. The effect of the orifice plate, as would be expected, is to restrict the flow into the exhaust manifold and hence increase the

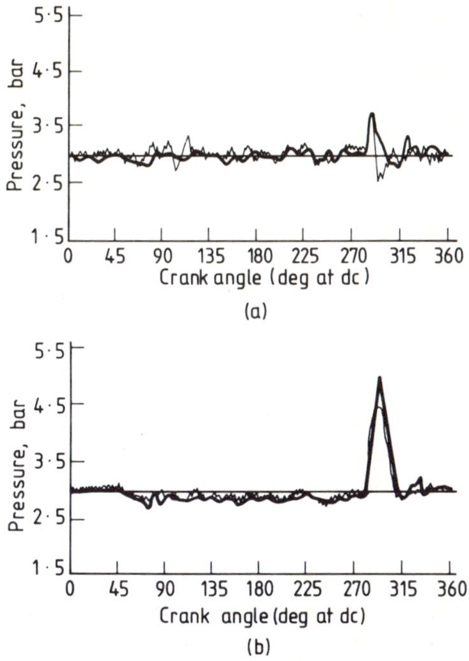

Fig. 59. Pressures in exhaust manifold just downstream of exhaust port for cylinder 2: (a) standard manifold; (b) with orifice plates fitted in each cylinder stub pipe.

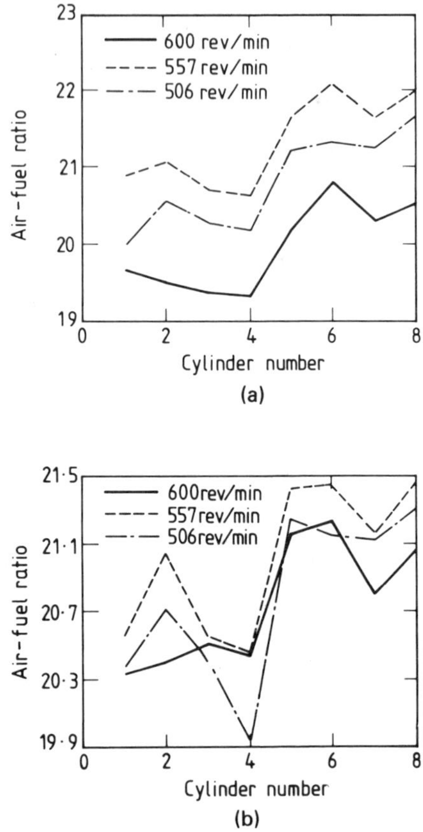

Fig. 60. Variation of trapped air : fuel ratio between cylinders of Wartsila–Wichmann two-stroke engines: (a) standard manifold; (b) with orifice plates fitted.

blowdown pulse (at around 270°); otherwise the pressure fluctuation in the manifold is lower than for the standard one. The surprising benefit was that the orifice plates reduced the fuel consumption on the engine by up to 4 g/kWh, down to 240 g/kWh at 506 rpm. The reason for this was found by examining the distribution of trapped air : fuel ratio with both the standard and modified manifolds. In Fig. 60(a) are shown the predicted values with the standard manifold, and there is a distinct variation of almost one air : fuel ratio between the two banks; Fig. 60(b) (for the same fuelling rate) indicates that the orifice plates increased the trapped air mass and improved the basic air : fuel ratio of the engine to about 20·7 : 1, from an original value of 19·8 : 1, and reduced the variation of air : fuel ratio between the banks. This is a substantial improvement and will result in a large increase in combustion efficiency: this explains the improvement in fuel consumption, even though the transmission efficiency of the exhaust manifold was lowered with the orifice plates.

These examples show that, while the simpler filling-and-emptying model would appear to be appropriate for evaluating the effects in large-volume constant-pressure manifolds, it is not capable of demonstrating some important features associated with them. It is always advisable to undertake a detailed analysis including wave effects even for these manifolds because there can be detrimental interactions occurring which would only come to light when the engine goes on the test bed.

5.2.3. *Pulse-converted Manifolds*

The basic concept of pulse converters was introduced in Section 4.7.3 of Chapter 12. Two types of pulse converter were discussed: the symmetrical, Sulzer type; and the asymmetrical, Birmann type. These devices are used in different ways when they are incorporated into manifolds, but both types have the basic aim of reducing interference between cylinders and reducing the number of entries required in the turbocharger turbine. The important parameter in designing a pulse converter is the ejector nozzle area ratio, ψ_N. Small values of ψ_N result in increased pulse converter effect but at the expense of larger throttling losses. The design of a pulse-converted manifold is the compromise between good flow rectification and excessive losses.

5.2.3.1. *Symmetrical (Sulzer-type) pulse converters.* The Sulzer type of pulse converter (see Fig. 18 of Chapter 12) is used at the turbine end of a pulse manifold. This means that the basic pulse manifold shape is retained right down to the turbine, and then the manifolds are brought together just upstream of the turbine. If this system is to work satisfactorily it is essential that the pulse converter aerodynamically isolates the individual manifolds from each other sufficiently to limit harmful interference between cylinders. Janota & Watson[67,68] (summarized in Ref. 22) describe the application of Sulzer pulse converters to three engines; two two-strokes and a four-stroke one, with one automotive and two medium-speed machines. They used a number of area ratios in all cases and it is apparent that the most appropriate value of ψ_N is dependent upon the particular engine and its duty. The best value of ψ_N for the automotive two-stroke engine was found to be 0·75, since this combined both acceptable exhaust interference and throttling losses; improvement of up to 6% was recorded in BSFC relative to the standard engine. On a medium-speed uniflow-scavenge two-stroke engine they found an improvement in fuel consumption of about 5% and a lowering of exhaust temperature of up to 80 K with a pulse converter of $\psi_N = 0{\cdot}92$. The medium-speed four-stroke engine operated best with a 'pulse converter' of $\psi_N = 1{\cdot}0$; this is effectively a junction with directional ejector characteristics but no throttling effects. This junction had a marginal effect on reducing the interference pulse, but the exhaust temperature was reduced by 60°C due to more efficient turbine operation, with the improved presentation of the exhaust gas energy at the turbine entry. This particular engine exhibited better fuel consumptions at high speed but poorer ones at low speeds.

Benson & Alexander[69] describe the application of pulse converters to automotive four-stroke engines. They show that the value of ψ_N has a large effect on the size of the interference pulse. The arrangement of the device is shown in Fig. 61, where it has been used to connect together two standard pulse manifolds (see Fig. 50). The diagram also shows the way in which smaller area ratios (i.e. smaller ψ_N) reduce the interfering pulses. It can be seen that the calculation method, based on the method of characteristics with the pulse converter simulated by quasi-steady loss coefficients (see Section 4.7.3 of Chapter 12), is well able to show the way in which the interference pulse is reduced. Benson & Alexander[69] (summarized in Horlock & Winterbone[22]) describe the introduction of various energy transmission parameters for the manifolds and show how these indicate the operation of the system. Perhaps the maximum benefit achieved in this study was an increase in the torque back-up of the engine: this was obtained by rematching the turbine size (A/R) to give the best performance. The best results achieved at full load are shown in Fig. 62. Benson & Alexander used three different sizes of pulse converter ($\psi_N = 0{\cdot}53$, $0{\cdot}63$ and $0{\cdot}83$) and three different turbines ($A/R = 0{\cdot}81$, $0{\cdot}96$ and $1{\cdot}31$). It was found that the smallest value of ψ_N resulted in high throttling losses, whereas the largest value allowed too much interference. In the case of turbine size, a small turbine will produce high boost at low speed but will tend to over-boost at higher speeds, and will also increase the exhaust back pressure, resulting in high exhaust temperatures. A larger turbine may not produce a high enough pressure in the exhaust to give a good level of available exhaust gas energy, leading to insufficient boost and a tendency for the engine to

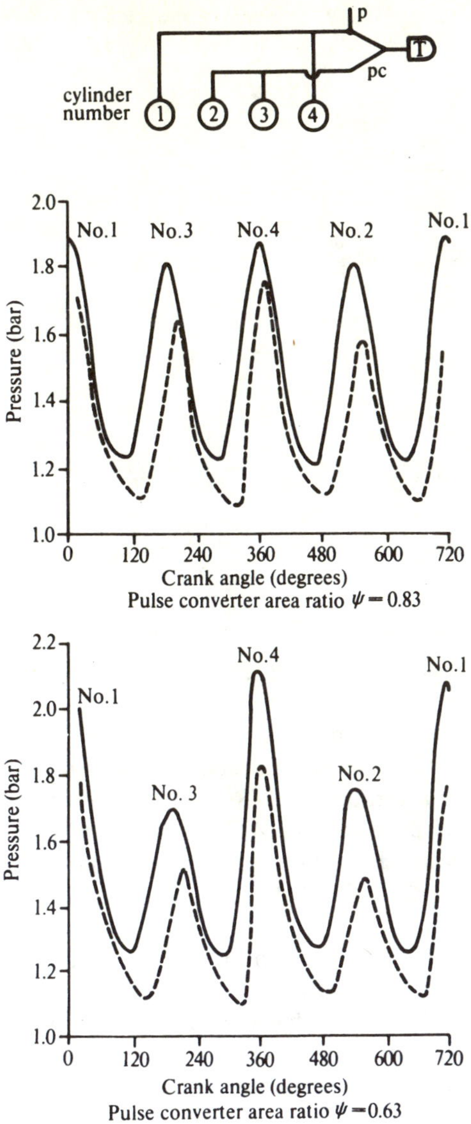

Fig. 61. Four-stroke engine fitted with pulse converter. ———, Calculated; - - - - - -, measured.

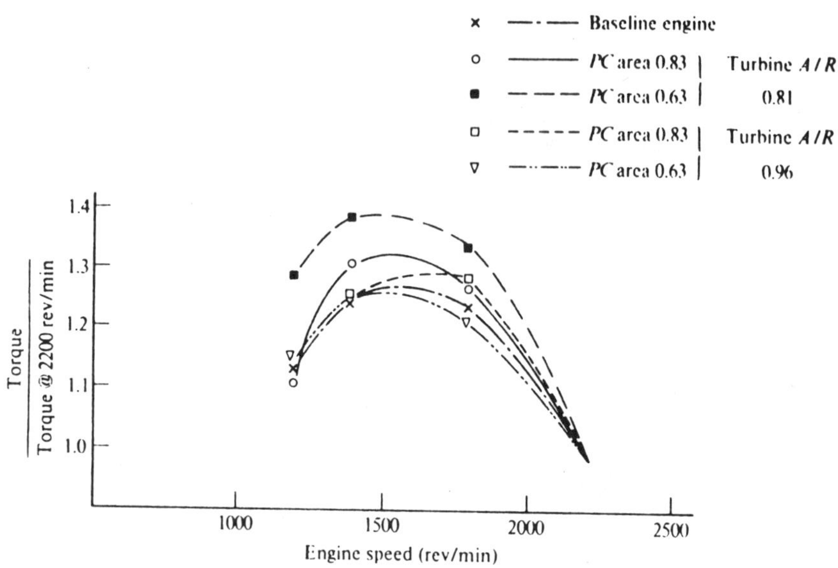

Fig. 62. Improvement in torque back-up obtained by fitting pulse converter.

smoke, particularly at low speed. The best combination of pulse converter and turbine for torque back-up was a value of $\psi_N = 0{\cdot}63$ and $A/R = 0{\cdot}81$, although an improvement might have been achieved with a slightly larger turbine because this one overboosted at 1800 rpm. The standard engine was fitted with a turbine of $A/R = 1{\cdot}75$; such a large turbine is necessary because it produces an A/R for each entry of about 0·85, and the device is sized (at low boosts) to swallow the individual pulses from each cylinder. The pulse converter produces a more favourable pattern of pulses at the turbine (effectively filling in the quiescent periods in the pulse pattern of the pulse manifold) and hence it uses the available area more effectively.

All the pulse converters described above have been used to connect two manifolds together. It is also possible to connect more than two manifolds through a multiple entry pulse converter, usually referred to as a multi-pulse device.

A further example of the use of pulse converters to improve engine performance is described by Borilla.[70] In this case he used both a large and small turbocharger and connected them both to the engine through a two-entry, two-exit pulse converter (Fig. 63). The pulse converter was used in this case to enable single entry turbines to be fitted, with a concomitant simplification in the pipework arrangement and a reduction in the number of valves required in the exhaust system. No predictions of the system performance are given but experimental results show, and Borilla states, that the pulse converter was sized for full load and speed operation (2200 rpm, 14 bar BMEP) and he also shows that it is not truly effective at 1200 rpm.

5.2.3.2. *Asymmetrical (Birmann-type) pulse converters.* These have been referred to as modular pulse converters or compact manifolds by the two companies that have used them extensively (SEMT and GEC–Ruston Diesels respectively). Both devices have the same basic aim, namely to enable the cylinders of a multi-cylinder engine to be connected to a simple exhaust manifold while limiting the

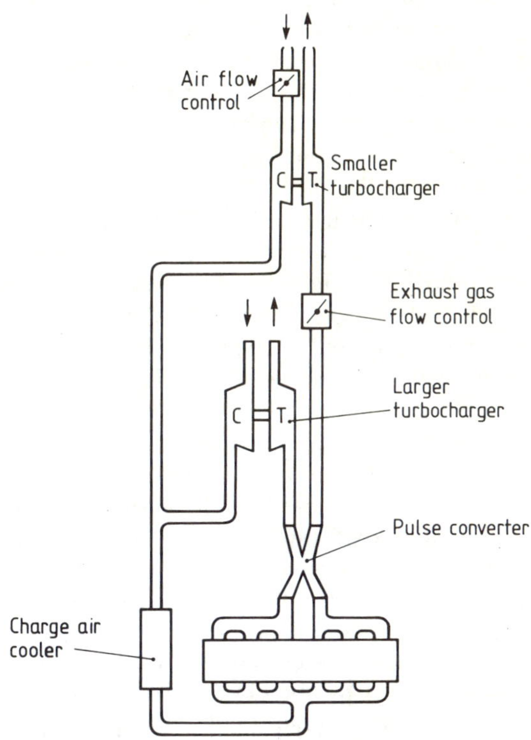

Fig. 63. Use of a pulse converter in a sequential turbocharging application.

interference between them. Birmann[71] originally proposed a manifold of this type in a patent, in which the exhaust system was in the form of a loop and circulating exhaust gas drew the exhaust flow from individual cylinders by a venturi (ejector) effect. This arrangement was never adopted, but the concept of asymmetrical pulse converters attached to cylinders was introduced by Curtil & Magnet[72] (SEMT). A similar system was developed by GEC–Ruston Diesels and this is described by Winterbone, Alexander, Nichols and others in Refs 56–58, 73 and 74. This description will concentrate on the author's experience.

As stated previously, the advantage of the pulse charging system, and hence the symmetrical pulse converter arrangement, is the small-volume exhaust manifold which preserves the integrity of the individual pulses of exhaust energy as they pass from the engine to the turbine. The disadvantages are the need for a lower efficiency multi-entry turbine (except if a pulse converter is used), and a complex manifold arrangement comprising a number of distinct pipes. The latter requirement results in the need to design the component parts of each manifold differently, reducing the interchangeability of components and increasing the stock of replacement parts. The asymmetrical pulse converter can overcome this problem since each component of the exhaust system can be identical: hence the term modular pulse converter. In Fig. 20 of Chapter 12 is shown the design of a junction for a compact manifold; Fig. 21 of Chapter 12 shows the variation in pressure loss for the junction over a range of Mach numbers and mass flow ratios. Alexander *et al.*[56] show the way in which these junctions can be put together to form a manifold (Fig. 64) for a 12-cylinder V-engine. The pulse manifold retains the usual, highly efficient, three-cylinder grouping of cylinders and hence requires a four-entry turbine (a multi-pulse converter could be fitted just upstream of the turbine to bring the individual manifolds together, but normally the available length limits the efficiency of the device). The constant-pressure exhaust manifold is a simple arrangement but results in unacceptable performance of the engine at low speed and load, with poor scavenge under these conditions due to the large valve overlap (>140° crank angle). The compact manifold results in the need for a twin-entry turbine (one entry per bank), but with a simplified arrangement of pipes compared to the pulse manifold. While benefits in exhaust pipe arrangement and layout can be achieved on 12-cylinder engines, even greater gains can be obtained with 16-cylinder engines,

Fig. 64. Comparison of manifolds for pulse and constant-pressure charging with a compact manifold.

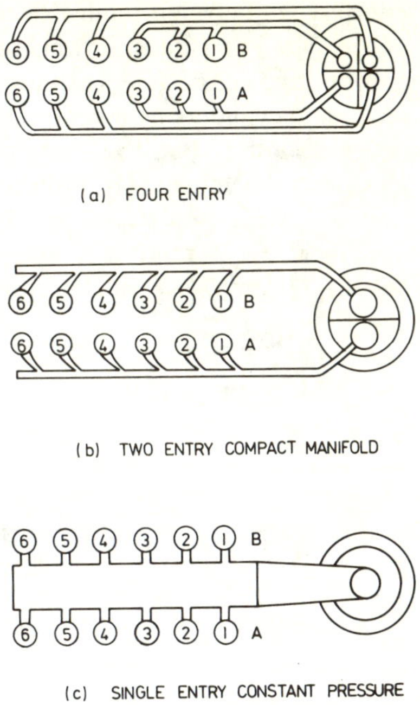

on which the cylinders are arranged in groups of two and connected to two four-entry turbochargers. The cross-sectional arrangements of a 16-cylinder engine with pulse charging and compact manifolds are compared in Fig. 65.

Experiments were performed on both 12- and 16-cylinder engines fitted with compact manifolds. The greatest gain was achieved on the latter because of the reduction in windage periods in the turbine, but even the performance of the 12-cylinder engine improved at high loads and speeds. Ref. 56 describes the results of some experiments with both systems; unfortunately a direct comparison is not possible, but Fig. 66 shows results for a 16-cylinder engine running at 900 rpm with a pulse manifold compared to those for the same engine at 1000 rpm with a compact manifold. At BMEPS greater than 9·7 bar the compact manifold is superior in respect of fuel consumption (3·5%), turbine inlet temperature (40°C) and thermal load; in fact, this has been achieved with an increase in power output of 11% (because of the higher speed). The results obtained for the 12-cylinder engine were similar but the gains achieved were not so great. This is not surprising

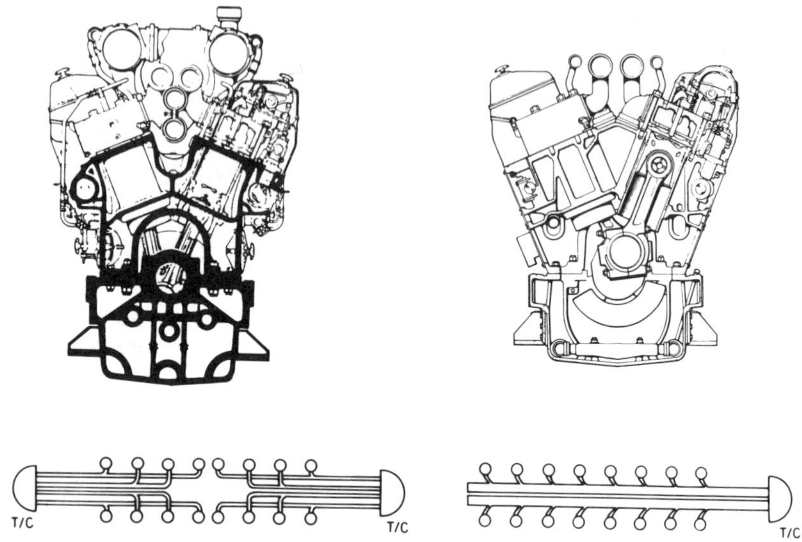

Fig. 65. Layout of engine with pulse charging and compact manifolds.

because the three-cylinder grouping in the exhaust system is the most effective one, with the best pulse utilization by the turbine. In all cases the performance of the engine with a compact manifold deteriorated at low load compared to the pulse charging system, because of the degradation of the pulse energy in the larger-volume manifold. Comparisons with a large-volume, constant-pressure manifold are not available.

Alexander *et al.*[56] modelled the compact manifold as a tapered pipe of area ratio equal to ψ_N, feeding into a constant-pressure junction. They compared the predictions with those for a parallel pipe feeding into the same pipe. It was apparent that the parallel pipe model was completely unrepresentative, underestimating the build-up of pressure in the manifold during blowdown, while the tapered pipe approach did simulate this effectively. However, neither model was adequate later in the scavenge process. The authors introduced a new model for the tapered pipe, based on one by Low, in which the values of properties in that section were based on the linear interpolation of pressure, mass flow and stagnation enthalpy. This scheme, while taking longer to calculate, is significantly more stable than that proposed in Section

3.2.2 of Chapter 12; this is because the change in Riemann variable occurring due to the change of area takes the calculation outside the stable step size.

In subsequent papers on asymmetric pulse converters, Winterbone *et al.* have used loss coefficients to model the junctions; these improve the simulation significantly. In Fig. 48 is shown a compact manifold for a six-cylinder engine, and the values of turbine work function, given in the boxes, show that a certain amount of energy is degraded by the

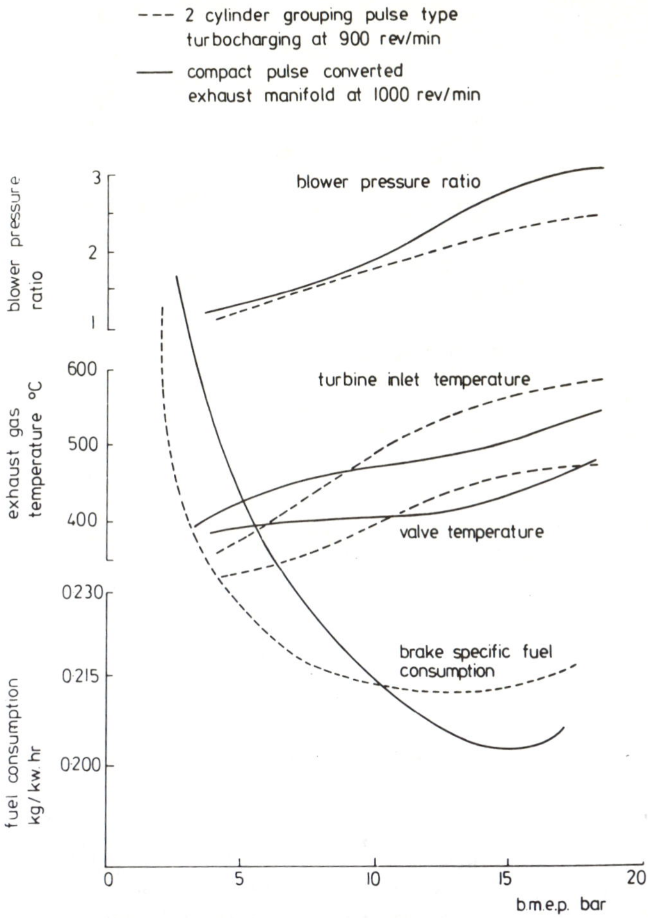

Fig. 66. Comparison of results for 16-cylinder V engine with conventional pulse charging and compact manifolds.

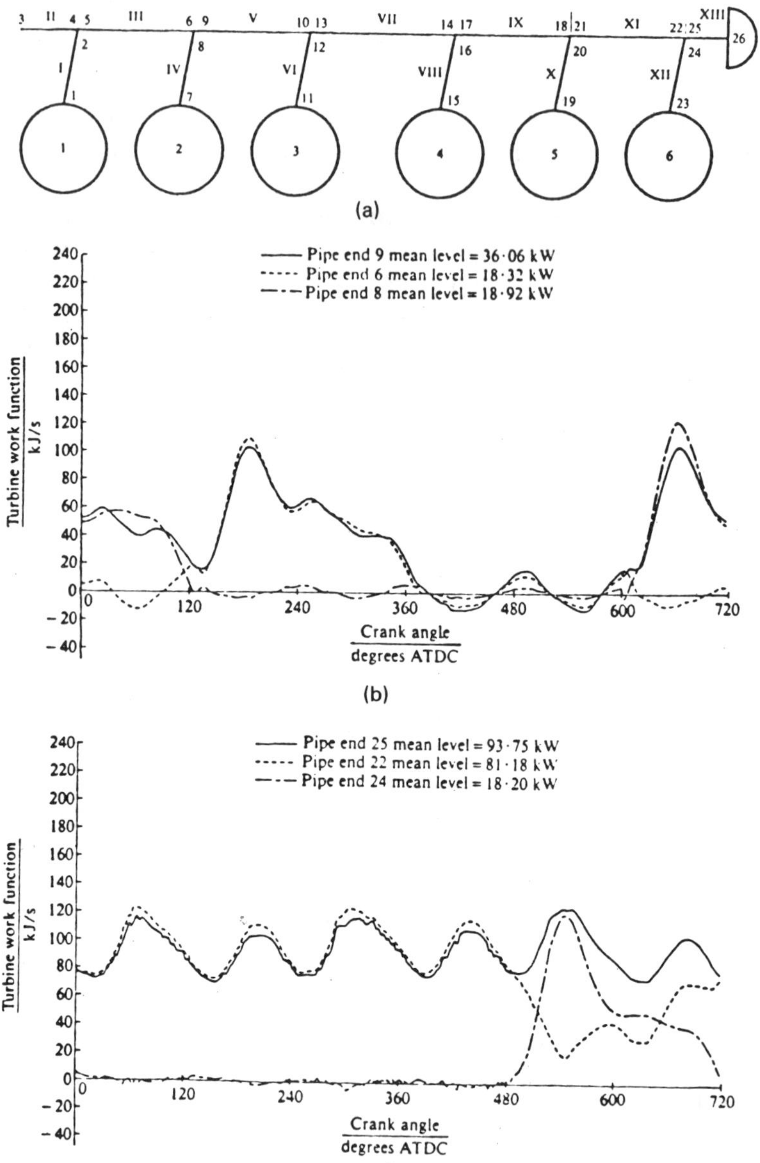

Fig. 67. Compact exhaust manifold: (a) schematic layout of junction; (b) variation of turbine work function with crank angle at three pipes of a pulse-converter junction (cylinder 2); (c) variation of turbine work function with crank angle at three pipes of a pulse-converger junction (cylinder 6).

junction (the amount of W_{is} leaving the junction is less than the sum of the quantities entering it). The layout of a compact manifold for a six-cylinder engine is shown in Fig. 67(a). The roman numerals indicate the pipes, while the arabic ones denote the pipe ends. In Fig. 67(b) is shown the variation of W_{is} in the three pipes downstream of cylinder 2 (viz. pipe ends 6, 8 and 9), while Fig. 67(c) shows a similar diagram for the junction of pipes 22, 24 and 25; that is, pipes at either end of the exhaust system. To simplify the analysis, assume that each cylinder releases $W_{is} = 19{\cdot}3$ kW into the exhaust system (see Fig. 48(a)); then at the first junction a total W_{is} of 38·6 kW would be expected and at the last junction the value should be 115·8 kW. The values calculated are only 36·1 kW and 93·8 kW respectively, which is 93·5% and 81% of the anticipated values. This loss of turbine work function is due to friction, heat transfer and pressure loss at the junctions. A similar calculation with constant-pressure junctions (neglecting the pressure loss) showed transmission efficiencies of 95·5% and 93·7% at the same junctions; hence the pressure loss brings about a significant reduction in turbine work function. The loss is a function of the position in the exhaust pipe, and the closer to the turbine the bigger the loss simply because the mean velocities are higher as all the exhaust gases join in the pipe. It can be seen that the major loss of turbine work function occurs between the side branch and the main manifold, which emphasizes the difference in loss coefficients in different flow directions.

To summarize, pulse converters have applications in medium- and low-speed engines. They can simplify the design of the engine and turbine, with benefits in performance. Methods of predicting the features found in pulse converters exist but these are currently dependent on empirical data.

6. THE PRESSURE EXCHANGER

In addition to devices which pressurize the air fed to the inlet manifold of an engine by centrifugal action or positive displacement, there is another which works purely on gas-dynamic mechanisms. This device is generically referred to as a pressure exchanger, but its more common name (registered by Brown Boveri and Co. Ltd) is the Comprex. A schematic view of a Comprex fitted to an engine is shown in Fig. 68. It can be seen that it consists of a rotor which is driven by a

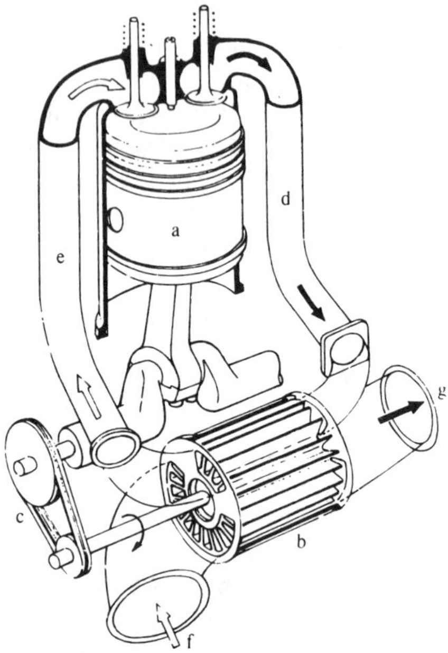

Fig. 68. Schematic of Comprex pressure exchanger. (a) engine. (b) cell wheel (rotor). (c) belt drive. (d) high pressure air. (e) high pressure air. (f) low pressure air. (g) low pressure gas.

belt from the engine crankshaft, and the stator which contains a series of air passages and pockets. The rotor itself consists of a number of axial passages which communicate on one side with the exhaust gas and on the other side the same passages are fed with air that will enter the engine. The times at which the exhaust and inlet gases can enter and leave the rotor are controlled by ports in the stator. The rotor is driven by the engine crankshaft but the speed of rotation is not rigidly defined by any necessity to synchronize the devices: the power provided by the engine crankshaft is only sufficient to rotate the rotor and does not do any work in compressing the gas. The compression of the air comes about purely by gas-dynamic action between the exhaust gas and the inlet air. Ideally the Comprex would operate with constant pressure levels on both the exhaust and intake sides; that is, the device is not dependent on wave action in the intake or exhaust manifolds and such wave action is detrimental to the performance of the machine.

The pressure exchanger is more similar to a turbocharger than a supercharger. This is because it uses the energy available in the exhaust gas to compress the air for delivery to the cylinder; the supercharger actually uses power from the engine crankshaft to perform that compression. Hence the Comprex should be compared with the turbocharger rather than the supercharger since it will achieve fuel economies which are comparable with the former device. There are a number of advantages of the Comprex over the turbocharger, the major one being its very rapid response to changes in engine load. It is well known that turbocharged engines suffer from a phenomenon referred to as turbocharger lag, which comes about from the mechanical inertia of the turbocharger and its inability to change speed (and hence operating point) rapidly under transient conditions. The Comprex does not suffer from such a limitation because the pressure ratio of this device is not directly related to the rotational speed of the device: it is much more closely related to the temperature (and energy) level in the exhaust system. Hence, a vehicle fitted with a Comprex pressure exchanger will behave like a naturally aspirated one, and accelerate with virtually no delay. The only inertia associated with the Comprex pressure exchanger is that of the gas, and this can be accelerated very rapidly in response to the changing fueling level. Comprex pressure exchangers have been applied mainly to diesel engines, where they can be used to overcome the relatively sluggish performance of such engines compared to gasoline-powered vehicles. In addition to improving the vehicle performance from the viewpoint of acceleration they also bring about benefits in fuel consumption and exhaust emissions.

The first pressure exchange devices were proposed as long ago as 1910 by Burghard. However, his proposal was backed by insufficient analytical capability and the device was extremely inefficient having been developed by an *ad hoc* empirical method. It was not further developed until a better understanding of compressible gas dynamics had been developed. The next major stage forward was when Seippel invented the so-called pressure exchanger as the high-pressure stage of a gas turbine arrangement. Developments in materials enabled the pressure exchanger in the gas turbine arrangement to be replaced later by a conventional high-pressure gas turbine stage and the Comprex again went into the background (it was at this time that the pressure exchanger was referred to as Comprex by Brown Boveri). It was not until about 1965–1970 that Brown Boveri seriously considered applying the Comprex as a means of supercharging diesel engines. The motivation was to use Comprex to provide an immediately available and ideal torque characteristic.

(a)

(b)

(c)

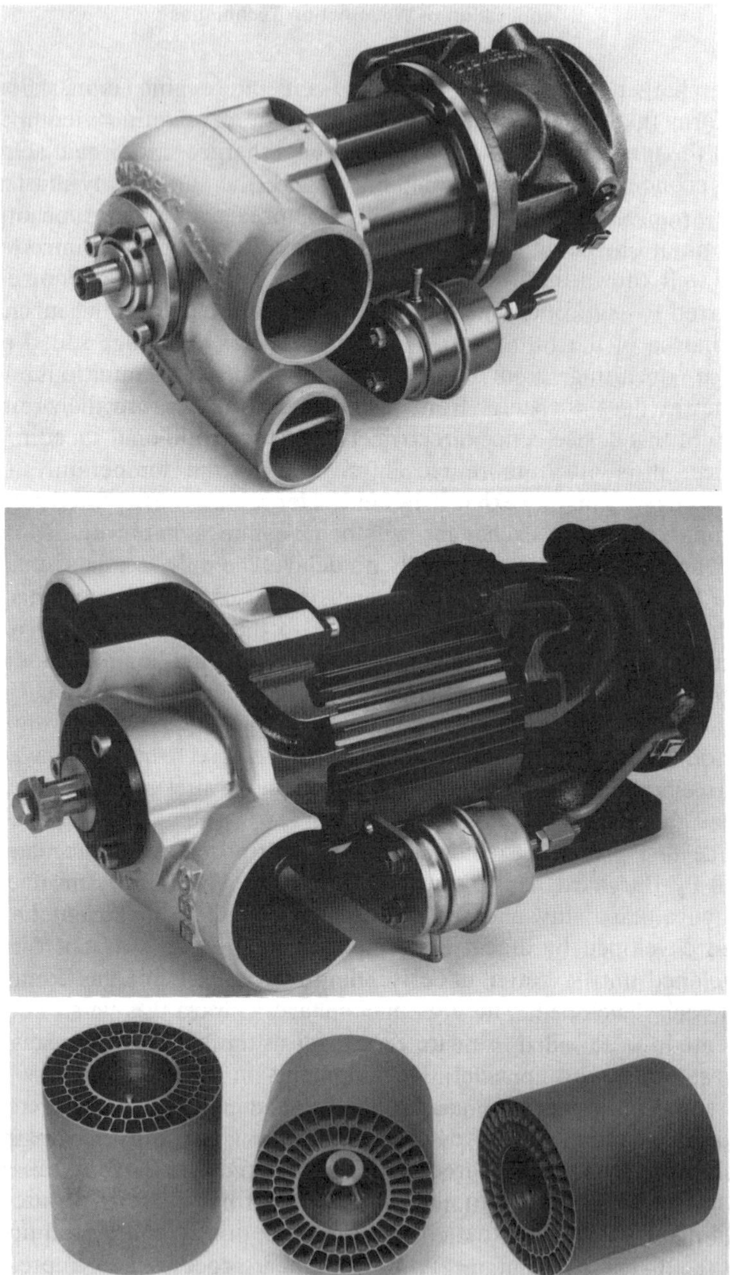

Fig. 69. Components of Comprex: (a) assembled device; (b) sectioned device showing rotor and stator passages; (c) rotor showing passage arrangement.

In Figure 69 is shown the construction and components of a Brown Boveri CX-series Comprex. Figure 69(a) is an external view of the device, showing the pressure cell for operating the waste-gate. Figure 69(b) is a sectioned view of a similar device, which clearly illustrates the rotor and stator passages, while Fig. 69(c) is a photograph of the rotor, depicting the cells. The two rows of cells have been introduced to reduce the characteristic whistle that had been associated with the early Comprexes: they do not alter the basic operating mechanisms of the device. Two other features are also apparent from this photograph: the ducts can be seen to be axial, and the thickness of the duct walls provides an appreciable blockage of the flow area available in the rotor. An extensive research project was required to achieve such thin rotor walls, which are down to 0·5 mm in mass production. A full description of the detailed design of the Comprex is given by Mayer.[75]

6.1. Principle of Operation

Although the pressure exchanger is a wave action device it operates most satisfactorily with constant-pressure inlet and exhaust manifolds. This is because the pressure exchanger generates the pressure waves it requires by rotation of the rotor. The basic principle of operation is that the sense of a wave can be modified by interaction with the end of a pipe; this has been demonstrated by the simple examples in Section 4 of Chapter 12 and this is shown graphically in Fig. 7 of Chapter 12. Hence, a pressure wave can be turned into a rarefaction wave, or simply reflected as a pressure wave—the sense of the reflection depends on the condition at the end of the rotor. The rotor consists of a number of cells (usually axially oriented in the case of the Comprex) which can be considered to be a number of pipes, the ends of which are open or closed depending on the position of the rotor. This will become clearer by considering an example.

The operating principles of Comprex and pressure exchangers are described in many references. Lutz & Scholz[76] and Azoury[77] give a good introduction to the basic features of the device; Azoury & Hai[78] describe a computer simulation of the flow in the exchanger, based on a method proposed by Spalding.[79] Pearson[80] gives a full description of the methods of simulating the individual processes in a pressure exchanger, and is a recommended text for those interested in the detailed analysis of the individual cell filling and emptying processes.

In Fig. 70, from Ref. 80 (based on Lutz & Scholz[76]), is shown a

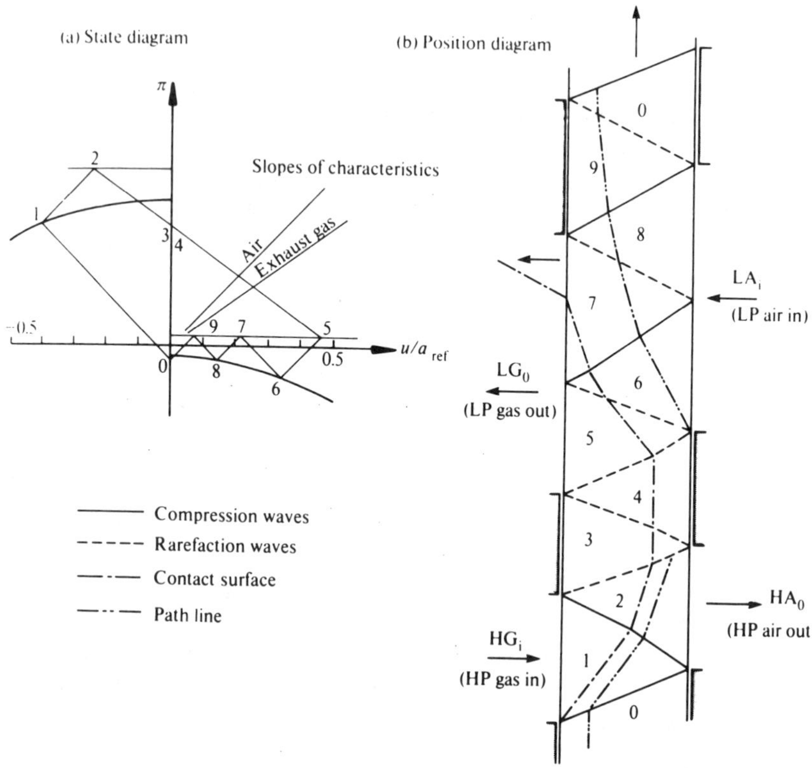

Fig. 70. Wave effects in Comprex.

developed view (the cylindrical surface has been opened onto the paper) of the rotor and stator of a Comprex. It is possible to evaluate the operation of the device by considering the cycle of a single rotor cell. The state diagram, Fig. 70(a), is derived from the method of characteristics, and indicates that the properties (the ratios of specific heats, κ) of the hot and cold gases are different: it is also used to evaluate the slopes of the lines on the position diagram. The position diagram, Fig. 70(b), shows exhaust gases entering and leaving on the left and air entering and leaving on the right. The cell has first been filled with air in a *low-pressure scavenging state*; the ports have been closed and the air is stagnant at state 0. One end of the cell is then opened to the *high-pressure gas inlet port* (HG_i). A pressure rise occurs in the air due to inflow of the driving gas. The opposite (or outlet) end of the cell is then opened to the *high-pressure air receiving port* (HA_0) and *high-pressure exchange* then follows in which compressed air at state 2 is displaced, to be delivered as output, by the inflow of further high-pressure driving gas. Both cell ends are then closed. Subsequently, the cell is opened to the 'low-pressure gas receiving port' (LG_0) to achieve blowdown of the gas from the cell to atmospheric pressure. Residual gases in states 5–7 are swept out to an 'exhaust duct' through port LG_0, by incoming air in states 6–8, entering through port LA_i in a low-pressure scavenge. An extended exhaust process (involving the complete traverse of the cell length by three waves) is used to provide a depression in the cells, which increases the amplitude of the wave generated by subsequent opening of the high-pressure gas inlet port (HG_i). An increased scavenging wave speed is thereby obtained.

This design (essentially that of the Brown Boveri Company) is most suitable for low pressure ratios, of the order 2:1. Further details of how the Comprex has been developed are given by Jenny & Naguib.[81] At pressure ratios of 3:1 and above, it is best to use a less extended exhaust process (two waves traversing the cells), leading to a cell pressure above atmospheric before the start of admission of high-pressure gas.

Consideration of the operating processes of the Comprex will show that the device is equivalent, in a single rotor, to the turbine and compressor of a turbocharger: it achieves this capability by utilizing the gas dynamics of the fluids. This has a number of effects on the characteristics of the device itself. First, the level of boost achieved from the Comprex is not primarily related to the rotational speed of the rotor (as in the case of a turbocharger) but more closely to the temperature ratio across the engine. Hence, it is possible to obtain a high boost at low engine speeds and this has a beneficial effect on the torque back-up that can be achieved by the engine. Second, since the boost achieved is not strongly dependent on the rotor speed it is possible for the response of the boost pressure to be extremely rapid—thus eliminating turbocharger lag.

The Comprex is driven by a belt from the crankshaft of the engine, which means that the ratio of rotor to engine speed is constant. For this to be acceptable the characteristics of the Comprex must be only weakly related to its rotational speed. Consideration will show that the device must have an optimum operating speed related to the natural

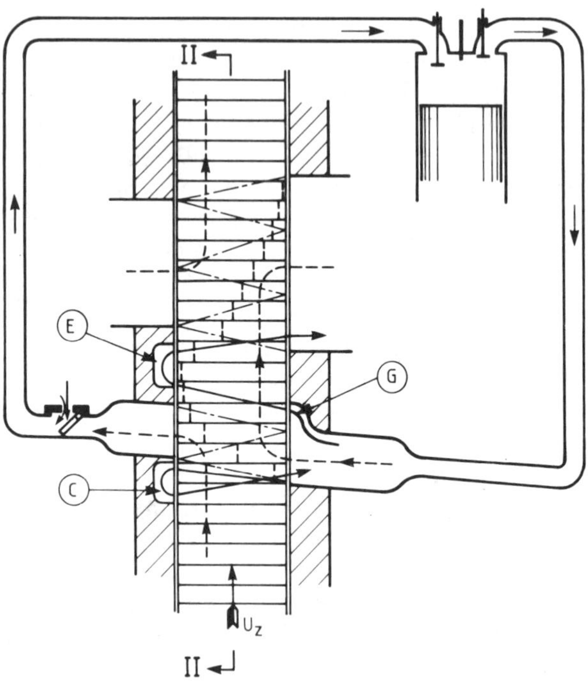

Fig. 71. Detuning pockets in Comprex; indicated by C, E and G.

frequency of wave travel in the rotor passages. This was a major shortcoming of the Comprex and it was not until 1960 that Waleffe, Jenny & Muller[82] patented a method of detuning the machine to reduce its dependence on rotational speed. The patent diagram which indicates how pockets C, E and G have been introduced in the stator is shown in Fig. 71. The effect of the pockets is to detune the wave action in the rotor by replacing the previously closed end by a partially open one. The theory of detuning pockets, and the detailed design of Comprex features, is described in detail in Pearson.[80]

6.2. Operating Characteristics of Pressure Exchangers

Comprexes have been fitted to a broad range of diesel engines. The original applications were large earthmoving and truck engines, where it was felt that the elimination of turbocharger lag would be a major benefit. It was found that development was slow and expensive and that the benefits achieved were not always as important as the engine manufacturer had expected. In the 1970s development was focused on automobile engines and many feasibility studies were performed on a range of engines. Brown Boveri fitted Comprexes to 50 Volkswagen vehicles and achieved extensive experience with the devices. Later, 750 Opel Senators were constructed with Comprexes fitted, and now Mazda has announced that Comprex will be used on their 2 litre engine.

A typical set of characteristics of a Comprex suitable for a 1·6 litre engine is shown in Fig. 72. The maximum pressure ratio achieved is about 2:1, and the maximum efficiency of compression (to be compared with the compressor efficiency with a turbocharger) is around 70%. The 'overall efficiency' of the device is quoted as up to 40%.

The very early Comprex devices suffered from a number of adverse characteristics, which included:

(a) limited operating speed range;
(b) low compression efficiency;
(c) poor mechanical integrity;
(d) excessive exhaust gas recirculation at part load and idling;
(e) harsh and unpleasant intake noise;
(f) awkward mechanical installation with belt drive;
(g) high price (relative to turbocharger).

Careful and detailed design by Brown Boveri has removed many of these limitations. Many of the design changes implemented are described by Mayer[75]. The speed range has been extended by the detuning pockets, the mechanical integrity has been improved by use of specially developed materials and the careful development of the rotor bearings, the compression efficiency has been raised by reducing the axial end clearances by careful design and the use of low-expansion alloys, and the intake and exhaust noise have been reduced by the use of a double row of staggered cells. In addition to these modifications, the latest Comprexes are fitted with exhaust waste-gates and inlet starting valves. The waste-gate enables the engine to be fitted with a Comprex which has a pressure ratio larger than required by the engine: the delivered pressure can be controlled by use of an exhaust waste-gate to give the appropriate characteristic from the engine. The inlet starting valve is important to reduce the inherent exhaust gas recirculation which occurs at low pressure ratios. This works by

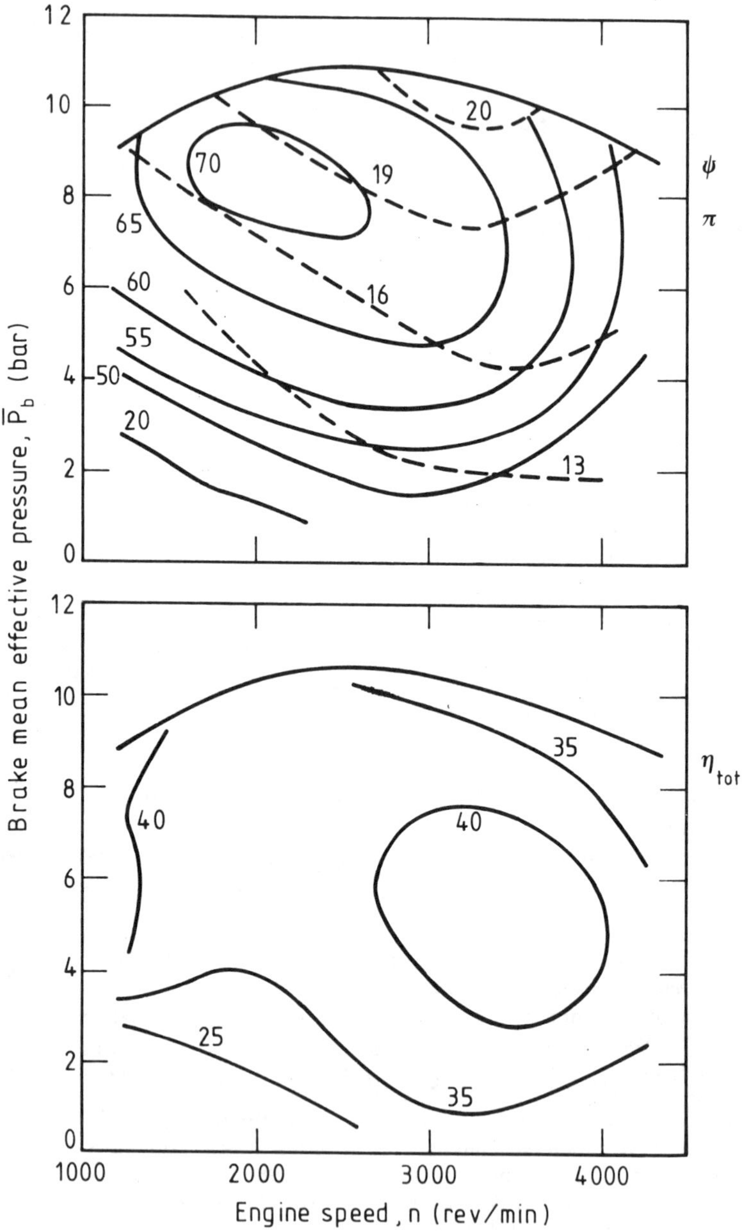

Fig. 72. Characteristics of 1·6 litre engine fitted with Comprex.

allowing the engine to breathe directly from the atmosphere at low levels of boost, the delivery from the Comprex being available only when the inlet manifold pressure rises above 0·15 bar gauge. A further shortcoming in the application of Comprex was the reliability and life of the belt drive. In early applications a breakage of the belt resulted in loss of boost and possible damage to the engine. Current Comprexes actually produce a torque on the rotor due to the stator port angles (see Pearson[80]) and will run even after the belt has broken. It is claimed that it is also possible to restart the Comprex even after the engine has been stopped; hence this is no longer a catastrophic failure.

REFERENCES

1. Broome, D., Induction ram. Part 1: The inertia and wave effects of ram. *Automobile Engr,* April 1969 130–33; Part 2: Inertial aspects of induction ram. ibid., May 1969, 180–84; Part 3: Wave phenomena and design of ram intake systems. Ibid., June 1969, 262–7.
2. Williams, T. J., Induction ramming of reciprocating engines. *Engineering,* 20 March 1959, 370–72.
3. Taylor, C. F., Livengood, J. C. & Tsai, D. H., Dynamics of the inlet system of a four-stroke single cylinder engine. *Trans. ASME,* **77** (1955) 1133–45.
4. Lighthill, M. J., *Waves in Fluids.* Cambridge University Press, 1978.
5. Cser, G., Some results of combined charging applications. Proc. Instn Mech. Engrs, Conf. on Turbocharging and Turbochargers, C64/78, London, 1978, 141–6.
6. Watson, N., Resonant intake and variable geometry turbocharging systems for a V8 diesel engine. Proc. Instn Mech. Engrs, Conf. on Turbocharging and Turbochargers, C40/82, London, 1982, 101–13.
7. Brands, M. C., Helmholtz tuned induction systems for turbocharged diesel engines. SAE paper 790069. Also in *Turbochargers and Turbocharged Engines.* SAE Publication SP 442, 1979, 139–47.
8. Vorum, P. C., Short pipe manifold design for four stroke engines. ASME paper 76-WA/DGP-4, 1976.
9. Vorum, P. C., Short pipe manifold design for four-stroke engines—2. *J. Engng Power, Trans. ASME,* **102**(4) (October 1980) 836–41.
10. Chen, A. & Gu, H.-Z., Simulation of a tuned induction system and the structural parameters analysis. Proc. Instn Mech. Engrs, Conf. on Turbocharging and Turbochargers, London, 1986.
11. Knecht, W. & Signer, M., Development of the Saurer D4KT-B diesel engine. SAE paper 810342, 1981.
12. Nikpour, B., Heat release and gas flow in a petrol engine. MSc Dissertation, University of Manchester, 1985.

13. Taylor, C. F., *The Internal Combustion Engine in Theory and Practice*, vol. 1. MIT Press, Cambridge, Massachusetts, 1977.
14. Ferguson, C. R., *Internal Combustion Engines, Applied Thermosciences.* John Wiley, New York, 1986.
15. Tabaczynski, R. J., Effects of inlet and exhaust system design on engine performance. SAE paper 821577, February 1982.
16. Fukutani, I. & Watanabe, E., An analysis of the volumetric efficiency characteristics of 4-stroke cycle engines using the mean inlet Mach no. SAE paper 790484; *Trans. SAE* **88** (1979) 1756–82.
17. Prosser, T. G., Induction ramming a motored high speed four-stroke reciprocating engine—influence of inlet port pressure waves on volumetric efficiency. *Proc. Instn Mech. Engrs,* **188** (1974) 577–84.
18. Adams, T. G., Effect of exhaust system design on engine performance. SAE paper 800319, Detroit, February 1980.
19. Hopkinson, B., The charging of two-cycle engines *Trans. NE Coast Inst. Engrs. Shipbldrs,* **30** (1914) 433–58.
20. Benson, R. S. & Whitehouse, N. D., *Internal Combustion Engines,* vols 1 & 2. Pergamon Press, Oxford, 1979.
21. Pearce, J. F., Hames, R. J. & Merrion, D. F., Two-stroke cycle diesel engine fuel economy improvement and emission reduction. SAE paper 770255, Detroit, February, 1977.
22. Horlock, J. H. & Winterbone, D. E., *The Thermodynamics and Gas Dynamics of Internal Combustion Engines,* vol. II. Oxford University Press, 1986.
23. Lilly, L. R. C. (ed.), *Diesel Engine Reference Book.* Butterworth, London, 1984.
24. Benson, R. S., *The Thermodynamics and Gas Dynamics of Internal Combustion Engines,* vol. I, eds J. H. Horlock & D. E. Winterbone, Oxford University Press, 1982.
25. Poloni, M., Winterbone, D. E. & Nichols, J. R., Comparison of unsteady flow calculations in a pipe by the method of characteristics and the two-step differential Lax–Wendroff method. *Int. J. Mech. Sci.,* **29**(5) (1987) 367–78.
26. Azuma, T., Yura, T. & Tokunaga, Y., Some aspects of constant pressure turbocharged marine diesel engines of medium and low speed. *J. Engng Power, Trans. ASME,* **105**(3) (1983) 697–711.
27. Annand, W. J. D. & Roe, G., *Gas Flow in the Internal Combustion Engine.* G. T. Foulis, Yeovil, 1974.
28. Abthoff, J., Bruggemann, H., Huttebraucker, D.-H. & Zeilinger, K., Daimler–Benz 2.3 litre, 16-valve high performance engine. SAE paper 841226, 1984.
29. Matsumoto, I. & Ohata, A., Variable induction systems to improve volumetric efficiency at low and/or medium engine speeds. SAE paper 860100, Detroit, February 1986.
30. Engleman, H. W., Design of a tuned intake manifold. ASME paper 73-WA/DGP-2, 1973.
31. Ohata, A. & Ishida, Y., Dynamic inlet pressure and volumetric efficiency of four-cycle four cylinder engine. SAE paper 820407, Detroit, 1982.

32. Chapman, M., Novak, J. M. & Stein, R. A., A non-linear acoustic model of inlet and exhaust flow in multi-cylinder internal combustion engines. ASME paper 83-WA/DSG-14, 1983.
33. Worth, D. R., Investigation of inlet manifold tuning of an automotive spark ignition engine. MSc thesis, Univ. Manchester, 1988.
34. Boucher, R. F. & Kitsios, E. E., Simulation of fluid network dynamics by transmission line modelling. *Proc. Instn Mech. Engrs,* **200** (1986) 21–9.
35. Auslander, D. M., Distributed system simulation with bilateral delay-line models. *Trans. ASME, J. Basic Engng,* June 1968, 195–200.
36. Soedel, W., Gas pulsations in compressor and engine manifolds. Short course text, Ray W. Herrick Laboratories, School of Mechanical Engineering, Purdue University, West Lafayette, Indiana, 1978.
37. Bingham, J. F. & Blair, G. P., An improved branched pipe model for multi-cylinder automotive engine calculations. *Proc. Instn Mech. Engrs,* **199,** No. D1, (1985) 65–77.
38. Bingham, J. F., Intake system design using a validated internal combustion engine computer model. Paper C25/87, Instn Mech. Engrs Conf. 'Computers in Engine Technology', Cambridge, UK, March 1986, pp. 111–22.
39. Takeyama, S., Ishizawa, S. & Yoshikawa, Y., Gas exchange simulation model for improving charging efficiency of four-valve internal combustion engine. Paper C28/87, Instn Mech. Engrs Conf. Computers in Engine Technology, Cambridge, UK, March 1986, 123–30.
40. Tosa, Y., Shimoda, K. & Oikawa, H., Calculation of two-dimensional unsteady flows in inlet pipe system and its application for V-8 resonant intake system. Int. Symp. *Flows in Internal Combustion Engines—III,* ASME, Miami Beach, 1985, pp. 63–70.
41. Chapman, M., Novak, J. M. & Stein, R. A., Numerical modelling of inlet and exhaust flows in multi-cylinder internal combustion engines. In *Flows in Internal Combustion Engines.* ASME, Phoenix, Arizona, November 1982, pp. 9-20.
42. Walter, G. A., & Chapman, M., Numerical simulation of the exhaust flow from a single cylinder of a two cycle engine. SAE paper 790243, Detroit, 1979.
43. Poloni, M., Winterbone, D. E. & Nichols, J. R., Calculation of pressure and temperature discontinuity in a pipe by the method of characteristics and the two-step differential Lax–Wendroff method. Int. Symp. *Flows in Internal Combustion Engines,* FED, Vol. 62. ASME Winter Annual Meeting, Boston, 13–18 December 1987.
44. Sierens, R., van Hove, W. & Snauwert, P., Comparison of measured and calculated gas velocities in the inlet channel of a single cylinder reciprocating engine. *Proc. Instn Mech. Engrs,* **198A**(1) (1984) 61–9.
45. Lakshminarayananan, P. A., Janakiraman, P. A., Gajendra Babu, M. K. & Murthy, B. S., Prediction of gas exchange processes in a single cylinder internal combustion engine. SAE paper 790359, June 1979.
46. Gajendra Babu, M. K. & Murthy, B., Simulation and evaluation of a four-stroke single cylinder spark-ignition engine. SAE paper 750687, *Trans. SAE,* **84** (1975) 1631.

47. Sato, A., Suenaga, K., Noda, M. & Maeda, Y., Advanced boost-up in Hino EP100 turbocharged and charge-cooled diesel engine. SAE paper 870298, Detroit, February 1987.
48. Blair, G. P. & Goulburn, J. R., An unsteady flow analysis of exhaust systems for multicylinder automobile engines. SAE paper 690469, 1969.
49. Takizawa, M., Uno, T., Oue, T. & Yura, T., A study of gas exchange process simulation of an automitive multi-cylinder internal combustion engine. SAE paper 820410, 1982.
50. Benson, R. S. & Baruah, P. C., Performance and emission predictions for a multi-cylinder spark ignition engine. *Proc. Instn Mech. Engrs* **191,** 32/77, (1977) 339–54.
51. Baruah, P. C., Benson, R. S. & Balouch, S. K., Performance and emission predictions for a multi-cylinder spark-ignition engine with exhaust gas recirculation. SAE paper 780663, 1978.
52. Baruah, P. C., Benson, R. S. & Gupta, H. N., Performance and emission predictions for a multi-cylinder spark-ignition engine with catalytic converter. SAE paper 780672, 1978.
53. Low, S. C. & Baruah, P. C., A generalized computer aided design package for I.C. engine manifold system. SAE paper 810498, 1981.
54. Herring, P., Sequential turbocharging of the MTU 1163 engine. *Trans. Instn Mar. Engrs,* **100** (1988) 145–56.
55. Benson, R. S. & Horlock, J. H., The matching of two-stroke engines and turbochangers. CIMAC, Copenhagen, 1962, p. 464.
56. Alexander, G. I., Nichols, J. R., Winterbone, D. E. & Sinha, S. K., Optimization of the design of an exhaust manifold system for improved turbocharger performance. Instn Mech. Engrs Conf. on Turbocharging and Turbochargers, C48/82, 1982, pp. 175–83.
57. Winterbone, D. E., Alexander, G. I. & Nichols, J. R., Developments in methods of considering wave action in pipes connected to IC engines. Int. Symp. *Flows in internal combustion engines—III,* ASME, Miami Beach, 1985, pp. 71–8.
58. Winterbone, D. E., Nichols, J. R. & Alexander, G. I., Efficiency of manifolds of turbocharged engines. *Proc. Instn Mech. Engrs,* **199** (D2) (1985) 137–49.
59. Haywood, R. L. E., *Equilibrium Thermodynamics.* John Wiley, Chichester, 1960.
60. Oldfield, S. G. & Watson, N., The influence of exhaust valve and port design on energy transfer to the turbocharger turbine. SAE paper 830151, 1983.
61. Büchi, A., French Patent No. 371281, 1905.
62. Payri, F., Desantes, J. M. & Boad, F., Prediction method of the operating conditions of a turbocharged diesel engine. Motorsympo. 86, April 1986, CSSR, pp. 22–24.
63. Benson, R. S. & Scrimshaw, K. H., An experimental investigation of non-steady flow in a radial gas turbine. *Proc. Instn Mech. Engrs,* **180,** part 3J, paper 23 (1965–66) 74–85.
64. Dale, A. & Watson, N., Vaneless radial turbocharger turbine performance. Instn Mech. Engrs Conf. on Turbocharging and Turbochargers, paper C110/86, London, May 1986, pp. 65–76.

65. Janota, M. S., Hallam, A. K., Brock, E. K. & Dexter, S. G., The prediction of diesel engine performance and combustion chamber component temperatures using digital computers. *Proc. Instn Mech. Engrs,* **182,** 3L (1967–68) 58–70.
66. Azuma, T., Tokunaga, Y. & Yura, T. Y., Characteristics of exhaust gas pulsation of constant pressure turbocharged diesel engines. *Trans. ASME, J. Engng Power,* **102** (October 1980) 827–35.
67. Janota, M. S. & Watson, N., Pulse converters—a method of improving the performance of the turbocharged diesel engine. *Proc. Instn Mech. Engrs,* **187,** 51/73 (1973) 635–47.
68. Watson, N. & Janota, M. R., Non-steady flow in an exhaust system with a pulse converter junction. Instn Mech. Engrs Conf. Internal Flows, Paper 25, Salford, 1971, pp. D17–D28.
69. Benson, R. S. & Alexander, G. I., The application of pulse converters to automotive four-stroke cycle engines. SAE paper 770034, Detroit, 1977.
70. Borilla, Y., Some aspects of performance optimization of the sequentially turbocharged highly-rated truck engine with turbochargers of unequal size and a pulse converter. Instn Mech. Engrs Conf. on Turbocharging and Turbochargers, paper C105/86, 1986, pp. 251–60.
71. Birmann, R., Exhaust energy converting means for internal combustion engines. US Patent 2406656, 1946.
72. Curtil, R. & Magnet, J., Exhaust pipe system for high pressure charging. Instn Mech. Engrs Conf. on Turbocharging and Turbochargers, C65/78, 1978, pp. 147–58.
73. Chan, C. L., Winterbone, D. E., Nichols, J. R. & Alexander, G. I., A detailed study of compact exhaust manifolds applied to automotive diesel engines. Proc Instn Mech. Engrs, Conf. on Turbocharging and Turbochargers, C113/86, 1986, pp. 269–81.
74. Winterbone, D. E., Nichols, J. R. & Alexander, G. I., The evaluation of the performance of exhaust systems equipped with integral pulse converters. 16th International Congress on Internal Combustion Engines (CIMAC), Oslo, June 1985.
75. Mayer, A., The Comprex supercharger—a simple machine for a highly complex thermodynamic process. *Brown Boveri Review,* **8** (1987) 422–30.
76. Lutz, T. W. & Scholz, R., Supercharging vehicle diesel engines by the Comprex system. *Proc. Instn Mech. Engrs,* (1968).
77. Azoury, P. H., An introduction to the dynamic pressure exchanger. *Proc. Instn Mech. Engrs,* **180**(1) (1965–66) 451–80.
78. Azoury, P. H. & Hai, S. M., Computerized analysis of dynamic pressure exchanger scavenge processes. *Proc. Instn Mech. Engrs,* **189** (1975) 149–58.
79. Spalding, D. B., A procedure for calculating the unsteady, one-dimensional flow of a compressible fluid, with allowance for the effects of heat transfer and friction. Report no. UF/TN/D/2, Imperial College, London, 1969.
80. Pearson, R. D., Pressure exchangers and pressure exchange engines. In *The Thermodynamics and Gas Dynamics of Internal Combustion Engines,* vol. II, ed. J. H. Horlock & D. E. Winterbone. Oxford University Press, 1986, Chapter 16.

81. Jenny, E. & Naguib, M., Development of the Comprex pressure-wave supercharger: in the tradition of thermal turbomachinery. *Brown Boveri Review*, **8** (1987) 416–21.
82. Waleffe, J., Jenny, E. & Muller, K., Brennkraftmaschine mit einem als Aufladegerat wirkenden. Drucktauscher Swiss Patent 378595, August 1960.

14

The Turbocharger

B. E. Walsham

Holset Engineering, Co. Ltd, Huddersfield, UK

&

D. E. Winterbone

UMIST, Manchester, UK

1. Introduction
2. The function of the turbocharger
3. Turbocharger construction
   3.1. Compressor wheels
   3.2. Compressor covers
   3.3. Bearing housings
   3.4. Bearing systems
   3.5. Turbine wheels
   3.6. Turbine housings
4. A general approach to turbocharger design
   4.1. General equations of turbomachinery
   4.2. The compressor
   4.3. The turbine
   4.4. The interspace between the casing and the rotor
   4.5. The turbine rotor
   4.6. Turbine efficiency (η_T) and degree of reaction (Λ)
   4.7. Streamline curvature methods
   4.8. Three-dimensional methods of analysis
5. Matching
6. Current research
   6.1. Research on wheels
   6.2. Research on compressor casings
   6.3. Turbine casings
   6.4. Pulsating flow
7. Testing facilities
   7.1. Compressor performance maps
   7.2. Turbine performance maps

8. Areas of special interest for research
 8.1. Aerodynamics: application of flow codes
 8.2. Pulsating flow
 8.3. Flow characteristics

9. The future
 9.1. Developments of conventional turbochargers
 9.2. Different air handling systems
 9.3. Ceramic rotor turbochargers

1. INTRODUCTION

Turbocharging was first proposed by Büchi, in patents, as long ago as 1905 and 1909, and has been used on large diesel engines for many decades. By the late 1950s, technology had advanced far enough for units to be used on truck diesel engines to make their power output competitive with the less economical gasoline engines. This application has grown such that very few truck engines above 6 litres in capacity are now made in a naturally aspirated form. Recently, the technique has come to prominence in the minds of the general public with the application of turbocharging to small automotive engines, and, in particular, the engines used in Formula 1 Grand Prix racing cars.

Turbocharging can produce two basic benefits in engine performance: it increases the power density of the engine and, particularly on diesel engines, can improve the specific fuel consumption. The actual effect required from a turbocharger installation depends very much on the application. Racing car designers go for maximum power output (although this has recently been modified by the overall fuel consumption limitation—note that lower specific fuel consumption with a higher power output may mean worse overall fuel consumption), while the designers of truck engines aim for modest power densities commensurate with long life and minimum fuel consumption, with military applications going for maximum power density, etc.

This chapter is divided roughly into four parts. The first section considers the construction of the turbocharger as a rotodynamic device. The next derives the simplified one-dimensional mathematical relationships that define the performance of the compressor and turbine. Then the interaction of the engine and turbocharger are considered in a section on matching. Finally, the detailed analysis of flows and additional areas of research are introduced. The transmission of exhaust gas energy from the engine to the turbine is described by Winterbone in Chapter 13, and methods for analysing this phenomenon are introduced there.

In addition to its direct role in engine systems, the turbocharger is an ideal piece of machinery for research. It can be used to provide aerodynamic data on the wheels and covers of radial compressor stages, and the radial inflow and axial turbine stages. It can also be adapted for use as a simple gas turbine or a gas turbine with a separate power turbine. The aerodynamic work can be as sophisticated as resources allow. The flow through a typical stage needs a fully three-dimensional flow analysis, with turbulence taken into account, and the experimental work really needs a laser anemometer to make meaningful measurements. Yet this wealth of possibilities comes from a mechanism that is neither expensive nor large and which is widely available!

A typical automotive turbocharger is shown in Fig. 1. It consists of a centrifugal compressor and radial inflow turbine, closely coupled on a shaft. In addition to these rotodynamic components there are casings for delivering the exhaust gas to the turbine and diffusing the air from

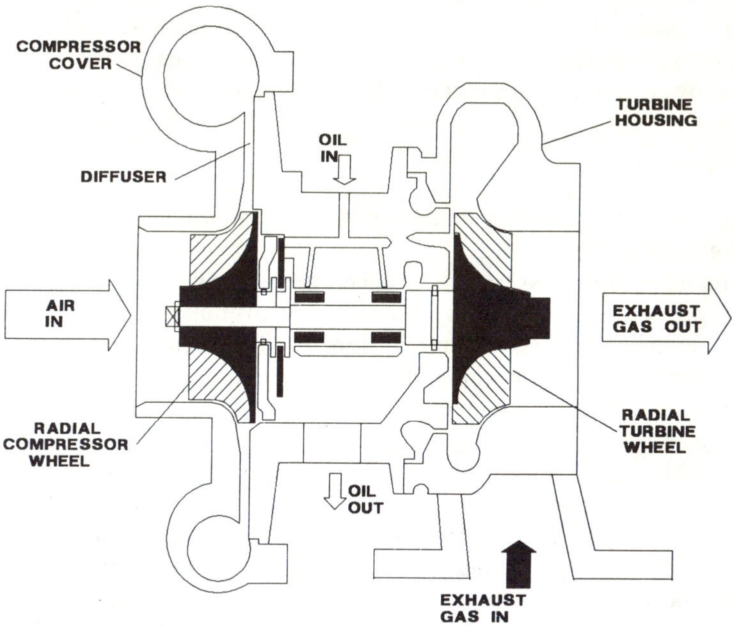

Fig. 1. Cross-section of automotive turbocharger.

the compressor. Between these casings is the bearing assembly, which is the central mounting point of all the components.

2. THE FUNCTION OF THE TURBOCHARGER

To make turbocharger research as meaningful as possible it is important to understand the function and duties of the machine. The turbocharger is not an integral part of an internal combustion engine, but has a symbiotic relationship with it. The power balance and thermal efficiency of the turbocharger are self-contained, yet the engine benefits from the addition of a turbocharger if the component efficiencies of the latter are high enough.

The need to increase the amount of air induced, or 'swallowed', by an engine is self-evident: the only way for an engine of a given size to produce more power is to burn more fuel and this, in turn, needs more air for its combustion. The limits on the amount of fuel burnt in an engine are:

(a) the allowable peak cylinder pressure;
(b) the production of too much smoke or other emissions;
(c) the temperature of some component reaches its limit; and
(d) a lower limit of the stoichiometric air : fuel ratio.

The reasons for the current dominance of the turbocharger in the field of pressure charging devices are fivefold:

(a) the turbocharger only delivers boost in response to the demand of the engine;
(b) the space required to fit a turbocharger is small;
(c) it has the capability of high pressure ratios;
(d) it matches the boost and airflow to the engine and avoids inefficient back-flow compression;
(e) it can have high component efficiencies.

However, the turbocharger has three distinct disadvantages, as follows:

(a) It takes time to go from one operating state to another in response to the demand of the engine—the well publicized 'turbo lag'.
(b) The swallowing capacity of a turbine (like any other flow restriction device) produces an expansion ratio that rises with

mass flow and so the turbine power (which is a function of mass flow and expansion ratio) rises at a rate greater than that of the mass flow—this can be shown to be far from ideal.

(c) The compressor stage of a turbocharger is not only limited by choke and poor efficiencies at high flow rates, but also suffers from an aerodynamic stall condition at low flows (referred to as *surge*), so the operating range of the compressor is limited and this has an effect on the allowable speed range of the engine.

Two major characteristics distinguish the operation of a turbocharger from that of a gas turbine. Firstly, the turbocharger does not have

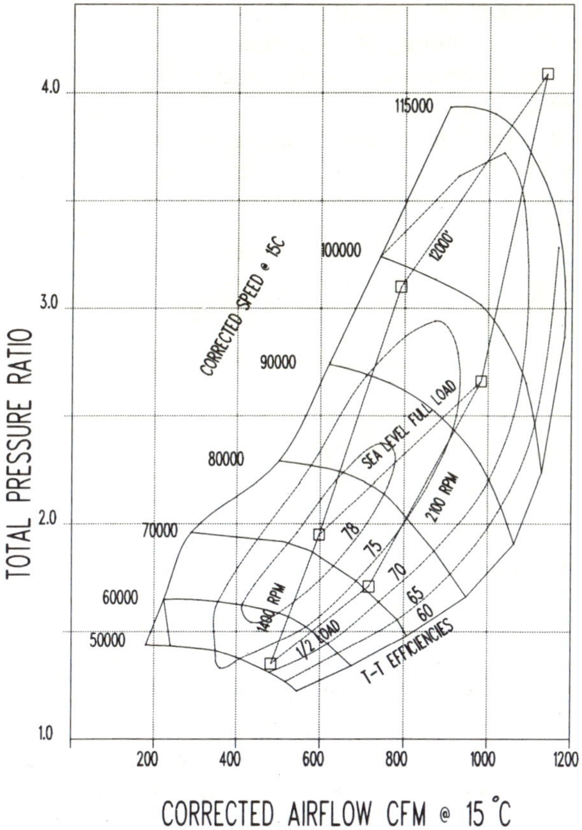

Fig. 2. Operating locus of truck engine on compressor map (CFM is volumetric flow rate in ft^3/min).

a single design point: in fact it has several, and the turbocharger is expected to operate over a wide range of mass flows and pressure ratios, as shown in Fig. 2, which indicates the running régime of a truck engine plotted on a compressor map. Secondly, the turbine stage has to accept a pulsating gas flow and convert this to shaft power with good efficiency. With these factors in mind, and recognizing that the aerospace industry has undertaken a great deal of research, that is relevant to its own applications, into the aerodynamics and structures of rotating machines, it becomes obvious that the real challenge of turbocharger research is in those fields that diverge from the aerospace applications. These are:

(a) the efficiency of compressors and turbines at mass flows and pressure ratios away from design conditions;
(b) the performance of vaneless diffusers and collectors in the compressor stage;
(c) the reduction of surge flow on the compressor stage;
(d) the enhancement, in general, of the performance of radial inflow turbine stages;
(e) the performance of turbine stages under pulsating flow conditions;
(f) the design and analysis of turbine inlet housings under pulsating flow;
(g) the design and development of highly efficient bearing systems; and
(h) the manipulation of turbine and compressor characteristics to modify the behaviour of the turbocharger/engine combination.

3. TURBOCHARGER CONSTRUCTION

The first task of this chapter is to familiarize the reader with the components of a turbocharger. Generalizing, there are two main types of turbocharger unit, which can be classified as 'big' and 'small'. The 'big' units are to be found on medium-speed and 'cathedral' diesel engines; i.e. engines for use in locomotives, power generating stations and ships. 'Small' turbochargers are used on road and agricultural vehicle engines, small boats, low-powered generator sets, etc. In terms of turbine and compressor wheel sizes, the break between the two sizes occurs at a diameter of about 150 mm.

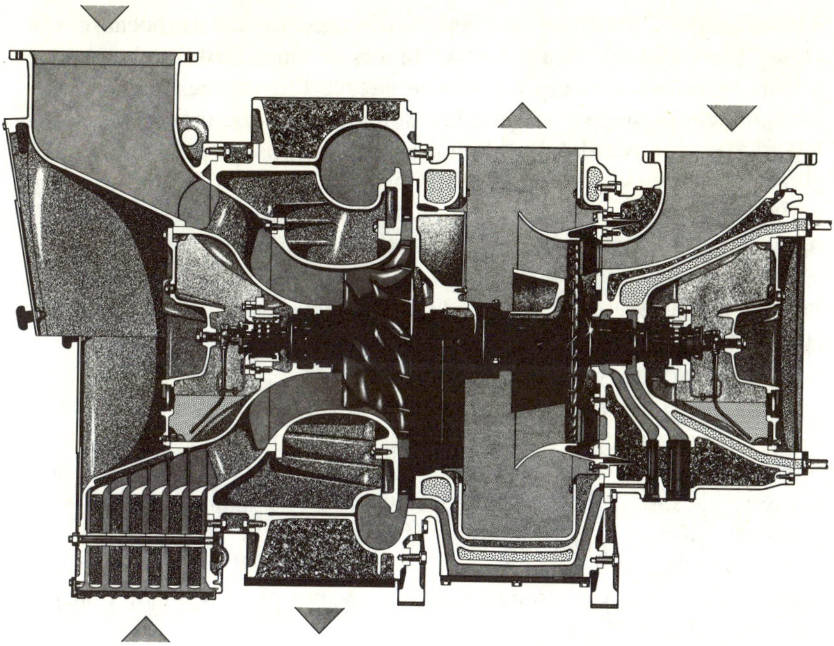

Fig. 3. Cross-section of large turbocharger for stationary or marine application (Brown Boveri VTR 454 . . . 714/E).

The reason for splitting the units into two groups becomes obvious from Figs 1 and 3. The cross-section of a typical small automotive turbocharger is shown in Fig. 1. Such devices are characterized by the use of inboard, fully floating, sleeve bearings (using the engine oil supply) and by the absence of vaned diffusers and nozzle rings. The units are designed for mass production at very low cost, and are characterized by the use of many cast parts, including both of the wheels and the three housings (compressor cover, turbine housing and bearing housing).

A typical cross-section of a 'big' turbocharger is shown in Fig. 3. There are many variants on this design: the bearings can be inboard or outboard, they can have a self-contained oil system or be fed from the engine oil supply, and the bearings themselves can be of the rolling element type or a variation of plain sleeve bearings. However, two items which are invariably found are (i) vaned diffusers in the compressor stage and (ii) nozzle rings as part of the (usually axial) turbine stage. The use of these two vaned stages is because the applications for which such turbochargers are intended do not require a 'wide' compressor map or 'torque back-up' from the engine. They are also those where stage efficiency is more important than turbocharger cost. A brief description of the component parts follows.

3.1. Compressor Wheels

During the 1980s, the design of compressors for turbochargers has developed from a simple, purely radial blade fibre design, to a complex blade design incorporating backsweep and non-radial inducers. This has come about, on the one hand, through the use of more sophisticated and better foundry practices and, on the other hand, through the development of more sophisticated stress and flow analysis packages. The compressor wheel is particularly sensitive to problems of low- and high-cycle fatigue, and the design procedure must take this into account. High-cycle fatigue should be borne in mind if there is a need to put structures, vanes or instrumentation just in front of the inducer!

The wheel is usually made from an aluminium investment casting, typically of C.355 aluminium. Other materials may be used for higher-temperature applications, such as KO1 aluminium, stainless steel or even titanium. If the wheel is not cast, e.g. for test purposes, low production numbers, very large size or to make use of a special material, then it may be cut from solid metal using a three- or five-axis milling machine; the number of axes required depends on the complexity of the wheel geometry.

3.2. Compressor Covers

As was mentioned above, typical compressor covers for small units consist of a vaneless diffuser followed by a scroll-type collector. The outer diameter of the package has been steadily reduced over the years by improved design methods to make the overall package size as small as possible. With larger units, or where the application requires a limited flow range, a similar overall construction will be used, but often some type of vaned diffuser will be incorporated: this will use flat, aerofoil or 'vane island' blades. The vanes may be cast into the diffuser structure, or fabricated and brazed from component parts. When considering the application of the turbocharger the following should be considered: vaned diffusers increase the production costs of the device, impart a higher stage efficiency, but result in a lower flow range below a pressure ratio of approximately 2·5 : 1.

Compressor covers are usually made from a sand-cast or gravity die-cast aluminium, and must display good long-term dimensional stability. In addition, they should have the ability to 'contain' the fragments of the compressor wheel after a burst failure. If the compressor wheel is made from stainless steel or titanium, it may be necessary to use a cast iron cover to achieve containment.

3.3. Bearing Housings

The bearing assembly of an automotive turbocharger is shown in Fig. 4. The role of the bearing housing is to be the base of the bearing system of the rotor, to distribute the lubricant and to locate the compressor and turbine housings. It might also be used to control the oil flow when the latter is additionally used for cooling the turbine end bearing, and it could also act as a heat sink to protect the bearings during running and 'hot shutdown'. The latter condition occurs when the engine is suddenly stopped after running at high temperature: there then follows a temperature rise in the bearing housing as the heat stored in the turbine wheel and housing flows back into the

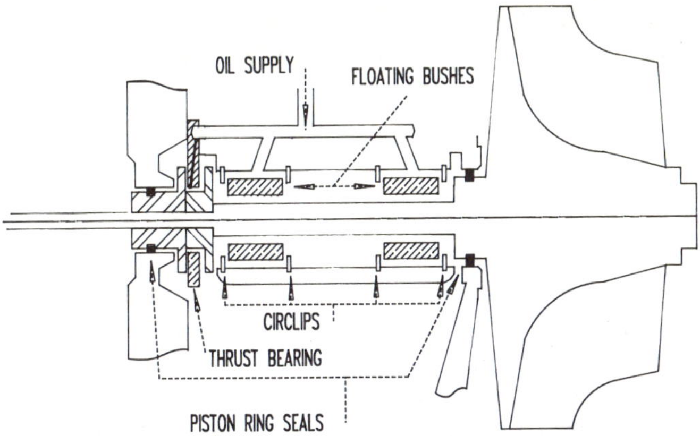

Fig. 4. Bearing system for automotive turbocharger.

bearing system. The bearing housing is usually a sand-casting of cast iron, or, exceptionally, of aluminium.

3.4. Bearing Systems

There is a choice of rolling element or plain bearings for the axial bearings of turbochargers, with various advantages for each type. Rolling element bearings will give a higher efficiency to the unit, particularly at part loads, but are more expensive and have a shorter life at the higher shaft speeds now encountered. Plain bearings, which are found in the majority of smaller automotive units, are usually of the floating sleeve variety. In these there is a clearance both between the shaft and bearing, and the bearing and its housing. In this design, the bearing is free to rotate at a speed which is lower than that of the rotor shaft, and this reduces its friction losses, with the additional advantage that the damping of both oil films allows stable running over a very wide range of speeds.

Rolling element or 'fixed' plain bearings, when used in small turbochargers units also require a second, external, squeeze film for damping purposes. In the larger units (with their more limited speed range) the second film is not required, but those with plain bearings often use internally lobed bearings to generate a higher load-carrying capacity.

The thrust loads, generated by unequal axial loads on the compressor and turbine wheels, are of a higher magnitude than the journal loads, and while rolling element journal bearings can be designed to accept these loads in the axial direction, plain bearings require a separate thrust bearing. Turbochargers are designed so that the net thrust is zero at some intermediate running point, which means that the thrust can be in either direction as the running point alters; hence the thrust system has to be double-sided. Current design practice is to make this bearing with tapered lands, a feature which increases the load capacity of a plain bearing.

Many designs of air bearings have been tried in turbochargers. These have the obvious advantages that come from divorcing the turbocharger from the oil supply of the engine. However, these have not yet reached production because the magnitude of the thrust loading results in a very large thrust bearing with a high friction loss.

3.5. Turbine Wheels

Turbine wheels have not changed in general style over the past two decades. Most radial wheels are investment cast in a high nickel alloy—usually Incocast 713C. However, there is now a movement towards ceramic wheels for use with very small (and shorter life) gasoline engines for passenger cars, and this trend could extend into other small turbochargers if the production processes improve and the material cost falls. Large turbochargers usually use axial turbines, although the changeover point between the two types is far from clear. Axial turbines offer a lower weight and moment of inertia, but at a significantly higher cost—mainly because the investment master used in the casting process is made of many parts as against the single part of the radial wheel.

Axial turbines in their smaller sizes are cast in one piece (known as a 'blisk') and, again, usually from Inco. 713. This process becomes impossible as the size increases, and the larger wheels are made from a cast slotted disc fitted with individual blades using 'fir-tree' roots or similar methods to attach the blades to the hub.

The unsteady nature of the inlet flow to the turbine rotor (because of cyclic pulses from the engine, and the turbine housing design) means that this component is very susceptible to high-cycle fatigue. Also, the relative specific strengths of the turbine and compressor wheels results in the former bursting before the compressor wheel (if this is not damaged by low-cycle fatigue), and so very detailed stress analysis is called for in its design.

3.6. Turbine Housings

Radial turbines without nozzle rings are fed from a snail type of housing, where the flow is fed in tangentially and then controlled in a centred vortex or other appropriate flow pattern before it enters the wheel. If it is used with a nozzle ring to control the flow out of the turbine housing, the shape of the housing is more of a collector type toroidal ring that feeds the nozzle ring entrance. The addition of a nozzle ring raises the efficiency of a turbine stage, but modifies the shape of the swallowing capacity curve so that the relationship between the mass flow and expansion ratio falls at lower mass flows and results in a lower boost at peak torque on an automotive engine. Hence these are only found on engines with applications in which the speed range is limited.

Axial turbines can be fed from either inboard or outboard of the turbocharger. The simplest installation arrangement is to feed the exhaust gases axially inwards from the back, into the stator ring, into a diffuser section and then turn them to allow the flow to exit at right angles to the shaft axis. However, in some installations the gas is fed in tangentially, as in the radial housing, and turned to feed the stator with the calculated amount of pre-whirl. The gas exits the wheel into a diffuser section and is then usually turned to give a radial outlet.

In a third type of installation, the inlet flow can be in a radial direction, then split and turned into the stator and then exited as before. Axial turbines also need stator rings, which have a similar effect on the turbine swallowing capacity as a nozzle ring on a radial machine. As most of the larger turbochargers which use axial wheels are not fitted to engines with wide speed or load ranges, this does not pose a problem.

4. A GENERAL APPROACH TO TURBOCHARGER DESIGN

As stated in the introduction to this chapter, the flow in the components of a turbocharger is extremely complex. Besides the normal features of flow through passages with varying cross-sectional areas and different curvatures, the turbocharger subjects the fluid to rotational fields, high levels of diffusion, and heat transfer. The calculation of the variation of gas properties in such a device is extremely complex, and currently at the limit of the computational techniques available. The basic features of turbocharger component aerodynamic design will be described by the use of one-dimensional methods for defining the gas velocities and relating these to the various gas properties. The approach will be further simplified by considering a 'mean' streamline; this can easily be seen to be a gross approximation in certain regions, e.g. the entry to the compressor inducer. At this stage a certain amount of empiricism will be introduced to account for deviations from one-dimensional flow.

Later in this chapter the various two- and three-dimensional flow régimes occurring in the turbocharger will be described, and methods of both measuring and calculating these effects will be discussed. In

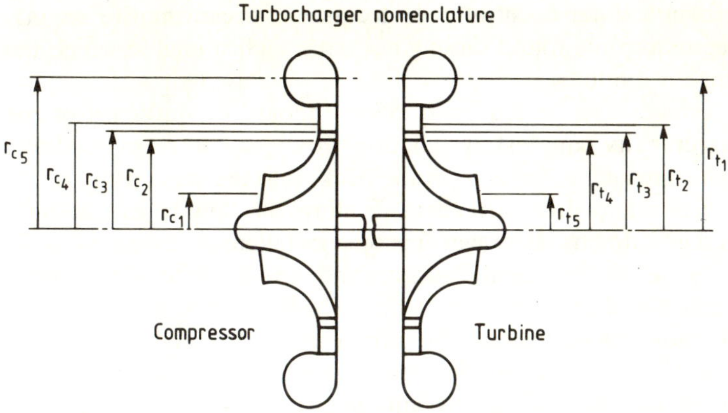

Fig. 5. Notation for analysis of turbocharger.

Fig. 5 is shown a general notation for the compressor and turbine sections. Note that r_{c1} and r_{t5} are mean radii at compressor inlet and turbine exit; a more detailed study would consider the effects at the roots and tips of the blades.

Further details of these approaches can be obtained from Cohen, Rogers & Saravanamutto,[1] Watson & Janota,[2] Sawyer[3] and Glassman.[4]

4.1. General Equations of Turbomachinery

The equation which governs the energy transfer in rotating aerodynamic machines is referred to as the Euler equation. This relates the power developed (or absorbed) by the component to the change in angular momentum of the fluid. Thus:

$$\mathscr{W} = \omega\tau = \omega\dot{m}(r_2 C_{\theta 2} - r_1 C_{\theta 1}) \tag{1}$$

where $\mathscr{W}$ is the work (negative shows work absorbed); τ is the torque on the rotor; ω is the angular velocity of the rotor; $\dot{m}$ is the mass flow rate of the fluid; r_1 is the mean radius at the rotor inlet; r_2 is the mean radius at the rotor exit; $C_{\theta 1}$ is the mean tangential fluid velocity at the rotor inlet; $C_{\theta 2}$ is the mean tangential fluid velocity at the rotor exit.

All turbomachinery can be analysed using this equation to evaluate the rotor torque. In addition to the Euler equation, which is essentially a momentum equation, it is also necessary to use the steady flow energy equation:

$$\dot{Q} - \dot{\mathscr{W}} = \dot{m}\{(h_2 + C_2^2/2 + gz_2) - (h_1 + C_1^2/2 + gz_1)\} \qquad (2)$$

and the continuity equation:

$$\dot{m} = \rho CF \qquad (3)$$

where C is the flow velocity and F is the flow area.

4.2. The Compressor

A compressor consists of three components; a rotor, a diffuser and a collector. Careful design of all these is necessary if the compressor is to operate efficiently. The rotor is used to impart velocity to the gas, and the diffuser then decelerates this gas to convert the kinetic energy into a rise in pressure, while the collector ensures uniform conditions downstream of the diffuser. These items will be considered in this section. In Fig. 6 are shown typical compressor maps for automotive turbochargers; both of these compressors have vaneless diffusers. It can be seen that the effect of backward-swept blades (see Section 4.2.1) is to broaden the mass flow range (i.e. between the surge and choke regions) of the compressor. This is particularly useful in automotive applications where the load–speed locus makes large demands on the size of the region of efficient operation of the compressor. Figure 7 is a compressor map for a large turbocharger: in this case a vaned diffuser has been used. The width of this map is significantly less than for automotive turbochargers, but the peak efficiency achieved is higher. In larger engines the width of the compressor map is not so important because the engine speed range is not as large as that of an automotive engine, and the periods away from high loads are usually a smaller proportion of operating time.

4.2.1. The Compressor Rotor

All the compressors used in turbochargers are centrifugal devices and are often referred to as radial compressors, although recently backward-swept blades have been introduced. (A backward-swept blade is one in which the rotor exit angle imparts a velocity component in the opposite direction to rotation). These single-stage radial wheels have the ability to generate a stage pressure ratio of up to 3·5 : 1, and

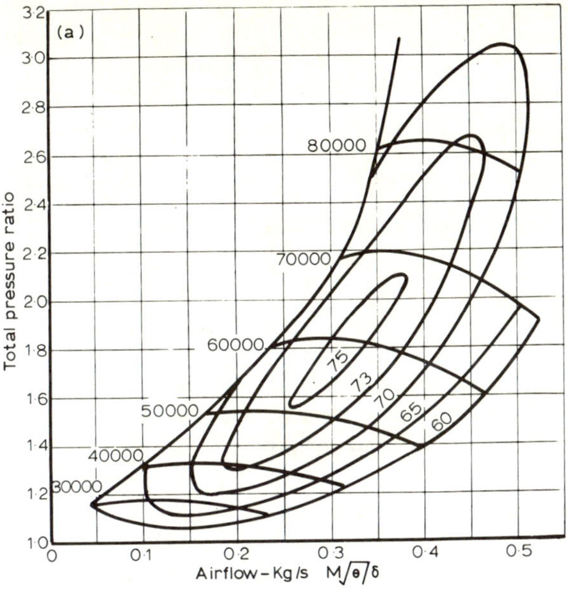

(a)

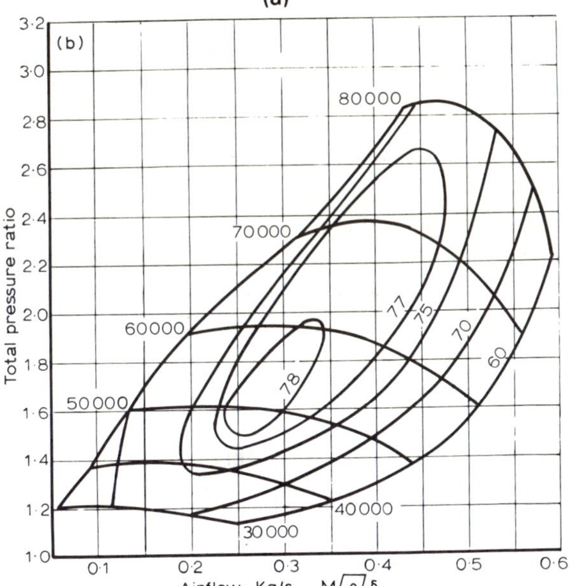

(b)

Fig. 6. Compressor maps for automotive turbocharger. (a) radial blades; (b) backward swept blades.

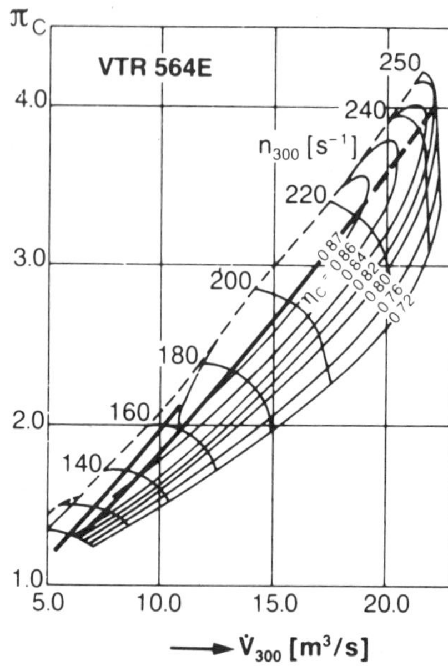

Fig. 7. Compressor map for large turbocharger with vaned diffuser (VTR 564E) ($\dot{n}_{300}$, turbocharger speed; $\dot{V}_{300}$, volumetric air flow; π_c, compressor pressure ratio (total-to-total); η_c, compressor efficiency (total-to-total).

are usually cast in an aluminium alloy. Only this type of wheel will be examined.

A radial compressor is one in which the fluid enters the wheel axially at its centre, also called the 'eye', and leaves almost tangentially at the outer, larger radius of the impeller. The motion of the gas through the wheel is achieved by centrifugal force.

It is possible to examine the performance of the unit by a one-dimensional analysis. This necessarily entails a large number of assumptions, but does give a basic design approach.

First it is necessary to draw the inlet and outlet vector diagrams for the rotor (Fig. 8). Using eqn (1):

$$-\mathcal{W} = \omega \dot{m}(r_{c2}C_{\theta 2} - r_{c1}C_{\theta 1}) \tag{4}$$

It is not usual to fit upstream guide vanes to generate pre-whirl for turbocharger compressors, so $C_{\theta 1} = 0$, and hence eqn (4) becomes:

$$-\mathcal{W} = \omega \dot{m} r_{c2} C_{\theta 2}$$

or

$$-\mathcal{W} = \dot{m} U_2 C_{\theta 2} \tag{5}$$

where, in Fig. 8, $U_2 = \omega r_{c2}$ is the rotor tip speed; C_2 is the absolute fluid velocity; C_{a2} is the axial fluid velocity; C_{r2} is the velocity of fluid in the radial direction; $C_{\theta 2}$ is the tangential velocity of the fluid at the rotor exit; and W_2 is the velocity of the fluid relative to the impeller.

Equation (5) allows for the case when U_2 is not equal to $C_{\theta 2}$ (see Fig. 8, where both cases are shown), but if the ideal case, with an infinite number of blades, is considered first, then these two velocities will be equal, and so:

$$-\mathcal{W} = \dot{m}U_2^2 \tag{6}$$

or

$$-\mathcal{W}/\dot{m} = U_2^2 \tag{7}$$

From eqn (2), for an adiabatic machine in which the change in elevation is negligible, then:

$$-\dot{\mathcal{W}}/\dot{m} = \Delta h_0 \tag{8}$$

where Δh_0 is the change in stagnation enthalpy across the machine.

Hence, for an ideal compressor, combining eqns (7) and (8) gives

$$\Delta h_0 = U_2^2 \tag{9}$$

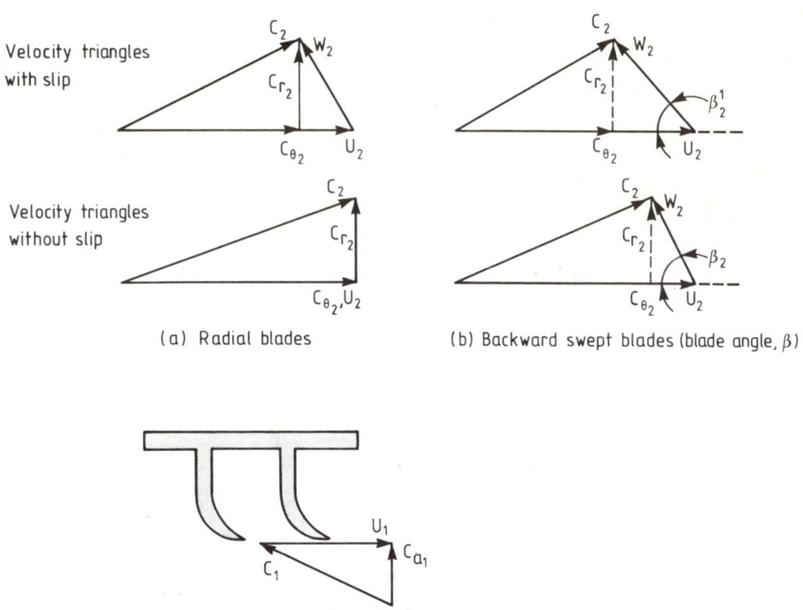

Fig. 8. Velocity vector diagram for compressor.

If the fluid does not leave the rotor in a radial direction, either because there are insufficient blades for perfect guidance, or because the blades are not radial at the rotor exit, then eqn (9) must be modified.

In the first case, with imperfect guidance but radial blades, it is necessary to introduce a modifier called the slip factor (σ). The simplest formulation of this is from Stantiz[5] who related it to the number of blades (n) by:

$$\sigma = 1 - \frac{0{\cdot}63\pi}{n} \tag{10}$$

where the definition of slip factor is

$$\sigma = C_{\theta 2}/U_2 \tag{11}$$

Therefore, eqn (9) becomes:

$$\Delta h_0 = C_{\theta 2} U_2 = \sigma U_2^2 \tag{12}$$

More comprehensive slip factors are available and Watson & Janota[2] quote results of a survey by Wiesner. Slip factors are a means of allowing for complex multi-dimensional flow fields in a simplified one-dimensional model. While increasing the number of blades will improve the guidance of the fluid and, in principle, the performance of the compressor, there is a compromise that has to be made to limit the increased flow losses due to friction. A typical turbocharger compressor has about eight full blades, and is often fitted with an equal number of splitter blades (which run about half-way through the blade passage).

In the case of non-radial blades, it is necessary to take account of the blade angle at the rotor exit (β_2) (see Fig. 8(b)), which gives

$$C_{\theta 2} = U_2 - C_{r2} \cot \beta_2 \tag{13}$$

and hence the work equation becomes

$$\frac{\dot{\mathscr{W}}}{\dot{m}} = U_2^2\left(1 - \frac{C_{r2}}{U_2} \cot \beta_2\right) \tag{14}$$

This equation shows that, if $\beta_2 > 90°$, the work requirement is increased, due to the larger change in angular momentum achieved (this is known as a forward-swept compressor wheel), whereas if $\beta_2 < 90°$, the work required is reduced and the wheel is known as a backward-swept type.

It is also possible to include a power input factor, ψ, which takes account of the frictional and windage losses. In this case eqn (12), for radial blades, becomes

$$\Delta h_0 = -\frac{\dot{\mathscr{W}}}{\dot{m}} = \psi \sigma U_2^2 \tag{15}$$

which, for an ideal gas, is

$$\Delta h_0 = C_p(T_{o3} - T_{o1}) \tag{16}$$

where T_{o3} is the stagnation temperature at the compressor diffuser exit and T_{o1} is that at the compressor entry.

Now, if the diffusion process is adiabatic, T_{o3} is equal to T_{o5} and hence the rise in stagnation temperature across the complete compressor (including both the diffuser and collector) can be defined as

$$T_{o5} - T_{o1} = \psi \sigma U_2^2 / C_p \tag{17}$$

It is possible to relate this change in stagnation temperature to the pressure ratio across the compressor, if an overall isentropic efficiency, η_C, is defined for the device, where:

$$\eta_C = (\Delta T_C)_{\text{isentropic}} / \Delta T_C = (T'_{o5} - T_{o1})/(T_{o5} - T_{o1}) \tag{18}$$

By definition,

$$\frac{p_{o5}}{p_{o1}} = \left(\frac{T'_{o5}}{T_{o1}}\right)^{\kappa/(\kappa-1)} = \left[1 + \frac{\eta_C \psi \sigma U_2^2}{C_p T_{o1}}\right]^{\kappa/(\kappa-1)} \tag{19}$$

4.2.2. The Compressor Diffuser

As stated previously, the gas leaving the rotor will contain a large amount of kinetic energy (equivalent to the velocity head, $C^2/2$): the purpose of the diffuser is to convert this energy into an increase in the pressure of the gas. The main advantage is the increased pressure ratio but, additionally, the engine is not capable of accepting a stream of high-velocity gas, and is much better able to operate from a reservoir of almost stagnant air to achieve an increase in the trapped mass in the cylinder.

The design of efficient diffusers is complicated and, in the case of the centrifugal compressor, this is exacerbated by the need to wrap the diffuser around the rotor and to deal with a swirl flow field. In all diffusers there is the compromise between having a large angle (rapid diffusion) with the possibility of separation along the walls, and a small angle (slow diffusion) with an increase in diffuser length and concomitantly greater friction losses. The length of the diffuser in the centrifugal compressor is necessarily limited by the general layout of the device.

Two basic types of diffuser can be used: either with or without guide vanes. Vaned diffusers are usually restricted to larger compressors and, hence, larger engines. This means that these are used on engines with reduced speed ranges, which often operate at fixed loads for a large proportion of their load cycle. The vaned compressor is well suited to this duty because it has a higher efficiency than its vaneless counterpart—but at the expense of a reduced mass flow range (see Fig. 7). This is caused by the incidence losses inherent in such a device when operating away from design point. The angles of the blades in the diffuser must be designed to minimize incidence losses as the gas passes from the rotor to the diffuser. In the simplest analysis the blade angle must equal that of the absolute velocity leaving the rotor (allowing for changes in velocity as the gas traverses the vaneless space between rotor and stator). As the operating point deviates from this design condition the angle of flow leaving the rotor changes and the difference between flow angle and blade angle increases: this results in a degradation of momentum as the flow is forced to change direction to travel along the blades. The effect can be readily envisaged by considering a compressor operating at fixed rotational speed but varying mass flow. The tangential velocity ($C_{\theta 2}$) of the gas leaving the rotor will be practically constant (if the slip factor does not vary significantly) but the radial velocity (C_{r2}) will reduce as the mass flow is reduced. This results in the angle of flow swinging closer to the tangential direction and hence, at some points in the range, the incidence losses will be high. Hence the efficiency of the compressor varies in the manner of the islands shown on compressor maps.

Most compressors on small turbochargers have vaneless diffusers because these are simpler, cheaper and have a broader mass flow range—but at the expense of maximum efficiency. A vaneless diffuser can be simply constructed using parallel walls because the flow area increases in proportion to the radius from the eye of the impeller. More complex designs can be achieved by profiling the side walls and these provide benefits in some areas of operation. The gas flow through the diffuser will follow a spiral, with the tangential component of velocity ($C_{\theta 2}$) obeying the equations of conservation of momentum, resulting in a free vortex, and the radial component (C_{r2}) obeying the continuity equation, resulting in a reduction in velocity as the flow area increases.

A mathematical approach to the performance of a vaneless diffuser is described by Dean,[6] who presents a map of diffuser performance. Some comments are also made on the matching of vaneless diffusers to a given compressor wheel, and their influence on the stage map. Work has also been attempted on variable geometry vaneless diffusers, but without success at this time.

4.2.3. The Compressor Collector

The purpose of this section of the casing is to provide uniform conditions downstream of the compressor diffuser. In this sense it is similar to the casing of the turbine, which is discussed in more detail in Section 4.3.2. The basic requirement of the casing is to preserve the free vortex flow that leaves the diffuser. This results in a casing defined by the equation $F_{\varphi}/r \propto \varphi$, where the constant of proportionality is related to the volume flow rate of the compressor. Watson & Janota[2] derive the relationship for the casing.

4.3. The Turbine

The radial inflow turbine is universally used in automotive turbocharger applications. Typical turbine maps for a small turbocharger are shown in Figs 9 and 10. The main advantages of these turbines are:

(a) a single-piece casting;
(b) more appropriate flow characteristics;
(c) lower manufacturing costs;
(d) simpler entry and exit ducting;
(e) small effect on efficiency of interactions between blade and cover;
(f) easier to maintain clearances in small wheel sizes.

In the larger turbochargers used for industrial, rail traction and marine applications the engine plus turbocharger systems, by virtue of their narrower speed–load operating ranges, can be designed to run under constant-pressure régimes and can use the high efficiency of an axial turbine. In addition to the efficiency advantages, the weight

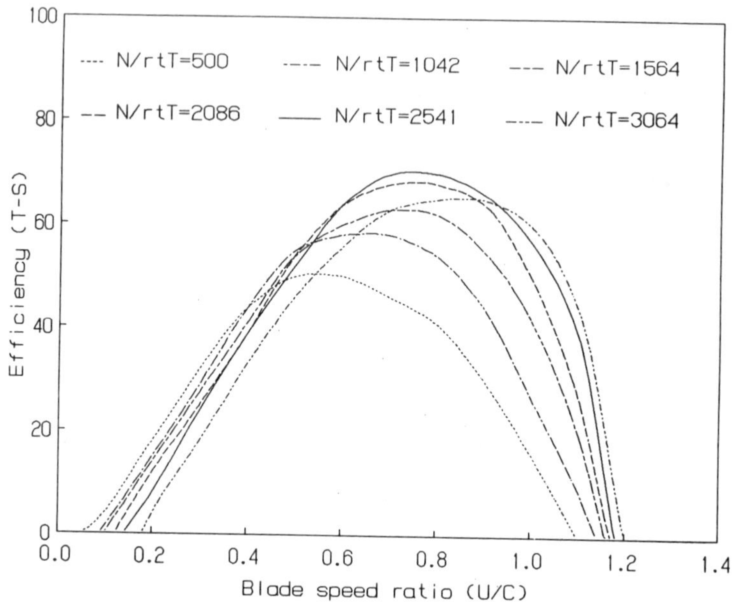

Fig. 9. Typical map for an inward flow radial turbine in an automotive size turbocharger (efficiency (total-to-static) against blade–speed ratio).

saving of an axial wheel relative to a radial one in these sizes, which can reach 1 m, is considerable. The 'changeover' size between radial and axial turbines is around a diameter of 250 mm, the smaller ones being radial and the larger axial.

This section will consider the design of the radial machine in some detail. Readers interested in the design of axial stages will find details in Sawyer[3] or in Watson & Janota.[2]

The basic one-dimensional steady state design approach is similar to that of the radial compressor, although, because of the pulsating nature of the flow, this approach is less reliable for turbines.

4.3.1. The Turbine Casing

Turbine casings can be either vaneless or vaned; the actual casing used depends very much on the particular application. In general, vaned casings produce a higher turbine efficiency than vaneless ones, but the width of the high-efficiency region (with respect to the turbine blade speed ratio, U/C) is usually significantly narrower. Most of the turbochargers fitted to automobile engines use vaneless casings, both

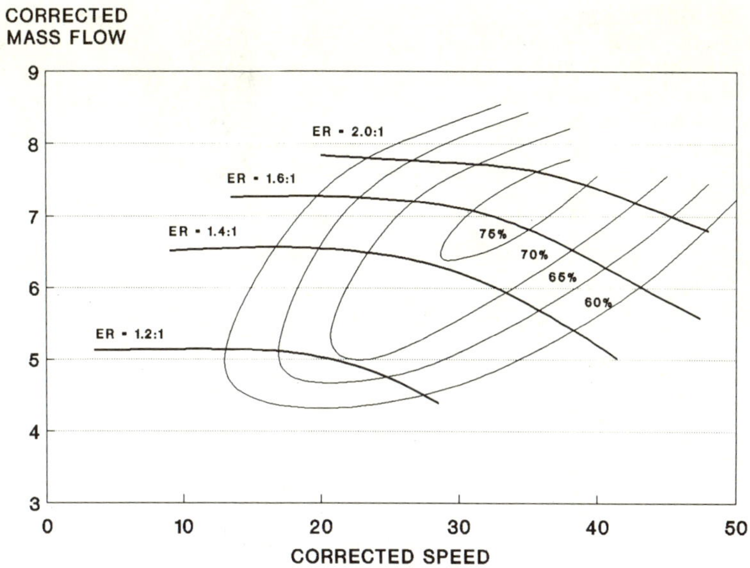

Fig. 10. Typical map for an inward flow radial turbine in an automotive size turbocharger (mass flow *vs* speed *vs* expansion ratio).

because of the broad operating spectrum expected in this application and for cheapness of production. Turbine casings are also available in single-entry and multiple-entry variants (Fig. 11). The latter are required to achieve separation of the gas pulses in the exhaust manifold to improve the scavenging of individual cylinders: this approach was proposed by Büchi in his patents. In Fig. 12 are shown two ways in which a twin-entry casing can be achieved. The gas flows

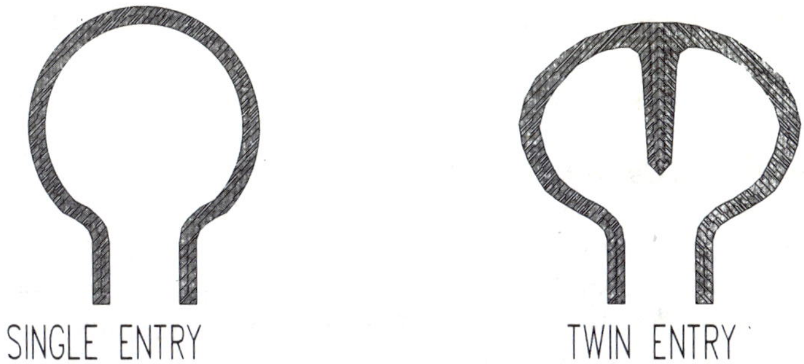

Fig. 11. Single and multiple entry turbine casings.

Fig. 12. Comparison between meridionally and circumferentially divided turbine casings. (a) circumferentially divided. (b) meridionally divided.

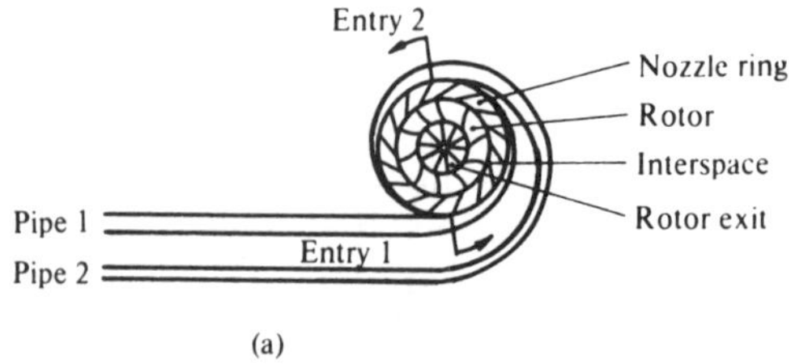

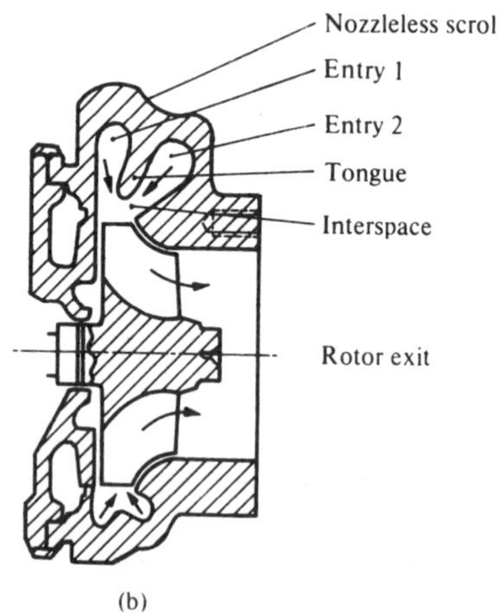

from the individual manifolds can be separated circumferentially or meridionally.

Under full admission conditions, when the flow through every entry of a multiple entry turbine is equal, the turbine operates in a similar manner to a single-entry one, except for minor effects as the flows from each entry mix and also increased wall friction. Most turbochargers for automotive use have meridionally divided casings because, on the whole, they are more effective than circumferentially divided ones under the conditions encountered. Circumferentially divided casings are often used for axial flow turbines, and also radial ones with vaned casings and these give good performance under full admission. However, if the flow through all entries is not equal the situation changes. There will be significant end of sector losses, and the nozzle exit angle will deflect as the flow varies. In addition, if the flow through a sector drops to a low value, the losses due to windmilling of the rotor will increase. The meridionally divided casing does not suffer from these losses but the mixing of the 'jets' leaving each entry reduces the flow efficiency. In the case when the pressure in one of the entries is low there can be some flow between entries, across the tongue. The effect of this is introduced in Section 5.2.1 of Chapter 13, where the interconnection path is indicated as a short pipe. Watson & Janota[2] and Benson[7] both show that the efficiency of the turbine is reduced by about 10 points (i.e. about 15%) under partial admission conditions. Benson shows that the characteristics of a twin-entry turbine are not equivalent for each entry: this has also been shown by Dale & Watson.[8]

Although vaneless casings are widely used on small automotive turbines with fixed geometry, the situation changes when variable-geometry turbocharging is considered. In these applications the casings usually revert to being single entry, even when adverse pulsation interference is possible, with the effective area being controlled either by rotating guide vanes or sliding plates which block the flow passages (see Fig. 13). These arrangements affect the turbine performance in different ways. The rotating guide vanes alter both the turbine effective area and the flow angle of the gas leaving the stator. To obtain optimum performance from the turbine the effective area and flow angle should be related in such a way as to minimize incidence loss when the flow enters the rotor. It is unusual for a simple mechanism to be able to achieve this relationship and, hence, the turbine performance is a compromise between an increase in available energy (brought about by restricting the flow area) and increased incidence losses (caused by the changing flow angle). In the other embodiment (used by Holset) the flow angle remains the same, because the guide vanes are fixed, but the width of the blade passage is reduced. Hence, this type of variable area mechanism simply changes the turbine effective area and increases the available energy in the gas without significantly altering the intrinsic efficiency curve of η_T against U/C.

The basic purpose of the turbine casing is to deliver the exhaust gas in a uniformly distributed manner to the turbine rotor. This means that an ideal casing will supply gas to the rotor with an equal velocity and equal incidence angle around the complete periphery of the casing. The corollary of this is that the angular momentum (per unit mass) and the pressure of the gas must also be uniform around the casing. The use of guide vanes (or nozzles) constrains the flow incidence angle to be similar around the periphery at all operating conditions, but it also results in an almost fixed incidence angle throughout the whole range of operation. This explains why a casing

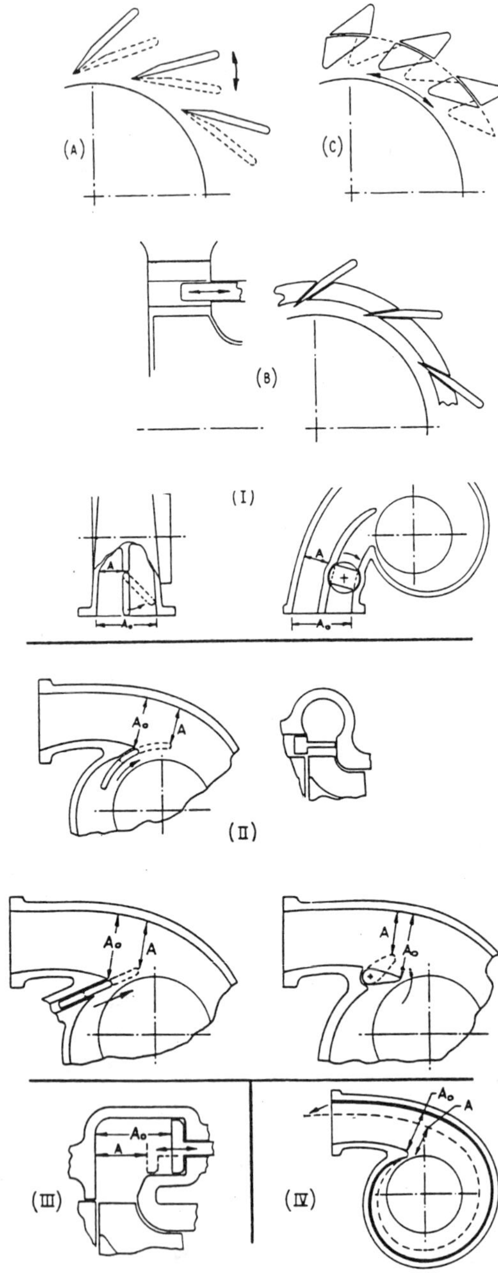

Fig. 13. Variable geometry arrangements for turbines.

with guide vanes can achieve a higher efficiency, and also why the region of high efficiency is more limited than its vaneless counterpart.

4.3.2. *Basic Geometry of Casing*

The basic volute-type scrolls for radial turbines are usually designed using incompressible flow theory based on constant angular momentum and uniform pressure. This should result in a constant angle of flow onto the rotor wheel. The following analysis will be limited to a single-entry turbine: it could be extended to a multi-entry device but the accuracy would be further reduced. Kastner & Bhinder[9] and Hussain, Ilyas & Bhinder[10] report work on the design of vaneless turbine volutes. They also assume a free vortex and incompressible flow. A typical volute for a vaneless radial turbine is shown in Fig. 14.

By definition of a free vortex,

$$rC_\theta = \text{constant} = K \tag{20}$$

If the flow is incompressible (a reasonable assumption in the casing under steady flow conditions) then the flow around the casing at any azimuth angle, φ, is

$$\dot{m}_\varphi = \rho F_\varphi C_{\theta\varphi} \tag{21}$$

and, for uniform mass flow from the casing to the rotor,

$$\dot{m}_\varphi = \dot{m}\varphi/2\pi \tag{22}$$

where $\dot{m}$ is the total mass flow entering the casing from the exhaust pipe, which is equal to the mass flow through the engine for a single-entry turbine. Thus the area of cross-section of the casing, at the azimuth angle φ, must be

$$F_\varphi = \dot{m}\varphi/2\pi\rho C_{\theta\varphi}. \tag{23}$$

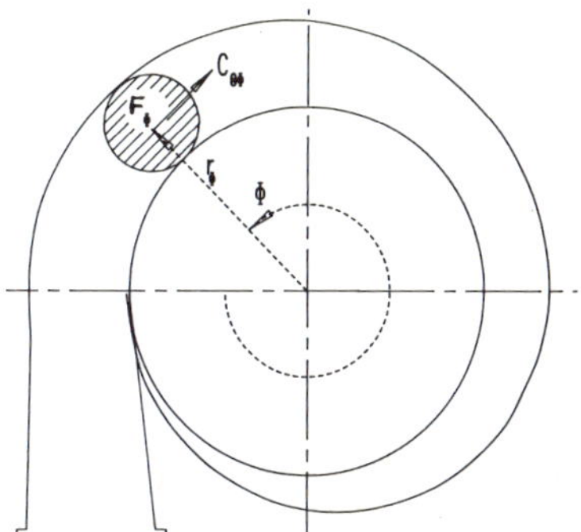

Fig. 14. Volute casing for inward flow radial turbine.

It is possible to relate the velocity at φ, $C_{\theta\varphi}$, to the strength of the vortex through eqn (20), so that eqn (23) becomes

$$F_\varphi = \dot{m}\varphi r_\varphi / 2\pi\rho K \tag{24}$$

Equation (24) introduces a concept which is used by one turbocharger manufacturer (Airesearch) for defining the size of its turbines, viz. A/r (or in this notation, F/r). This is a design feature that can be used to assess the size of turbocharger for any particular application. Rearranging eqn (24) gives

$$F_\varphi / r_\varphi = \dot{m}\varphi / 2\pi\rho K = K'\varphi \tag{25}$$

where $K' = \dot{m}/2\pi\rho K$, and is constant for any particular application.

Thus, the cross-sectional area of flow around the casing should vary in such a way that F_φ/r_φ is proportional to the azimuth angle, φ. Casings are often designed with F_φ proportional to the azimuth angle, φ, but in reality the value of r_φ also reduces significantly with azimuth angle, and as a result F_φ/r_φ is not constant and the ideal flow is not achieved. Care must also be taken when relating F/r values to those of turbine flow area (F). A turbine casing with an F/r value equal to 75% that of another casing does not have a flow area of 75% of the original casing because of movement of the centroid of the flow area. In Fig. 15 are shown two turbine casings which can be fitted to the

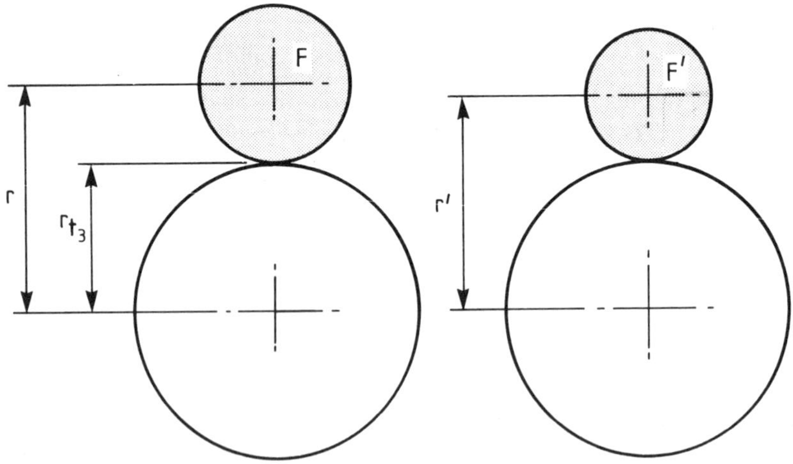

Fig. 15. Variation of F/r for two casings ($F/r = 1{\cdot}0$ in.; $F = 2{\cdot}37$ in.^{2}; $F'/r' = 0{\cdot}75$ in.; $F' = 0{\cdot}706F$.

same rotor. The F/r of the original casing is 1·0 in, and the rotor diameter, $r_1 = 1{\cdot}5$ in (Imperial units have been used because these are most commonly encountered with these devices); hence $(F/r)'$, for the smaller casing, is 0·75 in.

The value of F/r can be related to the rotor radius by

$$\frac{F}{r} = \frac{F}{r_1 + \sqrt{(F/\pi)}} \tag{26}$$

if the casing profile is circular.

This gives a turbine casing flow area (F) of 2·368 in^2. Substitution in eqn (26) gives $F'/F = 0{\cdot}706$, or an effective area for the new turbine of 1·672 in^2. If the casing was a meridionally divided twin-entry casing a similar analysis would be applied to each entry.

4.3.2.1. Turbine casing with guide vanes. If the casing is fitted with guide vanes (see Fig. 13) the flow leaving the casing is relatively constrained in its angle of exit. Mizumachi & Ariga[11] show that the angle of the flow leaving the blades can be estimated geometrically by a conformal transformation of the blades into a pseudo-axial turbine layout. Then the angle of flow is given by

$$\cos \alpha_3 = t/p \tag{27}$$

where t is the throat dimension between blades and p is the pitch of the blades.

This estimate is quite accurate for low expansion ratios, but Mizumachi & Ariga[11] show that at high expansion ratios (say, above 1·5) across the blades the flow deviates from the blade exit angle to a larger angle. They also show that the velocity coefficient Φ_N (defined as the ratio of actual absolute velocity to isentropic absolute velocity) of the blades varies with expansion ratio, but the actual effect was only about 5% over the whole range of expansion ratios considered (from 1·0 to 2·2).

4.3.2.2 Turbine casing without guide vanes. Vaneless turbines are commonly used in automotive applications and the casing design for such devices forms an important part of the overall turbine design. In such casings it is much more difficult to control the gas flow direction, especially at either end of the casing. However, it is initially worthwhile to attempt to evaluate a flow direction, α_3, similar to that defined for the vaned casing. This angle will not be constrained by the geometry of guide vanes, or similar devices, but will simply be the result of the areas of flow and the mass flow rates.

First, it will be assumed that the flow in the casing is a free vortex, in which case eqn (20) can be applied, giving

$$r_{t1}C_{\theta 1} = r_{t3}C_{\theta 3} = K \tag{28}$$

Also the continuity equation can be applied to the whole casing, and this states that the mass flow entering the turbine must flow out through the vaneless slot into the rotor. In this case,

$$\dot{m} = \rho_1 F_1 C_{\theta 1} = \rho_3 2\pi r_{t3} b_3 C_{r3} \tag{29}$$

where F_1 is the flow area of the casing normal to the flow vector in the casing. This gives

$$C_{r3} = \frac{F_1}{2\pi r_{t3} b_3}\left(\frac{\rho_1}{\rho_3}\right) C_{\theta 1} \tag{30}$$

or

$$C_{r3} = \left(\frac{F_1}{r_{t1}}\right)\left(\frac{\rho_1}{\rho_3}\right)\frac{1}{2\pi b_3} C_{\theta 3} \tag{31}$$

The actual angle of the flow leaving the casing is

$$\cot \alpha_3 = \frac{C_{r3}}{C_{\theta 3}} = \frac{F_1}{r_{t1}}\left(\frac{\rho_1}{\rho_3}\right)\frac{1}{2\pi b_3} \tag{32}$$

This is equivalent to the term derived by Mizumachi & Ariga[11] but, in this case, the coefficient of velocity (Φ_N) has been omitted. The definition of Φ_N is given in eqn (38).

Again, the term (F/r) appears in the equations and hence it can be seen that this ratio has a dominant effect on the angle of the flow entering the rotor. Watson & Janota[2] show how the inlet angle varies around the periphery of the turbine casing if the cross-sectional area (F_φ), rather than F_φ / r_φ, is varied linearly with azimuth angle.

4.4. The Interspace Between the Casing and the Rotor

On leaving the casing the gas must traverse the clearance to the rotor. It is reasonable to assume that, in the absence of losses, angular momentum is conserved and hence, both for varied and vaneless casings

$$r_{t3}C_{\theta 3} = r_{t4}C_{\theta 4} \tag{33}$$

Likewise, mass flow is conserved, and hence the continuity equation can be applied to give

$$\dot{m} = \rho_3 2\pi r_{t3} b_3 C_{r3} = \rho_4 2\pi r_{t4} b_4 C_{r4} \tag{34}$$

Combination of eqns (33) and (34) shows that for a vaneless turbine, if $b_3 = b_4$, the flow angle at the rotor is equal to that leaving the casing. In the case of the vaned casing this is not quite true because the total flow area through the guide vanes is less than the total circumferential area. In this case eqn (34) becomes

$$\dot{m} = \rho_3 t n b_3 C_{r3} \alpha_3 = \rho_4 2\pi r_{t4} b_4 C_{r4} \tag{35}$$

The angles of the flow leaving the casing are defined in Fig. 17 below: it should be noted that the angles are the complements of those in Mizumachi & Ariga[11] and Wallace,[12] but agree with those in Benson[13] and Watson & Janota.[2]

4.5. The Turbine Rotor

The turbine rotor is the component that produces the work from the turbine. For this reason it is usually designed prior to the casing and

Fig. 16. General arrangement of turbine end of turbocharger (turbine casing has been sectioned to show rotor).

the latter is fitted around it. A general arrangement of the turbine end of a turbocharger is shown in Fig. 16. This can be split into three parts; the casing, the rotor and the exit diffuser. The last item is usually omitted on turbocharger turbines because of space limitations, and the difficulty of designing such a device to operate under unsteady flow conditions. In this analysis it will be assumed that there is negligible diffusion of flow leaving the rotor exit; the benefit of an efficient diffuser would be to increase the turbine expansion ratio by utilizing the kinetic head in the exit stream to reduce the static pressure at outlet from the turbine.

The important equations for the design of the rotor are similar to those for the compressor.

From the Euler equation

$$\mathcal{W} = \omega \dot{m}(r_{t4}C_{\theta 4} - r_{t5}C_{\theta 5}) \tag{36}$$

In the ideal case at design point, where the angle of flow relative to the rotor (β_4') is equal to the geometric angle of the rotor blades at inlet (β_4), the conditions of flow leaving the casing are those entering the rotor. This means that $C_{\theta 4}$ in eqn (36) is the same as that calculated in eqn (33), and α_4 is the value obtained from eqn (32): the vector diagram is shown in Fig. 17(a) (from Benson[13]). In an actual rotor, with a finite number of blades, Rohlik[14] found that it was necessary for there to be some incidence (β_i), defined as $\beta_i = \beta_4' - \beta_4$. This is because the flow in the rotor blade passages is not one-dimensional and the flow vectors are deflected in the passages. Rohlik states that incidence as high as 40° can give optimum performance. The requirement for an incidence angle to give optimum conditions is an analogous effect to the slip factor that occurs in a compressor (see eqn (10)). The resultant enthalpy–entropy diagram for the case in which there is no incidence loss is given in Fig. 18(a), and it can be seen that the flow from the casing through the rotor takes place without any sudden 'shock' losses. If the turbine is operating away from design conditions, which is the more usual situation, the flow from the casing will not be inclined at the correct angle to enter the rotor without an incidence loss. This situation is depicted in Fig. 17(b), and Wallace,[12] supported by Benson,[13] proposes that the incidence loss takes place at *constant pressure*. Such a process, depicted in Fig. 18(b), causes increases in both entropy and enthalpy, and also results in a torque being developed by the change of momentum as the nozzle flow is deflected by the rotor blades.

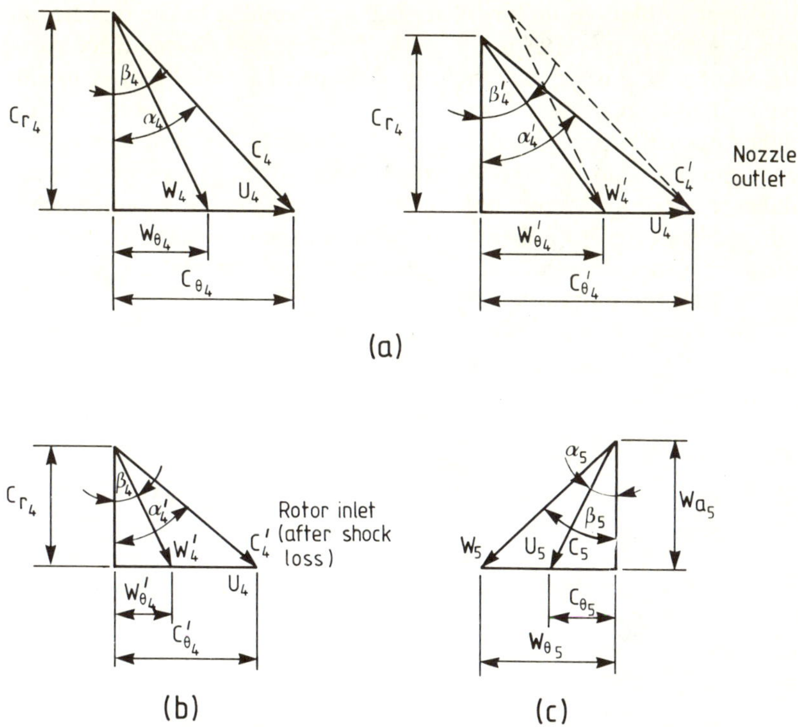

Fig. 17. Turbine vector diagrams. (a) design point operation at rotor entry; (b) off-design operation at rotor entry; (c) velocity diagram at rotor exit.

In analysing the performance of the complete turbine it is necessary to take account of three losses:

(a) the nozzle loss coefficient (ζ_N);
(b) the rotor incidence loss coefficient (ζ_I);
(c) the rotor frictional loss coefficient (ζ_{RB});

the last two are combined by Benson[13] to give the rotor overall loss coefficient (ζ_{RA}). These parameters can be related to velocity coefficients, and are also indicated on Fig. 18(b).

First, consider the nozzle loss coefficient (ζ_N). This is defined as

$$\zeta_N = \frac{h_4 - h_{4s}}{\frac{1}{2}C_4^2} \tag{37}$$

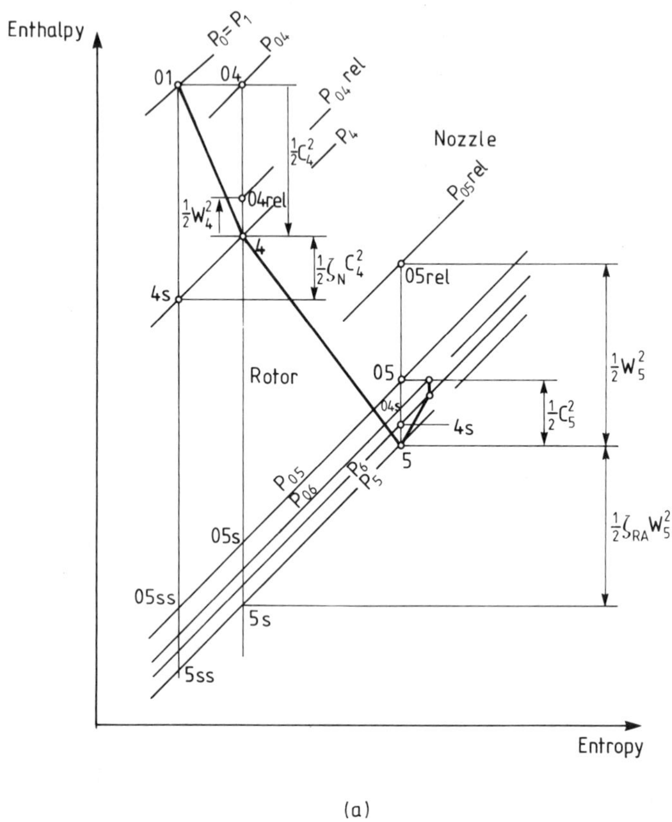

Fig. 18. H–S diagram for turbine. (a) design point operation.

and is indicated on Figs 18(a) and 18(b). The casing loss will occur in turbines both with and without nozzles, and results in a nozzle velocity coefficient given by

$$\Phi_N^2 = \frac{h_{01} - h_4}{h_{01} - h_{4s}} = \frac{1}{1 + \zeta_N} \tag{38}$$

The rotor incidence loss can be evaluated by applying the energy and continuity equations between the casing exit and the entry to the rotor. The energy equation is

$$h_4 + \frac{C_4^2}{2} = h_{4'} + \frac{C_{4'}^2}{2} + \dot{\mathscr{W}} \tag{39}$$

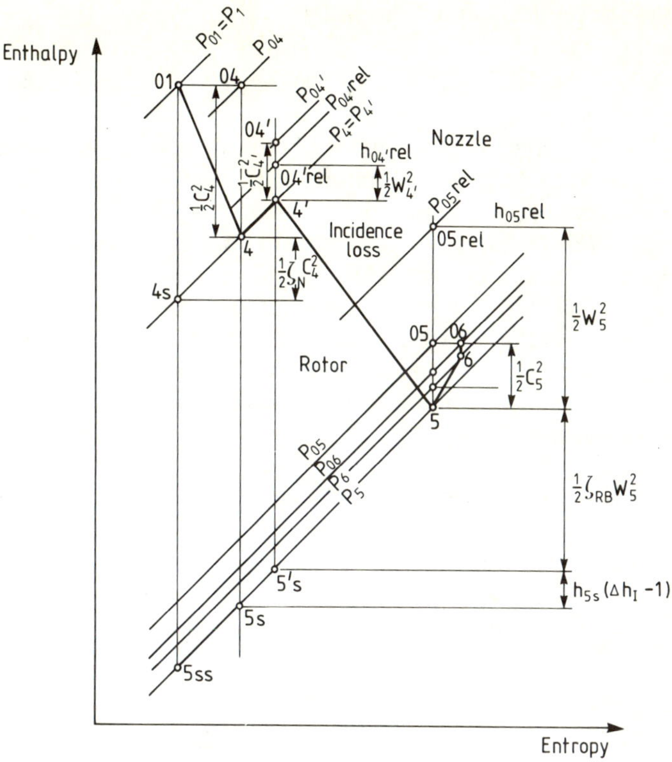

(b)

Fig. 18—*contd.* (b) off-design operation.

which, for a perfect gas, becomes

$$C_p T_4 + \frac{C_4^2}{2} = C_p T_{4'} + \frac{C_{4'}^2}{2} + \dot{\mathscr{W}} \tag{40}$$

where $\dot{\mathscr{W}}$ is the specific work done by the fluid changing direction at the entry to the rotor. This can be evaluated by application of the momentum equation, giving the rate of work output per unit mass as

$$\dot{\mathscr{W}} = \left(C_4 \sin \alpha_4 - C_4 \cos \alpha_4 \frac{T_{4'}}{T_4} \tan \beta_4 - U_4 \right) U_4 \tag{41}$$

If the continuity equation is applied, in the radial direction, then the gas velocity relative to the rotor is given by

$$W_{4'} = C_4 \cos \alpha_4 \frac{T_{4'}}{T_4} \sec \beta_4 \tag{42}$$

Equations (40), (41) and (42) can be combined to give an equation which may be solved for the temperature of the gas after incidence loss, i.e. the temperature of the gas entering the rotor. The ratio of the temperatures, $T_{4'}/T_4$, is equal to the ratio of enthalpies across the incidence process, and can be seen in Fig. 18(b) as $h_{4'}/h_4$.

The torque developed by the turbine rotor comes from two components; that developed by the gas flowing through the rotor without any incidence loss (as happens at design point) plus the torque produced by incidence loss. These two torques will now be derived.

The incidence torque, τ_I, is

$$\tau_I = \left(C_4 \sin \alpha_4 - C_4 \cos \alpha_4 \frac{T_{4'}}{T_4} \tan \beta_4 - U_4\right) r_{t4} \tag{43a}$$

and the rotor torque, τ_R, obtained from eqn (36), is

$$\tau_R = (r_{t4} C_{\theta 4} - r_{t5} C_{\theta 5}) = \left(C_{4'} \sin \alpha_{4'} - \frac{r_{t5}}{r_{t4}} C_5 \sin \alpha_5\right) r_{t4} \tag{43b}$$

The absolute velocity at exit from the rotor can be evaluated by considering the rotor passage geometry and by applying the energy equation along the mean streamline passing through the rotor passage. The energy equation, obtained by combining eqn (36) with

$$\mathcal{W} = \dot{m}(h_{04} - h_{05}) \tag{44}$$

gives

$$(U_4 C_{\theta 4} - U_5 C_{\theta 5}) = (h_{04} - h_{05}) \tag{45}$$

Then

$$2(h_{04} - h_{05}) = [U_4^2 + C_{\theta 4}^2 - (U_4 - C_{\theta 4})^2 - \{U_5^2 + C_{\theta 5}^2 - (U_5 - C_{\theta 5})^2\}] \tag{46a}$$

This can be rearranged to give

$$W_5^2 = W_4^2 + 2(h_4 - h_5) + U_5^2 - U_4^2 \tag{46b}$$

and then

$$C_5^2 = W_5^2 + U_5^2 - 2W_5 U_5 \sin \beta_5 \tag{47}$$

It is not possible to solve for these parameters explicitly and an estimate must first be made of the interspace pressure, which governs the subdivision of pressure ratio across the turbine. The various properties can then be calculated based on this estimate, and the accuracy can be assessed by checking the continuity of mass flow across the rotor.

Although these equations have allowed for incidence loss they have omitted the rotor friction loss. This can be included in the following manner. Basically, the overall loss coefficient of the rotor can be defined as

$$\zeta_{RA} = \frac{h_5 - h_{5s}}{\frac{1}{2}W_5^2} \tag{48}$$

and this gives the rotor velocity coefficient as

$$\Psi_A^2 = \frac{h_{05\text{rel}} - h_5}{h_{05\text{rel}} - h_{5s}} = \frac{1}{1 + \zeta_{RA}} \tag{49}$$

The rotor loss due to fluid friction is defined as

$$\zeta_{RB} = \frac{h_5 - h_{5s}}{\frac{1}{2}W_5^2} \tag{50}$$

with the velocity coefficient of

$$\Psi_B^2 = \frac{1}{1 + \zeta_{RB}} \tag{51}$$

The overall and rotor friction loss coefficients can be related to the incidence loss by

$$\zeta_{RA} - \zeta_{RB} = \frac{2h_{5s}(\Delta h_I - 1)}{W_5^2} \tag{52}$$

where Δh_I is the increase in enthalpy due to incidence loss and is defined as $\Delta h_I = h_{5'}/h_5$. In the case of a perfect gas, $\Delta h_I = T_{5'}/T_5$ and this is analogous to the temperature ratio due to loss in eqn (42). Care must be taken in using data from various sources because of differences in definition: this is shown here, where the loss coefficients have been related to different ends of the rotor.

Benson[14] describes methods of evaluating the loss coefficients for a small vaned turbocharger turbine. This paper includes a computer program which can be used to analyse gross performance data. It can also be used to extend the limited data normally available from turbine tests, especially if the compressor is used as the load. The results indicate that for a vaned casing the exit angle of the gas from the nozzles is almost constant, but reduces slightly with increasing absolute gas velocity, C_4. He also shows that the nozzle loss coefficient, ζ_N, reduces quite substantially with increasing nozzle Mach number, varying from 0·05 at $M_N = 0{\cdot}45$ to 0·14 at $M_N = 0{\cdot}2$. It is suggested by Benson that, for a turbine with radial blades ($\beta_4 = 0$), the incident loss can be evaluated as $Y_{SH} = \tan^2\beta_{4'} + (1 - \Delta h_I)$. Finally, Benson[13] relates ζ_{RB} to a parameter Z_{RN}, proposed by Futral & Wasserbauer,[15] in which $\zeta_{RB} = Z_{RN}(1 + (W_{4'}/W_5)^2)$, where the velocity ratio is actually converting the loss coefficient from the exit to inlet velocity. It has been found that Z_{RN} appears to be a function of turbine speed alone. The value of Z_{RN} varies significantly (from about 2 to 1) over the working speed range of the turbine tested (from 30 000 to 100 000 rpm). From this summary it is obvious that the one-dimensional techniques are very dependent on accurate empirical data. Benson recommends that the values of the various loss coefficients are calculated by analysis of turbine performance data: the values obtained in this manner can be used to extend the measurements and also for initial design of a geometrically similar device (see also Section 6.3.2, where it is shown that the peripheral variation of pressure and flow angle weakens further the ability of such a simple one-dimensional analysis to be used on a broad range of turbines). The only way to obtain more accurate synthesis of a turbine is to go to multi-dimensional analysis: such methods are described in Sections 4.7 and 4.8.

Wallace, Cave & Miles[16] have extended the use of the one-dimensional model to evaluate the characteristics of various turbines at both full- and part-load conditions. Wallace, with fellow workers,[17,18] has also applied the method to evaluating the unsteady flow in the turbine, as have Mizumachi, Yoshiki & Endoh.[19]

4.6. Turbine Efficiency (η_T) and Degree of Reaction (Λ)

In addition to the analysis given above, turbines are often defined by their isentropic efficiency (η_T) and the degree of reaction (Λ). The isentropic efficiency of a turbine is defined (from Fig. 18) as

$$\eta_T = \frac{h_{01} - h_{05}}{h_{01} - h_{5ss}}$$

$$= \frac{\text{actual work}}{\text{maximum work expanding to exhaust static pressure}} \quad (53)$$

This is related to speed of the turbine through the blade speed ratio (U/C), which is defined as

$$\frac{U}{C} = \frac{\text{rotor tip speed}}{\{2(h_{01} - h_{5ss})\}^{0\cdot 5}} \quad (54)$$

where the denominator is the gas velocity that would be achieved by expanding the flow across the turbine (stagnation to static) pressure ratio. This parameter is more appropriate to impulse turbines but is widely used in radial inflow turbines.

The degree of reaction of a turbine (Λ) relates the isentropic change of enthalpy in the rotor to the isentropic change of enthalpy across the whole turbine. Hence

$$\Lambda = \frac{h_{4s} - h_{5ss}}{h_{01} - h_{5ss}} \quad (55)$$

An impulse turbine, in which the pressure is all converted into velocity head in the casing, has a degree of reaction of zero ($\Lambda = 0$). A turbine in which the full change of enthalpy took place in the rotor would have $\Lambda = 1$. Watson & Janota[2] show that, if the degree of reaction is based on actual energy transfer, that is

$$\Lambda = \frac{\text{energy transfer due to change in static pressure in rotor}}{\text{total energy transfer in stage}} \quad (56)$$

then

$$\Lambda = \frac{h_4 - h_5}{h_{01} - h_{05}} = \frac{(h_{04} - C_4^2/2) - (h_{05} - C_5^2/2)}{(h_{01} - h_{05})} \quad (57)$$

This gives

$$\Lambda = 1 - \frac{C_4^2 - C_5^2}{2(h_{01} - h_{05})} \quad (58)$$

Manipulation of this equation, together with substitution of velocities, gives

$$\Lambda = 1 - \frac{C_{\theta 4}}{2U_4} \tag{59}$$

If the inflow is radial, i.e. $\beta_4 = 0°$, and there is no incident shock, i.e. $C_{\theta 4} = U_4$, then the degree of reaction will be 50%.

It can be shown that the isentropic efficiency of a turbine is related to its degree of reaction. An ideal impulse turbine (with $\Lambda = 0$) achieves its maximum efficiency at $U/C = 0{\cdot}5$, while a 50% reaction turbine (with $\Lambda = 0{\cdot}5$) achieves its maximum efficiency at $U/C = 0{\cdot}707$ ($\sqrt{2}/2$). Rohlik[13] shows how the value of U/C for maximum efficiency varies with specific speed for a radial turbine. It achieves the ideal value of $U/C = 0{\cdot}7$ for a dimensionless specific speed of 0·6, but at all other times the optimum U/C is lower. The degree of reaction also varies as the operating point of the turbine moves away from the design condition. Mizumachi *et al.*[19] show how Λ increases monotonically with U/C, and also how it changes around the periphery of a radial turbine with vaned casing.

Rohlik[14] also shows how the losses in a radial turbine are a function of the dimensionless specific speed. Dimensionless specific speed is defined as

$$N_s = NQ_5^{1/2}/H^{3/4} \tag{60}$$

and, in this case, N is the turbine speed (rad/s); Q_5 is the volume flow rate at turbine exit (m^3/s); and H is the ideal work based on inlet and exit stagnation pressures (J/kg).

In Fig. 19 it is shown how the distribution of the losses varies with the specific speed. A turbine with a low specific speed will have short blades (i.e. b_4 and b_5 will be small) but the blade passages (from entry to exit) will be long. This means that at low values of specific speed the stator and rotor viscous losses will be large, because of the high ratio of wall area to flow area. The clearance losses will also be large because the clearance will be a significant proportion of the blade height. As the specific speed increases the blades become higher (i.e. b at entry and exit becomes bigger) and the flow passages have more favourable viscous characteristics. However, the kinetic energy at exit to such a turbine design becomes large, and this loss then predominates.

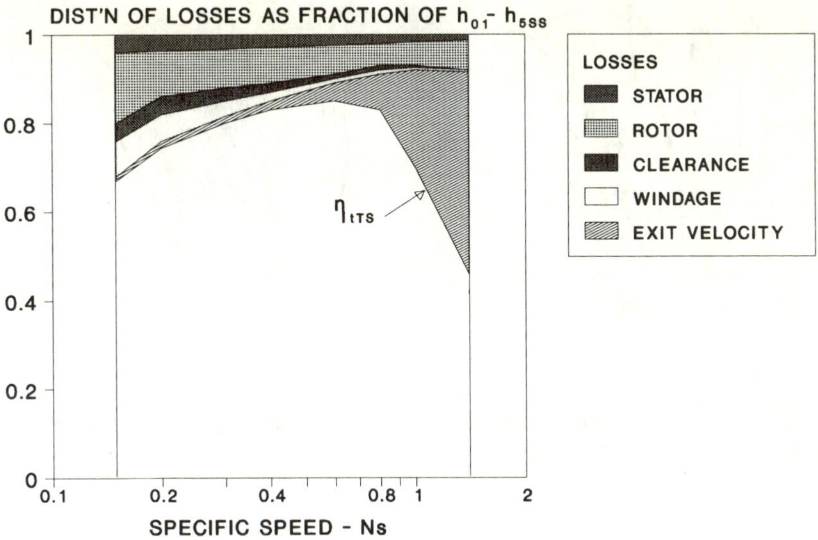

Fig. 19. Loss distribution in radial turbine as a function of specific speed.

4.7. Streamline Curvature Methods

The methods described above have been limited to a one-dimensional analysis of flow in either the turbine or compressor. In fact, the whole analysis has been performed by simply considering the entry and exit conditions at the turbine or compressor rotors: no account has been taken of the actual blade shape. To obtain solutions which are in good agreement with those measured on actual devices it has been necessary to introduce various empirical coefficients. The shortcomings of the one-dimensional approach lie in the inability to represent the complex flow field by a single streamline. Any one- or two-dimensional technique is unable to represent the full complexity of the fluid flow in the rotor, or even the much less complex casing. Katsanis[20] proposed an extension into two-dimensions of the methods used for rotor flow: this is referred to as the streamline curvature method.

The streamline curvature method is based on being able to define two-dimensional flow surfaces within the rotor geometry. Obviously two such families of non-planar surfaces can be defined: one being

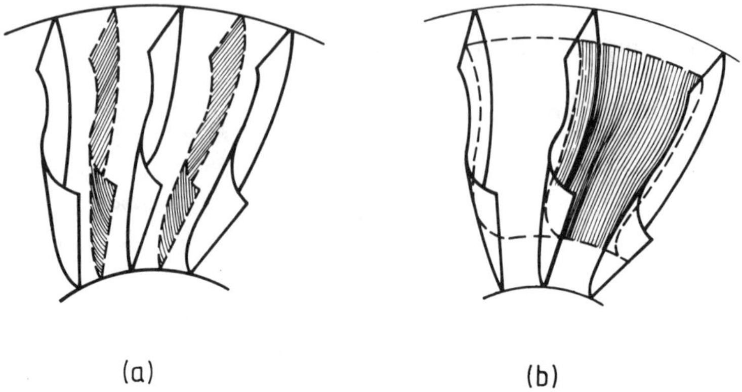

Fig. 20. Streamline curvature surfaces. (a) meridional; (b) blade to blade surface at mid-section.

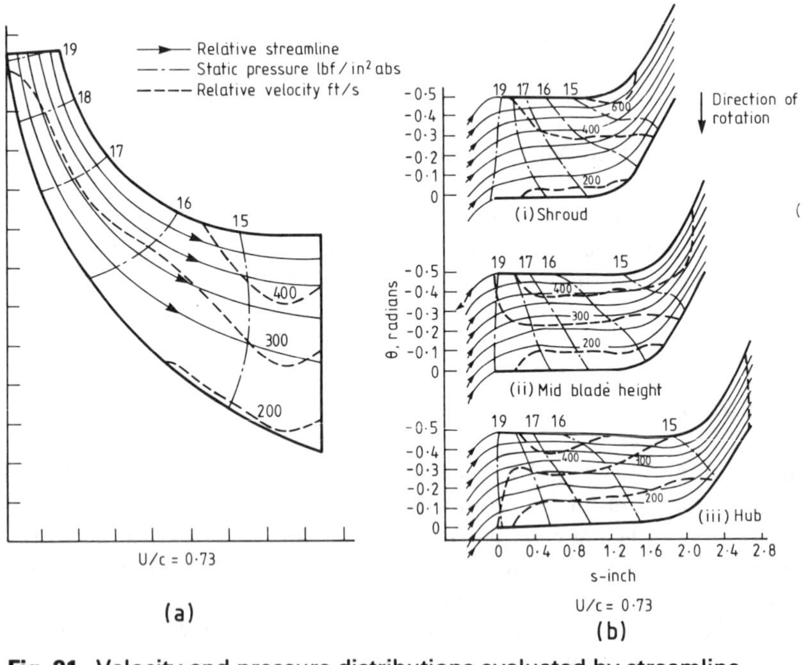

Fig. 21. Velocity and pressure distributions evaluated by streamline curvature method. (a) meridional surface; (b) blade to blade surfaces at various sections.

meridional in the blade space and the other passing from blade-to-blade. These surfaces are shown in Fig. 20. The meridional surfaces are often called *hub-to-shroud* and those orthogonal to them are called *blade-to-blade*. The equations of flow for such solutions are given by Mizumachi & Ariga,[11] who also present some results.

The advantage of this method over the one-dimensional approach is that it provides data on the velocity and pressure distribution in the rotor blade space. Typical results in Fig. 21 (from Benson, Cartwright & Woollatt[21]) show the distributions of pressure and velocity in both planes. These data could be used in the detailed design of blade shapes. The disadvantage of this method is that it constrains the flow to two orthogonal families of surfaces, and flow between these surfaces is not included in the solution. This immediately precludes many recirculation effects which are present in fully three-dimensional formulations of the flow equations, and in the rotor itself.

4.8. Three-dimensional Methods of Analysis

A number of three-dimensional analyses of flow in compressors and turbines are under development; most of these are for flow in the rotors. The calculations are normally restricted to individual blade passages only, and hence cannot allow for the distribution of conditions, e.g. pressure and velocity (both magnitude and direction), that occur around either a turbine or compressor casing. This work is discussed in more detail in Section 6, since it is in its basic research phase.

5. MATCHING

If the turbocharger is to deliver the required amount of air to the engine, the sizes of the compressor and turbine stages have to be matched both to the capacity of the engine and its power rating. If the engine is required to run over a limited speed range, this is no problem. However, in automotive and traction applications turbocharger matching can present difficulties because of the varying demand placed on the turbocharger as the load and speed of the engine change. In this case, the starting point of the calculation is usually taken as the design point rating of the engine. A knowledge of the following engine characteristics is required:

- engine swept volume (or bore and stroke)
- engine volumetric efficiency
- charge cooler characteristics (if fitted)
- engine power output
- engine speed
- predicted fuel consumption
- required air : fuel ratio
- the amount of heat lost to the cooling systems (sum of water, oil and air)

With these inputs, the required mass of air for the engine can be calculated:

$$\dot{m}_a = P_e b \varepsilon / 3600 \tag{61}$$

where m_a is the air mass flow (kg/s), P_e is the power developed (kW), b is the brake specific fuel consumption (kg/kWh), and ε is the air : fuel ratio.

The major engine parameter influencing the matching process is the swept volume of the engine; from this the volumetric flow rate (based on swept volume) can be evaluated:

$$\text{four-stroke engine: } V_s = \frac{\pi d^2 SnN}{48} \times 10^{-10} \quad \text{m}^3/\text{s} \tag{62a}$$

$$\text{two-stroke engine: } V_s = \frac{\pi d^2 SnN}{24} \times 10^{-10} \quad \text{m}^3/\text{s} \tag{62b}$$

where V_s is the volume swept by the engine (m^3/s), d is the bore (mm), S is the stroke (mm), n is the number of cylinders, and N is the engine speed rpm.

The actual volumetric flow of air through the engine is related to this by the volumetric efficiency, η_{vol}. giving:

$$\dot{Q}_a = \eta_{\text{vol}} V_s \quad \text{m}^3/\text{s} \tag{63}$$

where $\dot{Q}_a$ is the volume flow rate through the engine.

Now the density of the air *in the inlet manifold*, ρ_a, may be related to the mass and volumetric flow rates by:

$$\rho_a = \dot{m}_a / \dot{Q}_a \quad \text{kg/m}^3 \tag{64}$$

This is the density of air which must be delivered by the compressor. The inlet flow through the compressor should be calculated in units which are compatible with the compressor maps that are being used, and the temperatures and pressures at inlet to the compressor must be known. In the simplest case, using the conditions for which the compressor is mapped (see Section 7.1), the volumetric flow into the compressor is:

$$\dot{Q}_{in} = \dot{m}_a R T_1 / p_1 \quad \mathrm{m^3/s} \tag{65}$$

where T_1 is the air temperature at entry to the compressor (K), p_1 is the air pressure at entry to the compressor (N/m^2), and R is the gas constant (J/kgK).

The temperature of the air will rise as it is compressed, dependent on the pressure ratio and the compressor efficiency at that operating point. The required density of the manifold air is known (eqn (64)), so an iteration of pressure ratio and efficiency must be made.

With no charge cooler, the required equation is:

$$\text{temperature rise across compressor, } \delta T = \frac{T_1}{\eta_c}\left\{\left(\frac{p_2}{p_1}\right)^{(\kappa-1)/\kappa} - 1\right\} \tag{66}$$

where δT is the temperature rise across the compressor (°C), p_2 is the compressor outlet pressure (N/m^2), κ is the ratio of specific heats, and η_c is the compressor efficiency.

The density at the outlet of the compressor, ρ_2 can be calculated from the properties at that station:

$$T_2 = T_1 + \delta T, \qquad \rho_2 = p_2 / R T_2 \quad \mathrm{kg/m^3} \tag{67,68}$$

Comparison of this result with the required density enables a new estimate of compressor pressure ratio and efficiency to be made until the values converge.

If the engine is fitted with a charge cooler, then the characteristics of the latter must be known. There are three: the variation of pressure loss with flow; the variation effectiveness; and the inlet temperature of the coolant.

The effectiveness of a cooler is defined as:

$$e = \frac{(T_2 - T_2')}{(T_2 - T_{cool})} \tag{69}$$

where e is the cooler effectiveness, T_2' is the air temperature leaving the cooler (°C), and T_{cool} is the coolant inlet temperature (°C). Use of this equation together with eqns (67) and (68) again leads to an estimate of manifold density, and the calculation can be iterated to convergence.

The power required for the compressor stage can now be calculated as:

$$\dot{\mathscr{W}}_C = \dot{m}_a C_{pa} \delta T \tag{70}$$

where $\dot{\mathscr{W}}_C$ is the compressor work requirement (W), and C_{pa} is the specific heat of the air (J/kgK).

The inlet flow calculations are now completed and this allows a suitable compressor stage to be selected. The next stage of the matching calculation requires a knowledge of the engine. For a given running condition, the engine will divide the energy supplied by the fuel in three ways: part to power; part to the exhaust gas; and the rest to the combined cooling systems, external air, the oil system and the water system.

The following steady state equations can be constructed to define the exhaust temperature:

$$Q_f = \dot{m}_f PCV, \qquad \eta_e = 100/CV/b, \qquad \delta h = CVm_f(1 - \eta_e - Q_c) \tag{71–73}$$

$$\delta T_E = \frac{\delta h}{(C_{pe} m_f (1 + \varepsilon))} \tag{74}$$

$$T_3 = T_2 \text{ (or } T_2') + \delta T_E \tag{75}$$

where Q_f is the total heat supplied in the fuel (J/s); η_e is the engine thermal efficiency; CV is the fuel lower calorific value (J/kg); b is the brake specific fuel consumption (g/kWh); δh is the enthalpy rise across the engine; Q_c is the fraction of input heat transferred to the combined cooling systems (oil, water and air) within the engine; δT_E is the temperature rise across the engine (°C); C_{pe} is the specific heat of the exhaust gas (J/kgK); and T_3 is the exhaust gas temperature (K).

These equations have only two unknown quantities, the exhaust temperature and the total heat lost to the cooling systems. If no information is available on the latter, a value of 20% may be taken as typical of a modern diesel engine: if there is a set of data for the engine running at similar conditions, then the same equations can be iterated round the manifold temperatures, the fuel consumption and the air:fuel ratio so that an appropriate value for the heat lost term can be obtained.

From these calculations, the compressor power and speed, and the exhaust temperature are known. The work needed from the turbine stage can now be calculated:

$$\dot{m}_e = \dot{m}_a + \dot{m}_f \tag{76}$$

$$\delta T_T = \eta_T T_3\left(1 - \frac{1}{(p_3/p_4)}(\kappa_e - 1)/\kappa_e\right) \tag{77}$$

$$\dot{\mathscr{W}}_T = \dot{m}_e C_{pe} \delta T_T \tag{78}$$

where $\dot{m}_e$ is the mass flow of exhaust gas (kg/s), δT_T is the temperature drop across the turbine (°C), η_e is the total to static turbine efficiency, p_3 is the exhaust manifold pressure, p_4 is the pressure downstream of the turbine, κ_e is the ratio of specific heats of the exhaust gas, $\mathscr{W}_T$ is the turbine work (W), and C_{pe} is the specific heat of the exhaust gas (J/kgK).

The value of turbine efficiency will not be known at this stage of the calculation, so a figure (usually 75%) must be assumed until the actual running point is calculated.

The work seen at the compressor is the product of the turbine work and the mechanical efficiency of the turbocharger. This latter is difficult to find but, at an expansion ratio of over 2:1, a value of 97% is a good estimate. The two work equations, eqns (70) and (78), can now be equated and the value of expansion ratio can be obtained. This in turn allows $m_e\sqrt{T_3}/p_3$ to be evaluated and this, with $N/\sqrt{T_3}$, gives a full description of the turbine operating point.

To match this turbine point on an actual turbine map, it is convenient to use the 'swallowing capacity' curves of turbine performance (see Section 7.2) which, as in Fig. 46, show the mean operating lines of several turbine stages. With these, a match of $m\sqrt{T}/p$ and expansion ratio can be made, and the most appropriate size of turbine chosen. If there is no exact match, it must be decided whether to choose a smaller or larger size of turbine. A smaller size will increase the expansion ratio, thereby increasing the turbine work, and result in a higher boost pressure, air:fuel ratio and peak cylinder pressure, whereas a larger size will drop the boost and increase the exhaust temperature. The above series of calculations can be reworked until the exact match is achieved, including an accurate value for turbine efficiency, if known.

This calculation will be sufficient if the engine does not operate far from a single speed, but the majority of engines are required to run

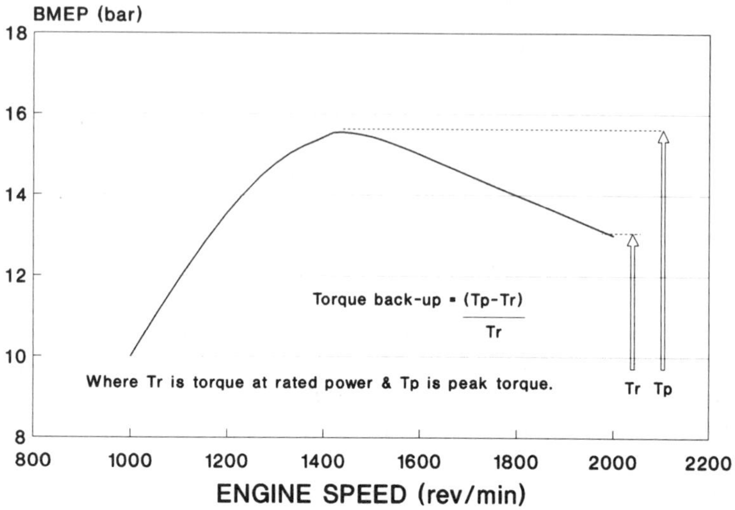

Fig. 22. Torque back-up for vehicle engine.

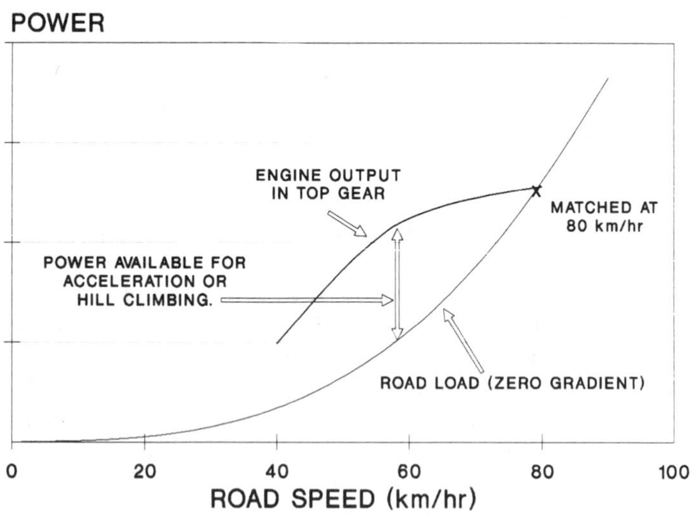

Fig. 23. Road load requirement of vehicle engine.

over a wide range of speeds and loads. This is particularly true of vehicle engines, which are required to generate the maximum torque back-up value (see Fig. 22).

In Fig. 23 is shown the road load requirement against speed of a vehicle on a level road and the power curve generated by the engine in a single gear. The margin between the two is the torque available for acceleration or for maintaining the speed on a gradient. The larger the value of back-up torque, the more convex is the power curve of the engine, and the fewer the number of gear changes that have to be made.

The net result of this is a requirement for the turbocharger–engine combination to deliver the maximum boost at lower engine speeds but, unfortunately, the natural characteristic of the turbine works against this. The expansion ratio of a turbine, as for a nozzle, rises with an increase in mass flow. The turbine work rises with both mass flow and expansion ratio, so when the engine speed falls and, consequently, the mass flow through the engine falls, it is impossible with a simple turbine to maintain the boost pressure at the level achieved at full load. Because the fuel injected per stroke of the engine must rise to give the required torque back-up as the speed falls, the engine must run at lower air:fuel ratios at low speed, with associated higher exhaust temperatures, and smoke and emission penalties. If, conversely, the turbocharger is matched at a lower speed, where the peak torque is to be developed, then the consequence of this is to develop a very high boost at high engine speeds, with associated high cylinder pressures.

To calculate what a particular match will deliver at a lower speed, the above calculations must be repeated. Choosing a speed and required power, the first estimate to be made is the new air:fuel ratio. This will be lower than the first, for the above reasons, and the whole calculation must be iterated to find the value the turbocharger will give. Continuing through the calculations, a new value of expansion ratio and $m\sqrt{T}/p$ for the turbine stage will be found. This must then be compared to the original swallowing capacity chart. If the value lies on the curve, then that is the required solution. If the value is above the curve, the value of air:fuel ratio must be reduced, and vice versa if the point is below the curve.

The problem of low boost at low speed is largely countered on trucks by the comparatively helpful characteristics of limited speed range, higher values of volumetric efficiency at low speeds, and the ability to run the compressor towards choke with lower efficiencies at high engine speed. However, with passenger cars the required speed range is so wide that other measures have to be used to enable the turbocharger to match over the full operating range. This has usually been achieved by fitting a waste-gate (see Fig. 24) to the turbocharger to by-pass some of the exhaust gas around the turbine at high speeds and loads.

Five different methods have been used to limit the boost pressure applied at the top speed of an engine while maintaining the boost at low speed. These are as follows:

(1) To use an orifice downstream of the turbine—see Fig. 25. The action of this is to increase the pressure ratio across the restriction orifice as airflow increases while the expansion ratio across the turbocharger turbine remains almost constant. The exhaust back-pressure of this system reaches a very high value at

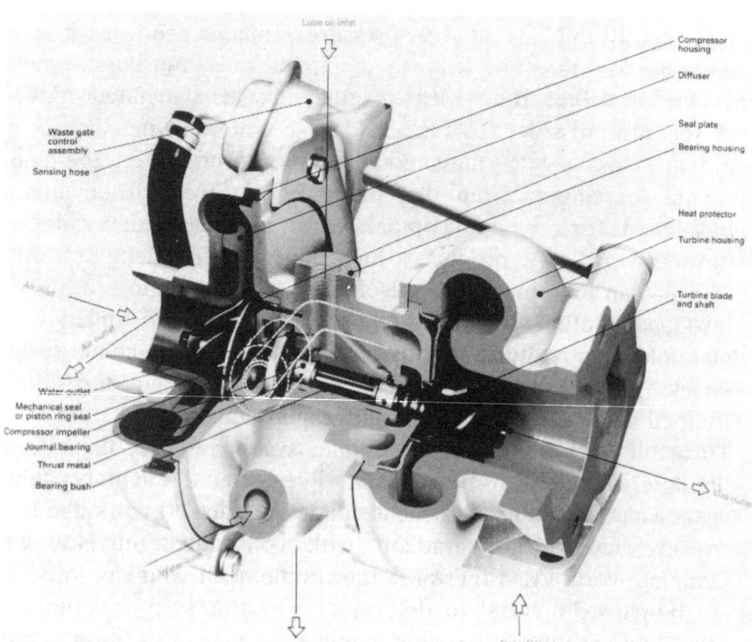

Fig. 24. Turbocharger with waste-gate for petrol engine use.

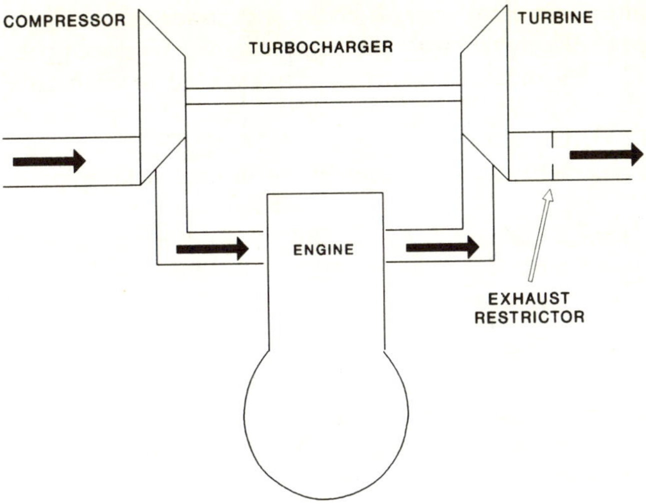

Fig. 25. Orifice plate downstream of turbine to widen operating range.

rated load, and this adverse pressure gradient produces a fuel consumption penalty.

(2) To use an orifice as a restrictor either upstream or downstream of the compressor—see Fig. 26. These generate an increasing pressure loss as the mass flow and boost pressure rises and isolate the engine from the higher boost levels. Their main problem is that as the compressor is using all the available turbine work, they produce a high inlet manifold temperature. This has an extremely adverse effect for petrol engines because it reduces the boost pressure which can be accepted for a given manifold temperature; the latter is restricted by combustion knock. In the case of a diesel engine it simply reduces the trapped air mass by lowering the trapped density.

(3) To use a blow-off valve in the inlet system so that the boost pressure is held sensibly constant—see Fig. 27. This system again causes some losses because the engine has to work against an adverse pressure gradient with consequent increase in pumping work. Also there are the problems of what to do with the by-passed air and to determine how this system could be fitted to a carburated gasoline engine.

(4) To use a waste-gate that by-passes the turbine stage when the boost reaches a limiting value—see Fig. 28. This system has

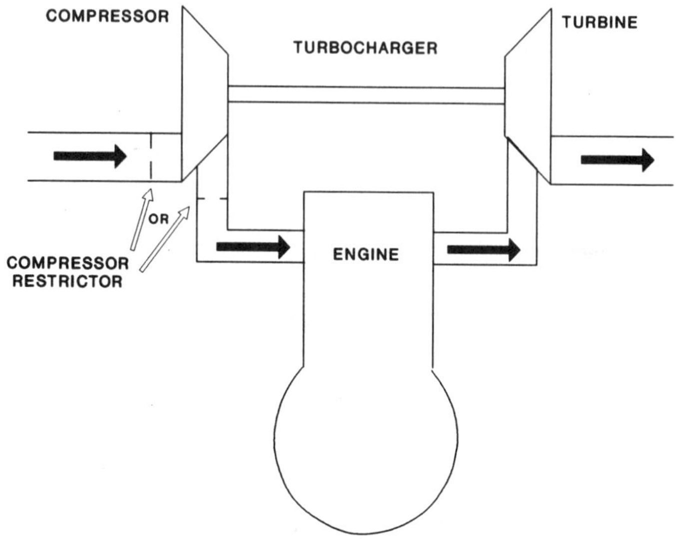

Fig. 26. Orifice plate upstream or downstream of compressor to improve turbocharger match.

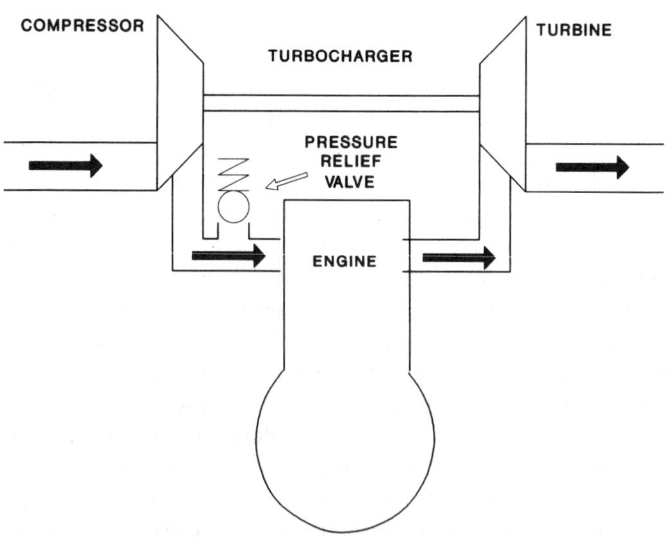

Fig. 27. Blow-off valve in inlet manifold to control boost.

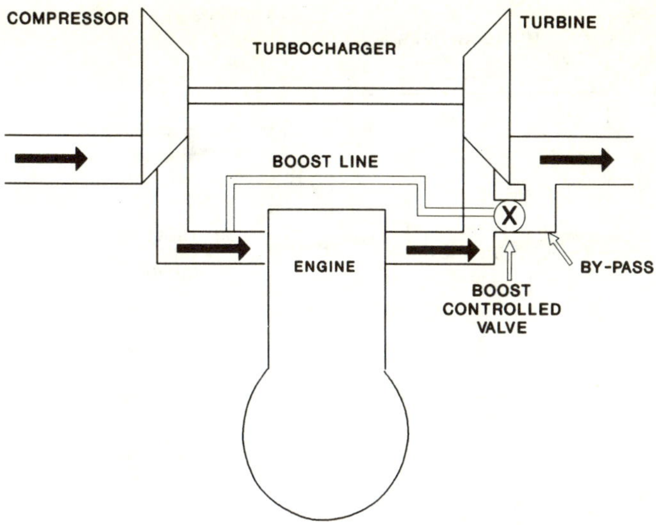

Fig. 28. Waste-gate on turbocharger to restrict boost level.

been universally used in passenger car applications. The engine is matched with a small turbine stage to allow boost to be developed at low engine speeds. When the engine speed rises, the boost is allowed to climb until the limiting value is reached. This limit may be detonation for a gasoline engine or peak cylinder pressure on a diesel engine. The exhaust gas is progressively by-passed round the turbine stage, maintaining a gently rising boost curve as the engine speed rises to its rated value. This system again gives a fuel consumption penalty because, when the valve is open, the exhaust pressure soon becomes greater than the inlet manifold pressure, with a consequent increase in pumping work.

(5) To vary the size of the turbine dynamically, so that it has a large swallowing capacity at high engine speeds and a small capacity at low engine speeds—see Fig. 29. This option of variable turbine geometry has been employed on wide speed range gasoline engines, both with and without a waste-gate. However, because of its mechanical complexity, cost and associated control systems, its main attraction is for engines with very high

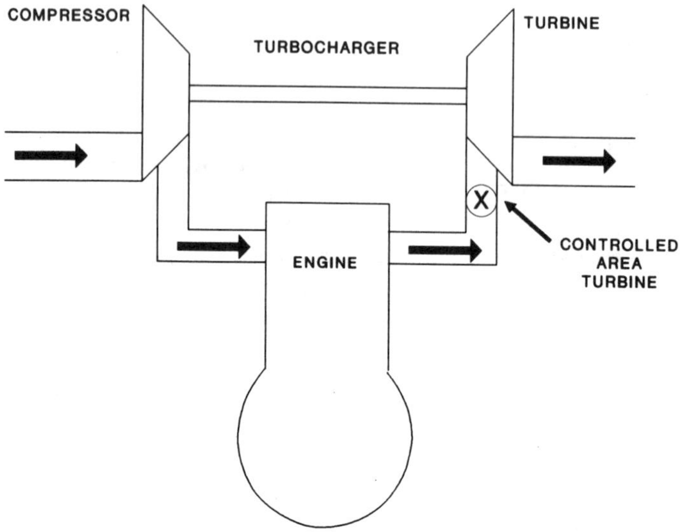

Fig. 29. Variable geometry turbine for turbocharger application.

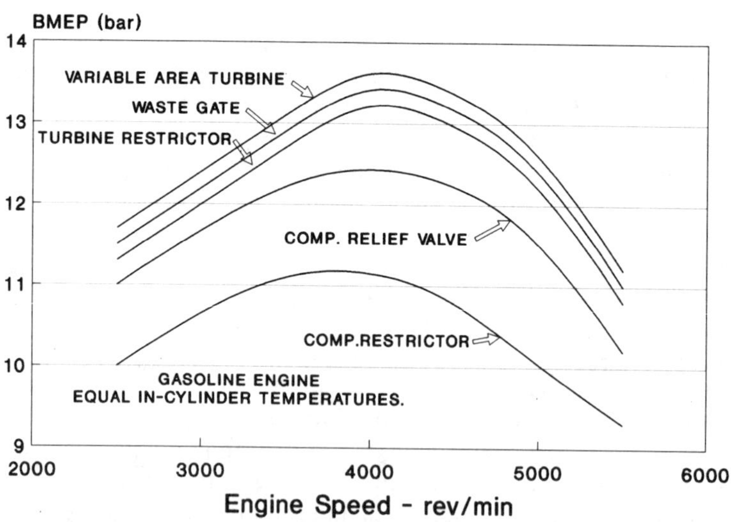

Fig. 30. Effect of using various boost restriction methods to achieve broad ranging match on gasoline engine.

load factors in their application, as this solution improves the fuel consumption of the engine at low speeds while maintaining the same value at the rated power point.

In Fig. 30 (from Baazari[22]) is shown the results of using each of these different control systems on a wide speed range gasoline engine. For each system, the BMEP has been set so that the engine has the same in-cylinder temperature before ignition. The compression ratio of the engine has also been changed to optimize the power and the values of this range from 5·7:1 for the compressor restriction results, increasing up to 7·6:1 for the variable geometry device. The consequence of this change would be to improve the fuel consumption of the better systems.

These, and other results, show that waste-gates and variable-geometry turbines are by far the most effective methods of controlling boost pressures. The choice between these two then becomes that of duty cycles and economics.

6. CURRENT RESEARCH

Research work on turbochargers can be split in several ways. The fundamental aerodynamic work is similar whether applied to a turbine or compressor stage, as was the case when considering the basic equations describing the devices in Section 4. Each stage can be split into a rotating part, the wheel, and a stationary part, the housing.

The compressor stage has been a favourite subject for research, and much more work has been done and published on this than the turbine stage. This is for two reasons: radial compressors are frequently used in the aerospace industry (where the investment in fundamental research work has been high); and the working environment is cooler and cleaner than for the turbine.

The turbine stage appropriate to small turbochargers has been somewhat neglected, because radial turbines are not frequently found outside the turbocharger industry. Also the operating environment of the turbine in an engine exhaust system is difficult to reproduce on a test rig, having both high temperatures and pulsating flow. In the past decade, more fundamental work on pulsating flow in turbines has been funded in the UK and so the current position is now improving.

6.1. Research on Wheels

6.1.1. Calculation of Flow in Centrifugal Machinery

For many years, the only practical way to calculate the flow going through a centrifugal machine was to use derivatives of the streamline curvature method (see Section 4.7), first proposed by Wu & Brown.[23] This work found most general application in the NASA suite of codes developed by Katsanis & McNally.[24] These programs were quasi-three-dimensional and produced an output of the blade loading in a wheel based on inviscid flow calculations (blade loading is defined by the pressure distribution curve acting across the blade, and the area between the curves represents the force acting in a tangential direction). In Fig. 31 is shown the output from such a code for a centrifugal compressor from Ref. 24.

With the advent of ever increasing computer power and speed, it has now become feasible to solve the full Navier–Stokes equations for three-dimensional viscous flow, and these codes are now finding general use in the design and analysis of stages. Various workers in many countries have been involved in the development of these codes, which are described in Refs 25–28. In Fig. 32 is shown a typical output

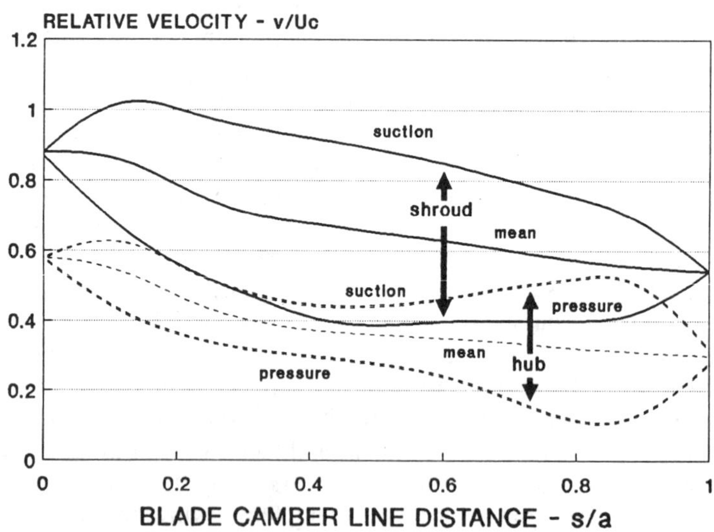

Fig. 31. Blade loading on compressor wheel based on inviscid streamline curvature method.

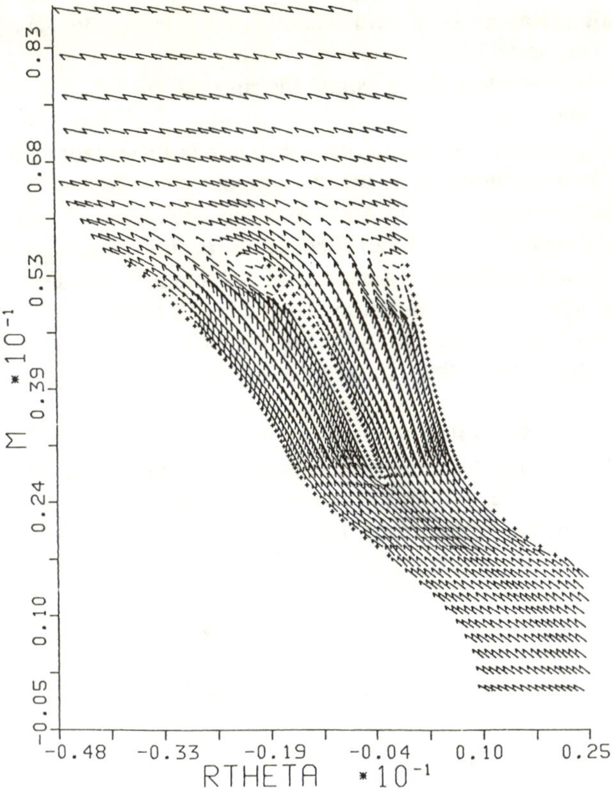

Fig. 32. Predictions of velocity vectors from three-dimensional computer code for compressor calculations from hub to shroud.

from the Wallitt code,[27] but, because of their three-dimensional form, the total outputs are very voluminous. The complexity of these codes precludes detailed discussion here, but the above references will enable the reader to pursue the subject further.

6.1.2. Experimental Work on Wheels

Until the advent of laser anemometry systems, experimental work on wheels was confined to the derivation of their internal flow performance from their overall performance and from external instrumentation. This led to the one-dimensional treatments, including gross loss coefficients, developed in Section 4. Many workers have attempted to use measured parameters at the wheel inlet, e.g. static pressure measurements along the shroud, and wheel exit to derive the wheel behaviour, but the small dimensions of the flow channels and the non-uniformity of the flow both spatially and temporally, made meaningful conclusions difficult.

Both laser–doppler and two-spot (time-of-flight) laser systems have been used on turbomachinery. The current state of technology of these measuring systems has resulted in laser–doppler anemometry being used in stationary components, i.e. vaned diffusers, nozzle rings and both compressor and turbine housings. Measurements inside the rotating components have been made using the two-spot systems, where their superior signal-to-noise ratio has enabled readings of flow vectors to be obtained closer than 1 mm from a wall (e.g. blade or hub surface).

The two-spot laser anemometry work by Eckhart & Krain[29–31] at DFVLR in Germany showed, for the first time, something of the real mechanism of flow inside a rotating wheel. With the commercially available laser systems, a two-spot laser (Polytech L2F) time of

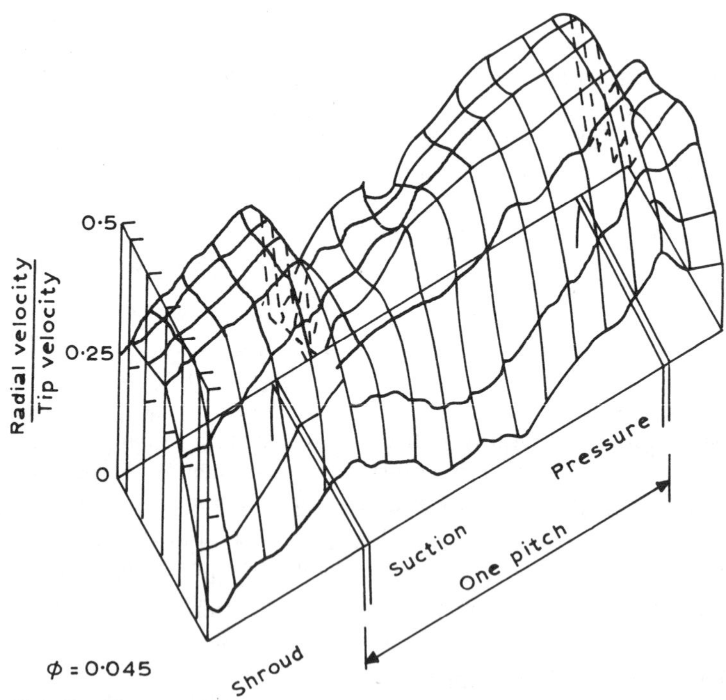

Fig. 33. Measurements from inside the compressor wheel by two-spot laser anemometer.

flight system can be used to measure flow vectors inside compressor wheels at rotational speeds over 100 000 rpm. In Fig. 33 (from Ref. 29) are shown typical results that are obtained from L2F studies in a compressor wheel, which enable a complete picture of the flow in a wheel or housing to be assembled. However, it is very time consuming, and hence expensive, to use laser anemometers in this way. To map the total characteristics of a stage would take many months of work. A more useful application for the technique is for validation of the three-dimensional viscous flow prediction programs, particularly in terms of turbulence modelling.

Similar techniques are being developed on the turbine, and critical experiments are aimed at getting the correct type and amount of seeding for the laser anemometer, which will still function when the turbine is running hot. If the working fluid is heated using a combustion chamber then the difficulties are obvious. Work done at Cranfield Institute of Technology has shown that such techniques are feasible, at least in the turbine housing.

The most comprehensive experimental study on compressor wheels has been that presented by Dean & Young[32]. The study pre-dated all the laser measurements, but showed a remarkable understanding of the flow. This work and most of the other published work on turbine stages has been taken under steady flow conditions. Investigations made under unsteady flow conditions are discussed in Section 6.4.

6.2. Research on Compressor Casings

Many results have been presented on the diffuser part of the compressor stage because 'conventional' instrumentation (by which is meant non-optical methods) can be used here. The experiments can be performed in a simplified way because the diffuser can be investigated without using a wheel. However, this simplicity can be a trap because investigations have now shown that the flow leaving a radial compressor wheel is very definitely a non-steady phenomenon. Hence not using a wheel and/or using 'conventional' instrumentation can produce misleading results which do not translate into the real environment.

The work on casings can be separated into two parts, in a similar manner to the analysis in Section 4; the vaned and the vaneless diffuser.

6.2.1. The Vaned Diffuser Stage

As a vaned diffuser has its largest impact on stage efficiencies at high pressure ratios (>3:1) and also restricts the flow range of a compressor, the majority of work has been done on gas turbine stages and on turbocharger compressors for large engines, where these characteristics are in keeping with the design aims. (The flow range of a compressor is defined as the distance between the surge and choke conditions at constant pressure ratio.) Much of the work presented has been of empirical form, where many tests are done, and the results presented as design charts. The most comprehensive studies of this type are those done by Sovran & Klomp[33] Runstadler & Dolan[34] and Dean & Young.[32]

A most comprehensive study, using a laser anemometer, was made by Jansen & Rautenberg[35] at Hannover Institute, where the result of the experimental work led to a non-uniform vane design, with incidence changing along its width to more closely match the angles of the gas leaving from the compressor wheel.

Another useful reference paper is by Fisher,[36] where the performance of different types of vaned diffusers is compared with the vaneless performance on two turbocharger compressors.

6.2.2. The Vaneless Diffuser Stage

The work by Dean & Young[32] gives a good general approach to the behaviour of the vaneless diffuser. It is possible to identify four major parameters in the design of a vaneless diffuser, namely:

(a) the radius ratio of the diffuser (effectively the area ratio).
(b) the inlet gas swirl angle;
(c) the inlet gas Mach number; and
(d) the friction width parameter, (fr/b), where f is the wall friction factor, r is the radius and b is the diffuser width.

Dean & Young show that an optimum value of $C_{\theta 2}/C_{r2} = 2$ exists.

Jansen & Rautenberg[35] and Senoo & Kinoshita[37] have worked on the onset of stall in a radial diffuser, and both found that stable operation could be ensured if $C_{\theta 2}/C_{r2}$ did not exceed 4.

6.3. Turbine Casings

Turbine casings, in a similar way to compressors, can be made with or without nozzles. The majority of current automotive turbochargers do not use nozzles for reasons of cost and performance. The benefits of using nozzles (higher maximum efficiency under steady state operation) are often not achieved under pulsating flow conditions, and the performance improvement (which manifests itself as a higher expansion ratio at peak torque) obtained by using a nozzleless housing is not well understood. As with compressors, nozzled casings are now found in applications where efficiency is of first importance, such as gas turbines and turbocharger stages used in applications with constant-pressure exhaust manifolds (see Chapter 13). In the past many small turbines were made with a nozzle ring, and this partly explains why there is much literature on such turbines.

6.3.1. Nozzled Turbine Casings

Little work has been done on the optimum design of nozzle rings for inward-flow radial turbines. In general, it is accepted that the gas exit angle from the blades is a constant and that production and cost constraints control the shape of the blades. The shape usually used is that of an uncambered aerofoil, but no work has been done on the optimization of the number of vanes. There is obviously a trade-off between gas guidance and friction losses, but this is usually arrived at experimentally.

Some work has been done in Japan by Mizumachi[38,39] and also by Mizumachi, Yoshiki & Endoh.[19] The last paper reports a comprehensive investigation of the performance of a turbine with nozzles under steady and unsteady flow conditions, with both full and partial admission (partial admission was achieved by blocking off one sector). The turbine was of 220 mm rotor diameter, and was loaded by an electrical dynamometer. Comprehensive results are given for turbine flow rate, degree of reaction, turbine work, turbine efficiency and pressure distributions around the casing. The steady flow tests will be considered here, and the unsteady flow results will be discussed in Section 6.4.

In the full admission case the pressure distribution around the casing was practically uniform. However, in the partial admission case (as would be expected) the pressure varies substantially: a typical pressure variation is shown in Fig. 34. At the end of the non-admission sector the pressure rises rapidly and reaches that of the centrifugal head, p_c, calculated from the rotational speed. This indicates that the gas is almost at rest relative to the rotor. When the rotor enters the open sector the pressure increases again as the gas from the nozzle enters the rotor and compresses the stationary gas still further. When the rotor passes around the admitting sector the static pressure falls as the

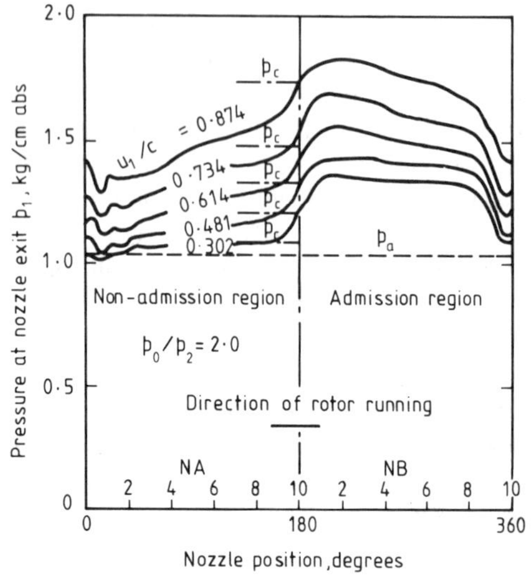

Fig. 34. Pressure variation around scroll of radial flow turbine under partial admission conditions: nozzled turbine.

gas in the rotor accelerates: by the end of the admission region the gas velocity has practically reached that existing during full admission. The effect of these changes in pressure around the periphery of the stator is to alter the effective degree of reaction at each azimuth angle. However, the mean degree of reaction of the admitting sector under partial admission is approximately equal to that under full admission, for the same conditions. Mizumachi *et al.* found that the flow rate under partial admission conditions was almost half that under full admission conditions over a broad range of operation, but the efficiency was substantially less (being about 81% under full and 62% under partial admission flows). (This is in broad agreement with the findings of Benson & Scrimshaw[40] for nozzled casings.)

Mizumachi *et al.* also developed two computer programs for simulating these results: an incompressible flow model and a compressible flow one. They found that the incompressible flow model did not accurately simulate the pressure variation in the partial admission case due to the intrinsically unsteady nature of the flow. The compressible flow model, based on method of characteristics, was capable of simulating pressure traces similar to Fig. 34. The flow in turbines operating under partial admission conditions is essentially unsteady and hence it is necessary to develop unsteady flow simulations of both rotor and stator to model partial admission effects—even if the entry conditions are not pulsating.

6.3.2. Nozzleless Turbine Casings

For steady state flow, the nozzleless housing has been analysed using both inviscid and viscous flow, single and multi-dimensional computer codes. While this will undoubtedly improve the steady state performance, it is questionable how accurate it will be with pulsating flow.

The housing flow is dominated by secondary flows and viscosity effects, and even when the input flow conditions are steady, these combine to produce a flow into the wheel that varies in both the axial and circumferential directions. This effect, in turn, has a major influence on the vibratory strain levels experienced by the rotor blades.

The assumption was made in Section 4.3.2 that the flow in the casing of a nozzleless turbine was basically a free vortex. Hussain, Ilyas & Bhinder[10] report an experimental investigation of a casing designed on this basis. Due to the necessity to retain a clearance between the tongue and the rotor tip it was not possible to obey the idealized equation for flow in such a casing, viz. eqn (25). This means that the value of F/r did not vary linearly with the azimuth angle, φ, and hence the idealized results could not be achieved. The important feature is how close to the ideal flow is that in a real turbine. To recapitulate, the ideal flow should result in the mass entering the turbine rotor being uniformly distributed around the casing, and the flow angle, α_3, should be equal around the casing. It should also be remembered that, in the ideal case, no allowance was made for loss in stagnation pressure due to frictional effects in the casing. Hussain *et al.* show that the blade flow angle α_3 was substantially constant over the main part of the periphery of the casing, but that a rapid increase in flow angle occurred in the vicinity of the tongue ($\varphi = 360°$). However, although the flow angle was relatively constant it was substantially different from that for which the casing was designed, namely $\alpha_3 = 20°$. The measured values varied between 10° and 18°; and they were almost independent of the inlet pressure over a reasonable range. It was also found that the tangential velocities, C_θ, were accurately calculated by the simple theory. The value of C_θ at the centroid of the casing increased around the periphery simply because the radius of the centroid of the flow, from the rotor axis, decreased as the flow passage became smaller. This is similar to the effect shown in Fig. 15. Hussain *et al.*[10] suggest that the large variation in flow angle around the volute might be due to an uneven mass flux in the radial direction relating to secondary flow effects. If this were the case then the flow into the rotor would not be uniform and loss coefficients of the type proposed by

Benson[13,14] would not be appropriate for predicting the loss of performance of that which occurs in such machines.

In summary, while one-dimensional methods enable an initial estimate to be made of the relevant geometry for the casing of a radial inflow turbine, the performance achieved will be substantially poorer than predicted from uniform flow. If such large variations in flow direction occur in an ideal casing under steady flow conditions then the variation in flow direction under unsteady flow is likely to be substantial. This will again create further losses as the incidence angles on the rotor become less favourable at various azimuth angles around the stator.

Lymberopoulos, Baines & Watson[41] describe calculations performed on both single- and twin-entry turbine casings using a quasi-three-dimensional technique for solving the equations of inviscid flow. First, the results for single-entry casings will be discussed. This code is able to predict many of the features measured experimentally, and shows that the flow transforms from a uniform velocity profile at the entry to the casing into a free vortex profile almost immediately on entering the scroll. This is not in agreement with Chapple, Flynn & Mulloy,[42] who showed a substantial transition region, and the difference could be due to differences in turbine geometry or neglect of viscosity in the predictions. In Fig. 35 is shown a comparison between prediction (by Lymberopoulos *et al.*[41]) and measurement for a rotor tested by Scrimshaw & Williams.[43] It can be seen, from Fig. 35(a), that the code predicts the variation in static pressure at the exit from the casing quite accurately. Also, Fig. 35(b) shows that the code is capable of showing the major features of variation of exit flow angle with azimuth angle. The positive exit flow angle (i.e. indicating a radial-outward component of flow) at the tongue is probably caused by an ejector effect as the inflowing gas passes the tongue. The computer program has not been able to predict the large negative flow angle at 40° azimuth angle: this is probably due to the low-momentum wake from the tongue being drawn into the rotor, and this is essentially a viscous effect which is not included in the program.

The characteristics of twin-entry, circumferentially divided, nozzled turbines under partial admission conditions are substantially poorer than those of the same turbine with full admission. This can be attributed to two factors; end-of-sector losses, and windage. The end-of-sector losses occur because the flow vectors at the ends of each sector are not constrained under partial admission conditions, and hence deviate towards the quiescent sector: this results in increased

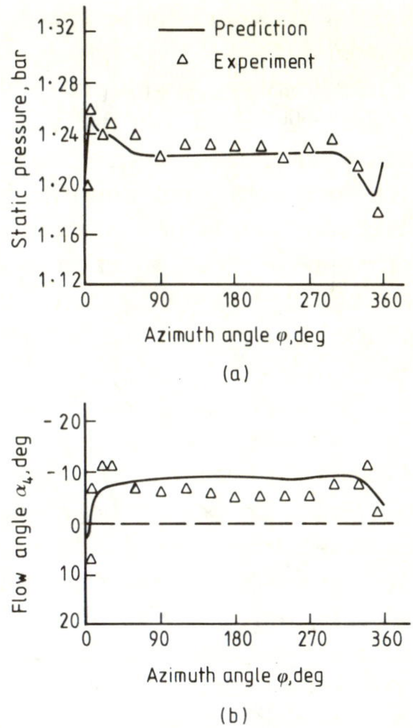

Fig. 35. Conditions at exit of nozzleless radial turbine casing. (a) static pressure as function of azimuth angle; (b) flow angle ($\alpha 4$) as function of azimuth angle; (△, experimental results; ———, predictions).

incidence losses. Windage occurs simply because the rotor generates a flow field which interacts with the quiescent sector and produces frictional effects.

The same effects will not so obviously occur with a meridionally divided twin-entry turbine, because there are no end-of-sector losses and windage is less apparent. However, Benson[7] shows that for such a turbine the variation of mass flow with pressure ratio is not the same for each entry. This is because the flow passages are not geometrically identical (even if the actual throat areas are the same) and hence the entries do not interact with each other in a reciprocal manner. Pischinger & Wunsche[44] and Dale & Watson[8] have both undertaken similar investigations of flow under partial admission conditions. Dale & Watson showed that, for their turbine, the mass flow through one entry was very close to half that of the total flow *if the pressure ratio was equal for both entries*. However, if the pressure in each entry is not equal then there is an interaction between the entries. In Fig. 36 (adapted from Dale & Watson[8]) it is shown how the entries interact with each other at a particular turbine speed. It can be seen that the

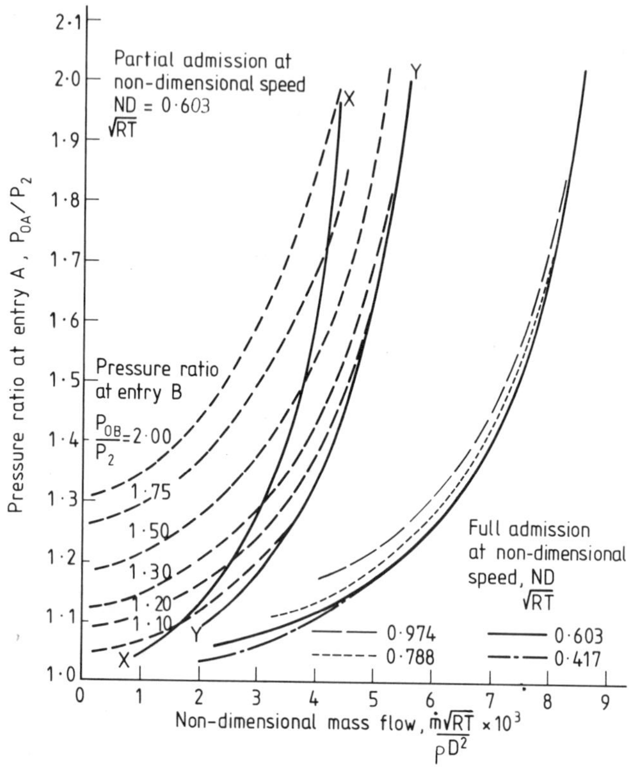

Fig. 36. Mass flow characteristics of a twin-entry turbine under steady flow conditions with both full and partial admission.

flow through entry 'A' is close to half that for both entries when there is equal pressure in each entry (line X–X). Line Y–Y on Fig. 36 is the flow through entry 'A' when there is no flow through entry 'B'. In addition, the diagram shows lines of constant expansion ratio across entry 'B': from these it is possible to evaluate the flow through entry 'B' for no flow through entry 'A' simply by drawing a line at constant pressure ratio p_{oA}. Note that there is no flow through entry 'A' if the expansion ratio in entry 'B' is 1·75, while that in entry 'A' is around 1·25.

While it is relatively straightforward to show how mass flow varies with different pressure ratios, Dale & Watson[8] explain that it is more difficult to do the same for the efficiency of each entry. It is usually assumed that the efficiency is the same as would be achieved for the instantaneous U/C for the entry, but Dale & Watson show that this assumption is incorrect. They evaluated the overall efficiency of the turbine as a unit based on the ratio of measured power output to the sum of the isentropic powers at each entry (see the definition of turbine work function in Chapter 13). They also evaluated a mean blade–speed ratio for the turbine, with $\bar{C}$ based on the mass average of the blade–speed ratios at each entry. Typical diagrams for the variation of the total-static efficiency (η_T) against mean blade–speed ratio are shown in Fig. 37. The different curves relate to various ratios of flow in each entry, and exhibit the surprising feature that the highest efficiency is not achieved at equal mass flows in each entry. In both cases the turbine is most efficient when the mass flow in entry 'A' is greater than that in entry 'B'. The diagram also shows that the efficiency is greatly reduced if the flow in either entry is zero, indicating the poor flow from the casing if only one entry is functioning. They point out that the lack of symmetry of effect on the efficiency has occurred in a casing which has been designed with identical (mirror-image) flow passages for each entry. It is suggested that the effect comes about because of the difference in angle (for each entry) through which the flow turns on entering the rotor.

No fully three-dimensional calculations have been made on flow through asymmetrical twin entry turbines, and Lymberopoulos *et al.*[41] state that their quasi-three-dimensional calculation method is not so accurate in this case because of the much larger influence of viscous

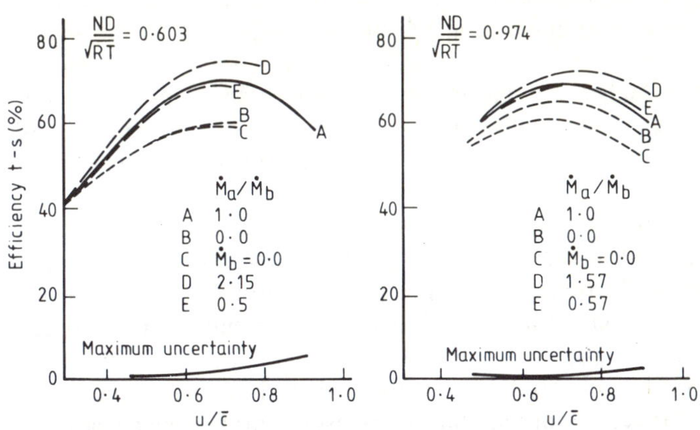

Fig. 37. Efficiency of a twin-entry turbine under steady flow conditions with both full and partial admission.

effects in such casings. These effects are dominant when the two flows from the meridionally divided casing mix just downstream of the separating wall. However, the results obtained did show some of the major features of such casings, and included flow from one entry to the other under conditions when the inlet pressures were substantially different. This effect has also been simulated in one-dimensional modelling of engine systems, by means of a short pipe connecting the two entries (or manifolds) of the turbine (see Section 5.2.1 of Chapter 13).

6.4. Pulsating Flow

The performance of turbocharger components is usually measured under steady flow conditions. This is quite appropriate for compressors because the pressure pulsations in the inlet manifold are normally very small percentages of the mean level. However, this is not the situation in the exhaust manifold where very large pressure pulsations can occur. Winterbone, in Chapters 12 and 13, describes methods of calculating these unsteady flows and also the effect of these flows on the turbocharger. As described in Chapter 13, the energy in the exhaust gas in the cylinder may be delivered to the turbine in two basic forms: approximately steady flow, by means of a constant pressure exhaust manifold; and in fluctuating form via a pulse manifold. In the first design the turbine is able to operate in steady flow but large losses in available energy occur during the blowdown period. In the second design the integrity of the pressure pulsations is retained right up to the turbine, which then has to operate under these pulsating flow conditions. This section will consider the second case in some detail: the constant pressure system can be analysed by a quasi-steady approach.

In Fig. 38 (from Horlock & Winterbone[45]) is shown the way in which the instantaneous efficiency of the turbine varies when presented with a fluctuating exhaust pressure, typical of an engine operating at low load. It can be seen that the efficiency drops to zero for long periods of the cycle, and varies substantially during the cycle. The variations are caused by the fluctuation of pressure which results in the gas velocity, C, varying during the cycle: this in turn causes U/C to fluctuate and hence the turbine efficiency also varies. In the case shown, the actual energy at the turbine is only about 15% of the

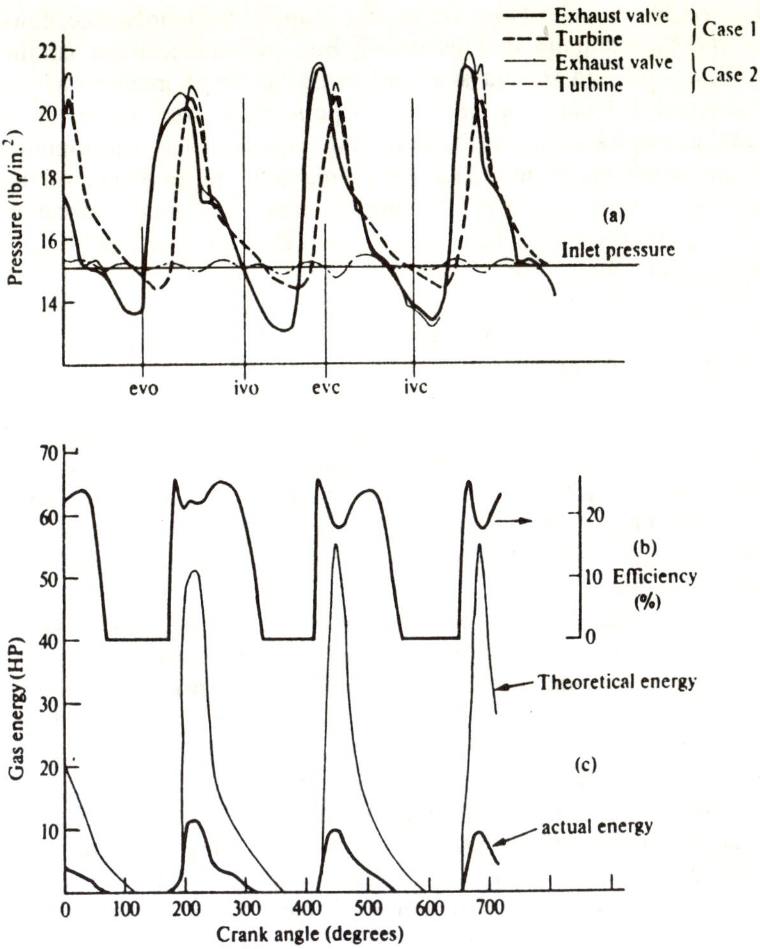

Fig. 38. Variation of pressure and turbine work under pulsating flow conditions, for an engine operating under low load.

theoretical energy in the gas. This case was analysed by assuming that the turbine operated in a quasi-steady manner, i.e. the instantaneous efficiency under pulsating conditions is the same as would exist under the equivalent steady flow situation. Experience shows that this is not true and the performance under pulsating flow can be substantially poorer than would be predicted from a quasi-steady assumption. An example of this is found when comparing the performance of nozzled and nozzleless turbines on an engine with a pulse manifold. The nozzled turbine will have a substantially better peak efficiency than the nozzleless one, and the quasi-steady operating locus might result in better average efficiency under engine operation. However, it is often found that the actual performance on the engine is substantially better with the nozzleless casing. This can be caused by the ability of the nozzleless casing to adapt to unsteady flow due to dynamic effects in the casing and rotor.

Benson[7] summarizes a number of studies into the effect of flow pulsation on turbine power output and mass flow. The picture is not completely clear, but it is apparent that the turbine does not appear to operate under quasi-steady conditions.

Wallace and fellow workers[17,18,46] and Benson & Scrimshaw[40] describe some early work on the measurement and calculation of turbines operating in an unsteady environment. In both cases the turbines considered had nozzled casings and hence the results will have been affected by this restriction on flow angle variation. Wallace, with others, also attempted to calculate the flow in the turbine under unsteady conditions.

A major problem in assessing turbine and compressor performance under unsteady flow conditions is measuring the torque being delivered by, or provided to, this component. Many people have used the turbocharger as a unit, and loaded the turbine by means of the compressor outlet pressure. This has a disadvantage, even in steady state, of being capable of providing turbine characteristics over an extremely limited range. Benson & Scrimshaw[40] overcame this problem, in part, by using two compressors: a small compressor was used for the low-load part of the turbine operation; while a larger compressor was used for the high-load part. Such a technique is relatively cheap and simple to set up but has the major deficiency of not providing information on the torque losses in the bearings. Under unsteady flow a further problem is introduced because it has to be assumed that the compressor is operating in a quasi-steady mode when attempting to evaluate the torque developed by the turbine.

A better method, used by Wallace,[17,18] Miyashita, Tomita & Ishihara,[47] Mizumachi *et al.*[19] and Dale & Watson[8] is to use a turbine dynamometer. A number of different designs of turbine dynamometer have been constructed. These are usually based upon hydraulic, air or, more recently, eddy-current machines. The design of suitable turbine dynamometers is extremely difficult because the torques produced by a typical automotive turbocharger turbine are very small (less than 10 Nm under full load conditions, at about 100 000 rpm). It is essential, if the device relies on torque reaction, that the fully floating portion of the dynamometer is extremely light and freely mounted. Some of the original turbine dynamometers based on hydraulic methods suffered difficulties of instability and were unreliable. The aero-dynamometer proposed by Miyashita *et al.*[47] does not appear to have been used by other workers, but they claim an accuracy of 0·05%. While the authors did use it for measuring flow under pulsating conditions, examination of the layout of the dynamometer suggests that the frequency response of the device would be quite low. The fully floating nature of that dynamometer enabled the friction of the bearing to be measured, and this was used as the load at low loads on the turbine. The most recently reported dynamometer is an eddy-current one which has been constructed by Dale & Watson. This appears to have an excellent frequency response, although it is somewhat limited in the maximum speed that can be achieved. Holset has developed a family of hydraulic dynamometers for turbocharger research. A schematic of one of these devices is shown in Fig. 39. Tests have shown that the smaller dynamometer, suitable for automotive-size turbochargers, is capable of running stably at speeds above 70 000 rpm: the torque resolution of this dynamometer is 0·04 Nm (about 1% of full load).

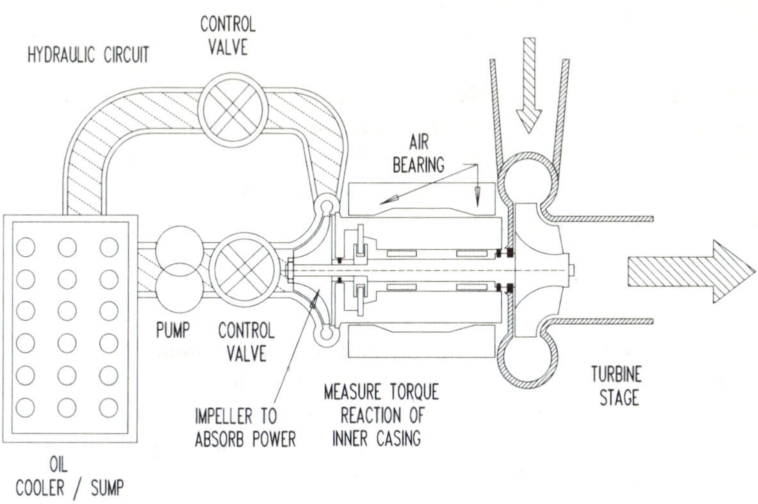

Fig. 39. Cross-section of hydraulic turbine dynamometer.

In summary, there is still room for a reliable, high-speed, high-accuracy turbine dynamometer to be developed. Up to now, these devices have been either unreliable or inaccurate.

6.4.1. Turbine Operation: Quasi-steady or Dynamic

As stated previously, some work has been undertaken on establishing whether turbochargers operate under quasi-steady conditions when subjected to pulsating inlet conditions or whether the operating locus leaves the steady state operating line. All of the results except those obtained by Dale & Watson relate to nozzled turbine casings. This should be borne in mind when examining the results reported.

Early work by Wallace & Blair[46] indicated that there could be discrepancies of about 25% between the power and mass flow predicted using quasi-steady methods and the measured values. They suggest that the quasi-steady flow hypothesis becomes less accurate as the pulse frequency is increased. The situation was worse when there was partial admission to the turbine, and the quasi-steady approach led to drastic underestimation of turbine performance. Wallace & Blair concluded that the swallowing capacity of the turbine was less under non-steady flow conditions than the equivalent steady flow ones.

Benson & Scrimshaw[40] tested a turbine using a pulse generator based on an engine. They measured the instantaneous pressure and temperature at entry to the turbine, and overcame the problem of predicting these values from those at the entry to a longer pipe (the method favoured by Wallace and fellow-workers). The turbine was loaded using a compressor. Benson & Scrimshaw showed that, under normal operating conditions in an engine, there would usually be flow in each sector (or scroll) of this circumferentially divided turbine (see Fig. 40). This explains why the mass flow and power evaluated using a quasi-steady approach for partial admission will grossly underestimate the unsteady values—particularly power. Usually, the steady state data for partial admission are obtained with flow in only one sector and no flow in the other sector. This exaggerates the end-of-sector (or inter-entry) and other losses that occur in the turbine. Benson & Scrimshaw show that the steady state torque under partial admission conditions is substantially less than 50% of that from the turbine under full admission (which would be expected from a twin-entry device). As stated above, the characteristics of a multi-entry turbine should be measured with different ratios of flow in each entry to give data that can be used with some confidence under partial admission conditions.

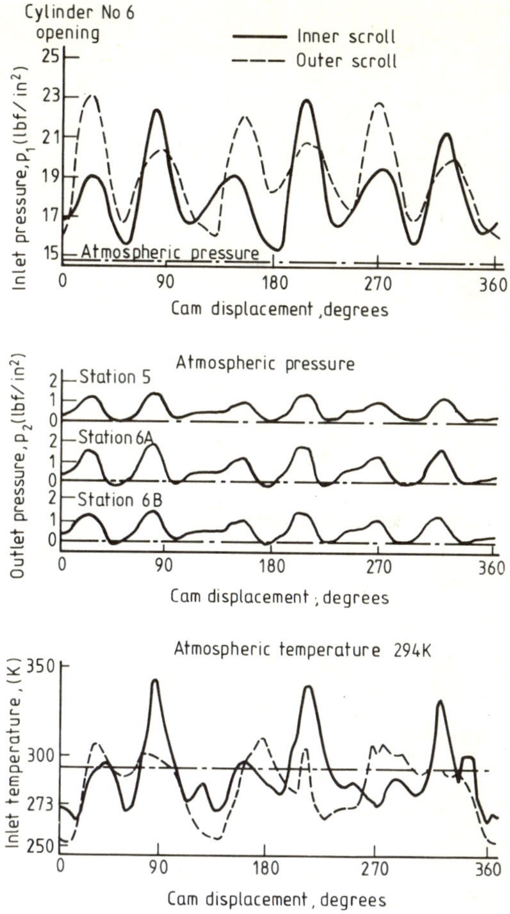

Fig. 40. Pressure variation in each entry of twin-entry turbine.

Benson & Scrimshaw also show that the accuracy of quasi-steady predictions is much better under full than partial admission conditions. In Fig. 41 it is shown that the quasi-steady approach underestimates the experimental results by around 7% for mass flow and 25% for power (of which part is because of the mass flow error). Benson & Scrimshaw also showed that pressure pulses pass through the turbine and result in fluctuating pressures downstream of the device: these affect the pressure ratio across the turbine itself and hence reduce the power output and mass flow. They found that the magnitude of all the unsteady effects was a function of pulse frequency and turbine speed.

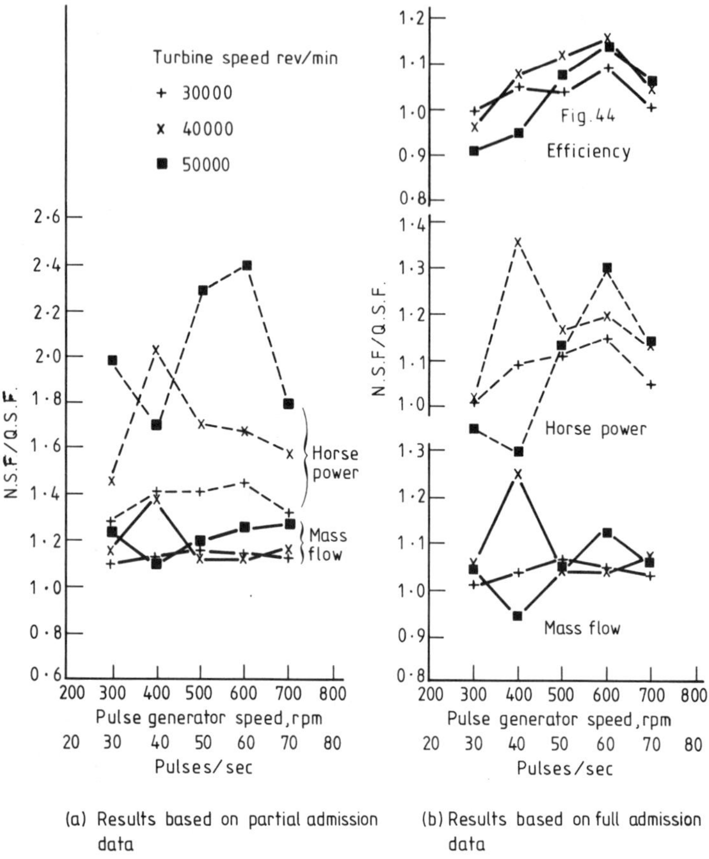

Fig. 41. Comparison of turbine parameters predicted using quasi-steady (QSF) assumption with those measured under non-steady flow (NSF).

Wallace, Adgey & Blair[17] performed experiments on a turbine similar to that of Benson & Scrimshaw, but in this case it had three entries. They used a turbine dynamometer as a load. In Fig. 42 it is shown that for a given mean mass flow the turbine produces a higher torque under pulsating conditions than steady ones. This is because the energy content of a pulsating flow is higher than the steady state equivalent, a feature which is recognized by pulse charging systems. A similar result was found by Miyashita *et al.*[46] Wallace *et al.*[17] also calculated the flow through the turbine, and its associated pipework.

They used homentropic flow in the pipes, neglecting friction and heat transfer, but included an entropy change term in the rotor interspace. In addition, they simplified the partial admission analysis. The simulation gave good agreement between pressures at various stations but there were some substantial differences when reflected waves occurred. The output of the turbine was evaluated using a work function similar to that in Chapter 13 (Section 5.1) and the instantaneous efficiency was also evaluated. Both of these parameters showed negative values in some regions due to the negative incidence occurring under unsteady flow. Even with this degree of sophistication the accuracy in predicting turbine torque was not good, with the program overestimating the measured value by about 50%. Wallace & Miles[18] extended this work and considered two computer simulations, a fully unsteady one and a quasi-steady one. The full wave action approach, based on method of characteristics, was similar to that in Ref. 17. The quasi-steady approach modelled the turbine in a similar manner to Section 4. All this work was performed at low mean pressure ratios (about 1·2) appropriate to pulse charging conditions. The results show the following:

(a) For a single-entry turbine both methods predict the experimental results quite well, although there is some phase error with the quasi-steady method due to the simulation of the turbine as a number of pipes. The power predicted was about 7% higher than that measured but this could be due to the neglect of friction and heat transfer in the simulation.
(b) For multi-entry turbines the prediction of pulse shapes, especially at the interspace between the stator and rotor, was good.

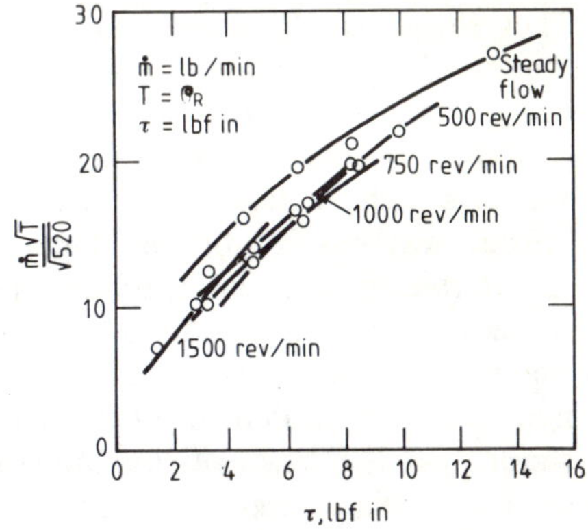

Fig. 42. Comparison between torque and mass flow characteristics with various frequencies of pulsating pressure.

The programs both accurately predicted mass flow, but the quasi-steady method overestimated power by about 11%, while the full pulsation model only overpredicted by about 5%.

(c) There is little effect of pulse frequency on turbine performance.

Mizumachi *et al.*[19] extended their steady flow study into unsteady flow and used a rotary pulse generator to produce these conditions. They also developed computer simulations, based on method of characteristics, for this flow régime for both full and partial admission (the program is very similar to that by Wallace and others, except for a slightly more sophisticated rotor model). They conducted experiments on one turbine at one speed (25 000 rpm) and two pulse frequencies (16·6 and 25 Hz). The computer program was able to predict accurately the pressure variations during full admission, with predicted mass flows about 5% higher than the measured values. (If the mass flows had been estimated using the mean expansion ratio the error would have been +30%.) They also found that, under pulsating conditions, flow reversals took place due to the centrifugal head developed by the rotor and negative torques occurred (similar to Wallace). The calculations were somewhat poorer in respect of turbine torque and the predictions were between 10 and 17% higher than the measured values. The quasi-steady method underestimated the torque by between 15 and 25%: this is simply because the pulse energy is substantially higher than the steady state energy at low mean expansion ratios (about 1·2).

Miyashita, Tomita & Ishihara[47] report unsteady flow tests on a turbine with pulsation frequencies over the range 20–50 Hz. The tests have been restricted to full admission and are essentially quasi-steady. They show that the mass flow is less under unsteady flow conditions than would be estimated from mean expansion ratios, and the largest effect is at 20 Hz. This observation is probably due to the non-linearity of the characteristics and is not a truly dynamic effect, although the variation with frequency might show some dynamics in the system.

The most recent work in this area is by Dale & Watson.[8] They report results of a twin-entry turbine operating under full admission conditions. The load was a high-speed eddy-current dynamometer capable of resolving the instantaneous torque, while the instantaneous mass flow was evaluated by a hot-wire anemometer. In Fig. 43 it is shown that the locus of instantaneous mass flow and pressure ratio does not follow the steady flow characteristic of the turbine, but

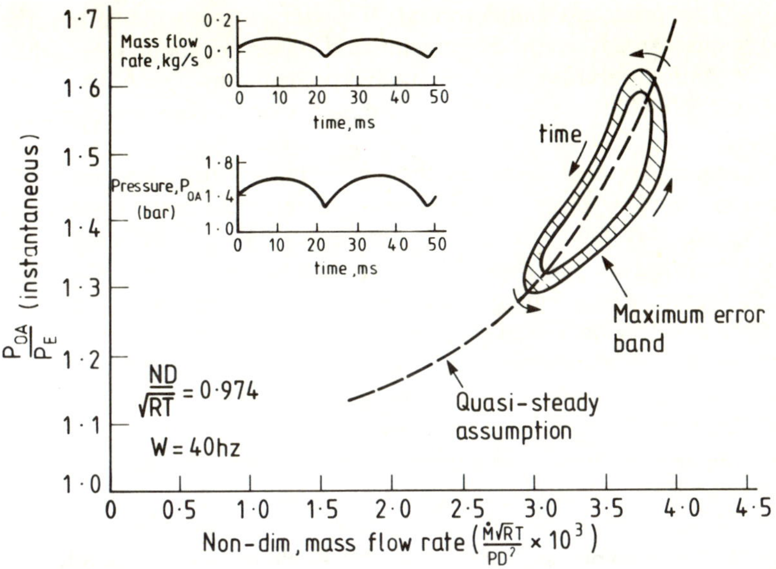

Fig. 43. Variation of instantaneous turbine operating point on swallowing capacity curve under unsteady conditions.

exhibits a hysteresis effect. This is due to the mass flow rate and pressure ratio becoming out of phase during the dynamic tests. The maximum deviation of mass flow rate from the steady flow line is about 6%, but the mean value is almost equal to the value appropriate to the mean pressure ratio. This is probably due to the capacity (or volume) of the turbine casing, which then acts as an integrating device and introduces a phase lag between the pressure ratio and mass flow.

Dale & Watson[8] also show a graph of instantaneous efficiency against U/C (Fig. 44). This again exhibits a hysteresis effect, with the instantaneous efficiency under unsteady conditions exceeding that under steady flow. This somewhat surprising result can easily be explained and is again the result of a phase lag. Since the instantaneous turbine power was related to the conditions at the entry to the device it is possible for the actual power measured by the torquemeter to be out of phase with it; a high torque can be the result of a previous pulse entering the turbine and a high efficiency would be calculated if the present entry conditions were a pressure trough. Fortunately, again, the average turbine efficiency over the cycle is not far removed

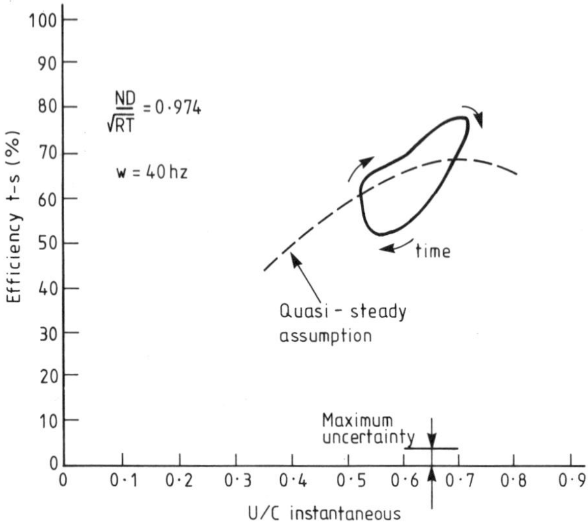

Fig. 44. Variation of instantaneous turbine efficiency under unsteady conditions.

from that which could have been evaluated by quasi-steady methods: the authors suggest that the error is within a few percent.

7. TEST FACILITIES

Theoretical research can, of course, be carried out on any size of turbomachinery, but when practical work is involved, then the effect of size must be taken into account. The simplest way to drive a turbocharger is as a gas turbine, when the compressor exit gas is passed through a combustion chamber and then used to drive the turbine. Unfortunately, this produces only one equilibrium point at any speed and the locus of these points produces one line, known as the self-drive line, which severely limits any experiments.

The better way to run the units for test is to supply compressed air from an external source, take it through a combustion chamber and then feed it into the turbine. By using a throttle on the compressor outlet flow, there are three controls; the compressor throttle, the throttle on the air supply to the combustion chamber and the fuel supply control. The only problem is that all the work produced by the compressor is now thrown away, so the running costs can become quite high! It is, of course, possible to omit the combustion chamber from the installation, if the inlet air temperature to the turbine can be kept high enough to avoid icing at the turbine exit. This can be achieved by inserting electrical heaters in the supply line: this has the additional benefit in that these produce clean hot air without any carbon particles, a necessity for any laser anemometry work on the turbine stage.

From a measurement point of view, the larger the components, the better. The smallest turbochargers have passages only a few millimetres in size, which makes any physical instrumentation very intrusive—and distorts the flow being measured. A further benefit of a large turbine is that the rotational speeds for a given pressure ratio are inversely proportional to the wheel diameter, so the speeds can be very high on a small unit (*c.* 250 000 rpm).

These factors make the experimenter choose the largest possible unit and the economist the smallest. As usual there is a good compromise, and this results in using units with wheel sizes between 100 and 150 mm in diameter. There is a secondary benefit from using these sizes, in that commercial units are available with both nozzled and nozzleless turbines and vaned and vaneless diffusers. This is because the applications for these sizes straddle the dividing line between automotive and medium-speed engines.

It is appropriate here to examine how the performance of a turbocharger is presented. Like any aerodynamic device, the governing parameter for both turbine and compressor stages is the Mach number, and all performance maps of a turbocharger reflect this, in spite of the misleading axes that are used.

7.1. Compressor Performance Maps

Compressor maps can be produced in three different forms, as well as in a variety of units. The three basic forms are shown in Figs 45(a), (b) and (c). In Fig. 45(a) corrected volumetric flow is plotted against pressure ratio, while Fig. 45(b) shows corrected mass flow against pressure ratio and Fig. 45(c) shows the term $m\sqrt{T}/p_0$ plotted against pressure ratio (where m is the mass flow, T is the absolute compressor inlet temperature and p_0 is the stagnation (or total) pressure at inlet).

It is important to understand the meaning of these axes, and their

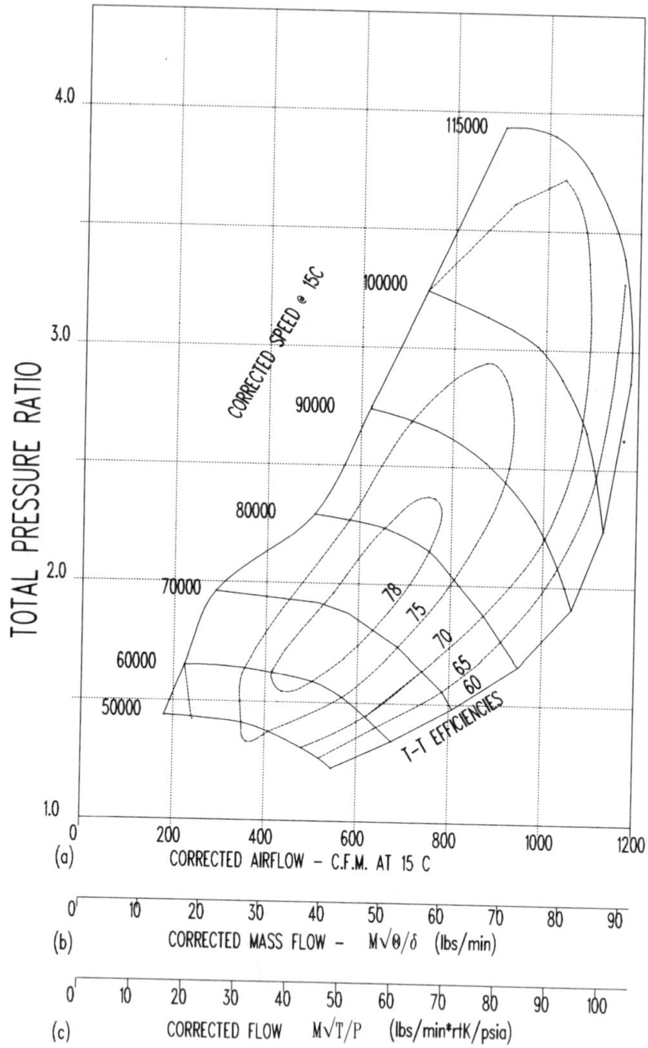

Fig. 45. Compressor characteristics. (a) corrected volumetric flow against pressure ratio; (b) corrected mass flow against pressure ratio; (c) $m\sqrt{T}/p_0$ against pressure ratio.

derivation follows. From fundamental relationships:

$$\text{density,} \quad \rho = p/RT \tag{79}$$

$$\text{volumetric flow,} \quad Q = m/\rho$$

$$= mRT/p \tag{80}$$

$$\text{velocity,} \quad V = Q/F \tag{81a}$$

$$= mRT/pF \tag{81b}$$

$$\text{speed of sound,} \quad a = \surd(\kappa RT) \tag{82}$$

$$\text{Mach number,} \quad M = V/a$$

$$= mRT/pF/\surd(\kappa RT) \tag{83}$$

Now, R is sensibly constant over the range of conditions encountered, and the area F is a constant for a given turbocharger component; so eqn (83) can be rewritten as:

$$M \propto m\surd T/p \tag{84}$$

In other words, the relationship $m\surd T/p$ is proportional to the axial flow Mach number in the inlet to the compressor wheel.

The other parameter on a compressor map that is related to Mach number is the speed of the rotor. On a map, it is usual to find either a speed quoted 'at 15°C' or some other value, or a speed given as a number $N/\surd T$. Both of these numbers have the same meaning. Consider the tip of the compressor inducer. Its tangential velocity, U, is:

$$U = r\omega \tag{84}$$

So the Mach number at the tip is

$$M = r\omega/\surd(\kappa RT) \tag{85}$$

Again, the terms which are constant for any given turbocharger can be neglected, giving

$$M \propto \omega/\surd T \tag{86}$$

So the term plotted as $N/\surd T$ is proportional to the tip Mach number and hence the intersection of the $m\surd T/P$ with the $N/\surd T$ line is a measure of the relative Mach number at the inducer tip, and *it is this value that produces flow similarity,* i.e. will reproduce the pressure ratio and efficiency of the base map.

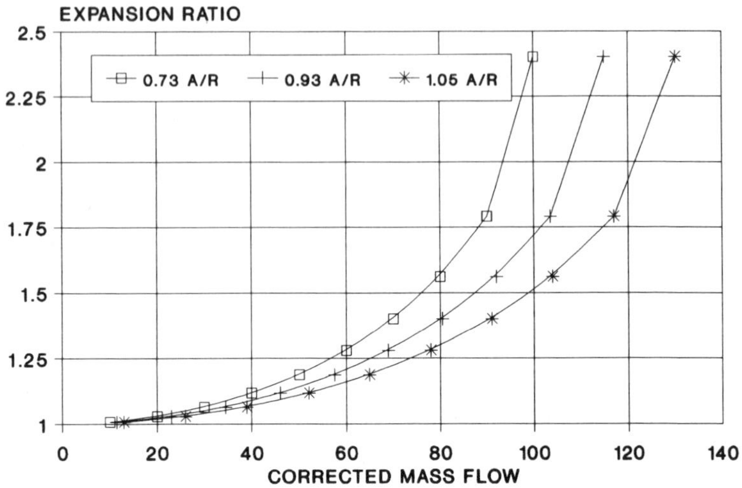

Fig. 46. Turbine characteristic—mass flow against expansion ratio.

All the other variants of the map still plot the same data, but try to present it in a more usable form. When the rotor speeds are given in apparently 'real' speeds, these are obtained by multiplying the $N/\sqrt{T}$ term by the square root of a particular reference temperature. It is thus important to multiply the measured speed by $\sqrt{(T_{ref}/T)}$ to obtain the map point.

The x-axis variations can be more difficult to understand. When a map displays a mass flow figure, then both the inlet temperature and pressure have been 'corrected' to some base condition. That is to say, if Φ is the ratio of temperatures and Ψ is the ratio of pressures, then the 'mass flow' is really $m\sqrt{\Phi}/\Psi$, and must be multiplied by the appropriate values of Φ and Ψ to obtain the corrected mass flow with the proper Mach number.

Maps are also still produced with a volume flow based x-axis, as they are the easiest to use for simple matching calculations. The derivation of the scale is interesting. Equation (81a) gives the flow velocity at the inlet in terms of the volume flow, Q, and the cross-sectional area. If this velocity is divided by the speed of sound and the constant terms are eliminated, then

$$\text{Mach number, } M \propto Q/\sqrt{T} \tag{87}$$

If this is multiplied by the square root of a reference temperature then a Mach number in terms of the inlet volumetric flow is obtained (Note that it is important to understand that this is volumetric flow at the *actual* inlet density.) To get on to the map from real figures only requires the measured flow to be multiplied by the square root of the ratio of the reference temperature/measured inlet temperature.

The other parameters shown on compressor maps, efficiencies and pressure ratios, have only one variation. Traditionally, both were given as total/total (or stagnation/stagnation) ratios, but there is now a move to change this to total to static, which will produce lower values of pressure ratio and efficiency—for the same stage!

7.2. Turbine Performance Maps

The situation is less confusing in the case of the turbine. A turbine performance map is usually presented as mass flow, expressed as $m\sqrt{T}/p$ against expansion ratio P/P_0 for a range of turbine speeds $N/\sqrt{T}$ (Fig. 46). However, just to confuse, the expansion ratio across a turbocharger turbine is always quoted as the total to static value—this being because any dynamic head leaving the turbine stage cannot be recovered. A second plot (see Fig. 9) shows a typical radial turbine map plotted with the efficiency results (again total/static) plotted on the y-axis against speed (corrected for Mach number). The term U/C is known as the blade–speed ratio, and was defined in Section 4.6. It will be noted from Fig. 9 which is for a typical radial inflow turbine that the maximum efficiency is at a U/C value of about 0·7. It was shown in Section 4·6 that the optimum efficiency for an ideal radial turbine occurs at a value of U/C of 0·707 and, in practice, it is difficult to achieve maximum efficiency at a significantly different figure.

8. AREAS OF SPECIAL INTEREST FOR RESEARCH

8.1. Aerodynamics: Application of Flow Codes

Many flow analysis codes are now available to the worker, with a wide range of complexity. These enable the flow in the wheel or cover of a compressor or turbine to be analysed. The codes are by no means complete, as the whole area of turbulence development in non-

Newtonian force fields has yet to be evaluated. Thus much work can be done to develop these codes, especially in conjunction with laser anemometry results which will enable the models to be refined. The flow fields in all the turbocharger components (even under steady flow conditions at inlet) have now been found to be essentially non-steady and, as computational power continues to increase, the application of these codes to non-steady flow is a future challenge.

All the codes outlined above are analysis codes and, although they allow the nature of the flow in a component to be quantified, they do not suggest ways in which the flow can be optimized: that area still remains a designer's art. A major development would be if such codes could be applied in an inverse way, so that a required flow pattern could be specified and the program then used to define a geometry that would achieve them. Some attempts have been made to do just that, but this also remains a major area for work in the future.

The non-steady aspects of the flow mean that only the upstream boundary conditions of the compressor inlet can be specified with any certainty, and major advances will come when the wheel code can be linked to the compressor cover code, so the whole stage can be analysed as a unit. This will need great computer power to run in a sensible time, but it is a requirement that will remain until it is achieved.

8.2. Pulsating Flow

Perhaps the biggest challenge for the future is the ability to measure accurately and predict the performance of the turbine stage under the pulsating conditions that exist on an engine. The upstream conditions for the stage are both unsteady and installation dependent. It is believed that better utilization of the engine exhaust energy will result from such analysis, and initial exploration of unsteady effects in the turbine stage show that design points for 'real' operations may differ significantly from the assumed steady conditions.[8]

8.3. Flow Characteristics

From an operational standpoint, the flow characteristics of both compressor and turbine may hold up the development of engines with higher BMEP. As discussed in Section 9 below, this is manifesting itself in a requirement for higher efficiencies, pressure ratios and 'map width' (the surge to choke range of airflow at a given pressure ratio) for compressors and higher efficiencies, expansion ratios and lower U/C ratios for maximum efficiency. In addition, on the turbine it is believed that a characteristic more suited to the demand of an engine might be achievable if the characteristics on a turbine map could be manipulated to more favourable shapes.

9. THE FUTURE

9.1. Developments of Conventional Turbochargers

The history of the diesel engine has been one of steadily increasing BMEP and this trend is expected to continue. The increase in BMEP is achieved by increasing the air flow and hence boost pressure. In addition to this increase in power, the demand for ever tighter emission control is leading to the use of higher values of trapped air:fuel ratios (i.e. weaker mixtures) in the engine, again necessitating a higher boost pressure. Currently, the commercial demand for boost pressure ratios is approaching the existing limit for single-stage operation, based on using commercially available alloys in both wheels. A further increase in boost pressure will only come about either by using more exotic materials or by using more than one stage in the compressor.

Extension of the pressure ratio capability of a single-stage unit is not the whole answer, as the operating range of a single-stage compressor (the flow margin between surge and choke) becomes narrower as the pressure ratio is increased. This is contrary to the demand placed on the compressor by the engine, so it is essential that any upward extension of the pressure ratio is accompanied by a wide range of flow.

As the BMEP of the engine rises, the turbocharger shaft work becomes greater as a proportion of the engine work, and so the influence of the turbomachinery on the engine fuel consumption increases.

The life expectancy required of a turbocharger is also rising, as is that of the engine. The combination of higher pressures, speeds, stresses and life present a very great challenge and, assuming that quality is not a problem, can only be attacked by advances in material properties. Current engine requirements and turbocharger technology allow turbochargers to be designed where the basic design parameters can be met and result in units where the lives of the turbine and compressor wheels are approximately equal. The changes in engine design which are currently being investigated are altering this.

A rising level of trapped air : fuel ratio leads to a smaller turbine flow parameter ($m\sqrt{T/P}$) being required, which leads to lower value of specific speed for the turbine—a new avenue for turbocharger design. The adoption of two-stage turbocharging changes the design match, particularly for the low-pressure turbine which requires a higher flow capacity wheel than standard. Both of these changes are leading the designers of turbocharger wheels into new fields, because the previous design technology for gas turbines was aimed primarily at narrow flow range machines.

9.2. Different Air Handling Systems

As engine output continues to rise, the turbomachinery and air handling systems become more important. This has resulted in the development of systems intended to make better use of the available exhaust energy (see Chapter 13).

9.2.1. Turbocompounding

Turbocompounding is the most prominent of the systems for achieving better utilization of exhaust gas energy. It is a method of adding a second turbine to the exhaust system which can transfer mechanical energy back into the crankshaft or to some accessory drive.[48–51] This leads to a completely new requirement from the turbomachinery, described by Tennant & Walsham.[52] The most significant change is to the design of the turbocharger turbine when used in conjunction with a downstream power turbine. For the same engine output, the compressor duty and the expansion ratio requirements from the turbine remain the same, but the inlet density to the turbine (and $m\sqrt{T/P}$ values) are reduced to a value that is 50–60% of the normal single-stage turbine. This in turn leads to a version of the turbine with much lower flow and specific speed to maintain efficiency.

The other characteristic of the turbocompound engine is that the boost developed along the full load torque curve does not droop as much as on a standard engine as the speed is reduced from the rated value. This translates into a wider operating range for the compressor, again posing a design problem.

9.2.2. Variable-geometry Turbines

Another development, being driven mainly by emissions legislation (particularly under transient loads), is that of a turbine with variable swallowing capacity. As was explained in Section 5, once one operating point is chosen with a standard turbocharger, the engine receives what the turbocharger delivers—not always what it requires—for the remainder of the running envelope. The most efficient way of correcting this is to be able to change dynamically the size of the turbine stage, so that the turbine work and thus airflow can be tailored to the requirements of the engine.

Various methods of achieving variable swallowing capacity turbines have been developed. The simplest in conception, if not in mechanical simplicity, durability and cost, are those utilizing a turbine stage with a nozzle ring incorporating pivoting vanes. By such a mechanism, changing the angle of the vanes changes both the throat area of the vanes and the incidence of the gas onto the turbine wheel.[53–55]

Another group involves modification of the critical area of a vaneless housing. This can be changed by modifying the 'tongue' of the housing, where the closed section ends, or by changing the whole side wall of the housing, which modifies the critical area and the volute cross-section.[55]

A third system[56,57] uses a nozzled turbine stage, where the nozzle ring is progressively blanked off to change the throat area of the nozzles.

All these systems result in some compromise of the aerodynamic design, because of the need for a robust design which will work under very adverse conditions. This ability to vary the airflow for the engine is becoming more critical as engine emissions legislation becomes more severe, and some form of dynamic control will probably be needed for the US emissions levels being proposed for 1994. This is particularly true under transient load conditions, where the ability to run with a very small turbine stage increases the air:fuel ratio during transients without the necessity to by-pass inefficiently the flow when the engine is running at rated power.

These types of turbines also bring about a compressor problem, as it now becomes possible to achieve a flat boost curve along the full load curve of the engine at different speeds and to achieve higher boost pressures at very low engine speeds. These possibilities are reflected in the need for a much wider compressor map from surge to choke.

9.3. Ceramic Rotor Turbochargers

Ceramic turbine rotors can potentially offer many advantages over the normal metallic component. First, the low material density reduces the inertia of the rotor system, giving a quicker response for acceleration, which is particularly desirable for automotive applications. Secondly, they will operate at high temperature (1000°C), often experienced in spark-ignition passenger car applications. This in turn reduces the need for the exotic and expensive materials used in metal wheels.

Against these advantages is the great difficulty in manufacture: this problem has retarded their introduction in production owing to the large number of failures that are experienced when overspeeding due to flaws or inclusions in the casting. A second difficulty is that of bonding the ceramic disc to a metallic shaft. Many methods have been developed, including shrink fitting, adhesive bonding and brazing and, because of the difficulties with all these techniques, discs with a continuous ceramic shaft have been tested.[58]

Materials must have a high tensile strength and low fatigue point and be hard enough to resist erosion from particles in the exhaust. Both silicon nitride and silicon carbide are the primary candidates, but the former is favoured.

The present situation (1989) is that small production quantities are being produced by a few manufacturers, but a high scrap rate has to be accepted and sophisticated inspection methods are essential.

REFERENCES

1. Cohen, H., Rogers, G. F. C. & Saravanamuttoo, H. I. H., *Gas Turbine Theory,* 3rd edition, London, Longman, 1987.
2. Watson, N. & Janota, M. S., *Turbocharging the Internal Combustion Engine*. Macmillan, London, 1982.
3. Sawyer, J. W., *Sawyer's Gas Turbine Engineering Handbook*; vol I, *Theory and Design*. Turbomachinery International Publ. Co., Connecticut, USA, 1985.
4. Glassman, A. J., Turbine design and application. NASA SP-290, 1972–75.
5. Stanitz, J. D., Some theoretical aerodynamic investigations of impellers in

radial and mixed-flow centrifugal compressors. *Trans. ASME,* **74** (1952) 473.
6. Dean, R. C., Jr & Young, L. R., The fluid dynamic design of advanced centrifugal compressors. In *Radial Flow Turbomachinery,* Fluid Dynamics Institute, Hanover, New Hampshire, USA, 1976, pp. 45–9.
7. Benson, R. S., *Thermodynamics and Gas Dynamics of Internal Combustion Engines,* vol. I. Oxford University Press, 1982.
8. Dale, A. & Watson, N., Vaneless radial turbocharger turbine performance. Instn Mech. Engrs Conf. on 'Turbocharging and Turbochargers, London, 1986, pp. 65–76.
9. Kastner, L. & Bhinder, F. S., A method for predicting the performance of a centripetal gas turbine. ASME 75-GT-65, 1975.
10. Hussain, M., Ilyas, M. & Bhinder, F. S., A contribution to designing a nozzleless volute casing for the inward flow radial gas turbine. *I. Mech. E Conference, Turbochargers and Turbocharging, London, 1982.* Paper C35/82, 1982.
11. Mizumachi, N. & Ariga, I., Radial inflow turbine. In *Sawyer's Gas Turbine Engineering Handbook*; vol. I, *Theory and Design.* Turbomachinery International Publ, Co., Connecticut, USA, 1985.
12. Wallace, F. J., Theoretical assessment of the performance characteristics of inward radial flow turbines. *Proc. Instn Mech. Engrs,* **172** (33) (1958) 931–52.
13. Benson, R. S., Prediction of performance of radial gas turbines in automotive turbochargers. ASME paper 71-GT-66, Houston, 1971.
14. Rohlik, H. E. & Kovskey, M. G., Recent radial turbine research at the NASA Lewis Research Center. ASME paper 72-GT-42, 1972.
15. Futral, S. M. & Wasserbauer, C. A., Off-design performance prediction with experimental verification for a radial-inflow turbine. NASA Technical Note, TN D-2621, 1965.
16. Wallace, F. J., Cave, P. R. & Miles, J., Performance on inward radial flow turbines under steady flow conditions with special reference to high pressure ratios and partial admission. *Proc. Instn Mech. Engrs,* **184**(1) (1969–70).
17. Wallace, F. J., Adgey, J. M. & Blair, G. P., Performance of inward radial flow turbines under non-steady flow conditions. *Proc. Instn Mech. Engrs,* **184**(1) (1969–70) 183–95.
18. Wallace, F. J. & Miles, J., Performance of inward radial flow turbines under unsteady flow conditions with full and partial conditions. *Proc. Instn Mech. Engrs,* **185**(77/71) (1970–71) 1091–105.
19. Mizumachi, N., Yoshiki, H. & Endoh, T., A study on performance of radial turbine under unsteady flow conditions. *Rep. Inst. Indust. Sci., Univ. Tokyo,* **28**(1) (1979) 1–77.
20. Katsanis, T., Use of arbitrary quasi-orthogonals for calculating flow distribution in the meridional plane of a turbomachine. NASA Technical Note TN D-2546, 1964.
21. Benson, R. S., Cartwright, W. G. & Woollatt, G., Calculations of the flow distribution within a radial turbine rotor. *Proc. Instn Mech. Engrs,* **184**(3G) (1969–70) 17–22.

22. Baazari. Z., Control systems for turbocharged petrol engines. MSc thesis, University of London, 1974.
23. Wu, C. H. & Brown, C. A., Method of analysis for compressible flow past arbitrary turbomachine blades on general surfaces of revolution. NASA TN 2407, 1951.
24. Katsanis, T. & McNally, W. D., Revised FORTRAN program for calculating velocities and streamlines on the hub-shroud mid-channel stream surface of an axial, radial or mixed flow turbomachine or annular duct—II. Programmers' Manual, NASA Technical Note D-8431, July 1977.
25. Bosman, C. & Highton, J., The computation of 3-dimensional, viscous, compressible flow. Proc. 1st. Int. Conf. on Numerical Methods in Laminar and Turbulent Flow, University of Swansea, March 1978.
26. Moore, J. & Moore, J. G., Calculation of three-dimensional viscous flow and wake development in a centrifugal compressor. In *Performance Prediction of Centrifugal Pumps and Compressors.* ASME Publication, New York, March 1980.
27. Wallitt, L., Numerical analysis of the three-dimensional viscous flow field in a centrifugal compressor. AGARD Conference Proceedings, May 1980.
28. Dawes, W. N., Application of full Navier–Stokes solvers to turbomachinery flow problems. In *Numerical Techniques for Viscous Flow Calculations in Turbomachinery Blading,* VKI Lecture Series 2, January 1986.
29. Eckardt, D., Detailed flow investigations within a high speed centrifugal compressor impeller. ASME Gas Turbine and Fluids Engineering Conference, New Orleans, 1976. SAE paper 76-FE-13.
30. Krain, H., A study on centrifugal impeller and diffuser flow. ASME paper 81-GT-9, 1981.
31. Eckardt, D., Flow-field analysis of radial and backswept centrifugal compressor impellers: Part I, Flow measurements using a laser velocimeter. In *Performance Prediction of Centrifugal Pumps and Compressors.* ASME, New York, March 1980.
32. Dean, R. C. & Young, L. R., The fluid dynamic design of advanced centrifugal compressors. Fluid Dynamics Institute, Dartmouth College, Hanover, USA, 1976.
33. Sovran, G. & Klomp, E. D., Experimentally determined optimum geometries and rectilinear diffusers with rectangular, conical or annular cross-section. In *Fluid Mechanics of Internal Flow.* Elsevier, Amsterdam, 1967.
34. Runstadler, P. W. & Dolan, F. X., *Further Data on the Pressure Recovery Performance of Straight Channel, Plain Divergence Diffusers at High Subsonic Mach numbers.* ASME Publication 73-FE-5, New York, 1973.
35. Janser, W. & Rautenberg, M., Design and investigations of a 3-D twisted diffuser for centrifugal compressors. ASME paper 82-GT-102; 27th Int. Gas Turbine Conf., London, April 1982.
36. Fisher, F. B., Developments of varied diffuser compressors for heavy duty diesel turbochargers. Paper C108/86, Instn Mech. Engrs Conf. on Turbocharging and Turbochargers, London, 1986.

37. Senoo, Y. & Kinoshita, Y., Influence of inlet flow conditions and geometries of centrifugal vaneless diffusers on critical flow angles for reverse flow. *Trans. ASME* **99** (March 1977).
38. Mizumachi, N., A study of radial gas turbines. *Rep. Inst. Indust. Sci. Univ. Tokyo,* **8**(1) (December 1958).
39. Mizumachi, N., A study of radial turbine nozzles at high expansion ratio. *J. Inst. Ind. Sci., Univ. Tokyo,* **19**(6) (June 1967).
40. Benson, R. S. & Scrimshaw, K. H., An experimental investigation of non-steady flow in a radial gas turbine. *Proc. Instn Mech. Engrs,* **180** (Part 3j) (1965–6), 74–85.
41. Lymberopoulos, N., Baines, N. C. & Watson, N., Flow in single and twin entry radial turbine volute. ASME paper 88-GT-59, 1988.
42. Chapple, P. M., Flynn, P. F. & Mulloy, J. M., Aerodynamic design of fixed and variable geometry nozzleless turbine casings. *J. Engng Power, Trans. ASME,* **102,** 141–7.
43. Scrimshaw, K. H. & Williams, T. J., Size effects in small radial turbines. ASME paper 84-GT-215, 1984.
44. Pischinger, F. & Wunsche, A., The characteristic behaviour of radial turbines and its influence on the turbocharging process. Proc. CIMAC, 1977.
45. Horlock, J. H. & Winterbone, D. E., *The Thermodynamics and Gas Dynamics of Internal Combustion Engines,* vol. II. Oxford University Press, 1986.
46. Wallace, F. J. & Blair, G. P., The pulsating flow performance of inward flow radial turbines. ASME paper 65-GTP-21, 1965.
47. Miyashita, T., Tomita, T. & Ishihara, D., Performance of inward radial flow turbine under unsteady flow conditions. *IHI Engng Rev.,* No. 1, Ishikawajima Harima Heavy Industries (January 1974).
48. Leroy, J. L. & Grosshans, G., New developments of turbocompound diesels. Paper C123/86. Instn Mech. Engrs Conf. on Turbocharging and Turbochargers, London, 1986.
49. Wallace, F. J., Tarabad, M. & Howard, D., The differential compound engine. SAE paper 670110, 1967.
50. Wilson, D. E., The design of a low specific fuel consumption turbocompound engine. SAE paper 860072, 1986.
51. Hoehne, J. & Kramer, S., Vehicle testing of Cummins turbocompound diesel engine. SAE paper 810073, 1981.
52. Tennant, D. H. C. & Walsham, B. E., The turbocompound diesel engine. SAE paper 890647, 1989.
53. Satih, H., Miyauchi, J. & Nakazawa, N., Development of variable geometry turbocharger for trucks and buses. GTSJ 83-Tokyo-IGTC-77.
54. McCutcheon, A. R. S. & Brown, M. R., Evaluation of a variable geometry turbocharger turbine on a commercial diesel engine. Paper C104/86. Instn Mech. Eng. Conf. on Turbocharging and Turbochargers, London, 1986.
55. Yokota, K., Hattori, H., Shimizu, M. & Furukawa, H., A high BMEP diesel engine with variable geometry turbocharger. Paper C119/86. Instn Mech. Engrs Conf. on Turbocharging and Turbochargers, London, 1986.

56. Berenyi, S. G. & Raffa, C. J., Variable area turbocharger for high output diesel engines. SAE paper 790064, 1979.
57. Franklin, P. C., Performance development of the Holset variable geometry turbine. SAE paper 890646.
58. Lasker, M. F. & Byrne, J. L., Experience with ceramic rotor turbochargers. Paper C106/86. Instn Mech. Engrs Conf. on Turbocharging and Turbochargers, London, 1986.

15

Atmospheric Pollution

J. H. Weaving

ERA Ltd, Dunstable, UK

& J-P. Pouille

Renault, Rueil Malmaison Cedex, France

1. The situation
2. Pollutant damage
3. Modelling of air pollution
   - 3.1. Description of urban model
   - 3.2. Chemical kinetic mechanism
   - 3.3. Input data
   - 3.4. Application of the model
   - 3.5. Regional transport model: PHOXA
   - 3.6. Description of the model
   - 3.7. Results for regional model
   - 3.8. Discussion and future research
4. Toxicity of exhaust gases
   - 4.1. Lead toxicity
   - 4.2. Carbon monoxide
   - 4.3. Oxides of nitrogen
   - 4.4. Ozone
   - 4.5. Particulate emissions
5. Atmospheric pollution regulations
   - 5.1. Passenger car regulations including light trucks: Europe
   - 5.2. Car/light truck emission standards: USA
   - 5.3. Car/light truck emission standards: Japan
   - 5.4. Diesel-engined vehicle regulations
   - 5.5. The measurement of emissions: petrol engines
   - 5.6. The measurements of pollutants: diesel engines
6. The spark-ignition engine: emissions
   - 6.1. Hydrocarbons
   - 6.2. Nitric oxide
   - 6.3. Carbon monoxide
   - 6.4. Emission control technologies: spark-ignition engines

6.5. The lean-burn engine approach
6.6. Exhaust gas recirculation
6.7. Ignition retardation

7. Diesel engine emissions
7.1 Comparison between diesel and spark-ignition engine exhaust emissions
7.2. Basic mechanisms of pollutant formation in diesel engines
7.3. Emissions from IDI diesel engines
7.4. Emissions from DI diesel engines
7.5. Fuel effects on diesel IDI exhaust emissions
7.6. Diesel particulate filters

1. THE SITUATION

The internal combustion engine does pollute the atmosphere, this mainly through its exhaust products. The major pollutants from SI engines are carbon monoxide, oxides of nitrogen and hydrocarbons. The diesel engine, generally a lower pollutant generator (see Section 7.1), also puts out a small amount of sulphur dioxide from the sulphur content of the fuel but, more importantly, from a regulation standpoint, it puts out particulate matter and smoke. A further pollutant, although non-toxic, is carbon dioxide, the main product of combustion, which contributes to the well known 'greenhouse' effect.

The general situation in the UK (1985) is shown in Fig. 1(a)–(d). In the USA, where atmospheric pollution effects were first appreciated, the situation is similar but also attended with particular blackspot problems. The worst of the blackspots is Los Angeles, where, due to temperature inversion, the topography and meteorology frequently form a basin in which complex chemical reactions take place in sunlight to form 'smog'. This 'smog' is an eye irritant and may be carcinogenic. Smog is termed a secondary pollutant and it is associated with ozone and nitrogen dioxide.

There is, therefore, a *prima facie* case for the severe regulations that have been introduced in the USA and followed in a modified form in Europe and Japan. Nevertheless, examination makes it clear that regulations have preceded research into the optimum requirements, thereby consuming some 5–10% more of the limited supply of fuel than is necessary and also, as a result of this, generated more carbon dioxide. Clearly scientific research is vital and some has been performed, which will be reviewed in the following sections.

2. POLLUTANT DAMAGE

It is first necessary to identify the target. From a health standpoint carbon monoxide, which is lethal in large doses, can be tolerated in the proportions found in cities by healthy people. However, those with lung or heart deficiencies suffer from the lack of oxygen fed to their bloodstream due to the carboxyhaemoglobin occupying oxygen sites. Smokers are particularly vulnerable as their blood is already heavily polluted. With CO therefore we have an urban problem and the target area will be at low level. NO and HC from unburnt fuel are only mildly harmful, but are important components in the production of the secondary pollutants, ozone and nitrogen dioxide. These secondary pollutants will be produced at a much higher level in the urban situation and will form immediately above, or immediately downwind of, towns such as Los Angeles, and will form a flume which may travel miles with little diffusion and dilution and could well be harmful to vegetation and even health if diffused to lower levels.

It will be apparent from the discussion above that it is essential to define air quality standards (AQS): this has been done in the USA and Europe and they are summarized in Table 1. It is clearly very difficult to decide objectively what is clean air, and such standards should be reviewed with increasing medical knowledge and research.

3. MODELLING OF AIR POLLUTION

Having decided on the standard for any pollutant which it is wished to control, it is necessary to link the sources of pollution with the AQS. This may be done with a mathematical model coupled with extensive monitoring points for pollutant measurements. It is this correlation that makes a mathematical model so useful. If the model, after validation, proves reasonable, an overall picture of pollution may be obtained, the effect of reducing the sources of emission may be evaluated and thereby regulations may be based on firmer foundations.

The above reasoning pinpoints the need for an urban model to ascertain the pollution situation in the various parts of major cities. The urban model would then itself be used as a source for a regional model. Needless to say, such an exercise is very expensive and the extent of the resolution has to be determined by the money available.

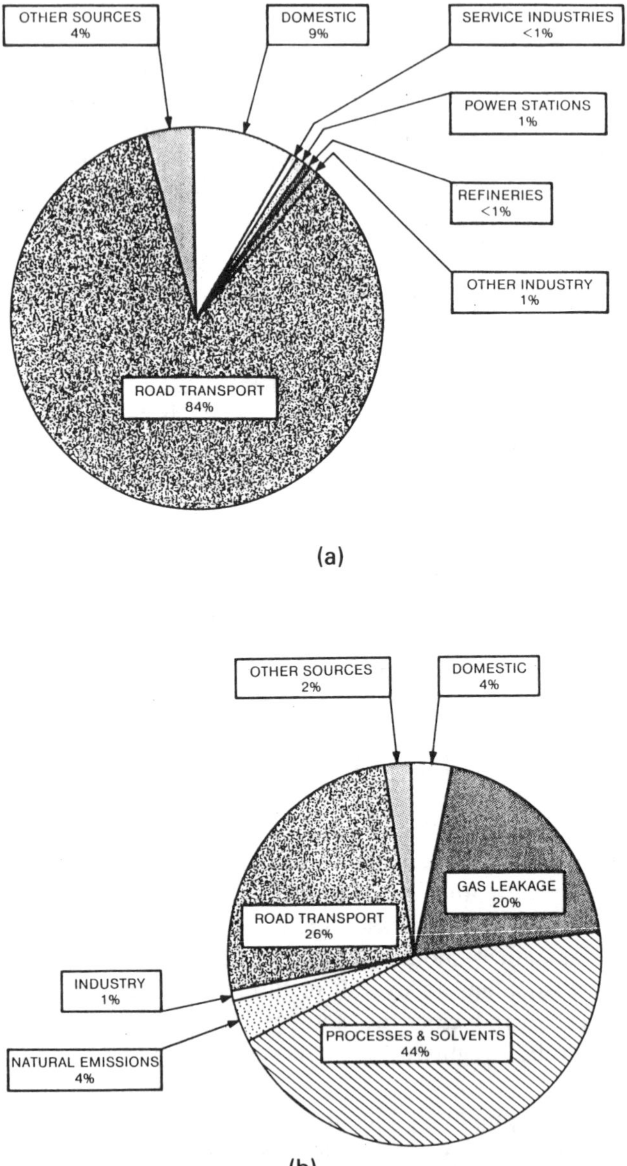

Fig. 1. (a) UK carbon monoxide emissions 1985 (total = 5·39 million tonnes). (b) UK volatile organic compound emissions 1985 (total 2·06 million tonnes).

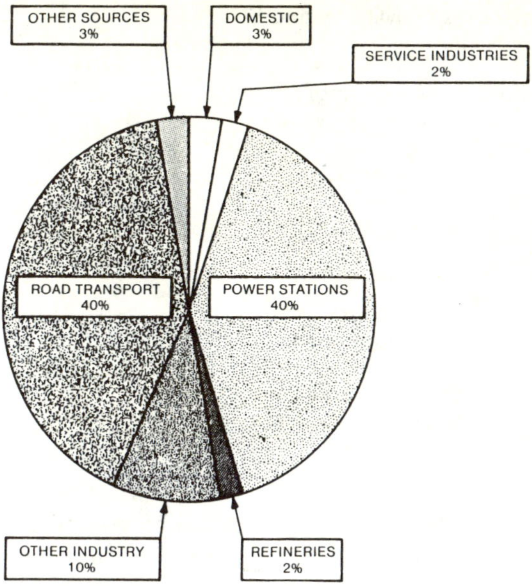

(c)

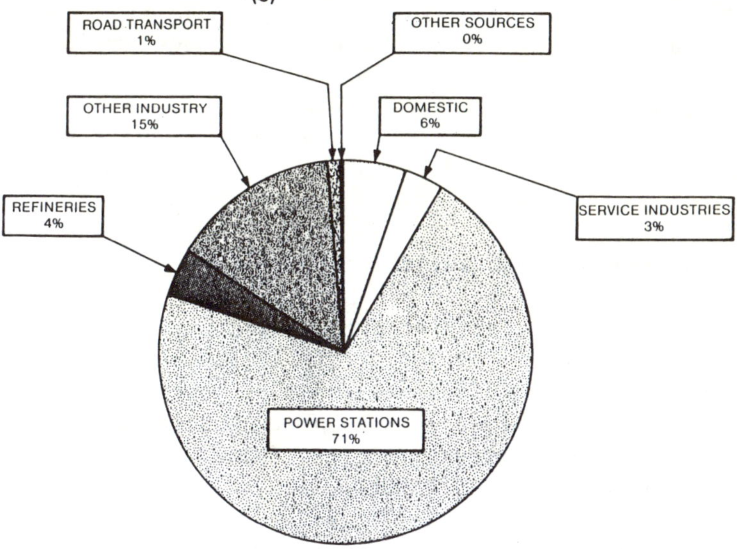

(d)

Fig. 1—*contd.* (c) UK nitrogen oxides emissions 1985 (total 1·84 million tonnes. (d) UK sulphur dioxide emissions 1985 (total 3·58 million tonnes).

Table 1. Air Quality Standards: CO, NO_2, O_3 and SO_2 for USA

Pollutant	Concentration ppm	Averaging/Time Frequency
CO	35	1 hour
	8·6	8 hour
NO_2	0·05	Annual mean
O_3	0·12	1 hour
SO_2	0·14	24 hour
	0·03	Annual mean

NO_2 is limited in Europe at 200 $\mu g/m^3$ and Lead at 2 $\mu g/m^3$ (Lead in the USA is limited to 1·5 $\mu g/m^3$).

There are several possible types of model, which can be divided into two broad categories, grid models and trajectory models. The former lays a grid over the projected area and calculates the pollution in each zone, while the latter follows a 'parcel' of air as it is blown by the wind through the target area. These modelling techniques constitute a subject in its own right and only a broad outline will be given here: the references should be used to follow up this very important area of research, which should give guidance on the optimum pollution control methodology.

3.1. Description of Urban Model

The mathematical model developed by Systems Applications Inc.[1] is of the grid or eulerian type. It gives a mathematical representation of the transport and chemical reactions over the city to which it is applied. It consists of a set of non-linear coupled partial differential equations which express the conservation of mass of each pollutant.

The computer program, which embodies the model, obtains a finite difference solution of the equations by the method of fractional steps. The output is the spatial and temporal variation of the ground cell concentration of each primary and each secondary pollutant. The inputs to the model are numerical representations of the meteorological and emission conditions as a function of the time of day. The initial concentration conditions and the inflow boundary conditions also have to be specified.

For the computation, the plan of the city is divided by a square grid (600 m). The mixing depth is an input and the atmosphere may be divided vertically into as many as six layers, giving a grid of cells (Fig. 2). Within each cell the conditions are assumed to be uniform. Time is also divided into small steps. The program simulates the behaviour of the atmospheric reactions by simulating the chemical and transport process in each grid cell for each time step. This results in an implicit time and space averaging of the concentrations of the pollutants.

The simulation program predicts the ground cell concentration (hourly averaged) of each pollutant in each square as a function of time. It also predicts the vertical concentration profile at each monitor station. The simulation package consists of four main programs:

(1) The atmospheric pollution simulation (APSP), which performs the task described above.
(2) The emission data preparation program, which takes the emission inventory data and prepares the emissions data file, from which the APSP takes the input.

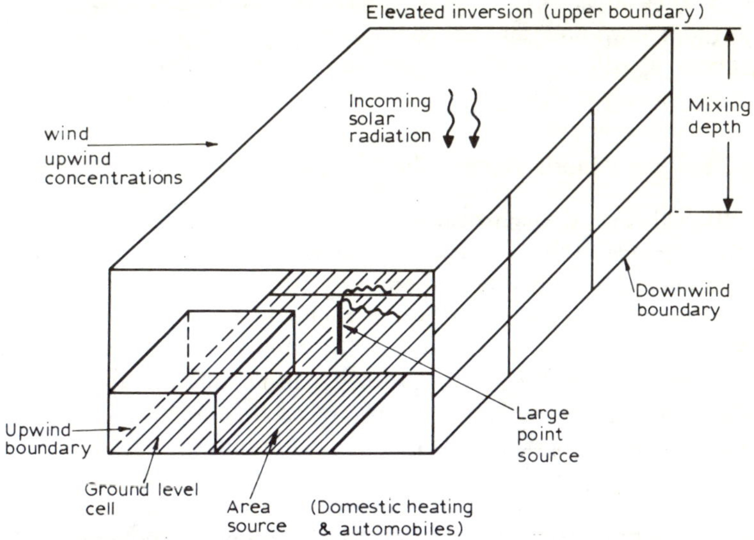

Fig. 2. Schematic diagram of the grid-based model showing its main components.

Table 2. The Simulation Package

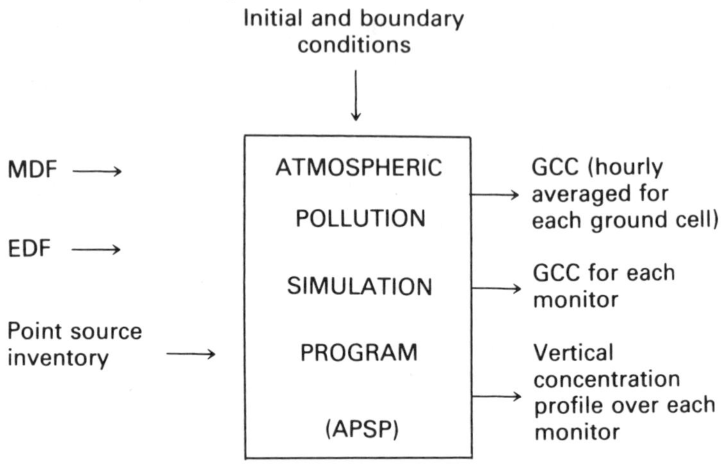

EDF, Emissions Data File; MDF, Meteorology Data File; GCC, Ground Cell Concentration.

(3) The data plotting program, which plots the results of the simulation together with the monitor station data.
(4) The meteorology data preparation program, which takes the meteorological data and prepares the meteorology data file.

3.2. Chemical Kinetic Mechanism

The original kinetic mechanism consists of 15 reactions involving 10 chemical species (Table 3). The mathematical representation is based on four differential equations to permit change of concentration with time of NO, NO_2, O_3 and hydrocarbons (HC).

The working equation used in the model for describing the conservation of species i is:

$$\frac{\partial\langle C_i\rangle}{\partial t}+\frac{\partial}{\partial x}(\bar{u}\langle C_i\rangle)+\frac{\partial}{\partial y}(\bar{v}\langle C_i\rangle)+\frac{\partial}{\partial z}(\bar{w}\langle C_i\rangle)=\frac{\partial}{\partial x}\left(K_h\frac{\partial\langle C_i\rangle}{\partial x}\right)+\frac{\partial}{\partial y}\left(K_h\frac{\partial\langle C_i\rangle}{\partial y}\right)+\frac{\partial}{\partial z}\left(K_v\frac{\partial\langle C_i\rangle}{\partial z}\right)+R_i(\langle C_i\rangle,\ldots,\langle C_N\rangle,T)+S_i(x,y,z,t),\ i=1,\ldots,N$$

Table 3. The Generalized Photochemical Kinetic Mechanism

$NO_2 + h\nu \rightarrow NO + O$	$HO_2 + NO \rightarrow OH + NO_2$
$O + O_2 + M \rightarrow O_3 + M$	$HO_2 + NO_2 \rightarrow HNO_2 + O_2$
$O_3 + NO \rightarrow NO_2 + O_2$	$HC + O \rightarrow aRO_2$
$O_3 + NO_2 \rightarrow NO_3 + O_2$	$HC + OH \rightarrow \beta RO_2$
$NO_3 + NO_2 \xrightarrow[H_2O]{} 2HNO_3$	$HC + O_3 \rightarrow \gamma RO_2$
$NO + NO_2 \xrightarrow[H_2O]{} 2HNO_2$	$RO_2 + NO \rightarrow NO_2 + \varepsilon OH$
$HNO_2 + h\nu \rightarrow OH + NO$	$RO_2 + NO_2 \rightarrow PAN^a$
$CO + OH \xrightarrow[O_2]{} CO_2 + HO_2$	

[a] Peroxyacetyl nitrate.

where $\langle C_i \rangle$ is the mean concentration for species i; $\bar{u}$, $\bar{v}$, $\bar{w}$ are the components for the mean wind vector; R_i is the rate of formation of species i; S_i is the rate of emission of species i; x, y, z are the co-ordinate axes; t is time; and K_h, K_v are the horizontal and vertical eddy diffusivities respectively.

The resultant N partial differential equations are solved over the urban area defined by the region:

$$X_w \leq x \leq X_e, \qquad Y_s \leq y \leq Y, \qquad H(x, y) \leq z \leq H(x, y, t)$$

where X_w, Y_e, Y_s and Y_n are the west, east, south and north boundaries of the airshed; $h(x, y)$ is the ground elevation above sea level; and $H(x, y, t)$ is the elevation above sea level of the assumed upper limit for vertical mixing or transport.

The equations are solved by the finite difference method of fractional steps.[1]

In this model hydrocarbons were divided into two classes, high reactivity and low reactivity. This involved three more equations using different rate constants appropriate to the reactions.

In an updated version of the SAI model greater care has been taken with the chemical reactions. This has been termed the 'carbon-bond mechanism'.[2]

Due to the very large number of hydrocarbons in petrol the number of possible reactions are too large to be practical, and so simplification is necessary. In the SAI model the hydrocarbons from a typical exhaust gas analysis, of the exhaust from a vehicle running on an urban cycle, are analysed into components. The concentrations of the hydrocarbons are then divided into chemical groups and an appropriate rate constant assigned. The groups chosen are single bonded

Table 4. The Carbon-bond Mechanism

$NO_2 + h\nu \rightarrow NO + O$	$OLE + O_3 \xrightarrow{O_2} (HC(O)O_2 + HCHO + OH)$
$O + O_2(+M) \rightarrow O_3(+M)$	$PAR + OH \xrightarrow{O_2} CH_3O_2 + H_2O$
$O_3 + NO \rightarrow NO_2 + O_2$	$PAR + O \xrightarrow{O_2} CH_3O_2 + OH$
$O + NO_2 \rightarrow NO + O_2$	$CAR + OH \xrightarrow{O_2} HC(O)O_2 + H_2O$
$O_3 + NO_2 \rightarrow NO_3 + O_2$	$CAR + h\nu \xrightarrow{2O_2} aHC(O)O_2 + aHO_2 + (1-a)CO$
$NO_3 + NO \rightarrow NO_2 + NO_2$	$ARO + OH \xrightarrow{O_2} CAR + CH_3O_2$
$NO_3 + NO_2 + H_2O \rightarrow 2HNO_3$	$ARO + O \xrightarrow{2O_2} HC(O)O_2 + CH_3O_2$
$HO_2 + NO_2 \rightarrow HNO_2$	$ARO + O_3 \xrightarrow{O_2} HC(O)O_2 + CAR + OH$
$NO_2 + OH \rightarrow HNO_3$	$ARO + NO_3 \rightarrow$ products (aerosols)
$HNO_2 + h\nu \rightarrow NO + OH$	$CH_3O_2 + NO \rightarrow NO_2 + CAR + HO_2$
$NO + OH \rightarrow HNO_2$	$HC(O)O_2 + NO \rightarrow NO_2 + CO_2 + HO_2$
$CO + OH \xrightarrow{O_2} CO_2 + HO_2$	$HC(O)O_2 + NO_2 \rightarrow PAN$
$HO_2 + NO \rightarrow OH + NO_2$	$CH_3O_2 + HO_2 \rightarrow CH_3OOH + O_2$
$HO_2 + HO_2 \rightarrow H_2O_2 + O_2$	$HC(O)O_2 + HO_2 \rightarrow HC(O)OOH + O_2$
$PAN\ HC(O)O_2 + NO_2$	
$H_2O_2 + h\nu \rightarrow OH + OH$	
$OLE + OH \xrightarrow{O_2} CAR + CH_3O_2$	

carbon atoms (paraffins, PAR), fast or relatively reactive double bonds (olefins, OLE), slow double bonds (aromatics, ARO) and finally carbonyl groups. Some 30 chemical reactions are considered in Table 4. An alternative method of handling hydrocarbons has been used by Derwent,[3] in which some 16 hydrocarbons which had the highest concentrations were selected and the appropriate kinetic mechanism for each adopted.

3.3. Input Data

The model requires comprehensive input data, including emissions from stationary point sources, vehicle emissions and domestic emissions which are treated as area sources.

The horizontal wind field near the ground is required and the mixing depth has to be measured. It is also important to have adequate data of the initial and boundary conditions. The radiation input has to be measured with a pyrometer.

3.4. Application of the Model

The SAI model has been used in several cities in the USA and also in the city of Turin in Italy.[4] A few results of the latter study are described.

The town was divided into three zones, central, urban and suburban, and a census of vehicles and the driving pattern in each were measured, and hence the emissions. Monitoring for validation and comparison was used from some five monitoring stations and a helicopter. Predictions for carbon monoxide, nitrogen dioxide and ozone are shown in Fig. 3 as it varies throughout the day due to traffic concentrations, radiation intensity and other variables.

Mathematical models are particularly valuable for predicting the effects of legislative proposals and, using the SAI model, the effect of

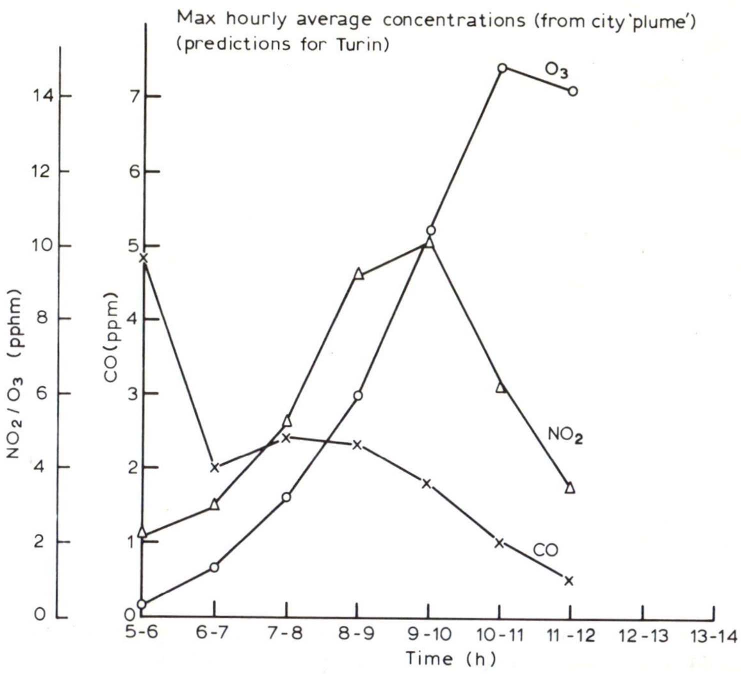

Fig. 3. Mathematical model predictions for ozone, nitrogen dioxide and carbon monoxide.

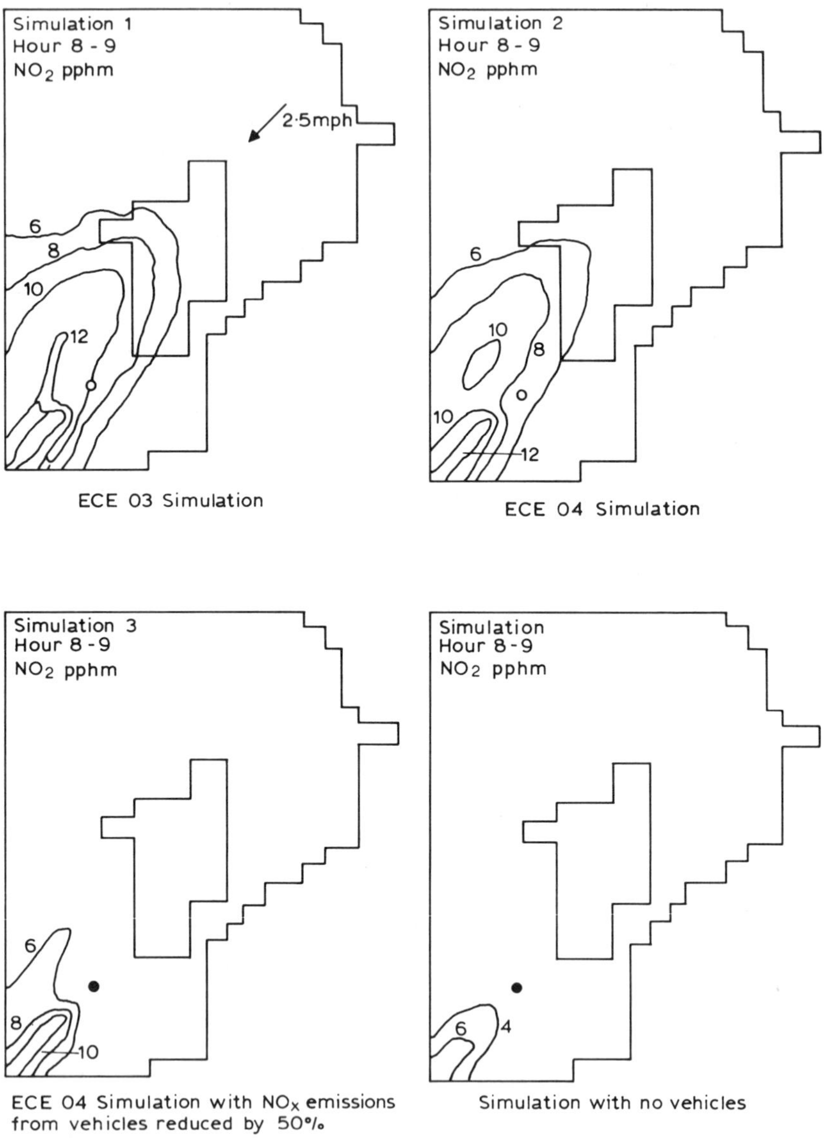

Fig. 4. Prediction of nitrogen dioxide emissions for vehicles tuned to ECE 03, ECE 04; also vehicles reduced by 50% and 'no vehicles'.

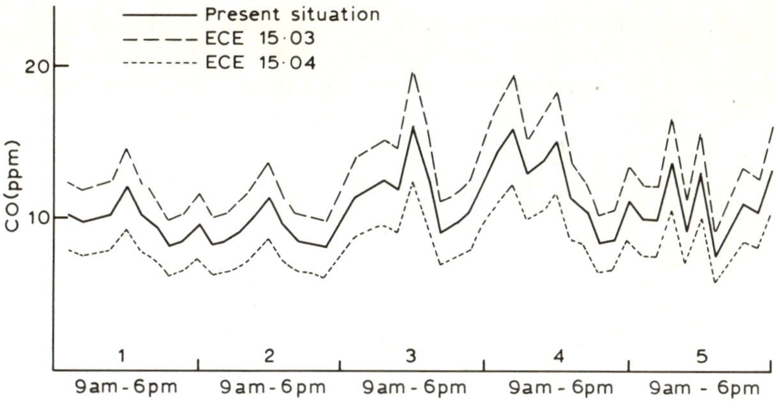

Fig. 5. Canyon Micromodel predictions for carbon monoxide emissions for vehicles tuned to 15·03 ECE Regulations.

the ECE 15.03 Regulations are compared in Fig. 4 with 15.04 for NO_2. It will be seen that there is a modest improvement. Indeed this is not necessarily so with a secondary pollutant, as the ratio of the HC to NO_x is important, and this is not addressed in the later European regulations where HC and NO_x are added together.

Carbon monoxide, a primary pollutant mainly from road transport (Fig. 5), will effectively be higher than that shown in Fig. 4, as these figures are the average in the cell while the 'target' will be at low level. A canyon micro-model was therefore used for such evaluation, and the predictions for five consecutive days, beginning on a Saturday are shown in Fig. 5.

3.5. Regional Transport Model: PHOXA

A 'regional' transport model has been used in Northern Europe, centred on the Netherlands, particularly to assess the effects of ozone and acid deposition (the PHOXA program). The SAI model, in modified form, was used for this work.[5]

3.6. Description of the Model

The PHOXA model is essentially a grid model, but because of the large territory that needed to be covered, uses simpler emission input data than was used for the urban model. As the emphasis was on 'ozone' and 'acid rain', the vehicle input was combined into an area source and considerable attention was given to the point source emissions from power stations for which a flume sub-model (puff-in-grid hybrid approach) was used to feed the grid system.

The SAI model (RTM-III) is based on the time-dependent, reactive species atmospheric diffusion equation for multiple layers, which can be written in a Cartesian co-ordinate system as follows:

$$\frac{\partial \bar{C}_i^j}{\partial t} + u\frac{\partial \bar{C}_i^j}{\partial x} + v\frac{\partial \bar{C}_i^j}{\partial y} = \frac{\partial}{\partial x}\left(K_h \frac{\partial \bar{C}_i^j}{\partial y}\right) + \frac{\partial}{\partial y}\left(K_h \frac{\partial \bar{C}_i^j}{\partial y}\right) + E_i^j + R_i^j + Q_i^j - S_i^j - W_i^j \quad (1)$$

where the subscript i refers to pollutant species as indicated in Table 5 and the superscript j denotes the vertical layer. In the present study, three layers are used, as shown in Fig. 6. A lower layer, defined as the mixed layer, is designated by $j = 1$; and two upper layers, within an elevated inversion layer, are designated by $j = 2$ and $j = 3$. Other symbols have conventional definitions: an overbar denotes the layer average, u and v are the horizontal wind velocities in the x and y directions, respectively, and K_h is the horizontal turbulent diffusivity. The reaction rate, pollutant source, and dry and wet deposition terms are denoted by R_i^j, Q_i^j, S_i^j and W_i^j, respectively. Entrainment/

Table 5. Species Ordering and Definitions Used in the RTM-III Model

Species number	Species name	Species description
1	NOX	Nitrogen oxides ($NO + NO_2$)
2	OX	Oxidants ($O_3 + NO_2$)
3	MGLY	Methyl glyoxal
4	ETH	Ethylene type carbon bonds
5	OLE	Olefinic type carbon bonds
6	PAR	Paraffinic type carbon bonds
7	ALD2	Higher aldehydes
8	FORM	Formaldehyde
9	XYL	Xylene and other multi-alkylbenzenes
10	TOL	Toluene and other mono-alkylbenzenes
11	PAN	Peroxyacetyl nitrate
12	CO	Carbon monoxide
13	PHEN	Aromatic phenols
14	HONO	Nitrous acid
15	SO2	Sulphur dioxide
16	SO4	Sulphates

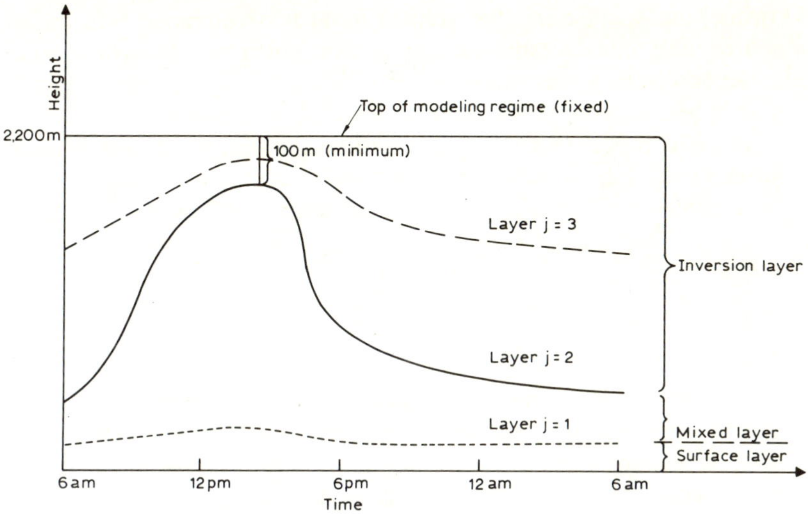

Fig. 6. PHOXA mathematical model (SAI) showing diurnal variation of vertical layers.

detrainment across the layer interface due to large-scale divergence and/or mixed layer variations and vertical diffusion is specified by E_i^j. The source term Q_i^j consists of a gridded array of area sources, the emissions of which are assumed to disperse instantaneously into the appropriate cells in the lower layer, and a set of major point sources, the emissions of which either enter the Gaussian puff sub-module (as a user option), or are dispersed within the appropriate vertical layer, as determined by the plume rise algorithm.

The chemical kinetics simulation package built into RTM-III uses a modified version of the carbon bond mechanism-IV (CBM-IV) to describe changes in the concentrations of various species due to their chemical reactions. The original CBM-IV mechanism (Whitten *et al.*[2]) was developed as a highly condensed mechanism suitable for inclusion in urban and regional air quality simulation models. As implemented in chemistry solution subroutines for such models, the mechanism contained 20 state species (those species that are tracked from one time step to another and whose mass is conserved) and nine steady state species (species whose steady state concentrations are calculated during solution at each time step). Because the computational time required to converge on a chemical solution at each time step is

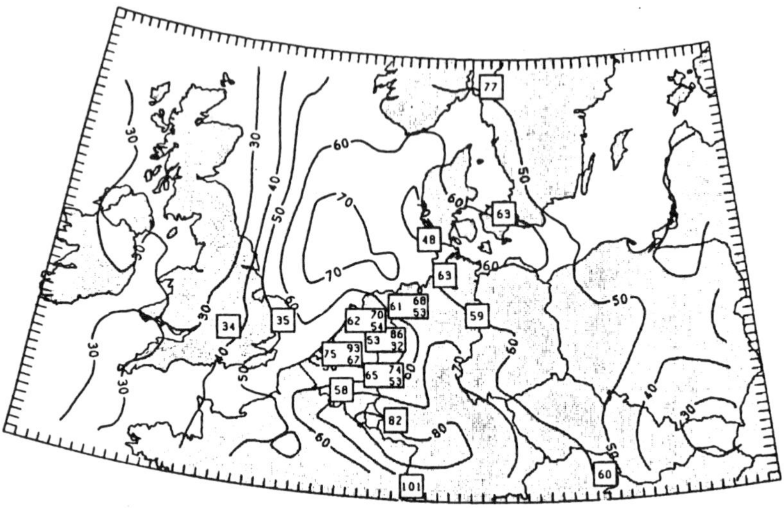

(a) 23 July

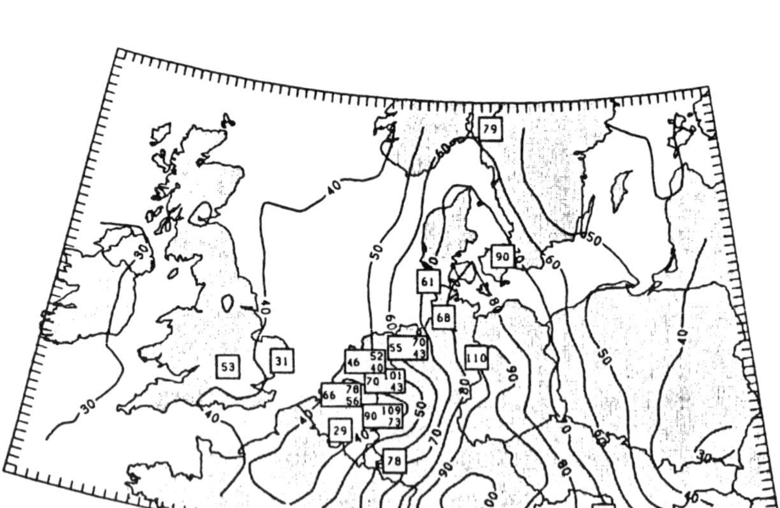

(b) 24 July

Fig. 7. PHOXA spatial maps illustrating predicted and measured ozone concentrations (ppb) at 1500 GMT on 23–26 July 1980.

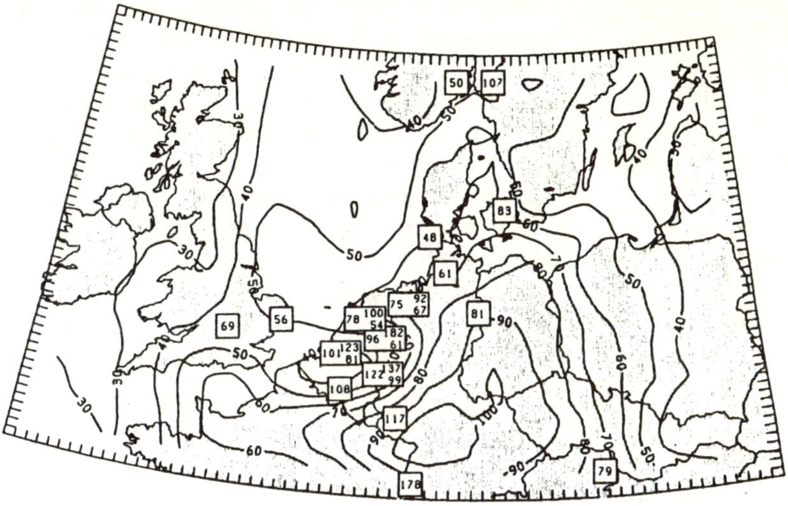

(c) 25 July

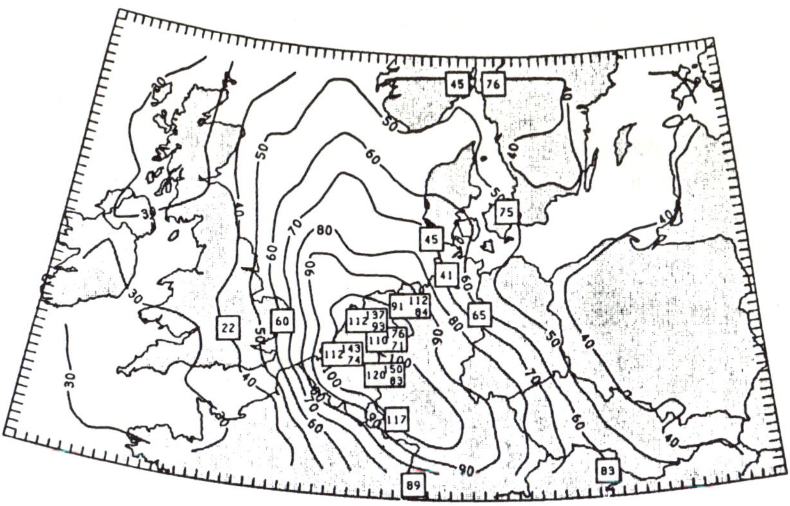

(d) 26 July

Fig. 7—*contd.*

strongly related to the number of state species, further condensation of the CBM-IV was performed to eliminate unimportant state species, thereby alleviating some computational burden in the RTM-III model.

The release of a significant portion of emissions at elevated levels coupled with possible sheared flow aloft (such as the nocturnal jet) calls for special treatment of major point sources. In the sulphate version of the regional transport model (RTM-II) and the current oxidant version (RTM-III), a point-source module embedded in a Gaussian puff framework is used to simulate the near-source transport and dispersion of a portion of the elevated point source emissions. This subgrid-scale treatment is terminated when the dimensions of the puff are commensurate with those of the grid cell, after which the emissions are released into the appropriate grid cells.

3.7. Results for Regional Model

The predicted mixed layer and measured ozone concentrations at 1500 GMT on 23–26 July 1980 are shown in Fig. 7. Results at 1500 GMT are displayed since this is the time when the highest ozone concentrations are predicted. In these figures, the numbers in the boxes are hourly average measured ozone concentrations. Note that some of the boxes also display a range of ozone values (i.e. minimum and maximum) for a cluster of sites within a block of grid cells. This method of displaying the observations was selected to accommodate visually all available measurements within the high-density Dutch ozone monitoring network, without completely obscuring the model prediction isopleths.

3.8. Discussion and Future Research

From a review of the results presented in Fig. 7, it is apparent that the model exhibits some skill in predicting ozone concentrations. Many features of the observed spatial pattern are reproduced by the model.

As indicated earlier, in future work it would be desirable to input the vehicle emissions more accurately so that the effect of existing and proposed regulations could be evaluated. This would require a combination of an urban model with a regional model.

4. TOXICITY OF EXHAUST GASES

Although this section may seem unnecessary for engineers, there are very good reasons for accentuating its importance. First, the engineer must be able to play his part in formulating regulations as much time and vast resources are at stake. Secondly, in designing engines and conducting research into pollution control he should know which pollutants are dangerous.

Probably the most conspicuous example of this is the question of lead in fuel. No additive is as effective in preventing 'knock'. As a result of eliminating it in the USA, principally because it poisons catalysts, octane numbers dropped from 98 RON to 91. This necessitated reducing compression ratios and running at stoichiometric fuel:air ratio with a consequent loss of between 5 and 10% in fuel consumption. There is thus a conflict between a suspected risk of poisoning against a known loss of a limited fossil fuel supply and an increase in the production of carbon dioxide, which is, as is well known, the principal pollutant contributing to the 'greenhouse' effect, with its consequent warming up of the Earth's atmosphere.

In engineering terms this translates into three-way catalyst control versus high compression lean-burn or stratified charge engines.

The motor industry has taken the pollution problem very seriously. In Europe the major toxicity work has been sponsored by the CCMC (Common Market Automobile Constructors) and in addition the various European governments have conducted important work. In England this has been performed principally at the Warren Spring Laboratory (DTI, Department of Trade and Industry) and AERE, Harwell; partly sponsored by the Department of Environment.

A summary of the present position follows, but for any investigation in depth considerable use of the references will be required.

4.1. Lead Toxicity

Tetra-ethyl lead and tetra-methyl lead, together or separately, are the additives used in petrol. In earlier days, before 1981, as much as 0·84 g per litre was permitted, although rarely used. EEC regulations limited this to 0·4 g per litre in 1981 and to 0·15 g per litre by 1985.

It is not easy for the general public to form a reliable view of the toxic nature due to the emotive propaganda on its dangers. Lead is

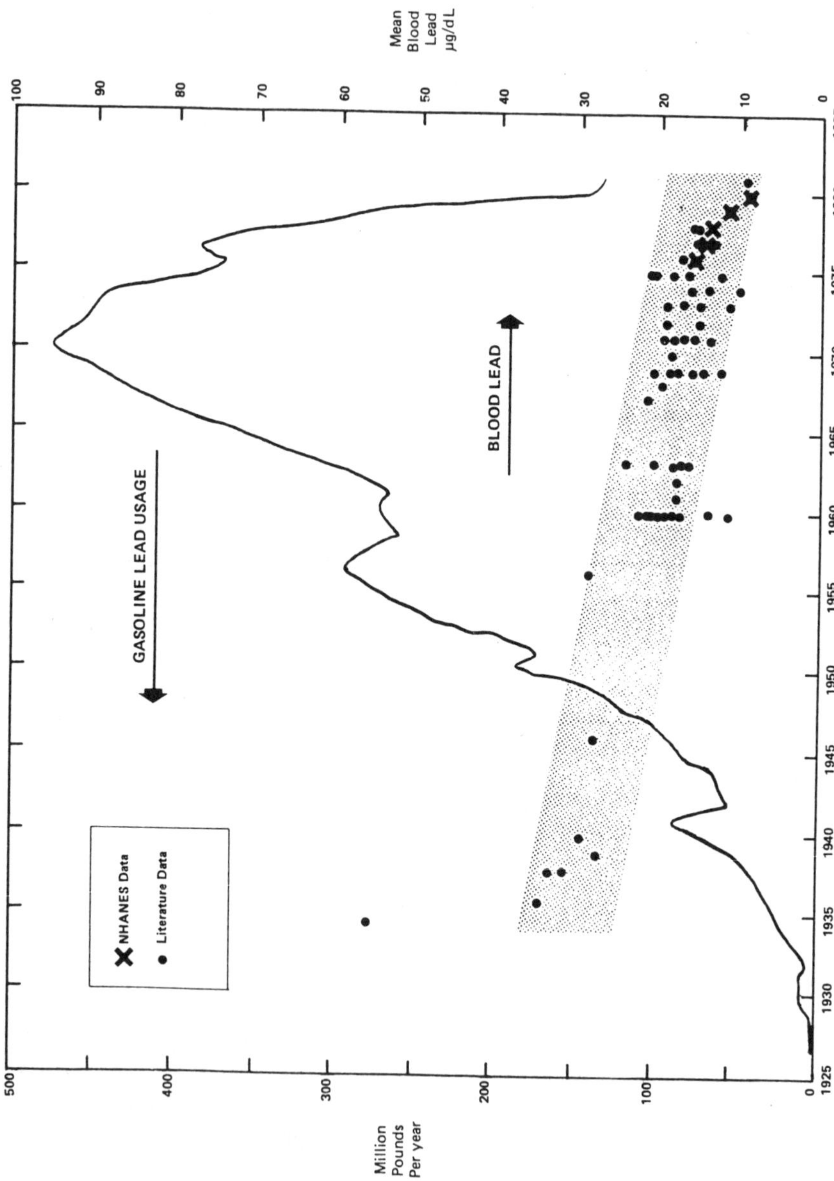

Fig. 8. US blood lead levels, 1930–82.

well known indeed to be toxic, but there will be a minimum concentration which will be harmless. What is this minimum? The criterion for measurement generally used is lead blood level in μg/dl. A study has been carried out in the USA (The Second National Health and Nutrition Examination Survey, NHANES II). From these results and others, a submission was made to the EPA relating the lead in fuel per annum to the blood levels (Fig. 8). It will be noted that although there is a drop over the years 1976–80, the correlation with lead used in petrol is weak, and during this period the NHANES report observed a reduction of 45% in lead in canned food. It is clear therefore that petrol is not the main factor affecting blood lead levels. The Lawther Report (Chairman, Professor Lawther) commissioned by the Department of Health and Social Security stated that the main environmental sources are, in descending order, food, water and air.[6]

A significant investigation by Harwell[7] was to ascertain the uptake of lead aerosols from the environment derived from motor vehicles. For this the lead additive was labelled with ^{203}Pb isotope and human volunteers inhaled diluted exhaust gas containing this ^{203}Pb aerosol from vehicles (which had catalysts to remove carbon monoxide); they also inhaled air from the side of busy motorways. As the lung uptake is very dependent on particle size, the size distribution was very carefully measured. It should be borne in mind that tetramethyl lead and tetraethyl lead, which are very toxic, are partly converted to lead oxide PbO and lead sulphate to $PbSO_4$ in the combustion chamber, and some of these compounds are deposited on the walls of the combustion chamber and are emitted by flaking. Most of these particles are too heavy to be airborne. As noted above, Harwell measured airborne lead particle size carefully. To quote from this report:

> The particle size, concentration and dispersion of ambient lead aerosols were assessed at a variety of roadside sites to determine the typical nature of airborne lead. The results indicate that lead emitted from cruising vehicles, as on a motorway, is mainly as very small discrete particles of about 0·03 μm diameter. Under urban driving conditions, with vehicles accelerating strongly, and when the aerosol had 'aged' as in car parks or in the countryside, a wider spectrum of particle size was found, with some lead attached to agglomerate particles of 0·2–2 μm diameter. Contrary to reports from the USA, very little lead was associated with particles greater than 5 μm.
>
> Correlation of measurements by various authors shows that roadside

concentrations are usually about 2 μg/m^3 for each 1000 petrol-driver vehicles per hour, but are higher on gradients and in poorly ventilated situations.

Lead concentrations decline rapidly with distance from a road. The main factor is upward dispersion of the plume rather than deposition of the lead; only about 10% of the airborne lead seems to be deposited within 100 m.

Dispersion calculations relating traffic flow to the concentration of airborne lead at various distances from a road suggest that, averaged over 24 hours, for 1000 Vh^{-1} there is a zone about 15 m wide on either side of the road in which the concentration exceeds 1 μg/m^3. For a very busy motorway carrying 4000 Vh^{-1} the equivalent zone is 100 m wide. Concentrations in urban houses are about 60% of those outside.

The researchers established the important relationship between the elevation of blood lead to a 24 h exposure to air lead as

$$\frac{\text{Elevation of blood lead } \mu\text{g } 100 \text{ m}}{24 \text{ h exposure to air lead } \mu\text{g/m}^3} = 2$$

From this work they estimate the proportion of lead uptake. To quote, 'lead uptake for the majority of urban and rural residents for whom airborne lead averages about 0·5 μg/m^3 indicates that the contribution of air lead to the total lead uptake is less than 10%'.

The researchers showed that all the lead in the blood was eventually excreted in the urine and faeces. A tiny part finds its way into the bones and is retained for a long time being only slowly discharged.

The emotive issue, which is still hotly argued, is whether a low level of blood lead can influence the intelligence of children. This concern seemed to be supported by a study by Needleman *et al.*[8] However considerable further work has since been performed which does not bear this out.

Such an investigation carried out at the University of Birmingham 1988 by Harvey *et al.*[9] was conducted on a random sample of 110 boys and 91 girls. Blood samples were taken from each child and they were subjected to some 10 different modes of test including IQ, reaction times and behaviour assessment by their school teachers. In discussion of the results they stated:

> Although these tests are extremely difficult both to perform and evaluate, it would appear that there is no good evidence that, with blood lead levels below 30 μg/dl, there is any impairment of intelligence.

In all health safety standards it is clearly desirable to err on the safe side, and it is the opinion of the author of this section, in company with the majority who have studied the evidence, that lead levels in fuel should be reduced to the level recommended by the EEC, namely 0·15 g per litre. This allows a 3 octane number improvement of petrol quality and helps refineries to save processing energy. A higher value would almost certainly be safe and would further conserve fuel and save more expensive fuel refining. When medical research throws up sufficiently reliable data it would seem logical that the situation should be reassessed.

4.2. Carbon Monoxide

Although this is a highly toxic gas in itself, at the low concentrations in the urban environment there is no danger to healthy people. However, smokers with pulmonary deficiencies who are exposed to the worst pollution areas are in increased danger. Additionally, because the affinity of haemoglobin in the blood for CO is many more times greater than its affinity for oxygen, CO hinders the transport of oxygen from the lungs to the rest of the body and, as a consequence, the heart has to pump more to compensate. Those who have weak hearts are put under additional strain. For these reasons air quality standards have been set (Table 1).

4.3. Oxides of Nitrogen

Nitric oxide (NO), the major oxide of nitrogen emitted by both diesel- and petrol-engined vehicles, is not highly poisonous but plays an important part in the complex reactions that take place in the environment, particularly with hydrocarbons, giving rise to ozone and nitrogen dioxide (NO_2) which are both highly toxic. Dose levels of 0·5 ppm for two hours showed little effect on healthy individuals but produced mild symptoms with asthmatic volunteers.[10]

Nitrogen dioxide with water vapour gives nitric acid, which is highly irritative to the respiratory airway system from the throat to the lungs. It also interferes with the oxygen-carrying capacity of the haemoglobin. It has been the subject of AQS (Table 1).

Additionally, with hydrocarbons nitric oxide is a primary pollutant in the formation of photochemical smog and contributes, with sulphur dioxide, to acid rain.[11]

4.4. Ozone

There is a not-inconsiderable background of ozone from natural causes to which the motor vehicle adds its burden through the reactions discussed in Section 3.2. Ozone causes chest discomfort, coughing, inflammation of the respiratory tracts, increased asthma attacks and reduces the pulmonary workings of the lungs.[12]

Ozone in sufficiently high concentration has been shown to cause severe damage to crops and is suspected to be a contributor to forest damage, although no satisfactory theory for forest damage is yet generally agreed.[13] AQS for ozone were established in the USA and were in fact revised when more data became available. Revisions of AQS are indeed most important for obvious reasons but, in fact, are rarely performed.

4.5. Particulate Emissions

Diesel engines are the main contributors to the emission of particles. When in high concentration, these are visible as black smoke. There have been smoke regulations since the early 1970s, introduced largely because of the socially unacceptable nuisance from smell and poor visibility. However, it was soon found that bound to the particles were hydrocarbons which were highly mutagenic and also some that were known carcinogens e.g. benzo(a)pyrene. In view of this, very severe regulations have been introduced in the USA and will soon be in force in Europe.

As with lead, it seems that fears have overtaken scientific investigation. Most of the early evidence considered was of a mutagenic character. Mutagenicity, that is the modification of the DNA, must be distinguished from carcinogenicity which is the production of tumours. The body has many defence mechanisms, which reject mutuated DNA.

4.5.1. Short-term Tests

The advantage of mutagenic tests, such as the Ames test, is that they are short term. They are useful for a first screening to alert one to the possibility of carcinogenic potential but should not be used as evidence for carcinogenicity. They are useful too for a quick comparison between two substances for possible carcinogenic activity. Unfortunately, in the early legislation in the USA, too much reliance was placed on the Ames test in formulating regulations.

The Ames test consists of extracting the hydrocarbons from the particles with dichloromethane (DCM) and subjecting the extract to strains of *Salmonella* bacteria which gave a high mutagenic response.[14] However, it has been shown by Garner *et al.* and further investigations that the ability of other commonly used solvents in this type of work to extract mutagenic material, such solvents as corn oil, gum arabic, bovine serum albumen and new-born calf serum is very limited.[15] This indicates that DCM as solvent is very much more potent than the natural fluids of the human lung. Dr Vostal of General Motors performed careful experiments on the macrophage clearance system present in mammals, using rats, and showed that these were effective in removing diesel particles from the lung.[16]

It is clear that although these short-term tests are useful indicators they cannot give confidence on the vital question as to whether diesel exhaust in the concentration as breathed in the atmosphere causes cancer. Recourse must be made to animal inhalation tests.

4.5.2. Animal Inhalation Tests

The best animal would of course be *Homo sapiens,* but it would be difficult to find volunteers. However, Professor P. Lawther did such an experiment in the very early days.[17] He compared the state of health of mechanics in a diesel-engined bus depot and service garage with a similar facility for the Underground tube trains. In this comparison he found no significant differences for persons examined over a period of 25 years. Professor Lawther concluded that diesel fumes, in the concentration experienced in this garage, were not carcinogenic. Some did not feel that this test was conclusive on the grounds that the engines were not running under typical road conditions. This is a matter of opinion and could only be confirmed by extensive biological analysis of the atmosphere in the depot compared with road load conditions.

A further human test is the detailed examination of smokers, who subject themselves to what could be regarded very loosely as dosing themselves with a very high concentration of 'exhaust gas'. We know that there is little doubt that cigarette smoke is carcinogenic.

In the above consideration we are left still with considerable uncertainty. It would appear that diesel exhaust at 'garage' concentration is not harmful, but probably at very high concentration it may well be.

In view of this uncertainty the European car manufacturers (CCMC) undertook a considerable inhalation investigation using rats and hamsters. At the Battelle Research Centre in Geneva, three engines were selected, a small passenger car IDI diesel engine, and two similar sized petrol engines—one tuned at stoichiometric air: fuel ratio with a three-way catalyst and the other with a normal 'European' tune to meet EEC Regulations.

Three dose levels were selected, the maximum the animals could tolerate without reducing their life-span. For the lowest dose a value

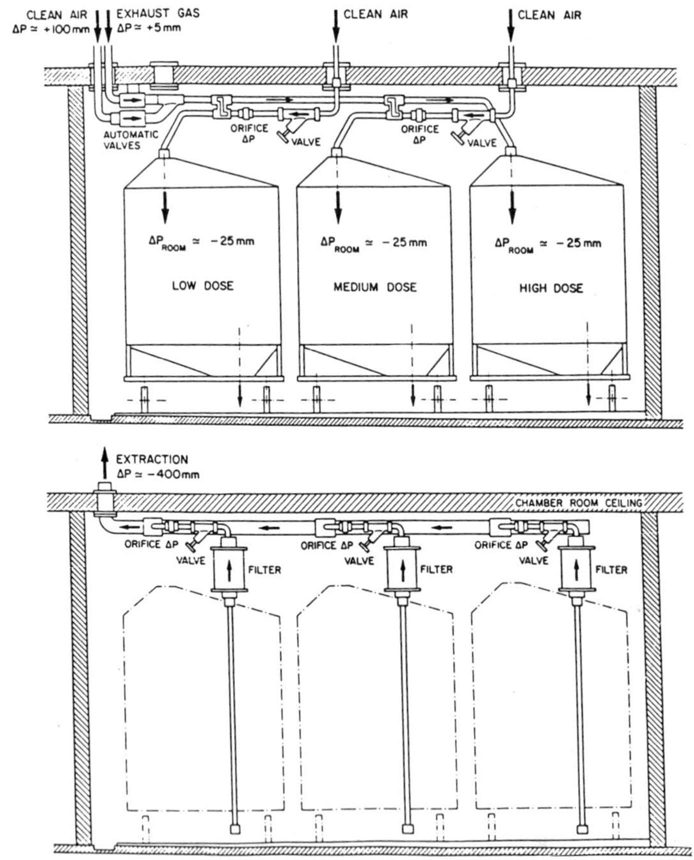

Fig. 9. Schematic diagram of engines, inhalation and pipework layout at the Battelle Laboratory (Battelle, Geneva).

of roughly 10 times urban curbside measurements and an intermediate dose were used in an endeavour to obtain a dose/response curve. A diagrammatic layout of the pipework and inhalation chambers is shown in Fig. 9. The exhaust was first discharged into a buffer tank to remove oscillations and from here diluted from a compressed air supply to the appropriate value. For the diesel engine only, the exhaust was divided and a portion passed through filters to remove the particulates before being conveyed to a group of animals. The engines were run for 16 hours per day for 5 days per week to a scheduled road-load cycle. The rest of the day was left for rest, servicing and feeding the animals. Each animal was kept in a separate stainless steel cage. In each group 144 rats and 208 hamsters were exposed to the various levels of pollution. The exhaust dose levels expressed as a percentage of total air plus exhaust volume for the diesel engine were:

High	8·2%
Medium	2·7%
Low	0·9%

The results of these tests were published[18] and the principal findings from a very detailed pathological examination were that with the hamsters the pollution concentrations produced less than 1% tumours for all the groups, which was not significantly different from the controls.

The rats, when exposed to gasoline exhaust, with catalyst or without, were clear of lung tumours (less than in the control group) as also were the filtered diesel subjects. There were, however, a significant number of tumours in the medium and heavy doses for the unfiltered diesel group. The low dose group had fewer tumours than in the controls.

For the pathological examination we quote from Ref. 18.

> At necropsy a statistically significant increase in lung weight was seen in both rats and hamsters exposed to diesel exhaust compared with controls. This finding was more marked in the rats where the increase was progressive with both duration of exposure and dose level. The increase was greatest at 30 months (after the end of the observation period) in high dose level males where the lung weight was 2·7 times the control and at 24 months in high dose level females (3·9 times control).
>
> The major histopathological finding was an increase in the incidence of primary lung tumours in rats exposed to the medium and high dose levels of diesel exhaust. The incidence of tumour-bearing rats in diesel exhaust exposed males is 3/72 (4%) and 16/71 (23%) for the medium and high

dose levels, respectively. In females the incidence in medium and high dose rats was 11/72 (15%) and 39/72 (54%), respectively. The incidence in control males was 3/140 (2%) and in females 1/142 (1%). The incidence of lung tumours is therefore dose-related and is higher in females than in males. There was no increase in the incidence of primary lung tumours in any other treatment group, including the low dose level of diesel exhaust emissions where the incidence was 1/72 (1%) for males and 0/72 (0%) for females.

It is not possible to transfer the results from rats to humans with certainty but it is, as is well known, the method used with new drugs and has proved to be quite reliable. It is usual with drugs, where a benefit may outweigh a risk, to consider it safe if the dosage is 10 times less than that which is safe for the animal. It is clear that this is the case for these inhalation tests. However, as there is no beneficial bonus for inhaling exhaust gas a further margin is desirable. That there is such a margin is apparent from the results of the medium dose. Clearly the animals were not overwhelmed as 85% did survive. A further safety factor, clear from the pathology, is that with the highly dosed animals the lung defences had been overwhelmed. As well as the carcinogens there is quite a high oxide of nitrogen content with an acidic aggravation as well as reducing the red blood cell oxygen retention. The overall safety factor is that very few people, if any, are subjected for 16 hours per day to diesel fumes.

In terms of mean concentration of soot particles from the unfiltered diesel exhaust, it was found that tumours only occurred in the rats exposed to at least 2200 $\mu g/m^3$, while Cuddihy, Griffith and McClellan reported that an estimate of diesel particulates for a major city is about 0·5 $\mu g/m^3$. It is believed, therefore, that the best toxicological opinion would be that no individual is at risk from atmospheric diesel particulates. However, as it has been found that tumours are possible with very high concentrations it would be prudent in a confined diesel environment, such as in coal mines or, to a lesser degree, tunnels, that ventilation should be adequate and particulate or catalytic traps should be used for vehicles in mines.

5. ATMOSPHERIC POLLUTION REGULATIONS

Pollution regulations were first introduced in California in the 1960s to deal with the severe 'smog' problem in Los Angeles, but it was not until 1970 that regulations were introduced into Europe. This regulation related to the type approval of new light duty vehicles equipped with spark-ignition or diesel engines and was negotiated under the auspices of the United Nations Economic Commission for Europe (ECE).

5.1. Passenger Car Regulations Including Light Trucks: Europe

Vehicle manufacturers have to meet these regulations, which have a considerable influence on engine design. In 1970 Regulation ECE R15 was adopted by the European Community (EEC) and published as a Community Directive (70/220/EEC). Initially these regulations applied to carbon monoxide and hydrocarbons only. Several amendments followed, increasing the severity, including the addition of oxides of nitrogen in 1977 (ECE 15.02). In the fourth amendment (ECE 15.04) severity was again increased, but hydrocarbon emissions and nitric oxide were added together to give motor manufacturers extra scope with their pollution control devices and in tuning their engines.

Needless to say, pollution regulations must be related to a standard driving pattern. Three major patterns have been defined for Europe, the USA and Japan. The ECE urban cycle is shown in Fig. 10(a). It has indeed been criticized for its low speed profile (averaging 18·7 km/h). The distance is 1·01 km and a complete test comprises four cycles, the first being from cold. A higher-speed test has been discussed.

To ascertain the quantity of pollutants the vehicles are run on a chassis dynamometer to the appropriate driving cycle and the pollutants from the exhaust are diluted and subsequently measured (Section 5.5).

As indicated above, it is usual for the ECE (Geneva) to recommend regulations and for the Community to adopt them. This applied to regulations up to 15·04 (1984) and these are quoted in Table 6 for a vehicle below 750 kg in weight. Figures are a little relaxed for heavier vehicles.

However, at the time of writing (1989) the European Community have overtaken the ECE with Directive 88/76/EEC (often referred to as the Luxembourg Accord), based on engine size and not on vehicle weight; namely vehicles with engines (i) over 2 litres, (ii) between 1·4 and 2·0 litres, and (iii) below 1·4 litres (see Table 7).

It will be noted that Phase 2 is shown for the smallest engine. Phase 1 was introduced with less severe figures, to enable the requirements to be met without the expense of catalysts.

Following hard on the heels of the above directive, has come yet a further directive from the EEC, due largely to German pressure. This directive 89/458/EEC refers only to cars with the small engines (below 1·4 litres) and limits carbon monoxide to 19 g/test and hydrocarbons plus nitric oxide to 5 g/test. These regulations will be introduced in 1992. These figures are extremely severe and will almost certainly mean that all cars will require catalysts, certainly in the short term. This directive also gives advance warning that carbon dioxide will be limited because of the greenhouse effect.

The ECE, through their committee WP29, are preparing an all-embracing regulation to cover the EEC Directive 88/76/EEC described above, and in addition a US regulation used by some European countries, as follows:

$$CO = 2{\cdot}1\ g/km$$
$$HC = 0{\cdot}25\ g/km$$
$$NO_x = 0{\cdot}62\ g/km$$

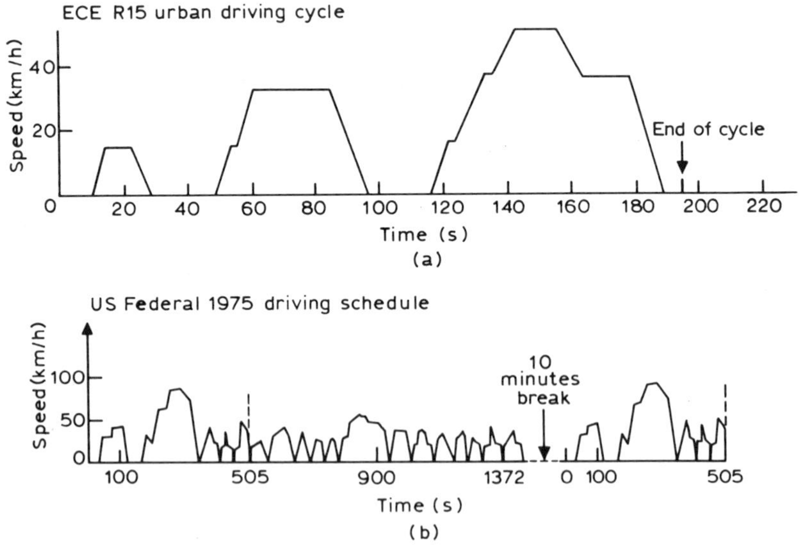

Fig. 10. Driving cycles: (a) ECE R15 urban driving cycle; (b) U.S. Federal 1975 driving schedule.

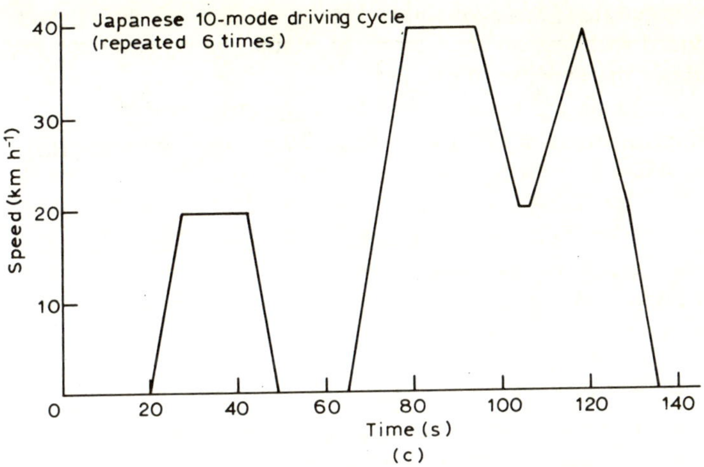

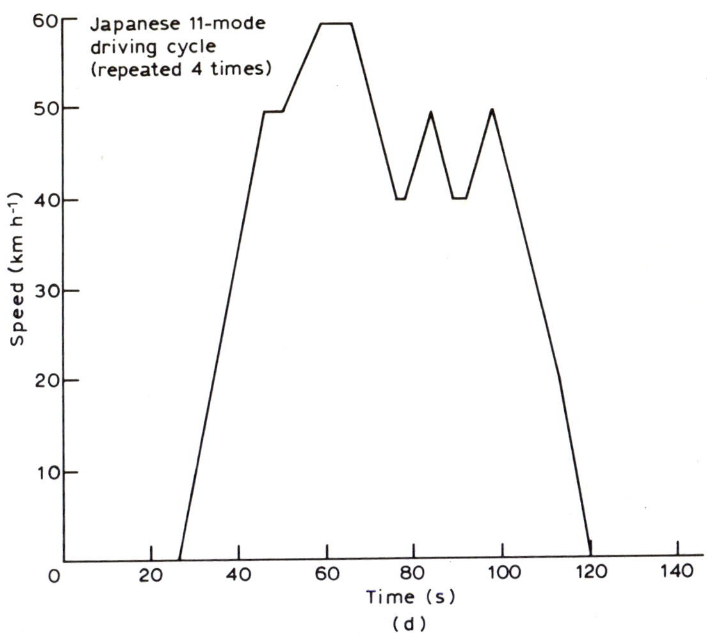

Fig. 10—*contd.* (c) Japanese 10-mode driving cycle; (d) Japanese 11-mode driving cycle.

Table 6. EEC Exhaust Emission Limits

Vehicle reference weight (kg)	ECE 15 (1971)		ECE 15.01 (1975)		ECE 15.02[a] (1976)	ECE 15-03 (1979)			15-04 (1984)	
	CO	HC	CO	HC	NO_x	CO	HC	NO_x	CO	$HC + NO_x$ combined
Below 750	100	8·0	80	6·8	10·0	65	6·0	8·5	58	19·0

Figures are in grams of pollutant per test.
[a] As 15.01 for CO and HC.

Table 7. Regulation 88/76/EEC Limits for Light Duty Vehicles

Cylinder capacity (litres)	Mass of carbon monoxide (g/test)	Combined mass of hydrocarbons and nitrogen oxides (g/test)	Mass of nitrogen oxides (g/test)
2·0	25	6·5	3·5
1·4–2·0	30	8	—
1·4 (phase 1)	45	15	6
1·4 (phase 2)	30	8	—

In this case the vehicle driving cycle is the US one. Particulate limits for diesel cars and trucks are proposed. The regulation will also include the previous 15.04 figures as it is believed that this will be retained or used by some non-European countries.

5.2. Car/Light Truck Emission Standards: USA

In the USA, the current driving cycle is 'The US Federal 1975 Driving Schedule' and is shown in Fig. 10(b): this represents a typical commuting drive into Los Angeles. There are three phases: a cold-start phase of 305 s duration with stops and starts reaching a maximum speed of 56 mph; a stabilized phase followed by a 10 minute break; then a final hot-start phase of a further 505 s duration. The average speed is 31·7 km/h and the distance 17·8 km. The accelerations are quite critical and have to be followed accurately for reliable results.

The US limits are shown in Table 8.

Table 8. Emissions (g/mile)

	HC	CO	NO_x	Particulates (diesel only)
Federal	0·41	3·4	1·0	0·2
California	0·41	7·0	0·4	0·2

5.3. Car/Light Truck Emission Standards: Japan

The Japanese 10- and 11-mode driving cycles are shown in Fig. 10(c) and 10(d). 10-mode cycle has a driven distance of 0·664 km per cycle; it is a hot-start 'stabilized' test in which the cycle is repeated six times to give a total distance of 3·98 km.

The 11-mode test cycle is repeated four times to give a total distance of 4·08 km per test. In this case the test is a cold-start procedure. The limits for the 10-mode cycle are shown in Table 9.

Table 9. Emission (g/km)

HC	CO	NO_x
0·39	2·7	0·48 (max.)
0·25	2·1	0·25 (mean)

5.4. Diesel-Engined Vehicle Regulations

5.4.1. Light Duty Diesel Engines: Europe

Passenger vehicles with diesel engines will have to conform to the same regulations as spark-ignition petrol engines but, in addition, have to meet a particulate emission regulation (88/436/EEC). The main reason for this regulation is the fact that the exhaust of diesels contains a very small quantity of known carcinogens (Section 4.4). The allowable emissions for all sizes of engines for Type Approval will be 1·1 g per test.

5.4.2. Heavy Duty Diesel Engines: Europe

There is a somewhat different regulation for heavy diesel-engined vehicles (over 3·5 tonne gross vehicle weight). This regulation was

Table 10. ECE Regulation 49: Emission Limits (g/bhp.h)

HC	CO	NO_x
2·6	10.4	13·4

promulgated in the USA and adopted in Europe within the ECE R49 Regulation in 1982. It has, however, been superseded in the USA from 1984 by the transient cycle (see below).

ECE R49 requires that engines are tested on the testbed rather than using a cumbersome dynamometer which would be required for a heavy commercial vehicle. The test cycle is a 13-mode cycle covering a range of loads from idle to full load and three speeds; idle, intermediate and rated. The ECE R49 proposal was formulated several years ago and was not in fact adopted by the European Commission. However, they have now agreed reduced emissions as shown in Table 10 (88/77/EEC).

5.4.3. Heavy Duty Diesel Engines: USA

The US Federal Heavy Duty Test Procedure is based on a transient cycle rather than the series of steady state running conditions of the European tests. It consists of two tests of 20 minute duration, the first from a cold start. As with the US petrol-engined vehicle cycle, the transient diesel test cycle is intended to represent the variable load and speed requirement of an engine to meet a typical road running condition for this type of heavy vehicle. The second cycle follows after allowing the engine to rest for 20 minutes before a restart. Details are given in Ref. 19.

5.5. The Measurement of Emissions: Petrol Engines

5.5.1. Passenger Cars

Except for specific research purposes, the emissions to be measured are those for which legislation has been promulgated; namely carbon monoxide, unburnt hydrocarbons and oxides of nitrogen. It is first necessary to trap the exhaust gases and then to analyse for each component. The system for trapping now used worldwide is the so-called Constant Volume Sampler System (CVS), which is defined as an SAE Recommended Code of Practice contained in the SAE

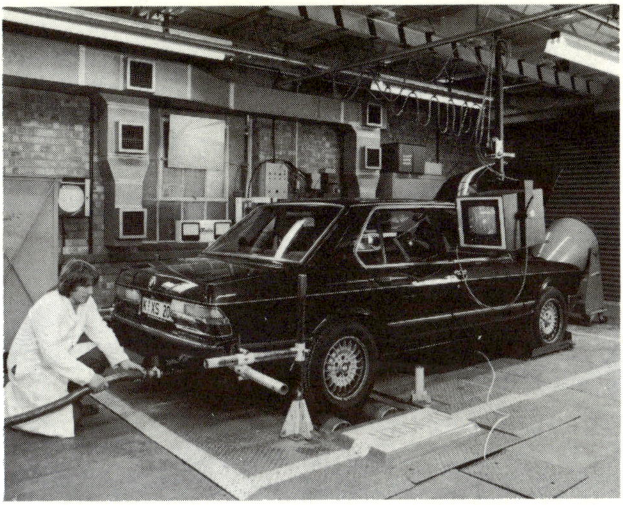

Fig. 11. Vehicle on chassis dynamometer under test for U.S. or European EEC Regulations.

Handbook (SAE J1094a). It is a complete vehicle test with the vehicle supported with its driving wheels on the rolls of a chassis dynamometer (Fig. 11). The rolls are connected to an absorption dynamometer and large flywheels of a size to represent adequately the inertia of the particular vehicle. The dynamometer is loaded in such a way as to simulate driving on the road as defined for the US Federal 1975 Drive Cycle (Section 5.2) The test driver drives the vehicle to this cycle with the help of the 'Drivers' Aid' outside the window. This plots the required speed pattern which he follows.

The total system is shown diagrammatically in Fig. 12. The whole of the exhaust is discharged into an airstream and conducted through a heat exchanger, that controls the temperature, to a 'Constant Volume Sampler' which is an apparatus that measures a constant volume of gas for the total duration of the test. In the older systems this was a two-lobe rotary pump (Roots Blower) but in newer designs the flow is measured by a critical flow venturi. By measuring the concentration of each pollutant component in this total volume of gas collected, the mass of each is ascertained. As the total time of the driving cycle is represented by a particular mileage (11·1 miles), the result is expressed in g/mile or g/km. To measure the concentration a small sample is drawn off and collected in a plastic sampling bag for analysis.

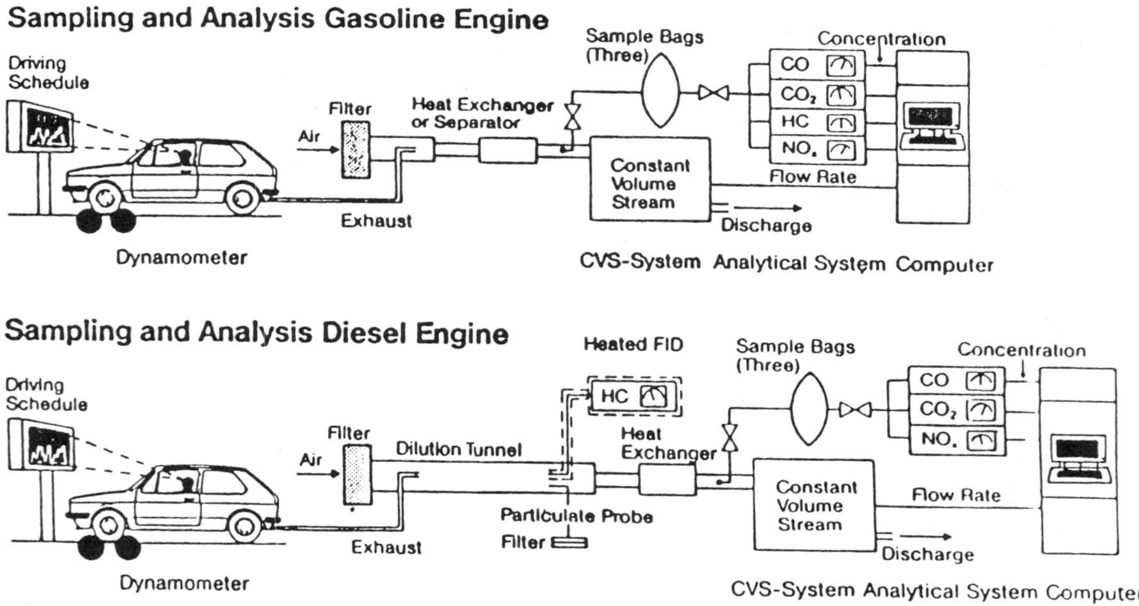

Fig. 12. Schematic arrangement of constant-volume sampling systems (CVS) for gasoline- and diesel-engined vehicles (reprinted with permission of the Society of Automotive Engineers).

The heat exchanger is necessary to prevent water condensation and to stabilize the temperature of the mixed exhaust gas and air for accurate measurement.

5.5.2. Gas Analysis

The bag samples of gas are passed at a predetermined rate through each of the analysers (Fig. 13). Carbon monoxide is analysed by non-dispersive Infra Red Analyser (NDIR), hydrocarbons by a Flame

25.95

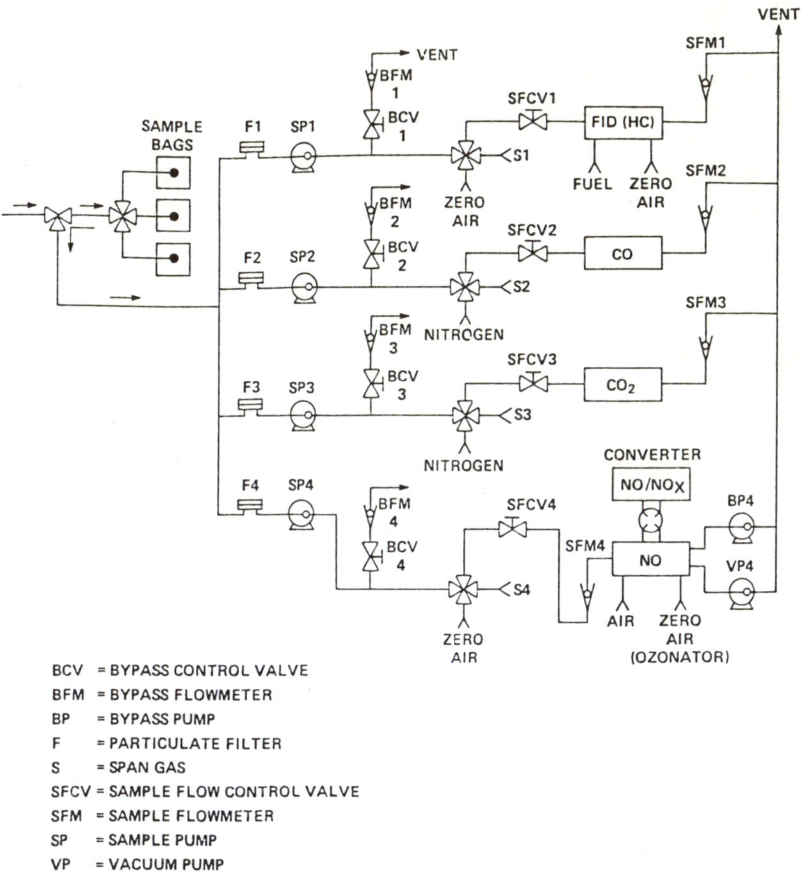

Fig. 13. Analytic system for sample bag measurement (reprinted with permission of the Society of Automotive Engineers).

Ionization Detector (FID) and oxides of nitrogen by means of a converter which converts any other oxides or ammonia into nitric oxide, which is then analysed by chemiluminescence (Fig. 14). Clearly, all these analyses work for convenience in a secondary manner, the output being analogue electrical signals, and therefore need to be 'zeroed' (with nitrogen) and calibrated with diluted gases of precise concentration (span gases).

It is also useful to measure oxygen and carbon dioxide.

5.5.3. Modal Analysis

The process is similar to the bag sampler analysis except that it includes a continuous sampling analysis of undiluted exhaust gas for HC, CO, NO_x, O_3 and CO_2 and the diluted gases. The diluted and undiluted sampling of CO_2 gives the dilution factor which, with the total diluted volume, will give the total vehicle exhaust volume and the modal mass emissions of each pollutant. The system is shown in Fig. 15.

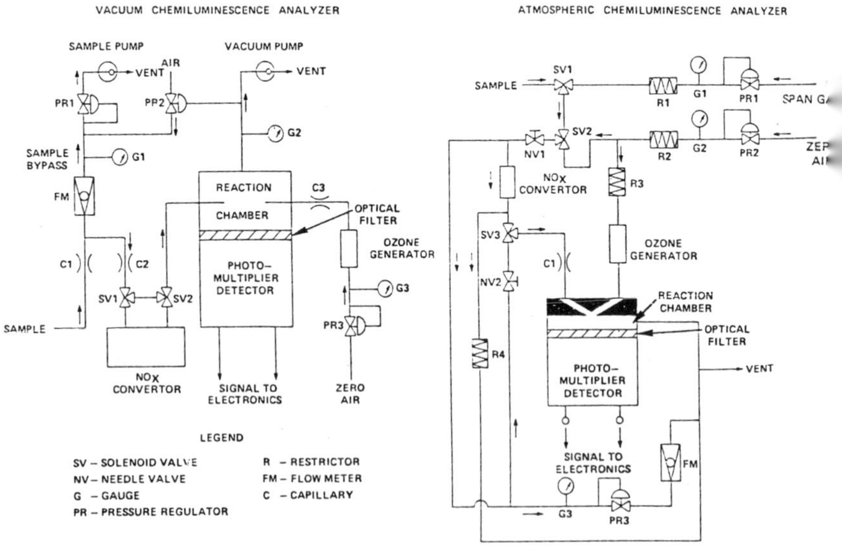

Fig. 14. Schematics of two chemiluminescent analysers: low-pressure type (left); atmospheric type (right) (reprinted with permission of the Society of Automotive Engineers).

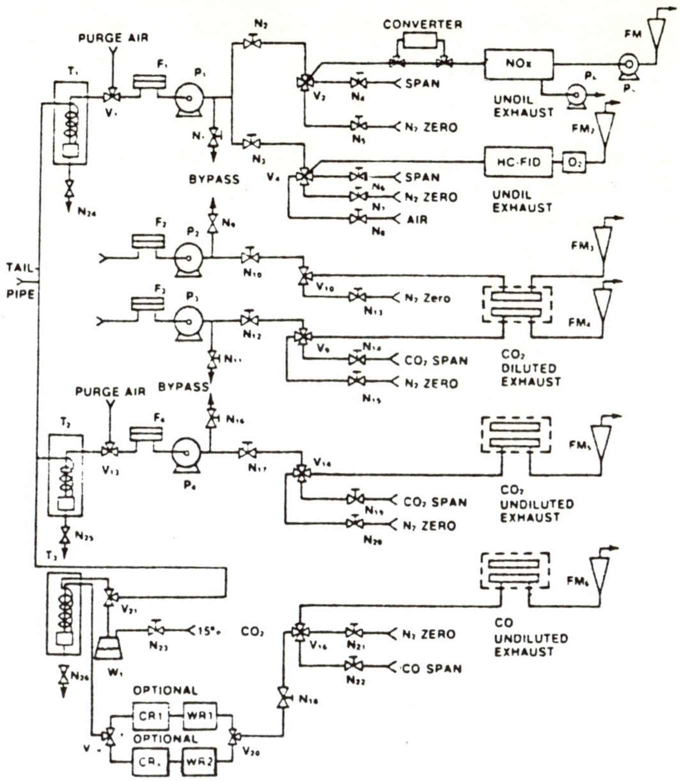

Fig. 15. Modal sampling and analysis train (reprinted with permission of the Society of Automotive Engineers).

5.6. The Measurement of Pollutants: Diesel Engines

5.6.1. Light Duty Engines

The test arrangements to comply with US Regulations have much in common with those for petrol engines (Fig. 13, lower diagram), the main difference being the case of a 'particulate tunnel' into which the exhaust is conducted. A specially designed probe and a prescribed gas flow is used to sample the gas and the particulates collected on a filter (Fig. 16(a)). Hydrocarbons are measured as with petrol engines by a flame ionization detector. It is, however, necessary to use an electrically heated sample line to prevent condensation. Particulates are closely related to black smoke. A smoke meter is shown in Fig. 16(b).

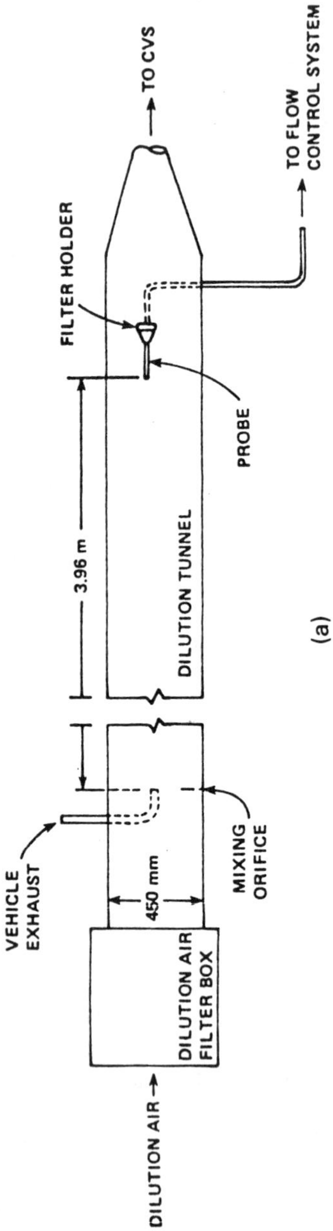

Fig. 16. (a) Dynamic dilution system.

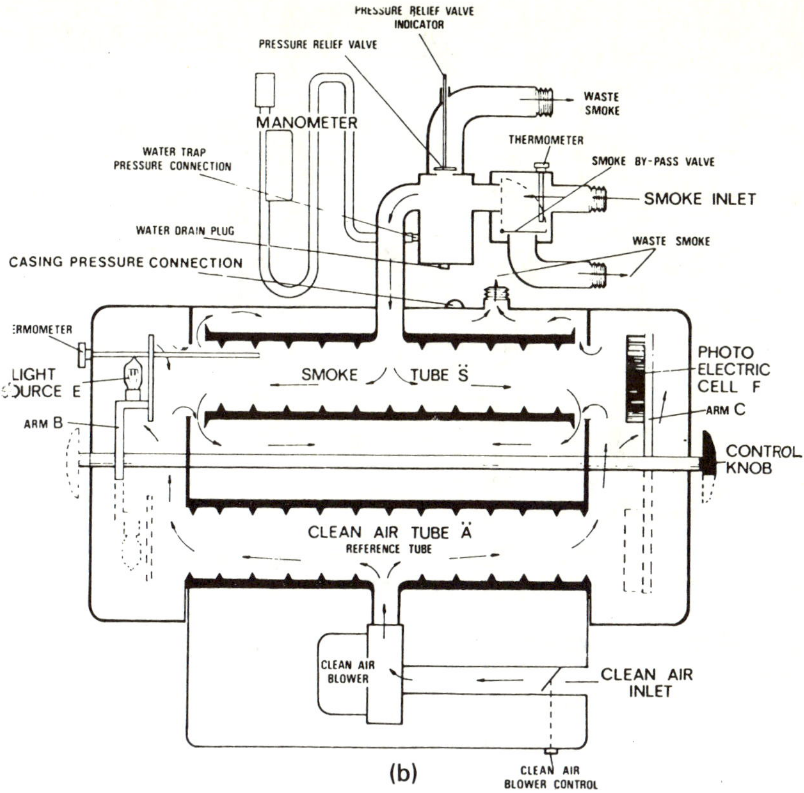

Fig. 16—*contd.* (b) Smoke opacity meter (reprinted with permission of the Society of Automotive Engineers).

5.6.2. Heavy Duty Engines

In the steady state 13-mode test procedure pollutants are measured with undiluted gas in accordance with SAE J1003 Recommended Practice (February 1984). Similar instruments as for petrol engines are used to measure HC, CO, CO_2 and NO_x but results are assessed in g/kWh. HC lines have to be heated. Particulates, for which severe regulations apply, are measured with a particulate tunnel as above.

For the transient testing required by the US Environmental Protection Agency (EPA) with a very severe particulate requirement (Section 7.6) a complex sampling arrangement is needed. Such a facility has been set up by Ricardo[19] and is shown in Fig. 17. The regulations stipulate that at the rated power of the engine the exhaust

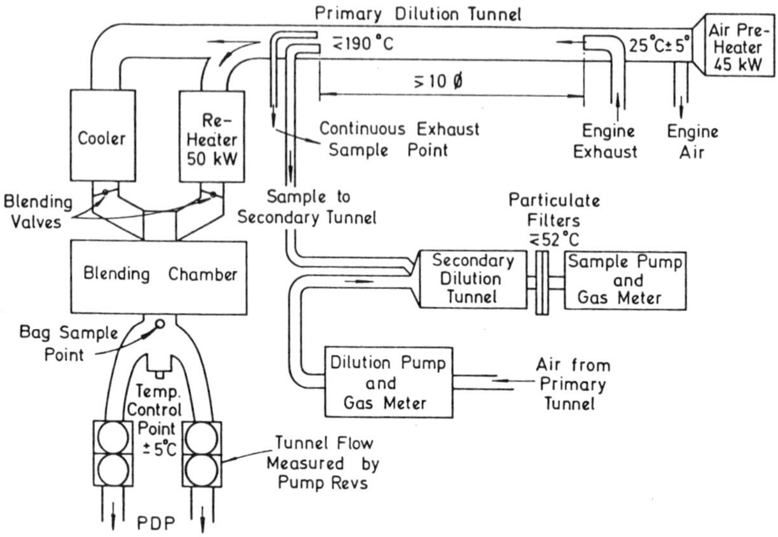

Fig. 17. Schematic of an FTP emissions sampling system (reprinted with permission of the Society of Automotive Engineers, 1984, 840348).

gas, when mixed with the dilution air to the particulate tunnel, shall not exceed 190°C. Indeed the temperature control is so closely specified that an electric heater and cooler are incorporated in the circuit, in addition to the air preheater.

6. THE SPARK-IGNITION ENGINE: EMISSIONS

The pollutants emitted from spark-ignition (SI) engines are very much influenced by the air : fuel ratio (A/F) of the petrol/air mixture. This is shown graphically in Fig. 18. The stoichiometric A/F ratio is shown dotted. As would be expected, with carbon monoxide (CO) and unburnt or partially burnt fuel (HC), as the mixture is weakened from stoichiometric, the HC and CO emissions are reduced to a minimum just beyond stoichiometric but remain constant until at about 20/1 HC and CO both increase due to poorer combustion. If misfire occurs HC goes up drastically.

The oxides of nitrogen (NO_x) picture is more complex. The collective term is used in the regulations as there is a small amount of nitrogen dioxide (NO_2) emitted. However, this is very small and it can be generally taken that the oxide emitted is nitric oxide (NO).

This is formed from the nitrogen and oxygen in the air at the very high temperature that exists in the combustion chamber. This will be considered in greater detail below but, broadly speaking, with rich mixtures there is competition for the oxygen and for weak mixtures the temperature is lower and less nitric oxide is formed. The peak occurs on the weak side of stoichiometric.

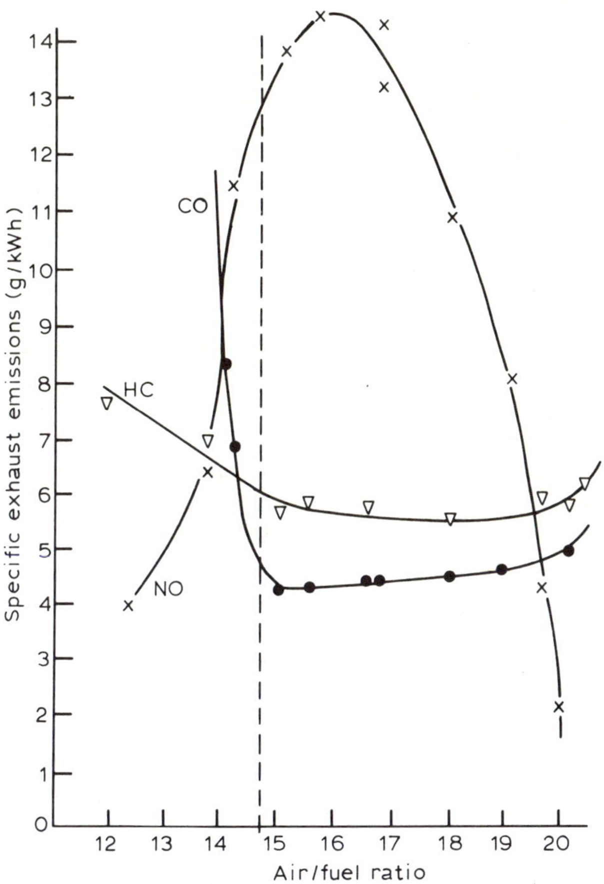

Fig. 18. Variation of emissions with air: fuel ratio.

6.1. Hydrocarbons

Fuel that escapes combustion arise from several causes. The most obvious is that which is retained at low temperature in crevices where it is forced by compression, namely the sparking plug and in the gap between piston and cylinder above the piston ring. During the expansion stroke this flows out but as temperatures have dropped it is only partially burned.[20] A second source is the oil layer on the cylinder walls which absorb fuel vapour during induction and compression and desorb it on the expansion stroke. A third source, even more difficult to quantify, is wall quenching: as a flame approaches a cool surface it is quenched as the mixture immediately adjacent to the wall is not at a high enough temperature to burn. This third source is difficult to quantify because although quench temperatures can be quantified under ideal controlled conditions, in the combustion chamber the flame may be normal to the surface, parallel to it or at any angle between. Another factor which confounds the whole HC emission problem is that the HC left in these low-temperature zones continue to oxidize even after discharge from the cylinder, so that when evaluating the measuring position it is very significant. The regulations for emission measurement specify—the end of the exhaust tail pipe—(Section 5).

Several research projects have helped to clarify the situation using 'bombs' and special single-cylinder engines. A significant contribution to the problem was made by Namazian & Heywood[21] who constructed a transparent single-cylinder engine with a square cylinder. This enabled high-speed photography to be utilized. The transparent section consisted of two parallel steel walls and two parallel quartz glass walls. They used graphite piston rings without oil. As the engine was uncooled it could only be run under combustion conditions for periods up to 50 s. These experiments demonstrated conclusively that during the expansion stroke, when the cylinder pressure drops, flow out of the crevices takes place. The major crevice was between piston and cylinder but crevice flow was also observed from the spark plug.

Ma *et al.*[22] endeavoured, with a series of experiments and model computation, not only to locate the sources but to follow the route through to the exhaust pipe and observe the extent of oxidation. In Figs 19 and 20 are shown time-resolved HC concentrations for a modern 2 litre OHC at 9:1 compression ratio. As has been observed by Tabaczynski, Heywood & Keck,[23] HC peak occurs in the manifold

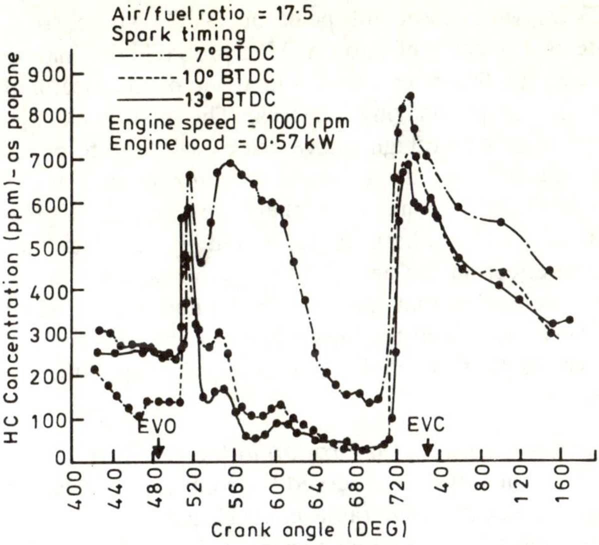

Fig. 19. Time-resolved HC samples in the exhaust manifold at 15·7 A/F ratio (from Institute of Mechanical Engineers, 1987, 336/87).

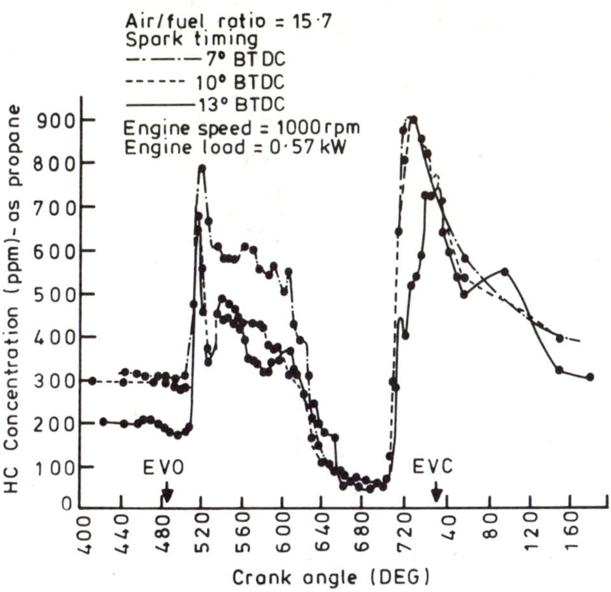

Fig. 20. Time-resolved HC samples in the exhaust manifold at 17·5 A/F ratio (from Institute of Mechanical Engineers, 1987, 336/87).

soon after the EVO, with a second peak at EVC. These, they considered, indicate two pockets of unburnt HC, probably originating from different sources, the first from crevices near the piston face, and the latter peak from the piston ring crevices. The first peak corresponds to the blowdown where high gas velocities will tend to purge the cylinder. When the pressure has dropped to atmospheric the last HC to be emitted is likely to come from the piston ring area. The authors consider that this hypothesis is supported by the fact that spark timing has little effect on the peaks but a significant effect in the mid-range HC, thus indicating that the peaks are unburnt fuel while the mid-range HC may come from the bulk where a small quantity has escaped oxidation and is thus affected by exhaust gas temperature, a function of spark timing.

6.1.1. Hydrocarbons Adsorption/Desorption in the Oil Film

This phenomenon was probably first observed by Wentworth[24] and has been shown to make a not-inconsiderable contribution to HC emissions. It is very sensitive to fuel composition as it depends upon the solubility of the fuel vapour in the oil, which is a function of Henry's constant (the larger the Henry's constant the lower the solubility). To verify this Dent[22] used two fuels; iso-octane, which has a high solubility, and iso-pentane which has a low one. The effect of HC emissions is shown in Fig. 21. The effect is further borne out by the reduction obtained by raising the cylinder wall temperature. From these results Ma *et al.* suggested that the development of fuels with

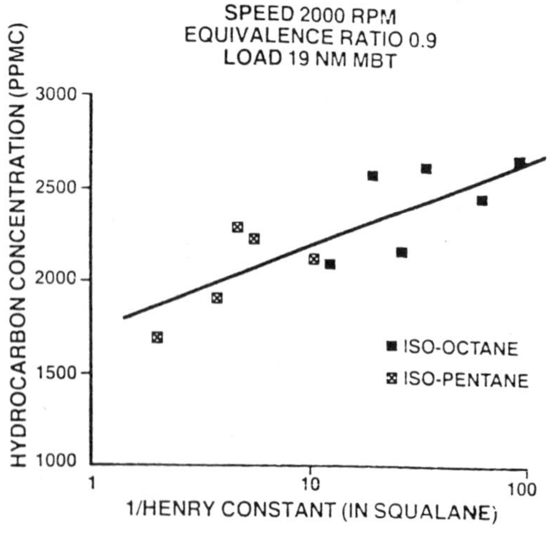

Fig. 21. Effect of fuel solubility in lubricating oil on HC emissions from a single-cylinder engine at 1000 rpm.

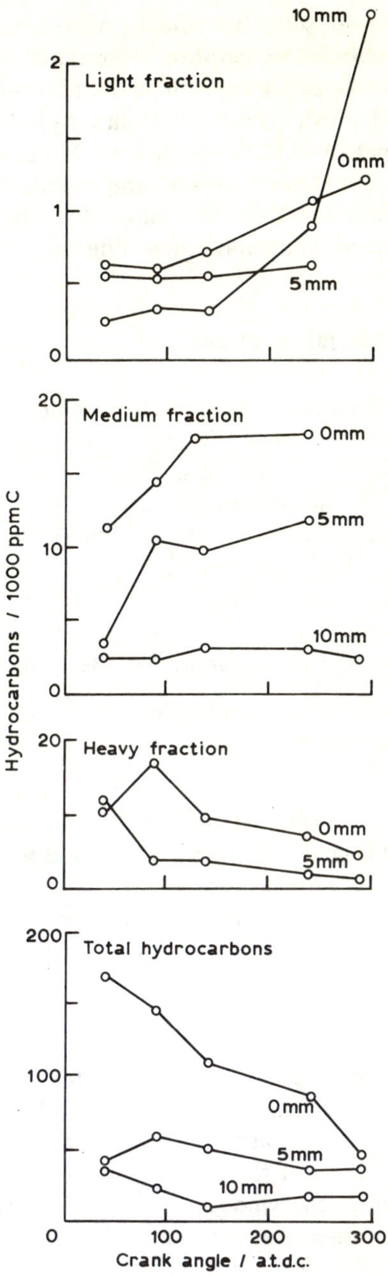

Fig. 22. Concentration of some hydrocarbons in the combustion chamber, when burning petrol (figures on curve show depth of probe insertion) (from Institute of Mechanical Engineers, 1987, 332/87).

lower solubility and an increase in cylinder wall temperatures were practical propositions to reduce HC emissions.

Yates[25] used a high-speed sampling valve in the disc combustion chamber of a single-cylinder (500 cc OHV) engine. Some results using petrol are shown in Fig. 22. It will be noted that concentration is highest near the wall and reduces rapidly. Yates concluded, with some caution, that this supported the oil absorption/desorption sources rather than wall quenching.

6.2. Nitric Oxide

Regulations for emission control (Section 5) quote NO_x in order to cover both nitric oxide (NO) and nitrogen dioxide (NO_2). However, by far the major pollutant of the oxides in IC engines is nitric oxide. The NO is formed from the nitrogen and oxygen in the combustion air at the very high temperatures present in the combustion chambers of both diesel and SI engines. The reactions are usually termed the extended Zeldovich reactions,[26] after the person who first called attention to the effect of atomic nitrogen and oxygen under high-temperature conditions. He proposed the following reactions:

$$O + N_2 \rightleftharpoons NO + N, \qquad N + O_2 \rightleftharpoons NO + O.$$

The rates of the reactions are an exponential function of temperature, and this unfortunately gives a basic problem with methods of reducing the formation of NO_x in a spark-ignition engine. As is well known, thermal efficiency is increased with increased maximum temperature. The situation is shown in simplified form in Fig. 23, where both efficiency and NO_x formation are plotted against maximum temperature. This is for the ideal constant-volume cycle and an average kinetic rate constant.[27] It makes clear, however, as so often in practice, that a compromise has to be made between efficiency and pollution. The consequence of this is that calculation and experiment have to be conducted as accurately as feasible.

The early classic work (1967) in this area was performed by Newhall[28] and Starkman,[29] who used the Zeldovich chain reactions as a basis of their calculations and spectrographic measurements of NO through windows in the combustion chamber of an experimental engine.

The Zeldovich reactions have been shown to be oversimplified for the complex interactions that take place in the SI combustion chamber

Fig. 23. Influence of maximum cycle temperature on air cycle efficiency and NO_x concentration.

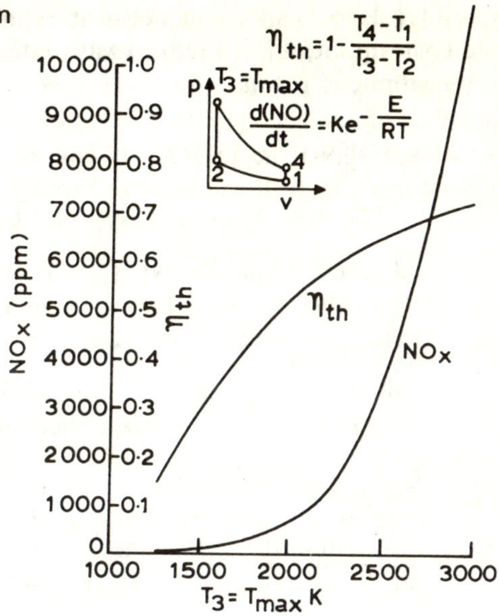

where both pressures and temperatures are high. Further, in addition to the nitrogen and oxygen in the air, there are innumerable reactions taking place due to the combustion of the fuel, giving the presence of many other important radicals.

For model calculations a reaction between the atomic nitrogen and the ubiquitous OH radical has been found necessary:

$$N + OH \rightleftharpoons NO + H$$

The three equations are usually termed the Extended Zeldovich reactions:

$$N + NO \rightleftharpoons N_2 + O + 75\ \text{kcal}$$

$$N + O_2 \rightleftharpoons NO + O + 31{\cdot}8\ \text{kcal}$$

$$N + OH \rightleftharpoons NO + H + 394\ \text{kcal}$$

Heywood[20] uses these equations to predict the rate of nitric oxide formation:

$$\frac{d[NO]}{dt} = k_1^+[O][N_2] + k_2^+[N][O_2] + k_3^+[N][OH] - k_1^-[NO][N] - k_2^-[NO][O] - k_3^-[NO][H] \quad (1)$$

where [ ] denotes species concentration in moles/cc, k_1^+ is the forward rate constant and k_1^- is the backward rate constant.

For atomic nitrogen,

$$\frac{d[N]}{dt} = k_1^+[O][N_2] - k_2^+[N][O_2] - k_3^+[N][OH] - k_1^-[NO][N] + k_2^-[NO][O] + k_3^-[NO][H] \quad (2)$$

Heywood assumes that as [N] is small compared with the concentration of the other significant species then eqn (2) can be equated to zero. Then, from eqns (1) and (2),

$$\frac{d[NO]}{dt} = 2k_1^+[O][N_2]\frac{1 - [NO]^2/(K[O_2][N_2])}{1 + k_1^-[NO]/(k_2^+[O_2] + k_3^+[OH])} \quad (3)$$

where $K = (k_1^+/k_1^-)(k_2^+/k_2^-)$.

Heywood further assumes that although NO is produced both in the flame front and in the post-flame zones the latter predominates due to its much larger volume and the fact that it is compressed as

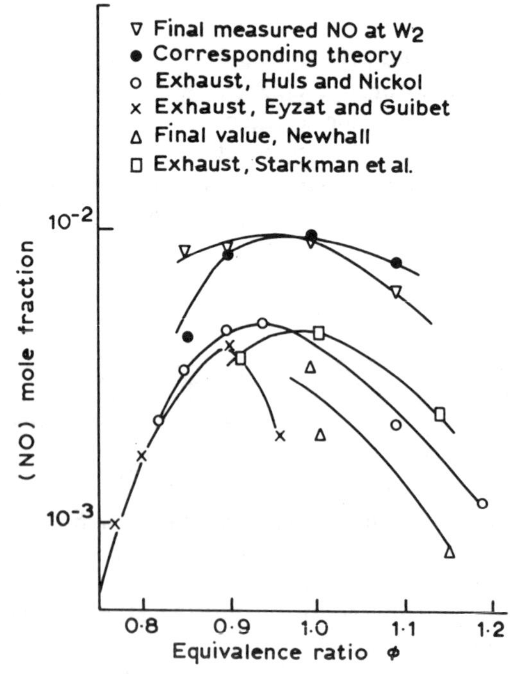

Fig. 24. Nitric oxide concentration versus equivalence ratio at constant speed and spark advance. Top two sets of points are the spectroscopically measured points and the corresponding theory. Experimental results of others for similar conditions are shown.

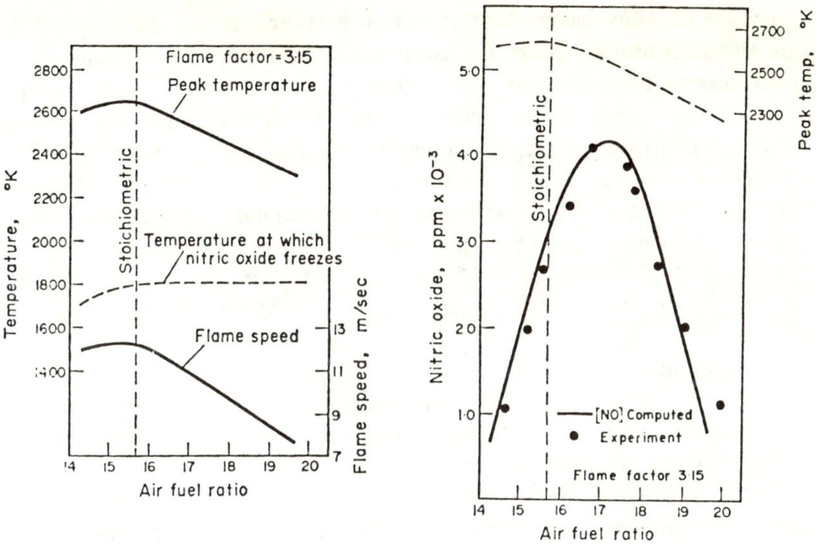

Fig. 25. Influence of air : fuel ratio on NO emissions, computed and experimental (Pergamon Press, 1975).

combustion proceeds. He then introduces equilibrium concentration values for the reactants, takes available data for rate constants and arrives at eqn (4) below:

$$\frac{\mathrm{d[NO]}}{\mathrm{d}t} = \frac{6 \times 10^{16}}{T^{1/2}} \exp\left(\frac{-69\,090}{T}\right) [\mathrm{O_2}]_e^{1/2} [\mathrm{N_2}]_e \quad \mathrm{mol/cm^3s} \qquad (4)$$

The relationship between NO_x formation rate and temperature will be seen from eqn (4) and the strong temperature dependence will be noted.

Lavoie[30] used a spectroscopic technique for measuring nitric oxide in a single-cylinder L-head CFR engine fitted with a number of quartz windows. The compression ratio was rather low at 4·5, which will produce a lower value of NO but nevertheless will be quite valid for the verification of the comparatively simple Extended Zeldovich reactions. Some results are shown in Fig. 24. Among his conclusions he states 'for the purposes of engine studies nitric oxide formation can be described adequately by an Extended Zeldovich mechanism in which all species of the C—O—H system are taken to be present in equilibrium concentrations.'

Lavoie, Heywood & Keck,[31] in a further investigation, used a somewhat extended chemistry and reached similar conclusions. This work was used by Benson, Annand & Baruah[32] and they applied the kinetics to a simplified two-zone (burnt and unburnt gas) mathematical model of combustion in a SI engine. Their results are shown in Fig. 25. It should be noted that in the calculated concentrations against air: fuel ratio agreement with experiment appears very good, but it should be borne in mind that the simplified model was adjusted to give a measured NO figure at maximum concentration. These adjustments were necessary with mathematical models (M.M.) at this time but it is not good practice unless a physical explanation can be deduced. This does not mean that the Extended Zeldovich mechanism is not adequate and these functions could be fed into the latest M.M.'s; see Spalding (Chapter 11) and others. Lavoie, Heywood & Keck[31] made an interesting comparison between equilibrium and actual NO (Fig. 26). This is of considerable interest as it shows the 'freezing' of the NO due to the slow reverse reactions in comparison with the engine speed. Also even the faster forward reactions do not keep pace to allow equilibrium values to be attained until 20° after TDC.

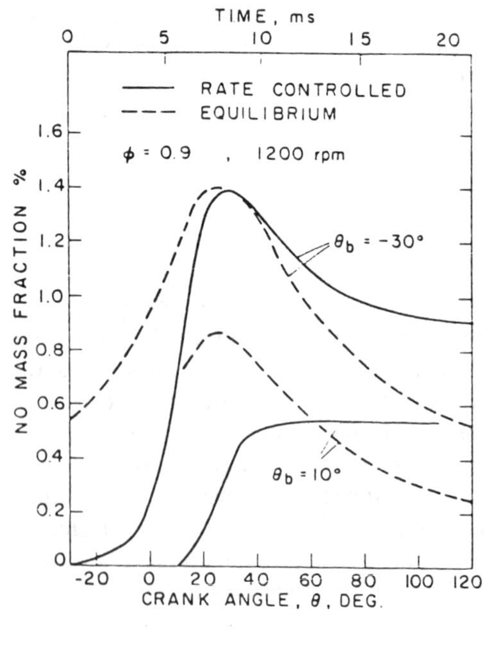

Fig. 26. Nitric oxide mass fractions as a function of time in the burnt gas for two elements which burn at different times. $\theta_b = -30°$ is the first part of the charge to burn; at $\theta_b = 10°$, one third of the charge is burnt (Gordon and Breach Science Publ.).

The above work should allow NO to be predicted for a particular engine configuration and lead to modifications to optimize between efficiency and NO production.

6.3. Carbon Monoxide

As will be seen from Fig. 18, carbon monoxide emissions from SI engines are primarily dependent on the air:fuel ratio or, in other words, the availability of oxygen to complete the combustion of the hydrocarbons in the fuel to carbon dioxide. Where this reaction fails to go to completion carbon monoxide is emitted; it follows that as one goes rich of stoichiometric CO values increase rapidly. At stoichiometric, although there is sufficient oxygen, due to imperfect mixing some carbon monoxide is still formed. In a diesel engine which runs with an excess of air CO is very low and a good SI stratified charge engine such as the Honda CVCC goes part way towards this desirable situation.

It will be noted that even at 20:1 CO emissions are not zero; a residual amount is emitted due at least partly to the hydrocarbons that escape complete combustion during the combustion period (see Section 6.1). These oxidize during the expansion and discharge period. Oxidation can be relatively slow and continues to take place in the exhaust port and pipe if the temperature is sufficiently high.

6.4. Emission Control Technologies: Spark-Ignition Engines

The previous sections have examined the major pollutants from SI engines. This section examines methods of control. As we have seen, emissions are very much influenced by the air:fuel (A/F) ratio. To summarize, Fig. 18 shows exhaust emissions for a typical engine. The stoichiometric A/F ratio is shown dotted. It will be noted that unburnt hydrocarbons and carbon monoxide, rich of stoichiometric, are high due to lack of sufficient oxygen. They flatten out just weak of stoichiometric, but finally turn up at about 20:1 A/F ratio due to slow burning and finally misfire. Nitric oxide is low when rich due to competition by the fuel for the oxygen, reaches a maximum just lean of stochiometric and falls rapidly toward 20:1 A/F ratio. This is due to a drop in the temperature of combustion which, as we have seen

(Section 6.2), is a controlling factor. Two main approaches can be made for emission reduction. Nitric oxide is the controlling factor and this can be reduced to harmless nitrogen by means of reducing agents which are present in the combustion process. The obvious one is carbon monoxide but hydrogen is also available, being derived under certain conditions from the fuel. To promote these reactions a catalyst is required.

The second method is to run at a very weak mixture ratio, 20:1, where due to the low temperature the nitric oxide produced is low. Both these methods are valid and will be considered below.

6.4.1. The Catalyst Method

The first experimental approach was to tune the SI engine to run slightly rich of stoichiometric, say 1–2% CO, to promote the reaction:

$$2CO + 2NO = 2CO_2 + N_2$$

This very adequately reduced the nitric oxide, and many catalytic materials in the transient metal groups were effective; that is, noble metals platinum, palladium, and among base metals nickel, cobalt and iron. With such a catalyst some excess carbon monoxide and unburnt hydrocarbons would remain in the exhaust due to the slightly rich condition. These could be effectively oxidized by following the reduction catalyst by an oxidation catalyst and adding air from a small pump between the two catalysts. Unfortunately, it was found that the nitric oxide then reappeared! Research showed that this was due to the fact that ammonia (NH_3) had been produced in the first catalyst due to the presence of hydrogen and this was subsequently oxidized in the second catalyst. Much research was then undertaken by catalyst firms, and UOP first developed the so-called 'three-way catalyst' which, by tuning the engine to run at stoichiometric air:fuel ratio, would reduce all three pollutants, NO, CO and HC, together. Such catalysts were further developed by Johnson & Matthey[33] and Englehart & Degussa.[34] They manage with one catalyst acting as both a reduction and oxidizing catalyst, with the A/F ratio oscillating slightly to each side of stoichiometric. The situation is shown in Fig. 27, and it will be noted that within a small band of equivalence ratio, shown dotted, an efficiency of conversion above 80% is achievable. This clearly requires a very precise control of equivalence ratio, which is achieved with an exhaust gas sensor (Section 6.4.5) with feedback to the mixture preparation system, i.e. the fuel injector system or

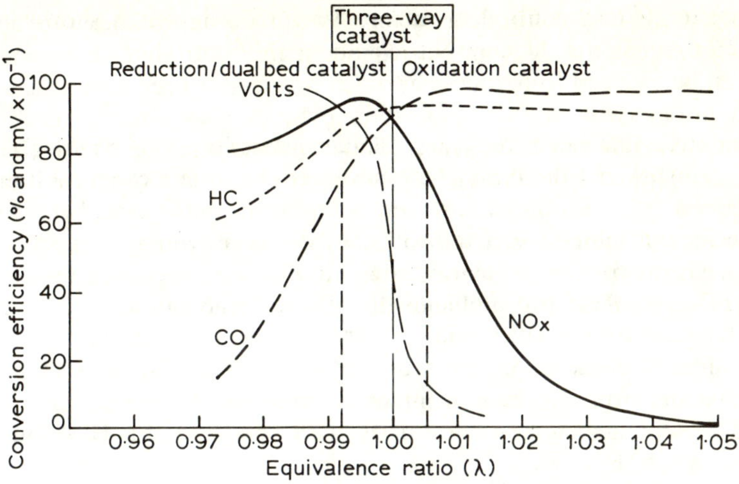

Fig. 27. Conversion of NO, CO and HC for a three-way catalyst as a function of air : fuel ratio. Voltage from exhaust gas sensor.

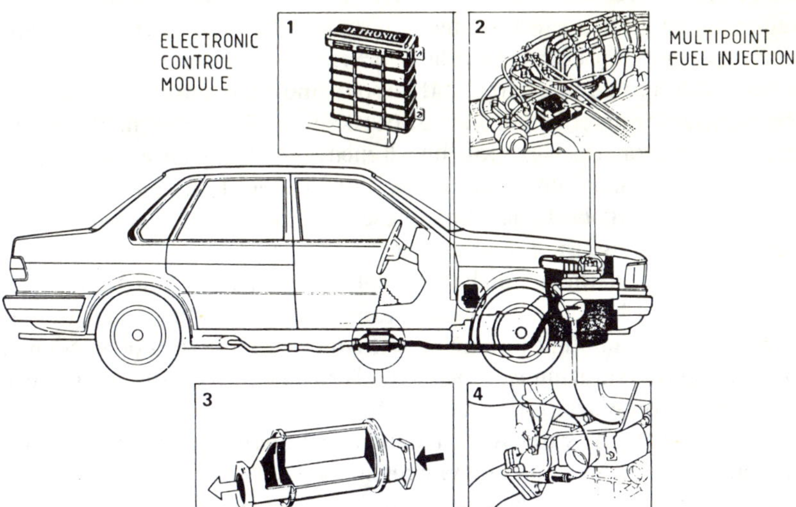

Fig. 28. Three-way catalyst system for a passenger car.

carburettor (Chapter 6). A typical passenger car layout is shown in Fig. 28.

6.4.2. Catalysts

The catalytic method of reducing pollutants described above may give the impression that these may be easily obtained from a chemical firm of manufacturers. Nothing, however, is further from the case, because of specific difficulties associated with the hostile environment in which they have to work in a vehicle, coupled with a severe limitation of space. Despite these two problems the catalytic reactions have to have very high conversion efficiencies, *c.* 90%, to meet the severe regulations already promulgated or proposed (Section 5). Because of all these factors over 20 years of intensive research has taken place to develop the highly efficient catalysts used today in the USA and increasingly in Europe.

In the forefront of catalyst research and development have been the Ford Motor Company in the USA[35] together with catalyst manufacturers Johnson & Matthey, Englehart & Degussa.

6.4.3. The Catalytic Reactor

The best catalytic materials for oxidation of CO and HC and the reduction of NO lie in the transition group of metals in the periodic table; namely, the precious metal group—platinum (Pt), palladium (Pd) and rhodium (Rh) and others lower down the table—iron, (Fe), nickle (Ni) and cobalt (Co), and copper and vanadium. Some of the metals may be eliminated due to attack by acidic components in the exhaust gas or due to possible toxicity. For instance, copper is attacked by small amounts of sulphur in the fuel and vanadium pentoxide is highly toxic. For these and attendant reasons, most catalysts today use the noble metals Pt, Pd and Rh.

Physical factors demand that the catalytic reactors shall not have a high pressure drop, as this causes loss of power from the engine. It must be strong enough to withstand high-frequency vibration induced by the engine and it must be sufficiently insulated to avoid setting fire to dry grass on which a vehicle may be stopped.

Beginning with size, a convenient parameter used by catalyst manufacturers is space velocity, defined as

$$V_s = \frac{\text{volume of gas flowing per hour}}{\text{catalyst volume}}$$

A typical figure for vehicle reactors is 50 000. This is a compromise and is much higher than would be used in stationary applications. The other important factors are to have a very large surface area for the catalytic material on which to expose the gas for rapid reaction and, finally, to have a sufficiently dense bed that the gas does not by-pass the catalyst surface in its rapid flow. Pelletted catalyst supports, as frequently used in industry, have been used in vehicles but have now largely given way to a matrix type (Fig. 29) made by extruding a refractory material, cordierite or similar, through a fine die. This produces a very large surface area, as matrices as fine as 600 per in^2 are produced. More commonly, 400 per in^2 are used to keep pressure drop to reasonable levels. It is then necessary to hold the matrix in a firm but cushioned outer shell, which is achieved by wrapping the matrix in a high-temperature-resistant mat and sandwiching it between two halves of a stainless steel shell compressing the mat.

The authors of this chapter have worked closely with Johnson & Matthey, and what follows in this particular section is largely a summary of two papers by authors from this firm.[36,37]

The surface area of the matrix, although large, can be greatly increased by coating it with γ alumina, which offers a very porous structure for molecular penetration. The internal surface area is typically 200-400 m^2g^{-1}. Scanning electron micrographs of a ceramic monolith substrate are shown in Fig. 30. The washcoated matrix is then dipped into a solution of the catalyst and finally baked at high

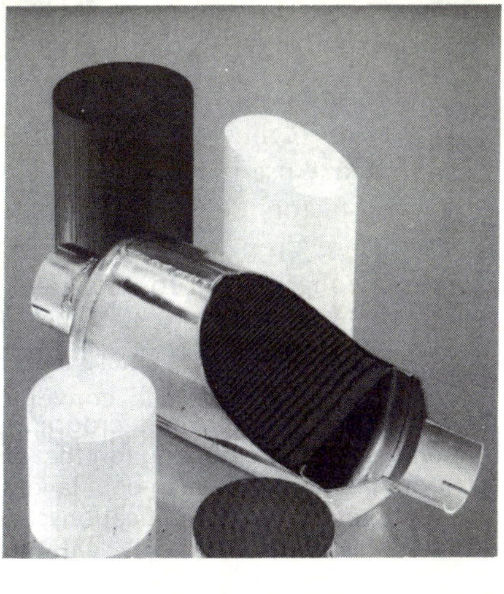

Fig. 29. Matrix type catalyst. Ceramic honeycomb supports and a cut-way catalytic converter (from Johnson and Matthey).

temperature, resulting in metal crystallites being highly dispersed on the porous alumina.

However, there are still problems due to the very high temperature which the catalysts have to withstand for a considerable period. This is worse in Europe, in countries where there are no speed limits, than in the USA. Unfortunately, γ alumina turns to α alumina at temperatures above 900°C, and in-service temperatures of 1000°C can be reached; hence the surface area is lost and the efficiency of the catalyst reduced. Much research has succeeded in minimizing this loss of surface area by means of stabilizers involving the addition of other oxides.

In place of a ceramic matrix a high-temperature metal matrix may be used. To withstand the temperatures a special material, Fecralloy, has been developed. This too has to be coated with porous washcoat and presents the same problems.[38]

The desirable characteristics of catalysts for automotive use are: (1) low light-off temperature; (2) high conversion efficiency; and (3) resistance to deterioration at high temperatures. Obviously low cost would be an additional advantage, but the difficulty of finding a base metal catalyst as good as a noble metal one is great, although the incentive is worthwhile; oxidation catalysts are normally platinum–palladium, while three-way catalysts need rhodium as well as platinum to give good reduction potential. Rh has a selective resistance to the

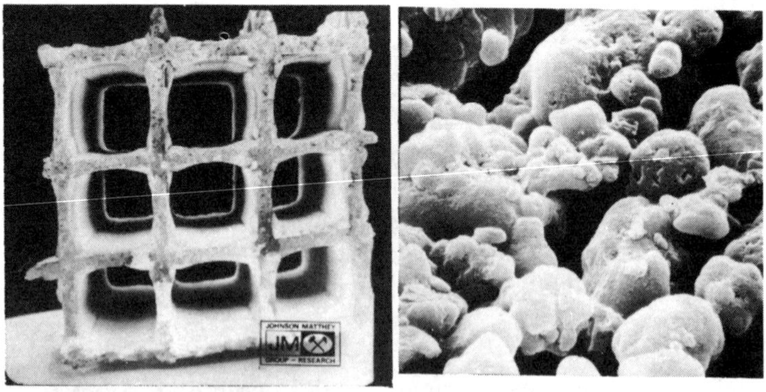

Fig. 30. Scanning electron micrographs of a wash-coated ceramic monolith substrate showing (on the left) the open-channel structure of the monolith with the wash-coat layer deposited on the channel walls, and (right) at higher magnification, the porous structure of the wash-coat (from Johnson and Matthey).

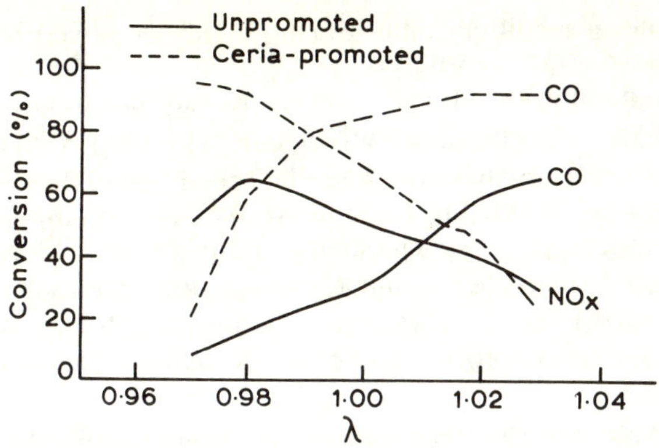

Fig. 31. Effect of ceria-promotion on a three-way catalyst.

formation of ammonia (NH_3) from the hydrogen present. Unfortunately, rhodium is sensitive to de-activation at high temperatures. The de-activation is thought to be due to reaction between rhodium and the γ alumina oxide state, which is difficult to eliminate. It has been shown that this interaction can be retarded by the incorporation of ceria into the the catalyst (Fig. 31). It has also been shown[39] that ceria in the presence of water vapour, which in engines is derived from the hydrogen component of the fuel, also increases considerably the conversion efficiency for both CO and NO. This is considered to be due to the water gas shift reactions:

$$H_2O + CO = CO_2 + H_2$$

Additionally, it has been shown that the activity of three-way catalysts can be improved in the region of stoichiometric air:fuel ratio by the addition of base metal promoters. Ceria has been discussed above but nickel is also helpful with ceria, but does not appear to give a great advantage. Base metal can add oxygen storage in the $\lambda = 1$ zone by being oxidized when slightly weak and giving up oxygen on reduction.

6.4.4. Oxidation Catalysts

For the high-compression lean-burn approach to pollution reduction, a catalyst may still be required, in some cases, to oxidize the unburnt hydrocarbons. Much depends on engine size: for cars with engines below 1·4 litres capacity it would have been quite feasible to meet the severe European regulations (88/76/EEC) Phase 1 and possibly the more stringent Phase 2, without a catalyst but with Directive 89/458/EEC an oxidizing or 3-way catalyst will be needed. For medium-sized engined vehicles (1·4–2·0 litres) a catalyst will be required to oxidize the unburnt hydrocarbons; in fact a three-way catalyst can be used with advantage. The three-way catalyst will oxidize when the engine is running on a weak mixture but when richened for acceleration (if a stoichiometric ratio or beyond is reached), it will also reduce the remnant nitric oxide. The same applies for vehicles with larger engines, but generally the three-way stoichiometric system is preferred as it is easier to achieve the low pollution figures required.

Oxidation catalysts do not need rhodium and a platinum–palladium mix gives good conversion,[40] and in vehicles with sophisticated engine management controls a feedback sensor may be dispensed with, thus giving a considerable saving.

6.4.5. Lead-tolerant Catalysts

In the USA lead was removed from most petrol as it poisoned catalysts. However, as noted above, there is a *prima facie* case for the retention of a small amount of lead in petrol to maintain higher octane quality and thus use more highly efficient engines and save fuel. To this end Johnson & Matthey, in co-operation with British Leyland (now Austin Rover), produced a lead-tolerant catalyst that was capable in a 1·7 litre car, of completing 50 000 km and still almost meeting the Luxembourg Agreement limit for CO with 28 g/km test and $HC + NO_2 - 10$ g/km test (Luxembourg limits $CO = 30$ g/km, $HC + NO_x = 8$ g test). The lead content of the fuel was 0·15 g/per litre.[41]

6.4.6. Oxygen Sensors

For three-way catalyst systems for gasoline engines (Section 6.4.1) the fuel system has to supply the engine with a stoichiometric mixture of fuel (gasoline) and air (A/F ratio 14·7). The zirconia cell which is sensitive to the small amount of oxygen present in the exhaust gas has a suitable characteristic for this purpose and is frequently termed a 'lambda' sensor. 'Lambda' is defined as the ratio of the actual A/F ratio divided by the stoichiometric A/F ratio, and for three-way catalyst this ratio has to be unity. The characteristics of the cell are shown in Fig. 27 and a cross-section of the cell itself in Fig. 32. It will be noted from the graph that if the mixture supplied to the engine goes rich a voltage of approximately 1·0 V is generated. This provides a

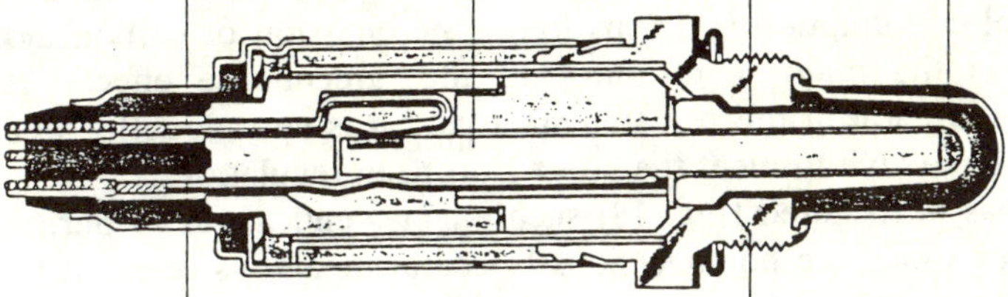

Fig. 32. Heated zirconia lambda sensor.

signal to the microprocessor which in turn provides a signal to the electrical control unit which controls the duration of opening of the fuel injector, thus restoring the ratio to $\lambda = 1$. This system is thus a 'feedback' control system. It will be noted that there is almost a step change for a small increase or decrease in the A/F ratio about stoichiometric, thus making the system sensitive. Although this type of cell will operate without electrical heat an electric heating element is usually incorporated for stability, quick operation from a cold start, and the reduction of poisoning from contaminants in the fuel. The principal developer of this type of cell for automotive use is the Bosch organization.

Titania (TiO_2) sensors are also being developed, which are claimed to have some advantage over the lambda sensor. These are in the form of thick-film semiconductors which change their electrical resistance with oxygen potential and temperature. Temperature is controlled by a heating element, and with an appropriate circuit an electrical voltage response to A/F can be obtained similar to that of the lambda cell. It has, however, not yet reached the production stage.

6.5. The Lean-burn Engine Approach

There are two different methods of obtaining very lean burn. The first is the use of the stratified charge system, treated in detail in Chapter 5, and the second is the so-called high compression lean-burn system, addressed in Chapter 1.

The advantage of lean-burn is that it is the only method of reducing the main pollutants CO, NO, and HC and at the same time improve the fuel consumption and thus lower the emission of carbon dioxide which is the major contributor to the 'greenhouse effect'. These engines are still in the research stage and it is very doubtful whether they will be able to meet the latest European regulations.

As has been noted (Fig. 18) such engines endeavour to burn weak mixtures which are homogeneous in character and as constant as one can achieve. If a large reduction in NO_x is required, the A/F ratio needs to be about 20:1. Unfortunately, as the mixture is weakened the rate of burning (heat release) becomes slower and, generally speaking, there is a loss of efficiency. However, by raising the compression ratio and increasing turbulence in the combustion chamber the speed of burning can be increased again, but unfortunately high compression ratios cause a greater tendency towards 'knock'. Knock is familiar to most people who can remember the low-octane fuel days during and following the Second World War when most engines 'knocked' at low speeds and high loads, namely when accelerating at full throttle, or climbing hills. This low speed knock does not damage the engine and dies out with high speed. However, 'high-speed knock' also occurs if the compression ratio is too high and the additional heat generated causes pre-ignition with subsequent piston failure. It is possible to prevent this knock by ensuring that the ignition setting is prevented from advancing into the sensitive areas, but clearly this requires much more precise ignition equipment than is commonly used. However, such equipment is indeed available and is being progressively introduced; this includes a knock sensor.

Because of the great reduction and ultimately elimination of lead from fuel the octane number (research method) of premium fuel is likely to drop from the present (1989) 98 octane to not more than 95, and this means a compromise in the compression ratio (CR) selected. A modest CR of 9:1 could be used with a 95 octane fuel with only a small retardation at low speeds on full load. At 12:1, about the optimum ratio for fuel economy, considerable retardation at low speeds and some retardation at high speeds is necessary for near full load operation. However, this may be still the preferred method for small- and medium-sized engines to meet the new proposed emission standards (see Section 5.2.2). The reason for preferring this method is that at part load, under cruising conditions, knocking is not a problem and some loss of power and economy near full load is tolerated. Having said that designers of lean-burn engines aim to utilize a weak homogeneous charge, this can lead to ignition problems necessitating high-energy sparks and it is desirable to help this situation by trying to produce a richer mixture in the vicinity of the spark plug; that is, a degree of stratification. This is achievable with high swirl, as developed in the combustion chamber due to Weslake and used in many Austin Rover engines (Fig. 33). The four-valve engine[42] is also good for lean burn due to its high turbulence, giving high flame speed.

The design details of lean-burn engines have to balance the concerns of nitrogen oxide and hydrocarbon emission and fuel economy. The situation is shown in Figs 34 and 35, in which pollutant concentrations are plotted against A/F ratio. The results shown are for an experimental engine running under cruise conditions with two types of cylinder head, one with a four-valve hemispherical combustion chamber (CC) and the other with a two-valve biscuit-shaped CC. It will be noted, however, that with the four-valve cylinder head the better filling gives an even worse NO_x situation near stoichiometric A/F, but combustion can be continued to a weaker air: air fuel ratio where low NO_x is obtained.

Due to the high induced turbulence with the four-valve cylinder head fast burning is achieved, with improved fuel consumption. The hydrocarbon emissions are numerically higher with the four-valve, but can be maintained to a considerably weaker A/F ratio. The balance of economy against pollution concentration is shown in Fig. 36. HC are

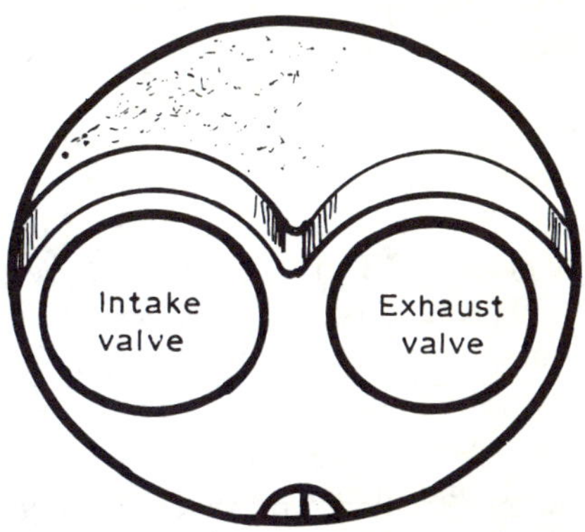

Fig. 33. Weslake combustion chamber.

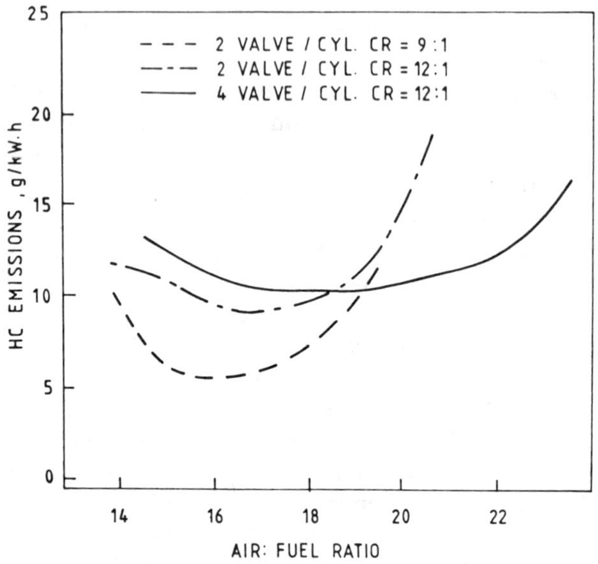

Fig. 34. Relationship between hydrocarbon emissions and air : fuel ratio for an experimental 500 cm^3, single-cylinder petrol engine (2000 rpm, 2 bar BMEP).

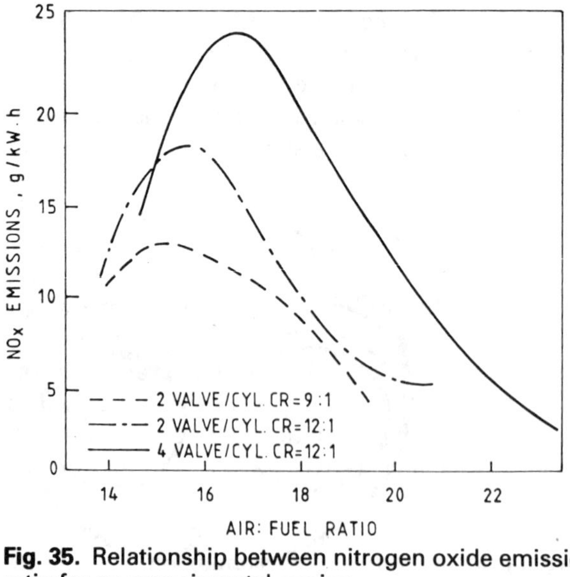

Fig. 35. Relationship between nitrogen oxide emissions and air : fuel ratio for an experimental engine.

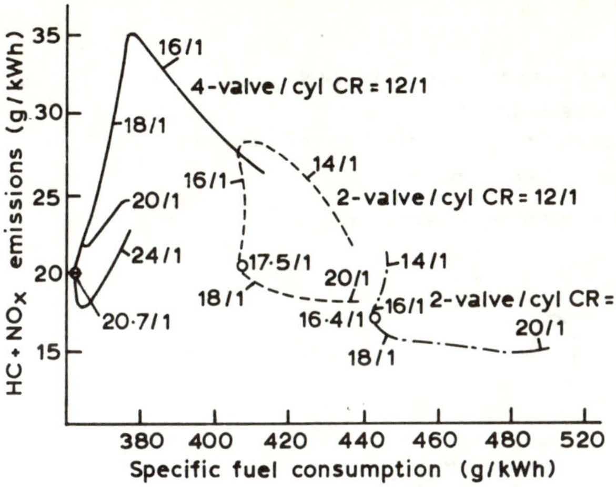

Fig. 36. Emissions and fuel consumption for an experimental engine.

added to NO_x as in the current EEC regulations. It will be seen that the lowest pollutants are obtained for the two-valve 9:1 CR engine, but this is at a considerable fuel penalty. A much improved fuel economy, with little sacrifice in pollution concentration, is obtained with the four-valve engine at 12:1 CR, provided that it can operate satisfactorily at about 20–22:1 A/F ratio.

The pollution reductions described above are for a single-cylinder research engine. For the practical multi-cylinder situation major problems are encountered because engine power is reduced at high A/F so that a mixture nearer to stoichiometric is necessary to achieve adequate vehicle acceleration and maximum speed. Despite some difficulties, lean-burn engines are of major interest to European manufacturers and are expected to be able to satisfy the prospective regulations (88/76/EEC) for the smallest engine sizes (below 1·4 litres).

Ricardo have made a valuable contribution for evaluating the emission potential of engines in this range. Single-cylinder research was performed on four types of combustion chamber and one of these, the High Ratio Compact Chamber (HRCC), was chosen for full pollution investigation on a vehicle.[43] Pollution results for the four combustion chambers are shown in Fig. 37. These curves were obtained by starting with a rich mixture and weakening off until misfire occurs with very high hydrocarbon emission. It will be noted that

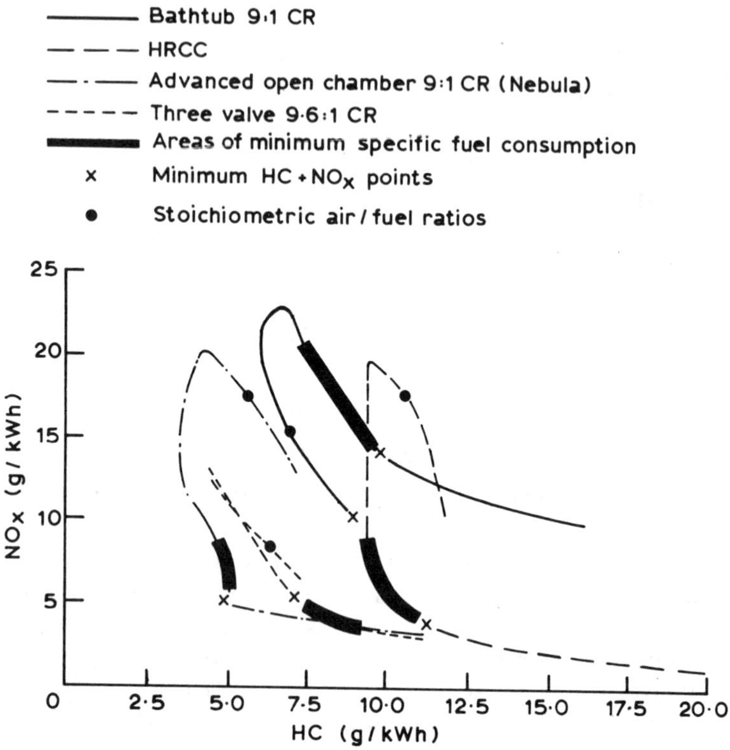

Fig. 37. NO_x–HC trade-off mixture response (2400 rpm, 2·5 bar BMEP; Ricardo) (from Institute of Mechanical Engineers).

minimum HC plus NO_x is obtained with the nebula combustion chamber, while the bathtub, although good on NO_x, is poor on HC. The best A/F ratio for the nebula chamber was 19·5.

The HRCC chamber (11:1 compression ratio) was incorporated in a production VW 1·5 litre engine and installed in a car with an oxidation catalyst for pollution and economy evaluation over a range of A/F ratios. Results are shown in Fig. 38. The Euromix fuel consumption is the arithmetic mean of the urban test and the 90 and 120 km/h tests.

It will be noted, somewhat surprisingly, that the fuel consumption remains fairly constant but that the HC + NO_x, as well as the CO, are well reduced as the mixture is leaned off.

These tests demonstrate well the potential of the lean-burn system.

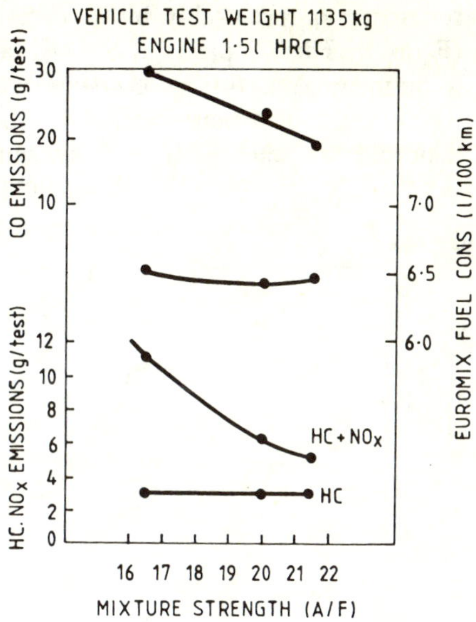

Fig. 38. Effect of air : fuel ratio on emissions and economy of a 1·5 litre car fitted with a Ricardo HRCC (11 : 1 CR) engine (from Institute of Mechanical Engineers).

6.6. Exhaust Gas Recirculation

Exhaust Gas Recirculation (EGR) is a simple method of reducing oxides of nitrogen. Carbon dioxide is a major constituent in the exhaust gases. There is also a reasonably large proportion of water vapour derived from the hydrogen in the fuel and oxygen in the combustion air. These two gases have high specific heats and these properties enable the exhaust gas to be used to reduce the temperature of the combustion process in the cylinder. The effect of EGR is shown in Fig. 39 for work on a single-cylinder research engine. It will be noted that there are two sets of curves, one for a 9:1 compression ratio (CR) engine (conventional) and one for a 12:1 CR lean-burn engine, both at part load. It will be observed that although NO_x decreases with increasing EGR, hydrocarbons tend to increase, particularly in the case of the conventional engine. Exhaust gas recirculation is therefore usually limited to 5–10%.

Exhaust gas can conveniently be induced into the inlet manifold by virtue of the fact that the pressure is below that pertaining in the exhaust pipe. However, the quantity has to be controlled either by a restriction or preferably by a special exhaust gas recirculating valve (Fig. 40). This simple type of valve allows the amount of exhaust gas to be metered according to engine requirements.

It consists of a pintle valve situated between the exhaust and inlet manifold and actuated by a diaphragm, one side of which is subject to

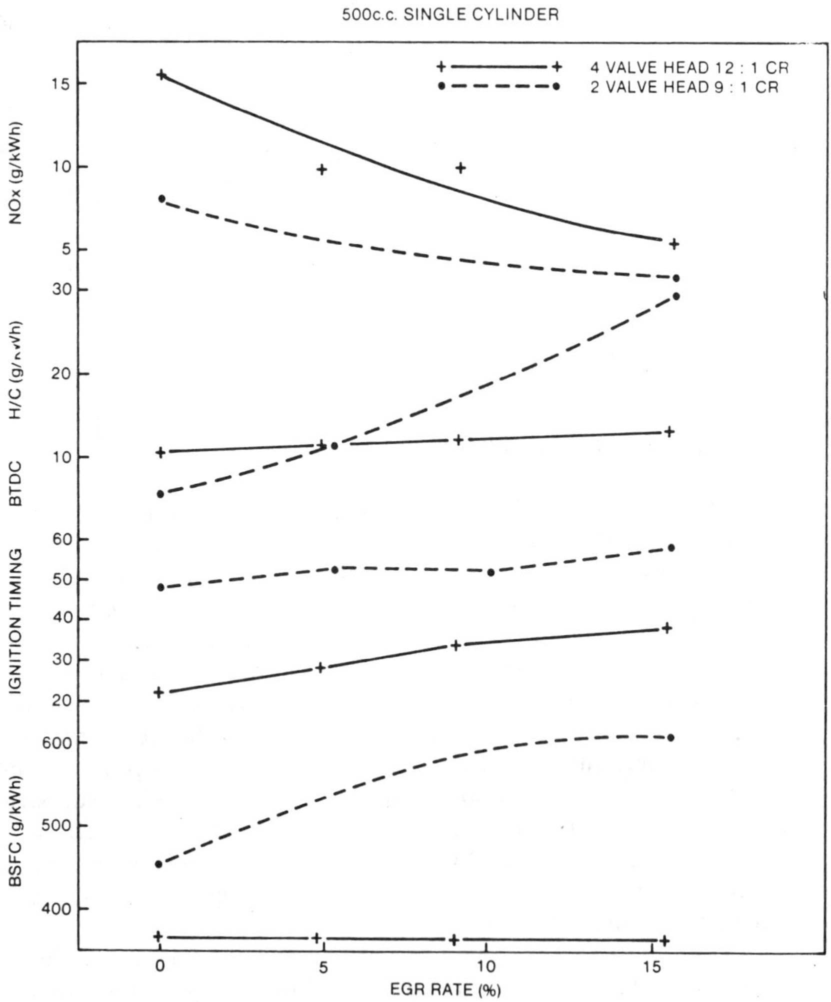

Fig. 39. Effect of EGR at 2000 rpm, 2·0 bar, 18 : 1 AFR, on NO_x HC emissions, BSFC and ignition timing.

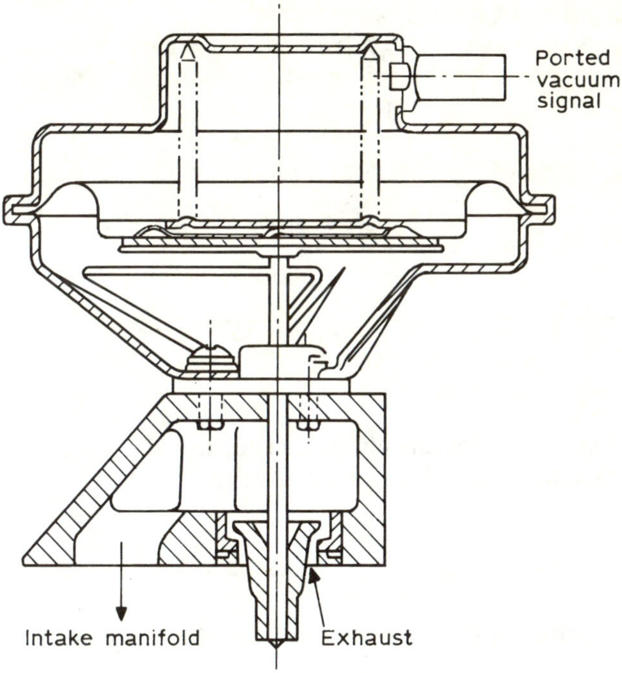

Fig. 40. An example of an exhaust gas recirculation valve.

the inlet manifold pressure but resisted by a calibrated spring to suit the application. With a complete engine management system the pressure on the diaphragm may be controlled to provide any desired flow characteristic for speed and load.

EGR can be useful for normal and lean-burn engines and also for diesels.

6.7. Ignition Retardation

As a last resort to meet regulations for nitric oxide, the retardation of the ignition is very effective. This works by reducing the temperature of combustion and hence the rate of formation of NO: this is

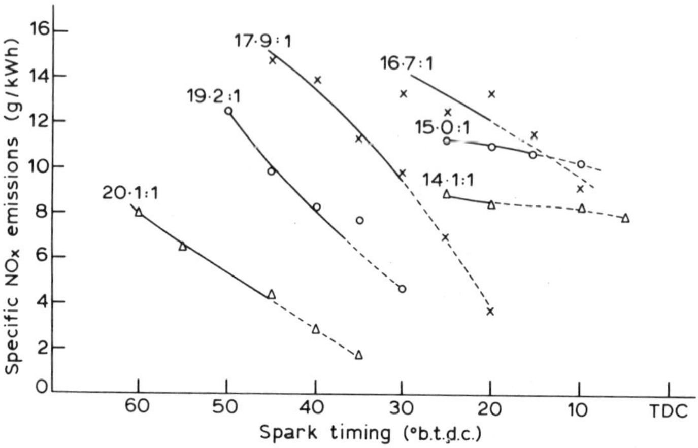

Fig. 41. Effect of ignition retardation.

demonstrated in Fig. 41. Unfortunately, it also reduces the efficiency of the engine and should be avoided if at all possible.

7. DIESEL ENGINE EMISSIONS

7.1. Comparison Between Diesel and Spark-ignition Engine Exhaust Emissions

Compared to spark-ignition engines, diesel engines not only have the advantage of low fuel consumption, but also of low pollutant emissions, as it can be seen in the example shown in Table 11.

Table 11. Comparison of Pollutant Emissions of Diesel and Spark-ignition Engines on Cold ECE Test

Passenger vehicle			Pollutant emissions on cold ECE test (g/test)		
Inertial weight (lb)	Engine displacement (litres)	Engine type[a]	HC	CO	NO_x
3000	2·0	SI	12	60	10
3000	2·0	IDI	2·5	6	4

[a] SI = spark-ignition engine; IDI = indirect injection diesel engine.

Table 12. Typical Examples of Particulate Emissions on U.S. F.T.P. 75 Cycle

	Particulate emissions on U.S. F.T.P. 75 cycle (g/mile)
SI engine without catalyst	0·05
SI engine with catalyst	0·015
IDI diesel engine	0·5–0·7

It will be noted that CO emissions, about ten times lower in the case of diesel engines as compared to SI engines, are generally far below the required standards.

Due to the principle of combustion in the diesel engine, the mixture of air and fuel is very heterogeneous. This causes particulates and smoke formation. Table 12 shows that, in this field, diesel engines have a disadvantage as compared to SI engines.

7.2. Basic Mechanisms of Pollutant Formation in Diesel Engines

The main physical and chemical processes which occur in the combustion chambers of all DI or IDI diesel engines are very similar.

Three phases are generally distinguished in a combustion cycle:

(1) *The delay period.* This period corresponds to the time lapse between the beginning of the injection and the beginning of the combustion. During this period, fuel is injected, partially evaporated and mixed with air. Some pre-reactions occur but do not give a significant increase in pressure.

(2) *Period of rapid combustion.* At the end of the delay period, almost all of the fuel evaporated and, mixed with air, burns very rapidly, producing a sharp increase of pressure and temperature. High temperatures in the presence of oxygen explain the formation of nitrogen oxides in this phase of combustion.

(3) *Third phase of combustion.* During this phase the combustion rate is limited by the mixing process between fuel and air, and by the rate of injection if all the fuel has not been injected at this time. The combustion is then much slower than during the previous phase. In the zones where fuel is not mixed with air,

high temperatures can produce the cracking of fuel molecules, and hence smoke and particulates formation.

The general mechanisms of formation of pollutants in diesel engines can be summarized as follows:

(1) *Nitrogen oxides* (NO_x) are formed when the temperature is high in presence of air, mainly during the rapid combustion period. Therefore an increase in the quantity of fuel injected during the delay period increases the maximum temperature and hence the NO_x formation.
(2) *Carbon monoxide* (CO) is formed when there is a lack of oxygen during combustion, either near full load, or sometimes also at low loads when there is a poor mixing of fuel with air (too large fuel droplets, insufficient swirl, etc.).
(3) *Hydrocarbons* (HC) are mainly due to fuel which is unburnt because of insufficient temperature: this is the case of fuel near the walls of the combustion chamber ('wall quenching'), and of fuel not yet burnt at the end of the combustion for lack of pressure and temperature ('bulk quenching').
(4) *Particulate emissions* are due to several phenomena:
    (a) *Nucleation*: formation of polyaromatic nuclei from condensation of hydrocarbons at high temperature and pressure in the absence of sufficient oxygen.
    (b) *Surface growth*: graphitic carbon makes a deposit around these nuclei which grows into small spheres of about 0·01–0·05 μm in diameter.
    (c) *Coagulation of these spheres* to form chain-like coagulates.
    (d) *Re-oxidation* in the presence of oxygen at high temperature. This re-oxidation is influenced by the gas speeds, which are very high in swirl-chamber engines.
    (e) *Adsorption of hydrocarbons* on the particulates, during the whole expansion stroke and continuing in the exhaust system.

Therefore particulates which are emitted have an equivalent diameter of 0·1–0·3 μm, and consist of two parts: the SOF (Soluble Organic Fraction), which contains mainly hydrocarbons; and carbon black and other insoluble matter.

As a function of load, the general shape of particulate emissions is

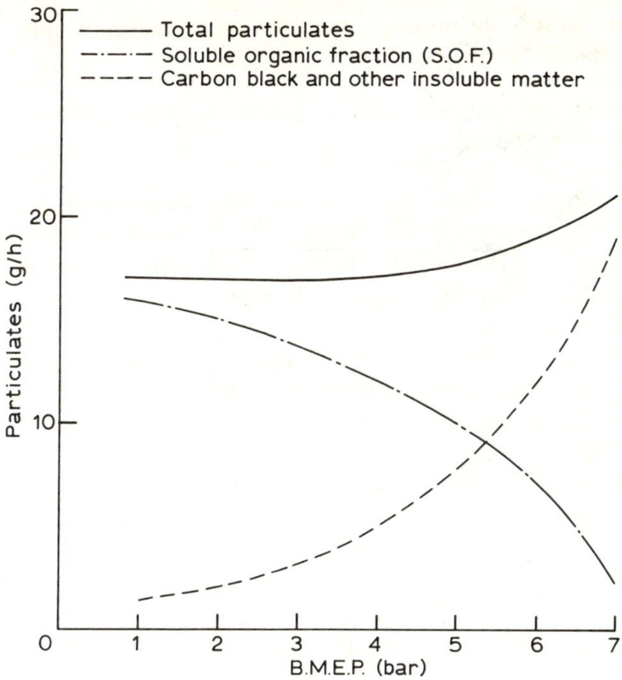

Fig. 42. Example of particulate composition as a function of load (2 litre IDI engine; 2500 rpm).

shown in Fig. 42. At high load, the particulates are mainly formed of carbon, by lack of oxygen during the combustion. At low loads, the particulates have a high SOF content due to adsorption of hydrocarbons remaining from an incomplete combustion.

Sometimes, if oil consumption is too high, particulates are formed from oil molecules: but, as these hydrocarbons are heavier than those in diesel fuel, they can be recognized by thermogravimetric analysis of particulates.

7.3. Emissions from IDI Diesel Engines

7.3.1. The Effect of the General Principle of IDI Diesel Engines on the Combustion Process

Combustion takes place in two separate combustion chambers, namely: the *pre-chamber,* where the fuel is injected; and the *main*

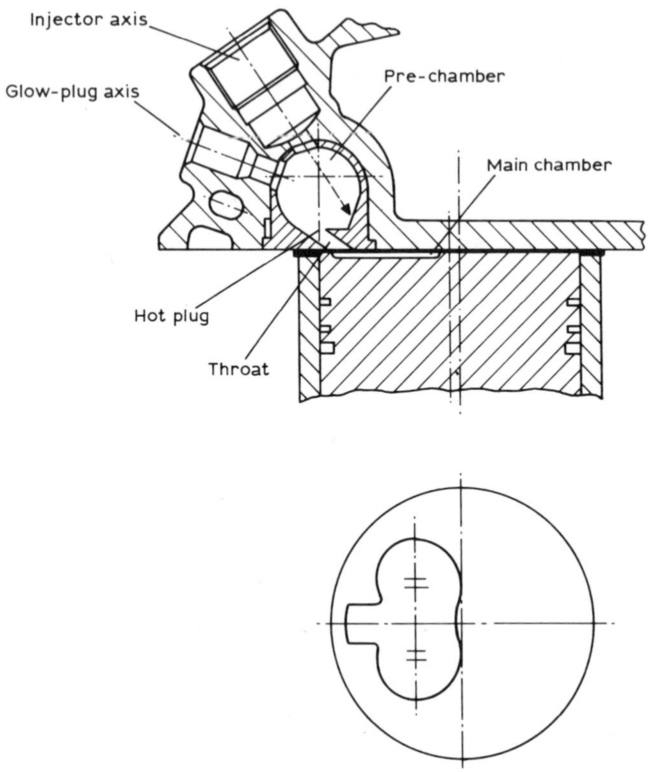

Fig. 43. IDI diesel engine.

chamber, situated above the top of the piston and communicating with the pre-chamber through a narrow passage (throat) (see Fig. 43).

The objectives of this design are the following:

(a) *During the compression stroke,* the air passing through the throat has a high velocity, and that allows good mixing of fuel and air with a pintle nozzle injector. This type of injector, with a single hole and a relatively low injection pressure (15–25 MPa), cannot by itself provide a good mixing of fuel and air without aerodynamic assistance. Conversely, the big advantage of this type of injector is to have a variable hole area as a function of needle lift, so that the quantity of fuel injected during the delay period can be low: consequently, noise and NO_x emissions can be better controlled.

(b) *During the expansion stroke,* the hot gases of the pre-chamber are expelled at high speed into the main chamber in a partially burnt state. This allows a very good use of air (relative equivalence ratios can reach 0·85) for a wide range of engine speeds.

7.3.2. NO_x Emissions

NO_x emissions (Fig. 44) increase with increasing of speed because of: the increase of gas temperatures; and the increase of injection advance (since injection duration increases with engine RPM, injection advance has to be increased to avoid combustion too late in the cycle).

When progressively increasing the load, the NO_x emissions show a maximum and then decrease because the temperature and oxygen concentration vary in opposite ways.

7.3.3. HC Emissions

HC emissions (Fig. 45), at least at low and moderate values of engine speed, have a relatively constant value, independent of load and speed.

On some swirl chamber engines, HC emissions can show an increase at low loads:

(a) at low speed, because of too slow combustion ('bulk quenching' during the expansion stroke);

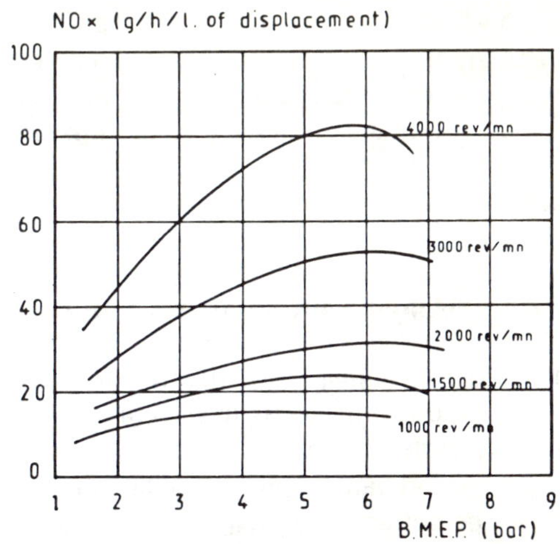

Fig. 44. NO_x emissions of IDI diesel engine.

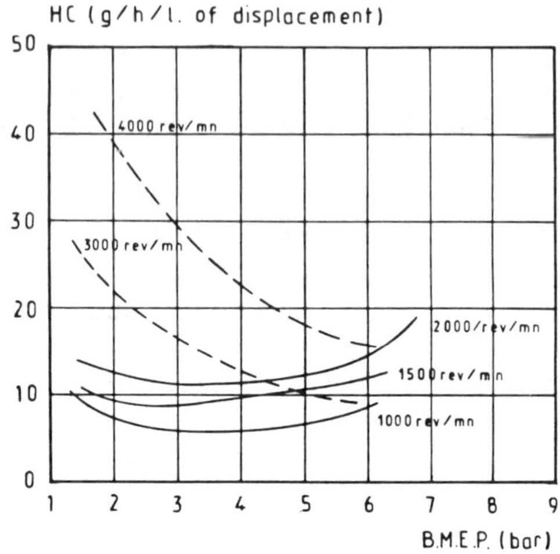

Fig. 45. HC emissions of IDI diesel engine.

(b) at high speed, because the swirl in the pre-chamber is high, and the fuel going out of the nozzle has a relatively low kinetic energy—it is supposed that the fuel is pushed by the swirl to the wall of the pre-chamber and cannot burn (wall quenching).

7.3.4. Local Gas Composition in the Swirl Chamber

Some idea of gas composition inside the swirl chamber can be given from sampling experiments (see Fig. 46).

The fuel spray, not injected towards the centre of the chamber is, furthermore, deviated by the swirl. Therefore the fuel : air ratio is very high near the walls, generating CO and HC formation. Near the centre of the pre-chamber, however, the mixture is relatively lean, and the presence of oxygen in a high-temperature zone explains the high concentration of NO_x in this area.

7.4. Emissions from DI Diesel Engines

In most of these engines, the fuel is injected by a multi-hole nozzle in order to distribute the fuel in several zones of the combustion chamber (see Fig. 47).

As soon as the needle lift begins, practically the whole area of holes is available for the passage of fuel, which is injected at a high pressure (35–80 MPa). Therefore, although the ignition delay is generally shorter than in IDI engines, the quantity of fuel injected during the delay period is much higher. As a consequence, the noise and the NO_x emissions are also much higher (see Fig. 48).

HC and particulate emissions are similar to those of IDI engines. Their variation as a function of load and speed is, of course,

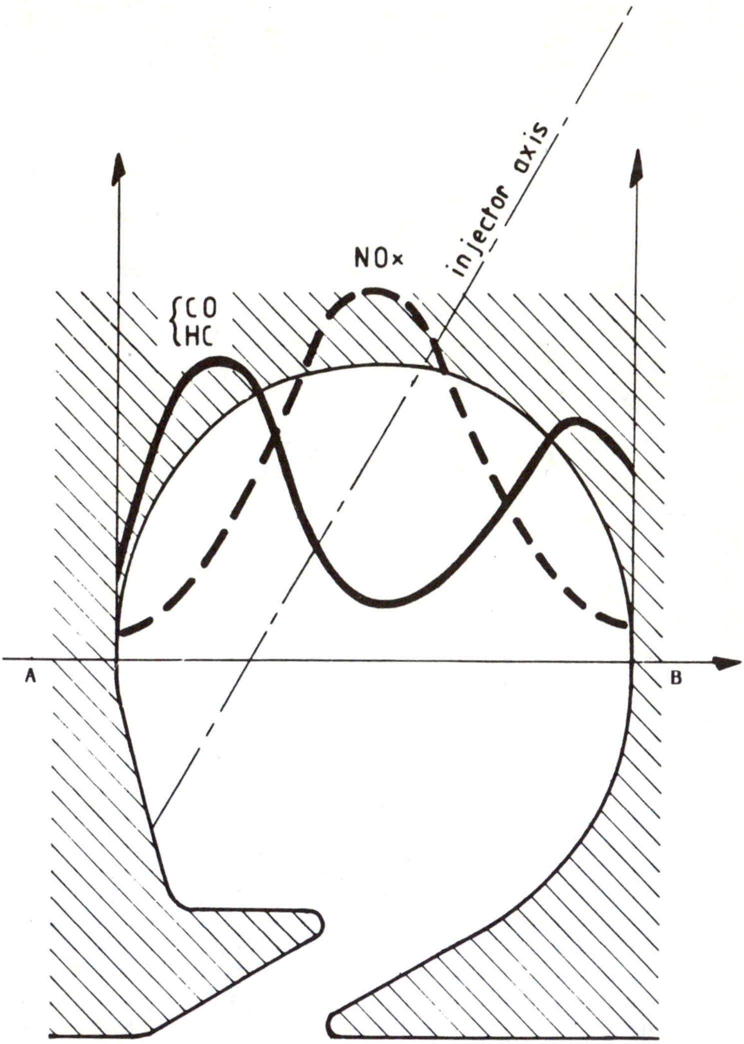

Fig. 46. Local gas composition in swirl chamber of IDI diesel engine along the A–B line.

Fig. 47. DI diesel engine.

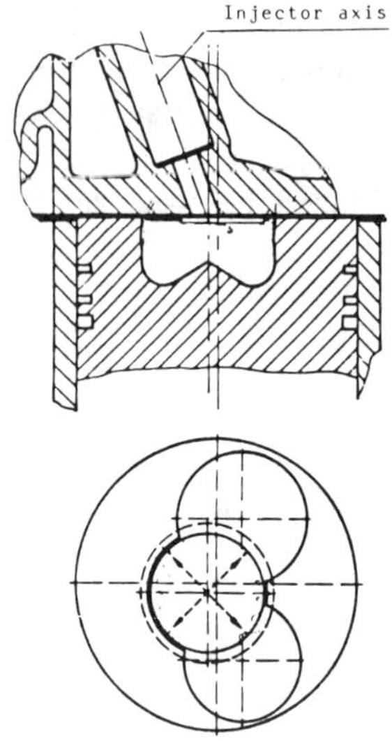

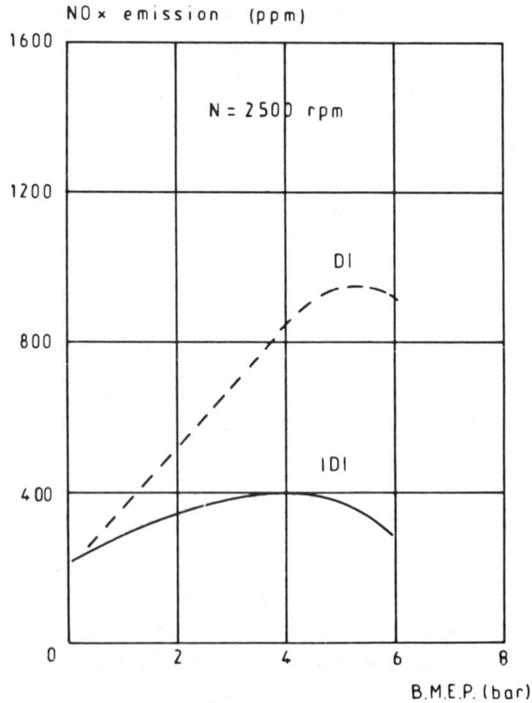

Fig. 48. Comparison between DI and IDI engine NO_x emissions.

dependent on engine design (on intake-port-generated swirl, on combustion chamber shape, on injection jet divergence angle, etc.).

7.5. Fuel Effects on Diesel IDI Exhaust Emissions

Although distillation properties have a slight effect on combustion (for instance 90% distillation temperature), diesel IDI exhaust emissions are chiefly influenced by fuel composition. Aromatic content and cetane number are the most significant parameters of fuel composition. Their influence can be more or less important from one engine to another; examples are shown in Figs 49 and 50.

NO_x emissions and fuel consumption are almost unchanged, but HC, CO and particulate emissions are generally strongly affected by fuel composition.

7.6. Diesel Particulate Filters

7.6.1. General Considerations on Dimensions and Quantity of Particulates

Particulates emitted by diesel engines have dimensions generally inferior to 1 μm, with a mean value of about 0·2 μm.

The speed at which such particulates would fall under the influence of their own weight can be obtained from the following formula:

$$6\pi\eta r u = \tfrac{4}{3}\pi r^3(\rho - \rho_0)g$$

in which ρ is the density of particulates (about 1000 kg/m^3); ρ_0 is the density of air (0·67 kg/m^3 at 300°C); η is the air viscosity ($29{\cdot}5 \times 10^{-6}$ da P at 300°C); g is the acceleration due to gravity (9·81 m/s^2); and r is the particulate radius ($0{\cdot}1 \times 10^{-6}$ m).

From this formula one can calculate

$$u = 7{\cdot}4 \times 10^{-7}\ \text{m/s} = 6{\cdot}4\ \text{cm/day}$$

This extremely low value explains why particulates cannot be separated from exhaust gases by centrifuging and settling.

A diesel particulate filter on the exhaust line has been tested as a method of reducing particulate emissions. If it is supposed that the trapping efficiency of the filter is about 70%, and that the particulate

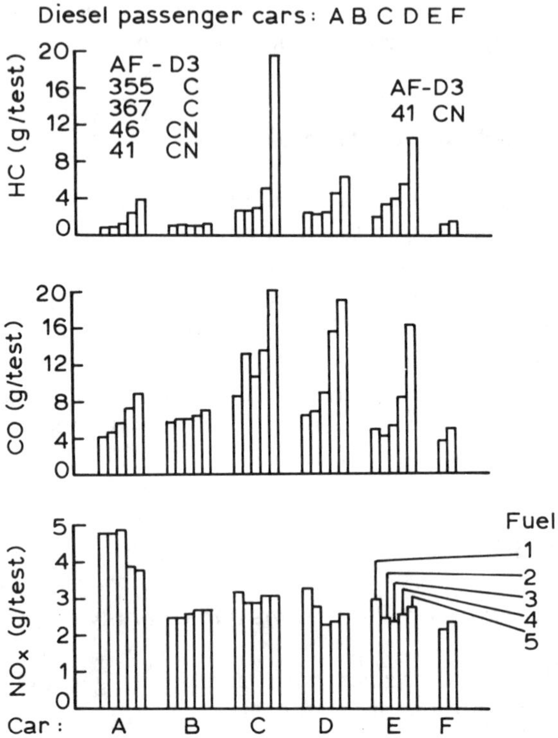

Effect of cetane number and 85% point on gaseous emissions of passenger vehicles (ECE-R 15/04 test results)

A, B, C, D, E : IDI/Swirl chamber | F : IDI/Prechamber

Fuel	Temperature of 85% point	Cetane number
1	320	53
2	355	49, 5
3	367	49, 5
4	322	46
5	325	41

Fig. 49. Influence of fuel composition on diesel IDI exhaust emissions (from Institute of Mechanical Engineers, 1986, C 306/86).

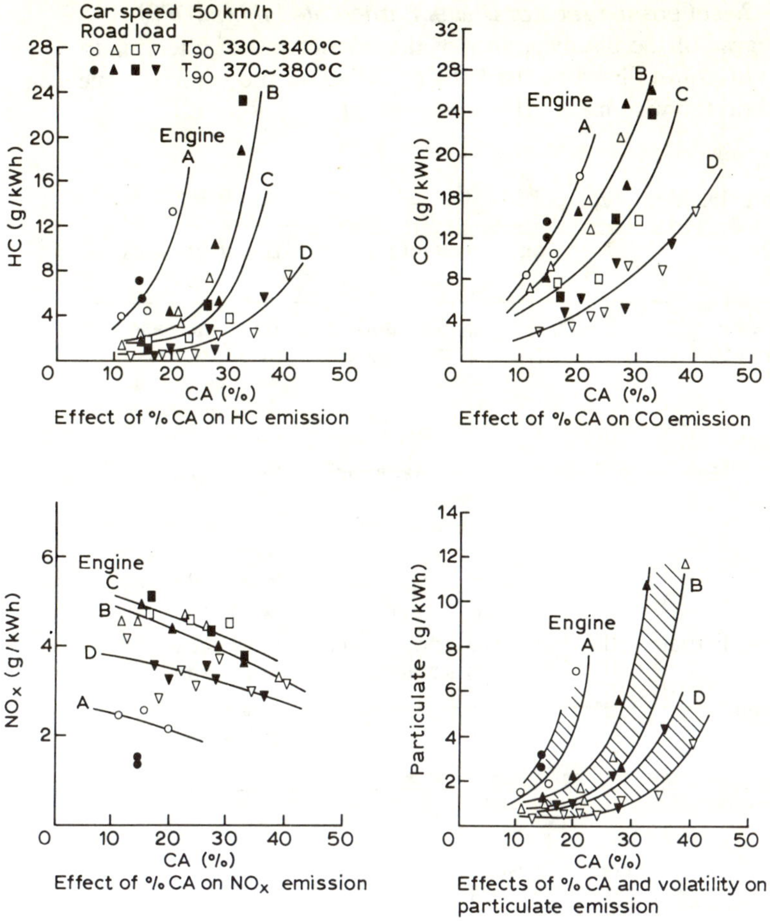

Fig. 50. Influence of fuel composition on diesel IDI exhaust emissions (%CA = aromatic ring carbon content) (reprinted with permission of the Society of Automotive Engineers, 1986, 860619).

emissions are about 0·2 g/km, the filter will collect about 14 g of particulates each 100 km. To avoid clogging of the filter, which generally begins to occur when it contains more than 20 g of particulates, it is necessary to regenerate the filter (by burning the particulates) each 140 km.

7.6.2. Different Types of Diesel Particulate Filters (DPF)

Because of the disadvantages of the other types of filters, automotive manufacturers have focused their studies on the ceramic honeycomb and on the wire mesh DPFs.

Table 13. Main Types of DPF

DPF type	Trapping efficiency	Advantages	Disadvantages
Ceramic honeycomb (Fig. 51)	60–90%	good trapping efficiency, low cost	rapid increase of back-pressure with clogging, thermal shock resistance
Wire mesh (Fig. 52)	30–60%	thermal shock resistance, slow clogging	low trapping efficiency, decreasing at high speed and when filter is clogged
Ceramic foam	30–75%	thermal shock resistance	high back-pressure
Pellets	30%	trapping of HC	low trapping efficiency, wear

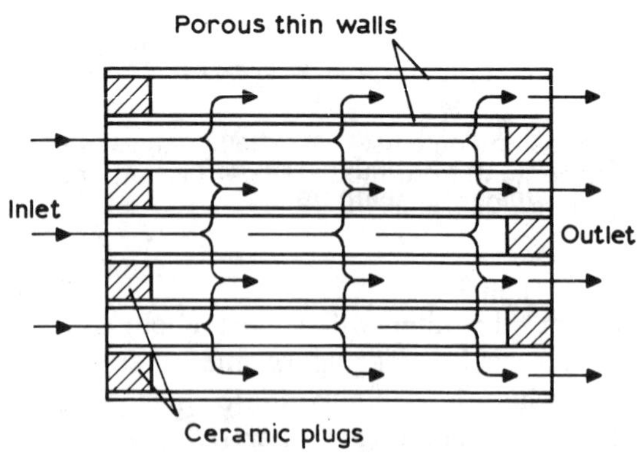

Fig. 51. Schematic of a ceramic honeycomb diesel particulate filter.

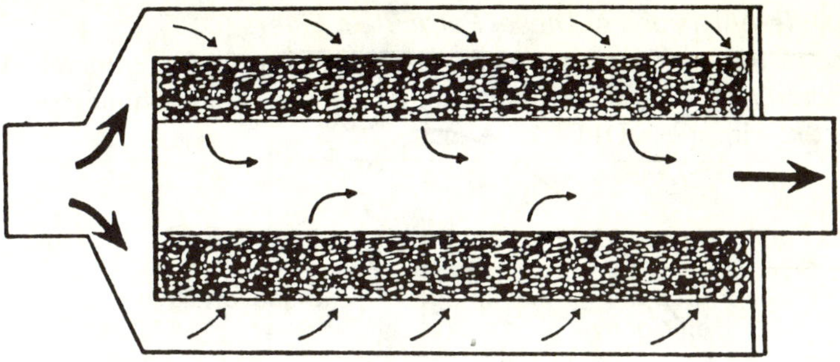

Fig. 52. Schematic of a wire-mesh diesel particulate filter.

7.6.3. Regeneration of a DPF

The burning of carbon deposits on a DPF needs a minimum temperature and oxygen content.

In Table 14 are given typical temperature levels needed as a function of the catalytic action of the filter and of the SOF content of the particulates.

One cannot be sure that these temperatures will be obtained during car use, as can be seen on the speed–load diagram of Fig. 53.

One must not forget, on the other hand, that if the quantity of carbon collected is too high, there is a risk of overheating of the DPF during regeneration.

Several methods have been tested to increase the temperature of exhaust gases in the DPF:

(a) Mounting of the DPF as close as possible to the exhaust ports.

Table 14. Typical Regeneration Temperatures of Diesel Particulate Filters

DPF with catalytic wash-coat[a]		SOF content of particulates		Regeneration temperature (°C)
No	Yes	Low	High	
×		×		600
	×	×		500
	×		×	350

[a] Catalytic action can also be provided by fuel additives.

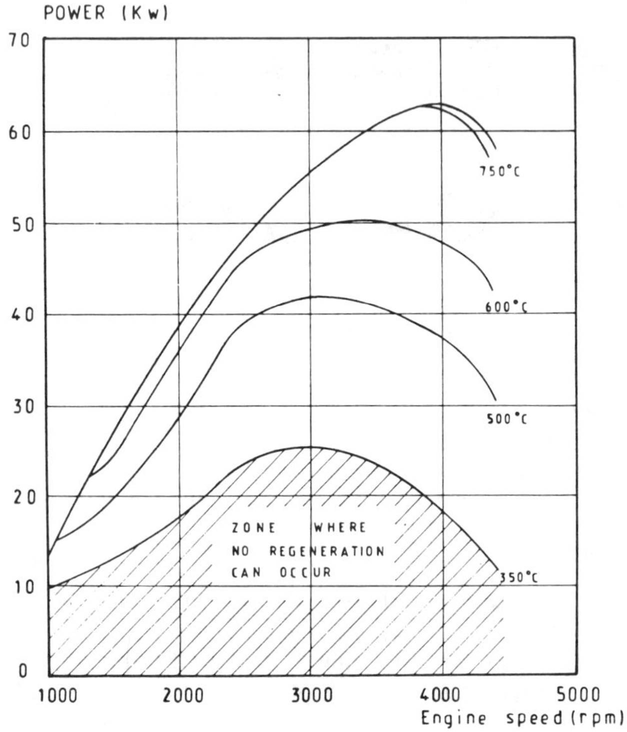

Fig. 53. Typical exhaust gas temperatures of an IDI diesel engine.

(b) Throttling the intake manifold (this technique does not increase the temperature enough at low load and low speed conditions).

(c) Use of a burner: this technique allows a good regeneration of the filter at low load and low speed, but it necessitates a precisely controlled air and fuel supply to obtain a stable combustion for different exhaust gas flow levels. The combustion in the burner also generates HC and CO emissions, increases the fuel consumption, and is a possible cause of fire hazard. Because of the great complexity of this system it does not seem realistic to apply it to light duty vehicles.

Although several regeneration techniques have been worked on, none has yet proved to be satisfactory.

REFERENCES

1. Reynolds, S. D., Roth, P. M. & Seinfield, J. H., Mathematical modelling of photo-chemical air pollution. *Atm. Environ.* **7** (1973) 1033–61.
2. Whitten, G. Z., Hugo, M. & Killus, J. P. The carbon-bond mechanism for photochemical smog. *Environ. Sci. Technol.*, **14** (1980) 690.
3. Derwent, R. G., A computer modelling study of the relationship between urban exposure to nitrogen dioxide and motor vehicle exhaust emissions. In *Air Pollution by Nitrogen Oxides*. Studies in Environmental Science, **21**. Elsevier, 1982, p. 309.
4. Weaving, J. H. & Gaddo, P. P., Urban scale mathematical model for pollutants in a European city. In *Air Pollution by Nitrogen Oxides*. Studies in Environmental Science, **21,** Elsevier, 1982, p. 289.
5. Builtjes, P. J. H., Stern, R. M. & Reynolds, S. D., PHOXA: the use of a photochemical dispersion model for several episodes in north-western Europe. In *Air Pollution Modelling and its Application,* **VI**. Plenum, 1988.
6. Lawther, P. J., Lead and the environment. Report of DHSS Working Party on Lead and Health. HMSO, London, 1980.
7. Chamberlain, A. C. *et al.*, Investigations into lead from motor vehicles. HL 78/4122(c10) AERE, Harwell, 1978. HMSO, London, c10.
8. Needleman, H. L., Gunnoe, C., Leviton, A., Reed, R., Peresie, H., Maher, C. & Barrett, P., Deficits in psychologic and classroom performance of children with elevated lead levels. *New England J. Med.*, **300** (1979) 689–95.
9. Harvey, P. G. *et al.*, Relationship between blood-lead, behaviour, psychometric and neuropsychological test performance in young children. *Br. J. Develop. Psych.*, **6** (1988) 145–56.
10. Kulle, T. J., Effects of nitrogen dioxide on pulmonary function. In *Air Pollution by Nitrogen Oxides*. Studies in Environmental Sciences, **21**. Elsevier, Amsterdam, 1982, p. 477.
11. Mellanby, K. (ed) *Air Pollution, Acid Rain and the Environment* Watt Committee Report No. 18. Elsevier, London, 1988.
12. Veninga, H. E., Comparison of biological effects of NO_2, O_3 using mice. Effects of nitrogen dioxide on pulmonary function. In *Air Pollution by Nitrogen Oxides*. Studies in Environmental Science, **21**. Elsevier, Amsterdam, 1982, p. 441.
13. Blank, L. W. & Roberts, T. M., Ecological effects of nitrogen oxides and hydrocarbons on Central European forests. Paper C352/87, Instn Mech. Engrs Conf. on Vehicle Emissions, 1987.
14. Kier, L. D., Yamasaki, E. & Ames, B. N., Detection of mutagenic activity in cigarette smoke condensate. *Proc. Natl Acad. Sci., USA,* **71** (1974) 4159–63.
15. Garner, R. C. & Campbell, J., Nitrated polycyclic aromatic hydrocarbons: potent bacterial mutagens and stimulators of DNA repair synthesis in cultured human cells. *Carcinogenesis,* **2**(6) (1981) 559–65.
16. Vostal, J. J. & Garg, B. D., Studies on the lymphatic transport of

particulates in the lungs of rats and guinea pigs exposed to diluted diesel exhaust. International Symposium on Health Effects of Diesel Engine Emissions, EPA, Cincinnati, 1979.
17. Waller, R. E. & Lawther, P. J., Trends in lung cancer in London in relation to exposure to diesel fumes. International Symposium on Health Effects of Diesel Engine Emissions, EPA, Cincinnati, 3–5 December 1979.
18. Brightwell, J., Fouillet, X., *et al.*, Neoplastic and functional changes in rodents after chronic inhalation of engine exhaust emissions. *Conf. Carcinogenic and Mutagenic Effects of Diesel Engine Exhaust.* Elsevier, 1986; Brightwell, J., Fouillet, X., Cassano-Zoppi, A.-L. *et al.*, Tumours of the respiratory tract in rats and hamsters following chronic inhalation of engine exhaust emissions. *J. Appl. Toxicol.*, **9**(1) (1989) 23–31.
19. Clark, C. A., May, M. P. & Challen, B. J., Transient testing of diesel engines. SAE paper 840348, 1984.
20. Heywood, J. B., *Internal Combustion Engine Fundamentals.* McGraw-Hill, New York, 1988, pp. 596–620.
21. Namazian, M. & Heywood, J. B., Flow in piston ring crevices of an SI engine. SAE paper 820088; *Trans. SAE,* **91** (1982) 261.
22. Ma, T. H., Wood, W. A., Tanesar, A., Brown, P., Dent, J. C. & Lambert, N. in Tracing the sources of hydrocarbon emissions in engines—a review of several research programmes made over the past five years. Paper C326/87, Instn Mech. Engrs Conf. on Vehicle Emissions and their Impact on European Air Quality, 1987, p. 229.
23. Tabaczynski, R. J., Heywood, J. B. & Keck, J. C., Time-resolved measurements of hydrocarbon mass flow rate in the exhaust of a spark-ignition engine. SAE paper 720112; *Trans. SAE,* **81** (1972) 379.
24. Wentworth, J. T., The piston and ring variables affect exhaust hydrocarbon emissions. SAE paper 680109; *Trans. SAE,* **77** (1968) 402.
25. Yates, D. A., Hydrocarbon sampling from the combustion chamber of a lean-burn Engine. Paper C332/87. Instn Mech. Engrs Conf. on Vehicle Emissions and their Impact on European Air Quality, 1987.
26. Zeldovich, Y. B., The oxidation of nitrogen in combustion explosions. *Acta Phys.-chim., USSR,* **21** (1946) 577–628.
27. Gruden, D., Wurster, W. & Brachert, Th. F., Brake efficiency and NO_x emission of the spark ignition engine. In *Air Pollution by Nitrogen Oxides.* Studies in Environmental Science, **21**. Elsevier, Amsterdam, 1982, p. 803.
28. Newhall, H. K. & Shahed, S. M., Kinetics of nitric oxide formation in high-pressure flames. In *Proc. 13th Int. Symp. Combustion.* The Combustion Institute, 1971, pp. 381–90.
29. Starkman, E. S., Steward, H. E. & Zvonow, V. A., Investigation into formation and modification of exhaust emission precursors. SAE paper 690020, 1969.
30. Lavoie, G. A., Spectroscopic measurement of nitric oxide in spark-ignition engines. *Combust. Flame,* **15** (1970) 97–108.
31. Lavoie, G. A., Heywood, J. B. & Keck, J. C., Experimental and theoretical investigation of nitric oxide formation in internal combustion engines. *Combust. Sci. Technol.,* **1** (1970) 313–26.

32. Benson, R. S., Annand, W. J. D. & Baruah, P., A simulation model including intake and exhaust systems for a single cylinder four-stroke cycle spark ignition engine. *Int. J. Mech. Sci.*, **17** (1975) 97–124.
33. Evans, W. D. J. & Wilkins, A. J. J., Single bed, three-way catalyst in the European environment. SAE paper 852096, 1985.
34. Koberstein, E., Pletka, H. D. & Volker, H., Control of nitrogen oxide emissions from gasoline engines. In *Air Pollution by Nitrogen Oxides*. Studies in Environmental Science, **21**. Elsevier, Amsterdam, 1982.
35. Williamson, W. B., Gandhi, H. S., Szpilka, M. E. & Deakin, A., Durability of automotive catalysts for European applications. SAE paper 852097, 1985.
36. Cooper, B. J., Durability of platinum-containing automotive exhaust control catalysts. *Platinum Metals Rev.*, **27**(4) (1983) 146.
37. Cooper, J., Evans, W. D. J. & Harrison, B., Catalyst preparation, performance and durability. *Proc. Symp. 'Catalysis in Automotive Pollution Control'*, ed. A. Crucq & A. Frennet. Brussels, Elsevier, 1987.
38. Nonnenmann, M., Metal supports for exhaust gas catalysts. SAE paper 850131: *Trans. SAE* (1985) 814.
39. Harrison, B., Diwell, A. F. & Hallett, C., Promoting platinum metals by ceria. *Platinum Metals Rev.*, **32**(2) (1988) 73–83.
40. Evans, W. D. J., Searles, R. A. & Wilkins, A. J. J., Catalytic exhaust emission control—progress in application of catalysts in Europe. Paper C347/87, Inst Mech. Engrs Conf. on Vehicle Emissions, p. 284.
41. Harrison, B., Taylor, J. R., Diwell, A. F. & Salathiel, A., Lead species in vehicle exhaust: a thermodynamic approach to lead tolerant catalyst design. SAE paper 830268, 1983.
42. Weaving, J. H., Contribution to Watt Committee Report No. 18. Approaches to exhaust emission reduction. *Ibid* Ref. 11, pp. 91–3.

16

Instrumentation for Engine Flows

C. Arcoumanis

Department of Mechanical Engineering, Imperial College of Science, Technology and Medicine, London, UK

M. Yianneskis

School of Engineering, King's College London, University of London, London, UK

D. R. Williams, C. A. Baker & D. A. Greenhalgh

Harwell Laboratory, UK Atomic Energy Authority, Oxfordshire, UK

1. Introduction
2. Definition of mean flow and turbulence as appropriate to an internal combustion engine
3. Overview of flow measurement techniques
   3.1. Flow visualization
   3.2. Hot-wire anemometry
   3.3. Laser–Doppler anemometry
4. Scalar measurement techniques
   4.1. Laser Rayleigh scattering
   4.2. Laser Raman scattering
5. CARS diagnostic measurements in IC engines
   5.1. Introduction
   5.2. CARS spectroscopy overview
   5.3. Spectroscopy of other molecules
   5.4. Measurement of in-cylinder fuel: air ratios

1. INTRODUCTION

In order to improve combustion efficiency and reduce pollutants, previous chapters have highlighted the need for accurate experimental measurements under realistic engine conditions. First, to determine the state of the gas, we need measurements of pressure, temperature and the instantaneous volume of the combustion chamber. Secondly, for determining and predicting the fluid motion and combustion process, we need the velocity pattern of the air and the magnitude of both turbulence intensity and scale, together with the distribution and characterization of the fuel state for both diesel and spark-ignition engines.

Since the instrumentation available in the past was not sufficiently adequate to characterize the flow with the time- and space-resolution required, time and resources have been allocated to the development of laser-based techniques.

This chapter describes the instrumentation presently available in the research departments of the major engine manufacturers and universities involved in engine research to characterize qualitatively and quantitatively the flow in gasoline and diesel engines, and provides a few representative examples of their application to research and production engines. The discussion allows the reader to appreciate the advantages and limitations of each technique but it is not intended to substitute for the numerous review papers and textbooks available on the subject (see the Bibliography at the end of the chapter).

For the state of the gas, accurate instantaneous volume determination depends on having good temporal resolution by a shaft encoder, which is also needed for triggering electronic instrumentation at particular crank angles in the engine cycle. Pressure measurement, though still by no means perfect, has been greatly improved in commercially available pick-ups; these need frequent calibration, preferably under engine conditions. These techniques are well known and reference is made to them in Chapter 1, so they will not be described here.

The measurement of instantaneous gas temperature is difficult. Thermocouples, due to their thermal inertia, may not follow faithfully the gas fluctuations of temperature. For a motored engine fine platinum wire resistance thermometry may be used and by using two diameters of wire, a correction may be applied for lag. However, these are very delicate and would not survive in a fast gas stream. Optical spectral methods seem to offer the best alternative and these are described in the third section of this chapter.

For the fluid dynamics aspects, two well tried methods are now available—hot wire anemometry (HWA) and laser–Doppler anemometry (LDA). Both methods have certain advantages and are described below in Sections 3.2 and 3.3 respectively.

In characterizing the combustion process in a particular engine it is desirable to obtain a general concept of the whole flow by some visual means. This gives an overall picture and enables precise instrumentation such as HWA and LDA to be used where it is most useful. Several methods have been used, usually involving high speed photography, ciné and still, and are noted in Section 3.1.

The advent of lasers has opened up many useful instrumentation techniques involving light scattering.[94] Two such processes are Raman techniques and Rayleigh scattering.

When a laser beam interacts with a gas containing droplets or particles, photons are scattered (Mie scattering). The intensity of the scattering depends on the size of the particles and the gas density while the frequency of the scattered light depends on the velocity of the particles.

In other applications, in the absence of particles, the laser beam is used to interact with the gas molecules themselves. When the photons are scattered at the incident wavelength there is no loss of energy, namely elastic scattering (Rayleigh scattering); or if due to interaction with the molecules of the gas, the emission radiation frequency of the scattered light is shifted up or down, that is non-elastic scattering; the process is termed 'Raman scattering'. If the frequency is lower than the incident beam it is termed 'Stokes', and if higher 'anti-Stokes'. The scattered photons may be collected and the spectral frequency characteristics measured with a spectrometer from which some gas species concentrations may be determined. In particular, the characteristic of the nitrogen spectrum may be used to determine temperature. This is the principle used in the Coherent Anti-Stokes Raman Spectrometry (CARS) described in Section 5 of this chapter. It is also possible with Rayleigh scattering to estimate the gas temperature by determining the total gas concentration; that is, the density from which the temperature may be calculated from the gas laws.

The past few years have seen significant improvements in the understanding of the quantitative relationship between engine geometry and in-cylinder flow: see, for example, the reviews of Arcoumanis and Whitelaw,[1] Dyer,[2] Gosman,[3,4] and Heywood.[5] Proper acknowledgement, however, has also to be given to the contributions made by the non-laser classical techniques,[6] which have provided, and continue to provide, complementary information, thus allowing maximum benefit to be obtained. The automotive engineer is now better equipped to evaluate new design concepts within previously unthinkable short times.

2. DEFINITION OF MEAN FLOW AND TURBULENCE AS APPROPRIATE TO AN INTERNAL COMBUSTION ENGINE

There is a significant number of engine designs that have reached production, and each has aimed at influencing the flow before and after ignition in a way that could improve combustion. Despite the peculiarities of the various engine types and geometries, there are certain flow characteristics that are common in reciprocating engines:

(a) unsteady or non-stationary flow;
(b) turbulent flow at all engine speeds and for all inlet port/cylinder geometries;
(c) three-dimensional flow;
(d) bulk flow in phase with engine speed;
(e) local flow which varies from cycle to cycle;
(f) time scales associated with bulk flow variations of the same order as the turbulent time scales;
(g) large- and small-scale motions present simultaneously in the engine cylinder at the time of ignition.

It becomes clear from the above that the flow is complex and its measurement requires carefully matched instrumentation. The processing of the instantaneous information at a point, which is the only measurable quantity, is not straightforward and statistical analysis is necessary to define the local flow field. The separation of the mean flow from turbulence in an engine cycle becomes, therefore, a matter of definition and is important because different phases of combustion are preferentially affected by different flow characteristics. From the computational point of view, the distinction is also important since it is necessary that definitions of experimentally measured quantities have a modelling counterpart (see Chapter 11).

In stationary turbulent flows, the conventional statistical approach is to resolve the instantaneous flow velocity $U(x, y, z, t)$ or any other property into a time-averaged component $\bar{U}(x, y, z)$ and a fluctuating component $u(x, y, z, t)$. In non-stationary flows, as in reciprocating engines, an equivalent definition is required and the one most appropriate to periodic flows is ensemble-averaging over a statistically adequate number of engine cycles (data points). Similarly to stationary flows, the instantaneous velocity $U(x, y, z, t)$, which for a specified direction at a certain crank angle θ in engine cycle i becomes $U(\theta, i)$, can be decomposed into two components

$$U(\theta, i) = \bar{U}_e(\theta) + u_e(\theta, i) \quad (1)$$

where $\bar{U}_e(\theta)$ is the ensemble-averaged velocity at crank angle θ over i cycles; that is,

$$\bar{U}_e(\theta) = \lim\{(1/N)\, \Sigma\, U(\theta, i)\} \quad (2)$$

and $u_e(\theta, i)$ is the ensemble-averaged turbulence intensity

$$u_e(\theta, i) = \lim\{(1/N)[\Sigma\, \{(U(\theta, i) - \bar{U}_e(\theta)\}^2]^{1/2}\} \quad (3)$$

An important implication of this definition is that $\bar{U}_e(\theta)$ becomes a function of crank angle alone, which implies that the mean flow at a particular crank angle does not vary from one cycle to the next. Flow visualization studies, however, have shown that this may not always be true. And the question arises about whether this is due to the random nature of turbulence or whether there are engine-specific phenomena that are responsible for the observed cycle-to-cycle variations. If the mean flow varies, then the argument can be made that according to the ensemble-average analysis any variations in the mean flow will be included in $u_e(\theta, i)$, resulting in overestimation of the true in-cycle turbulence intensity. In this case, the turbulence-producing capability of an engine may not be properly estimated from ensemble-averaged velocity data. The difficulty lies with the method of determining the mean velocity in individual cycles. If there was a physically sound method of separating mean flow from turbulence, then the difference between the mean velocity $\bar{U}(\theta, i)$ in cycle i and the previously defined ensemble-averaged velocity $\bar{U}_e(\theta)$ over many cycles could have resolved the cycle-to-cycle variability in cycle i, i.e.

$$\hat{U}(\theta, i) = \bar{U}(\theta, i) - \bar{U}_e(\theta) \quad (4)$$

This approach, which effectively corresponds to a decomposition of the instantaneous velocity as

$$U(\theta, i) = \bar{U}_e(\theta) + \hat{U}(\theta, i) + u(\theta, i) \quad (5)$$

assumes that the flow in general consists of an ensemble-average component, an individual cycle mean velocity and a third component which represents the fluctuations in time about the cycle-resolved mean rather than the ensemble-averaged mean; in this case $u(\theta, i)$ is different from $u_e(\theta, i)$.[7]

A number of attempts have been made to estimate the magnitude of the mean velocity during individual engine cycles with both continuous and intermittent signals. Lancaster,[8] for example, attempted to eliminate the non-stationary velocity component by subtracting the ensemble-averaged velocity from the instantaneous velocity and time-averaging the difference over a time interval Δt:

$$U(\theta, i) = \bar{U}_e(\theta) + \bar{U}_{\Delta t}(i) + u(\theta, i) \tag{6}$$

where $\bar{U}_{\Delta t}(i)$ is the time-averaged mean velocity for the ith cycle over a time interval Δt (corresponding to a crank angle window $\Delta\theta$) during which turbulence is assumed to be stationary; that is,

$$\bar{U}_{\Delta t}(i) = \lim(1/\Delta t)\left[\int_t^{t+\Delta t} \{U(\theta, i) - \bar{U}_e(\theta)\}\, \mathrm{d}t\right] \tag{7}$$

In the absence of cycle-to-cycle flow variations $\bar{U}_{\Delta t}(i)$ which corresponds to $\hat{U}(\theta, i)$ of eqn (5), becomes zero and this definition is equivalent to eqn (1). Nevertheless, the choice of time interval Δt requires physical justification and can influence the results. The window size $\Delta\theta$ which corresponds to Δt is also an important factor in the ensemble-average analysis but for different reasons. Since certain measurement techniques such as laser–Doppler anemometry (LDA) provide information which is not continuous but random in nature, and depend on the arrival of a seeding particle at the measurement volume, the window over which data are collected should be of finite width around the selected angle θ. The larger the window is, the better from the statistical point of view, but at the same time this is undesirable in flows with large temporal gradients, since it introduces errors in the estimate of the mean velocity and turbulence intensity (crank angle broadening[9]). The only approach which does not give a single value over the finite window is smoothed ensemble analysis[10]; this uses a cubic-spline smoothing routine to pass a smooth curve through data which have been ensemble-averaged over 1° CA windows, thus eliminating crank angle broadening.

Alternative ways for estimating the mean velocity in individual engine cycles have been proposed, and include the moving window approach[11] and filtering techniques.[12,13] In the former method, a crank angle window is selected on the basis of being larger than the turbulent microscale but smaller than the characteristic period of piston motion (say 10° CA) and, using heavy seeding and fast sampling, high data rates can be achieved ($\geqslant$2 measurements per degree crank angle) to resolve most of the turbulence frequencies and scales. The evaluation of the time-varying mean velocity within the selected window requires some sort of smoothing (i.e. cubic-spline) which is controlled through a pre-determined sum-of-squares-error and, therefore, by the range of frequencies that are included in the cycle-resolved mean velocity.[14] With the filtering approach, the instantaneous velocities pass through a filter where, depending on the filter setting, information can be obtained about the mean flow (low-pass filter) or turbulence (high-pass filter). Since the method is based on selection of a cut-off frequency which is crank angle dependent, it involves some arbitrariness and requires physical justification which is difficult to provide despite some first attempts.[15]

An alternative data reduction scheme has been proposed to resolve the relationship between flow characteristics before ignition near the spark plug, or in the unburned region ahead of the propagating flame, with some combustion parameter during individual engine cycles.[16,17] It is based on conditional sampling and identifies groups of engine cycles of similar behaviour that can be further treated using ensemble-averaging. The difficulty in selecting the most representative conditioning parameter is that those which can be most easily measured, such as peak cylinder pressure, flame speed and flame arrival time, are time- and space-averaged quantities and resolution may be partly lost; however, simultaneous sampling of a number of these parameters minimizes the integration effect. A limiting factor in conditional ensemble-averaging is that only the most probable events are statistically represented.

At present there is no consensus among research workers about the proper method of data reduction in engine flows. The argument against ensemble-averaging is that since the flow characteristics in individual cycles influence combustion in that same cycle, the cycle-resolved mean velocity and turbulence intensity have to be determined at any cost (even with the uncertainties associated with the arbitrary selection of a smoothing parameter or cut-off frequency). On the other hand, the argument in favour of ensemble-averaging is that before combustion starts any variations in the flow from one cycle to the next are simply due to turbulence, since no other physical mechanism has yet been identified which can give rise to variations in the instantaneous flow.† After ignition, the observed variations in the direction

† Swirl centre precession has been identified as a mechanism that, under certain conditions, gives rise to quasi-periodic variations in the position of zero swirl velocity around the cylinder axis with frequencies proportional to engine speed.[18–20]

and speed of the propagating flame front may be taken into account by conditioning the data upon flame arrival at the measurement location and then ensemble-averaging them within small crank angle windows to resolve the flow behaviour.[17,21]

Ensemble-averaging is unambiguous in its implementation and allows comparison of data from different engines and direct compatibility between measurements and calculations; in addition, at present only ensemble-averaged results can provide boundary conditions and validation data for the development and refinement of multi-dimensional computer codes, which solve the ensemble-averaged equations of motion.[3,4,22]

3. OVERVIEW OF FLOW MEASUREMENT TECHNIQUES

The measurement of the flow in the induction and exhaust systems and inside the cylinder during the whole engine cycle may be achieved by hot-wire anemometry (HWA) or laser–doppler anemometry (LDA). The former has the advantage of giving continuous records but has the disadvantages that it is an indirect method, since it measures heat transfer, does not readily give the direction of flow and needs calibration. As the wires are delicate, they will not stand combustion and can therefore only be used on a motored engine.[8,23–25]

It can provide cycle-resolved velocities and turbulent frequencies during the compression stroke and is useful for measuring flow velocities in manifolds and around valves during induction.[26,27]

LDA has the great advantage over HWA in that it provides a direct measurement of velocity, is non-intrusive and is independent of fluid properties. A disadvantage is that it requires optical access.

Laser-based techniques are also finding increasing application as scalar measuring devices. The fuel concentration before combustion has been measured by Rayleigh scattering,[28–30] laser-excited fluorescence,[31] Raman scattering[32–34] and holographic interferometry.[35] Imaged Rayleigh scattering, and two-dimensional flow visualization based on Mie scattering[36–38] have provided very interesting information about the turbulent flame thickness and structure, and valuable insight for the development of computer codes for predicting premixed combustion.

There is also an increasing demand for spatially resolved information on diesel sprays. The need to understand fuel atomization, vaporization and its mixing with air can only be satisfied by simultaneous measurements of droplet size, concentration and velocity; this type of data can validate available spray models.[39,40] Point measurements have relied on LDA and interpretation of signal visibility and/or maximum amplitude, as droplet diameter requires careful consideration of the light attenuation by the surrounding droplet field. The development of frequency-based rather than intensity-based techniques such as phase Doppler anemometry (see, e.g., Ref. 41) shows promise for accurate droplet sizing in multi-hole or hollow-cone sprays, but less so in full-cone sprays where the presence of the liquid core precludes meaningful measurements close to the nozzle. Some information about the dense core length and its break-up may be obtained with single laser beam deflection.[42,43]

3.1. Flow Visualization

A wide range of flow visualization methods has been used to study the flows in engine cylinders before or during combustion. The principle of the techniques used is to introduce particles, as tracers, into the induction system, and the flow is illuminated by a light source to allow still or high-speed photography to be obtained. This requires optical access to the combustion chamber, which can be provided by quartz or sapphire windows in firing engines and Plexiglass windows in motored engines. In the former case, the seeding particles (e.g. Al_2O_3 or TiO_2) have to withstand combustion without loss in their scattering efficiency,[44] in addition to the requirements for motored engines which are for the particles to follow the flow faithfully, to be in sufficient concentration in the region of illumination and to scatter enough light: scattering intensity is directly related to particle diameter. Since the quality of the source of illumination and of the flow tracers have improved significantly in recent years, a renewed interest has developed in flow visualization techniques and, in particular, in laser sheet methods[45] as means of obtaining quickly and at a reasonable cost information about the bulk flow in engine cylinders. However, the requirement for optical access necessitates significant modifications to the engine which can be expensive and may influence the flow. Due to the unsteady nature of the cylinder flow, synchronization of the source of illumination with the crank angle window of interest may be required, which can be provided by either a mechanical chopper or, more frequently, by an acousto-optic modulator in front of the laser beam; the latter offers advantages in controlling with accuracy the duration and frequency of the light pulses, but limits laser operation to the single-line mode.

One of the earliest flow visualization studies in motored engines was that of Lee,[46] who used feathers in the intake air stream as tracer particles and still photography through a glass cylinder. Other researchers used grids of tufts[47] or paper flags,[48] which were inserted in the cylinder during induction to indicate flow directions. Kent & Eaton[49] used helium-filled bubbles around 2 mm in diameter as tracers to study the induction process in a motored engine, and recorded the patterns from two angles of observation using stereoscopic photography. Since the bubbles are destroyed during compression the technique can only be used for induction studies. Marko *et al.*[50] and Arcoumanis *et al.*[18] used phenolic microballoons (10–40 μm) to mark the flow, which scatter light adequately due to their size but also, being hollow, are light enough to follow the flow faithfully, at least at low engine speeds (Fig. 1). Marko *et al.*[50] employed a multiple-exposure video technique to study the flow in an engine where air was

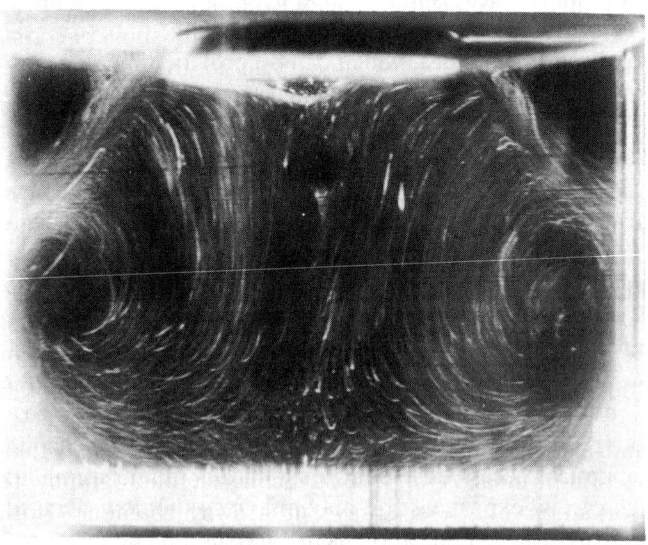

Fig. 1. Flow structure during induction stroke obtained with laser sheet photography using microballoons as tracers.[18]

replaced by freon gas, thus allowing running at low speeds but at dynamically similar conditions to those in production engines.

The difficulties involved with visualization of air flows have motivated several researchers to use scaled water models, to simulate the induction stroke, and to use polystyrene or acrylic beads as tracers. Willis *et al.*[51] provided considerable insight into the mechanism of swirl generation during intake by photographing neutral density tracer particles suspended in water flowing through a model engine with a transparent piston. Eckchian & Hoult[52] carried out a detailed study of the flow structures generated by an axisymmetric intake port/valve in a transparent cylinder, while Kent *et al.*[53] resolved differences in large-scale flow structures and velocity levels caused by changes in the orientation of the intake port. Khalighi & Huebler[54] used a transient water analogue model with a transparent single-cylinder engine to study the flow during the intake and exhaust strokes: by pulsing the illuminating source in a coded manner it was possible to resolve the flow direction from the streaks of the tracer particles.

During combustion, other techniques which respond to the density and derivatives of density of the fluid, such as Schlieren photography, shadowgraphy and interferometry, have been used to study spray development and flame propagation. Shadowgraphy and high-speed photography are particularly useful in studying combustion phenomena and their variation from one engine cycle to the next. In the Schlieren method, a knife edge or aperture is used to partially obstruct the light, and the technique is sensitive to the first spatial derivative of density, whereas shadowgraphy is sensitive to the second derivative of density and there is no obstruction. The simplicity and reduced cost of shadowgraphy has made it popular in engine applications. Witze & Vilchis,[55] for example, have used high-speed shadowgraph photography to examine the influence of mixture motion (swirl) on flame propagation, while Smith used microshadowgraphs (see Fig. 23 of Chapter 8) to examine the influence of engine speed on the turbulent flame structure in a spark-ignition engine. Colour Schlieren photography was used by Ishikawa & Daily[56] to investigate the intake, compression and exhaust processes in a single-stroke engine with a square cross-section. Tindal *et al.*[57] used Schlieren photography to visualize the formation of diesel sprays in a swirl chamber. Both Tindal *et al.*[57] and Mattavi *et al.*[58] using simple cinematography, digitized the recorded images for computer processing and subsequent analysis with densitometer-type devices.

The uncertainties associated with the use of microballoons to visualize the flow at realistic engine speeds, as discussed previously, have encouraged researchers to use smoke particles as tracers, which are of sub-micron size. Wakisaka *et al.*,[59] for example, placed oil on a nichrome wire which, when heated, allowed oil to vaporize, thus permitting the path of the oil smoke to be visualized by high-speed photography. Eaton & Reynolds[60] also used smoke particles to observe the flow field in a single-valve transparent engine motored at speeds up to 500 rpm; the required sheet illumination was provided by a 10 W copper vapour laser.

Finally, another type of flow tracer can be provided in the form of a high-voltage discharge, the motion of which through the flow field can be photographed. Quantitative information can be obtained by measuring the time between the initiation of the discharge at a point in the flow and the response of a voltage-sensitive or ionization probe at a known distance away from the source. The calibration of the spark discharge/time-of-flight technique has been reported to be independent of the fluid temperature, pressure and local turbulence levels.[61] The technique has been used by Ohigashi *et al.*[62] to study the flow in an engine cylinder under motored and firing conditions.

3.2. Hot-wire Anemometry

3.2.1. Description of the Technique

A hot-wire anemometer is essentially a device that measures the change in electrical resistance or current that occurs in a fine, heated wire as it loses heat to the surroundings; in this case a turbulent gas flow in an engine. Hot-wire anemometers comprise very fine wires suspended between two prongs. The wire diameter is chosen as a compromise between mechanical strength, which is better with coarse wires, and spatial resolution and response, which are better with fine wires (e.g. platinum–irridium or platinum–rhodium wires of 5–10 μm diameter are normally used in engine applications). If the wire is heated by supplying electrical power to it, the heat loss from the cylindrical wire varies with flow velocity perpendicular to the cylinder, V, according to King's law

$$q = I^2R \simeq a + b(\rho V)^n \tag{8}$$

where I and R are the applied current and the resistance of the wire respectively, and a, b and n are constants determined by calibration.

However, the response of the wire is strongly influenced by the properties of the fluid being measured (see, e.g., Ref. 7). King's law can also be written as

$$Nu = A + B\,Re^n \tag{9}$$

where A and B are calibration constants and Nu and Re are the Nusselt and Reynolds number, respectively, of the flow past the wire.

Hot-wire probes cannot withstand combustion temperatures, having a higher temperature limit of around 600°C. Therefore they have been mainly used to measure the gas flow in motored engines, ever since the first application to engine flows by Semenov.[63]

A hot wire responds to mass flow; hence the pressure, p, and temperature, T, of the fluid must be known to deduce the velocity. In engine applications, p and T vary during a cycle as well as between cycles and are normally monitored, by means of a thermocouple and a pressure transducer, to allow the estimation of V. The wire can be operated either at constant current, I, or at constant resistance, R (or, equivalently, temperature), so that either R or I, respectively, is a measure of V. In constant-temperature anemometry, a feedback control circuit adjusts the current through the sensor to maintain a constant resistance (temperature). The anemometer output is a voltage which is a non-linear function of the pressure and temperature of the gas, and of the velocity component perpendicular to the wire.

A hot wire indicates only the magnitude of the flow velocity; a single-wire probe cannot indicate flow direction, and a multi-wire probe is often used to resolve the velocity direction. However, multi-wire probes are more obtrusive and single wires are normally employed to minimize interference with the flow. In single-wire probes, the anemometer output is related to two orthogonal velocity components perpendicular to the wire. By varying the orientation of the wire, different pairs of velocity components can be measured, e.g. a wire parallel to the cylinder axis will respond to the radial and tangential components, whereas a wire in line with the cylinder radius will respond to the axial and tangential components. Williams & Tindal[64] used shielded probes to determine the direction and unshielded probes to determine the magnitude of the measured velocities. Extreme care is necessary, though, when using shielded probes to identify flow directions. Wakisaka *et al.*[59] used the spark-discharge technique to obtain the flow directions in a motored engine, in conjunction with hot-wire anemometry for the flow magnitudes.

The main limitation of the application of hot-wire anemometry

(HWA) to engine flows is the need for calibration of the wires at the pressures and temperatures encountered in engines. The hot-wire probes must be calibrated before (and sometimes after) each engine test in a simple flow rig. The calibration equation is of the form of eqn (8) or eqn (9). Use has been made of temperature-compensated hot wires so that q conforms to eqn (8) by varying the wire temperature as the gas temperature varies during the engine cycle.[24] In addition, the operation of the hot-wires under unsteady conditions must be accounted for. Arnold *et al.*[65] calibrated their wires in a steady gas stream and checked their dynamic response under unsteady conditions using a shock tube: this test showed that the measured values were on average within 5% of those calculated from the steady flow calibration. It has been established that as the spatial and temporal velocity gradients of the flow in the latter part of the compression stroke are small and the response of the hot wires is normally fast, the errors arising from calibration in steady flow streams can be neglected for that part of the cycle.

The convective heat transfer must be separated from the radiative and conductive heat transfer to permit accurate velocity measurements. Radiation effects will not be negligible for high flow temperatures, and the heat loss by conduction to the prongs will be significant when the wire aspect ratio (l/d) is relatively small, and errors will be introduced. For engine applications $l \simeq 250\text{–}400\ \mu m$ and $d \simeq 5\text{–}10\ \mu m$, and therefore relatively small aspect ratios of around 50 are common. Horvatin & Hussmann[66] attempted to account for this effect by considering a simplified heat balance comprising the entire hot-wire probe; in this way a correlation between flow velocity and voltage drop across the wire could be established for a certain probe if the geometrical and operating conditions of the wire were known. The determination of the latter is, unfortunately, not without considerable difficulties which can only be overcome by calibration. Although the hot wire can be calibrated for the range of velocities to be measured, it is not possible to calibrate over the full range of temperatures and pressures encountered in engines. An increase in pressure can be accounted for relatively easily, because it affects only the density of the gas, whereas the temperature affects the viscosity and thermal conductivity as well and it is more difficult to account for. The output of the hot wire must then be corrected by calibrating at the appropriate engine temperature and pressure conditions.[67] Hassan & Dent[68] suggested corrections to the estimated velocities by a factor

$(T_f/T_g)^{0.5}$, where T_f is the film temperature (average of the absolute gas and wire temperatures) and T_g is the absolute gas temperature, to take account of the operational temperatures. Tindal *et al.*[48] calibrated their hot wires under steady flow conditions, with velocities similar to those encountered in engines and temperatures up to 450°C and pressures up to 15 bar, and found that over a wide range of temperatures and pressures the relationship between the Nusselt number and Reynolds number could be best represented by a relationship of the form:

$$Nu(T_f/T_g)^{0.5} = A + B\,Re^n \tag{10}$$

The velocity values obtained with this correction equation are similar to values recorded with laser anemometry techniques and far lower than those obtained with correction methods previously used, and indicated clearly that normal correction methods result in overestimation of the velocities.

Another problem in HWA is probe interference with the flow in the swept and clearance volume, and probe impingement on the moving piston. Probe interference cannot be eliminated, but it can be minimized by careful design and positioning of the probe. Stationary wires attached to the cylinder head can be used to measure in the piston bowl and clearance volume at TDC, but moving wires, synchronized with the piston motion, must be used for measurements in most of the swept volume. Horvatin & Hussmann[66] used water-cooled stationary probes inserted through the cylinder head, which plunged into holes in the piston crown when the piston approached the cylinder head. Arnold *et al.*[65] used both stationary and moving wires to measure in the inlet port, cylinder and combustion chamber of a motored engine. The moving wires were inserted through the cylinder head and were translated by a traversing mechanism driven from the piston of an adjoining cylinder and plunged into the cylinder and combustion chamber. Tindal *et al.*[67] used a similar method in a single-cylinder engine, employing a positive displacement cam mechanism to traverse the probes in synchronization with the piston motion. Williams & Tindal[64] used small hot-wire probes fixed on the piston crown to measure in the swept volume.

The most significant advantage of hot wires is their fast response to velocity fluctuations, as they are analogue devices and real-time measurements can be obtained; for this reason they are very well suited for the recording of velocity fluctuations during an engine cycle, and such data can be used to determine cycle-to-cycle velocity variations in engine flows. The very high frequency response enables the determination of time correlations and turbulence scales. A characterization of the eddy size distribution within a turbulent flow is given by the time scales of turbulence defined by:

$$L_t = \int_0^T \frac{1}{T}\left(\int_0^T \frac{u(\tau)u(t+\tau)}{u^2}\,\mathrm{d}t\right)\mathrm{d}\tau \tag{11}$$

where t denotes time, T is the period of measurement and τ is the correlation time. There is no need to calibrate the wires in order to determine turbulence time scales, as the measurement of turbulence frequencies is independent of the temperature and pressure of the gas.[69] The temporal (L_t) and spatial (L_x) integral scales of turbulence can be related by:

$$L_x = L_t U \tag{12}$$

when U is constant and $U \gg u'$; however, these conditions are not generally met in engine flows near the end of the compression stroke. L_x is useful in computing turbulent diffusion. The microscales of turbulence can also be determined using procedures similar to that described above. The advent of 'flying' hot-wire anemometry, in which a probe is traversed at great speed through a flow field, has enabled measurements in recirculating flows in engine cylinders to be obtained and turbulence length scales to be measured (see, e.g., Ref. 70). The probe is traversed through the flow so that it moves at a speed significantly faster than that of the gas, in order to maintain directional sensitivity and to 'freeze' the turbulence; however, probe interference can still be significant and wire calibration is still far from perfect.

A significant advantage of hot-wire anemometry is that signal acquisition and processing equipment is relatively simple and less expensive than that for LDA. In addition, the signal-to-noise ratio for HWA is very high, no seeding of the flow is necessary, and access to the flow can be obtained with relative ease. At present, most applications of hot-wire anemometry to engine flows are concerned with steady flows in the curtain area of the inlet valve, or with the determination of length scales in the manner described above.

3.2.2. Applications

A large number of applications of hot-wire anemometry to measure engine flows have been reported and only a representative sample is outlined here. Tindal and co-workers[48,67,71] have studied the flow in single- and multi-cylinder direct-injection diesel engines, and in particular the effect of inlet port design and squish and swirl on the flow patterns. In spark-ignition engines, Lucas[72] investigated the effect of squish on turbulence and flame propagation. Measurements of the turbulence length scales in engines have yielded values of around 0·2–3·5 mm for the macroscale, depending on the shape of the combustion chamber, and 0·3–0·7 mm for the microscale.[23] Witze[73] found that the time scale was inversely proportional to engine speed, and that the mean velocity and turbulence intensity varied linearly with engine speed; the intensity during induction and compression was approximately 50%. Catania & Mittica[25] found that the microtime scale of turbulence was almost uniform during induction and compression, and that the turbulence intensity and its degree of anisotropy in general increase with the level of swirl. Winsor & Patterson[74] found that the turbulence was isotropic, increased linearly with engine speed, and decayed during the compression stroke, and suggested that cyclic variation was primarily a result of the turbulence in the cylinder. Lancaster[8] found that the turbulence near TDC of compression is isotropic, the turbulence intensity is a linear function of the intake volume flow rate and that the turbulence scale is dependent on the intensity. Witze[73] reported that the turbulence in the compression stroke is not homogeneous, but could not determine whether the spatial variation of turbulence is large enough to affect mixing and combustion. More recent applications of hot-wire anemometry to engine flows have been concerned with the measurement of steady[26] and unsteady[27] flows around the intake valves of reciprocating engines; the unsteady flow results exhibited many similarities to the steady ones, and the results were more sensitive to valve lift than to flow unsteadiness. Ahmed & Gosman[75] also used hot wires to obtain measurements of the steady flow in an axisymmetric port and cylinder, which were compared with numerical predictions of the flow.

In conclusion, the use of hot wires for engine measurements is advantageous in that the continuous output and fast response enables cycle-to-cycle variations and turbulence scales to be measured, but must be balanced against the drawbacks of interference with the flow and imperfect calibration.

3.3. Laser–Doppler Anemometry

3.3.1. Description of the Technique

Laser–Doppler anemometry (LDA) has been widely used in recent years for engine flow measurements as it is an unobtrusive technique,

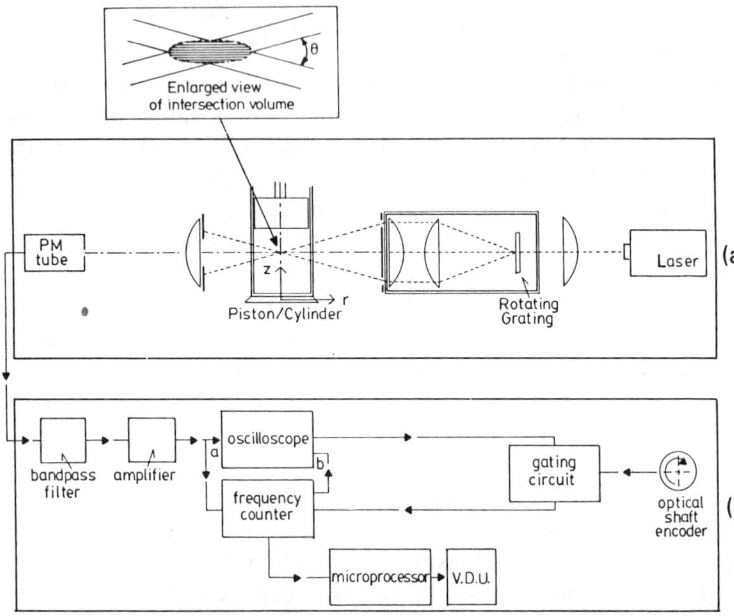

Fig. 2. Schematic diagram of a laser–Doppler anemometer (velocimeter) for the measurement of in-cylinder flows: (a) optical system; (b) signal processing system.

causing no disturbance in the flow field, and requires no calibration. Typical optical and signal processing systems of a laser anemometer are shown in Figs 2(a) and (b). The beam from a laser is split into two beams of equal intensity which are collimated and then focused to cross at a point inside the flow field. At their intersection volume a set of interference fringes is formed and as micron-sized particles suspended in the flow cross the fringe pattern, light is scattered which is collected by a photomultiplier. The frequency of the scattered light is directly proportional to the velocity of the particle:

$$U = \frac{f_D \lambda}{2\sin(\theta/2)} \tag{13}$$

where λ is the wavelength of the laser light, f_D is the Doppler frequency of the scattered light and U is the velocity component of the particle in the direction perpendicular to the fringes. Provided that the aerodynamic diameter of the particle is small enough so that it can follow the flow faithfully, the flow velocity component in a direction perpendicular to the bisector of the angle between the two beams can be measured every time a particle crosses the intersection volume, the dimensions of which are, typically, 50–100 μm diameter and 500–1000 μm length. Different velocity components can be measured either separately by rotating the plane of the beams, or simultaneously by using two or three pairs of beams of different wavelengths crossing at the same point so that their bisectors are mutually orthogonal.

The three main considerations in the application of laser anemometry to engine flows are the introduction of particles into the flow to scatter the laser light ('seeding' the flow), the ability to resolve the direction of the velocity measured, and to obtain optical access to the flow field. Particles are normally introduced by seeding the intake

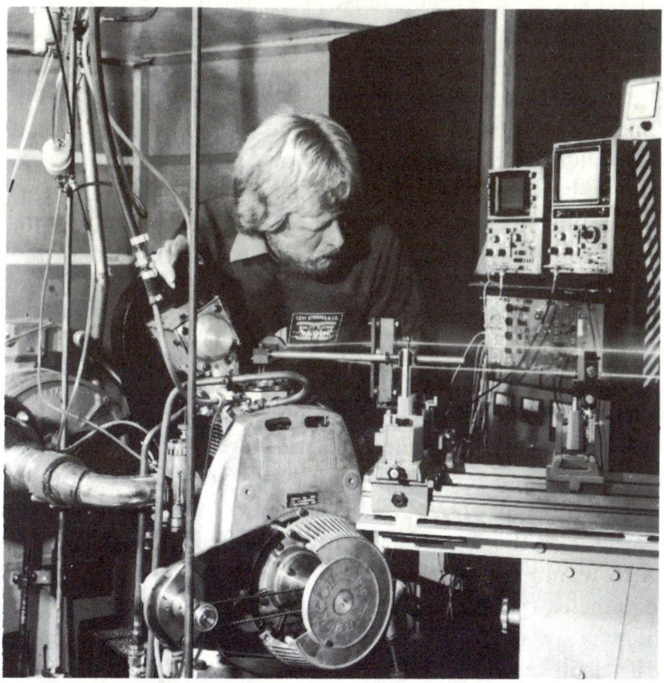

Fig. 3. The Harwell laser–Doppler anemometer in use with the British Leyland experimental stratified charge cylinder head (see Chapter 5). Optical access is through a quartz window in the head which is fitted to a special motoring rig.

airstream with droplets produced by a blast atomizer or, for firing engine applications, with solid particles such as TiO_2 powder produced by a fluidized bed (e.g. Ref. 44). The directional ambiguity is resolved by using a frequency-shifting device such as a Bragg cell or a rotating diffraction grating: these devices impose a higher frequency on one laser beam with respect to the other and thus the sign of the velocity measured will be determined by whether the measured frequency is above or below the shift frequency.

Optical access to the cylinder has been generally obtained either through transparent extensions to the cylinder barrel, or through glass windows fitted on the cylinder head or barrel or on the piston crown. The first method is the most advantageous due to the extensive access that can be obtained and the use of the anemometer in the forward scatter mode. However, transparent cylinders that can withstand high temperatures are expensive as they must be made from either fused silica or sapphire. Fibre optics are well suited for engine applications using lasers, as most of the optical system can be located away from the engine. Durst & Krebs[76] and Obokata *et al.*[77] have reported the development and application of miniaturized fibre-optic probes which could be fitted through an injector or spark plug or other similar access hole in an engine; however, the problems associated with misalignment of the optics due to engine vibration are severe and have not been fully resolved yet. Other potential problems are the effects of the fibres on signal intensity and beam polarization.

The signals obtained at the photomultiplier output, the Doppler 'bursts', are first amplified and band-pass filtered to remove the low-frequency 'pedestal' frequency and any high-frequency noise that may be present. The doppler bursts are then processed to measure their frequency; this is most frequently carried out with a frequency counter. The frequencies of the signals with amplitudes above a preset threshold are counted and often the frequency of the first 10 or so cycles is compared with the frequency of the first 16 cycles and only if they are equal to within a preset limit is the measurement considered valid. The frequencies (velocities) are paired with the corresponding crank angle information derived from a shaft encoder connected to the crank shaft. Either all the data is stored for subsequent processing, or the signals are 'gated', so that only those occurring within a preset crank angle window are used to build ensemble averages from many engine cycles. Photon-correlation is also sometimes used for signal processing (see, for example Refs 78 & 79) as lower levels of scattered light are needed to obtain the velocity information, and optimization of the light level is often difficult at high engine speeds and compression ratios if limited optical access is available. Detailed descriptions of typical laser anemometers for engine flow measurements are given by Wigley & Glanz,[80] Wigley & Hawkins[81] and Suen & Yianneskis.[82]

3.3.2. Applications

As with HWA applications, a number of LDA studies have been concerned with steady flows. A review of the fluid mechanics of engines, based at least partly on LDA measurements, is given by Arcoumanis & Whitelaw.[1] A comprehensive comparison of steady and unsteady flows through engine axisymmetric inlet ports and cylinders has been reported by Bicen *et al.*[83]; it was shown that the flow in the inlet port and up to the valve exit is insensitive to confinement, the proximity of the piston or flow unsteadiness. Arcoumanis *et al.*,[19] compared steady and unsteady flows through a helical port and, in general, confirmed the above finaings. However, it has been shown[94] that the quasi-steady assumption is not valid during the early and the late parts of the induction stroke. Tindal *et al.*[79] made use of the demonstrated quasi-steadiness of the flow in the port and valve passage and investigated the steady flows through both axisymmetric and helical ports using a liquid-flow rig and a transparent port and cylinder made of acrylic plastic. The refractive index of the working fluid was matched to that of the acrylic test section so that optical access to the entire flow space was obtained without beam refraction at the walls of the port and cylinder. Measurements of the steady flows through helical and directed ports (see, e.g., Refs. 81 & 84) have shown that the flows through the two types of port are substantially different.

Due to the three-dimensionality and complexity of the engine flow processes, model engines of idealized geometries have been used widely to study the effect of individual parameters such as compression ratio, valve shape, etc. on the flow patterns and to identify the more fundamental aspects of the flows (see, for example, Refs 85–87).

Measurements in motored production engines have been reported by a number of researchers (see, e.g., Refs 10, 19, 69, 73, 88–90). The flow field in an engine prior to combustion is affected by the inlet port and valve geometry during induction, and by the piston crown shape during the compression stroke. The determination of the flow field before injection in diesel engines or ignition in petrol engines is therefore important for the design of the cylinder and piston head, and much research has been aimed in this direction. In spark-ignition engines, for example, the combustion–turbulence interaction is, of course, of significant interest to the engine designer, but the state of mixing at the time of ignition—which is related to the turbulence structure of the air–fuel mixture—is of paramount importance. There is evidence[62] that the flow velocities during compression are little affected by whether or not combustion has occurred in previous cycles. Thus measurements in motored engines will be representative of the conditions existing in a firing engine prior to combustion, and they can be obtained more easily. The flow patterns in a piston bowl, determined from LDA measurements in a motored idealized engine, are shown in Fig. 4.

Significant insight into engine flows has been obtained through cycle-resolved measurements, since, as already mentioned in Section 2, there is no general agreement as to the origin of the velocity variations from cycle-to-cycle. Fansler & French,[78] for example, obtained cycle-resolved measurements in a motored engine and, using both low- and high-pass filtering to analyse the data, found the existence of random fluctuations of high frequency and evidence of organized fluctuations of low frequency during the engine cycle, which can be expected to be related to the quasi-periodic precession of the swirl centre at frequencies proportional to the engine speed.[20] Typical variations of the instantaneous (single-cycle), ensemble and low-pass filtered velocities are shown in Fig. 5.

Measurements of turbulence length scales have been obtained in engines using primarily two kinds of LDA techniques. Multi-point techniques deduce the integral length scale by correlating measurements obtained at two points or by using an elongated probe volume (see, for example, Ref. 91). The lateral turbulence integral length scale at x_1 is approximated by:

$$L_g = \int_0^{x_2^*} R(x_1, x_2)\,\mathrm{d}x_2$$

where x_2^* is the first point at which $R(x_1, x_2) = 0{\cdot}0$. Alternatively, the scanning LDA technique—similar in approach to the flying hot wire described in Section 3.2.1 above—can be used: the probe volume is traversed rapidly through the flow, 'freezing' the turbulence[92]: longitudinal and transverse autocorrelation functions are calculated from

Fig. 4. Flow patterns in idealized cylindrical and re-entrant piston-bowls; (a), (b) without swirl; (c), (d) with swirl.[87]

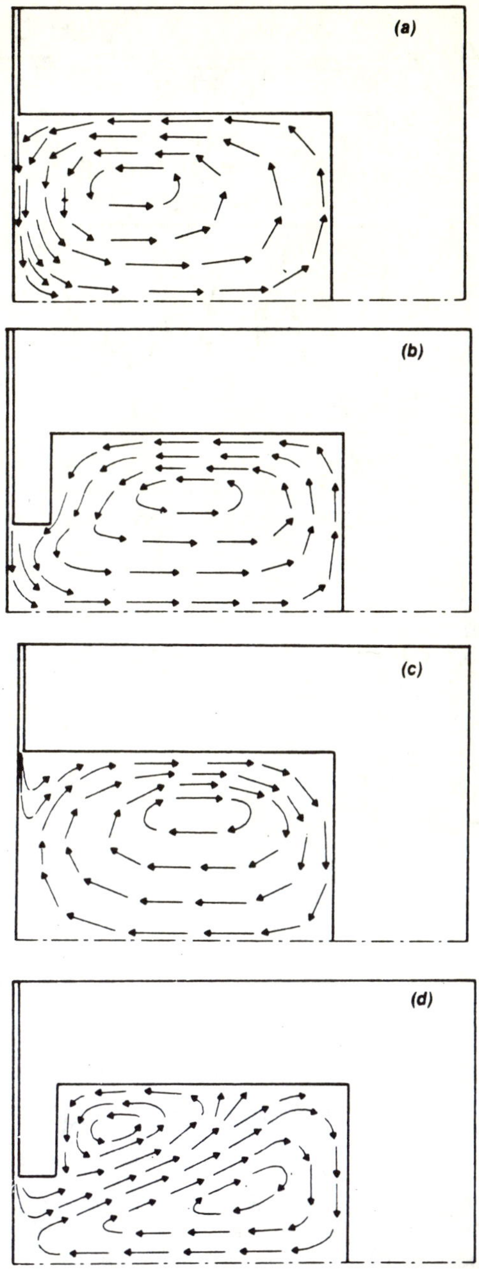

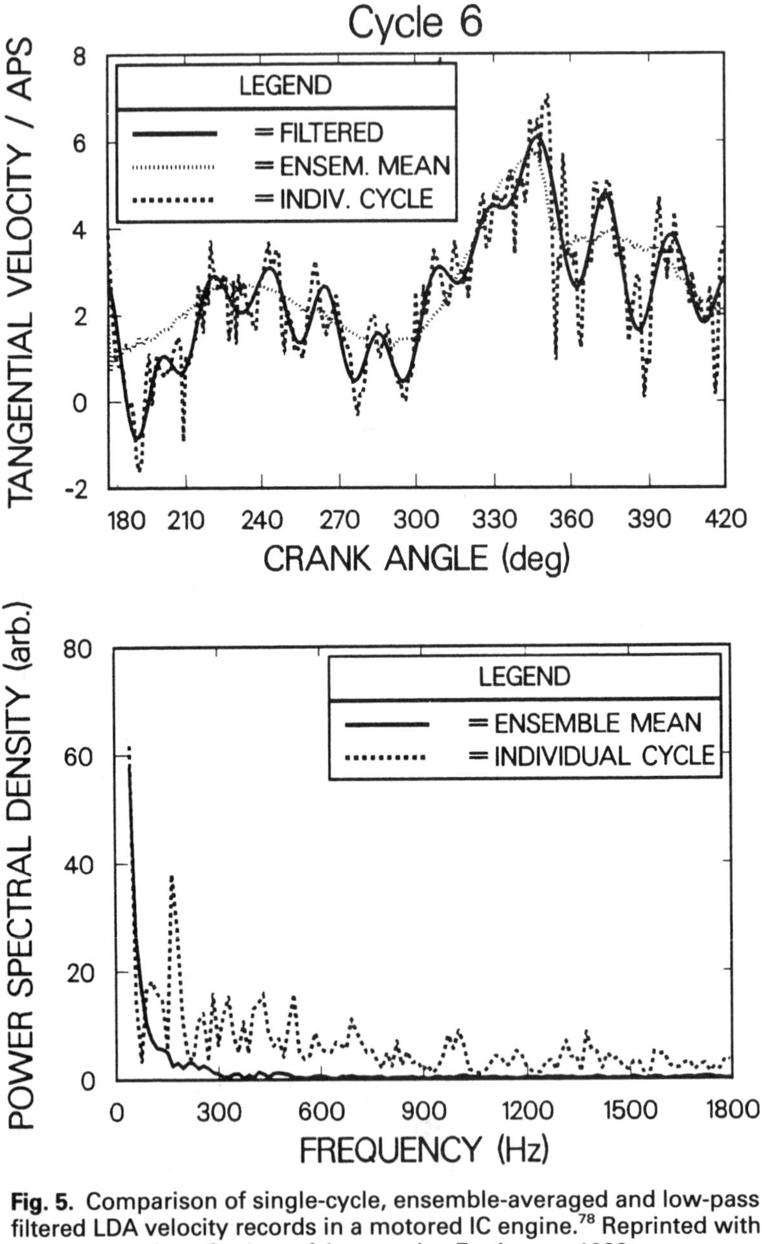

Fig. 5. Comparison of single-cycle, ensemble-averaged and low-pass filtered LDA velocity records in a motored IC engine.[78] Reprinted with permission of the Society of Automotive Engineers, 1988.

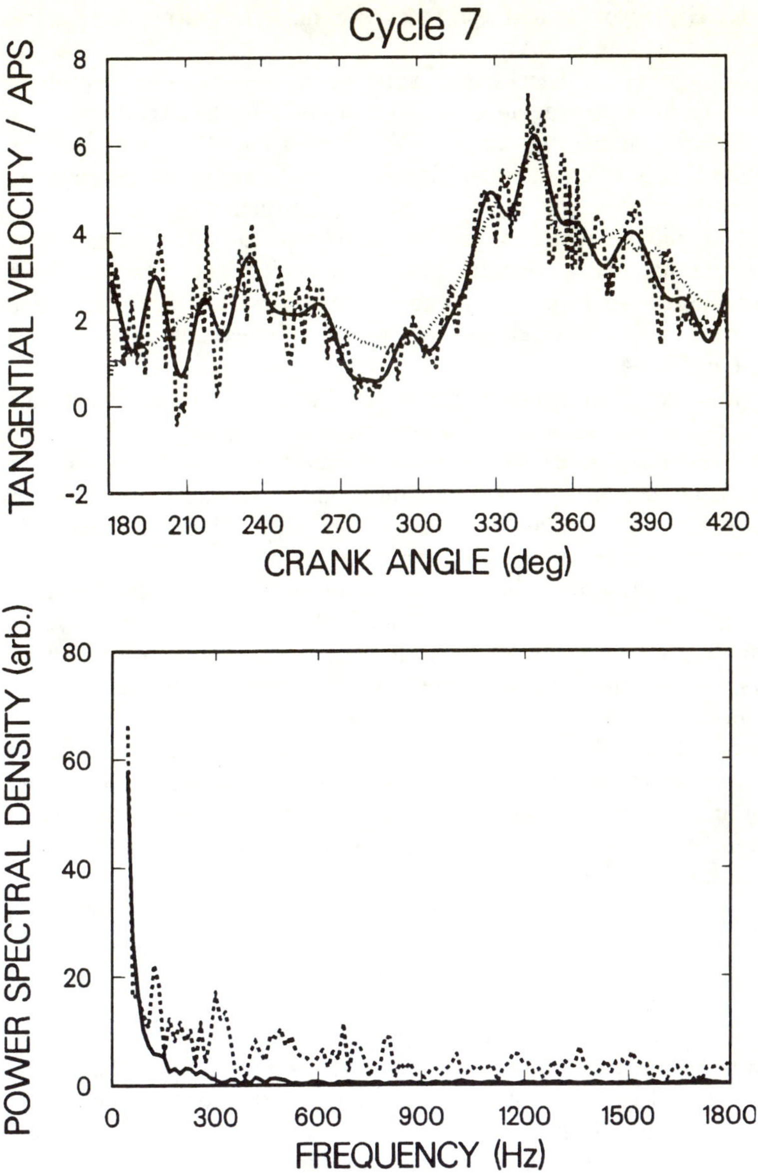
Cycle 7
TANGENTIAL VELOCITY / APS
8
6
4
2
0
-2
180 210 240 270 300 330 360 390 420
CRANK ANGLE (deg)
POWER SPECTRAL DENSITY (arb.)
80
60
40
20
0
0 300 600 900 1200 1500 1800
FREQUENCY (Hz)

Fig. 5—*contd.*

the velocity records, and estimates are made of turbulence length scales.

Measurements of diesel engine sprays and, in general, of two-phase flows in engines provide an interesting problem to the experimenter as the velocities of the two phases (i.e. gas and liquid droplet) must be identified. It is often necessary to measure the spray region near the nozzle, which is very difficult due to the obstruction of the laser light by the liquid droplets. A number of two-phase flow measurement techniques have been developed in recent years,[93] but so far they have found limited applications in engines. Most related work has been concerned with measurements of sprays in quiescent environments (see, e.g., Ref 43).

Measurements in firing engines warrant special attention due to the additional difficulties imposed by the combustion process. It has been found that fused silica windows are preferable to sapphire ones for laser beam access, as the condensation on the surface of the window can be prevented relatively easily with silica windows. Errors in velocity measurements due to laser beam refraction at the flame front have been shown to be insignificant, especially if the optical path of the beams is minimized.[44] The data rate is reduced under firing conditions and some improvement may be possible if laser line optical filters are used to remove visible emissions from the flame. Velocity measurements in firing engines are normally correlated with the position of the flame front, which is located by means of flame-detection techniques. The standard technique is the ionization probe; when ionized gases in the flame reach a pair of electrodes with a large voltage across them, a current passes across the gap between the electrodes. Another method employs the deflection of a laser beam passing through the combustion chamber by the refractive index gradients within the flame.[2] The visible luminosity technique is also sometimes used, employing the natural emissions of the flame to detect its arrival. Foster & Witze[95] have compared the three flame detection techniques and concluded that the ionization probe would seem to be the preferred technique, provided that the perturbations of the flow field by the probe can be tolerated. The optical techniques have the advantage that they do not interfere with the flow, but errors may occur due to the length of the line-of-sight—which should therefore be minimized—and due to irregularly-shaped flame fronts and variations in intensity with location.

4. SCALAR MEASUREMENT TECHNIQUES

The two laser-based techniques that will be covered in more detail below are Rayleigh scattering and Raman scattering. In each case, the basic principles will first be discussed, followed by description of a typical measurement system, and a few representative examples will be presented of their application to engine flows. From the various techniques based on Raman scattering, only spontaneous Raman scattering will be discussed briefly here while the various versions of CARS (Coherent anti-Stokes Raman Spectroscopy) are exclusively covered in Section 5.

4.1. Laser Rayleigh Scattering

4.1.1. Basic Principles

Rayleigh scattering refers to elastic scattering of electromagnetic radiation as a result of photon interaction with the gas atoms and molecules or with small particles (less than 50 nanometres). Elastic scattering means that the scattered light is at the same wavelength as the incident (laser) light which implies that, during this interaction, there is no energy change in the particle's internal state. The relative size of the scattering centre is determined in comparison with the wavelength of the incident light; small particles (molecules, atoms) are those whose characteristic length (i.e. radius) is much smaller than the laser wavelength. The intensity of the Rayleigh scattered light is expressed in terms of an angular scattering cross-section σ_{Ri} which, for isotropic molecular species i, is related to the index of refraction, i.e.

$$\sigma_{Ri} = (4\pi^2/\lambda^4)\{(n_i - 1)/N_i\}^2(\sin^2 \theta_s) \qquad (14)$$

where λ is the wavelength of incident laser light, n_i is the index of refraction of species i, N_i is the number density of species i, and θ_S is the scattering angle.

From eqn (14) it is clear that the Rayleigh scattering cross-section is a maximum at a scattering angle of 90° in a plane perpendicular to the electric field vector and decreases significantly at higher wavelengths due to the λ^{-4} dependence. When instead of a single gas there is a mixture of M gases in the scattering volume, the effective scattering cross-section is the sum of the mole fraction weighted molecular cross-sections. The total scattered intensity will, therefore, include contributions from all the constituent molecular species and is given by

$$I_{SC} = KI_L N\Omega(\Delta x) \sum X_i \sigma_{Ri} \tag{15}$$

where I_{SC} is the total scattered intensity, K is the calibration constant of the optical system, I_L is the intensity of incident laser light, N is the total number density (i.e. number of molecules per unit volume), Ω is the solid angle of collection optics, Δx is the length of the scattering volume, X_i is the mole fraction of species i ($\Sigma x_i = 1$), and σ_{Ri} is the Rayleigh scattering cross-section of species i.

Equation (15) is based on the assumption that the gas mixture is optically thin so that the incident intensity is the same for each individual molecule. For an ideal gas, the relationship between pressure P, number density N and temperature T is given by

$$P = (N/N_0)R_g T \tag{16}$$

where N_0 is Avogadro's number, and R_g is the universal gas constant. Substituting eqn (16) into eqn (15), the total scattered intensity I_{SC} is given by

$$I_{SC} = (K'/T) \sum X_i \sigma_{Ri} \tag{17}$$

where

$$K' = (KI_L P N_0 \Omega \, \Delta x)/R_g$$

For low-velocity flows of nearly constant static pressure, K' is a constant and therefore variations in the Rayleigh scattering intensity can result from temperature and/or concentration variations. For two-component constant temperature flows, variations in the scattered light intensity are attributed to variations in the mole fraction (concentration) of the major component. For constant effective scattering cross-section, variations in the scattered light intensity are attributed to variations in density/temperature; this is the case in some reacting flows where the variation in the effective cross-section from reactants to intermediates and then to products is less than 10%. There is also the case of both temperature and scattering cross-section varying as in a gasoline engine during compression, which requires additional information to be provided (i.e. pressure) to allow decomposition of the instantaneous Rayleigh signal into its two time-varying components.

Information about concentration or temperature can also be obtained, in principle, from the spectral shape of the Rayleigh line (rather than intensity). For laser incident light, the Doppler-broadened Rayleigh line due to the molecular translational motion has a Gaussian profile. Deviations from this shape are indicative of variations in concentration (multi-component mixture) and in temperature (single-component ideal gas). In practice, however, due to the extremely narrow line-width and line distortion from background light, it is rather difficult to obtain accurate temperature or concentration measurements from Rayleigh scattering line shapes.

4.1.2. Measurement System

In Fig. 6 is shown the schematic of a typical laser Rayleigh scattering (LRS) system for engine applications. The technique provides a directionally sensitive signal, the wavelength of which is the same with that of the incident, single laser beam (elastic scattering) and its intensity proportional to the number density of the molecules in the cylindrical scattering volume and their scattering cross-section. The intensity of the Rayleigh signal can be interpreted as density/temperature or fuel concentration in 'tailored' fuel/air gaseous mixtures. The most important components of an LRS system are listed below, together with explanatory comments:

(1) *Laser*: Ar-ion CW† lasers have been used extensively to obtain information about fuel concentration (see, e.g. Refs. 28 & 30); He–Ne lasers of much smaller power can also be used in conjunction with photon correlators.[96]

(2) *Optics*: there is need for careful matching between the light transmission and collecting systems due to the directional sensitivity of the scattered light which exhibits maximum intensity at 90° to the direction of beam propagation and its plane of polarization.

(3) *Photodetector*: due to the low Rayleigh scattering cross-section (relative to Mie scattering), very sensitive low-noise photodetectors are required with internal or external signal amplification.

(4) *Particle/noise filters*: in addition to the interference and electronic frequency filters used to minimize stray light and white noise, respectively, software and/or hardware particle filtering is required for on-line and off-line discrimination between Mie and Rayleigh scattered light.

† CW refers to continuous wave as distinct from pulsed laser systems.

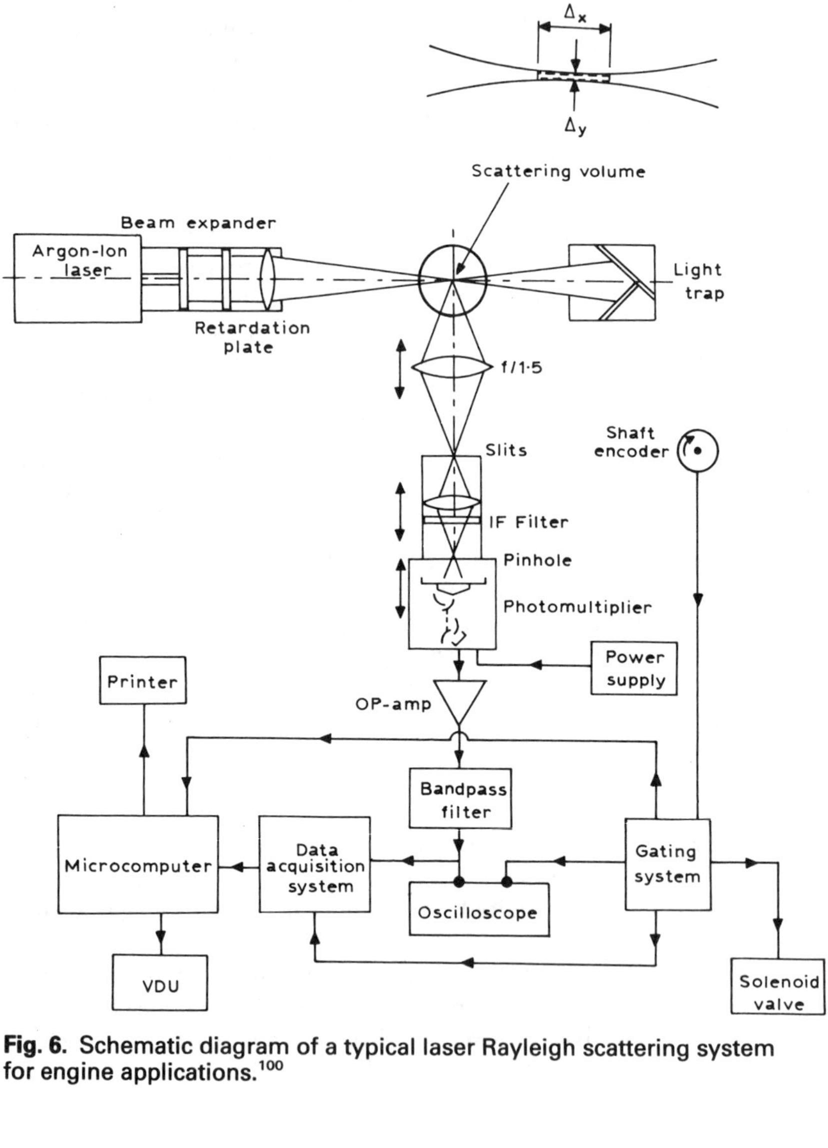

Fig. 6. Schematic diagram of a typical laser Rayleigh scattering system for engine applications.[100]

As an intensity-based technique, LRS is sensitive to interference from light of the same or different wavelength reaching the photodetector or internal variations in the response of the system. The various error sources associated with LRS are listed below:

(1) *System's response*: fluctuations in the laser power output; fluctuations in the input voltage to the photodetector; drift in the electronics; electronic shot noise; limited response time of photomultiplier; dark noise in the photomultiplier; digitization error in the analog-to-digital converter.
(2) *Background radiation*: ambient light; scattered light from optics; scattered light from entrance/exit windows and cylinder walls; flame radiation (chemiluminescence); variation in the scattering cross-section of reactants, intermediates and products.
(3) *Particle interference*: impurities in the air/gas stream; fuel droplets; combustion products (i.e. particulates).

Due to the difficulty of eliminating particles from the scattering volume or its vicinity before, during or after combustion, software particle filtering has been introduced[28,97] to discriminate between Mie and Rayleigh scattered light. The criteria for software filtering are the following:

(1) The intensity of scattered light from particles is usually higher than the maximum intensity of the fuel/air mixture (*amplitude filtering*).
(2) The initial rise in signal amplitude appears greater for particle scattering than molecular scattering (*initial rise filtering*).
(3) The rate of fall in signal amplitude appears greater for particle scattering than molecular scattering (*transient filtering*).

The need for adequate signal-to-noise ratio requires careful selection of the gaseous fuel on the basis of its scattering cross-section. In premixed flames, as in gasoline engines, careful matching is needed between the scattering efficiency of reactants, intermediates and products. For example, propane gas, which is a good scatterer of light, exhibits at stoichiometric conditions a 23% difference in the scattering cross-section between reactants and products[98] and, therefore, only a lean mixture of propane and air can be used. Alternatively, a correction factor should be calculated and applied to the Rayleigh signal as a function of the degree of reaction, fuel type and equivalence ratio.[99] In diffusion flames, as in diesel engines, there is need to match the scattering cross-section of the fuel to that of air and this requires premixing the fuel with hydrogen in order to reduce the total effective scattering cross-section of the mixture.

4.1.3. Applications

There are relatively few applications of Laser Rayleigh Scattering (LRS) to engine flows, and this is attributed to the sensitivity of the technique to light scattering from windows, optics and moving solid boundaries (i.e. valves, piston) and the need for a particle/droplet free flow. However, the main limitation of the technique is that it cannot identify individual components of a gaseous mixture. The applications are divided into two groups dealing with motored and firing engines, respectively.

The difficulty of characterizing the diesel spray in an operating engine has prompted attempts to simulate it by a transient gaseous jet.[28,29,100] In an idealized motored engine, Freon-12 was injected through a nozzle inserted in the axisymmetric inlet/exhaust valve near TDC and its mixing with quiescent or swirling air was examined by LRS through a transparent cylinder wall. Freon-12 was selected due to its high scattering cross-section and its high density relative to air. Initially, the engine was motored without compression and this allowed direct interpretation of the intensity of the Rayleigh signal as Freon concentration (mole fraction). Under compression, however, the time-varying pressure requires that the signal be analysed at the instantaneous pressure at which scattering takes place.[29] Calibration made with different Freon/air mixtures (Fig. 7) showed that signal intensity varies linearly with pressure. The instantaneous and ensemble-averaged Freon concentration during injection was quantified as a function of downstream distance from the nozzle (Fig. 8) using the system of Fig. 6 and found to exhibit a very sharp rise, followed by a relatively flat plateau whose level decreases with downstream distance and a relatively sharp fall. The degree of mixing with air was also quantified as a function of nozzle geometry, compression ratio, induction swirl and piston–bowl geometry and provided insight into the relative importance of the various parameters.[101]

LRS was also used in a motored spark-ignition engine to provide time- and space-resolved measurements of vapour Freon-12 or Freon-113 concentration in the combustion chamber during the induction stroke, for both gaseous and liquid Freon injected into the intake

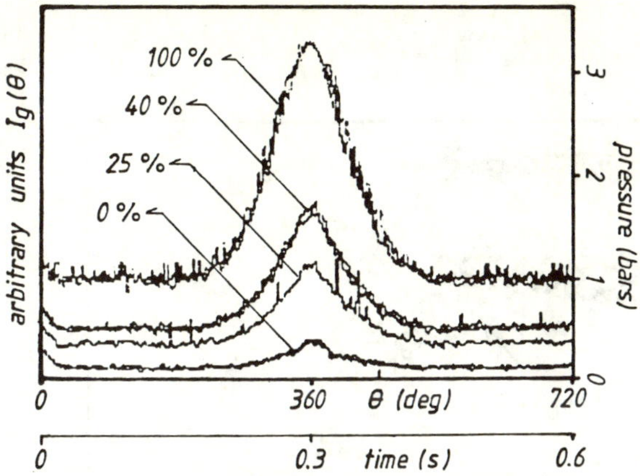

Fig. 7. Calibration of different Freon/air mixtures during compression in a model engine.[29]

port.[30] The results showed significant differences in the space-resolved Freon concentration, with a strong dependence on the location of the fuel injector and the intake valve lift.

To gain insight into the mechanism of turbulent flame propagation, a technique using imaged multi-point LRS has been developed and applied to an operating spark-ignition engine to measure turbulent flame thickness.[36] The scattered light from the propagating flame front was imaged through a flat window in the cylinder head and a lens system onto a multichannel analyser. A typical density (intensity) profile through a flame is shown in Fig. 9, with the flame thickness defined as the distance between the intercepts of the preflame and postflame densities with the straight line coinciding with the average maximum density gradient. The mean flame thickness was found to increase with engine speed and the turbulent flame front to become more wrinkled. It has been observed that islands of unburned mixture can be engulfed behind the flame front and this is more pronounced at the higher engine speeds.

Another area where LRS is expected to find application in engine flows is the measurement of the space- and time-resolved temperature in spark-ignition engines. The technique may also be capable of resolving mixture non-homogeneities in the vicinity of the spark plug before ignition, provided that a high-power laser is available to allow good signal-to-noise ratios to be obtained.

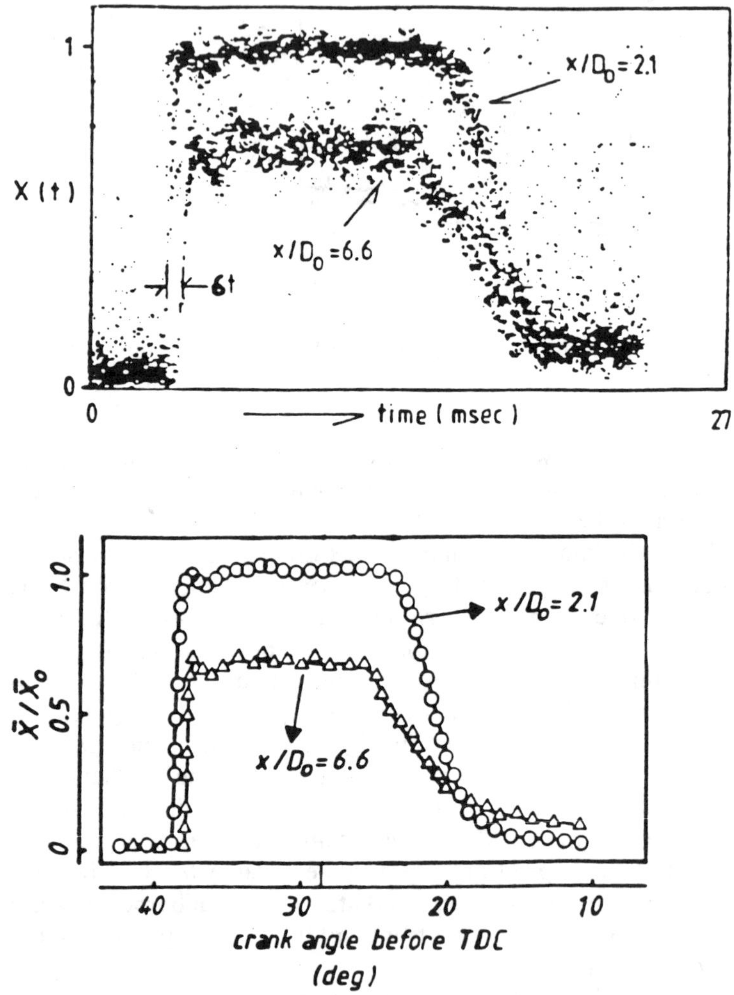

Fig. 8. Instantaneous and ensemble-averaged Freon mole concentration (injection schedule) in a model engine.[29]

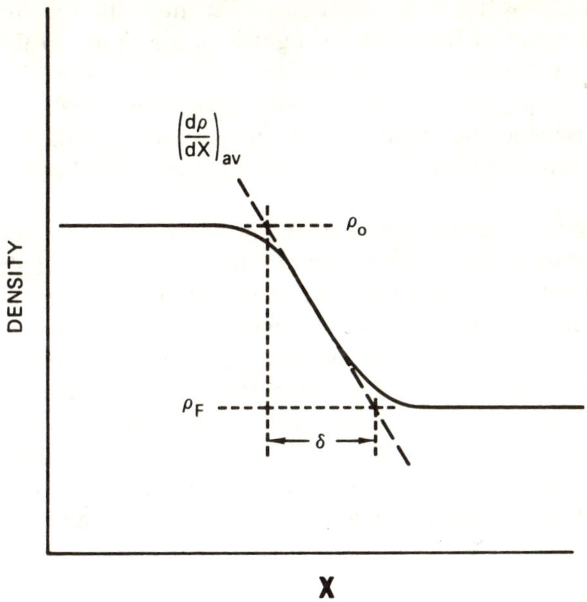

Fig. 9. Density–intensity profile through a propagating flame in the Sandia research engine, obtained with laser Rayleigh scattering.[132]

4.2. Laser Raman Scattering

4.2.1. Basic Principles

As discussed previously, Rayleigh scattering cannot provide information about individual components of a gaseous mixture consisting of more than two gases. The requirement for a technique capable of quantifying the concentration of specific components has led to the development of laser Raman scattering.

Raman scattering refers to the inelastic scattering of light as a result of photon interaction with molecules. Contrary to Rayleigh scattering, the energy of the Raman scattered light is shifted from that of the incident light by an amount equal to the rotational or vibrational energies of the interacting molecule; if the shifted frequency is lower than the frequency of the incident light, the scattering process is referred to as Stokes scattering and, if it is higher, anti-Stokes scattering. Since each molecule has different rotational and vibrational line spacings, it is possible to spectrally resolve the scattered light and identify the type of molecule with which the light interacts. On the other hand, the intensity of the scattered light is proportional to the number density of the interacting molecules, similarly to Rayleigh scattering. Since the populations of the vibrational and rotational molecular levels depend on temperature, it is also possible to determine the temperature of the gas from Raman scattering (Section 5).

Although spontaneous Raman scattering offers advantages in being able to identify individual major species and in that it requires a single laser which can operate at any visible wavelength without the need of being tuned to resonances of the molecules being probed, it is a very weak scattering process. In relative terms, rotational Raman scattering cross-section is about 100 times smaller than Rayleigh cross-section, while vibrational cross-sections are about 1000 times smaller. Despite its lower scattering efficiency, vibrational Raman scattering offers advantages over rotational in that the spectra of the molecular species are well displaced from the exciting laser line and interferences between them are rare.[102,103]

4.2.2. Measurement System

The schematic of a typical laser Raman system for engine applications is shown in Fig. 10. The focusing system is the same with that of Rayleigh scattering but in this case the power of the laser light is even more important since it determines the type of information that can be obtained. While Rayleigh scattering cross-sections allow time-resolved measurements to be obtained with CW Argon-ion lasers, for the same power output (<5 W) Raman scattering can only provide time-averaged information unless the technique of Birch *et al.*[104] is used, which allows time-resolved information to be obtained from short-duration Raman scattered light and application of photon correlation. The more widely used approach is to induce scattering by a high-power pulsed laser,[33] which allows instantaneous concentration measurements within the 5–20 ns period of illumination; however, the temporal variation of the species concentration cannot be measured as a result of the low repetition rates of pulsed lasers.

The light collecting system is also similar to that for Rayleigh scattering, except that a spectrometer is needed to disperse the scattered light, which is then focused on the cathode of a thermo-electrically cooled photomultiplier and measured by a photon counter.[32] As with all previous measurement techniques, a shaft

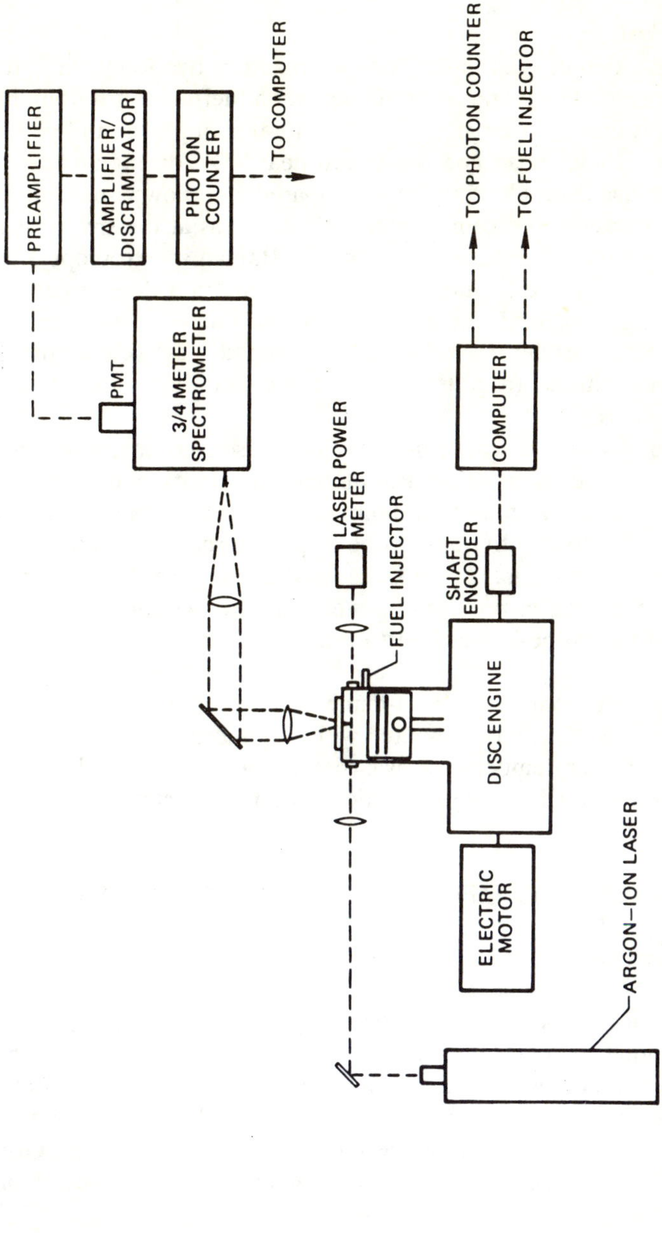

Fig. 10. Schematic diagram of a typical laser Raman spectroscopy system for engine applications.[32]

encoder with good resolution is required to allow measurements to be obtained within pre-determined crank angle windows.

4.2.3. Applications

Spontaneous Raman scattering has been used by Johnston[32] to quantify the ensemble-averaged pre-combustion fuel/air distribution in the engine cylinder; a large spatial variation along the mid-plane between the cylinder head and the piston near TDC during and after fuel injection has been identified and compared with flow visualization results. The transient pre-combustion fuel: air ratios at the spark plug electrode gap have been measured with CW Raman spectroscopy and correlated with mixture ignitability measurements in a direct-injected stratified charge engine.[33] In addition to equivalence ratios, mean vapour temperatures have also been determined[34] as well as time-resolved temperatures from the ratio of the anti-Stokes to Stokes vibrational intensities.[36]

Spontaneous Raman scattering has been used to measure pre-combustion hydrocarbon concentration measurements and compare with those measured with a gas sampling probe and a flame ionization detector in order to identify any probe-related aerodynamic effects in the probe measurements. The results showed large differences between the concentration profiles obtained with and without the probe as a result of the probe-induced perturbations to the air motion.

Pulsed Raman spectroscopy was used to measure gas temperatures and CO concentrations in the post-flame gases at various crank angle degrees after TDC and compare with equilibrium CO concentrations.[106] The gas temperature was determined by comparing measured and theoretical nitrogen vibrational Raman spectra.

5. CARS DIAGNOSTIC MEASUREMENTS IN IC ENGINES

5.1. Introduction

Coherent Anti-Stokes Raman Spectroscopy (CARS) is now an established technique for in-cylinder IC engine thermometry of gases. The technique was first demonstrated on a research Otto engine in 1979 by Stenhouse *et al.*[107] and has spread to more extended measurements on similar engines[108] and, more recently, to a production Otto engine.[109,110] The technique has also been extensively deployed on other industrial combustion devices, such as furnaces[111–13] and jet combustors.[114–18] As well as thermometry the technique is increasingly being developed for species concentration measurements in gases. For IC engine studies it offers a useful combination of advantages which include the following:

(a) it is non-invasive to the combustion space (apart from windowing);
(b) it is spatially precise;
(c) measurements are made in a single 15 ns laser pulse—thus flow fluctuations (turbulence) are frozen;
(d) it is insensitive to background luminosity or fluorescence emissions;
(e) unlike mechanical devices, laser beams are not irreparably damaged by violent flow transients or particles.

5.2. CARS Spectroscopy Overview

Both the theory and application of CARS spectroscopy have received wide attention in a number of reviews.[115–17] The essentials of the CARS process are illustrated in Fig. 11. Two lasers provide beams of frequencies ω_1 and ω_s, which are focused into the combustion space to generate a signal beam at a frequency ω_{as}. To generate the signal efficiently the beams must be geometrically combined in the medium to achieve phase or momentum matching. The resultant signal is a coherent 'laser-like' beam at a frequency $\omega_{as} = 2\omega_1 - \omega_s$. The laser at ω_1 is termed the 'pump laser' and is usually spectrally narrow. The laser at ω_s, usually broadband, is termed the 'Stokes' laser due to its 'red shift' with respect to the pump laser. The generated CARS signal beam at ω_{as} is termed 'anti-Stokes' since it has a 'blue shift' from the pump wavelength.

Phase matching requires that the pump and Stokes lasers are properly vectored so that the generated CARS signal is in phase from all points within the intersection volume of the lasers. This ensures that the CARS signal grows coherently and in a specific direction. The phase-matching geometry illustrated in Fig. 11 is commonly known as 'BOXCARS'.[117] However, for engine experiments, a simple collinear or similar[118] arrangement is employed in which the beams are overlaid using dichroic mirrors. Most practical applications of CARS utilize the

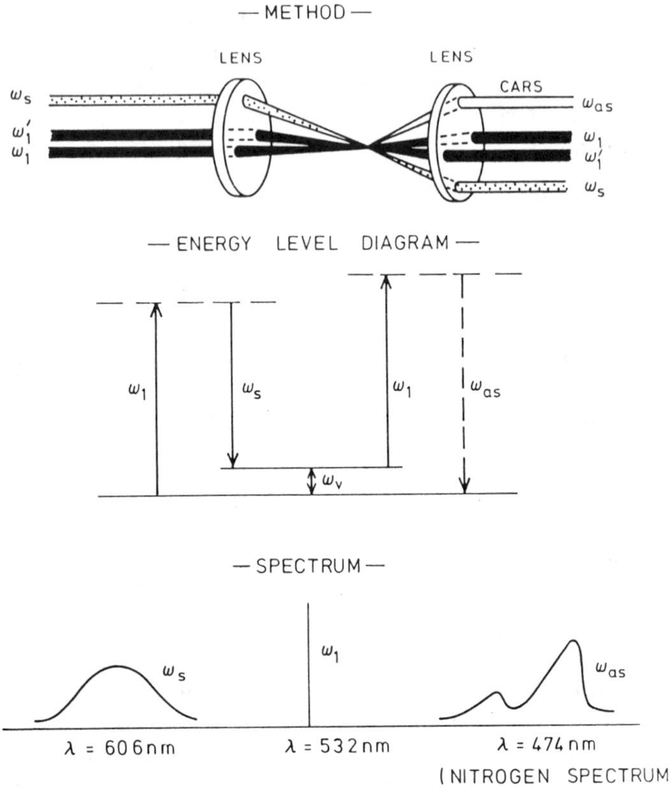

Fig. 11. Schematic of CARS process.

'broadband' technique where the Stokes laser, normally a dye laser, is designed to operate with a 150 cm^{-1} (3–4 nm) bandwidth, because a broadband spectral source generates a complete Raman spectrum of a single species. This is illustrated in the lower part of Fig. 11. CARS signals are generated when a Raman resonance is driven by the difference frequency of the input lasers. Suitable choice of the difference frequency between ω_1 and ω_s allows particular molecular Raman resonances to be probed. For CARS spectroscopy it is convenient to use a Q-switched Nd:YAG laser with a 10–20 Hz repetition rate. Its output is frequency doubled to 532 nm (ω_1) and one third of this, typically up to 40 millijoules at 532 nm, is used to optically pump a broadband dye laser to provide the Stokes source

(ω_s). The difference frequency ($\omega_1 - \omega_s$) required to drive the nitrogen resonance is 2330 cm^{-1}; this means that ω_s must be centred at 607·3 nm. The generated CARS signal is then at 473·4 nm. The spectral properties of the generated anti-Stokes beam are recorded using a spectrograph and a multi-channel detector.

In practice, the IC engine is situated between the laser focusing and recollimating lenses. For an air-fed engine the probed gas is predominantly nitrogen and the nitrogen CARS spectrum is recorded. This spectrum changes shape according to the local environment of the gas, and important influences on spectral shape are temperature, pressure and contributions from other molecular species present. This is the basis for CARS thermometry.

5.2.1. CARS Thermometry

To enable a temperature to be derived from a single laser pulse the whole of the nitrogen Q-branch spectrum must be generated and recorded as a result of that pulse. The main problems with single-pulse broadband CARS are noise in the spectrum arising from the dye laser[119] and Poisson noise arising in the multi-channel detector.[120] These noise contributions can be minimized by adding together spectra from several similar engine firings. CARS nitrogen Q-branch spectra are strongly temperature dependent, and theoretical spectra from 500 K to 2000 K at 1 bar pressure are shown in Fig. 12. At low temperatures the spectrum arises from transitions between the molecular vibrational levels $V = 0$ and $V = 1$ for every rotational level (quantum number J). Each rotational 'side-band' is shifted by approximately $\alpha_e J(J+1)$ from the $J = 0$ level; α_e is the vibrational–rotational coupling constant and is of order 0·02 cm^{-1}. A typical broadband CARS system has a resolution of only 1–2 cm^{-1}, so many of these rotational 'side-bands' are not fully resolved. However, at high temperatures, where high J states are populated, the spacing between lines is just sufficient to produce some structure in the observed spectrum. Generally, the effect of increasing temperature is to broaden the $V = 1 \leftarrow V = 0$ Q-branch. At high temperatures, say above 1000 K, new bands appear at lower Raman shifts; these correspond to $V = 2 \leftarrow V = 1$, $V = 3 \leftarrow V = 2$ etc. transitions.

Temperature analysis of a CARS spectrum is performed by analysing its overall shape. The most effective method is to model spectra and least squares fit them to experimental spectra with temperature as the principal variable.[109–11,117,121,122]

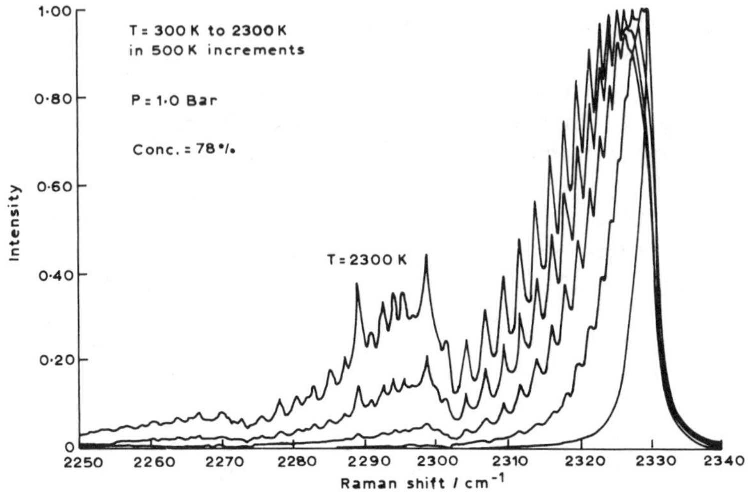

Fig. 12. Effect of temperature on the spectral shape of the nitrogen Q branch at 1 atmosphere.

This possibility arises because the CARS spectrum of nitrogen is due to transitions (or resonances) caused by changes in the vibrational quantum level of the molecule. A diatomic molecule such as nitrogen is essentially a pair of coupled masses which vibrate only at specific frequencies, i.e. quantum states. Associated with each vibrational band is a large number of rotational bands corresponding to the molecules rotating in different rotational states, which are also quantized.

The statistical population of molecules in the various vibrational–rotational states is described by the Boltzmann principle. This states that the quantum states are populated according to the expression $\exp(E_{J,V}hc/KT)$, where $E_{J,V}$ is the energy of the Vth vibrational and Jth rotational state, h is Planck's constant, c is the speed of light, k is the Boltzmann constant and T is the absolute temperature. However, the measurement of temperature by spectral shape analysis is not straightforward because the intensities of the transitions are also determined by (i) oscillation strength, (ii) the degeneracy of the states and (iii) their nuclear spin statistics. Furthermore, each transition has a line width Γ, which is inversely proportional to the 'mean' lifetime of a molecule in a given state. Lifetimes are determined by the mean time between rotationally inelastic collisions and, for a given temperature, are inversely proportional to pressure.

During engine studies, the gas pressures are normally substantially above ambient. A simplistic view would suggest that only the line widths of individual transitions need be scaled as the pressure increases. This approach is, however, inadequate when the individual transitions broaden sufficiently to overlap. At this point a new phenomenon—'motional narrowing'—becomes important.

Consider a hypothetical molecule with two energy states. In a rotationally inelastic collision, the CARS effect sees essentially a change in frequency as the molecule changes rotational states, but no change in phase. This process is shown in Fig. 13 for the hypothetical two-level molecule. At low pressure, molecular lifetimes τ are long and the transitions narrow. As pressure increases, τ decreases and the transitions broaden in a linear fashion. However, when τ is small the molecule switches rapidly between its quantum states and the spectrum collapses to a single sharp line. Under these circumstances, the switching time, or mean time between collisions, is comparable to the inverse of the frequency difference between the spectral lines.

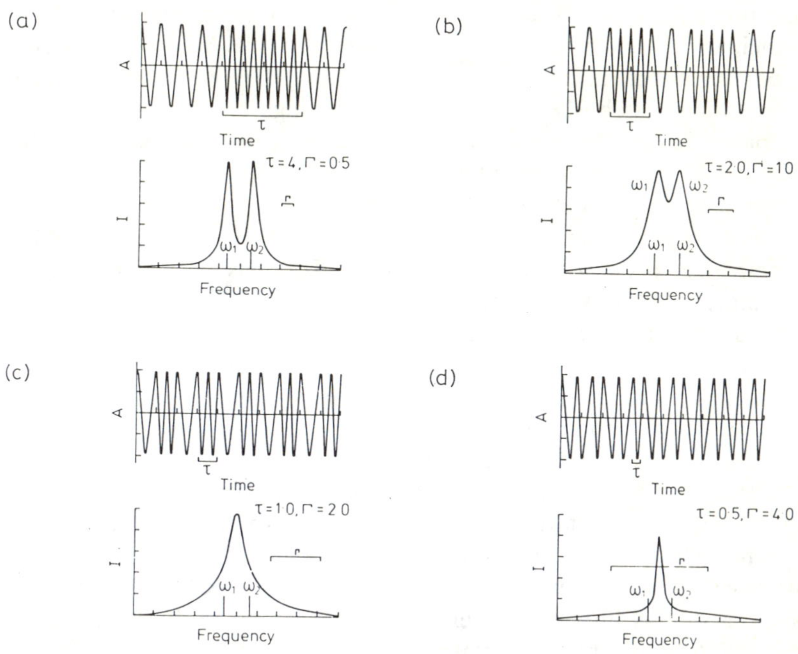

Fig. 13. Diagrammatic representation of the phenomenon of motional narrowing.

Nitrogen behaves in this fashion at high pressure, but here the calculation is very complex because up to 50 rotational states and four vibrational bands can be involved.

Calculation of a nitrogen CARS spectrum at a particular temperature and pressure requires (1) a model which describes the relaxation rates between individual states (for nitrogen, some 2500 exchanges are possible) and (2) a method of computing the spectrum on the basis of the model.

The first practical formulation was found by Hall & Greenhalgh[123] and is based on the Gordon rotational diffusion model. More recently, experimental data, particularly from Sandia National Laboratories[124] and also from in-cylinder engine measurements at Harwell have shown that this model is inadequate. Furthermore, complex effects arising from the coherence properties of the pump laser used in CARS[125] have further complicated spectral calculation. A complete mechanism which enables fast, general and rigorous calculation of spectra has now been developed by Greenhalgh & Hall.[126] Also, a new model which describes rotational relaxation rates for a wide range of temperatures and pressures has been developed.[120] These developments, which have principally been conducted to support the UKAEA's underlying programme for nuclear research and the IEA programme, have substantially established high-temperature, high-pressure CARS thermometry as an effective technique. An example of the success of this work is engine research given in a recent paper presented to the Autotech Conference[109] (see Figs 14 and 15).

Recently the computation time needed for the generation of CARS spectra has been considerably reduced. This has been achieved by solving integrations for convolution of the laser line shape and the 'instrument function', arising from the detection system, on to the spectrum in closed form analytical formulations rather than by numerical methods. There is no complete analytical solution for the laser cross-coherence terms, so the integrations used are part analytical and part numerical, with coarse grid spacing.

Large sets of CARS spectral data can now be processed using an interpolation procedure. This is based on a library of precalculated spectra which is computed to cover a range of temperatures and pressures greater than those expected in the CARS data. The grid spacing of this library is carefully chosen to ensure minimization of errors introduced by interpolation.

Such libraries must be calculated for each series of consecutive

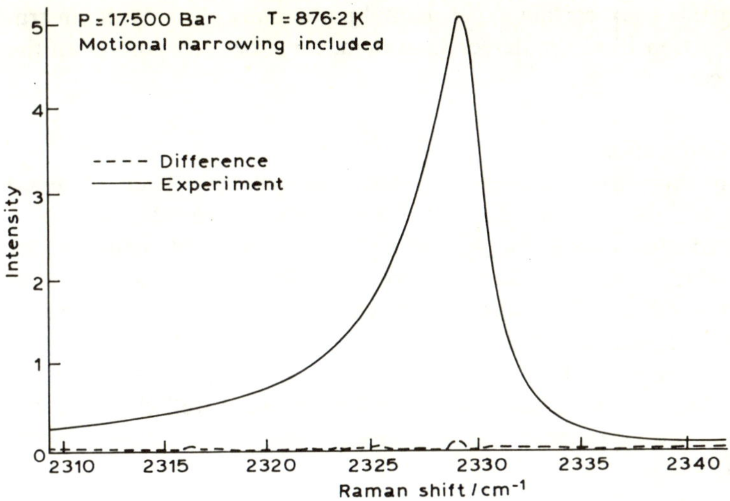

Fig. 14. Pre-combustion CARS spectrum of nitrogen from a firing IC engine.

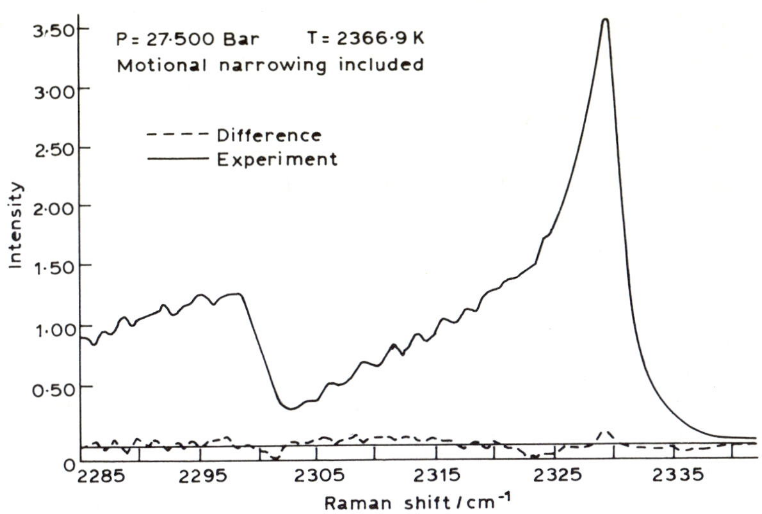

Fig. 15. Post-combustion CARS spectrum of nitrogen from a firing IC engine.

experiments since certain experimental parameters, notably the instrument function from the detection system, must be incorporated in the library calculations.

5.2.2. Calibration

Good quality CARS nitrogen spectra have recently been obtained from a high-temperature/high-pressure furnace. This consisted of a windowed pressure cell totally contained in a windowed circulating air oven, which could be temperature controlled over the range of 300–870 K. The pressure cell was charged with pure nitrogen at pressures between 1 and 25 bar.

Several aspects of the equipment were carefully investigated to ensure that the gas conditions within the cell were well characterized and that the oven and cell windows did not distort the observed spectra. These were:

(1) *Oven and cell window quality*: flat, striation-free, fused silica windows were used for the oven and sapphire windows for the cell. The sapphire windows selected were 'C' cut—i.e. the window faces were normal to the 'C' axis of the crystal—to ensure that birefringence effects on the spectra due to the windows did not occur.

(2) *Make-up gas preheating*: as the cell leaked very slightly at the high-temperature and high-pressure limits of its range, a 1·5 m long coil of thin-walled stainless steel tubing was incorporated into the gas feed line. This coil was contained within the oven and ensured that any make-up gas entering the cell was at the correct temperature.

(3) *Thermocouple measurements*: thermocouples monitored the circulating air temperature and the make-up gas temperature. A third, small-diameter, thermocouple was introduced into the pressure cell itself. It was formed into a loose spiral with its measuring tip approximately 2 mm from the CARS measurement volume. This thermocouple was far more sensitive to the temperature of the gas in the cell than to radiation heating from the cell walls.

(4) *Equilibration times*: after a step change in oven temperature or cell pressure the system was given sufficient time to allow temperature equilibrium between the cell and the gas within it to be attained.

Spectra were obtained from the cell at temperatures of 295, 466, 589, 737 and 865 K. At each temperature, well averaged spectra were taken at pressures of 1, 2, 5, 10 and 20 bar. The current CARS model has been tested using this database and, on the whole, good agreement between CARS and thermocouple temperatures has been achieved. There is, however, some divergence between the results at the highest temperatures and pressures, and the current model is being refined using this data.

5.3. Spectroscopy of Other Molecules

In addition to measuring gas temperature from the structure of the nitrogen spectrum, CARS can be used to measure the concentration of several species which are of direct interest to engine combustion studies, to investigate if possible some of the reactions in both normal and 'knocking' combustion. This is achieved by measuring the efficiency or strength of the generated CARS signal. However, as well as species number density, the CARS signal depends, non-linearly, on laser pulse energy and other optical factors such as window transmission. Consequently, this method is unreliable unless the measurement is made relative to another species in the system.

Most species of interest generate CARS signals at well separated wavelengths so that to change from the study of nitrogen to, say, oxygen requires a change in the dye of Stokes laser wavelength and a readjustment of the centre frequency of the spectrometer/detector used to analyse the signal. Consequently, simultaneous generation of CARS from two species would normally require two dye lasers.

Preliminary results have been obtained for hydrogen, water, carbon dioxide and hydrocarbons.

5.3.1. Hydrogen

The spectrum of hydrogen is extremely simple and may therefore be readily computed. However, as with all species there exists an uncertainty in the precise pressure-broadened line widths; uncertainty in these quantities can lead to erroneous CARS concentration measurements.

Calibration spectra gained for a few conditions of temperature and pressure from a test cell will provide a basis for more accurate interpretation of spectra from the engine and hence provide valuable information on pre-combustion fuel chemistry.

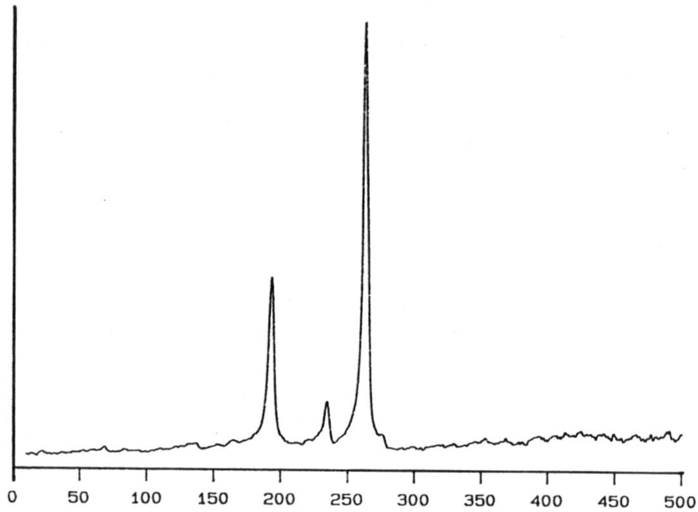

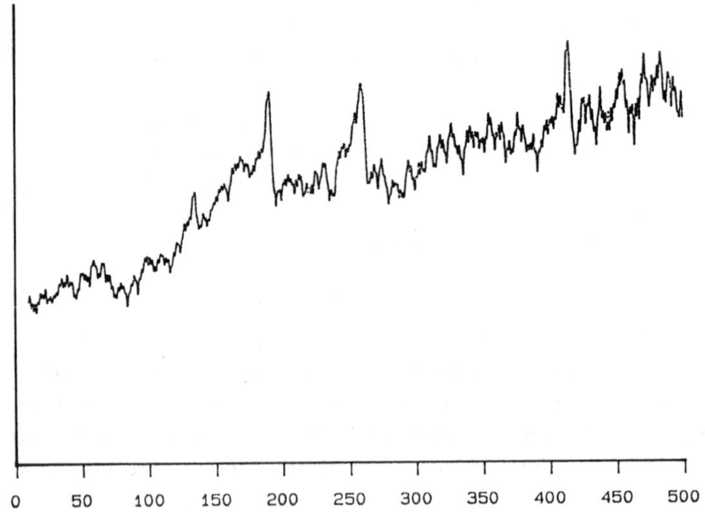

Fig. 16. Hydrogen spectra from a firing IC engine (intensity *v* channel): (a) CARS H_2 spectrum from 1% injection into air intake; (b) CARS H_2 spectrum, no artificial H_2 injection.

In Fig. 16 is shown a spectrum from the engine where 1% of H_2 was bled into the carburettor intake. The spectral shape is due to the populations of the $J=1$, $J=2$ and $J=3$ levels. The lower trace is a spectrum from the engine at 30° BTDC with no H_2 addition. The features due to H_2 at channels 193 ($J=3$) and 257 ($J=1$) are still clearly visible.

5.3.2. Water

Water is one of the major combustion products and therefore there is a considerable incentive to use CARS H_2O spectroscopy to determine the concentration of residual exhaust gases present in the compressed charge before combustion. The CARS spectroscopy of water vapour is by no means as extensively investigated as nitrogen, but is nonetheless one of the most studied CARS spectra to date. High-pressure, high-temperature spectra of water vapour have been recorded in the laboratory and compared with computer predictions by Greenhalgh *et al.*[127] and have also been obtained from practical combustion systems.[128]

In Fig. 17 is shown the spectrum of water vapour at 1 bar, which may be compared with the lower trace taken at 60° BTDC; the same dispersion-shaped signal is clearly visible. This latter spectrum is from water vapour at high temperature and high pressure.

5.3.3. Carbon Dioxide

There have been three significant studies of the quantitative CARS spectroscopy of CO_2.[129–31] In particular, the work of Papineau & Pealat has shown that accurate simulation of the CO_2 spectrum is possible both at room temperature[130] and also at higher temperatures.[131]

In Fig. 18 is shown the spectrum of cold CO_2 in air at 1 bar. As can be seen from the lower trace, CO_2 is readily observable at 60° BTDC. In post-combustion spectra from hot gas mixtures CO_2 is not visible. However, a new signal tentatively assigned to the $J=7 \leftarrow 5$ pure rotational line of H_2 appears. This particular line is known to be the strongest of H_2 at ~2000 K. Some spectral modulations are visible at lower channel numbers. They may well be due to hot CO_2 but the spectroscopy involved is exceedingly complex.

5.3.4. Hydrocarbons

Initial measurements were made in a stationary engine with cylinder 2 charged with propane. A very strong resonant CARS signal was

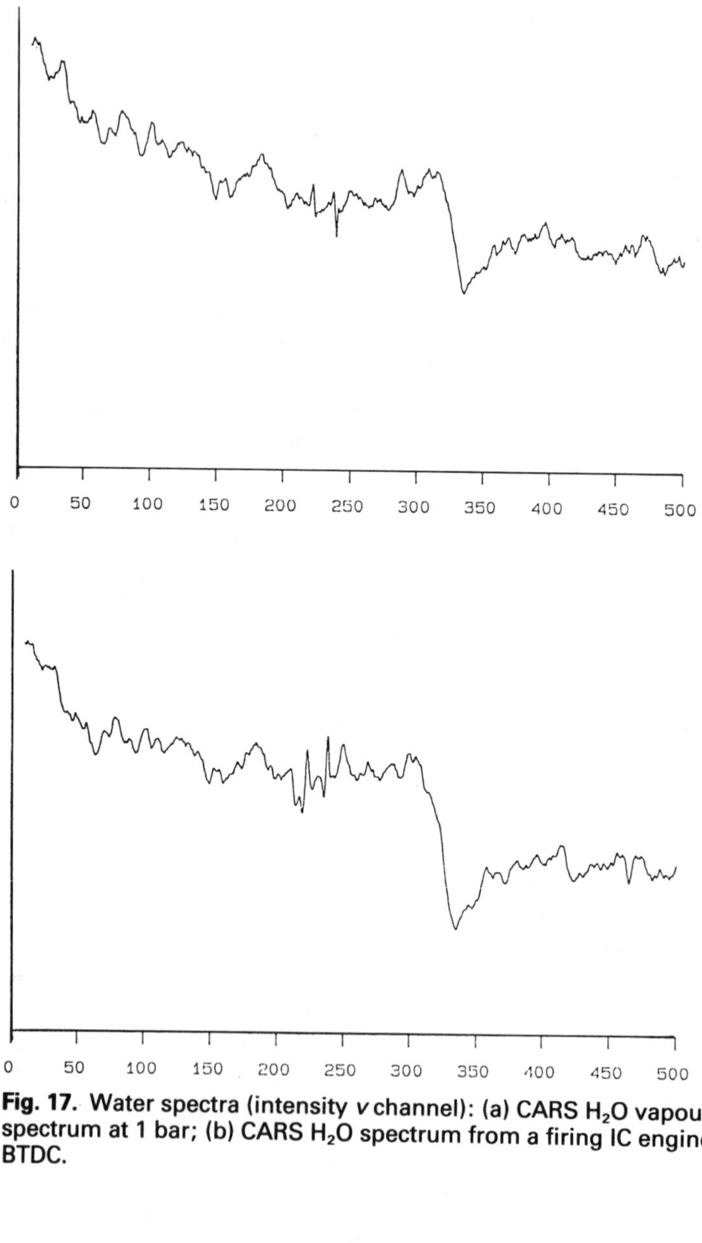

Fig. 17. Water spectra (intensity *v* channel): (a) CARS H_2O vapour spectrum at 1 bar; (b) CARS H_2O spectrum from a firing IC engine at 60° BTDC.

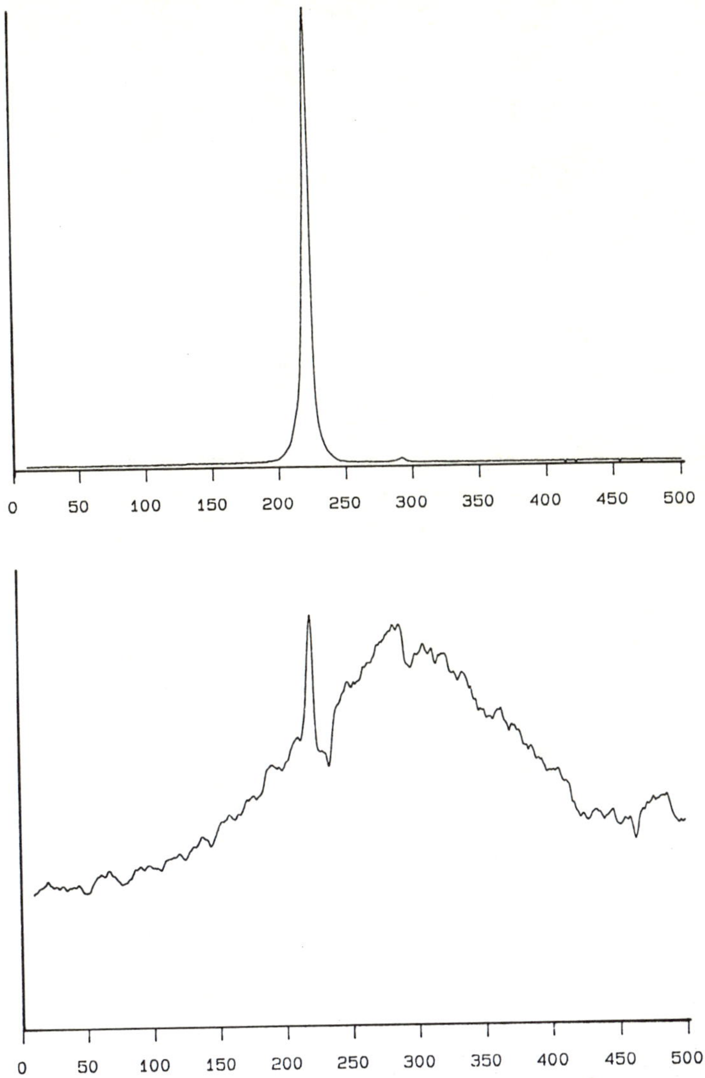

Fig. 18. Carbon dioxide spectra (intensity *v* channel: (a) CARS CO_2 spectrum at 1 bar; (b) CARS CO_2 spectrum from a firing IC engine at 60° BTDC.

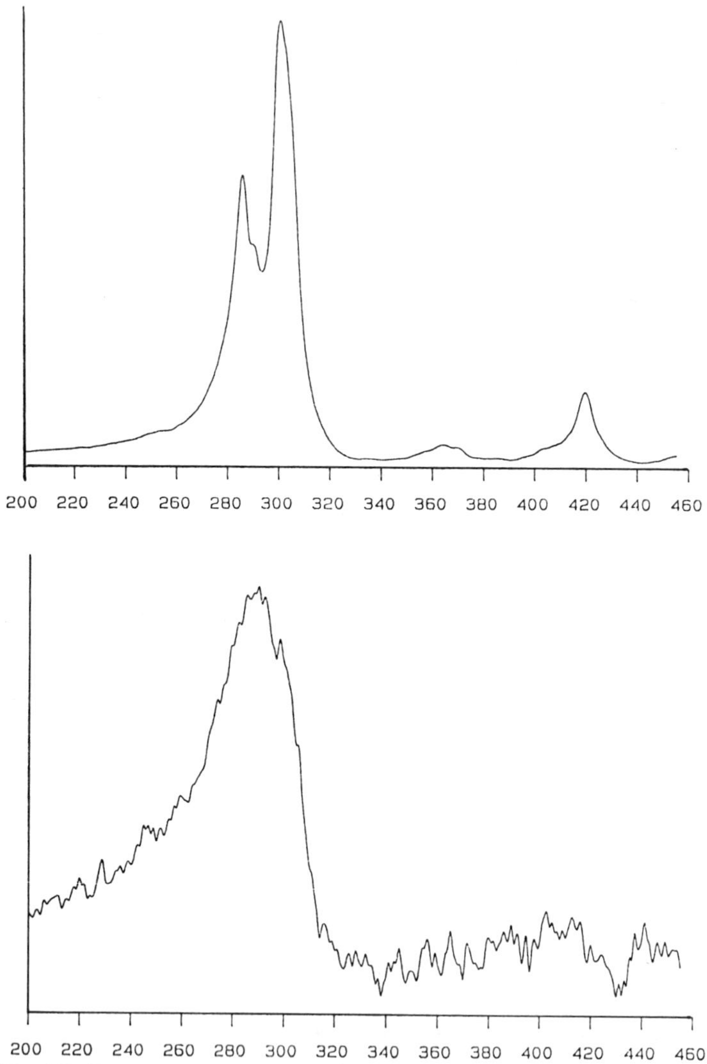

Fig. 19. Propane spectra (intensity *v* channel): (a) CARS propane spectrum at 1 bar; (b) CARS propane spectrum from a firing IC engine at 40° BTDC.

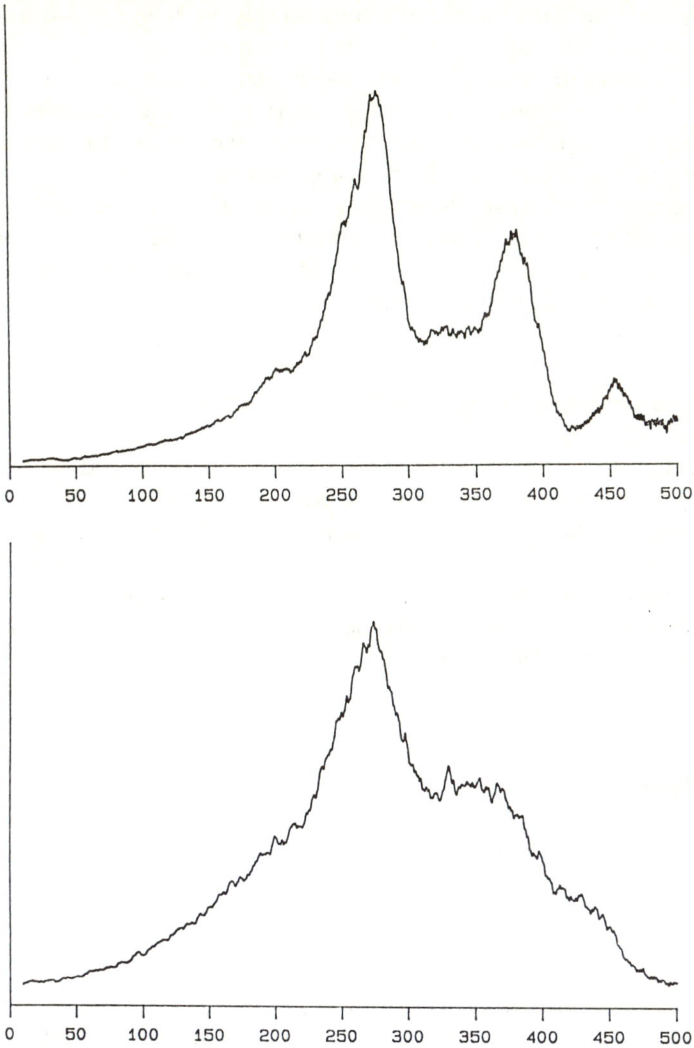

Fig. 20. Iso-octane spectra (intensity *v* channel): (a) CARS iso-octane vapour spectrum at 1 bar; (b) CARS iso-octane vapour spectrum from a firing IC engine at 10° BTDC.

generated (Fig. 19). The engine was then fuelled with propane and run at 1400 rpm; strong signals were obtained up to flame front arrival. The propane spectrum broadened considerably as temperature and pressure increased as can be seen in the lower trace.

Similar behaviour was observed when the engine was run at 3000 rpm using iso-octane fuel. In Fig. 20 is shown the spectrum of room temperature iso-octane, and the lower trace shows the severe broadening of the spectrum as TDC is approached.

These experiments have shown that it is possible to detect these species in a firing engine. However, measurement of the concentration from the observed signals is not yet possible and will involve further development, depending upon the particular species.

5.4. Measurement of In-cylinder Fuel : Air Ratios

To make fuel : air ratio measurements it is necessary to produce and detect CARS signals generated simultaneously from both nitrogen and hydrocarbons in the optical probe volume. Since the two signals are generated by the interactions of two different dye laser (Stokes) beams with a common pump beam, the overall intensities (i.e. spectral areas) of the Stokes beams must also be measured to normalize the intensities of the CARS signals.

REFERENCES

1. Arcoumanis, C. & Whitelaw, J. H., Fluid mechanics of internal combustion engines—a review. *Proc. Instn Mech. Engrs,* **201** (No. C1) (1987).
2. Dyer, T. M., New experimental techniques for in-cylinder engine studies. SAE paper 850396, 1985.
3. Gosman, A. D., Multidimensional modelling of cold flows and turbulence in reciprocating engines. SAE paper 850344, 1985.
4. Gosman, A. D., Computer modelling of flow and heat transfer in engines: progress and prospects. Proc. JSME Symp. on Diagnostics and Modelling of Combustion in Reciprocating Engines, Tokyo, 1985.
5. Heywood, J. B., Fluid motion within the cylinder of internal combustion engines. The 1986 Freeman Scholar Lecture. *J. Fluids Engng,* **109** (1987) 3.
6. Amann, C. A., Classical combustion diagnostics in engine research. SAE paper 850395, 1985.

7. Witze, P. O., A critical comparison of hot-wire anemometry and laser–Doppler anemometry for IC engine applications. SAE paper 800132, 1980.
8. Lancaster, D. R., Effects of engine variables on turbulence in a spark-ignition engine. SAE paper 760159, 1976.
9. Morse, A. P., The effect of crank-angle broadening on laser Doppler measurements in reciprocating engines. Report FS/77/8, Mech. Engng Dept, Imperial College, 1977.
10. Rask, R. B., Laser Doppler anemometer measurements in an internal combustion engine. SAE paper 790094, 1979.
11. Ball, W. F., Pettifer, H. F. & Waterhouse, C. N. F., Laser Doppler velocimeter measurements of turbulence in a direct-injection diesel combustion chamber. Paper C52/83 in *Proc. Int. Conf. on Combustion in Engineering,* vol. 1. Mechanical Engineering Publications, London, 1983.
12. Liou, T. M., Hall, D. A., Santavicca, D. A. & Bracco, F. V., Laser Doppler velocimetry measurements in valved and ported engines. SAE paper 840375, 1984.
13. Hall, M. J., Bracco, F. V. & Santavicca, D. A., Cycle-resolved velocity and turbulence measurements in an IC engine with combustion. SAE paper 860320, 1986.
14. Rask, R. B., Laser Doppler anemometer measurements of mean velocity and turbulence in internal combustion engines. Report 4839, General Motors, 1984.
15. Bracco, F. V., Structure of flames in internal combustion engines. Symp. on Calculations of Turbulent Reactive Flows, ASME Winter Meeting, New Orleans, 1986.
16. Swords, M. D., Kalghatgi, G. T. & Watts, A. J., An experimental study of ignition and flame development in a spark ignition engine. SAE paper 821220, 1982.
17. Witze, P. O. & Martin, J. K., Cyclic-variation bias in spark ignition engine turbulence measurements. In *Laser Anemometry in Fluid Mechanics,* ed. R. J. Adrian *et al.* Ladoan-Instituto Superior Tecnico, 1986.
18. Arcoumanis, C., Green, H. G. & Wong, K. Y., Flow structure visualization in reciprocating engines. In *Flow Visualization IV,* ed. C. Veret. Hemisphere, New York, 1987.
19. Arcoumanis, C., Vafidis, C. & Whitelaw, J. H., Valve and in-cylinder flow generated by a helical port in a production diesel engine. *J. Fluids Engng,* **109** (1987) 270–6.
20. Arcoumanis, C., Hadjiapostolou, A. & Whitelaw, J. H., Swirl center precession in engine flows. *Trans. SAE*; SAE paper 830370, 1987.
21. Martin, J. K., Witze, P. O. & Borgnakke, C., Combustion effects on the preflame flow field in a research engine. SAE paper 850122, 1985.
22. Ahmadi-Befrui, B., Arcoumanis, C., Bicen, A. F., Gosman, A. D., Jahanbasksh, A. & Whitelaw, J. H., Calculations and measurements of the flow in a motored model engine and implications for open-chamber direct-injection engines. Proc. Symp. on Three-Dimensional Turbulent Shear Flows, AIAA/ASME Meeting, St. Louis, 1981.

23. Dent, J. C. & Salama, N. S., The measurement of turbulence characteristics in an internal combustion engine cylinder. SAE paper 750886, 1975.
24. Tindal, M. J., Williams, T. J. & El Khafaji, A. H. A., Gas flow measurements in engine cylinders. SAE paper 740719, 1974.
25. Catania, A. E. & Mittica, A., Induction system effects on the small scale turbulence in a high-speed diesel engine. ASME Conf. on Energy Sources, Dallas, 1987.
26. Khalighi, B., El Tahry, S. H. & Kuziak Jr., W. R., Measured steady flow velocity distributions around a valve/seat annulus. SAE paper 860462, 1986.
27. El Tahry, S. H., Khalighi, B. & Kuziak, W. R., Unsteady flow velocity measurements around the intake valve of a reciprocating engine. SAE paper 870593, 1987.
28. Arcoumanis, C., Green, H. G. & Whitelaw, J. H., Velocity and concentration measurements in a model diesel engine. *Exp. Fluids,* **3** (1985) 368–75.
29. Green, H. G., Characteristics of steady and unsteady jets. Ph.D. thesis, University of London, 1986.
30. Kadota, T., The application of laser Rayleigh scattering to the local mixture strength measurements in SI engine during intake stroke. SAE paper 872151, 1987.
31. Cole, J. B. & Swords, M. D., Measurement of concentration fluctuations in an internal combustion engine. *J. Phys. D: Appl. Phys.,* **13** (1980) 733.
32. Johnston, S. C., Precombustion fuel: air distribution in a stratified charge engine using laser Raman spectroscopy. SAE paper 790433, 1979.
33. Johnston, S. C., Raman spectroscopy and flow visualization study of stratified charge engine combustion. SAE paper 800136, 1980.
34. Johnston, S. C., An experimental investigation into the application of spontaneous Raman scattering to spray measurements in an engine. Proc. ASME Symp. on *Fluid Mechanics of Combustion Systems,* Boulder, 1981.
35. Dent, J. C., Supplementary comments: potential applications of holographic interferometry to engine combustion research. In *Combustion Modeling in Reciprocating Engines,* ed. J. N. Mattavi & C. A. Amann. Plenum Press, New York, 1980.
36. Smith, J. R., Instantaneous temperature and density by spontaneous Raman scattering in a piston engine. AIAA-80-1359, AIAA 13th Fluid and Plasma Dynamics Conf., 1980.
37. Baritaud, T. A. & Green, R. M., A 2-D flame visualization technique applied to the IC engine. SAE paper 860025, 1986.
38. Zur Loye, A. O. & Bracco, F. V., Two-dimensional visualization of premixed-charge flame structure in an IC engine. SAE paper 870454, 1987.
39. Bracco, F. V., Modelling of diesel sprays. SAE paper 850394, 1985.
40. Watkins, A. P., Gosman, A. D. & Tabrizi, B., Calculation of 3-D spray motion in engines. SAE paper 860468, 1986.

41. Bachalo, W. D. & Houser, M. J., Phase/doppler spray analyser for simultaneous measurements of drop size and velocity distributions. *Opt. Engr,* **23**(5) (1984) 583.
42. Yule, A. J., Mo, S. L., Tham, S. Y. & Aval, S. M., Diesel spray structure. Proc. *ICLASS*-85, London, 1985.
43. Arcoumanis, C., Cossali, E., Paal, G. & Whitelaw, J. H., Transient characteristics of single-hole diesel sprays. SAE paper 890314, 1989.
44. Witze, P. O. & Baritaud, T. A., Particle seeding for Mie scattering measurements in combustion flows. Sandia Report no. 985-8912, 1986.
45. Hentschel, W. & Stoffregen, B., Flow visualization with laser light-sheet techniques in automotive research. In *Flow Visualization IV,* ed. C. Veret. Hemisphere, New York, 1986, pp. 230–4.
46. Lee, D. W., A study of air flow in an engine. Report No. 653, NACA 25th Annual Report, 1939.
47. Huebner, K. H. & McDonald, A. T., Experimental determination of airflow patterns in piston engines with induction swirl. SAE paper 720026, 1972.
48. Tindal, M. J., Williams, T. J. & Aldoory, M., The effect of inlet port design on cylinder gas motion in direct injection diesel engines. ASME Winter Meeting, Phoenix, 1982.
49. Kent, J. C. & Eaton, A. R., Stereo photography of helium-filled bubbles for 3-D fluid motion studies in an engine cylinder. *Appl. Opt,* **21** (1982) 904.
50. Marko, K. A., Li, P., Rimai, L. & Ma, T., Flow field imaging for quantitative cycle resolved velocity measurements in a model engine. SAE paper 860022, 1986.
51. Willis, D. A., Meyer, W. E. & Birnie, C., Mapping of airflow patterns with induction swirl. SAE paper 660093, 1966.
52. Ekchian, A. & Hoult, D. P., Flow visualization study of the intake process of an internal combustion engine. *Trans. SAE,* **88** (1979); SAE paper 790095.
53. Kent, J. C. *et al.,* Effect of intake port design on in-cylinder flow and burn rate. *Trans. SAE,* paper 872153.
54. Khalighi, B. & Huebler, M. S., A transient water analog of a dual-intake-valve engine for intake flow visualization and full-field velocity measurements. SAE paper 880519, 1988.
55. Witze, P. O. & Vilchis, F. R., Stroboscopic laser shadowgraph study of the effect of swirl on homogeneous combustion in spark-ignition engine. SAE paper 810226, 1981.
56. Ishikawa, N. & Daily, J. W., Observation of flow characteristics in a model IC engine cylinder. SAE paper 780230, 1978.
57. Tindal, M. J., Williams, T. J. & Hartcombe, A. T., Fuel spray formation in an experimental swirl chamber. SAE paper 870451, 1987.
58. Mattavi, J. N., Groff, E. G. & Matekunas, F. A., Turbulence flame motion and combustion chamber geometry—their interactions in a lean-combustion engine. Instn Mech. Engrs, Conference paper C100/79, 1979.
59. Wakisaka, T., Hamamoto, Y., Ohigashi, S. & Hashimoto, M., Measure-

ments of air swirl and its turbulence characteristics in the cylinder of an internal combustion engine. Paper C91/79, Instn Mech. Engrs Conf., 1979.
60. Eaton, A. R. & Reynolds, W. C., High-speed photography of smoke marked flow in a motored axisymmetric engine. ASME Conf. on Energy Sources, Dallas, 1987.
61. Tippelman, G., A new method of investigation of swirl ports. SAE paper 770404, 1977.
62. Ohigashi, S., Hamamoto, Y. & Tanabe, S., Swirl—its measurement and effect on combustion in a diesel engine, Instn Mech. Engrs Conf. on Air Pollution in Transport Engines, Solihull, 1971.
63. Semenov, E. S., Studies of turbulent gas flow in piston engines. NACA Technical Translation F-97, 1958.
64. Williams, T. J. & Tindal, M. J., Gas flow studies in direct injection diesel engines with re-entrant combustion chambers. SAE paper 800027, 1980.
65. Arnold, M. J., Tindal, M. J. & Williams, T. J., Measurement of induction gas velocities in a reciprocating engine cylinder. SAE paper 720115, 1972.
66. Horvatin, M. & Hussmann, A. W., Measurement of air movements in internal combustion engine cylinders. *DISA Information,* no. 8 (1969) 13.
67. Tindal, M. J., Brown, P. G. & Kyriakides, S. C., An investigation of swirl and turbulence in the cylinders of direct injection diesel engines. Paper C127/82, Proc. Instn Mech. Engrs Conf. on *Diesel Engines for Passenger Cars and Light Duty Vehicles,* 1982.
68. Hassan, H. & Dent, J. C., The measurement of air velocity in a motored internal combustion engine using a hot-wire anemometer. *Proc. Instn Mech. Engrs,* **185** (1971) 583.
69. Haghgooie, H., Kent, J. C. & Tabaczynski, R. J., Turbulent time-scale measurements in a spark-ignition engine using hot wire anemometry and fast response ion probes. In *Flows in Internal Combustion Engines.* ed. T. Uzkan, 1982, pp. 1–8.
70. Dinsdale, S., Roughton, A. & Collings, N., Length scale and turbulence intensity measurements in a motored internal combustion engine. SAE paper 880380, 1988.
71. Tindal, M. J. & Williams, T. J., An investigation of cylinder gas motion in the direct injection diesel engine. SAE paper 770405, 1977.
72. Lucas, G. G. & Anton, R., The effect of squish on charge turbulence and flame propagation in an SI engine. Instn Mech. Engrs, Conference paper C87/79, 1979.
73. Witze, P. O., Measurements of the spatial distribution and engine speed dependence of turbulent air motion in an IC engine. SAE paper 770220, 1977.
74. Winsor, R. E. & Patterson, D. J., Mixture turbulence—a key to cyclic combustion variation. *Trans. SAE,* **82** (1973) 363–83.
75. Ahmed, A. M. Y. & Gosman, A. D., Measurements and multidimensional prediction of flow in an axisymmetric valve/port assembly. SAE paper 870590, 1987.

76. Durst, F. & Krebs, H., Adaptive optics for IC engine measurements. In *Experimental Measurements and Techniques in Turbulent Reactive and Non-reactive Flows,* vol. AMD 66. ASME Winter Meeting, New Orleans, 1984.
77. Obokata, T., Hanada, N. & Kurabayashi, T., Velocity and turbulence measurements in a combustion chamber of SI engine under motored and firing operations by LDA with fiber-optic pick-up. SAE paper 870166, 1987.
78. Fansler, T. D. & French, D. T., Cycle-resolved laser velocimetry measurements in a re-entrant-bowl-in-piston engine. SAE paper 880377, 1988.
79. Tindal, M. J., Cheung, R. S. & Yianneskis, M., Velocity characteristics of steady flows through engine inlet ports and cylinders. SAE paper 880383, 1988.
80. Wigley, G. & Glanz, R., A laser anemometer for engine flow studies, Proc. 2nd Int. Symp on *Applications of Laser Anemometry to Fluid Mechanics,* Lisbon, 1984.
81. Wigley, G. & Hawkins, M. G., Three dimensional velocity measurements by laser anemometry in a diesel engine under steady flow conditions. SAE paper 780060, 1978.
82. Suen, K. O. & Yianneskis, M., A Laser-anemometry system for the investigation of in-cylinder flows. Proc. I. Mech. E. Seminar on Experimental Methods in Engine Research and Development, 1988.
83. Bicen, A. F., Vafidis, C. & Whitelaw, J. H., Steady and unsteady airflow through the intake valve of a reciprocating engine. *J. Fluids Engng,* **107** (1985) 413.
84. Coghe, A., Brunello, G. & Tassi, E., Effects of intake ports on the in-cylinder air motion under steady flow conditions. SAE paper 880384, 1988.
85. Morse, A. P., Whitelaw, J. H. & Yianneskis, M., The influence of swirl on the flow characteristics of a reciprocating piston–cylinder assembly. *J. Fluids Engng,* **102** (1980).
86. Morse, A. P., Whitelaw, J. H. & Yianneskis, M. The flow characteristics of a piston-cylinder assembly with an off-centre, open port. *Proc. Instn Mech. Engrs,* **194** (1980) 291–9.
87. Arcoumanis, C., Bicen, A. F. & Whitelaw, J. H., Squish and swirl–squish interaction in motored model engines. *J. Fluids Engng,* **105** (1983) 105–12.
88. Morse, A. P. & Whitelaw, J. H., Measurements of the in-cylinder flow of a motored 4-stroke reciprocating engine. *Proc. R. Soc.,* **377** (1981) 309–29.
89. Fansler, T. D., Laser velocimetry measurements of swirl and squish flows in an engine with a cylindrical piston bowl. SAE paper 850124, 1985.
90. Suen, K. O., Tindal, M. J., Yianneskis, M. & Paul, G. R., The development of swirl in the cylinder of a direct injection diesel engine. Proc. Second Int. Conf. on *Laser Anemometry—Advances and Applications,* Glasgow, 1987.

91. Fraser, R. A. & Bracco, F. V., Cycle-resolved LDV integral length scale measurements in an IC engine. SAE paper 880381, 1988.
92. Glover, A. R., Hundleby, G. E. & Hadded, O., The development of scanning LDA for the measurement of turbulence in engines. SAE paper 880379, 1988.
93. Yianneskis, M., Velocity, particle sizing and concentration measurement techniques for multi-phase flow. *Powder Technol,* **49** (1987) 261–9.
94. Arcoumanis, C. & Whitelaw, J. H., Are steady flow inlet boundary conditions valid for engine cylinder calculations? Proc. Congress on Modeling of Internal Combustion Engines, Valencia, (1987).
95. Foster, D. E. & Witze, P. O., A comparison of flame detection techniques for premixed-charge combustion in spark ignition engines. *Exp. Fluids,* **6** (1988) 283.
96. Haumann, J., Wu, G. & Leipertz, A., Low power laser Rayleigh probe for flow mixing studies. *Exp. Fluids* **5** (1987).
97. Green, H. G. & Whitelaw, J. H., Scalar field measurements in reciprocating engines. *AIAA J. Propulsion and Power,* **3** (1987) 350–8.
98. Dandekar, K. V., Velocity and density measurements in premixed turbulent flames. Ph.D. thesis, Cornell University, 1982.
99. Namer, I. & Schefer, R. W., Error estimates for Rayleigh scattering density and temperature measurements in premixed flames. *Exp. Fluids* **3** (1985) 1–10.
100. Arcoumanis, C., Green, H. G. & Whitelaw, J. H., The application of laser Rayleigh scattering to a reciprocating model engine. *Trans. SAE,* **2** (1984) 949; SAE paper 840376.
101. Wong, K. Y., Gaseous simulation of diesel sprays. PhD Thesis, University of London, 1989.
102. Eckbreth, A. C., Recent advances in laser diagnostics for temperature and species concentration in combustion. *18th Int. Symp. on Combustion,* The Combustion Institute, 1981, pp. 1471–88.
103. Pitts, W. M. & Kashiwagi, T., The application of laser-induced Rayleigh light scattering to the study of turbulent mixing. *J. Fluid Mech.,* **141** (1984) 391–429.
104. Birch, A. D., Brown, D. R., Dodson, M. G. & Thomas, J. R., The turbulent concentration field of a methane jet. *J. Fluid Mech.,* **88** (1978) 431.
105. Johnston, S. C. & Green, R. M., Engine hydrocarbon concentration measurements: a comparison between a sampling probe and Raman spectroscopy. SAE paper 820499, 1982.
106. Zur Loye, A. O. & Santavicca, D. A., Temperature and concentration measurements in an internal combustion engine using laser Raman spectroscopy. AIAA-83-1551, 1983.
107. Stenhouse, I. A., Williams, D. R., Cole, J. B. & Swords, M. D., CARS on an internal combustion engine. *Appl. Opt.,* **18** (1979) 3819–25.
108. Kajiyama, K., Sajiki, K., Kataoka, H., Maeda, S. & Hirose, C., N_2 CARS thermometry in a diesel engine. SAE paper 821036, 1982.
109. Greenhalgh, D. A., Williams, D. R. & Baker, C. A., CARS thermometry in a firing production petrol engine. Proc. 'Autotech' Conference, Instn Mech. Engrs, Birmingham, 1985.

110. Greenhalgh, D. A. & Williams, D. R., Modelling of CARS N_2 spectra for in-cylinder engine research. Report AERE-R12222, Harwell, 1986.
111. Greenhalgh, D. A. & Porter, F. M., CARS temperature measurements in a 30 kW oil-fired furnace. Report AERE-R12221, Harwell, 1986.
112. Ferrario, A. & Malvicini, C., Real time temperature and concentration measurements in a semi-industrial furnace with CARS spectroscopy. *NATA ASI Series, Ser. B,* **19** (1985) 89–105.
113. Alden, M. & Wallin, S., CARS experiments in a full-scale (10 × 10 m) industrial coal furnace. *Appl. Opt.,* **24** (1985) 3434–7.
114. Eckbreth, A. C., Dobbs, G. M., Stufflebeam, J. H. & Tellex, P. A., CARS temperature and species measurements in an augmented jet engine exhaust. *Appl. Opt.,* **23** (1984) 1328–39.
115. Druet, S. & Taran, J.-P. E., Coherent anti-Stokes Raman spectroscopy. In *Lasers,* ed. C. B. Moore, Academic Press, New York, 1979.
116. Nibler, J. W. & Knighten, G. V., Coherent anti-Stokes Raman spectroscopy. In *Raman Spectroscopy of Gases and Liquids,* ed. A. Weber. Springer-Verlag, Berlin, 1979.
117. Hall, R. J. & Eckbreth, A. C., Coherent anti-Stokes Raman spectroscopy (CARS): applications to combustion diagnostics. In *Laser Applications,* Vol. 5, ed. R. K. Erf. Academic Press, New York, 1984.
118. Davis, L. C., Marko, K. A. & Rimai, L., Angular distribution of coherent Raman emission in degenerate four-wave mixing with pumping by a single diffraction coupled laser beam: configurations for high spatial resolution. *Appl. Opt.,* **20** (1981) 1685.
119. Greenhalgh, D. A. & Whittley, S. T., Mode noise in broadband CARS spectroscopy. *Appl. Opt.,* **24** (1985) 907–13.
120. Greenhalgh, D. A. & Porter, F. M., CARS applications in chemical reactors, combustion and heat transfer. Proc. 1st Laser Sci. Conf. Dallas, Texas, November 1985.
121. Greenhalgh, D. R., England, W. A. & Porter, F. M., The application of CARS of turbulent combustion thermometry. *Combust. Flame,* **49** (1983) 171.
122. Porter, F. M., A study of temperature measurements using coherent anti-Stokes Raman spectroscopy. Ph.D. thesis, University of Surrey, 1985.
123. Hall, R. J. & Greenhalgh, D. A., *Opt. Commun,* **35** (1980) 69.
124. Koszykowski, M. L., Farrow, R. L. & Palmer, R. E., Calculation of collisionally narrowed coherent anti-Stokes Raman spectra. *Opt. Letts.* **10** (1985) 479.
125. Teets, R. E., Accurate convolutions of coherent anti-Stokes Raman spectra. *Opt. Letts,* **19** (1984) 226.
126. Greenhalgh, D. A. & Hall, R. J., A closed form solution for the CARS intensity convolution. *Opt. Commun,* **57** (1986) 125–8.
127. Greenhalgh, D. A., Hall, R. J., Porter, F. M. & England, W. A., *J. Raman Spectroscopy* **15** (1984) 71.
128. Rahn, L. A. & Palmer, R. E., *J. Opt. Soc. Am.* (B), **3** (1986) 1164.
129. Hall, R. J. & Stufflebeam, J. H., *Appl. Opt.,* **23** (1984) 4319.
130. Papineau, N. & Pealat, M., *J. Chem. Phys.,* **79** (1982) 5758.
131. Paineau, N. & Pealat, M., *Appl. Opt.,* **24** (1985) 3007.

132. Smith, J. R., Turbulent flame structure in a homogeneous-charge engine. SAE paper 820043, 1982.

BIBLIOGRAPHY

Drain, L. E. (1980). *The Laser Doppler Technique,* John Wiley & Sons Ltd.
Durst, F., Melling, A. & Whitelaw, J. H. (1981). *Principles and Practice of Laser–Doppler Anemometry,* 2nd edn. Academic Press, London.

Index

Air dilution, 16–18
Air entrapment, 338–40
Air motion, 18–19
  in-cylinder, 30–1
Air utilization, diesel engines, 373–9
Air-valve carburettor. *See* Carburettors
Anemometry, 165–8, 235–7, 806–20
Atmospheric pollution, 707–93
  damage by, 709
  diesel-engine emissions and, 776–90
  exhaust gases, 725–34
  modelling of, 709–24
  spark-ignition engine emissions, 748–76
Atomization, 220–3
Autoignition, 322–3

Bend impaction modelling, 195
Birmann pulse converter, 589–93
BL stratified charge engine, 140
  see also Stratified charge engines
Blowdown, and two-stroke engines, 103–18
  flow equations, 107
  pressure fluctuation, 115
  simulation schemes, 114–15
  sonic discharge phase, 109–12
  subsonic discharge phase, 112–14
Burning, 318–320
  different engine speeds, 319–20
  compression, ahead of flame, during, 318–19
  laminar, velocity, 291–5
  mass, rate, 307–10
  turbulent, 290–1
    velocity, 302–5
Burnt-gas
  disturbance, diesel engine, 373–9
  self re-entrainment, 340–9

Carbon dioxide, CARS, 843, 845
Carbon monoxide, 729, 759
Carburettor pipe boundary conditions, 491–3
Carburettors, and droplet formation, 183–6
CARS, 832–48
  calibration, 840–1
  carbon dioxide, 843, 845
  hydrocarbons, 843, 846–8
  hydrogen, 841–3
  thermometry, 835–40
  water, 843–4
Cars, pollution by, 740–5
  Europe and, 735–8
  Japan and, 739
  USA and, 738
Catalyst methods, 760–7
  lead-tolerant, 766

Catalyst methods—*contd.*
oxidation, 765–6
oxygen, 766–7
reactor, 762–5
Central fuel injection systems, 186–8
Centrifugal machinery flow, wheels, 670–1
Ceramic rotor turbocharger, 702
Chemical kinetics, 408–9
air pollution monitoring, and, 714–16
Closed-end pipe boundary conditions, 471
Cloud of flame mass, 358–63
Coefficient of variation, 9
Coherent anti-Stoke Raman spectroscopy. *See* CARS
Combustion
compression, ahead of flame, 318–19
computer simulation of, 385–444
cyclic variations, 317–18
diesel engine, 333–84
main, 52–62
gasoline engines, 287–331
initial burning, 317–18
NO_x emission, 55–9
particle emission, 52–5
small diesel engines, 32–63
spark-ignition engines, 1–32
stratified charge engine modelling, 161–3
turbulent flow, 279–81
see also other specific engines
Combustion chamber design, 25–30
turbulence, and, 281–4
Compressor pipe boundary conditions, 493–4
Compressor, of turbocharger, 628–34
casings, 673–4
collector, 635
covers, 622–3
diffuser, 633–5
performance maps, 693–7
rotor, 628–33
wheels, 622
Computer simulation
accessibility, 408
Computer simulation—*contd.*
advantages of, 387
boundary conditions, 391–2, 398
convergence, 403–5
definition, 386
differential equations and, 389–90
display and, 407
distributed reaction representation, 408–24
economy of, 405–6
finite-volume equation, 393–4
history of, 387–8
interpolation formulae, 396–7
mass conservation equation, 398–9
numerical accuracy of, 406
physical realism and, 406–7
physics of, 388–9
spatio-temporal discretization, 392–3
two-fluid model turbulent, 424–5
ways to solve equations, 399–403
Constant-pressure manifolding, 581–9
Crankcase compression type two-stroke engines, 65–101
induction control, 77–82
noise prediction and, 82–84
performance of, 69–77
scavenging
design, 86–95
efficiency, 84–6
wave action, 67–8
Curtis Wright rotary engine, 145

Degree of reaction, 652–4
Diesel engines, 32–63
direct injection, 35–46
direct, pollutants, 782–5
fuel injection processing, 215–42
ignition delay, 42–52
indirect injection, 35–46
indirect, pollutants, 782–5
main combustion in, 52–62
pollutants, measurement, 745–8
swirl generation, 37–42
vehicle regulations, pollution, 739–40

Diesel-engine combustion
air utilization, 373–9
burnt-gas
disturbance, 373–9
self re-entrainment, 340–9
equivalence ratio, 365–9
flame mass cloud, 358–63
free-spray
flame, 337–55
formation, 349
in-cylinder swirl, 354–5
initial stage of, 337–8
low fuel consumption, 379–81
low particulate emission, 379–81
main stage, 354–5
nitrogen oxides and, 365–9
piston cavity chamber, 355–81
ROAE in-cylinder, 363–5
spray, 334–7
spray flame air entrapment, 338–40
spray turbulence, 349
squish flow, 370–2
swirl, 335–7
swirl interference, 370–2
temperature distribution, 365–9
wall impingement, 350–3
Diesel-engine manifolding, 576–99
constant-pressure, 581–9
pulse, 576–81
pulse-converted, 589–99
Diesel flame. *See* Spray; Diesel spray
Diesel particulate filters, 785–90
regeneration of, 789–90
types of, 788
Diesel spray, 372–3
aerodynamics of, 223–31
break-up of, 221–3
visualization, 218–20
see also Spray
Diffraction, 205, 233–235
Direct injection processes, 35–37
Distributed-reaction computer representation, 408–424
grid-fineness problems, 415–16
mixture fraction, 412–13
reaction rate, 411
SCRS, 409–10
Distributed-reaction computer representation—*contd.*
steadily propagating laminar flame 413–15
turbulent fluctuation problems, 416–20
two-phase problems, 420–3
Drop evaporation, 188–92
motionless drop, 188–91
moving drop, 191–2
Droplet formation, 177–88
basics of, 177–8
carburettors, 183–6
central injection systems, 186–8
diameter prediction, 180–2
diffraction, 205
fuel systems, 183–8
halography, 205
Hinze equation, 182–3
impaction methods, 204
Ingebo & Foster equation, 182
laser-doppler technique, 205
multi-point fuel injection systems, 188
Nukiyama & Tanasawa equations, 181–2
photographic techniques, 204
Prandtl equation, 182
Rosin & Rammler distribution, 179–80
Sauter mean, 178–9
size distribution and, 179–83, 204–5
Droplet size, 46–7
distribution diagnostics, 231–7

Eddy break-up model, 419–20
EGR. *See* Exhaust gas recirculation
Engine exhaust systems, 502–23
Engine flow instrumentation, 795–856
Engine-intake tuning, 502–23
Engines, research, 2–5
see also under specific types
Engine speed
burning and, 319–20
ignition and, 46

Engine starting and ignition delay, 47–50
Entropy discontinuity, 475
Evaporation. *See* Drop evaporation
Exhaust gases, 725–34
  dilution, 19–22
  recirculation (EGR), 51–2, 57–9, 773–5
Exhaust lead, 116–17
Exhaust manifold
  calculations, 560–66
  tuning, 521–3
Exhaust pulse, 117–18

Film removal, 205
Finite difference
  equation, 160
  grid, 158–9
  technique, 461–8
    Lax–Wendroff technique, 462–6
    RCM, 467
    Runge–Kutta solution and, 467
Finite-volume equations, 393–9
  boundary conditions, 398
  interpolation formulae, 396–7
  linear-seeming form, 394–6
  mass conservation equation, 398–9
  solution, 399–403
    quality, 403–8
Flame
  front, 164
  speed, stratified charge engines, 163
  straining, 313–16
Flammability limits, and gasoline combustion, 296–8
Flow symmetry ratio, 89–92
Flow technique, 802–820
  see also under specific aspects
Flow visualization, 803–6
Fluctuation velocity, 419
Fluid, flow, in reciprocating engines, 385–444
Ford Proco
  open combustion chamber engine, 153–4
  stratified charge engine, 144
Frauenhofer diffraction. *See* Diffraction
Fuel, and, ignition delay, 43–4
Fuel injection
  boundary condition, 491–3
  components, 215–18

Gasoline engines
  burning velocity in, 302–5
  combustion in, 287–331
  combustion control, 289–90
  engine performance, and combustion, 317–26
  flammability limits, 296–8
  ignition and, 310–17
  laminar burning velocity, 291–5
  laminar flame straining, 295–6
  mass burning rate, 307–10
  steady-state turbulent burning, 298–9
  turbulent burning, 290–1, 298–310
  turbulent explosion bombs, 306–7
  unsteady-state turbulent burning, 299–302
Glow plug, 50–1
Grid fineness, 415–16
Guide vanes, 643–4

Heat transfer, to walls, 59–62
High ratio compact chamber, 25–30
Hinze equation, 182–3
Holography, 205, 232–3
Homentropic finite waves, 451–3, 457–60
Honda CVCC engine, 140–143
Hot-wire anemometry, 165–8, 806–11
Hydra engine, 2–5
Hydrocarbon emission, 750–4
  adsorption/desorption oil film, 752–4
Hydrocarbons, and CARS, 843, 845
Hydrogen, CARS, 841–5

Ignition, auto-, 322–6

Ignition delay, 42–52
engine
speed, 46
starting, 47–50
exhaust gas recirculation, 51–2
fuel and, 43
glow plug and, 50–1
size of droplets and, 46–7
swirl, 46–7
Ignition, and gasoline combustion, 310–16
flame straining, 313–16
spark and, 310–13
spark-assisted propagation, 313–16
Ignition retardation, 19, 775–6
IMEP, 7–9
Impaction, 204
In-cylinder
air motion, 30–1
swirl, 358–63
Indirect injection processes, 35–7
Induction control, on two-stroke engines, 77–82
Ingebo & Foster equation, 182
Inlet manifold calculation, 530–60
linearized, 531–42
non-linear, 542–55
variable geometry, 555–60
Inlet valves, 208
Ionization probes, 15
Intake manifold tuning, 502–10
momentum effects, 505–10
resonators, 510–16
wave effects, 504–5
Intake process, 275–8
Isobaric–isochoric model, uniflow scavenging, 121–30
analysis of results, 125–30
boundary conditions, 121–3
individual phases in, 123–5

Jet-based (phenomenological) scavenging models, 130–5
analysis of results, 133–5
governing equations, 132–3

Karlovitz stretch factor, 296
Knock, 322–6
Kolmogorov. *See* Microscale

Lambda sensor, 766–7
Laminar burning velocity, 291–5
Laminar flame
speed, 418–19
straining, 295–6
Laser–doppler anemometry, 165–8, 205, 235–7, 811–20
applications of, 815–20
Laser Raman scattering, 829–32
applications of, 832
Laser Rayleigh scattering, 821–9
applications of, 826–9
Law of Thermodynamics, 7–9
Lax–Wendroff technique, 462–6
LDA. *See* Laser–doppler anemometry
Lead-tolerant catalyst, 766
Lead toxicity, 725–9
Lean-burn engine, 767–73
Lewis number, 295
Linear waves, 449–50, 456

Macroscale, and turbulence, 262–7
Manifold calculations
exhaust, 560–6
inlet, 530–60
turbocharged engines, 567–70
Manifolds, mixture transport, 192–203
bend impaction models, 195
design, 200–3
droplet behaviour of, 193–5
theory of, 192
wall-film models, 195–6

Manifold tuning, 502–23
exhaust, 521–3
intake, 502–10
resonators, 510–15
Mass burning rate, 307–10
Matching, turbocharger, 657–69
Mean droplet volume diameter, 178–9

Mean flow, 798–802
Microscale, and turbulence, 267–74
Mixture preparation, and spark-ignition engines, 173–211
  droplet formation, 177–88
  engine requirements and, 175–7
  evaporation, droplets, 188–92
  inlet-valve studies, future, 208
  manifold
    design, 201
    studies, future of, 208
    transport, 192–7
  studies into, 197–200
Motionless drop, 188–91
Moving drop, 191–2
Multi-pipe junction, boundary conditions, 482–91
Multi-point fuel injection, 188

Nakiyama & Tanasawa equation, 181–2
Naturally aspirated two-stroke engines, 524–6
  wave action, manifold systems, 527–66
Nitric oxide, 164–5, 754–9
  emission, 754–9
Nitrogen oxides, 148–50, 379–81
  air pollution, and, 729–30
  emission of, 55–9
Noise prediction, two-stroke engines, 82–4
Non-homentropic finite waves, 453–6
  path line equations, 455–6
  wave equations, 455
Nozzle boundary conditions, 471–3
Nozzled turbine casings, 675–6
Nozzleless turbine casings, 677–82
Nusselt number, 191–2

One-dimensional flow equation, 449–56
Open-end pipe boundary conditions, 469–70
Optical measurements, 9–15
Otto cycle. *See* Spark-ignition engines

Oxygen sensors, 766–7
Ozone, 730

Particle emission, diesel engines, 52–5, 379–81, 730–4
  filters, diesel, 785–90
Performance mapping, turbocharger
  compressor, 693–7
  turbine, 697
Phenomenological scavenging models, 130–5
Photography, 204, 207
  diesel spray, and, 219–20
PHOXA model, 719–24
Pipe boundary conditions, 469–97
  carburettor, 491–3
  closed-end, 471
  compressor, 493–4
  cross-section change and, 474–5
  entropy discontinuity, 475
  fuel injection, 491–3
  multi-pipe junction, 482–91
  turbine, 495–7
  valve and port boundaries, 476–81
Piston-cavity chamber, 355–81
Piston movement, 159–60
Pollution. *See* Atmospheric pollution
Porsche stratified charge engine, 143
Prandtl equation, 182, 191
Pressure correction, 160–1
Pressure exchangers, 599–609
  operating characteristics, 606–9
  operating principle, 603–6
Pulsating flow, turbochargers, 682–92, 698
  quasi-steady/dynamic turbine operation, 686–92
Pulse-converted diesel-engine manifolds, 589–99
  Birmann type, 593–9
  Sulzer type, 589–93

Raman scattering. *See* Laser Raman scattering
Raman spectrometry. *See* CARS
Rate of heat release, 229–31

Rayleigh scattering. *See* Laser Rayleigh scattering
Reaction rate, 411
Reciprocating internal combustion engines, 243–85, 385–444
  wave action
    technology, 501–614
    theory, 445–99
Regional transport air pollution modelling, 719–24
Regulations, air pollution, 734–48
  cars, 735–9
  diesel engines, 739–40, 745–8
  petrol engines, 741–4
Research engines, 2–5
Resonators, 510–15
Reynolds number, 191, 223
  see also Turbulent flow
Rosin & Rammler distribution, 179–80
Runge–Kutta solution, 467

Sauter mean diameter, 178–9
SCAN combustion simulation, 168–70
Savenging
  design for, 86–95
  efficiency, 84–6
  equation for, 66
  modelling of, 95–7
  two-stroke engines, 523–4
  uniflow. *See* Uniflow scavenging
SCE. *See* Stratified charge engine
Schmidt number, 192
Sonic discharge, and blow-down, 109–12
  subsonic, 112–14
Spark-assisted propagation, 313–16
Spark ignition, 310–13
  SCE, and, 138–40
  see under specific types
Spark-ignition engines, 1–32
  air-motion characteristics, 22–5
  analysis methods, 5–15
  capacity, 34–5
  combustion chamber, design, 25–30
Spark-ignition engines—*contd.*
  combustion characteristics, 15–22
  emission, 748–76
  in-cylinder air motion, 30–1
  ionization probes, 15
  measurement methods, 5–15
  mixture preparation, 173–211
  optical measurement, 9–15
  pressure data, 7–9
  research, 2–5
  small diesel, combustion, 32–63
  stroke : bore ratio, 34–5
  swirl, 22–4
  tumble, 24–5
Spatio-temporal discretization, 392–3
Speed, of engine, and gasoline burning, 319–20
Spray, 334–7
  break-up, 221–3
  diagnostic methods, 215–42
  flame, and air entrapment, 338–40
  turbulence, 349
Squish flow, 370–2
  see also Turbulent flow
Steadily propagating laminar flame, 413–15
Steady-state turbulent burning, 298–9
Stefan's equation, 188–9
Stratified charge engines, 137–71
  achievement, 140–5
  BL, 140
    investigation, 149–53
  combustion, behind flame front, 164
  combustion modelling, 161–3
  Curtis Wright rotary, 145
  design of, 145–6
  finite difference equations, 160
  flame front, 164
  flame speed prediction, 163
  Ford Proco, 144, 153–4
  GM, 146
  Honda, 140–3
  Mathematical
    modelling, 155–70
    representation, 155–8
  model validation, 165–8
  nitric oxide, 164–5

Stratified charge engines—*contd.*
  open-chamber, 143–5
  Porsche, 143, 146–9
  piston movement, 159–60
  pre-chamber design, 140–3
  pre-charge combustion system, 146–53
  pressure correction equations, 160
  SCAN combustion, 168–70
  Texaco, 143–4
  theory of, 138–40
Streamline curvature methods, 655–7
Stroke: bore ratio, 34–5
Subsonic discharge, and blowdown, 112–14
Sulzer pulse converter, 589–93
Supercharged engines, 526–7
Supercharged manifold wave action, 566–99
  diesel engines, 576–99
  exhaust, transmission efficiency, 570–6
Swirl, 23, 37–42, 370–2, 431–4
  spray, and, 335–7
  turbulence, 281–4

Tanasawa–Tesima distribution, 180
Taylor microscale, 247–8
Temperature, CARS and, 835–41
Texaco stratified charge engines, 143–4
Top dead centre, 25
Tracer mixing, 205–7
Tumble, 23–4
Tuned intake manifolds. *See* Intake manifold tuning
Turbine
  efficiency, 652–4
  performance maps, 697
Turbine casings, turbocharger, 636–40
  geometry, 641–3
  guide vanes, 643–4
Turbine pipe boundary conditions, 495–7
Turbines, of turbocharger, 635–45
  casings for, 675–82
Turbines, of turbocharger—*contd.*
  housings, 625–6
  rotor, 645–52
  variable-geometry, 701–2
  wheels, 625, 670–3
Turbocharged manifold wave action. *See* Supercharged manifold wave action
Turbochargers, 615–705
  aerodynamics of, 697–8
  air handling systems research, 700–2
  bearings
    housings, 623–4
    systems, 624
  casing and rotor interspace, 644–5
  ceramic rotor, 702
  compressor, 628–35
    covers, 622–3
    wheels, 622
  construction, 620–6
  conventional, future of, 699–702
  degree of reaction, 652–4
  design of, 626–57
  function of, 618–20
  matching, 657–69
  streamline curvature, 655–7
  test facilities, 692–7
  three-dimensional, analysis, 657
  turbine, 634–44
    efficiency, 652–4
    housing, 625–6
    rotor, 645–52
    wheels, 625
Turbocompounding, 700–1
Turbomachinery equations, 627–8
Turbulence, 798–802
  scales, 262–74
    energy spectra, 269–74
    macro-, 262–7
    micro-, 267–74
Turbulent burning, gasoline engines, 290–1
  steady-state, 298–9
  unsteady-state, 299–302
  velocity, 302–5
Turbulent explosion bombs, 306–7

Turbulent flow, 243–85
basics of, 243–5
compression process and, 278–81
intake process, 275–8
mean flow field, 249–55
modelling, 274–84
parameters, 246–9
scales, 262–74
swirl, 281–4
Turbulent fluctuation, 416–18
Two-fluid turbulent combustion model, 424–35
one-dimensional representation of, 426–8
two-dimensional representation of, 431–5
Two-stroke engines
blowdown, 103–18
crankcase compression type, 65–101
pipework, 523–7
naturally aspirated, 524–6
supercharged, 526–7, 566–99
uniflow scavenge process, 118–35

Uniflow scavenging, two-stroke engines, 118–35
basics of, 119–21
isobaric–isochoric model, 121–30
jet-based model, 135
port/valve arrangement, 105–7
Unsteady-state turbulent burning, 299–302
Urban model, of air pollution, 712–14

Valve and port pipe boundary, 476–81
Vane diffuser, 673–4
Vaneless diffuser, 674
Velocity distribution diagnostics, 231–7
diffraction, 233–5
holography, 232
laser-doppler, 235–7
results of, 237–40
Volumetric efficiency, 516–21

Wall-film modelling, 195–6
Wall-film rate measurements, 205–7
film removal methods, 205
photographic methods, 207
tracer mixing methods, 205, 207
Wall impingement, spray flame, 350–3
Wave action theory, 445–99
compressible flow calculations, 456–69
crank case compression engines, 67–8
finite difference techniques, 461–8
homentropic finite, 451–3, 457–60
linear, 449–50, 456
non-homentropic finite, 453–6, 460–1
pipe boundary conditions, 469–97
reciprocating engine, 501–614
Weber number, 183
Weslake combustion chamber, 769–71